D1030678

THE
Bacterial
Spore

FROM MOLECULES TO SYSTEMS

THE
Bacterial
Spore
FROM
MOLECULES
TO SYSTEMS

EDITED BY

ADAM DRIKS
Stritch School of Medicine, Loyola University Chicago, Maywood, IL

PATRICK EICHENBERGER
Center for Genomics & Systems Biology, New York University, New York, NY

ASM PRESS

Washington, DC

Library of Congress Cataloging-in-Publication Data

Names: Driks, Adam, editor. | Eichenberger, Patrick (Professor of biology), editor.
Title: The bacterial spore : from molecules to systems / edited by Adam Driks and Patrick Eichenberger.
Description: Washington, DC : ASM Press, [2016] | ?2016 | Includes bibliographical references and index.
Identifiers: LCCN 2016012264 (print) | LCCN 2016014762 (ebook) | ISBN 9781555816759 (print : alk. paper) | ISBN 9781555819323 ()
Subjects: LCSH: Bacterial spores. | Sporeforming bacteria.
Classification: LCC QR79 .B328 2016 (print) | LCC QR79 (ebook) | DDC 579--dc23 LC record available at http://lccn.loc.gov/2016012264

Printed in the United States of America

10 9 8 7 6 5 4 3 2 1

Address editorial correspondence to: ASM Press, 1752 N St., N.W., Washington, DC 20036-2904, USA.

Send orders to: ASM Press, P.O. Box 605, Herndon, VA 20172, USA.
Phone: 800-546-2416; 703-661-1593. Fax: 703-661-1501.
E-mail: books@asmusa.org
Online: http://estore.asm.org

Contents

Contributors

ESTHER R. ANGERT
Department of Microbiology, Cornell University, Wing Hall, Ithaca, NY 14853

ASHLEY R. BATE
Center for Genomics and Systems Biology, Department of Biology, New York University, New York, NY 10003

CASEY B. BERNHARDS
U.S. Army Edgewood Chemical Biological Center, 5183 Blackhawk Rd., E3835, Aberdeen Proving Ground, MD 21010

RICHARD BONNEAU
Center for Genomics and Systems Biology, Department of Biology, New York University, New York, NY 10003

JOEL A. BOZUE
U.S. Army Medical Research Institute of Infectious Diseases, Division of Bacteriology, 1425 Porter Street, Fort Detrick, MD 21702

CRISTINA N. BUTTERFIELD
Division of Environmental and Biomolecular Systems, Institute of Environmental Health, Oregon Health & Science University, 3181 S. W. Sam Jackson Park Rd., Portland, OR 97239

GARETH COOPER
Krebs Institute, Department of Molecular Biology & Biotechnology, University of Sheffield, Sheffield S10 2TN, United Kingdom

CHRISTOPHER K. COTE
U.S. Army Medical Research Institute of Infectious Diseases, Division of Bacteriology, 1425 Porter Street, Fort Detrick, MD 21702

ADAM DRIKS
Department of Microbiology and Immunology, Stritch School of Medicine, Loyola University Chicago, Maywood, IL 60153

DAVID DUBNAU
Public Health Research Center of New Jersey Medical School, 225 Warren Street, Newark, NJ 07103

PETER DÜRRE
Institut für Mikrobiologie und Biotechnologie, Universität Ulm, 89069 Ulm, Germany

JONATHAN DWORKIN
Department of Microbiology & Immunology, College of Physicians and Surgeons, Columbia University, New York, NY 10032

PATRICK EICHENBERGER
Center for Genomics and Systems Biology, Department of Biology, New York University, New York, NY 10003

PATRICIA FAJARDO-CAVAZOS
Department of Microbiology & Cell Science, University of Florida, Space Life Sciences Laboratory, 505 Odyssey Way, Rm. 201, Exploration Park at Kennedy Space Center, Merritt Island, FL 32953

MICHAEL Y. GALPERIN
National Center for Biotechnology Information, National Library of Medicine, National Institutes of Health, Bethesda, MD 20894

ELIZABETH A. HUTCHISON
Department of Biology, SUNY Geneseo, 1 College Circle, Geneseo, NY 14454

RACHELE ISTICATO
Department of Biology, Federico II University of Naples, Naples, 80126 Italy

PAUL KEIM
Center for Microbial Genetics and Genomics, Northern Arizona University, 1298 S. Knoles Drive, Northern Arizona University, Flagstaff, AZ 86011-4073

SUNG-WOO LEE
Division of Environmental and Biomolecular Systems, Institute of Environmental Health, Oregon Health & Science University, 3181 S. W. Sam Jackson Park Rd., Portland, OR 97239

JIHONG LI
Department of Microbiology and Molecular Genetics, University of Pittsburgh School of Medicine, Pittsburgh, PA 15219

INES MANDIC-MULEC
University of Ljubljana, Biotechnical Faculty, Department of Food Science and Technology, Vecna pot 111, 1000 Ljubljana, SIovenia

HEATHER MAUGHAN
Ronin Institute (Headquartered in Montclair, NJ), Mildmay, ON Canada N0G 2J0

BRUCE A. McCLANE
Department of Microbiology and Molecular Genetics, University of Pittsburgh
School of Medicine, Pittsburgh, PA 15219

DAVID A. MILLER
Department of Microbiology, Medical Instill Development, 201 Housatonic Ave.,
New Milford, CT 06776

NICOLAS MIROUZE
UMR1319 Micalis, Bat. Biotechnologie (440), I.N.R.A., Domaine de Vilvert,
78352 Jouy-en-Josas Cedex, FR

ANNE MOIR
Krebs Institute, Department of Molecular Biology & Biotechnology, University
of Sheffield, Sheffield S10 2TN, United Kingdom

WAYNE L. NICHOLSON
Department of Microbiology & Cell Science, University of Florida, Space Life
Sciences Laboratory, 505 Odyssey Way, Rm. 201, Exploration Park at Kennedy
Space Center, Merritt Island, FL 32953

RICHARD T. OKINAKA
Center for Microbial Genetics and Genomics, Northern Arizona University,
1298 S. Knoles Drive, Northern Arizona University, Flagstaff,
AZ 86011-4073

DANIEL PAREDES-SABJA
Departamento de Ciencias Biológicas, Universidad Andrés Bello, Santiago,
920-8640, Chile

DAVID L. POPHAM
Department of Biological Sciences, Life Sciences I – MC0910, 970 Washington
St. SW, Virginia Tech, Blacksburg, VA 24061

EZIO RICCA
Department of Biology, Federico II University of Naples, Naples, 80126 Italy

MAHFUZUR R. SARKER
Department of Biomedical Sciences, College of Veterinary Medicine,
Department of Microbiology, College of Science, Oregon State University,
Corvallis, OR 15219

PETER SETLOW
Department of Molecular, Microbial and Structural Biology, University of
Connecticut Health Center, Farmington, CT 06030-3305

POLONCA STEFANIC
University of Ljubljana, Biotechnical Faculty, Department of Food Science and
Technology, Vecna pot 111, 1000 Ljubljana, SIovenia

BRADLEY M. TEBO
Division of Environmental and Biomolecular Systems, Institute of Environmental
Health, Oregon Health & Science University, 3181 S. W. Sam Jackson Park Rd.,
Portland, OR 97239

NANCY TWENHAFEL
Pathology Division, United States Army Medical Research Institute of Infectious
Diseases, Frederick, MD 21702

JAN DIRK VAN ELSAS
Department of Microbial Ecology, Centre for Ecological and Evolutionary
Studies, University of Groningen, Linneausborg, Nijenborgh 7, 9747AG
Groningen, The Netherlands

SUSAN WELKOS
U.S. Army Medical Research Institute of Infectious Diseases, Division of
Bacteriology, 1425 Porter Street, Fort Detrick, MD 21702

Preface

If bacteria are really the cause of anthrax, then we must hypothesize that they can go through a change in life history and assume a condition which will be resistant to alternate drying and moisture. What is more likely, and what has already been indicated by Prof. Cohn, is that the bacteria can form spores which possess the ability to reform bacteria after a long or short resting period.

Robert Koch, in *Beiträge Biol Pflanzen* 7:277–310, 1876

In an earlier generation, biochemists regarded spores as inert, resistant little bodies which had few or no enzymes. With what he knows of biochemistry today, an informed student, aware of the spore's potential for rapid germination from an inert state and its resistance to adverse external conditions, would postulate a far more elaborate composition for a spore.

Arthur Kornberg, James A. Spudich, David L. Nelson,
and Murray P. Deutscher, in *Annu Rev Biochem* 37:51–78, 1968

In no organism are questions of cell fate, cell differentiation, and morphogenesis more accessible to experimental investigation than in the spore-forming bacterium Bacillus subtilis.

Patrick Stragier and Richard Losick in
Annu Rev Genet 30:297–241, 1996

The first two publications reporting that some bacteria have the ability to form dormant spores appeared in 1876 as back-to-back articles in the plant journal *Beiträge zur Biologie der Pflanzen*. Remarkably, each report was authored by a founding father in the field of microbiology. The first paper, which described sporulation in the bacterium *Bacillus subtilis*, was written by Ferdinand Cohn, who is now remembered as the first bacterial taxonomist. The second paper, contributed by Robert Koch, focused on the etiology of anthrax, revealing the role played by spores of *Bacillus anthracis* in causing the disease. Koch's discoveries on spores (and his following work on tuberculosis), along with studies from his archrival Louis Pasteur, established the basis for the germ theory of disease. Pasteur himself stumbled on spores on many occasions: sometimes as a nuisance, when his famous experiments to disprove spontaneous generation of bacteria were challenged by Felix Pouchet (it was later understood that the hay infusions that Pouchet had used for his experiments were contaminated by spores that were resistant to boiling); sometimes more auspiciously, when Pasteur developed a vaccine against anthrax. The study of spore formation is thus as old as microbiology itself. Major conceptual and technological advances in the field of microbiology have impacted the study of sporeformers ever since. Similarly, many techniques and hypotheses pioneered in spore-forming bacteria continue to influence microbial research at large.

Molecular mechanisms of sporulation were studied in detail during the second half of the 20th century. Progress was significantly aided by the realization, from the work of Spizizen and Anagnostopoulos, that *B. subtilis* cells were naturally competent, thus opening the door for manipulations of the *B. subtilis* genome. A combination of genetics-, biochemistry-, and electron microscopy-based approaches led to a deep understanding of spore structure and a clear description of the different stages of the spore formation process. As an example, a review by Piggot and Coote from 1976 (i.e., a century after Cohn and Koch) listed over 30 genetic *spo* loci, i.e., genes or operons, whose mutation affected the efficiency of sporulation without disturbing the bacterium's ability to grow. Each locus was named using roman numerals referring to the stage of blockage in the sporulation cycle of the corresponding mutants. These stages were originally defined by Ryter, Schaeffer, and collaborators, who used electron microscopy to investigate the morphology of sporulating cells.

Biochemical and cytological analysis of sporulation mutants played a key role in cementing the position of *B. subtilis* as an excellent model system to investigate basic developmental mechanisms. Importantly, spores formed by *B. subtilis* and related bacteria are endospores, in the sense that spore maturation must occur inside a mother cell. The mother cell and the forespore (i.e., the cell that will develop into a spore) both result from an asymmetric division of the sporulating cell near the cell pole. Thus, sporulating cells provide a relatively simple system for the study of cellular differentiation. More specifically, how can the asymmetric division of a progenitor cell be exploited to produce two daughter cells that will experience distinct fates? Many studies during that period focused on understanding the mechanisms of intercellular communication and the cell-specific activation of the sporulation sigma factors that control the parallel lines of gene expression established in the forespore and the mother cell.

In *The Bacterial Spore: From Molecules to Systems* (which is published exactly 140 years after the founding papers by Cohn and Koch), we have attempted to summarize this vast amount of knowledge. One of our goals was to integrate several decades of research about the process of spore formation in the model organism *B. subtilis* with other efforts to characterize medically, evolutionarily, and ecologically relevant species of spore-forming Firmicutes (primarily *Bacillaceae* and clostridia). The publication of the full genome sequence of *B. subtilis* in 1996 represented a milestone, which was later followed by the release of a multitude of complete genome sequences of other spore-forming bacteria. Comparative genomics (chapter 1) led to a broader understanding of evolution in spore-forming species (chapters 2 and 4), while the surge of metagenomics is likely to enrich our view of the ecology of spore-forming species (chapter 3).

Our book is subtitled *From Molecules to Systems* to indicate our desire to provide a link between classic molecular genetics approaches (chapters 8 to 11) and more recent efforts that view the spore as a system (chapters 5 and 6). As an illustration of the systems biology of spores, transcriptomic studies have revealed that several hundred genes are expressed during sporulation. Strikingly, this is an order of magnitude higher than the number of sporulation genes estimated from genetic screens. Furthermore, a bridge between traditional methods and novel technologies is provided by fluorescence microscopy-based methods, which have become very common in recent years (chapter 7).

Among pathogenic sporeformers, *B. anthracis* has always occupied a central position (chapters 12 through 14); however, pathogenic *Clostridium* species (chapters 15 and 16) have also garnered considerable attention including, for instance, *Clostridium difficile*, a bacterium that has recently caused nosocomial infections in alarming numbers. In contrast to being agents of infectious diseases, spore-forming bacteria can also be used for practical applications. For example, in an attempt to generate new vaccine delivery systems, it was shown that spore surfaces can be useful for antigen display (chapter 17). In addition, spores are excellent candidates for use in bioremediation, in particular the removal of metal contaminants (chapter 18).

In spite of this vast trove of information, it goes without saying that an enormous amount remains to be discovered; each chapter in this book defines important open questions for future research. As just one example, one of the most intriguing gaps in our knowledge of spore-forming bacteria is a detailed understanding of their natural ecology. We still do not know the exact roles that the diverse stress responses produced by these bacteria play in their incompletely characterized ecosystems. This is especially true of sporulation, the lifestyle that so powerfully caught the attention of Robert Koch and Ferdinand Cohn many years ago. We hope that the open questions posed in these chapters are as intriguing to our readers as are the detailed answers, and that they will inspire continued study for many more years to come.

We wish to thank all of the extremely skilled people at ASM Press who have been instrumental in bringing this book forward. In particular, we wish to acknowledge our colleagues Ellie Tupper and Eleanor Riemer, whose efforts have been especially important.

General Overview of the *Bacillaceae* and Clostridia

I

The Bacterial Spore: From Molecules to Systems
Edited by P. Eichenberger and A. Driks
© 2016 American Society for Microbiology, Washington, DC
doi:10.1128/microbiolspec.TBS-0015-2012

Michael Y. Galperin[1]

Genome Diversity of Spore-Forming *Firmicutes*

1

Later the *Bacillus* filaments begin to prepare for spore formation. In their homogenous contents strongly refracting bodies appear. From each of these bodies develops an oblong or shortly cylindrical, strongly refracting, dark-rimmed spore.

> Ferdinand Cohn. 1876. Untersuchungen über Bacterien. IV. Beiträge zur Biologie der Bacillen. *Beiträge zur Biologie der Pflanzen*, vol 2, p 249–276. (Studies on the biology of the bacilli. In: Milestones in Microbiology: 1546 to 1940. Translated and edited by Thomas D. Brock. Prentice-Hall, Englewood Cliffs, NJ, 1961, p 49–56).

BACTERIAL SYSTEMATICS FROM GRAM STAIN TO 16S rRNA

The taxonomy of spore-forming Gram-positive bacteria has a long and colorful history. In 1872, 35 years after Christian Ehrenberg provided the initial description of *Vibrio subtilis* (and also *Vibrio bacillus*), Ferdinand Cohn assigned it to the genus *Bacillus* and family *Bacillaceae*, specifically noting the existence of heat-sensitive vegetative cells and heat-resistant endospores (see reference 1). Soon after that, Robert Koch identified *Bacillus anthracis* as the causative agent of anthrax in cattle and the endospores as a means of the propagation of this organism among its hosts. In subsequent studies, the ability to form endospores, the specific purple staining by crystal violet-iodine (Gram-positive staining, reflecting the presence of a thick peptidoglycan layer and the absence of an outer membrane), and the relatively low (typically less than 50%) molar fraction of guanine and cytosine in the genomic DNA have been used as diagnostic characteristics of the phylum *Firmicutes* (low-G+C Gram-positive bacteria).

Remarkably, neither of these traits proved to be a clear-cut predictor of the organism's membership in the *Firmicutes*. Many members of the phylum (lactic acid bacteria, listeria, staphylococci) do not form endospores, some *Firmicutes* stain Gram-variable or even Gram-negative, and some, like *Symbiobacterium thermophilum*, have the G+C content of >60%, which is more typical for the *Actinobacteria*.

Obviously, microorganisms can be classified by a variety of parameters, including the cell shape, staining pattern, spore formation, relation to oxygen, nutritional requirements, the ability to use CO_2 and fix nitrogen, salt tolerance, and, last but not least, pathogenicity. For many years, "numerical" taxonomy based on a combination of such parameters seemed the only way to impose some order onto the enormous diversity of microbial life. In many respects, it was successful, and the early descriptions of many bacterial species, genera, and families were later upheld by molecular techniques.

[1]National Center for Biotechnology Information, National Library of Medicine, National Institutes of Health, Bethesda, MD 20894.

However, the deep-level systematics of bacteria required different approaches. The breakthrough came from the studies by Carl Woese and colleagues, who sought to base bacterial classification on the evolutionary history of the respective organisms and used the similarity of 16S rRNA sequences as a universal measure of the evolutionary proximity of the organisms (2–5). The 16S rRNA-based phylogenetic classification of bacteria has become universally accepted (see Table 1 for a list of resources) and proved so successful that it is sometimes hard to imagine that it is only 25 years old. In the case of Gram-positive bacteria, the 16S rRNA sequences were found to share conserved oligonucleotide signatures, lending molecular biology support for the Gram-staining results (3). Furthermore, 16S rRNA-based trees mostly agreed with phylogenetic trees based on other popular markers, such as ribosomal proteins, DNA gyrase subunit GyrB, RNA polymerase subunits, and others (6, 7). Subsequent more detailed analyses led to the recognition of the substantial differences between the low- and high-G+C Gram-positive bacteria, which had been assigned to two different phyla, the *Firmicutes* and the *Actinobacteria*, respectively. The recent genome-based studies confirmed the absence of a close relationship between the members of *Firmicutes* and *Actinobacteria*, justifying their separation into two

different phyla. This chapter discusses the taxonomy of spore-forming *Firmicutes*, aiming to show that the importance of bacterial systematics goes beyond simply reflecting the current state of knowledge on the relatedness of various taxa within the phylum. Modern taxonomy strives to reflect the evolutionary history and serves as a guiding tool for further genome-sequencing projects and also for comparative genomics studies, which combine to provide a better understanding of bacterial physiology, including the sporulation processes in diverse members of the *Firmicutes*.

FIRMICUTES AS A SEPARATE EARLY-DIVERGING PHYLUM

In the absence of a reliable fossil record, any speculations on the timing of the divergence of the major bacterial phyla are bound to remain controversial. In addition, the bacterial phylogenetic tree appears to have a starlike topology with all major phyla diverging at approximately the same time; only tentative groupings of some phyla ("superphyla") have been put forward (8–10). However, in the case of the *Firmicutes*, there is a general consensus that they had diverged from other bacterial phyla at a relatively early stage (8, 11, 12). The evolution of *Firmicutes* obviously included

Table 1 Principal data sources on bacterial (*Firmicutes*) taxonomy

Data resource, reference, URL[a]	Comment
Taxonomy databases	
Approved Lists of Bacterial Names (103), http://www.ncbi.nlm.nih.gov/books/NBK814/	A list of validly published bacterial species names, last updated in 1989
Prokaryotic Nomenclature Up-to-date at the German Collection of Microorganisms and Cell Cultures (DSMZ), http://www.dsmz.de/bacterial-diversity/	An updated listing of validly published bacterial species names and nomenclature changes
Bergey's Manual of Systematic Bacteriology (32), http://www.bergeys.org/outlines.html	The official bacterial systematics from the Bergey's Trust
ITIS - Catalogue of Life (104), http://www.catalogueoflife.org/annual-checklist/	Integrated Taxonomic Information System, a partnership of North American government agencies
List of Prokaryotic Names with Standing in Nomenclature (42), http://www.bacterio.net/	A constantly updated listing of validly published species names, includes bacterial classification and *Candidatus* organisms
NCBI taxonomy database (41), http://www.ncbi.nlm.nih.gov/taxonomy	A hierarchical database of all organisms that have nucleotide sequences deposited in GenBank
The Taxonomic Outline of Bacteria and Archaea (31), http://www.taxonomicoutline.org/	Text-based bacterial taxonomy files (in PDF), updated in 2007
Taxonomic Outline of the Phylum *Firmicutes* (105), http://www.bergeys.org/outlines/bergeys_vol_3_outline_linked.pdf	Text-based *Firmicutes* taxonomy files (in PDF), updated in 2009
16S rRNA databases	
The SILVA database (106), http://www.arb-silva.de/	A constantly updated 16S rRNA-based Tree of Life
Greengenes (107, 108), http://greengenes.lbl.gov/	Includes an improved classification of uncultivated bacteria
Ribosomal Database Project (109), http://rdp.cme.msu.edu/	A constantly updated 16S rRNA-based Tree of Life

[a]The resources are listed in alphabetical order.

numerous events of lateral gene transfer to and from representatives of other phyla, which is why certain gene families are shared by *Firmicutes* with *Fusobacteria*, *Thermotogae*, and other groups (13, 14). Still, the core set of well-conserved informational genes and their protein products show much higher similarity within the members of the phylum than to any organisms from other phyla. The phylogenetic trees built from ribosomal proteins and/or RNA polymerase subunits are typically consistent with the 16S rRNA-based trees and show confident clustering of various *Firmicutes* members (9, 15–18). The unity of *Firmicutes* is also supported by other means of comparative genome analysis, including dinucleotide frequencies, codon usage, presence of simple sequence repeats, and distribution of insertion and deletions in highly conserved proteins (19–21), reviewed in references 11 and 22.

Much of the discussion of the *Firmicutes* evolution focuses on the absence of the outer membrane. Gupta and several other researchers argued that the presence of a single cytoplasmic membrane must be an ancient feature, unifying *Firmicutes* with *Actinobacteria*, *Mollicutes*, *Thermotogae*, and/or *Archaea*—and potentially *Chloroflexi*—into a single group *Monodermata* (11, 19, 23) or *Posibacteria* (24, 25) that was ancestral to Gram-negative bacteria (*Didermata* or *Negibacteria*). While severely criticized by others, see e.g., reference 26 and the discussion in reference 25, these ideas helped in highlighting the big—and still unresolved—questions of the early evolution of *Bacteria*. In any case, the distinct identity of the *Firmicutes* as a separate early diverging phylum is not being disputed by anyone. Even the highly controversial (and generally dismissed by other microbiologists) classification of bacteria proposed by Cavalier-Smith retained the *Firmicutes* as a separate class of *Teichobacteria*, based on the presence of teichoic acid in their cell walls (24).

THE EVOLVING SYSTEMATICS OF *FIRMICUTES*

While the existence of *Firmicutes* as a separate phylum is no longer a matter of contention, the systematics within the phylum is still very much in flux. Just in the past several years, one class (*Mollicutes*) has been removed from the *Firmicutes*, three new taxa have been elevated to the class level, a number of new taxa—at the genus, family, and order level—have been described, and some previously characterized species have been reassigned to new taxa.

As mentioned above, the classification of *Firmicutes*, as well as the assignment of new isolates to various taxa within this phylum, is based primarily on the 16S rRNA similarity patterns, the thickness of bacterial cell walls, and several additional traits. The reliance on the cell wall as a key diagnostic feature recently led to a noteworthy conflict between the taxonomic and phylogenetic approaches. The *Mollicutes* (mycoplasmas), which fall within the *Firmicutes* in most 16S rRNA and ribosomal protein-based trees (9, 15, 18, 27) and share with *Firmicutes* a number of common traits (28–30), had been previously considered a distinct class-level lineage within the *Firmicutes* (31). However, because *Mollicutes* lack the peptidoglycan cell wall, they have been reassigned to a separate phylum, the *Tenericutes* (32).

The removal of *Mollicutes* left the *Firmicutes* as a paraphyletic group with just two classes, *Bacilli* and *Clostridia*, both having a typical Gram-positive cell wall (31, 32). However, another group of *Firmicutes*, the family *Erysipelotrichaceae*, was found to contain an unusual type of peptidoglycan and share with *Mollicutes* a number of traits, which prompted its elevation to the class level (29, 32, 33). The genomes of two representatives, *Erysipelothrix rhusiopathiae* and *Eubacterium cylindroides*, have been sequenced (29), and two dozen other genomes are in the pipeline. Genome analysis is expected to shed light on the cellular physiology of these interesting bacteria, which include several spore-forming species that had been previously misassigned to the *Clostridium* genus (Table 2).

Two more lineages of *Firmicutes* have also been elevated to the class level, forming classes *Thermolithobacteria* and *Negativicutes* (32, 34, 35). The class *Thermolithobacteria* currently includes just two species, both appear to be asporogenous (34), and so far neither of them has been the subject of a genome-sequencing project. The last class, *Negativicutes*, unifies bacteria that stain Gram-negative (i.e., do not retain the Gram stain); in some carefully studied cases, they were seen surrounded by two membranes and had a thin cell wall (36). Nevertheless, based on 16S rRNA-, ribosomal proteins-, RpoB-, GyrB-, and DnaK-based trees, these bacteria are legitimate members of the *Firmicutes* phylum (18, 35, 37). Several representatives of this group have been shown to form endospores (36, 38, 39). The current classification of *Negativicutes* includes a single order *Selenomonadales* with two families, *Acidaminococcaceae* and *Veillonellaceae*, with sporeformers found only in the latter one (Table 2). It has been recently argued that elevation of Gram-negative *Firmicutes* to a separate class was not justified and that they should be left as a separate order within the class *Clostridia*, consistent with the 16S rRNA- and

Table 2   Distribution of spore-forming bacteria among *Firmicutes*

		Spore-forming members		
Class, order[a]	Family[a]	Fraction[b]	Complete genomes[c]	Example (GenBank entry or reference)
Bacilli				
Bacillales	*Alicyclobacillaceae*	+++	2 (3)	*Kyrpidia tusciae* (CP002017)
	Bacillaceae	++	32 (73)	*Bacillus subtilis* (CP000922)
	Listeriaceae	−	− (28)	
	Paenibacillaceae	+++	6 (10)	*Paenibacillus polymyxa* (CP000154)
	Pasteuriaceae	+++	−	*Pasteuria penetrans* (110)
	Planococcaceae	+	1 (1)	*Solibacillus silvestris* (AP012157)
	Sporolactobacillaceae	++	−	*Sporolactobacillus inulinus* (AFVQ00000000)
	Staphylococcaceae	−	− (42)	
	Thermoactinomycetaceae	+++	−	*Desmospora activa* (111)
	Other	+	− (3)	*Tuberibacillus calidus* (112)
Lactobacillales	*Aerococcaceae*	−	− (1)	
	Carnobacteriaceae	−	−	
	Enterococcaceae	−	− (10)	
	Lactobacillaceae	−	− (44)	
	Leuconostocaceae	−	− (10)	
	Streptococcaceae	−	− (99)	
Clostridia				
Clostridiales	*Caldicoprobacteraceae*	+++	−	*Caldicoprobacter oshimai* (113)
	Catabacteriaceae	−	−	
	Christensenellaceae	−	−	
	Clostridiaceae	+++	25 (43)	*Clostridium botulinum* (CP000727)
	Defluviitaleaceae	+++	−	*Defluviitalea saccharophila* (114)
	Eubacteriaceae	−	− (9)	
	Gracilibacteraceae	−	−	
	Heliobacteriaceae	++	1 (1)	*Heliobacterium modesticaldum* (CP000930)
	Lachnospiraceae	++	− (7)	*Anaerostipes butyraticus* (115)
	Oscillospiraceae	−	− (1)	
	Peptococcaceae	++	11 (15)	*Desulfitobacterium hafniense* (CP001336)
	Peptostreptococcaceae	++	1 (11)	*Clostridium difficile* (AM180355)
	Ruminococcaceae	+	− (9)	*Sporobacter termitidis* (116)
	Syntrophomonadaceae	+	− (2)	*Pelospora glutarica* (117)
	Other	+	3 (7)	*Symbiobacterium thermophilum* (AP006840)
Halanaerobiales	*Halanaerobiaceae*	−	−	
	Halobacteroidaceae	−	− (4)	
Natranaerobiales	*Natranaerobiaceae*	−	− (1)	
Thermoanaero-bacterales	*Thermoanaerobacteraceae*	++	11 (13)	*Carboxydothermus hydrogeniformans* (CP000141)
	Thermodesulfobiaceae	−	− (2)	
	Other	+	2 (13)	*Mahella australiensis* (CP002360)
Erysipelotrichia				
Erysipelotrichales	*Erysipelotrichaceae*	+	− (2)	*Clostridium ramosum*
Negativicutes[d]				
Selenomonadales	*Acidaminococcaceae*	−	− (2)	
	Veillonellaceae	+	− (5)	*Pelosinus fermentans* (57, 58)
Thermolithobacteria				
Thermolithobacterales	*Thermolithobacteraceae*	−	−	

[a]Taxonomy is according to the List of Prokaryotic Names with Standing in Nomenclature (42) and the NCBI Taxonomy database (41); see Table 1 for the URLs.
[b]The distribution of sporeformers among the experimentally characterized members of the respective family is indicated as follows: +++, all (or nearly all) characterized members of the family produce spores; ++, a significant fraction of species are sporeformers; +, the family includes some sporeformers; −, no known sporeformers in the family.
[c]The number of spore-forming species with completely sequenced genomes in the respective family (according to the RefSeq database [56] as of November 1, 2012); the total number of completely sequenced genomes is given in parentheses.
[d]See reference 18 for a discussion on whether the order *Selenomonadales* deserves to be placed in the separate class *Negativicutes* as opposed to the class *Clostridia*.

protein-based phylogenetic trees (18). In any case, this interesting group of *Firmicutes* is being intensively studied, and its systematics is likely to change in the future. Here, we just refer to these bacteria as members of the *Selenomonadales*.

Although the families *Bacillaceae* and *Clostridiaceae* were established and described in detail many years ago, orders *Bacillales* and *Clostridiales* were created only in 1953 by André-Romain Prévot, and the classes *Bacilli* and *Clostridia* were codified only recently to accommodate the rapidly growing number of newly described Gram-positive bacteria. Indeed, more than half of the families listed in Table 2 have been described only in the past 5 to 10 years. In other instances, selected genera have been elevated to the family level after molecular analysis showed that they were only distantly related to other genera in the same family.

It should be noted that, although bacterial taxonomy may seem to be in a constant flux, it is generally stable at the (intermediate) levels of family and order; many families of *Bacilli* and *Clostridia* that were described in the early 20th century are still recognized as such. Genera are somewhat less stable, because the description of new species often reveals new groupings and leads to the subdivision of a single genus into two or three new genera. This inevitably leads to the name change, sometimes affecting well-known and often-used organisms, such as *Bacillus* (now *Lysinibacillus*) *sphaericus*, *Bacillus* (now *Geobacillus*) *stearothermo-*

philus, and many others (Table 3). That said, *Clostridium difficile* still retains its name, even though molecular data revealed that it—along with several related *Clostridium* spp.—clearly falls outside the family *Clostridiaceae* and probably belongs in the family *Peptostreptococcaceae* (40). This has led to a recent proposal to rename it *Peptoclostridium difficile* and assign new names to 77 other former *Clostridium* species (18). Fortunately, such resources as the NCBI Taxonomy database (41) and the List of Prokaryotic Names with Standing in Nomenclature (42) keep track of the name changes and allow searches using the old names.

SPORE-FORMING AND ASPOROGENOUS *FIRMICUTES*

The ability to form spores depends on a conserved set of at least 60 genes, mutations of which interrupt the sporulation process at various steps and decrease the fraction of sporeformers by several orders of magnitude or even render the cells completely asporogenous (43–47). Accordingly, the ability to form spores is easily lost, and many lineages contain both spore-forming and non-spore-forming members. Some other lineages do not include any spore-forming members, suggesting that the ability to form spores was either lost very early in the history of that lineage or was absent from its ancestors (see below). As an example, the current classification subdivides the class *Bacilli* into two orders,

Table 3 Recent renaming of some well-known sporeformers

Old name	New name (GenBank genome entry)[a]
Bacilli	
Bacillus acidocaldarius	*Alicyclobacillus acidocaldarius* (CP001727)
Bacillus brevis	*Brevibacillus brevis* (AP008955)
Bacillus globisporus	*Sporosarcina globispora*
Bacillus haloalkaliphilus	*Alkalibacillus haloalkaliphilus* (AKIF00000000)
Bacillus pantothenticus	*Virgibacillus pantothenticus*
Bacillus polymyxa	*Paenibacillus polymyxa* (CP000154)
Bacillus sphaericus	*Lysinibacillus sphaericus* (CP000817)
Bacillus stearothermophilus	*Geobacillus stearothermophilus*
Bacillus tusciae	*Kyrpidia tusciae* (CP002017)
Clostridia	
Clostridium fervidum	*Caloramator fervidus*
Clostridium lentocellum	*Cellulosilyticum lentocellum* (CP002582)
Clostridium thermoaceticum	*Moorella thermoacetica* (CP000232)
Clostridium thermohydrosulfuricum	*Thermoanaerobacter thermohydrosulfuricus*
Clostridium thermosaccharolyticum	*Thermoanaerobacterium thermosaccharolyticum* (CP002171)
Desulfotomaculum orientis	*Desulfosporosinus orientis* (CP003108)
Thermoanaerobacter tengcongensis	*Caldanaerobacter subterraneus* (AE008691)
Thermoanaerobium brockii	*Thermoanaerobacter brockii* (CP002466)

[a]GenBank accession number of the genome sequence, if available.

Bacillales and *Lactobacillales*, which include, respectively, nine and six families (Table 2). There are no (known) spore-forming representatives within *Lactobacillales* and in two families of *Bacillales*, *Listeriaceae*, and *Staphylococcaceae*; the remaining seven families of *Bacillales* contain both spore-forming and apparently asporogenous members. There are four recognized orders in the class *Clostridia*; two of them (*Clostridiales* and *Thermoanaerobacterales*) include sporeformers, whereas the other two (*Halanaerobiales* and *Natranaerobiales*) do not (Table 2).

It must be noted that the information on whether a particular organism forms spores is somewhat biased; the presence of spores in the culture positively identifies the bacterium as a sporeformer, whereas the absence of the visible spores is not sufficient to label the organism as asporogenous. Indeed, the absence of spores in a studied sample could be due to the specific isolation and cultivation conditions; the organism's inability to form spores cannot be ascertained without a specific concerted effort to detect spore formation under a variety of growth conditions. Thus, even when light or electron microscopy and/or a heat resistance test indicate the absence of spores in the culture, there remains a distinct possibility that proper conditions for the organism's sporulation have not yet been found.

As an example, the thermophilic, facultatively chemolithoautotrophic anaerobe *Thermincola ferriacetica* has been observed to forms spores (48). In contrast, its close relative *Thermincola carboxydiphila* has not been seen to do that (49), but there has been no special effort to detect spore formation in its culture (E.A. Bonch-Osmolovskaya, personal communication). The third member of the genus, *Thermincola potens*, had its genome sequenced without microbiological characterization of the organism (50). Thus, there is no easy way to predict whether *T. potens* is a sporeformer, even though its genome appears to encode all essential sporulation proteins (44). Hopefully, the perspective of using *T. potens* in microbial fuel cells (51, 52) would lead to a better microbiological description of this interesting organism.

Spore formation is a particularly interesting trait for the members of the *Selenomonadales*. Since these bacteria stain Gram negative, many newly characterized members of this group have only been checked for sporulation with the use of light microscopy. Nevertheless, the reports on the inability of many members of the *Selenomonadales* to form spores appear to be correct and are supported by the available genomic data. Sporulation has been observed in some representatives of *Veillonellaceae*, such as *Sporomusa sphaeroides*

and *Acetonema longum* (36, 38, 39), but seems to be restricted to just a handful of genera (Table 2) that belong to a separate branch of the phylogenetic tree (18, 53). Further, even within that branch there are reported non-spore-forming species, such as, for example, *Sporomusa paucivorans* (54). In such cases, a recent loss of the ability to sporulate seems very likely.

Summing up, a number of taxa within the phylum *Firmicutes* contain both spore-forming and asporogenous members. The apparently independent loss of the ability to form endospores in distinct lineages of this phylum suggests that, despite providing a clear evolutionary advantage when it comes to surviving environmental challenges, sporulation comes with its own costs. Accordingly, adaptation of many members of the phylum (lactic acid bacteria, staphylococci, and others) to their specific (e.g., nutrient-rich) ecological niches apparently included a loss of their ability to form spores.

COVERAGE OF THE *FIRMICUTES* DIVERSITY BY GENOME-SEQUENCING PROJECTS

Phylogenetic Diversity

The medical, environmental, and industrial importance of many Gram-positive bacteria fueled a sustained effort in genome sequencing of numerous members of the *Firmicutes* phylum. By the end of October 2012, almost five hundred complete genomes of various *Firmicutes* had been available in the public databases (see http://www.ebi.ac.uk/genomes/bacteria.html or http://www.ncbi.nlm.nih.gov/genome/browse/) there were also several hundred partially sequenced genomes and over 3,000 genome sequencing projects (see http://www.genomesonline.org/) (55, 56). The majority of the sequenced genomes came from well-characterized genera, such as *Bacillus*, *Clostridium*, *Staphylococcus*, and *Streptococcus*, with multiple complete genomes of various strains of *Bacillus subtilis* and the human pathogens *Bacillus anthracis*, *Clostridium botulinum*, *Clostridium difficile*, *Listeria monocytogenes*, *Staphylococcus aureus*, *Streptococcus pneumoniae*, and *Streptococcus pyogenes* and the insect pathogen *Bacillus thuringiensis*. However, owing largely to the efforts of the Genomic Encyclopedia of Bacteria and Archaea (GEBA) project (45), there is now a fairly detailed genomic coverage of the *Firmicutes* diversity, with complete genomes available for representatives of all (currently recognized) orders—and most families—of

Bacilli and *Clostridia*, as well as several representatives of *Negativicutes* and *Erysipelotrichia* (see http://www.ncbi.nlm.nih.gov/genomes/MICROBES/microbial_taxtree.html).

So far, complete genomes of sporeformers have come exclusively from the representatives of *Bacilli* and *Clostridia*. As mentioned above, most members of *Erysipelotrichia* are asporogenous. However, *Clostridium ramosum*, *Clostridium innocuum*, and *Clostridium spiroforme*, recently reassigned to this class, are sporeformers, whose draft genome sequences are already available. Among *Selenomonadales*, complete genomes are currently available for several non-spore-forming representatives, such as *Acidaminococcus intestini*, *Acidaminococcus fermentans*, *Megasphaera elsdenii*, *Selenomonas ruminantium*, and *Veillonella parvula*, and just one sporeformer, *Sporomusa ovata* (133). However, draft genomes of six metal-reducing strains of the sporeformer *Pelosinus fermentans* have been sequenced and assembled into 65, 76, 98, 134, 844, and 887 contigs, respectively (57, 58). In addition, a genome-sequencing project of another sporeformer, *Acetonema longum*, has been brought to the level of 296 contigs. There is also a draft genome of the non-spore-forming *Thermosinus carboxydivorans* (59). Remarkably, each of these unfinished genomes encodes orthologs of more than 60 key sporulation proteins of *B. subtilis*, confirming the overall unity of the sporulation machinery in all *Firmicutes*. Our recent study found that the set of sporulation proteins encoded in the unfinished genomes of *A. longum*, *P. fermentans*, and *T. carboxydivorans* was essentially the same as that encoded in most clostridial genomes, lending credence to the suggestion that *Selenomonadales* are just very unusual members of the *Clostridia* (18).

Obviously, for comparative purposes, it would be advantageous to have at least two representative genomes from each major lineage (e.g., at the genus level), but the existing genome coverage has already allowed some meaningful comparisons, see e.g., references 17, 44, 46, 60–62.

Ecological Diversity

The breadth of the genomic coverage of the *Firmicutes* is also reflected in the variety of ecological niches inhabited by already sampled organisms. The ability of spores to survive environmental challenges, such as heat, desiccation, presence of organic solvents and oxidizing agents, and UV irradiation, as well as predation by protozoa (63–65), helped spore-forming *Firmicutes* colonize a wide variety of diverse habitats. Sporeformers inhabit most aquatic and terrestrial habitats,

both aerobic and anaerobic, and have been found in a variety of environments, including deep in the ocean (e.g., *Oceanobacillus iheyensis* and *Geobacillus kaustophilus* [66–68]) and in the Earth's crust ("*Candidatus*[1] Desulforudis audaxviator" [69]).

In the past several years, complete genome sequences have become available for a number of Gram-positive extremophiles (Table 4), including acidophiles, alkaliphiles, thermophiles, psychrophiles, and halophiles. Again, the existing genome coverage, while far from exhaustive, has already allowed some interesting comparative analyses (67, 70). Still, genome sequences remain to be determined from many extremophilic sporeformers, such as, for example, *Psychrobacillus* spp. that can grow even at −2°C (71, 72).

It is important to note that the ability to grow at extreme conditions is not exclusive for sporeformers; there are many extremophiles in other phyla, as well as non-spore-forming extremophiles among *Firmicutes*. For example, no spores were observed in the culture of the obligately halophilic and alkaliphilic (growth at >3 M NaCl and pH >8.3) thermophilic bacterium *Natranaerobius thermophilus*, a member of *Clostridia* isolated from a soda lake in Egypt (73). Although its genome encodes a nearly complete set of sporulation genes, it lacks both *dpaB* and *etfA* genes that code for two subunits of clostridial dihydrodipicolinate reductase, an essential sporulation enzyme (44, 74, 75). The extreme thermophiles *Ammonifex degensii* and *Caldicellulosiruptor* spp. are also nonsporeformers (76–80).

Model Organisms

Aside from the popular model organism *B. subtilis*, spore-forming *Firmicutes* are probably most famous for the diseases that they cause, which include anthrax, food poisoning, infectious diarrhea, enterocolitis, gas gangrene, tetanus, and various kinds of bacteremia. However, their importance is not limited to pathogenicity. Such organisms as *B. thuringiensis* are being actively used for pest control, and there are now more than two dozen completely sequenced genomes of various strains of *B. thuringiensis* that differ in their insecticidal activity. Clostridia are being studied for their potential use in production of biofuel (*C. acetobutylicum*) and/or wood processing (*C. cellulolyticum*, *C. cellulovorans*, *C. clariflavum*, *C. thermocellum*), and

---

[1]The *Candidatus* name is used for incompletely described organisms, including those that have not been cultivated (and deposited in internationally recognized culture collections).

Table 4 Genome sequencing of extremophilic sporeformers

Organism name	Growth conditions	GenBank accession number, reference
Acidophiles		
Alicyclobacillus acidocaldarius	pH_{opt} 3.5(pH 2–6)	CP001727 (118, 119)
Sulfobacillus acidophilus	pH_{opt} 1.8(pH 1.6–2.3)	CP002901 (83, 120); CP003179 (121, 122)
Alkaliphiles		
Alkaliphilus metalliredigens	pH_{opt} 9.5(pH 7.5–11)	CP000724 (123)
Bacillus halodurans	pH_{opt} 9.5(pH 7.0–11)	BA000004 (70, 124)
Bacillus pseudofirmus	pH_{opt} 8.5–10.6(pH 7.5–11.4)	CP001878 (125, 126)
Halophiles		
Oceanobacillus iheyensis	3.6 M NaCl(0%–21%)	BA000028 (66, 67)
Virgibacillus halodenitrificans	4.2 M NaCl(2%–25%)	ALEF00000000 (127, 128)
Thermophiles		
Alicyclobacillus acidocaldarius	T_{opt} 65°C(45–70°C)	CP001727 (118, 119)
Thermoanaerobacter mathranii	T_{opt} 70–75°C(50–75°C)	CP002032 (129)
Psychrophiles		
Bacillus weihenstephanensis	T = 4°C(4–35°C)	CP000903 (130)
UV-resistant strains		
Bacillus pumilus SAFR-032	UV_{254} <1 kJ/m²	CP000813 (131, 132)

genome sequencing is increasingly being used to analyze the encoded hydrolases and their ability to degrade cellulose, lignin, and other components of plant cell walls (81).

In physiological terms, spore-forming *Firmicutes* include both autotrophs and heterotrophs, many of which have been used as model organisms for biochemical and biophysical studies and have completely sequenced genomes. Chemolithoautotrophs include a variety of hydrogen- or formate-oxidizing bacteria that grow by reducing sulfur, sulfate, or nitrate (69, 82). Other strains grow by oxidizing minerals, including ferrous iron (83). A number of sporeformers are capable of utilizing carbon monoxide. As its name suggests, *Carboxydothermus hydrogenoformans* produces molecular hydrogen (84), whereas *Clostridium ljungdahlii* can use CO/H_2 and CO_2/H_2 mixtures (85). The family *Heliobacteriaceae* includes phototrophic members that use anaerobic anoxygenic photosynthesis as a source of energy; they are also able to fix nitrogen (86–88). The photosynthetic reaction centers of *Heliobacillus mobilis* and *Heliobacterium modesticaldum* represent some of the most primitive photosynthetic systems and are being studied to understand the mechanisms and evolution of photosynthesis. *Bacillus methanolicus* is a sporeformer that can utilize methanol as its sole carbon and energy source; unfinished genome sequences of two of its strains (7 and 12 contigs, respectively) were released in 2012 (89). Further studies are bound to find new unexpected applications of sporeformers, such as the above-mentioned use of *T. potens* in microbial fuel cells (51).

GENOMICS OF SPORULATION

Diagnostic Sporulation Genes

Comparative analyses of diverse *Firmicutes* genomes identified a core set of sporulation genes that are conserved in (nearly) all sporeformers (43, 44, 46, 47, 90) and could be considered a "sporulation genomic signature" (91). Incidentally, most of these genes are also essential for sporulation: the respective mutations affect sporulation in *B. subtilis*, *C. acetobutylicum*, and/or other model organisms, resulting in the decrease of spore count by two orders of magnitude or more. Unfortunately, the attempts to identify tell-tale sporulation genes proved unsuccessful; there was not a single gene that would be present in all sporeformers and absent in all asporogens (44).

One of the best indicators of the spore-forming ability is Spo0A, the master regulator of sporulation. Spo0A is a transcriptional regulator that combines the two-component receiver domain with a specific type of the helix-turn-helix DNA-binding domain, which so far has been seen only among the members of *Firmicutes* (92). However, Spo0A is also encoded in the genomes of several unequivocally non-spore-forming organisms, such as *Caldicellulosiruptor* spp. (77–79), *Exiguobacterium* spp. (93), *Macrococcus caseolyticus* (94), and many others (44) (see Table 5). Still, Spo0A can be used as a molecular marker for evaluating the abundance of sporeformers in natural environments (134).

In addition to *spo0A*, other apparently sporulation-specific genes are occasionally found in the genomes of nonsporeformers (Table 5) and can be seen even outside

Table 5 Distribution of sporulation genes in some non-spore-forming bacteria[a]

Taxonomy: class, order	spo0A	spmA	spo0M	spoIIM	spoIIIAA	spoIVA	spoIVFA	spoVAC	spoVG	spoVR	spoVS	sspF
Class *Bacilli*												
Order *Bacillales*												
Bacillus selenitireducens	+	–	–	–	–	–	–	–	+	–	+	–
Macrococcus caseolyticus	+	–	(–)	–	–	–	–	–	+	–	+	–
Exiguobacterium sibiricum	+	–	+	v	–	–	–	–	+	–	+	–
Class *Clostridia*												
Order *Clostridiales*												
Clostridiales genomosp. BVAB3	+	–	–	–	–	–	–	–	+	–	–	–
Eubacterium rectale	+	–	–	–	+	+	+	+	+	–	–	+
Eubacterium eligens	+	–	–	+	+	+	+	+	+	–	–	+
Roseburia hominis	+	–	–	–	–	–	–	+	+	–	–	+
Oscillibacter valericigenes	+	+	–	–	+	+	+	+	+	–	+	+
Ethanoligenens harbinense	+	+	–	–	+	+	+	+	+	+	+	+
Ruminococcus albus 7	+	–	–	–	+	+	+	+	+	–	+	+
Syntrophomonas wolfei	+	+	–	+	+	+	+	+	–	–	+	+
Thermaerobacter marianensis	+	+	–	+	+	+	+	+	+	+	+	+
Order *Halanaerobiales*												
Acetohalobium arabaticum	+	+	–	+	+	+	+	+	–	+	+	+
Halanaerobium praevalens	+	–	–	–	–	–	–	–	+	+	+	–
Halothermothrix orenii	+	+	–	+	+	+	+	+	+	–	+	+
Order *Thermoanaerobacterales*												
Ammonifex degensii KC4	+	+	–	+	+	+	+	+	–	+	+	+
Caldicellulosiruptor saccharolyticus	+	+	–	+	+	+	+	+	+	–	+	+
Other phyla												
Actinobacteria	–	–	+	+	+	–	–	–	–	–	+	–
Cyanobacteria	–	+	+	+	+	–	–	–	–	+	–	–
Proteobacteria	–	+	+	+	+	–	–	+	–	+	–	–
Chloroflexi	–	–	+	+	+	–	–	–	–	+	+	–
Spirochetes	–	–	–	–	–	–	–	–	+	–	–	–
Thermotogae	–	–	–	–	–	–	–	–	–	–	+	–
Euryarchaeota	–	–	+	+	–	–	–	–	+	+	–	–

Sporulation genes

[a]Based on the data from references 44 and 62.

of the phylum *Firmicutes* (43, 44, 62). On the other hand, owing to the phenomenon of nonorthologous gene displacement (i.e., the ability of unrelated or distantly related proteins to perform the same function), some essential sporulation genes can be missing in certain genomes. A good example is the ability of the electron transfer flavoprotein EtfA to catalyze the oxidation of dihydrodipicolinate into dipicolinate, a universal component of the developing spore (75). This activity underlies replacement of the *dpaA* and *dpaB* genes by *etfA* in such spore-forming clostridia as *C. acetobutylicum*, *C. botulinum*, and *C. perfringens* (44, 75). A nonorthologous gene displacement of *spoIIQ* or *spoIVFA* has been proposed to explain the absence of these genes in clostridia (44).

What Is the Minimal Genome Size of a Sporeformer?

A recent genome comparison of spore-forming and asporogenous *Firmicutes* revealed a certain degree of correlation between the ability of the bacterium to form spores and its genome size. Most *Firmicutes* whose completely sequenced genomes have sizes of more than 3 million base pairs (Mbp) were found to be sporeformers (44). The few exceptions among *Clostridia* included *Oscillibacter valericigenes* (4.7 Mbp), *Eubacterium limosum*, and *Eubacterium rectale* (4.5 and 3.6 Mbp, respectively), *Butyrivibrio proteoclasticus* (3.8 Mbp), and certain representatives of *Enterococcus*, *Lactobacillus*, *Roseburia*, and *Ruminococcus* genera. Among *Bacilli*, the only exceptions were *Bacillus selenitireducens* (3.6 Mbp), *Exiguobacterium* spp. (3.0 Mbp), and *Listeria* spp. (3.0 Mbp). This correlation also showed up in the observation that all cultured *Firmicutes* with genome sizes of less than 2.3 Mbp were asporogenous; these included the majority of lactobacilli and streptococci (44). It was concluded that, at least among free-living *Firmicutes*, sporulation was a property of relatively gene-rich bacteria.

Among the free-living representatives of classes *Bacilli* and *Clostridia*, the lower boundary of genome size for sporeformers was estimated at 2.8 Mbp in *Anoxybacillus flavithermus* (17) and 2.3 Mbp in *Thermoanaerobacter mathranii*, respectively (44).

However, since the ability to form spores appears to depend on a just a few dozen genes (43, 44, 46), it is quite likely that there exist uncharacterized free-living spore-forming *Firmicutes* with much smaller genome sizes. Such organisms are likely to be found in relatively stable environments, such as the open sea or the Earth's crust, that have not yet been sufficiently sampled. Be-

sides, organisms with relatively small genomes are likely to have limited metabolic capabilities and therefore can be expected to be fastidious and harder to cultivate.

Indeed, there already is an example of sporeformers whose genome sizes are much smaller than 2.3 Mbp. These organisms are referred to as unculturable segmented filamentous bacteria and are closely related to the genus *Clostridium*, forming a proposed genus "*Candidatus* Arthromitus" in the *Clostridiaceae* family (95, 96). Over the years, these bacteria have been seen attached to the intestinal walls of many animals, including mice, rats, cats, dogs, and chickens; a dedicated study using light microscopy expanded the range of hosts of "*Ca.* Arthromitus" to include human, monkeys, domestic fowl, toad, and carp (97, 98). Although "*Ca.* Arthromitus" spp. have not yet been cultivated outside of the mammalian hosts, they have been shown to form spores (99), and the ability of the spores to survive treatment with 3% chloroform has been used to obtain a (nearly) pure culture of these bacteria, suitable for genome sequencing (96).

The 1.5- to 1.6-Mbp genome sequences of three strains of "*Ca.* Arthromitus" spp. proved to be much smaller than those of any free-living sporeformers, owing largely to the apparent loss of genes responsible for amino acid biosynthesis, carbohydrate and nucleotide metabolism, and energy conservation (96, 100, 101). At the same time, "*Ca.* Arthromitus" spp. encoded a relatively large number of sporulation proteins, most likely comprising a nearly minimal core set of essential sporulation genes (44), see references 96 and 101. We should not exclude the possibility of eventually finding sporeformers with even smaller genimes—and probably even more dependent upon the host for the supply of essential nutrients.

Evolution of Sporulation

The presence of a conserved set of sporulation genes that confers the ability to form endospores in two different, early diverging branches of *Firmicutes*, *Bacilli*, and *Clostridia*, not to mention the *Selenomonadales* (*Negativicutes*), strongly suggests that it was already present in their common ancestor. An alternative explanation, horizontal transfer of numerous (at least sixty, probably many more) sporulation genes from one branch to another after their separation, sounds extremely unlikely. However, the assumption that sporulation ability is an ancestral feature would indicate that the various asporogenous lineages within the *Firmicutes* phylum (see Table 2) lost (most of) their sporulation genes relatively late in their evolution, after the separation from the closely related spore-forming lineages. For saprophytic and fastidious bacteria (listeria, staphylococci, streptococci, etc.), the loss of sporulation genes could have been part of a systemic genome compaction in the course of their adaptation to the relatively nutrient-rich ecological niches. Indeed, representatives of these lineages not only have significantly smaller genomes than their spore-forming relatives, but they also encode fewer biosynthetic pathways (e.g., references 61 and 102). This genome compaction, accompanied by the loss of metabolic genes, is particularly evident in the case of *Erysipelothrix rhusiopathiae*, which lacks the genes coding for the enzymes of the tricarboxylic acid cycle, fatty acid biosynthesis, synthesis of biotin, riboflavin, pantothenate, thiamine, and folate and a number of amino acids (29). Given this scale of genome compaction, the complete loss of sporulation genes in *Erysipelothrix* is hardly surprising.

Several years ago, a study of an *Anoxybacillus flavithermus* strain isolated from a super-saturated silica solution revealed a genome sequence that was one-third shorter than that of *B. subtilis* and, accordingly, encoded 33% fewer proteins (17). This finding prompted an examination of gene conservation among the 20 bacillar, 5 clostridial, and 6 mollicute genomes sequenced by that time, which led to a somewhat paradoxical conclusion that the common ancestor of all *Firmicutes* might have encoded just 1,318 protein families (17). Taking into account paralogy and "orphan" reading frames, that number would still correspond to fewer than 2,000 genes. This relatively small genome was then dramatically expanded during the evolution of *Bacillaceae* lineage and somewhat contracted in the *Anoxybacillus*/*Geobacillus* branch (17). Such reconstructions are necessarily tentative and depend strongly on the available set of complete genomes. Still, they suggest that the genome of the common ancestor of *Firmicutes* could have been reasonably close to the smallest genomes of modern free-living spore formers, such as *Thermoanaerobacter mathranii* and *A. flavithermus*.

Two new classes, *Tissierellia* and *Limnochordia*, have been proposed, the first one to unify a number of organisms with previously obscure status (135) and the second one to accommodate a single cultured isolate (and some uncultured ones) with very unusual properties (136, 137). Both new classes include spore-forming representatives, further supporting the idea that the common ancestor of all firmicutes was able to form spores. New genomic sequences continue to pour in, due to the efforts to characterize the microbial dark matter (138) and also to the human microbiome project, which revealed a surprising abundance of the spore-forming firmicutes in the human gut (139).

CONCLUDING REMARKS

Spore-forming members of the phylum *Firmicutes* are extremely diverse in their biochemical, physiological, and ecological properties and range from obligate parasites to free-living phototrophs and chemolithotrophs. Many sporeformers attract wide interest, either as model organisms, or because of the diseases they cause, or because of their potential use in biotechnology, bioremediation, or insect control. Despite their diversity, all sporeformers share a common heritage and encode very similar sets of sporulation genes that they likely inherited from a common ancestor of all *Firmicutes*. The ability to form spores has been lost numerous times in a variety of lineages of the *Firmicutes*, usually in the course of adaption to life in nutrient-rich conditions. This indicates that the ability to survive environmental challenges by forming endospores comes with certain strings attached, which could be a reason why it is not found anywhere outside the *Firmicutes* phylum.

While this chapter was in preparation, the worldwide genome-sequencing efforts brought us complete and draft genomes of many new members of the *Firmicutes*, including some spore-forming species (55, 56, 133). These sequence data keep feeding the new science of phylogenomics, which uses comparative genomics data to reconstruct bacterial evolution and aims at a better understanding of the entire phenomenon of life.

Acknowledgments. I thank Boris Belitsky and Eugene Koonin for helpful comments. This work was supported by the NIH Intramural Research Program at the National Library of Medicine.

Citation. Galperin MY. 2013. Genome diversity of spore-forming *Firmicutes*. Microbiol Spectrum 1(2):TBS-0015-2012.

References

1. **Drews G.** 1999. Ferdinand Cohn, a founder of modern microbiology. *ASM News* **65:**547.

2. **Olsen GJ, Woese CR, Overbeek R.** 1994. The winds of (evolutionary) change: breathing new life into microbiology. *J Bacteriol* **176:**1–6.

3. **Woese CR, Stackebrandt E, Macke TJ, Fox GE.** 1985. A phylogenetic definition of the major eubacterial taxa. *Syst Appl Microbiol* **6:**143–151.

4. **Woese CR.** 1987. Bacterial evolution. *Microbiol Rev* **51:**221–271.

5. **Woese CR, Kandler O, Wheelis ML.** 1990. Towards a natural system of organisms: proposal for the domains Archaea, Bacteria, and Eucarya. *Proc Natl Acad Sci USA* **87:**4576–4579.

6. **Ludwig W, Schleifer KH.** 1994. Bacterial phylogeny based on 16S and 23S rRNA sequence analysis. *FEMS Microbiol Rev* **15:**155–173.

7. Ludwig W, Strunk O, Klugbauer S, Klugbauer N, Weizenegger M, Neumaier J, Bachleitner M, Schleifer KH. 1998. Bacterial phylogeny based on comparative sequence analysis. *Electrophoresis* 19:554–568.

8. Wolf YI, Rogozin IB, Grishin NV, Tatusov RL, Koonin EV. 2001. Genome trees constructed using five different approaches suggest new major bacterial clades. *BMC Evol Biol* 1:8. doi:10.1186/1471-2148-1-8.

9. Yutin N, Puigbo P, Koonin EV, Wolf YI. 2012. Phylogenomics of prokaryotic ribosomal proteins. *PLoS One* 7:e36972 doi:10.1371/journal.pone.0036972.

10. Wagner M, Horn M. 2006. The Planctomycetes, Verrucomicrobia, Chlamydiae and sister phyla comprise a superphylum with biotechnological and medical relevance. *Curr Opin Biotechnol* 17:241–249.

11. Gupta RS. 2000. The natural evolutionary relationships among prokaryotes. *Crit Rev Microbiol* 26:111–131.

12. Lake JA, Skophammer RG, Herbold CW, Servin JA. 2009. Genome beginnings: rooting the tree of life. *Philos Trans R Soc Lond B Biol Sci* 364:2177–2185.

13. Mira A, Pushker R, Legault BA, Moreira D, Rodriguez-Valera F. 2004. Evolutionary relationships of *Fusobacterium nucleatum* based on phylogenetic analysis and comparative genomics. *BMC Evol Biol* 4:50. doi:10.1186/1471-2148-4-50.

14. Zhaxybayeva O, Swithers KS, Lapierre P, Fournier GP, Bickhart DM, DeBoy RT, Nelson KE, Nesbo CL, Doolittle WF, Gogarten JP, Noll KM. 2009. On the chimeric nature, thermophilic origin, and phylogenetic placement of the *Thermotogales*. *Proc Natl Acad Sci USA* 106:5865–5870.

15. Ciccarelli FD, Doerks T, von Mering C, Creevey CJ, Snel B, Bork P. 2006. Toward automatic reconstruction of a highly resolved tree of life. *Science* 311:1283–1287.

16. Beiko RG. 2011. Telling the whole story in a 10,000-genome world. *Biol Direct* 6:34. doi:10.1186/1745-6150-6-34.

17. Saw JH, Mountain BW, Feng L, Omelchenko MV, Hou S, Saito JA, Stott MB, Li D, Zhao G, Wu J, Galperin MY, Koonin EV, Makarova KS, Wolf YI, Rigden DJ, Dunfield PF, Wang L, Alam M. 2008. Encapsulated in silica: genome, proteome and physiology of the thermophilic bacterium *Anoxybacillus flavithermus* WK1. *Genome Biol* 9:R161.

18. Yutin N, Galperin MY. 9 July 2013. An genomic update on clostridial phylogeny: Gram-negative spore formers and other misplaced clostridia. *Environ Microbiol* 15:2631–2641.

19. Gupta RS. 1998. Protein phylogenies and signature sequences: a reappraisal of evolutionary relationships among archaebacteria, eubacteria, and eukaryotes. *Microbiol Mol Biol Rev* 62:1435–1491.

20. Karlin S, Mrazek J, Campbell AM. 1997. Compositional biases of bacterial genomes and evolutionary implications. *J Bacteriol* 179:3899–3913.

21. Mrazek J, Guo X, Shah A. 2007. Simple sequence repeats in prokaryotic genomes. *Proc Natl Acad Sci USA* 104:8472–8477.

22. Karlin S, Campbell AM, Mrazek J. 1998. Comparative DNA analysis across diverse genomes. *Annu Rev Genet* 32:185–225.

23. Sutcliffe IC. 2010. A phylum level perspective on bacterial cell envelope architecture. *Trends Microbiol* 18:464–470.

24. Cavalier-Smith T. 2002. The neomuran origin of archaebacteria, the negibacterial root of the universal tree and bacterial megaclassification. *Int J Syst Evol Microbiol* 52:7–76.

25. Cavalier-Smith T. 2006. Rooting the tree of life by transition analyses. *Biol Direct* 1:19. doi:10.1186/1745-6150-1-19.

26. Woese CR. 1998. Default taxonomy: Ernst Mayr's view of the microbial world. *Proc Natl Acad Sci USA* 95:11043–11046.

27. Falah M, Gupta RS. 1997. Phylogenetic analysis of mycoplasmas based on Hsp70 sequences: cloning of the *dnaK* (*hsp70*) gene region of *Mycoplasma capricolum*. *Int J Syst Bacteriol* 47:38–45.

28. Wolf M, Muller T, Dandekar T, Pollack JD. 2004. Phylogeny of *Firmicutes* with special reference to *Mycoplasma* (*Mollicutes*) as inferred from phosphoglycerate kinase amino acid sequence data. *Int J Syst Evol Microbiol* 54:871–875.

29. Ogawa Y, Ooka T, Shi F, Ogura Y, Nakayama K, Hayashi T, Shimoji Y. 2011. The genome of *Erysipelothrix rhusiopathiae*, the causative agent of swine erysipelas, reveals new insights into the evolution of firmicutes and the organism's intracellular adaptations. *J Bacteriol* 193:2959–2971.

30. Zhao Y, Davis RE, Lee IM. 2005. Phylogenetic positions of 'Candidatus Phytoplasma asteris' and *Spiroplasma kunkelii* as inferred from multiple sets of concatenated core housekeeping proteins. *Int J Syst Evol Microbiol* 55:2131–2141.

31. Garrity GM, Lilburn TG, Cole JR, Harrison SH, Euzeby J, Tindall BJ. 2007. The *Bacteria*: phylum *Firmicutes*: class *Mollicutes*, p 317–332. *In* Garrity GM (ed), *The Taxonomic Outline of Bacteria and Archaea, release 7.7*, http://www.taxonomicoutline.org/.

32. Ludwig W, Schleifer K-H, Whitman WB. 2009. Revised road map to the phylum *Firmicutes*, p 1–8. *In* De Vos P, Garrity GM, Jones D, Krieg NR, Ludwig W, Rainey FA, Schleifer K-H, Whitman WB (ed), Bergey's Manual of Systematic Bacteriology, 2nd ed, **vol 3**. *The Firmicutes*. Springer, New York, NY.

33. Verbarg S, Rheims H, Emus S, Fruhling A, Kroppenstedt RM, Stackebrandt E, Schumann P. 2004. *Erysipelothrix inopinata* sp. nov., isolated in the course of sterile filtration of vegetable peptone broth, and description of *Erysipelotrichaceae* fam. nov. *Int J Syst Evol Microbiol* 54:221–225.

34. Sokolova T, Hanel J, Onyenwoke RU, Reysenbach AL, Banta A, Geyer R, Gonzalez JM, Whitman WB, Wiegel J. 2007. Novel chemolithotrophic, thermophilic, anaerobic bacteria *Thermolithobacter ferrireducens* gen. nov., sp. nov. and *Thermolithobacter carboxydivorans* sp. nov. *Extremophiles* 11:145–157.

35. Marchandin H, Teyssier C, Campos J, Jean-Pierre H, Roger F, Gay B, Carlier JP, Jumas-Bilak E. 2010. *Negativicoccus succinicivorans* gen. nov., sp. nov., isolated from human clinical samples, emended description of the family *Veillonellaceae* and description of *Negativicutes* classis nov., *Selenomonadales* ord. nov. and *Acidaminococcaceae* fam. nov. in the bacterial phylum *Firmicutes*. *Int J Syst Evol Microbiol* 60:1271–1279.

36. Tocheva EI, Matson EG, Morris DM, Moussavi F, Leadbetter JR, Jensen GJ. 2011. Peptidoglycan remodeling and conversion of an inner membrane into an outer membrane during sporulation. *Cell* 146:799–812.

37. Izquierdo JA, Goodwin L, Davenport KW, Teshima H, Bruce D, Detter C, Tapia R, Han S, Land M, Hauser L, Jeffries CD, Han J, Pitluck S, Nolan M, Chen A, Huntemann M, Mavromatis K, Mikhailova N, Liolios K, Woyke T, Lynd LR. 2012. Complete genome sequence of *Clostridium clariflavum* DSM 19732. *Stand Genomic Sci* 6:104–115.

38. Möller B, Ossmer R, Howard BH, Gottschalk G, Hippe H. 1984. *Sporomusa*, a new genus of gram-negative anaerobic bacteria including *Sporomusa sphaeroides* spec. nov. and *Sporomusa ovata* spec. nov. *Arch Microbiol* 139:388–396.

39. Kane MD, Breznak JA. 1991. *Acetonema longum* gen. nov. sp. nov., an H$_2$/CO$_2$ acetogenic bacterium from the termite, *Pterotermes occidentis*. *Arch Microbiol* 156:91–98.

40. Collins MD, Lawson PA, Willems A, Cordoba JJ, Fernandez-Garayzabal J, Garcia P, Cai J, Hippe H, Farrow JA. 1994. The phylogeny of the genus *Clostridium*: proposal of five new genera and eleven new species combinations. *Int J Syst Bacteriol* 44:812–826.

41. Federhen S. 2012. The NCBI Taxonomy database. *Nucleic Acids Res* 40:D136–D143.

42. Munoz R, Yarza P, Ludwig W, Euzéby J, Amann R, Schleifer KH, Glockner FO, Rossello-Mora R. 2011. Release LTPs104 of the All-Species Living Tree. *Syst Appl Microbiol* 34:169–170.

43. Onyenwoke RU, Brill JA, Farahi K, Wiegel J. 2004. Sporulation genes in members of the low G+C Gram-type-positive phylogenetic branch (*Firmicutes*). *Arch Microbiol* 182:182–192.

44. Galperin MY, Mekhedov SL, Puigbo P, Smirnov S, Wolf YI, Rigden DJ. 2012. Genomic determinants of sporulation in *Bacilli* and *Clostridia*: towards the minimal set of sporulation-specific genes. *Environ Microbiol* 14:2870–2890.

45. Wu D, Hugenholtz P, Mavromatis K, Pukall R, Dalin E, Ivanova NN, Kunin V, Goodwin L, Wu M, Tindall BJ, Hooper SD, Pati A, Lykidis A, Spring S, Anderson IJ, D'Haeseleer P, Zemla A, Singer M, Lapidus A, Nolan M, Copeland A, Han C, Chen F, Cheng JF, Lucas S, Kerfeld C, Lang E, Gronow S, Chain P, Bruce D, Rubin EM, Kyrpides NC, Klenk HP, Eisen JA. 2009. A phylogeny-driven genomic encyclopaedia of Bacteria and Archaea. *Nature* 462:1056–1060.

46. de Hoon MJ, Eichenberger P, Vitkup D. 2010. Hierarchical evolution of the bacterial sporulation network. *Curr Biol* 20:R735–R745.

47. Stragier P. 2002. A gene odyssey: Exploring the genomes of endospore-forming bacteria, p 519–526. *In* Sonenshein AL, Hoch JA, Losick R (ed), *Bacillus subtilis and Its Closest Relatives: From Genes to Cells*. ASM Press, Washington, DC.

48. Zavarzina DG, Sokolova TG, Tourova TP, Chernyh NA, Kostrikina NA, Bonch-Osmolovskaya EA. 2007. *Thermincola ferriacetica* sp. nov., a new anaerobic, thermophilic, facultatively chemolithoautotrophic bacterium capable of dissimilatory Fe(III) reduction. *Extremophiles* 11:1–7.

49. Sokolova TG, Kostrikina NA, Chernyh NA, Kolganova TV, Tourova TP, Bonch-Osmolovskaya EA. 2005. *Thermincola carboxydiphila* gen. nov., sp. nov., a novel anaerobic, carboxydotrophic, hydrogenogenic bacterium from a hot spring of the Lake Baikal area. *Int J Syst Evol Microbiol* 55:2069–2073.

50. Byrne-Bailey KG, Wrighton KC, Melnyk RA, Agbo P, Hazen TC, Coates JD. 2010. Complete genome sequence of the electricity-producing 'Thermincola potens' strain JR. *J Bacteriol* 192:4078–4079.

51. Wrighton KC, Thrash JC, Melnyk RA, Bigi JP, Byrne-Bailey KG, Remis JP, Schichnes D, Auer M, Chang CJ, Coates JD. 2011. Evidence for direct electron transfer by a gram-positive bacterium isolated from a microbial fuel cell. *Appl Environ Microbiol* 77:7633–7639.

52. Carlson HK, Iavarone AT, Gorur A, Yeo BS, Tran R, Melnyk RA, Mathies RA, Auer M, Coates JD. 2012. Surface multiheme c-type cytochromes from *Thermincola potens* and implications for respiratory metal reduction by Gram-positive bacteria. *Proc Natl Acad Sci USA* 109:1702–1707.

53. Strompl C, Tindall BJ, Jarvis GN, Lunsdorf H, Moore ER, Hippe H. 1999. A re-evaluation of the taxonomy of the genus *Anaerovibrio*, with the reclassification of *Anaerovibrio glycerini* as *Anaerosinus glycerini* gen. nov., comb. nov., and *Anaerovibrio burkinabensis* as *Anaeroarcus burkinensis* [corrig.] gen. nov., comb. nov. *Int J Syst Bacteriol* 49(Pt 4):1861–1872.

54. Hermann M, Popoff M-R, Senbald M. 1987. *Sporomusa paucivorans* sp. nov., a methylotrophic bacterium that forms acetic acid from hydrogen and carbon dioxide. *Int J Syst Evol Microbiol* 37:93–101.

55. Pagani I, Liolios K, Jansson J, Chen IM, Smirnova T, Nosrat B, Markowitz VM, Kyrpides NC. 2012. The Genomes OnLine Database (GOLD) v.4: status of genomic and metagenomic projects and their associated metadata. *Nucleic Acids Res* 40:D571–D579.

56. Pruitt KD, Tatusova T, Brown GR, Maglott DR. 2012. NCBI Reference Sequences (RefSeq): current status, new features and genome annotation policy. *Nucleic Acids Res* 40:D130–D135.

57. Bowen De León K, Young ML, Camilleri LB, Brown SD, Skerker JM, Deutschbauer AM, Arkin AP, Fields MW. 2012. Draft genome sequence of *Pelosinus fermentans* JBW45, isolated during in situ stimulation for Cr(VI) reduction. *J Bacteriol* 194:5456–5457.

58. Brown SD, Podar M, Klingeman DM, Johnson CM, Yang ZK, Utturkar SM, Land ML, Mosher JJ, Hurt RA Jr, Phelps TJ, Palumbo AV, Arkin AP, Hazen TC, Elias DA. 2012. Draft genome sequences for two metal-reducing *Pelosinus fermentans* strains isolated from a Cr(VI)-contaminated site and for type strain R7. *J Bacteriol* 194:5147–5148.

59. Sokolova TG, Gonzalez JM, Kostrikina NA, Chernyh NA, Slepova TV, Bonch-Osmolovskaya EA, Robb FT. 2004. *Thermosinus carboxydivorans* gen. nov., sp. nov., a new anaerobic, thermophilic, carbon-monoxide-oxidizing, hydrogenogenic bacterium from a hot pool of Yellowstone National Park. *Int J Syst Evol Microbiol* 54:2353–2359.

60. Rasko DA, Altherr MR, Han CS, Ravel J. 2005. Genomics of the *Bacillus cereus* group of organisms. *FEMS Microbiol Rev* 29:303–329.

61. Makarova KS, Koonin EV. 2007. Evolutionary genomics of lactic acid bacteria. *J Bacteriol* 189:1199–1208.

62. Rigden DJ, Galperin MY. 2008. Sequence analysis of GerM and SpoVS, uncharacterized bacterial 'sporulation' proteins with widespread phylogenetic distribution. *Bioinformatics* 24:1793–1797.

63. Horneck G, Klaus DM, Mancinelli RL. 2010. Space microbiology. *Microbiol Mol Biol Rev* 74:121–156.

64. Setlow P. 2007. I will survive: DNA protection in bacterial spores. *Trends Microbiol* 15:172–180.

65. Klobutcher LA, Ragkousi K, Setlow P. 2006. The *Bacillus subtilis* spore coat provides 'eat resistance' during phagocytic predation by the protozoan *Tetrahymena thermophila*. *Proc Natl Acad Sci USA* 103:165–170.

66. Lu J, Nogi Y, Takami H. 2001. *Oceanobacillus iheyensis* gen. nov., sp. nov., a deep-sea extremely halotolerant and alkaliphilic species isolated from a depth of 1050 m on the Iheya Ridge. *FEMS Microbiol Lett* 205:291–297.

67. Takami H, Takaki Y, Uchiyama I. 2002. Genome sequence of *Oceanobacillus iheyensis* isolated from the Iheya Ridge and its unexpected adaptive capabilities to extreme environments. *Nucleic Acids Res* 30:3927–3935.

68. Takami H, Takaki Y, Chee GJ, Nishi S, Shimamura S, Suzuki H, Matsui S, Uchiyama I. 2004. Thermoadaptation trait revealed by the genome sequence of thermophilic *Geobacillus kaustophilus*. *Nucleic Acids Res* 32:6292–6303.

69. Chivian D, Brodie EL, Alm EJ, Culley DE, Dehal PS, DeSantis TZ, Gihring TM, Lapidus A, Lin LH, Lowry SR, Moser DP, Richardson PM, Southam G, Wanger G, Pratt LM, Andersen GL, Hazen TC, Brockman FJ, Arkin AP, Onstott TC. 2008. Environmental genomics reveals a single-species ecosystem deep within Earth. *Science* 322:275–278.

70. Takami H, Nakasone K, Takaki Y, Maeno G, Sasaki R, Masui N, Fuji F, Hirama C, Nakamura Y, Ogasawara N, Kuhara S, Horikoshi K. 2000. Complete genome sequence of the alkaliphilic bacterium *Bacillus halodurans* and genomic sequence comparison with *Bacillus subtilis*. *Nucleic Acids Res* 28:4317–4331.

71. Abd El-Rahman HA, Fritze D, Sproer C, Claus D. 2002. Two novel psychrotolerant species, *Bacillus psychrotolerans* sp. nov. and *Bacillus psychrodurans* sp. nov., which contain ornithine in their cell walls. *Int J Syst Evol Microbiol* 52:2127–2133.

72. Krishnamurthi S, Ruckmani A, Pukall R, Chakrabarti T. 2010. *Psychrobacillus* gen. nov. and proposal for reclassification of *Bacillus insolitus* Larkin & Stokes, 1967, *B. psychrotolerans* Abd-El Rahman et al., 2002 and *B. psychrodurans* Abd-El Rahman et al., 2002 as *Psychrobacillus insolitus* comb. nov., *Psychrobacillus psychrotolerans* comb. nov. and *Psychrobacillus psychrodurans* comb. nov. *Syst Appl Microbiol* 33:367–373.

73. Mesbah NM, Hedrick DB, Peacock AD, Rohde M, Wiegel J. 2007. *Natranaerobius thermophilus* gen. nov., sp. nov., a halophilic, alkalithermophilic bacterium from soda lakes of the Wadi An Natrun, Egypt, and proposal of *Natranaerobiaceae* fam. nov. and *Natranaerobiales* ord. nov. *Int J Syst Evol Microbiol* 57:2507–2512.

74. Zhao B, Mesbah NM, Dalin E, Goodwin L, Nolan M, Pitluck S, Chertkov O, Brettin TS, Han J, Larimer FW, Land ML, Hauser L, Kyrpides N, Wiegel J. 2011. Complete genome sequence of the anaerobic, halophilic alkalithermophile *Natranaerobius thermophilus* JW/NM-WN-LF. *J Bacteriol* 193:4023–4024.

75. Orsburn BC, Melville SB, Popham DL. 2010. EtfA catalyses the formation of dipicolinic acid in *Clostridium perfringens*. *Mol Microbiol* 75:178–186.

76. Huber R, Rossnagel P, Woese CR, Rachel R, Langworthy TA, Stetter KO. 1996. Formation of ammonium from nitrate during chemolithoautotrophic growth of the extremely thermophilic bacterium *Ammonifex degensii* gen. nov. sp. nov. *Syst Appl Microbiol* 19:40–49.

77. Rainey FA, Donnison AM, Janssen PH, Saul D, Rodrigo A, Bergquist PL, Daniel RM, Stackebrandt E, Morgan HW. 1994. Description of *Caldicellulosiruptor saccharolyticus* gen. nov., sp. nov: an obligately anaerobic, extremely thermophilic, cellulolytic bacterium. *FEMS Microbiol Lett* 120:263–266.

78. Bredholt S, Sonne-Hansen J, Nielsen P, Mathrani IM, Ahring BK. 1999. *Caldicellulosiruptor kristjanssonii* sp. nov., a cellulolytic, extremely thermophilic, anaerobic bacterium. *Int J Syst Bacteriol* 49:991–996.

79. Miroshnichenko ML, Kublanov IV, Kostrikina NA, Tourova TP, Kolganova TV, Birkeland NK, Bonch-Osmolovskaya EA. 2008. *Caldicellulosiruptor kronotskyensis* sp. nov. and *Caldicellulosiruptor hydrothermalis* sp. nov., two extremely thermophilic, cellulolytic, anaerobic bacteria from Kamchatka thermal springs. *Int J Syst Evol Microbiol* 58:1492–1496.

80. Blumer-Schuette SE, Ozdemir I, Mistry D, Lucas S, Lapidus A, Cheng JF, Goodwin LA, Pitluck S, Land ML, Hauser LJ, Woyke T, Mikhailova N, Pati A, Kyrpides NC, Ivanova N, Detter JC, Walston-Davenport K, Han S, Adams MW, Kelly RM. 2011. Complete genome sequences for the anaerobic, extremely thermophilic plant biomass-degrading bacteria *Caldicellulosiruptor hydrothermalis*, *Caldicellulosiruptor kristjanssonii*, *Caldicellulosiruptor kronotskyensis*, *Caldicellulosiruptor*

owensensis, and *Caldicellulosiruptor lactoaceticus*. *J Bacteriol* **193**:1483–1484.

81. Demain AL, Newcomb M, Wu JH. 2005. Cellulase, clostridia, and ethanol. *Microbiol Mol Biol Rev* **69**: 124–154.

82. Klenk H-P, Lapidus A, Chertkov O, Copeland A, Glavina del Rio T, Nolan M, Lucas S, Chen F, Tice H, Cheng JF, Han C, Bruce D, Goodwin L, Pitluck S, Pati A, Ivanova N, Mavromatis K, Daum C, Chen A, Palaniappan K, Chang YJ, Land ML, Hauser LJ, Jeffries CD, Detter JC, Rohde M, Abt B, Pukall R, Göker M, Bristow J, Markowitz V, Hugenholtz P, Eisen JA. 2011. Complete genome sequence of the thermophilic, hydrogen-oxidizing *Bacillus tusciae* type strain (T2^T) and reclassification in the new genus, *Kyrpidia* gen. nov. as *Kyrpidia tusciae* comb. nov. and emendation of the family *Alicyclobacillaceae* da Costa and Rainey, 2010. *Stand Genomic Sci* **5**:121–134.

83. Li B, Chen Y, Liu Q, Hu S, Chen X. 2011. Complete genome analysis of *Sulfobacillus acidophilus* strain TPY, isolated from a hydrothermal vent in the Pacific Ocean. *J Bacteriol* **193**:5555–5556.

84. Wu M, Ren Q, Durkin AS, Daugherty SC, Brinkac LM, Dodson RJ, Madupu R, Sullivan SA, Kolonay JF, Haft DH, Nelson WC, Tallon LJ, Jones KM, Ulrich LE, Gonzalez JM, Zhulin IB, Robb FT, Eisen JA. 2005. Life in hot carbon monoxide: the complete genome sequence of *Carboxydothermus hydrogenoformans* Z-2901. *PLoS Genet* **1**:e65. doi:10.1371/journal.pgen.0010065.

85. Kopke M, Held C, Hujer S, Liesegang H, Wiezer A, Wollherr A, Ehrenreich A, Liebl W, Gottschalk G, Durre P. 2010. Clostridium ljungdahlii represents a microbial production platform based on syngas. *Proc Natl Acad Sci USA* **107**:13087–13092.

86. Asao M, Madigan MT. 2010. Taxonomy, phylogeny, and ecology of the heliobacteria. *Photosynth Res* **104**: 103–111.

87. Sattley WM, Madigan MT, Swingley WD, Cheung PC, Clocksin KM, Conrad AL, Dejesa LC, Honchak BM, Jung DO, Karbach LE, Kurdoglu A, Lahiri S, Mastrian SD, Page LE, Taylor HL, Wang ZT, Raymond J, Chen M, Blankenship RE, Touchman JW. 2008. The genome of *Heliobacterium modesticaldum*, a phototrophic representative of the Firmicutes containing the simplest photosynthetic apparatus. *J Bacteriol* **190**: 4687–4696.

88. Tang KH, Yue H, Blankenship RE. 2010. Energy metabolism of Heliobacterium modesticaldum during phototrophic and chemotrophic growth. *BMC Microbiol* **10**:150. doi:10.1186/1471-2180-10-150.

89. Heggeset TM, Krog A, Balzer S, Wentzel A, Ellingsen TE, Brautaset T. 2012. Genome sequence of thermotolerant *Bacillus methanolicus*: features and regulation related to methylotrophy and production of L-lysine and L-glutamate from methanol. *Appl Environ Microbiol* **78**:5170–5181.

90. Paredes CJ, Alsaker KV, Papoutsakis ET. 2005. A comparative genomic view of clostridial sporulation and physiology. *Nat Rev Microbiol* **3**:969–978.

91. Abecasis AB, Serrano M, Alves R, Quintais L, Pereira-Leal JB, Henriques AO. 2013. A genomic signature and the identification of new sporulation genes. *J Bacteriol* **195**:2101–2115.

92. Galperin MY. 2006. Structural classification of bacterial response regulators: diversity of output domains and domain combinations. *J Bacteriol* **188**:4169–4182.

93. Farrow JA, Wallbanks S, Collins MD. 1994. Phylogenetic interrelationships of round-spore-forming bacilli containing cell walls based on lysine and the non-spore-forming genera *Caryophanon*, *Exiguobacterium*, *Kurthia*, and *Planococcus*. *Int J Syst Bacteriol* **44**: 74–82.

94. Kloos WE, Ballard DN, George CG, Webster JA, Hubner RJ, Ludwig W, Schleifer KH, Fiedler F, Schubert K. 1998. Delimiting the genus *Staphylococcus* through description of *Macrococcus caseolyticus* gen. nov., comb. nov. and *Macrococcus equipercicus* sp. nov., and *Macrococcus bovicus* sp. no. and *Macrococcus carouselicus* sp. nov. *Int J Syst Bacteriol* **48**: 859–877.

95. Snel J, Heinen PP, Blok HJ, Carman RJ, Duncan AJ, Allen PC, Collins MD. 1995. Comparison of 16S rRNA sequences of segmented filamentous bacteria isolated from mice, rats, and chickens and proposal of 'Candidatus *Arthromitus*'. *Int J Syst Bacteriol* **45**:780–782.

96. Kuwahara T, Ogura Y, Oshima K, Kurokawa K, Ooka T, Hirakawa H, Itoh T, Nakayama-Imaohji H, Ichimura M, Itoh K, Ishifune C, Maekawa Y, Yasutomo K, Hattori M, Hayashi T. 2011. The lifestyle of the segmented filamentous bacterium: a non-culturable gut-associated immunostimulating microbe inferred by whole-genome sequencing. *DNA Res* **18**:291–303.

97. Klaasen HL, Koopman JP, Poelma FG, Beynen AC. 1992. Intestinal, segmented, filamentous bacteria. *FEMS Microbiol Rev* **8**:165–180.

98. Klaasen HL, Koopman JP, Van den Brink ME, Bakker MH, Poelma FG, Beynen AC. 1993. Intestinal, segmented, filamentous bacteria in a wide range of vertebrate species. *Lab Anim* **27**:141–150.

99. Chase DG, Erlandsen SL. 1976. Evidence for a complex life cycle and endospore formation in the attached, filamentous, segmented bacterium from murine ileum. *J Bacteriol* **127**:572–583.

100. Prakash T, Oshima K, Morita H, Fukuda S, Imaoka A, Kumar N, Sharma VK, Kim SW, Takahashi M, Saitou N, Taylor TD, Ohno H, Umesaki Y, Hattori M. 2011. Complete genome sequences of rat and mouse segmented filamentous bacteria, a potent inducer of th17 cell differentiation. *Cell Host Microbe* **10**:273–284.

101. Sczesnak A, Segata N, Qin X, Gevers D, Petrosino JF, Huttenhower C, Littman DR, Ivanov II. 2011. The genome of th17 cell-inducing segmented filamentous bacteria reveals extensive auxotrophy and adaptations to the intestinal environment. *Cell Host Microbe* **10**: 260–272.

102. Makarova K, Slesarev A, Wolf Y, Sorokin A, Mirkin B, Koonin E, Pavlov A, Pavlova N, Karamychev V, Polouchine N, Shakhova V, Grigoriev I, Lou Y, Rohksar

D, Lucas S, Huang K, Goodstein DM, Hawkins T, Plengvidhya V, Welker D, Hughes J, Goh Y, Benson A, Baldwin K, Lee JH, Diaz-Muniz I, Dosti B, Smeianov V, Wechter W, Barabote R, Lorca G, Altermann E, Barrangou R, Ganesan B, Xie Y, Rawsthorne H, Tamir D, Parker C, Breidt F, Broadbent J, Hutkins R, O'Sullivan D, Steele J, Unlu G, Saier M, Klaenhammer T, Richardson P, Kozyavkin S, Weimer B, Mills D. 2006. Comparative genomics of the lactic acid bacteria. *Proc Natl Acad Sci USA* **103**:15611–15616.

103. Skerman VBD, McGowan V, Sneath PHA. 1980. Approved lists of bacterial names. *Int J Syst Bacteriol* **30**: 225–420.

104. Bisby FA, Roskov YR, Orrell TM, Nicolson D, Paglinawan LE, Bailly N, Kirk PM, Bourgoin T, Baillargeon G (ed). 2009. *Species 2000 & ITIS Catalogue of Life: 2009 Annual Checklist Taxonomic Classification.* Species 2000, Reading, UK.

105. Ludwig W, Schleifer K-H, Whitman WB. 2009. Taxonomic outline of the phylum Firmicutes, p 15–17. *In* De Vos P, Garrity GM, Jones D, Krieg NR, Ludwig W, Rainey FA, Schleifer K-H, Whitman WB (ed), Bergey's Manual of Systematic Bacteriology, 2nd ed, **vol 3**. *The Firmicutes.* Springer, New York, NY.

106. Quast C, Pruesse E, Yilmaz P, Gerken J, Schweer T, Yarza P, Peplies J, Glöckner FO. 2013. The SILVA ribosomal RNA gene database project: improved data processing and web-based tools. *Nucleic Acids Res* **41**: D590–D596.

107. DeSantis TZ, Hugenholtz P, Larsen N, Rojas M, Brodie EL, Keller K, Huber T, Dalevi D, Hu P, Andersen GL. 2006. Greengenes, a chimera-checked 16S rRNA gene database and workbench compatible with ARB. *Appl Environ Microbiol* **72**:5069–5072.

108. McDonald D, Price MN, Goodrich J, Nawrocki EP, DeSantis TZ, Probst A, Andersen GL, Knight R, Hugenholtz P. 2012. An improved Greengenes taxonomy with explicit ranks for ecological and evolutionary analyses of bacteria and archaea. *ISME J* **6**:610–618.

109. Cole JR, Wang Q, Cardenas E, Fish J, Chai B, Farris RJ, Kulam-Syed-Mohideen AS, McGarrell DM, Marsh T, Garrity GM, Tiedje JM. 2009. The Ribosomal Database Project: improved alignments and new tools for rRNA analysis. *Nucleic Acids Res* **37**:D141–D145.

110. Starr MP, Sayre RM. 1988. *Pasteuria thornei* sp. nov. and *Pasteuria penetrans* sensu stricto emend., mycelial and endospore-forming bacteria parasitic, respectively, on plant-parasitic nematodes of the genera *Pratylenchus* and *Meloidogyne. Ann Inst Pasteur Microbiol* **139**: 11–31.

111. Yassin AF, Hupfer H, Klenk HP, Siering C. 2009. *Desmospora activa* gen. nov., sp. nov., a thermoactinomycete isolated from sputum of a patient with suspected pulmonary tuberculosis, and emended description of the family *Thermoactinomycetaceae* Matsuo et al. 2006. *Int J Syst Evol Microbiol* **59**:454–459.

112. Hatayama K, Shoun H, Ueda Y, Nakamura A. 2006. *Tuberibacillus calidus* gen. nov., sp. nov., isolated from a compost pile and reclassification of *Bacillus naganoensis* Tomimura et al. 1990 as *Pullulanibacillus nagano-*

ensis gen. nov., comb. nov. and *Bacillus laevolacticus* Andersch et al. 1994 as *Sporolactobacillus laevolacticus* comb. nov. *Int J Syst Evol Microbiol* **56**:2545–2551.

113. Yokoyama H, Wagner ID, Wiegel J. 2010. *Caldicoprobacter oshimai* gen. nov., sp. nov., an anaerobic, xylanolytic, extremely thermophilic bacterium isolated from sheep faeces, and proposal of *Caldicoprobacteraceae* fam. nov. *Int J Syst Evol Microbiol* **60**:67–71.

114. Jabari L, Gannoun H, Cayol JL, Hamdi M, Fauque G, Ollivier B, Fardeau ML. 2012. Characterization of *Defluviitalea saccharophila* gen. nov., sp. nov., a thermophilic bacterium isolated from an upflow anaerobic filter treating abattoir wastewaters, and proposal of *Defluviitaleaceae* fam. nov. *Int J Syst Evol Microbiol* **62**:550–555.

115. Eeckhaut V, Van Immerseel F, Pasmans F, De Brandt E, Haesebrouck F, Ducatelle R, Vandamme P. 2010. *Anaerostipes butyraticus* sp. nov., an anaerobic, butyrate-producing bacterium from *Clostridium* cluster XIVa isolated from broiler chicken caecal content, and emended description of the genus *Anaerostipes. Int J Syst Evol Microbiol* **60**:1108–1112.

116. Grech-Mora I, Fardeau M-L, Patel BKC, Ollivier B, Rimbault A, Prensier G, Garcia JL, Garnier-Sillam E. 1996. Isolation and characterization of *Sporobacter termitidis* gen. nov., sp. nov., from the digestive tract of the wood-feeding termite *Nasutitermes lujae. Int J Syst Evol Microbiol* **46**:512–518.

117. Matthies C, Springer N, Ludwig W, Schink B. 2000. *Pelospora glutarica* gen. nov., sp. nov., a glutarate-fermenting, strictly anaerobic, spore-forming bacterium. *Int J Syst Evol Microbiol* **50**:645–648.

118. Wisotzkey JD, Jurtshuk P Jr, Fox GE, Deinhard G, Poralla K. 1992. Comparative sequence analyses on the 16S rRNA (rDNA) of *Bacillus acidocaldarius, Bacillus acidoterrestris,* and *Bacillus cycloheptanicus* and proposal for creation of a new genus, *Alicyclobacillus* gen. nov. *Int J Syst Bacteriol* **42**:263–269.

119. Darland G, Brock TD. 1971. *Bacillus acidocaldarius* sp. nov., an acidophilic thermophilic spore-forming bacterium. *J Gen Microbiol* **67**:9–15.

120. Qi H, Chen H, Ao J, Zhou H, Chen X. 2009. Isolation and identification of a strain of moderate thermophilic and acidophilic bacterium from deep sea. *Acta Oceanol Sin* **31**:152–158.

121. Anderson I, Chertkov O, Chen A, Saunders E, Lapidus A, Nolan M, Lucas S, Hammon N, Deshpande S, Cheng J-F, Han C, Tapia R, Goodwin LA, Pitluck S, Liolios K, Pagani I, Ivanova N, Mikhailova N, Pati A, Palaniappan K, Land M, Pan C, Rohde M, Pukall R, Göker M, Detter JC, Woyke T, Bristow J, Eisen JA, Markowitz V, Hugenholtz P, Kyrpides NC, Klenk H-P, Mavromatis K. 2012. Complete genome sequence of the moderately thermophilic mineral-sulfide-oxidizing firmicute *Sulfobacillus acidophilus* type strain (NAL^T). *Stand Genomic Sci* **6**:293–303.

122. Watling HR, Perrot FA, Shiers DW. 2008. Comparison of selected characteristics of *Sulfobacillus* species and review of their occurrence in acidic and bioleaching environments. *Hydrometallurgy* **93**:57–65.

123. Ye Q, Roh Y, Carroll SL, Blair B, Zhou J, Zhang CL, Fields MW. 2004. Alkaline anaerobic respiration: isolation and characterization of a novel alkaliphilic and metal-reducing bacterium. *Appl Environ Microbiol* 70: 5595–5602.

124. Nielsen P, Fritze D, Priest FG. 1995. Phenetic diversity of alkaliphilic *Bacillus* strains: proposal for nine new species. *Microbiology* 141:1745–1761.

125. Sturr MG, Guffanti AA, Krulwich TA. 1994. Growth and bioenergetics of alkaliphilic *Bacillus firmus* OF4 in continuous culture at high pH. *J Bacteriol* 176: 3111–3116.

126. Janto B, Ahmed A, Ito M, Liu J, Hicks DB, Pagni S, Fackelmayer OJ, Smith TA, Earl J, Elbourne LD, Hassan K, Paulsen IT, Kolsto AB, Tourasse NJ, Ehrlich GD, Boissy R, Ivey DM, Li G, Xue Y, Ma Y, Hu FZ, Krulwich TA. 2011. Genome of alkaliphilic *Bacillus pseudofirmus* OF4 reveals adaptations that support the ability to grow in an external pH range from 7.5 to 11.4. *Environ Microbiol* 13:3289–3309.

127. Yoon JH, Oh TK, Park YH. 2004. Transfer of *Bacillus halodenitrificans* Denariaz et al. 1989 to the genus *Virgibacillus* as *Virgibacillus halodenitrificans* comb. nov. *Int J Syst Evol Microbiol* 54:2163–2167.

128. Lee SJ, Lee YJ, Jeong H, Lee HS, Pan JG, Kim BC, Lee DW. 2012. Draft genome sequence of *Virgibacillus halodenitrificans* 1806. *J Bacteriol* 194:6332–6333.

129. Larsen L, Nielsen P, Ahring BK. 1997. *Thermoanaerobacter mathranii* sp. nov., an ethanol-producing, extremely thermophilic anaerobic bacterium from a hot spring in Iceland. *Arch Microbiol* 168:114–119.

130. Lechner S, Mayr R, Francis KP, Pruss BM, Kaplan T, Wiessner-Gunkel E, Stewart GS, Scherer S. 1998. *Bacillus weihenstephanensis* sp. nov. is a new psychrotolerant species of the *Bacillus cereus* group. *Int J Syst Bacteriol* 48:1373–1382.

131. Gioia J, Yerrapragada S, Qin X, Jiang H, Igboeli OC, Muzny D, Dugan-Rocha S, Ding Y, Hawes A, Liu W, Perez L, Kovar C, Dinh H, Lee S, Nazareth L, Blyth P, Holder M, Buhay C, Tirumalai MR, Liu Y, Dasgupta I, Bokhetache L, Fujita M, Karouia F, Eswara Moorthy P, Siefert J, Uzman A, Buzumbo P, Verma A, Zwiya H, McWilliams BD, Olowu A, Clinkenbeard KD, Newcombe D, Golebiewski L, Petrosino JF, Nicholson WL, Fox GE, Venkateswaran K, Highlander SK, Weinstock GM. 2007. Paradoxical DNA repair and peroxide resistance gene conservation in *Bacillus pumilus* SAFR-032. *PLoS One* 2:e928. doi:10.1371/journal.pone.0000928.

132. Link L, Sawyer J, Venkateswaran K, Nicholson W. 2004. Extreme spore UV resistance of *Bacillus pumilus* isolates obtained from an ultraclean spacecraft assembly facility. *Microb Ecol* 47:159–163.

133. Poehlein A, Gottschalk G, Daniel R. 2013. First insights into the genome of the Gram-negative, endospore-forming organism *Sporomusa ovata* strain H1 DSM 2662. *Genome Announc* 1(5) doi:10.1128/genomeA.00734-13.

134. Wunderlin T, Junier T, Roussel-Delif L, Jeanneret N, Junier P. 2013. Stage 0 sporulation gene A as a molecular marker to study diversity of endospore-forming Firmicutes. *Environ Microbiol Rep.* doi:10.1111/1758-2229.12094.

135. Alauzet C, Marchandin H, Courtin P, Mory F, Lemee L, Pons JL, Chapot-Chartier MP, Lozniewski A, Jumas-Bilak E. 2014. Multilocus analysis reveals diversity in the genus *Tissierella*: description of *Tissierella carlieri* sp. nov. in the new class *Tissierellia* classis nov. *Syst Appl Microbiol* 37:23–34.

136. Watanabe M, Kojima H, Fukui M. 2015. *Limnochorda pilosa* gen. nov., sp. nov., a moderately thermophilic, facultatively anaerobic, pleomorphic bacterium and proposal of *Limnochordaceae* fam. nov., *Limnochordales* ord. nov. and *Limnochordia* classis nov. in the phylum *Firmicutes*. *Int J Syst Evol Microbiol* 65: 2378–2384.

137. Watanabe M, Kojima H, Fukui M. 2016. Complete genome sequence and cell structure of Limnochorda pilosa, a Gram-negative spore former within the phylum Firmicutes. *Int J Syst Evol Microbiol* doi:10.1099/ijsem.0.000881.

138. Rinke C, Schwientek P, Sczyrba A, Ivanova NN, Anderson IJ, Cheng JF, Darling A, Malfatti S, Swan BK, Gies EA, Dodsworth JA, Hedlund BP, Tsiamis G, Sievert SM, Liu WT, Eisen JA, Hallam SJ, Kyrpides NC, Stepanauskas R, Rubin EM, Hugenholtz P, Woyke T. 2013. Insights into the phylogeny and coding potential of microbial dark matter. *Nature* 499:431–437.

139. Sankar SA, Lagier JC, Pontarotti P, Raoult D, Fournier PE. 2015. The human gut microbiome, a taxonomic conundrum. *Syst Appl Microbiol* 38:276–286

The Bacterial Spore: From Molecules to Systems
Edited by P. Eichenberger and A. Driks
© 2016 American Society for Microbiology, Washington, DC
doi:10.1128/microbiolspec.TBS-0020-2014

Patricia Fajardo-Cavazos[1]
Heather Maughan[2]
Wayne L. Nicholson[1]

Evolution in the *Bacillaceae*

2

The family *Bacillaceae* (domain Bacteria; kingdom *Bacteria*; phylum *Firmicutes*; class *Bacilli*; order *Bacillales*) (Fig. 1) is a globally dispersed and phenotypically heterogeneous group of bacteria (1, 2). It therefore follows that this family possesses a considerable evolutionary history for scientists to unravel. In recent years, the issue of evolution in the *Bacillaceae* has been probed from two directions. Some researchers have taken an ecological approach to understand the organisms' adaptive evolution to particular environmental niches; others have pursued a laboratory-based approach in which single laboratory strains are directed to evolve under particular conditions determined by the experimenter. Both approaches have yielded new insights. As in so many other facets of the biology of the *Bacillaceae*, most of our knowledge has been derived from the intense study of relatively few members of the family, most notably *Bacillus subtilis* (3, 4). Much less information has been obtained concerning the vast majority of the *Bacillaceae*, prompting the present examination of diversity and evolution within this ubiquitous family. The present article builds upon previous reviews of the topic (5–10). Additional information on the genomic diversity of spore-forming *Firmicutes* and on the ecology of the *Bacillaceae* can be found in chapters by Galperin and by Mandic-Mulec et al. in this volume.

SHORT HISTORY OF THE TAXONOMY OF THE *BACILLACEAE*

An extensive and excellent review on the taxonomy of the *Bacillaceae* was published recently (10), so the topic will be covered here only briefly. Two representatives of the *Bacillaceae*, *Bacillus anthracis* and *Bacillus subtilis*, hold the distinction of being among the first systematically characterized bacteria. Ferdinand Cohn discovered sporulation in *B. subtilis* (11) and Robert Koch developed his postulates of infectious disease using *B. anthracis* (12); both bacteria continue to be studied intensely to this day (4, 13–15). The discovery and description of these bacteria marked the first step toward classification of what we now know as the family *Bacillaceae*. The initial proposal of the genus *Bacillus* by Cohn in 1872 (11), based on consideration of the cell shape, was followed by the adoption of the general term "bacillus" (derived from the Latin, meaning "little staff") to describe any rod-shaped bacterium.

Major advances in methodologies for bacterial taxonomy during the late 19th and early 20th century, including the development and use of various histological stains and metabolic tests, led to a system of taxonomic classification based on a combination of morphological, phenotypic, and metabolic characteristics, termed

[1]Department of Microbiology & Cell Science, University of Florida, Space Life Sciences Laboratory, Exploration Park at Kennedy Space Center, Merritt Island, FL 32953; [2]Ronin Institute (Headquartered in Montclair, NJ), Mildmay, ON Canada N0G 2J0.

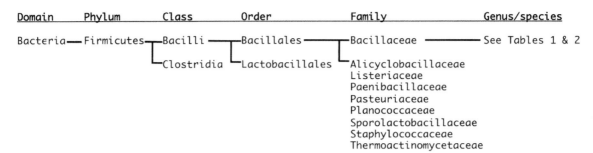

Figure 1 Taxonomic position of the *Bacillaceae* reconstructed from 16S rDNA data posted in the Ribosomal Database Project Release 10 (http://rdp.cme.msu.edu/) (177).

"phenetics." Based on their shared phenetic characteristics, the family *Bacillaceae* was first proposed in 1920 (16) and in 1923 was incorporated into the first edition of *Bergey's Manual of Determinative* (later *Systematic*) *Bacteriology* (17). The family *Bacillaceae* was originally defined as "aerobic or facultatively anaerobic chemo-organotrophic Gram-positive rods that form endospores." From that point, as with all other bacterial families, the taxonomy of *Bacillaceae* has been shaped and reshaped numerous times as more knowledge has accumulated.

In the late 20th century, a fundamental shift occurred in our understanding of the evolutionary relatedness of microorganisms with the advent of molecular phylogenetics, also known as "cladistics" (18). Grouping of the *Bacillaceae* by the use of 16S rRNA gene sequence comparison has gained increasing importance in taxonomy (19, 20) to the point that species and genera are now regularly defined *de facto* based on 16S rDNA identities of >97.5% and >95%, respectively (21, 22). However, cladistic methods based on 16S rDNA sequences result in groupings that often differ from those generated by traditional phenetic taxonomy (10), and phenotypes once considered characteristic of the *Bacillaceae* now appear outside the family. Even one of the major defining features of the *Bacillaceae*, the endospore, has lost its importance in 16S rDNA-based taxonomy, and numerous endospore formers have been placed into entirely different families, such as the *Paenibacillaceae, Alicyclobacillaceae, Pasteuriaceae*, or *Sporolactobacillaceae*. Conversely, in the family *Bacillaceae* there are exceptions to every benchmark characteristic originally listed in Bergey's manual, and now strict anaerobes (*B. infernus*), nonsporeformers (*Falsibacillus, Marinococcus, Saccharococcus* spp.), non-rod-shaped cells (*Marinococcus, Saccharococcus* spp.), and Gram-negative-staining bacteria (*Filobacillus, Halobacillus, Lentibacillus* spp.) are included in

the family (10, 23). This disparity between cladistic and phenetic taxonomy persists because opposing forces strengthen each approach. The goal of cladistics is to create a phylogeny that consistently reflects evolutionary relationships. However, for practical purposes, it is often useful to identify a microorganism as distinct from close relatives, based on its impact on human activities. For example, *B. anthracis, Bacillus cereus*, and *Bacillus thuringiensis* have long been considered as three separate species based on phenetics, despite the fact that by 16S rDNA cladistics they form a single species (24).

The recent application of affordable, high-throughput whole-genome sequencing technologies to the taxonomy of the *Bacillaceae* is yielding phylogenomic classifications that deviate even further from those that take into consideration phenotypic characteristics of historical and practical importance. Traditional phylogenetic tree-based models assumed that evolution is predominantly due to the vertical transmission of genetic information and that horizontal gene transfer played a lesser role. New network-based models allow the evolution of horizontally transferred genes to be tracked. These studies indicate that reticulated phylogenetic networks are favored over strictly bifurcating phylogenetic trees (482). In fact, reconstruction of evolution for groups of closely related bacteria and archaea, including *55 Bacillus* species, revealed an extremely rapid and highly variable flux of genes in evolving microbial genomes and demonstrated that extensive gene loss and horizontal gene transfer are the two dominant evolutionary processes (483). Support of this general observation has come from phylogenetic analysis of 29 *Geobacillus* species (484) and two strains of *B. megaterium* (485). Network-based approaches to taxonomy are now being incorporated into, and strengthening, current polyphasic approaches which rely on cladistic, phenetic, and ecological considerations.

PHENOTYPIC DIVERSITY AND EVOLUTION

The bulk of our knowledge regarding phenotypic diversity in *Bacillaceae* comes from laboratory studies of species in pure culture. Early phenetic studies (based on Gram-positive rods that sporulate) showed that their metabolic capabilities were vast. Species of the *Bacillaceae* were found to be quite diverse in their (i) sites of isolation, (ii) abilities to grow anaerobically, (iii) cell wall and membrane lipid compositions, (iv) production of fermentation products from a variety of sugars, (v) hydrolysis of casein, gelatin, and starches, (vi) utilization of citrate and/or propionate, (vii) salt tolerance, (viii) motility, (ix) spore shape and placement, and (x) growth ranges of temperature and pH (23). The phenotypic diversity exhibited by the family *Bacillaceae* has resulted from approximately 2 billion years of evolution (25, 26). The basal position of the *Firmicutes* in the bacterial tree of life suggests that *Bacillaceae* is an ancient family (2) that may have ancestrally been entirely thermophilic (27). Exciting new genome data collected from members of this family have shown that, although over 19,000 gene families are found within 19 species, only 814 orthologous genes are shared between all of those species' genomes (28). This small number of shared "core" genes is even more surprising when we consider that the average number of genes in each genome was almost 5,000. These core genes are predominantly housekeeping genes that include the majority of those found to be essential when knocked out in *B. subtilis* (29). The widely varying "dispensable" genome highlights the diverse environments that are encountered by *Bacillaceae*. They have been found in nearly every environment on Earth (Tables 1 and 2), and new members are continually being discovered from the exploration of increasingly novel environments such as arsenic-rich Mono Lake (30), Tunisian oilfields (31), subterranean brine (32), and rock wall paintings (33).

Bacillus spp. were first isolated from the human gastrointestinal (GI) tract in the early 20th century (*B. alcalophilus*, *B. badius*, *B. flexus*) (34, 35). More recently, sampling for the occurrence of *Actinobacteria* in dried fecal samples from 10 volunteer donors led to the discovery that 81 of 124 isolates in fact belonged to the *Bacillaceae* (36). *Bacillus clausii* and *Bacillus licheniformis* were found in the majority of donors, with *Lysinibacillus*, *Ureibacillus*, *Oceanobacillus*, *Ornithinibacillus*, and *Virgibacillus* spp. also found at lower frequencies (36). Many of the isolates exhibited antimicrobial activity against one or more strains of *Clostridium difficile*, *Clostridium perfringens*, *Listeria monocytogenes*, and *Staphylococcus aureus*,

which would be important for exclusion of competitors in the GI tract. Additional experiments measuring the adherence of the isolates to cultured human cells indicated that many of these strains have the potential for colonization of the GI tract, and were probably not simply passing through (36).

The information collected from environmental sampling is substantial, but deriving environmental preferences and global patterns of distribution of *Bacillaceae* from such data is still a difficult endeavor. As is the case with prokaryotes in general (37), obstacles in uncovering biogeographical trends include: (i) the lack of a generally accepted definition of what constitutes a microbial species ([38]; see also http://www.bacterio.cict.fr/); (ii) the lack of standardized descriptions of environments; (iii) difficulties in cataloging groups with such great diversity inherent to large populations with rapid growth rates; and (iv) the limited knowledge about the set of physiological activities exhibited by *Bacillaceae* in their natural habitats (39). Nevertheless, metagenomic and bioinformatic approaches have enabled the identification and quantification of *Bacillaceae* within various microbial communities. Tamames et al. (37) analyzed the complete set of 16S rDNA sequences obtained from environmental samples stored in the GenBank database and assigned them into a hierarchical classification of environments. The results support a cosmopolitan distribution of the *Bacillaceae* and actually found the genus *Bacillus* to be the second most abundant genus in the data set, after *Pseudomonas* (37). It should be noted that the numbers and variety of *Bacillaceae* represented in metagenomic data sets are likely to be underestimates, because commonly used DNA extraction kits are usually unable to efficiently lyse spores (40).

How do *Bacillaceae* come to occupy this diversity of environments? For the spore-forming genera, the spore is an ideal vehicle for widespread dispersal via air or water currents (6, 41, 42), and doubtless the spore has played a major role in establishing the ubiquity of the family. Indeed, spore-forming bacteria have been shown to dominate communities of bacteria isolated from air (43–45), regardless of time of year (46). As DNA-based studies continue to address microbial dispersal through the atmosphere (43), we will learn more about the aerial dispersal of *Bacillaceae* into novel environments.

However, as pointed out by Priest (20), the resistance, longevity, and widespread dispersal of spores make it uncertain whether spores found in environmental samples actually live in and contribute to the

Table 1   Species in the family *Bacillaceae*, genus *Bacillus*[a]

Species	Type strain	Source	Contains ComK (NCBI)	Reference
acidiceler	DSM 18954	Forensic specimen	—[b]	178
acidicola	DSM 14745	Acidic peat bog	—	179
acidiproducens	DSM 23148	Vineyard soil	—	180
aeolius	DSM 15084	Shallow marine hot spring	—	181
aerius	JCM 13348	High-altitude air sample	—	182
aerophilus	JCM 13347	High-altitude air sample	Yes	182
agaradhaerens	DSM 8721	Unknown	—	183
aidingensis	DSM 18341	Salt lake	—	184
akibai	JCM 9157	Industrial isolate	—	185
alcalophilus	DSM 485	Human feces	—	35
algicola	CIP 107850	Brown alga	—	186
alkalinitrilicus	NCCB 100120	Soda soil	—	187
alkalisediminis	DSM 21670	Soda pond sediment	—	188
alkalitelluris	DSM 16976	Sandy soil	—	189
altitudinis	DSM 21631	High-altitude air sample	—	182
alveayuensis	DSM 19092	Deep-sea sediment	Yes	190
amyloliquefaciens	DSM 7	Soil	Yes	191
anthracis	ATCC 14578	Causative agent of anthrax	Yes	11
aquimaris	DSM 16205	Tidal flat seawater	—	192
arsenicus	DSM 15822	Siderite concretion	—	193
aryabhattai	DSM21047	High-altitude air sample	—	194
asahii	JCM 12112	Soil	—	195
atrophaeus	DSM 7264	Soil	Yes	196
aurantiacus	DSM 18675	Soda lake	—	197
azotoformans	DSM 1046	Soil	—	198
badius	DSM 23	Intestinal tract of child	—	34
barbaricus	DSM 14730	Experimental wall painting	—	33
bataviensis	DSM 15601	Soil	—	199
beijingensis	DSM 19037	Ginseng root	—	200
benzoevorans	DSM 5391	Soil	—	201
beveridgei	DSM 22320	Soda lake	—	30
bogoriensis	LMG 22234	Soda lake	—	202
boroniphilus	DSM 17376	Soil	—	203
butanolivorans	DSM 18926	Soil	—	204
canaveralius	ATCC BAA-1493	Spacecraft assembly facility	—	205
carboniphilus	DSM 17613	Air fallout on witness plate	—	206
cecembensis	JCM 15113	Glacier	—	207
cellulosilyticus	DSM 2522	Industrial isolate	—	185
centrosporus	DSM 8445	Distinguished from *B. brevis*	—	208
cereus	DSM 31	Air from a cowshed	Yes	209
chagannorensis	DSM 18086	Soda lake	—	210
chungangensis	KCTC 13566	Sea sand	—	211
cibi	DSM 16189	Jeotgal, traditional Korean fermented food	—	212
circulans	DSM 11	Sewage	—	213
clarkii	DSM 8720	Unknown	—	183
clausii	DSM 8716	Unknown	—	183
coagulans	DSM 1	Evaporated milk	Yes	214
coahuilensis	NRRL B-41737	Desert lagoon	Yes	215
cohnii	DSM 6307	Soil	—	216
cytotoxicus	DSM 22905	Food poisoning cases	Yes	217
decisifrondis	DSM 17725	Soil	—	218
decolorationis	DSM 14890	Biodeteriorated mural painting	—	219

(Continued)

Table 1 *(Continued)*

Species	Type strain	Source	Contains ComK (NCBI)	Reference
drentensis	DSM 15600	Soil	—	199
endophyticus	DSM 13796	Inner tissue of cotton plant	—	220
endoradicis	LMG 25492	Inner tissue of soybean root	—	221
farraginis	DSM 16013	Milking apparatus	—	222
fastidiosus	DSM 91	Soil	—	223
firmus	DSM 12	Soil	—	224
flexus	DSM 1320	Intestinal tract of child	—	34
foraminis	LMG 23174	Alkaline groundwater	—	225
fordii	DSM 16014	Raw milk	—	222
fortis	DSM 16012	Milking apparatus	—	222
fumarioli	DSM 18237	Fumarole	—	226
funiculus	DSM 15141	Activated sludge	—	227
galactosidilyticus	DSM15595	Raw milk	—	228
galliciensis	DSM 21539	Seahorse feces	—	229
gelatini	DSM 15865	Gelatin	—	230
gibsonii	DSM 8722	Soil	—	183
ginsengi	DSM 19038	Ginseng root	—	200
ginsengihumi	DSM 18134	Soil	—	231
graminis	DSM 22162	Coastal dune plant	—	232
halmapalus	DSM 8723	Soil	—	183
halochares	DSM 21373	Solar saltern	—	233
halodurans	DSM 497	Soil	—	183
hemicellulosilyticus	DSM 16731	Industrial isolate	—	185
hemicentroti	DSM 23007	Sea urchin	—	234
herbersteinensis	DSM 16534	Medieval wall painting	—	235
horikoshii	DSM 8719	Soil	—	183
horneckiae	NRRL B-59162	Spacecraft assembly facility	—	236
horti	DSM 12751	Soil	—	237
humi	DSM 16318	Soil	—	238
hwajinpoensis	DSM 16206	Sea water	—	239
idriensis	DSM 19097	Blood of patient with neonatal sepsis	—	240
indicus	DSM 15820	Aquifer	—	241
infantis	DSM 19098	Blood of patient with neonatal sepsis	—	240
infernus	DSM 10277	Deep subsurface	—	242
isabeliae	CIP 108578	Sea salt evaporation pond	—	243
isronensis	DSM 21046	High-altitude air sample	—	194, 244
jeotgali	DSM 18226	Traditional Korean fermented food	—	244
koreensis	DSM 16467	Willow rhizosphere	—	245
korlensis	NRRL B-51302	Sand soil	—	246
kribbensis	DSM 17871	Soil	—	247
krulwichiae	DSM 18225	Soil	—	248
lehensis	DSM 19099	Soil	—	249
lentus	DSM 9	Soil	—	250
licheniformis	DSM 13	Soil	Yes	251
litoralis	DSM 16303	Tidal flat	—	252
locisalis	DSM 18085	Hypersaline and alkaline lake	—	253
luciferensis	DSM 18845	Volcanic soil	—	254
luteolus	DSM 22388	Salt field	—	255
macauensis	DSM 17262	Water	—	256
mannanilyticus	DSM 16130	Industrial isolate	—	185
marisflavi	DSM 16204	Tidal flat seawater	—	192
marmarensis	DSM 21297	Mushroom compost	—	257

(Continued)

Table 1 Species in the family *Bacillaceae*, genus *Bacillus*[a] *(Continued)*

Species	Type strain	Source	Contains ComK (NCBI)	Reference
massiliensis	CIP 108446	Cerebrospinal fluid	—	258
megaterium	DSM 32	Soil	Yes	259
methanolicus	DSM 16454	Waste water	Yes	260
methylotrophicus	NCCB 100236	Rice rhizosphere	Yes	261
mojavensis	DSM 9205	Desert soil	Yes	262, 263
muralis	DSM 16288	Mural painting	Yes	264
murimartini	DSM 19154	Wall mural	—	265
mycoides	DSM 2048	Soil	Yes	266
nanhaiensis	DSM 23009	Oyster	—	267
nanhaiisediminis	JCM 16507	Sea sediment	—	268
nealsonii	DSM 15077	Spacecraft assembly facility	—	269
neizhouensis	DSM 19794	Sea anemone	—	270
niabensis	DSM 17723	Cotton-waste composts	—	271
niacini	DSM 2923	Soil	—	272
novalis	DSM 15603	Soil	—	199
oceanisediminis	JCM 16506	Marine sediment	—	273
odysseyi	DSM 18869	Mars Odyssey spacecraft	—	274
okhensis	JCM 13040	Saltpan	—	275
okuhidensis	DSM 13666	Spa water	—	276
oleronius	DSM 9356	Hindgut flora of termite	—	277
oshimensis	DSM 18940	Soil	—	278
panaciterrae	DSM 19096	Soil	—	279
patagoniensis	DSM 16117	South American saltbush rhizosphere	—	280
persepolensis	DSM 21632	Hypersaline lake	—	281
plakortidis	DSM19153	Sea sponge	—	265
pocheonensis	DSM 18135	Soil	—	282
polygoni	JCM 14604	Indigo balls	—	283
pseudalcaliphilus	DSM 8725	Soil	—	183
pseudofirmus	DSM 8715	Soil	Yes	183
pseudomycoides	DSM 12442	Soil	Yes	284
psychrosaccharolyticus	DSM 6	Lowland marsh soil	—	285
pumilus	DSM 27	Soil	Yes	286
purgationiresistans	DSM 23494	Water from drinking-water treatment plant	—	287
qingdaonensis	JCM 14087	Sea salt	—	288
rigui	JCM 16348	Fresh water from wetland	—	289
ruris	DSM 17057	Raw milk	—	290
safensis	DSM 19292	Spacecraft assembly facility	Yes	291
salarius	DSM 16461	Salt lake	—	292
saliphilus	DSM 15402	Mineral pool	—	293
schlegelii	DSM 2000	Lake sediment	—	294
selenatarsenatis	DSM18680	Effluent drain of glass factory	Yes	295
selenitireducens	DSM 15326	Lake	—	296
seohaeanensis	DSM 16464	Solar saltern	—	297
shackletonii	DSM 18868	Volcanic soil	—	298
siamensis	KCTC 13613	Salted crab	Yes	299
simplex	DSM 1321	Soil	Yes	286
siralis	DSM 13140	Silage	—	300
smithii	DSM4216	Cheese	Yes	301
soli	DSM 15604	Soil	—	199
solisalsi	JCM 14863	Soil	—	302
sonorensis	DSM 13779	Soil	Yes	303
sporothermodurans	DSM 10599	UHT-treated milk[c]	—	304

(Continued)

Table 1 (*Continued*)

Species	Type strain	Source	Contains ComK (NCBI)	Reference
stratosphericus	JCM 13349	High-altitude air sample	—	182
subterraneus	DSM 13966	Deep subsurface aquifer	—	305
subtilis	DSM 10	Hay	Yes	306
taeanensis	DSM 16466	Solar saltern	—	307
tequilensis	LMG 25326	Ancient tomb shaft	Yes	308
thermoamylovorans	LMG 18084	Palm wine	—	309
thermocloacae	DSM 5250	Sewage sludge	—	310
thermolactis	DSM 23332	Raw milk	—	311
thioparans	CECT 7196	Wastewater treatment culture system	—	312
thuringiensis	DSM 2046	Flour moth caterpillars	Yes	313
trypoxylicola	KCTC 13244	Guts of beetle larvae	—	314
tusciae	DSM 2912	Geothermal pond	—	315
vallismortis	DSM 11031	Soil	—	316
vedderi	DSM 9768	Bauxite-processing tailing pond mud	—	317
vietnamensis	JCM 11124	Fish sauce	—	318
vireti	DSM 15602	Soil	—	199
wakoensis	JCM 9140	Industrial isolate	—	185
weihenstephanensis	DSM 11821	Milk	Yes	319
xiaoxiensis	DSM 21943	Soil	—	320

[a]All strains are type strains. Abbreviations: ATCC, American Type Culture Collection (www.atcc.org); CCM, Czech Collection of Microorganisms (http://web.natur.cuni.cz/fccm/colleczе.htm#ricp); CECT, Collección Española de Cultivos Tipo (Spanish Collection of Type Cultures (www.cect.org); CGMCC, China General Microbiological Culture Collection (http://english.im.cas.cn/rh/ss/200910/t20091009_44841.html); CIP, Collection of the Institut Pasteur (http://cip.pasteur.fr/rech-bacteries-gb.html); DSM, Deutsche Sammlung von Mikroorganismen und Zellculturen (DSMZ) (German Collection of Microorganisms and Cell Cultures (www.dsmz.de); JCM, Japan Collection of Microorganisms (www.jcm.riken.jp); KCTC, Korea Collection for Type Cultures (http://kctc.kribb.re.k:/English/ekctc.aspx); LMG, the Belgian Coordinated Collection of Microorganisms, BCCM/LMG Culture Collection (http://bccm.belspo.be/about/lmg.php); NBRC, NITE Biological Resource Center (http://www.nbrc.nite.go.jp/NBRC2/NBRCDispSearchServlet); NCCB, Netherlands Collection of Bacteria (http://www.cbs.knaw.nl/About/nccb.aspx); NRRL, Northern Regional Research Laboratory, United States Department of Agriculture (http://nrrl.ncaur.usda.gov/). For an up-to-date environmental distribution of prokaryotes, the reader is directed to consult the EnvDB database (http://metagenomics.uv.es/envDB).
[b]—, genome data unavailable.
[c]UHT, ultra high temperature.

environment, or have just accidentally accumulated; for example, see reference 47. Moreover, DNA-based studies are complicated by the relative stability of DNA in the environment, such that DNA from long-dead cells may contribute erroneously to abundance estimates (48). In addition, searches for microbes in increasingly novel or extreme environments, and the creation of new species often based on a single isolate (10), can often leave it unclear whether a newly described isolate actually lives in a particular habitat or was just an adventitious spore landing in that location (20). Nonetheless, in cases where phenotypes measured in the laboratory are consistent with the sampling site, we can be relatively confident that species are active members in the habitat from which they were isolated; for example, *Halobacillus*, *Salibacillus*, and *Salimicrobium* spp. are isolated from saline environments, and *Geobacillus* spp. are isolated from thermal environments. Also, experiments demonstrating the growth of vegetative cells in soils, in animal GI tracts, and on plant roots suggest that *Bacillaceae* are not necessarily just passersby at their sites of isolation (49).

SPORULATION AND GENETIC EXCHANGE

Although the ubiquity of *Bacillaceae* speaks to their physiological diversity, one of the most striking variable phenotypes is the ability to form spores. Early species descriptions required spore formation for inclusion in the *Bacillaceae*, but newer taxonomic outlines that rely more on cladistic than phenetic data have redefined the family to include nonsporeformers, as mentioned above. This raises the question of whether these lineages were ancient sporeformers that have lost the ability to sporulate, or whether sporulation has evolved independently in multiple lineages of *Firmicutes*. Within the currently accepted 16S-based taxonomic outline, the non-spore-forming lineages do not form a tight cluster but are instead dispersed within spore-forming lineages (50), indicating that sporulation was gained or lost multiple times. But given (i) the developmental complexity of sporulation, (ii) the large numbers of genes involved in sporulation and germination, and (iii) their widely scattered locations on the chromosome, it is difficult to imagine gain events being so numerous; thus, it is most parsimonious to hypothesize

Table 2 Genera and species in the family *Bacillaceae*, excluding *Bacillus* spp.[a]

Genus	Species	Type strain	Source	Carries ComK (NCBI)	Reference
Aeribacillus	*pallidus*	DSM 3670	Hot spring	—[b]	321
Alkalibacillus	*filiformis*	DSM 15448	Mineral pool	—	322
	flavidus	KCTC 13258	Solar saltern	—	323
	haloalkaliphilus	DSM 5271	Salt lake	—	324
	halophilus	DSM 17369	Hypersaline soil	—	325
	salilacus	DSM 16460	Salt lake	—	324
	silvisoli	DSM 18495	Forest soil	—	326
Allobacillus	*halotolerans*	LMG 24826	Shrimp paste	—	327
Amphibacillus	*fermentum*	DSM 13869	Sediment from alkaline lake	—	328
	jilinensis	JCM 16149	Sediment from alkaline lake	—	329
	sediminis	KCTC 13120	Lake sediment	—	330
	tropicus	DSM 13870	Sediment from alkaline lake	—	328
	xylanus	JCM 7361	Manure/rice straw compost	—	331
Anaerobacillus	*alkalidiazotrophicus*	NCCB 100213	Alkaline soil	—	332, 333
	alkalilacustris	DSM 18345	Sediment from alkaline lake	—	333
	arseniciselenatis	DSM 15340	Mono Lake	—	333
	macyae	DSM 16346	Gold mine	—	333
Anoxybacillus	*amylolyticus*	DSM 15939	Geothermal soil	—	334
	ayderensis	NCCB 100050	Hot spring	—	335
	contaminans	DSM 15866	Contaminated gelatin	—	230
	eryuanensis	KCTC 13720	Hot spring	—	336
	flavithermus	DSM 2641	Manure	Yes	337
	gonensis	NCCB 100040	Hot spring	—	338
	kamchatkensis	DSM 14988	Geothermal environment	—	339
	kestanbolensis	NCCB 100051	Hot spring	—	335
	mongoliensis	DSM 19169	Alkaline hot spring	—	340
	pushchinoensis	DSM 12423	Manure	—	337
	rupiensis	DSM 17127	Hot spring	—	341
	salavatliensis	DSM 2262	Well pipeline sediment	—	342
	tengchongensis	KCTC 13721	Hot spring	—	336
	thermarum	DSM 17141	Thermal mud	—	343
	voinovskiensis	JCM 12111	Hot spring	—	344
Aquisalibacillus	*elongatus*	DSM 18090	Saline lake	—	345
Bacillus	see Table 1				
Cerasibacillus	*quisquiliarum*	DSM 15825	Decomposing kitchen refuse	—	54
Falsibacillus	*pallidus*	KCTC 13200	Forest soil	—	346, 347
Filobacillus	*milensis*	DSM 13259	Sediment near hydrothermal vent	—	348
Geobacillus	*caldoxylosilyticus*	DSM 12041	Soil	Yes	349, 350
	debilis	DSM 16016	Soil	—	351
	jurassicus	DSM 15726	High-temperature oil field	—	352
	kaustophilus	DSM 7263	Pasteurized milk	Yes	353
	lituanicus	DSM 15325	High-temperature oil field	—	354
	stearothermophilus	DSM 22	Spoiled canned corn	—	355
	subterraneus	DSM 13552	Oil field	—	353
	tepidamans	DSM 16325	Beet sugar factory	—	356
	thermantarcticus	DSM 9572	Geothermal soil	—	357, 358
	thermocatenulatus	DSM 730	Unknown	—	353, 359, 360
	thermodenitrificans	DSM 465	Soil	Yes	353, 361, 362
	thermoglucosidans	DSM 2542	Soil	Yes	358, 363
	thermoleovorans	DSM 5366	Unknown	Yes	364
	toebii	DSM 14590	Hay compost	—	365
	uzenensis	DSM 13551	Oil field	—	353

(Continued)

Table 2 *(Continued)*

Genus	Species	Type strain	Source	Carries ComK (NCBI)	Reference
	vulcani	DSM 13174	Sediment near hydrothermal vent	—	366, 367
Geomicrobium	*halophilum*	DSM 21769	Soil	—	368
Gracilibacillus	*boraciitolerans*	DSM 17256	Soil	—	369
	dipsosauri	DSM 11125	Desert iguana	—	370, 371
	halophilus	DSM 17856	Saline soil	—	372
	halotolerans	DSM 11805	Salt lake	—	371
	lacisalsi	DSM 19029	Salt lake	—	373
	orientalis	CCM 7326	Salt lake	—	374
	saliphilus	DSM 19802	Salt lake	—	375
	thailandensis	JCM 15569	Fermented fish	—	376
	ureilyticus	JCM15711	Saline-alkaline soil	—	377
Halalkalibacillus	*halophilus*	DSM18494	Soil	—	378
Halobacillus	*aidingensis*	JCM12771	Salt lake	—	379
	alkaliphilus	DSM18525	Salt lake	—	380
	campisalis	KCTC 13144	Solar saltern	—	381
	dabanensis	JCM12772	Salt lake	—	379
	faecis	KCTC 13121	Mangrove area	—	382
	halophilus	DSM2266	Salt marsh soil	Yes	383
	karajensis	DSM14948	Saline soil	—	384
	kuroshimensis	DSM18393	Deep-sea methane seep	—	385
	litoralis	DSM10405	Saline sediment	—	383
	locisalis	DSM16468	Solar saltern	—	386
	mangrovi	CCM7397	Black mangrove	—	387
	profundi	DSM18394	Deep-sea methane seep	—	385
	salinus	JCM 11546	Salt lake	—	388
	salsuginis	DSM21185	Subterranean brine	—	32
	seohaensis	KCTC 13145	Solar saltern	—	389
	trueperi	DSM10404	Saline sediment	—	383
	yeomjeoni	DSM17110	Solar saltern	—	390
Halolactibacillus	*alkaliphilus*	NBRC 103919	Alkaline lake	—	391
	halophilus	DSM17073	Marine organisms	—	392
	miurensis	DSM17074	Marine organisms	—	392
Lentibacillus	*halodurans*	DSM18342	Salt lake	—	393
	halophilus	JCM 12149	Fish sauce	—	394
	jeotgali	KCTC 13300	Salt-fermented seafood	—	395
	juripiscarius	JCM 12147	Fish sauce	—	396
	kapialis	JCM 12580	Fermented shrimp paste	—	397
	lacisalsi	DSM16462	Saline lake	—	398
	persicus	DSM22530	Saline lake	—	399
	salarius	DSM16459	Saline sediment	—	400
	salicampi	JCM 11462	Salt field	—	401
	salinarum	KCTC 13162	Solar saltern	—	402
	salis	DSM16817	Salt lake	—	403
Marinococcus	*halophilus*	DSM20408	Salted mackerel	—	404
	halotolerans	DSM16375	Saline soil	—	405
	luteus	KCTC 13214	Salt lake	—	406
Natronobacillus	*azotifigens*	NCCB100215	Soda soil	—	407
Oceanobacillus	*caeni*	KCTC 13061	Wastewater	—	408
	chironomi	DSM 18262	Chironomid egg mass	—	409
	iheyensis	DSM 14371	Deep-sea sediment	Yes	410
	kapialis	KCTC 13177	Fermented shrimp paste	—	411

(Continued)

Table 2   Genera and species in the family *Bacillaceae*, excluding *Bacillus* spp.[a] *(Continued)*

Genus	Species	Type strain	Source	Carries ComK (NCBI)	Reference
	kimchii	DSM23341	Mustard kimchi	—	412
	locisalsi	KCTC 13253	Solar saltern	—	413
	neutriphilus	JCM 15776	Activated sludge	—	414
	oncorhynchi	JCM 12661	Skin of rainbow trout	—	415
	picturae	DSM 14867	Deteriorated mural painting	—	416, 417
	profundus	DSM 18246	Deep-sea sediment core	—	418
	soja	JCM 15792	Soy sauce production equipment	—	419
Ornithinibacillus	*bavariensis*	DSM 15681	Pasteurized milk	—	420
	californiensis	DSM 16628	Coastal surface sediment	—	420
	contaminans	DSM 22953	Human blood sample	—	421
	scapharcae	JCM17314	Dead ark clam	Yes	422
Paraliobacillus	*quinghaiensis*	DSM 17857	Saline lake sediment	—	423
	ryukyuensis	DSM 15140	Decomposing marine alga	—	424
Paucisalibacillus	*globulus*	LMG 23148	Potting soil	—	425
Piscibacillus	*halophilus*	DSM 21633	Salt lake	—	426
	salipiscarius	JCM 13188	Fermented fish	—	427
Pontibacillus	*chungwhensis*	DSM 16287	Solar saltern	—	428
	halophilus	DSM 19796	Sea urchin	—	429
	litoralis	DSM 21186	Sea anemone	—	430
	marinus	DSM 16465	Solar saltern	—	431
	yanchengensis	NRRL B-59408	Salt field soil	—	432
Pullulanibacillus	*naganoensis*	DSM 10191	Soil	—	433, 434
Saccharococcus	*caldoxylosilyticus*	DSM 12041	Soil	—	349
	thermophilus	DSM 4749	Beet sugar factory	—	435
Salimicrobium	*album*	DSM 20748	Solar saltern	—	404, 436
	flavidum	KCTC 13260	Solar saltern	—	437
	halophilum	DSM 4771	Solar saltern	—	436
	luteum	KCTC 3989	Solar saltern	—	436
	salexigens	DSM 22782	Salted hides	—	438
Salinibacillus	*aidingensis*	JCM 12389	Neutral saline lake	—	439
	kushneri	JCM 12390	Neutral saline lake	—	439
Salirhabdus	*euzebyi*	LMG 22839	Saltern	—	440
Salsuginibacillus	*halophilus*	CGMCC 1.7653	Soda lake	—	441
	kocurii	DSM 18087	Soda lake sediment	—	442
Sediminibacillus	*albus*	DSM 19340	Salt lake	—	443
	halophilus	DSM 18088	Salt lake	—	444
Sinobaca	*qinghaiensis*	DSM 17008	Saline soil	—	445, 446
Streptohalobacillus	*salinus*	DSM 22440	Saline soil	—	447
Tenuibacillus	*multivorans*	NBRC 100370	Saline soil	—	448
Terribacillus	*aidingensis*	CGMCC 1.8913	Salt lake sediment	—	449
	goriensis	DSM 18252	Coastal seawater	—	450, 451
	halophilus	KCTC 13937	Field soil	—	452
	saccharophilus	KCTC 13936	Field soil	—	452
Thalassobacillus	*cyri*	DSM 21635	Salt lake	—	453
	devorans	DSM 16966	Hypersaline habitat	—	454
	hwangdonensis	KCTC 13254	Tidal flat sediment	—	455
	pellis	DSM 22784	Salted hides	—	456
Tuberibacillus	*calidus*	DSM 17572	Compost	—	433
Virgibacillus	*alimentarius*	JCM 16994	Traditional Korean food	—	457

(Continued)

Table 2 Genera and species in the family *Bacillaceae*, excluding *Bacillus* spp.[a] *(Continued)*

Genus	Species	Type strain	Source	Carries ComK (NCBI)	Reference
	arcticus	DSM 19574	Arctic permafrost	—	458
	byunsanensis	KCTC 13259	Solar saltern	—	459
	campisalis	KCTC 13727	Solar saltern	—	460
	carmonensis	DSM 14868	Deteriorated mural painting	—	416
	chiguensis	CGMCC 1.6496	Saltern	—	461
	dokdonensis	DSM 16826	Seawater	—	462
	halodenitrificans	DSM 10037	Solar saltern	—	463, 464
	halophilus	KCTC 13935	Soil	—	465
	kekensis	DSM 17056	Salt lake	—	466
	koreensis	KCTC 3823	Salt field	—	417
	marismortui	DSM 12325	Water sample from Dead Sea	—	416, 467
	necropolis	DSM 14866	Biodeteriorated mural painting	—	416
	olivae	DSM 18098	Olive-processing wastewater	—	468
	pantothenticus	DSM 26	Soil	—	469, 470
	proomii	DSM 13055	Soil	—	469
	salarius	DSM 18441	Salt lake	—	471
	salexigens	DSM 11483	Saline soil	—	416, 472
	salinus	DSM 21756	Salt lake sediment	—	473
	sediminis	KCTC 13193	Salt lake	—	474
	siamensis	JCM 15395	Fermented fish	—	475
	soli	DSM 22952	Mountain soil	—	476
	subterraneus	DSM 22441	Saline soil	—	477
	xinjiangensis	DSM 19031	Salt lake	—	478
Vulcanibacillus	*modesticaldus*	DSM 14931	Deep-sea hydrothermal vent	—	479

[a]All strains are type strains. Abbreviations as in Table 1.
[b]—, genome data unavailable.

that sporulation has likely been lost several times during the evolution and divergence of *Bacillaceae* lineages (51). This hypothesis must be qualified by the potential for misdiagnosing sporeformers as nonsporeformers, because we know little or nothing about the spore-forming capability of most species of *Bacillaceae* in their natural habitats (28, 52, 53). In addition to the presence or absence of spore formation, the shape and placement of the developing spore are known to be variable and were early diagnostic tools for species classification. Spores may begin forming in the terminal, subterminal, or central region of the cell and may take on an ellipsoid or spherical shape. Unusual spore phenotypes have also been discovered, including the cherrylike spores of *Cerasibacillus* (the "cherry" *Bacillus*) (54) and the budding out of sporelike resting cells in the filamentous *Bacillus* sp. NAF001 (55). Additional genome data, along with comparative studies in the laboratory and the field, will no doubt reveal even greater diversity in spore formation.

From the evolutionary perspective, the widespread distribution of spores can be viewed as a driver of evolution within the *Bacillaceae*. When nutrients become exhausted in a particular environment, spore-forming species respond by producing metabolically dormant endospores that exhibit a high degree of resistance to environmental insults (41, 42). Spores can be transported by water or wind to distant sites, and during these sojourns spore DNA is subjected to various mutagenic damages induced by exposure to heat, desiccation, oxidants, or solar UV radiation (41). Subsequent germination of spores in new environments exposes them to novel environmental pressures that select for mutants exhibiting higher fitness, leading to subsequent local adaptation of the population.

Horizontal gene transfer can also accelerate evolution, and the *Bacillaceae* are replete with mechanisms that result in the exchange of genes. Phage-mediated genetic exchange via either generalized or specialized transduction has been demonstrated in numerous members of the family (*B. anthracis*, *B. cereus*, *B. subtilis*,

Bacillus megaterium, Bacillus pumilus, B. thuringiensis, Geobacillus stearothermophilus) (56–63). Conjugative transfer of plasmids and transposons among the *Bacillaceae* exhibits special promiscuity (see, for example, reference 64); on the other hand, fewer examples of conjugative transfer of chromosomal DNA markers have been documented (65). DNA-mediated transformation within the *Bacillaceae* was first discovered in a few strains of *B. subtilis* (66). However, genetic competence for transformation was not originally considered to be widespread among the *Bacillaceae*, perhaps because relatively few species had been tested, and the conditions for inducing competence in the laboratory were historically arrived at empirically (66–68). This view is changing, partly as a result of recent whole-genome sequencing efforts, which have uncovered that competence (*com*) genes are widespread among the *Bacillaceae*. For example, a simple search of the NCBI protein database for homologues to ComK, the master competence regulator first identified in *B. subtilis* (69), revealed the presence of ComK homologues in no fewer than 42 species of *Bacillaceae* (Tables 1 and 2), as well as numerous ComK homologues widely distributed throughout the families *Listeriaceae, Planococcaceae, Staphylococcaceae, Streptococcaceae*, and the class *Clostridia* (data not shown). Future genome-sequencing projects will no doubt reveal many more competence gene homologues.

But does the existence of competence genes in a bacterial genome necessarily mean that the bacterium can be transformed? Many competence genes probably evolved for the uptake of DNA as a nutrient. Indeed, current evidence supports the evolution of competence as a means of obtaining food from the environment in times of nutrient stress, with the occasional recombination of some incoming DNA into the chromosome by the already-present DNA replication/repair proteins, resulting in genetic transformation (70, 71). In support of the notion that competence genes found in bacterial genomes are in fact functional, recent work has shown that strains of *B. subtilis* and *B. cereus* that are not transformable in the laboratory, but that contain an apparently complete set of competence genes, can be induced to become transformable by transient overexpression of the *comK* master regulator gene under control of an IPTG (isopropyl-β-D-thiogalactopyranoside)-inducible promoter carried on an unstable plasmid (72–74). Therefore, it appears that some members of the *Bacillaceae* are indeed capable of developing competence under certain laboratory conditions. These observations bolster the possibility that transformation in the *Bacillaceae* may be much more widespread in the environment than previously believed.

THE PROBLEM OF ANCIENT SPORES

Not only are spore-forming *Bacillaceae* capable of widespread spatial transport, they can in a sense travel across long spans of time, as bacterial spores are probably the longest-lived cellular structures known. Numerous reports have documented the isolation of ancient spores, ranging in age from hundreds to millions of years, from diverse environmental samples such as dried soil in herbarium collections, ancient soils or lake sediments, permafrost soils, ice cores, or the guts of fossilized bees trapped in 25-million-year-old amber. (For an exhaustive list up to 1994, see reference 75.) More recently, in the year 2000, the discovery of *Virgibacillus* spp. isolate 2-9-3 was reported. This sporeformer was recovered from a halite (NaCl) brine inclusion within a 250-million-year-old primary salt crystal obtained from the subterranean Salado Formation near Carlsbad, New Mexico (76).

Such reports prompt the question: what is the limit on spore longevity? Reports of extraordinary spore longevity must first pass the inevitable hurdle of scientific critique; the more extreme the claim, the more severe the skepticism. Most criticisms rely on three basic arguments. First, it is argued that the organisms recovered may represent recent contaminants arising either from ineffective surface sterilization of the ancient sample, or from spores that have recently entered the ancient sample via microscopic fissures. While there is no way of completely eliminating this possibility, reports of rigorous sample selection criteria and sterilization control have been published for scientific scrutiny in the peer-reviewed literature (77–80).

A second criticism of the validity of claims of ancient spores relies on the argument that biological molecules, particularly DNA, cannot survive for such extended periods of time (80, 81). Often quoted as substantiating evidence for this contention is a landmark review by Lindahl (82) describing the various weaknesses in the chemical structure of DNA that render it susceptible to spontaneous degradation. This particular criticism overlooks several factors. (i) The bulk of DNA sensitivity studies were performed on naked DNA in aqueous solution, not on DNA within spores; in fact, Lindahl stressed that DNA packaged in spores is well protected from degradation (82). A substantial literature documents the fact that DNA in bacterial spores is indeed in a much more protected and long-lived state than naked DNA, due to well-described molecular mechanisms for spore DNA damage protection and repair (41, 42, 83). (ii) It is currently unknown what effect the medium in which spores were trapped (e.g., soil, rock, ice, amber, brine) would exert on either slowing or accelerating the rate of molecular degradation. For example, it has been suggested that one factor limiting the longevity of spores entrapped in brine inclusions is the radioactive isotope ^{40}K naturally present in brine, which may result in a radiation environment leading to the rapid fragmentation of DNA and accumulation of lethal "hits" in spores. The question whether the radiation environment prevailing inside subterranean brine inclusions limits spore longevity prompted modeling studies by two groups. One group concluded that the radiation environment was insufficient to fully inactivate spores after 250 Ma (84), while the other group concluded that spores exposed to ambient radiation in halite brine could not survive 250 Ma (85). Using published spore thermal inactivation kinetics, one of us (W.L.N.) concluded that spore populations of mesophilic species stood a poor chance of survival for millions of years, whereas thermophilic spore populations might be predicted to survive such time scales with rather high probabilities (86). In order to test the putative protective effect of NaCl brine on spore longevity, we recently measured the thermal inactivation kinetics of *B. subtilis* 168 spores, suspended either in distilled water or in saturated NaCl, and extrapolated the resulting decimal reduction times

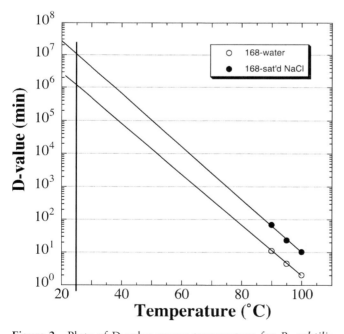

Figure 2 Plots of D value versus temperature for *B. subtilis* 168 spores suspended in water (open circles) or in saturated NaCl (filled circles). The lines are best-fit extrapolations of the data to ambient temperature (25°C; thick vertical line). Data are averages of triplicate determinations that varied by less than 5%.

(D values) to an ambient temperature of 25°C (Fig. 2). The resulting data clearly showed that spores in saturated NaCl were inactivated ~10-fold more slowly than spores in water. However, extrapolation of the curves to 25°C indicated that spores in water and brine exhibited D values of ~1.1×10^6 min (~2 years) and ~1.05×10^7 min (~20 years), respectively (Fig. 2), both of which are much too short to result in significant survival of a spore population for 250 million years.

A third criticism against the existence of ancient spores has come from molecular phylogenetic analyses, using either 16S rDNA sequence comparisons (76, 78) or comparisons of the deduced amino acid sequences of conserved protein-coding genes such as *recA* and *splB* (87). In all cases tested, sequences from the putative ancient sporeformers turned out to share considerable similarity to those of extant species (*Lysinibacillus sphaericus* in the case of the amber isolate; *Virgibacillus marismortui* in the case of the brine isolate). These observations have been used to argue that, because the DNA sequences of the "ancient" sporeformers closely matched those of "modern" bacteria, the isolates themselves must also be "modern" (81, 88). This third criticism highlights a fundamental problem with assigning absolute ages to microbes based on cladistics. In the absence of a microbial fossil record that can be linked to the geologic record, bacterial phylogenetic trees are constructed from the DNA of extant species; thus only relative, not absolute, ages for bacteria can be inferred (89). In an attempt to compensate for this deficit, various algorithms have been devised to transform percentages of nucleotide or amino acid similarity into "evolutionary rates" for bacteria. Such rates are valid only for the particular system under study and are much more accurate when supported by a true fossil record. Very few microbial systems exist that meet these criteria. For example, studies using the aphid endosymbionts *Buchnera* spp. are able assign molecular evolutionary rates with a rather high accuracy because they are externally supported by, and calibrated to, the aphid fossil record (90). But, in the absence of a reliable fossil record, "no gene, nor class of sites within genes, can serve as a reliable molecular clock for bacteria," and, furthermore, "no single evolutionary rate can be applied across diverse bacterial lineages or over broad evolutionary periods" (89). In particular, free-living, spore-forming *Bacillaceae* are subjected to a "feast or famine" existence and while in the spore state may not experience a single chromosomal replication event for years, decades, or centuries at a stretch. Thus, it is probably impossible to calculate a reliable molecular evolutionary rate for sporeformers. Due to the very fact of dormancy, it was thought that sporeformers may exhibit a lower overall rate of evolution than do enteric bacteria (87), but this would be very sensitive to the amount of time spores spend dormant, as spore-forming and non-spore-forming lineages could have equal rates of evolution even if sporeformers spent 10 to 30% of their time as dormant spores and if substitutions were equally as likely to occur during DNA replication or via other mutational processes (51). Indeed, an attempt has been made to test the hypothesis that spore-forming bacteria evolve more slowly than related non-spore-forming bacteria by comparing inferred rates of amino acid substitution between those two groups, and no difference in rates was detected (91). However, the time scale in this study was constructed from evolutionary events not related to the *Firmicutes*, again illustrating the difficulties and drawbacks of estimating evolutionary rates for the *Bacillaceae* in the absence of a fossil record for these organisms.

Indeed, the very concept of "modern" versus "ancient" bacteria can itself be called into question. Due to the cyclic geological processes of weathering, sedimentary deposition, compression, and uplift, soil is constantly formed from rock by erosion and ultimately compressed back into rock. Thus, it is reasonable to presume that subsurface endolithic spore-forming bacteria of indeterminate ages are constantly being exposed in soil at the Earth's surface, being distributed over the surface on dust particles and in water, and ultimately reentering the soil and bedrock by percolation and sedimentary deposition. Given these cyclic conditions, is a newly isolated bacterium "modern," thus a standard for comparison with a putative "ancient" bacterium, just because we humans obtained it from the environment yesterday?

THE ECOLOGICAL APPROACH TO STUDYING EVOLUTION IN *BACILLACEAE*

The evolution of any bacterial species depends on its environment: which abiotic and biotic factors influence survival and reproduction? Identifying ecological components of fitness and the ways in which they contribute to local adaptation is incredibly difficult in bacteria, for the same reasons as discussed in the context of biogeography above. Despite these limitations, advances in technology are enabling us to reach a more profound understanding of the ecology and evolution of the *Bacillaceae*. Ecological approaches to understanding evolution in *Bacillaceae* can be categorized into two basic methodologies: (i) sampling natural populations for culture and analysis in the laboratory, and

(ii) describing natural populations based on genetic data obtained *in situ*. Most studies have combined these approaches in different ways to address questions of adaptation to abiotic and biotic factors.

In its simplest form, the evolution of prokaryotes can be viewed as a clonal process; i.e., haploid bacteria undergo binary fission, variability within a population arises from random spontaneous mutations, and horizontal gene transfer does not occur. In cases that have been studied, some bacteria such as *Escherichia coli* and *Salmonella enterica* serovar Typhimurium tend to conform to this general "clonal" pattern of evolution (92, 93). However, studies of natural populations of *B. subtilis* in Sonoran desert soils revealed evidence of extensive genetic exchange (94), distinct from the "clonal" paradigm observed in enteric bacteria and suggesting that horizontal gene transfer and recombination were driving genetic variability in natural *B. subtilis* populations. The results from these early studies in many way presage the discovery of the dominance of horizontal gene transfer during bacterial evolution, as demonstrated using whole-genome, network-based comparisons (483).

The most impressive studies addressing the ecology of *Bacillaceae* have been performed at two field sites: Evolution Canyons I and II in Israel, and Death Valley in California, USA. Evolution Canyons I and II are two east-west running canyons separated by ~40 km, in which the north-facing slopes receive less direct sunlight than the south-facing slopes. Although the opposing slopes of each canyon are separated by only tens to hundreds of meters, the resulting environments of the two slopes differ markedly (95). In the Evolution Canyons, researchers have isolated strains of *Bacillus simplex* at locales known to differ in their abiotic characteristics such as UV intensity, temperature, and soil composition (96, 97). Measurements of survival under UV, mutation rates, and high-temperature growth in the laboratory showed strong correlations between the environment at the site of isolation and the phenotype of the bacterial isolate, indicating local adaptation to prevailing conditions. These were not merely physiological adaptations: *B. simplex* strains isolated from the same slope of Evolution Canyons I and II, at geologically distant sites, were found to be more genetically similar than strains isolated from opposite slopes within the same canyon (96).

Local adaptation to particular environments was also found in strains within the *B. subtilis*-*B. licheniformis* clade isolated from shaded versus sun-exposed areas of Death Valley, CA (98). It was demonstrated that ecotypes with more warm-adapted membrane fatty acid profiles were isolated more frequently from sites with greater solar exposure. The researchers used genetic data obtained from these strains to test the "ecotype formation" model of evolution (99) and found that, indeed, many "hidden" ecotypes were present in the strain samples (98). The existence of these hidden ecotypes suggests that the diversity within *Bacillaceae* is even greater than previously imagined. As seen in previous sections on taxonomy and biogeography, whole-genome sequence comparisons are also providing new insights into various aspects of evolutionary ecology. Pan-genome analyses allow for the identification of the "dispensable" subsets of genes in a phylogenetic clade that are important for adaptation to different niches (486). Such analyses have been applied to clades at levels of resolution ranging from extremely broad (Class Bacilli) (487) to extremely narrow (*B. subtilis* subsp. *spizizenii*) (488) and are helping to constrain various evolutionary models of ecotype formation.

Although there is much to be learned by the ecological studies just described, many researchers have also taken high-throughput approaches to identify potential interactions between bacterial species. Shank et al. (100) screened for interactions between *B. subtilis* and numerous soil isolates of multiple bacterial families to identify which species interactions resulted in the production of biofilm matrices. Interestingly, the microbes that were most able to drive the expression of biofilm matrices were other members of the genus *Bacillus* (100). A genomic approach to identifying species interactions was used by Hooper et al. (101), assuming that genomes sharing horizontally transferred transposase genes must have coexisted in nature at some point in time. Although in the case of *Bacillus* most of the inferred interactions tended to be with other *Bacillus* species, many interactions were also proposed between *Bacillus* species and *Escherichia*, *Shigella*, *Clostridium*, *Klebsiella*, *Salmonella*, *Yersinia*, and *Coxiella* spp. (101). Yet another study used a high-throughput automated literature-mining approach to identify potential coexisting species (102). Similar to the results using genomic transposases, *Bacillus* species were often "found" together or with *Clostridium*, *Enterococcus*, *Listeria*, *Lactobacillus*, *Pseudomonas*, *Staphylococcus*, and *Vibrio* spp. (102). These high-throughput approaches agree in their findings that *Bacillus* species are likely to interact with each other in nature, where in combination they may express particular phenotypes enhancing their colonization and survival in different niches.

Some of the most fascinating windows into the ecology of the *Bacillaceae* have come from simply examining the biology of organismal consortia. One recent example is the identification of *Bacillus* spp. as mutualists of termites (103). Not only did the *Bacillus* spp. help the termites maintain their cellulose-rich diet, they also helped suppress "weed" fungi from invading the fungus gardens grown by the termites for nitrogen (103). *Bacillus* spp. were abundant within the fungus gardens and the termite guts, resulting in a constant source of beneficial bacteria for the termite. Although the presence of *Bacillus* spp. would have been detected in termite guts and fungus gardens by using purely DNA-based approaches, subsequent laboratory experiments were essential to identify these interactions as beneficial or detrimental. The inability to culture many species has typically hindered such experiments in the past, causing us to rely on DNA-based studies, but new approaches to culturing "unculturable" microbes show promise (104).

EXPERIMENTAL EVOLUTION

Evolution can be observed in real time by using laboratory populations of bacteria owing to their short generation times, large population sizes, and ease of experimental manipulation. Important evolutionary parameters, including mutation rates, effective population size, fitness trajectories, and quantitative measures of specific traits, are relatively easy to measure in bacterial populations (105, 106). As usual, most laboratory evolution studies in the *Bacillaceae* have centered on the model laboratory organism *B. subtilis*.

The first laboratory evolution experiments date back to the 1970s, when Graham and Istock began to investigate the role of genetic exchange during *B. subtilis* evolution in soil by cultivating pairs of genetically marked strains in presterilized soil microcosms. They first observed genetic exchange of markers between the two input strains, leading to new recombinants in addition to the parental strains (107). In subsequent experiments, they observed that certain recombinant phenotypes eventually came to dominate the mixed-soil populations, thus demonstrating that not only do genetic exchange and recombination occur in *B. subtilis* growing in soil, but the process can lead to a rapid increase in fitness (108, 109). As noted in the previous section, extensive genetic exchange was later shown by Istock et al. (94) to occur also in natural Sonoran desert soil populations of *B. subtilis*.

In the year 2000, we began to take a laboratory approach to explore the role that stringent or relaxed selection for sporulation plays in the evolution of spore-forming bacteria. We designed a long-term evolution experiment using genetically marked congenic strains of

B. subtilis. Five batch cultures were propagated in a complex nutrient broth-based liquid medium that induced sporulation (Sporulating, or S populations), and five cultures in a similar medium that repressed sporulation with 1% (wt/vol) glucose (Repressed, or R populations). At daily intervals we simply diluted the five R populations 1:100 into fresh repressing medium, but subjected the five S populations to heat shock (80°C, 10 min) to select for spores before diluting them 1:100 into fresh sporulation medium (110). We continued this regimen for 6,000 generations, stored frozen aliquots of each culture at 50-generation intervals, and tracked the evolution of these cultures by using a variety of methods ranging from simple observations and phenotypic tests to population genetic and whole-genome techniques (110–117).

We observed a number of interesting events during the evolution of these populations. First, we noted that very rapidly, within 100 to 200 generations, both S and R populations began to exhibit a large diversity of colony morphotypes (which subsequently declined in later generations), and we observed sporulation-defective (Spo⁻) colonies arising in the R populations. These observations suggested to us that the populations were undergoing rapid diversification, consistent with a rise in mutation rate. By performing fluctuation tests, we were able to confirm that indeed the mutation rate in the evolving populations was increasing by as much as 1 to 2 orders of magnitude (110). At present, the exact mechanism of increased mutation rate in the evolving cultures is unknown but is likely similar or identical to that underlying the well-studied phenomenon of stationary-phase mutagenesis observed in *B. subtilis* cells subjected to nonlethal stress (118–125).

Competition experiments showed that, over the course of their evolution, both S and R populations gained in fitness over the ancestral strain, in part due to their faster exponential growth rates and shorter lag periods in their respective media (110). In addition, over the course of evolution, both S and R populations accumulated auxotrophic mutations (110), whereas only R populations accumulated mutations in *spo* genes rendering cells Spo⁻ (112). In short, both S and R populations generated diversity via an increase in mutation rate and evolved higher fitness in their respective environments by accumulating mutations that (i) decreased the lag time involved in adjusting to fresh medium, (ii) increased their exponential growth rates, and (iii) inactivated biosynthetic pathways not needed for growth in complex medium (110). In the case of R populations, mutations accumulated that inactivated the sporulation pathway, as

sporulation was no longer being actively selected for in these cells.

We became intensely interested in pinpointing and understanding the changes in the *B. subtilis* genome responding to stringent or relaxed selective pressure for sporulation, and we were greatly aided in this effort by the fact that, in the early years of the 21st century, genomic-scale technologies had become increasingly affordable for single laboratories. We thought that the propagation of *B. subtilis* for several thousand generations in a relatively predictable environment might result in "reductive evolution" or "genomic erosion," i.e., loss of large dispensable regions of the genome (126). We tested this notion by labeling chromosomal DNA from all 10 evolved strains at generation 6,000, probing DNA microarrays made from ancestral strain 168, and looking for spots with decreased fluorescent intensity (114). No loss of genomic DNA was detected in any of the five S populations or three of the R populations. But a deletion of ~10 kbp was detected in two of the R populations, corresponding to loss of the nonessential *ppsC* and *ppsD* genes of the *ppsABCDE* operon encoding the lipopeptide antibiotic plipastatin (114). Deletion of these two genes had also been reported in *B. subtilis* laboratory strain PY79 during its descent from ancestral laboratory strains 168 and W23 (127). We showed that the *ppsCD* deletions likely occurred by recombination between direct repeat regions embedded within the operon (128). Thus, most of the mutations leading to the observed adaptive phenotypes over 6,000 generations were not due to gross genomic losses, but more likely due to small changes in the genome, such as single-nucleotide polymorphisms (SNPs) or small insertion/deletions (InDels) (128).

In nature, *Bacillaceae* must be able to deploy a large host of functions in order to cope with a constantly changing environment, a concept known as phenotypic plasticity. Using DNA microarrays, we observed that descendants from R populations in particular had lost plasticity in their ability to alter their global transcription patterns in response to growth in the sporulation-inducing environment (128). This finding, coupled with the greater fitness of R populations in sporulation-repressing media, indicated that the populations were evolving from generalists into niche specialists (128).

About 1,000 to 2,000 generations into the evolution experiment, we were fortunate to actually witness an evolutionary sweep as it occurred. In four of the five R populations, a distinct small-colony morphotype arose and swept through the populations within a few hundred generations (115). In addition to their small-colony morphologies, all of the new variants grew to a higher cell density in sporulation-repressing medium and exhibited altered phenotypes such as cell filamentation and/or loss of motility (115). In order to perform more detailed analysis, we isolated a strain from the postsweep event in one of the R populations (strain WN716), and, going back to our frozen stock collection, we also isolated a strain typical of the same population before the sweep (strain WN715).

Pairwise competition experiments proved that postsweep strain WN716 had indeed gained a large competitive advantage over both the ancestor and the presweep strain WN715; in mixed culture, postsweep strain WN716 grew at a faster exponential growth rate and continued to grow in postexponential phase, while presweep strain WN715 dramatically lost viability during this phase (115). We used microarrays to compare the transcriptome profiles in early postexponential-phase cultures of strains WN715 and WN716 cultures to identify differentially transcribed genes. We found that the operons encoding pyruvate dehydrogenase and purine and pyrimidine biosynthetic pathways were transcribed at higher levels in postsweep strain WN716, as well as numerous genes involved in adaptation to stress. The data are consistent with WN716's continued growth and its response to some (at that time, unidentified) stress in the postexponential phase (115). At the same time, we found decreased transcription in strain WN716 of numerous operons encoding autolysins, flagella and chemotaxis functions, membrane-associated transporters and cytochromes, sporulation initiation, competence, extracellular enzymes, and antibiotic production. These data were consistent with the filamentous and nonmotile phenotype of strain WN716, as well as with a defect in its activation of transition-state functions in favor of continued growth in the postexponential phase (115). We were intrigued by the observation that in postsweep strain WN716 transcription of the *alsSD* operon, responsible for acetoin fermentation, was among the most severely downregulated, and that as a result strain WN716 had switched its fermentation pattern from the production of acetoin to acetate, with concomitant lowering of the medium pH from 7.0 to ~4.5 (115). This observation suggested that acid stress was the environmental condition to which WN716 had evolved resistance and which presweep strain WN715 lacked. Taken together, the data suggested that the increased fitness of WN716 was not due to a single mutation, but that a number of changes must have occurred during its evolution in order to result in its complex alteration of multiple phenotypic traits.

In order to understand the exact nature of the adaptive changes that had occurred during the evolution of strain WN716, we embarked on a genome-sequencing project in which we performed comparative sequence analysis of the genomes of the ancestral strain, presweep strain WN715, and postsweep strain WN716 (116). We found a total of 34 SNPs and +1 insertions in coding regions of known annotated genes, and 11 SNPs in intergenic regions, that had occurred in the short time (~460 generations) that had elapsed during the population sweep between WN715 and WN716. One of these mutations was found to be a nonsense mutation predicted to inactivate the extracytoplasmic factor (ECF) sigma factor SigW, which controls a small regulon whose members are involved in resistance to membrane-damaging agents and bacteriocins, and another mutation was found in the *alsR* gene, the positive transcriptional regulator of acetoin fermentation (116). Although the loss of motility and chemotaxis in strain WN716 likely contributed to its increased fitness, and multiple motility/chemotaxis operons were downregulated, we failed to find any mutation in the motility-specific sigma factor SigD (115, 116).

We directly tested whether the inactivation of these pleiotropic regulators was indeed responsible for the increased fitness of strain WN716 by constructing strains carrying insertional knockout mutations of the *alsR*, *sigW*, and *sigD* genes singly and in all possible combinations in ancestral strain WN624, then testing their relative fitnesses in pairwise competition experiments versus the ancestor (117). Each knockout mutation tested was observed to cause an incremental increase in fitness, and the triple *alsR sigW sigD* mutant displayed the highest degree of fitness increase compared with ancestral strain WN624 (relative fitness of 1.130 + 0.013). However, the triple mutant still did not achieve the dramatic increase in fitness displayed by strain WN716 itself (relative fitness of 1.322 + 0.003) (117). Therefore, using the genomic approach, we were able to uncover some, but not all, genetic factors responsible for the increase in WN716's fitness.

LABORATORY DOMESTICATION IN *BACILLACEAE*

In the laboratory evolution experiment above, we saw that continual propagation of a microbe under a specific set of conditions will lead to its genetic adaptation toward optimization of growth under those conditions. In fact, the very act of bringing environmental isolates of bacteria into the laboratory for either practical use or detailed study inevitably begins an evolutionary pathway in which the actions of the experimenter serve as agents of selection, a process termed domestication. In recent years, *B. subtilis* has become a rather well-studied example of certain aspects of the domestication process, mainly owing to its century-long heritage as an experimental organism and the existence in stock collections of numerous *B. subtilis* strains that have undergone varying degrees of domestication (127, 129).

Some hallmarks of bacterial domestication are genome decay, loss of pathways, horizontal acquisition of genomic elements, and accumulation of mutations that provide a growth advantage in nutrient-rich environments (130). In many cases, selection is imposed by the experimenter, consciously or unconsciously, in striving for strains that "behave" in culture (such as those producing homogeneous, nonspreading, nonsticky colonies that grow rapidly in rich medium and are easily manipulated). It has been documented that laboratory domestication in *B. subtilis* has led to the adventitious loss of some "wild" traits such as biofilm and fruiting body formation, surface swarming, and production of toxins (129, 131, 132). In other instances, experimenters have deliberately mutated or deleted genomic regions in the course of strain development for particular research applications, including deletion of resident prophages (133) or inactivation of individual genes or entire metabolic pathways; see reference 4 for a sampling of examples too numerous to mention here. In short, the process of evolution continues even after strains are moved from the environment into the laboratory. By bringing *Bacillaceae* into the laboratory for more detailed study, we have undoubtedly learned a great deal more about their basic biology than we could have otherwise, but in so doing our experimental workhorses have become domesticated and have lost traits that are critical to understanding their original adaptation to life in the wild. Furthermore, during their propagation in separate laboratories around the world, even "benchmark" laboratory strains like *B. subtilis* 168, SMY, JH642, etc. have embarked on their own distinct evolutionary trajectories. Recent whole-genome sequence comparisons of the same strains obtained from different laboratories found that the number of nucleotide changes between strains to be few, but that some of these changes occur in genes that can lead to significant differences in isolate-specific phenotypes (134).

DIRECTED EVOLUTION OF *BACILLACEAE*

Directed evolution is a method by which an experimenter (i) uses artificial means to generate diversity within a gene or pathway of interest, (ii) selects or screens for mutants exhibiting improvement in the desired trait, and (iii) amplifies and characterizes the desired mutant. The procedure can be reiterated any number of times. This approach traces its origins to the early domestication of microorganisms. Historically domesticated microbes such as the fungi and bacteria used in production of bread, cheese, beer, wine, and various fermented meat, dairy, and vegetable products (130) can be thought of as the results of directed evolution. Domesticated *Bacillaceae* are not commonly encountered in Western culture; however, various strains of *Bacillus* spp. (*B. cereus*, *B. subtilis*, *B. licheniformis*, *B. pumilus*) are widely used in preparation of traditional Asian fermented foods (e.g., natto, kimchi, douchi, jeotgali, thua nao, cheongkukjang, doenjang, kochujang) (135, 136) as well as traditional African fermented condiments (e.g., bikalga, okpehe, soumbala) (137).

The application of scientific principles to the improvement of traditional fermentation processes constituted the origins of industrial microbiology (for an excellent popular review of the field from antiquity up to 1981, the reader is referred to reference 138). Members of the *Bacillaceae* (for example, *Bacillus brevis*, *Bacillus circulans*, *B. licheniformis*, *B. subtilis*, *B. thuringiensis*, *G. stearothermophilus*, and numerous taxonomically unidentified *Bacillaceae*) are widely used as important workhorses for the production of industrially important products. The details of these systems have been thoroughly reviewed (139–143). Particularly important to the molecular biotechnology industry is the availability of host organisms that are well characterized, Generally Regarded As Safe (GRAS), and easily amenable to physiologic, genetic, molecular biological, and genomic manipulation. Of all the *Bacillaceae*, *B. subtilis* possesses all of these qualities in spades; thus, it has become the premier Gram-positive microbe used in industrial biotechnology (143).

Up to the present day, industrial microbiologists continue to take advantage of the traditional strategies of domestication, i.e., searching among environmental isolates for individual strains exhibiting desirable traits and then applying "classical" directed evolution methods for strain improvement. Such methods usually consist of some form of generalized mutagenesis followed by screening among the resulting population for mutants with enhanced yield of the desired product. Throughout the 20th century, this basic procedure became gradually more subtle and powerful with advances in technology, and progress was greatly accelerated by the application of techniques for genetic transfer (transformation and transduction) to strain improvement. For example, the enzyme α-amylase is widely used in the food, brewing, and detergent industries (144). To maximize amylase production in *B. subtilis*, a traditional strain development program, utilizing a combination of (i) screening *Bacillus* spp. strains obtained from environmental sources, (ii) nitrosoguanidine mutagenesis, and (iii) transformation of the desirable genes into a single strain, resulted in an increase of amylase production from 11 U/ml to 2,530 U/ml (145).

With the advent of recombinant DNA technology starting in the late 1970s and continuing until the present, traditional industrial microbiology has morphed into the new field of molecular biotechnology (142, 146), and directed evolution has become an integral part of this field. The experimenter no longer needs to use random mutagenesis of whole organisms in order to generate the diversity upon which natural selection acts. Enzymes, regulatory DNA circuitry, and indeed entire metabolic pathways can be rationally designed *in silico* and constructed in any variety of permutations by using *in vitro* site-directed mutagenesis, random PCR mutagenesis, DNA shuffling, and synthetic biology techniques. The number of ways in which mutant populations can be screened *en masse* for the desired phenotype or property is limited only by the experimenter's imagination. The final construct can then be easily delivered and optimally expressed within an industrially amenable host organism. Within the *Bacillaceae* are found numerous examples in which the modern directed-evolution paradigm has been used to improve the production levels or industrial properties (e.g., thermal or pH stability) of various industrially important enzymes such as α-amylase (147), laccases (148, 149), acidophilic enzymes (150), proteases (151), endoglucanases (152, 153) and xylanases (154). In addition, *B. subtilis* itself has also been exploited as a platform for industrial-scale production of many important products including vitamins such as biotin (155) and riboflavin (141, 156); D-ribose used in the synthesis of antiviral and anticancer drugs (157); lipopeptide antibiotics such as surfactin (158) and iturin (159); biopolymers such as poly-γ-glutamic acid (136) and polyhydroxybutyrate (PHB) (160); or lactic acid for synthesis of the bioplastic polylactic acid (PLA) (161), to name but a few.

From the above discussion it can be seen that examples of the use of directed evolution in molecular biotechnology abound. However, the very same principles can be used to better understand the fundamental properties of microorganisms. For example, it is well established that prokaryotes, taken as a whole, can grow under an incredible range of physical extremes (temperature, pressure, pH, salinity, etc.), but that any individual microbe can only exist within a limited subset of this range. Hence, the microbial world consists (for example) of psychrophiles, mesophiles, thermophiles, and hyperthermophiles, each with its own characteristic range of growth temperatures. What are the underlying cellular and molecular mechanisms that dictate at which temperature, pressure, pH, or osmotic strength a prokaryote can grow? This question not only has implications for the biology of extremophiles on Earth, but also impacts studies of the physical limits at which life can exist in the universe. Two studies have recently been undertaken to explore the directed evolution of *B. subtilis* to growth under the harsh conditions prevailing on the planet Mars, which has been an exploration target of high interest to the life detection community throughout the late 20th and early 21st centuries (162, 163).

Directed Evolution to UV Resistance

Mars suffers a high flux of biologically harmful (>190 nm) UV radiation owing to its thin, CO_2-dominated atmosphere (164). On Earth, in contrast, the interaction of solar radiation with our relatively dense oxygen-rich atmosphere causes the accumulation of stratospheric ozone, which only allows the passage of longer-wavelength UV (>290 nm). Studies were undertaken asking the question whether vegetatively growing *B. subtilis* cells could evolve higher resistance to UV if subjected to periodic episodes of selection by exposure to polychromatic UV (200 to 400 nm) (165). Serial exposure of cells to UV over the course of 700 generations indeed led to a statistically significant 4.4-fold increase in resistance of vegetative cells to polychromatic (200 to 400 nm) UV and monochromatic (254 nm) UV, respectively (165). Vegetative cells of the UV-adapted strains also exhibited significantly increased resistance to X rays, desiccation, and hydrogen peroxide, all of which treatments cause damage to DNA. In addition, spores of the UV-adapted strains were also slightly more resistant to 254-nm UV than were spores of the ancestor (166). Taken together, the data suggest that the cells may have evolved a generalized upregulation of DNA repair capacity, but at present the molecular mechanism for this effect is unknown.

Directed Evolution to Growth at Low Pressure

Earth's mean atmospheric pressure at sea level is 1,013 mbar and ranges from ~250 mbar at the top of

Mt. Everest to ~1,055 mbar at the Dead Sea. In contrast, the average atmospheric pressure at the Martian surface is ~7 mbar and ranges from ~1 mbar at the top of Olympus Mons to ~10 mbar at the bottom of Hellas Basin (167). Therefore, the atmospheric pressure on Mars is more than 2 orders of magnitude lower than that of Earth. Just as high pressures encountered in the ocean depths limit microbial growth to specialized piezophiles, low pressure has also been shown to be a factor limiting the growth of microorganisms normally adapted to grow at Earth-normal surface pressure. To date, the growth of several *Bacillaceae* (*B. licheniformis*, *Bacillus nealsonii*, *B. megaterium*, *B. pumilus*, and *B. subtilis*) has been shown to be inhibited at pressures below 25 mbar, 2 to 3 times above the highest pressure on the surface of Mars (168). Growth inhibition at low pressure was not merely caused by a lowering of the partial pressure of O_2, as the medium used contained both a fermentable substrate (glucose) and an alternative terminal electron acceptor (nitrate) and four of the five species were facultative anaerobes (168).

We wondered if terrestrial bacteria could evolve the ability to grow at lower pressure, so we propagated *B. subtilis* WN624 (*amyE::spc*) in liquid medium for 1,000 generations at 50 mbar, a pressure just above its inhibitory growth pressure (169). During the course of evolution, the population exhibited a stepwise evolution to better growth at 50 mbar, and we isolated a strain from the 1,000-generation culture, strain WN1106, which could grow to ~2.5-fold higher cell density at 50 mbar than the ancestral strain WN624 (169). In pairwise competition experiments, strain WN1106 readily outcompeted the ancestral strain at 50 mbar, but the two strains were competitively neutral when cocultivated at Earth-normal pressure (1,013 mbar). The growth advantage of WN1106 over ancestral strain WN624 at 50 mbar was not due to its better growth under oxygen limitation, because both strains grew equally under oxygen-limited conditions (169). To identify genes whose expression responded to low pressure, we isolated RNA from both ancestral strain WN624 and low-pressure-evolved strain WN1106 grown at either 1,013 mbar or 50 mbar, then compared their expression levels in all combinations by using microarrays (170). One set of transcripts seen at higher levels specifically in low pressure-evolved strain WN1106 cultivated at low pressure belonged to the *des-desKR* cluster of genes. The *des* gene encodes the major membrane fatty acid desaturase Des, while *desK* and *desR* encode the two-component kinase and response regulators, respectively, that regulate *des* transcription in response to changes in membrane fluidity

brought about by temperature shifts (171). We showed by both microarray analysis and quantitative reverse transcription PCR that *des-desKR* transcription in low-pressure-adapted strain WN1106, but not ancestral strain WN624, was strongly induced at low pressure (172). However, oxygen deprivation did not affect *des-desKR* transcript levels of either strain. Pairwise competition experiments showed that strain WN1106 was significantly more fit at low pressure than the ancestor and that its fitness advantage was significantly reduced (but not completely abolished) by a Δ*des::kan* knockout mutation (172). Piezophiles are known to respond to changes in pressure by altering the fatty acid composition of their membranes (173). However, we found no differences in the membrane fatty acid compositions of the ancestral and low-pressure-evolved strains grown at a variety of low pressures (172).

In order to identify mutations which may have occurred during the evolution of *B. subtilis* to enhanced growth at low pressure, we determined and compared the whole-genome sequences of ancestral strain WN624 and low pressure-evolved strain WN1106 (489). Strain WN1106 carried amino acid-altering mutations in the coding sequences of only seven genes (*fliI*, *parC*, *ytoI*, *bacD*, *resD*, *walK*, and *yvlD*) and a single 9-nucleotide in-frame deletion in the *rnjB* gene (489). We are currently investigating the possible role of all eight mutant genes in evolution to enhanced low-pressure growth.

CONCLUDING REMARKS

From their first appearance on our planet some 2 billion years ago, the *Bacillaceae* have evolved a dramatic diversification of capabilities and have conquered a majority of niches on Earth. Evolution of the spore has doubtless made a major contribution to their global success. The spore has also been seriously considered as an ideal package for the transmission of DNA beyond Earth. Since the 1960s, spores of *Bacillaceae* have been deliberately sent into low Earth orbit on various short- and long-duration missions (41, 174), and even into lunar orbit on the Apollo 16 and 17 missions (175). Furthermore, it is almost certain that spores have been launched as inadvertent contaminants of spacecraft and, to date, probably (i) have reached the surface of the Moon and Mars; (ii) are orbiting Jupiter and Saturn; and (iii) are on the verge of leaving our solar system entirely on the spacecraft Voyagers 1 and 2 and Pluto New Horizons (176). Seen from this perspective, it is enticing to envision that *Bacillaceae* may well be the first terrestrial colonizers of extraterrestrial environments, thus extending life into undreamed-of niches. What a boon to evolutionary and environmental microbiologists of the future!

Acknowledgments. We thank Changwoo Cha and Grant Boyadjian for technical assistance, and Ralf Moeller for helpful discussion. This work was supported in part by a grant from the NASA Astrobiology: Exobiology and Evolutionary Biology program (NNX08AO15G) to W.L.N. and P.F.-C.

Citation. Fajardo-Cavazos P, Maughan H, Nicholson WL. 2014. Evolution in the *Bacillaceae*. Microbiol Spectrum 2(5): TBS-0020-2014.

References

1. Ciccarelli FD, Doerks T, von Mering C, Creevey CJ, Snel B, Bork P. 2006. Toward automatic reconstruction of a highly resolved tree of life. *Science* 311:1283–1287.

2. Wu D, Hugenholtz P, Mavromatis K, Pukall R, Dalin E, Ivanova NN, Kunin V, Goodwin L, Wu M, Tindall BJ, Hooper SD, Pati A, Lykidis A, Spring S, Anderson IJ, D'haeseleer P, Zemla A, Singer M, Lapidus A, Nolan M, Copeland A, Han C, Chen F, Cheng JF, Lucas S, Kerfeld C, Lang E, Gronow S, Chain P, Bruce D, Rubin EM, Kyrpides NC, Klenk HP, Eisen JA. 2009. A phylogeny-driven genomic encyclopaedia of Bacteria and Archaea. *Nature* 462:1056–1060.

3. Graumann PE. 2007. Bacillus: *Cellular and Molecular Biology*. Caister Academic Press, Hethersett, UK.

4. Sonenshein AL, Hoch JA, Losick R (ed). 2002. Bacillus subtilis *and Its Closest Relatives: From Genes to Cells*. ASM Press, Washington, DC.

5. Nicholson WL. 2002. Roles of *Bacillus* endospores in the environment. *Cell Mol Life Sci* 59:410–416.

6. Nicholson WL. 2004. Ubiquity, longevity, and ecological roles of *Bacillus* spores, p 1–15. *In* Ricca E, Henriques AO, Cutting SM (ed), *Bacterial Spore Formers: Probiotics and Emerging Applications*. Horizon Scientific Press, Norfolk, UK.

7. Priest FG. 1993. Systematics and ecology of *Bacillus*, p 3–16. *In* Sonenshein AL, Hoch JA, Losick R (ed), Bacillus subtilis *and Other Gram-Positive Bacteria: Biochemistry, Physiology, and Molecular Genetics*. ASM Press, Washington, DC.

8. Slepecky RA, Leadbetter ER. 1994. Ecology and relationships of endospore-forming bacteria: changing perspectives, p 195–206. *In* Piggot PJ, Moran CP Jr, Youngman P (ed), *Regulation of Bacterial Differentiation*. American Society for Microbiology, Washington, DC.

9. Wiedenbeck J, Cohan FM. 2011. Origins of bacterial diversity through horizontal genetic transfer and adaptation to new ecological niches. *FEMS Microbiol Rev* 35: 957–976.

10. Logan NA, Halket G. 2011. Developments in the taxonomy of aerobic, endospore-forming bacteria, p 1–31. *In* Logan NA, De Vos P (ed), *Endospore-Forming Soil Bacteria*. Springer-Verlag, Heidelberg, Germany.

11. Cohn F. 1872. Untersuchungen über Bacterien. *I Beitr Biol Pfl* 1:127–224.

12. Koch R. 1876. Die Ätiologie der Milzbrand-krankheit, begründet auf die Entwicklungsgeschichte des *Bacillus anthracis*. *Beitr Biologie Pflanz* 2:277–310.

13. Beierlein JM, Anderson AC. 2011. New developments in vaccines, inhibitors of anthrax toxins, and antibiotic therapeutics for *Bacillus anthracis*. *Curr Med Chem* 18: 5083–5094.

14. Hudson MJ, Beyer W, Boehm R, Fasanella A, Garofolo G, Golinski R, Goossens PL, Hahn U, Hallis B, King A, Mock M, Montecucco C, Ozin A, Tonello F, Kaufmann SHE. 2008. *Bacillus anthracis*: balancing innocent research with dual-use potential. *Int J Med Microbiol* 298:345–364.

15. Graumann P. 2007. Bacillus: *Cellular and Molecular Biology*. Caister Academic Press, Hethersett, United Kingdom.

16. Winslow CE, Broadhurst J, Buchanan RE, Krumwiede C, Rogers LA, Smith GH. 1920. The families and genera of the bacteria: final report of the Committee of the Society of American Bacteriologists on Characterization and Classification of Bacterial Types. *J Bacteriol* 5: 191–229.

17. Bergey DH, Harrison FC, Breed RS, Hammer BW, Huntoon FW. 1923. *Bergey's Manual of Determinative Bacteriology: A Key for the Identification of Organisms of the Class Schizomycetes*. Williams & Wilkins, Baltimore, MD.

18. Woese CR, Fox GE. 1977. Phylogenetic structure of the prokaryotic domain: the primary kingdoms. *Proc Natl Acad Sci USA* 74:5088–5090.

19. Ash C, Farrow JAE, Wallbanks S, Collins MD. 1991. Phylogenetic heterogeneity of the genus *Bacillus* revealed by comparative analysis of small subunit ribosomal RNA sequences. *Lett Appl Microbiol* 13: 202–206.

20. Priest FG. 1993. Systematics and ecology of *Bacillus*, p 3–16. *In* Sonenshein AL, Hoch JA, Losick R (ed), Bacillus subtilis *and Other Gram-Positive Bacteria: Biochemistry, Physiology, and Molecular Genetics*. American Society for Microbiology, Washington, DC.

21. Schloss PD, Handelsman J. 2005. Introducing DOTUR, a computer program for defining operational taxonomic units and estimating species richness. *Appl Environ Microbiol* 71:1501–1506.

22. Stackebrandt E, Goebel BM. 1994. A place for DNA-DNA reassociation and 16S ribosomal RNA sequence analysis in the present species definition in bacteriology. *Int J Syst Bacteriol* 44:846–849.

23. De Vos P, Garrity GM, Jones D, Kreig NR, Ludwig W, Rainey FA, Schleifer K-H, Whitman WB (ed). 2009. *Bergey's Manual of Systematic Bacteriology*, 2nd ed, vol 3. *The Firmicutes*. Springer-Verlag, New York, NY.

24. Maughan H, Van der Auwera G. 2011. *Bacillus* taxonomy in the genomic era finds phenotypes to be essential though often misleading. *Infect Genet Evol* 11:789–797.

25. David LA, Alm EJ. 2011. Rapid evolutionary innovation during an Archaean genetic expansion. *Nature* 469: 93–96.

26. Moreno-Letelier A, Olmedo G, Eguiarte LE, Martinez-Castilla L, Souza V. 2011. Parallel evolution and horizontal gene transfer of the *pst* operon in Firmicutes from oligotrophic environments. *Int J Evol Biol* **2011**: 781642. doi:10.4061/2011/781642

27. Hobbs JK, Shepherd C, Saul DJ, Demetras NJ, Haaning S, Monk CR, Daniel RM, Arcus VL. 2012. On the origin and evolution of thermophily: reconstruction of functional precambrian enzymes from ancestors of *Bacillus*. *Mol Biol Evol* **29**:825–835.

28. Alcaraz LD, Moreno-Hagelsieb G, Eguiarte LE, Souza V, Herrera-Estrella L, Olmedo G. 2010. Understanding the evolutionary relationships and major traits of *Bacillus* through comparative genomics. *BMC Genomics* **11**: 332. doi:10.1186/1471-2164-11-332.

29. Kobayashi K, Ehrlich S, Albertini A, Amati G, Andersen K, Arnaud M, Asai K, Ashikaga S, Aymerich S, Bessieres P, Boland F, Brignell S, Bron S, Bunai K, Chapuis J, Christiansen L, Danchin A, Debarbouille M, Dervyn E, Deuerling E, Devine K, Devine S, Dreesen O, Errington J, Fillinger S, Foster S, Fujita Y, Galizzi A, Gardan R, Eschevins C, Fukushima T, Haga K, Harwood C, Hecker M, Hosoya D, Hullo M, Kakeshita H, Karamata D, Kasahara Y, Kawamura F, Koga K, Koski P, Kuwana R, Imamura D, Ishimaru M, Ishikawa S, Ishio I, Le Coq D, Masson A, Mauel C, Meima R, Mellado R, Moir A, Moriya S, Nagakawa E, Nanamiya H, Nakai S, Nygaard P, Ogura M, Ohanan T, O'Reilly M, O'Rourke M, Pragai Z, Pooley H, Rapoport G, Rawlins J, Rivas L, Rivolta C, Sadaie A, Sadaie Y, Sarvas M, Sato T, Saxild H, Scanlan E, Schumann W, Seegers J, Sekiguchi J, Sekowska A, Seror S, Simon M, Stragier P, Studer R, Takamatsu H, Tanaka T, Takeuchi M, Thomaides H, Vagner V, van Dijl J, Watabe K, Wipat A, Yamamoto H, Yamamoto M, Yamamoto Y, Yamane K, Yata K, Yoshida K, Yoshikawa H, Zuber U, Ogasawara N. 2003. Essential *Bacillus subtilis* genes. *Proc Natl Acad Sci USA* **100**:4678–4683.

30. Baesman SM, Stolz JF, Kulp TR, Oremland RS. 2009. Enrichment and isolation of *Bacillus beveridgei* sp. nov., a facultative anaerobic haloalkaliphile from Mono Lake, California, that respires oxyanions of tellurium, selenium, and arsenic. *Extremophiles* **13**:695–705.

31. Mnif S, Chamkha M, Labat M, Sayadi S. 2011. Simultaneous hydrocarbon biodegradation and biosurfactant production by oilfield-selected bacteria. *J Appl Microbiol* **111**:525–536.

32. Chen Y-G, Zhang Y-Q, Liu Z-X, Zhuang D-C, Klenk H-P, Tang S-K, Cui X-L, Li W-J. 2009. *Halobacillus salsuginis* sp nov., a moderately halophilic bacterium from a subterranean brine. *Int J Syst Evol Microbiol* **59**:2505–2509.

33. Taubel M, Kampfer P, Buczolits S, Lubitz W, Busse HA. 2003. *Bacillus barbaricus* sp. nov., isolated from an experimental wall painting. *Int J Syst Evol Microbiol* **53**:725–730.

34. Batchelor MD. 1919. Aerobic spore-bearing bacteria in the intestinal tract of children. *J Bacteriol* **4**:23–34.

35. Vedder A. 1934. *Bacillus alcalophilus* n. sp.; benevens enkele ervaringen met sterk alcalische voedingsbodems. *Antonie Leeuwenhoek* **1**:143–147.

36. Hoyles L, Honda H, Logan NA, Halket G, La Ragione RM, McCartney AL. 2012. Recognition of greater diversity of *Bacillus* species and related bacteria in human faeces. *Res. Microbiol* **163**:3–13.

37. Tamames J, Abellán JJ, Pignatelli M, Camacho A, Moya A. 2010. Environmental distribution of prokaryotic taxa. *BMC Microbiol* **10**:85. doi:10.1186/1471-2180-10-85.

38. Euzéby JP. 1997. List of bacterial names with standing in nomenclature: a folder available on the Internet. *Int J Syst Bacteriol* **47**:590–592.

39. Philippot L, Andersson SG, Battin TJ, Prosser JI, Schimel JP, Whitman WB, Hallin S. 2010. The ecological coherence of high bacterial taxonomic ranks. *Nat Rev Microbiol* **8**:523–529.

40. Lakay FM, Botha A, Prior BA. 2007. Comparative analysis of environmental DNA extraction and purification methods from different humic acid-rich soils. *J Appl Microbiol* **102**:265–273.

41. Nicholson WL, Munakata N, Horneck G, Melosh HJ, Setlow P. 2000. Resistance of *Bacillus* endospores to extreme terrestrial and extraterrestrial environments. *Microbiol Mol Biol Rev* **64**:548–572.

42. Setlow P. 2006. Spores of *Bacillus subtilis*: their resistance to and killing by radiation, heat and chemicals. *J Appl Microbiol* **101**:514–525.

43. Womack AM, Bohannan BJ, Green JL. 2010. Biodiversity and biogeography of the atmosphere. *Philos Trans R Soc Lond B Biol Sci* **365**:3645–3653.

44. DeSantis TZ, Brodie EL, Moberg JP, Zubieta IX, Piceno YM, Andersen GL. 2007. High-density universal 16S rRNA microarray analysis reveals broader diversity than typical clone library when sampling the environment. *Microb Ecol* **53**:371–383.

45. Brodie EL, DeSantis TZ, Parker JP, Zubietta IX, Piceno YM, Andersen GL. 2007. Urban aerosols harbor diverse and dynamic bacterial populations. *Proc Natl Acad Sci USA* **104**:299–304.

46. Merrill L, Dunbar J, Richardson J, Kuske CR. 2006. Composition of *Bacillus* species in aerosols from 11 U.S. cities. *J Forensic Sci* **51**:559–565.

47. Sass AM, McKew BA, Sass H, Fichtel J, Timmis KN, McGenity TJ. 2008. Diversity of *Bacillus*-like organisms isolated from deep-sea hypersaline anoxic sediments. *Saline Syst* **4**:8. doi:10.1186/1746-1448-4-8.

48. Nielsen KM, Johnsen PJ, Bensasson D, Daffonchio D. 2007. Release and persistence of extracellular DNA in the environment. *Environ Biosafety Res* **6**:37–53.

49. Earl AM, Losick R, Kolter R. 2008. Ecology and genomics of *Bacillus subtilis*. *Trends Microbiol* **16**:269–275.

50. Ludwig W, Schleifer K-H, Whitman WB. 2009. Revised road map to the phylum *Firmicutes*, p 1–13. *In* De Vos P, Garrity GM, Jones D, Kreig NR, Ludwig W, Rainey FA, Schleifer K-H, Whitman WB (ed), Bergey's Manual of Systematic Bacteriology, 2nd ed, **vol 3**. *The Firmicutes*. Springer, New York, NY.

51. Maughan H. 2007. Rates of molecular evolution in bacteria are relatively constant despite spore dormancy. *Evolution* **61**:280–288.

52. Earl AM, Losick R, Kolter R. 2007. *Bacillus subtilis* genome diversity. *J Bacteriol* **189**:1163–1170.

53. Singh AH, Wolf DM, Wang P, Arkin AP. 2008. Modularity of stress response evolution. *Proc Natl Acad Sci USA* **105**:7500–7505.

54. Nakamura K, Haruta S, Ueno S, Ishii M, Yokota A, Igarashi Y. 2004. *Cerasibacillus quisquiliarum* gen. nov., sp. nov., isolated from a semi-continuous decomposing system of kitchen refuse. *Int J Syst Evol Microbiol* **54**:1063–1069.

55. Ajithkumar VP, Ajithkumar B, Mori K, Takamizawa K, Iriye R, Tabata S. 2001. A novel filamentous *Bacillus* sp., strain NAF001, forming endospores and budding cells. *Microbiology* **147**:1415–1423.

56. Thorne CB. 1968. Transducing bacteriophage for *Bacillus cereus*. *J Virol* **2**:657–662.

57. Lovett PS, Bramucci D, Bramucci MG, Burdick BD. 1974. Some properties of the PBP1 transduction system in *Bacillus pumilus*. *J Virol* **13**:81–84.

58. Yasbin RE, Young FE. 1974. Transduction in *Bacillus subtilis* by bacteriophage SPP1. *J Virol* **14**:1343–1348.

59. Thorne CB. 1978. Transduction in *Bacillus thuringiensis*. *Appl Environ Microbiol* **35**:1109–1115.

60. Vary PS, Garbe JC, Franzen M, Frampton EW. 1982. MP13, a generalized transducing bacteriophage for *Bacillus megaterium*. *J Bacteriol* **149**:1112–1119.

61. Ruhfel RE, Robillard NJ, Thorne CB. 1984. Interspecies transduction of plasmids among *Bacillus anthracis*, *B. cereus*, and *B. thuringiensis*. *J Bacteriol* **157**:708–711.

62. Welker NE. 1988. Transduction in *Bacillus stearothermophilus*. *J Bacteriol* **170**:3761–3764.

63. Zahler SA. 1993. Temperate bacteriophages, p 831–842. *In* Sonenshein AL, Hoch JA, Losick R (ed.), *Bacillus subtilis and Other Gram-positive Bacteria: Biochemistry, Physiology, and Molecular Genetics*. American Society for Microbiology, Washington, DC.

64. Bertram J, Strätz M, Dürre P. 1991. Natural transfer of conjugative transposon Tn*916* between gram-positive and gram-negative bacteria. *J Bacteriol* **173**:443–448.

65. Torres OR, Korman RZ, Zahler SA, Dunny GM. 1991. The conjugative transposon Tn*925*: enhancement of conjugal transfer by tetracycline in *Enterococcus faecalis* and mobilization of chromosomal genes in *Bacillus subtilis* and *E. faecalis*. *Mol Gen Genet* **225**:395–400.

66. Spizizen J. 1958. Transformation of biochemically deficient strains of *Bacillus subtilis* by deoxyribonucleate. *Proc Natl Acad Sci USA* **44**:1072–1078.

67. Anagnostopoulos C, Spizizen J. 1961. Requirements for transformation in *Bacillus subtilis*. *J Bacteriol* **81**:741–746.

68. Bott KF, Wilson GA. 1968. Metabolic and nutritional factors influencing the development of competence for transfection of *Bacillus subtilis*. *Bacteriol Rev* **32**:370–378.

69. van Sinderen D, Luttinger A, Kong L, Dubnau D, Venema G, Hamoen L. 1995. *comK* encodes the competence transcription factor, the key regulatory protein for competence development in *Bacillus subtilis*. *Mol Microbiol* **15**:455–462.

70. Finkel SE, Kolter R. 2001. DNA as a nutrient: novel role for bacterial competence gene homologs. *J Bacteriol* **183**:6288–6293.

71. Redfield RJ. 2001. Do bacteria have sex? *Nat Rev Genet* **2**:634–639.

72. Mirończuk AM, Kovács Á, Kuipers OP. 2008. Induction of natural competence in *Bacillus cereus* ATCC14579. *Microb Biotechnol* **1**:226–235.

73. Duitman EH, Wyczawski D, Boven LG, Venema G, Kuipers OP, Hamoen LW. 2007. Novel methods for genetic transformation of natural *Bacillus subtilis* isolates used to study the regulation of the mycosubtilin and surfactin synthetases. *Appl Environ Microbiol* **73**:3490–3496.

74. Nijland R, Burgess JG, Errington J, Veening JW. 2010. Transformation of environmental *Bacillus subtilis* isolates by transiently inducing genetic competence. *PLoS One* **5**:e9724. doi:10.1371/journal.pone.0009724

75. Kennedy MJ, Reader SL, Swierczynski LM. 1994. Preservation records of microorganisms: evidence of the tenacity of life. *Microbiology* **140**:2513–2529.

76. Vreeland RH, Rosenzweig WD, Powers DW. 2000. Isolation of a 250 million-year-old halotolerant bacterium from a primary salt crystal. *Nature* **407**:897–900.

77. Cano RJ, Borucki MK, Higby-Schweitzer M, Poinar HN, Poinar GO, Pollard KJ. 1994. *Bacillus* DNA in fossil bees: an ancient symbiosis. *Appl Environ Microbiol* **60**:2164–2167.

78. Cano RJ, Borucki MK. 1995. Revival and identification of bacterial spores in 25-million-year-old to 40-million-year-old Dominican amber. *Science* **268**:1060–1064.

79. Rosenzweig WD, Woish J, Peterson J, Vreeland R. 2000. Development of a protocol to retrieve microorganisms from ancient salt crystals. *Geomicrobiol J* **17**:185–192.

80. Vreeland RH, Rosenzweig WD. 2002. The question of uniqueness of ancient bacteria. *J Indust Microbiol Biotechnol* **28**:32–41.

81. Graur D, Pupko T. 2001. The permian bacterium that isn't. *Mol Biol Evol* **18**:1143–1146.

82. Lindahl T. 1993. Instability and decay of the primary structure of DNA. *Nature* **362**:709–715.

83. Nicholson WL, Schuerger AC, Setlow P. 2005. The solar UV environment and bacterial spore UV resistance: considerations for Earth-to-Mars transport by natural processes and human spaceflight. *Mutat Res* **571**:249–264.

84. Nicastro AJ, Vreeland RH, Rosenzweig WD. 2002. Limits imposed by ionizing radiation on the long-term survival of trapped bacterial spores: beta radiation. *Int J Radiat Biol* **78**:891–901.

85. Kminek G, Bada JL, Pogliano K, Ward JF. 2003. Radiation-dependent limit for the viability of bacterial spores in halite fluid inclusions and on Mars. *Radiat Res* **159**:722–729.

86. Nicholson W. 2003. Using thermal inactivation kinetics to calculate the probability of extreme spore longevity: implications for paleomicrobiology and lithopanspermia. *Orig Life Evol Biosph* 33:621–631.

87. Maughan H, Birky CW, Nicholson WL, Rosenzweig WD, Vreeland RH. 2002. The paradox of the "ancient" bacterium which contains "modern" protein-coding genes. *Mol Biol Evol* 19:1637–1639.

88. Nickle DC, Learn GH, Rain MW, Mullins JI, Mittler JE. 2002. Curiously modern DNA for a "250 million-year-old" bacterium. *J Mol Evol* 54:134–137.

89. Kuo CH, Ochman H. 2009. Inferring clocks when lacking rocks: the variable rates of molecular evolution in bacteria. *Biol Direct* 4:10. doi:10.1186/1745-6150-4-35.

90. Ochman H, Elwyn S, Moran NA. 1999. Calibrating bacterial evolution. *Proc Natl Acad Sci USA* 96:12638–12643.

91. Maughan H. 2007. Rates of molecular evolution in bacteria are relatively constant despite spore dormancy. *Evolution* 61:280–288.

92. Ochman H, Selander RK. 1984. Evidence for clonal population structure in *Escherichia coli*. *Proc Natl Acad Sci USA* 81:198–201.

93. Selander RK, Beltran P, Smith NH. 1991. Evolutionary genetics of *Salmonella*., p 25–57. *In* Selander RK, Clark AG, Whittam TS (ed), *Evolution at the Molecular Level*. Sinauer Associates, Sunderland, MA.

94. Istock CA, Duncan KE, Ferguson N, Zhou X. 1992. Sexuality in a natural population of bacteria-*Bacillus subtilis* challenges the clonal paradigm. *Mol Ecol* 1:95–103.

95. Sikorski J. 2012. A glimpse into microevolution in nature: adaptation and speciation of *Bacillus simplex* from "Evolution Canyon," p 225–231. *In* Kolter R, Maloy S (ed), *Microbes and Evolution: The World That Darwin Never Saw*. ASM Press, Washington, DC.

96. Sikorski J, Nevo E. 2005. Adaptation and incipient sympatric speciation of *Bacillus simplex* under microclimatic contrast at "Evolution Canyons" I and II, Israel. *Proc Natl Acad Sci USA* 102:15924–15929.

97. Sikorski J, Nevo E. 2007. Patterns of thermal adaptation of *Bacillus simplex* to the microclimatically contrasting slopes of "Evolution Canyons" I and II, Israel. *Environ Microbiol* 9:71–726.

98. Connor N, Sikorski J, Rooney AP, Kopac S, Koeppel AF, Burger A, Cole SG, Perry EB, Krizanc D, Field NC, Slaton M, Cohan FM. 2010. Ecology of speciation in the genus *Bacillus*. *Appl Environ Microbiol* 76:1349–1358.

99. Lowry DB. 2012. Ecotypes and the controversy over stages in the formation of new species. *Biol J Linnean Soc* 106:241–257.

100. Shank EA, Klepac-Ceraj V, Collado-Torres L, Powers GE, Losick R, Kolter R. 2011. Interspecies interactions that result in *Bacillus subtilis* forming biofilms are mediated mainly by members of its own genus. *Proc Natl Acad Sci USA* 108:E1236–E1243.

101. Hooper SD, Mavromatis K, Kyrpides NC. 2009. Microbial co-habitation and lateral gene transfer: what transposases can tell us. *Genome Biol* 10:R45. doi:10.1186/gb-2009-10-4-r45.

102. Freilich S, Kreimer A, Meilijson I, Gophna U, Sharan R, Ruppin E. 2010. The large-scale organization of the bacterial network of ecological co-occurrence interactions. *Nucleic Acids Res* 38:3857–3868.

103. Mathew GM, Ju YM, Lai CY, Mathew DC, Huang CC. 2012. Microbial community analysis in the termite gut and fungus comb of *Odontotermes formosanus*: the implication of *Bacillus* as mutualists. *FEMS Microbiol Ecol* 79:504–517.

104. Stewart EJ. 2012. Growing unculturable bacteria. *J Bacteriol* 194:4151–4160.

105. Kassen R. 2002. The experimental evolution of specialists, generalists, and the maintenance of diversity. *J Evol Biol* 15:173–190.

106. Elena SF, Lenski RE. 2003. Evolution experiments with microorganisms: the dynamics and genetic bases of adaptation. *Nat Rev Genet* 4:457–469.

107. Graham JB, Istock CA. 1978. Genetic exchange in *Bacillus subtilis* in soil. *Mol Gen Genet* 166:287–290.

108. Graham JP, Istock CA. 1979. Gene exchange and natural selection cause *Bacillus subtilis* to evolve in soil culture. *Science* 204:637–639.

109. Graham J, Istock C. 1981. Parasexuality and microevolution in experimental populations of *Bacillus subtilis*. *Evolution* 35:954–963.

110. Maughan H, Callicotte V, Hancock A, Birky CW, Nicholson WL, Masel J. 2006. The population genetics of phenotypic deterioration in experimental populations of *Bacillus subtilis*. *Evolution* 60:686–695.

111. Maughan H, Nicholson WL. 2004. Stochastic processes influence stationary-phase decisions in *Bacillus subtilis*. *J Bacteriol* 186:2212–2214.

112. Maughan H, Masel J, Birky C, Nicholson W. 2007. The roles of mutation accumulation and selection in loss of sporulation in experimental populations of *Bacillus subtilis*. *Genetics* 177:937–948.

113. Masel J, Maughan H. 2007. Mutations leading to loss of sporulation ability in *Bacillus subtilis* are sufficiently frequent to favor genetic canalization. *Genetics* 175:453–457.

114. Maughan H, Birky CWJ, Nicholson WL. 2009. Transcriptome divergence and the loss of plasticity in *Bacillus subtilis* after 6,000 generations of evolution under relaxed selection for sporulation. *J Bacteriol* 191:428–433.

115. Maughan H, Nicholson WL. 2011. Increased fitness and alteration of metabolic pathways during *Bacillus subtilis* evolution in the laboratory. *Appl Environ Microbiol* 77:4105–4118.

116. Brown CT, Fishwick LK, Chokshi BM, Cuff MA, Jackson JM, Oglesby T, Rioux AT, Rodriguez E, Stupp GS, Trupp AH, Woollcombe-Clarke JS, Wright TN, Zaragoza WJ, Drew JC, Triplett EW, Nicholson WL. 2011. Whole-genome sequencing and phenotypic analysis of *Bacillus subtilis* mutants following evolution under conditions of relaxed selection for sporulation. *Appl Environ Microbiol* 77:6867–6877.

117. Nicholson WL. 2012. Increased competitive fitness of *Bacillus subtilis* under nonsporulating conditions via inactivation of pleiotropic regulators AlsR, SigD, and SigW. *Appl Environ Microbiol* **78**:3500–3503.

118. Debora BN, Vidales LE, Ramírez R, Ramírez M, Robleto EA, Yasbin RE, Pedraza-Reyes M. 2011. Mismatch repair modulation of MutY activity drives *Bacillus subtilis* stationary-phase mutagenesis. *J Bacteriol* **193**:236–245.

119. Pybus C, Pedraza-Reyes M, Ross CA, Martin H, Ona K, Yasbin RE, Robleto E. 2010. Transcription-associated mutation in *Bacillus subtilis* cells under stress. *J Bacteriol* **192**:3321–3328.

120. Vidales LE, Cárdenas LC, Robleto E, Yasbin RE, Pedraza-Reyes M. 2009. Defects in the error prevention oxidized guanine system potentiate stationary-phase mutagenesis in *Bacillus subtilis*. *J Bacteriol* **191**:506–513.

121. Robleto EA, Yasbin R, Ross C, Pedraza-Reyes M. 2007. Stationary phase mutagenesis in *B. subtilis*: a paradigm to study genetic diversity programs in cells under stress. *Crit Rev Biochem Mol Biol* **42**:327–339.

122. Ross C, Pybus C, Pedraza-Reyes M, Sung HM, Yasbin RE, Robleto E. 2006. Novel role of *mfd*: effects on stationary-phase mutagenesis in *Bacillus subtilis*. *J Bacteriol* **188**:7512–7520.

123. Pedraza-Reyes M, Yasbin RE. 2004. Contribution of the mismatch DNA repair system to the generation of stationary-phase-induced mutants of *Bacillus subtilis*. *J Bacteriol* **186**:6485–6491.

124. Sung HM, Yasbin RE. 2002. Adaptive, or stationary-phase, mutagenesis, a component of bacterial differentiation in *Bacillus subtilis*. *J Bacteriol* **184**:5641–5653.

125. Sung HM, Yeamans G, Ross CA, Yasbin RE. 2003. Roles of YqjH and YqjW, homologs of the *Escherichia coli* UmuC/DinB or Y superfamily of DNA polymerases, in stationary-phase mutagenesis and UV-induced mutagenesis of *Bacillus subtilis*. *J Bacteriol* **185**:2153–2160.

126. McCutcheon JP, Moran NA. 2012. Extreme genome reduction in symbiotic bacteria. *Nat Rev Microbiol* **10**:13–26.

127. Zeigler DR, Prágai Z, Rodriguez S, Chevreux B, Muffler A, Albert T, Bai R, Wyss M, Perkins JB. 2008. The origins of 168, W23, and other *Bacillus subtilis* legacy strains. *J Bacteriol* **190**:6983–6995.

128. Maughan H, Birky CWJ, Nicholson WL. 2009. Transcriptome divergence and the loss of plasticity in *Bacillus subtilis* after 6,000 generations of evolution under relaxed selection for sporulation. *J Bacteriol* **191**:428–433.

129. McLoon AL, Guttenplan SB, Kearns DB, Kolter R, Losick R. 2011. Tracing the domestication of a biofilm-forming bacterium. *J Bacteriol* **193**:2027–2034.

130. Douglas GL, Klaenhammer TR. 2010. Genomic evolution of domesticated microorganisms. *Annu Rev Food Sci Technol* **1**:397–414.

131. Branda SS, Gonzalez-Pastor JE, Ben-Yehuda S, Losick R, Kolter R. 2001. Fruiting body formation by *Bacillus subtilis*. *Proc Natl Acad Sci USA* **98**:11621–11626.

132. Gonzalez-Pastor JE. 2012. Multicellularity and social behaviour in *Bacillus subtilis*, p 351–375. *In* Graumann P (ed), *Bacillus: Cellular and Molecular Biology*, 2nd ed. Caister Academic Press, Wymondham, United Kingdom.

133. Yasbin RE, Fields PI, Andersen BJ. 1980. Properties of *Bacillus subtilis* 168 derivatives freed of their natural prophages. *Gene* **12**:155–159.

134. Srivatsan A, Han Y, Peng J, Tehranchi AK, Gibbs R, Wang JD, Chen R. 2008. High-precision, whole-genome sequencing of laboratory strains facilitates genetic studies. *PLoS Genet* **4**:e1000139. doi:10.1371/journal.pgen.1000139.

135. Hosoi T, Kiuchi K. 2004. Production and probiotic effects of natto, p 143–153. *In* Ricca E, Henriques AO, Cutting SM (ed), *Bacterial Spore Formers: Probiotics and Emerging Applications*. Horizon Bioscience, Wymondham, United Kingdom.

136. Kang SE, Rhee JH, Park C, Sung MH, Lee I. 2005. Distribution of poly-gamma-glutamate (gamma-PGA) producers in Korean fermented foods, Cheongkukjang, Doenjang, and Kochujang. *Food Sci Biotechnol* **14**:704–708.

137. Ouoba LII, Thorsen L, Varnam AH. 2008. Enterotoxins and emetic toxins production by *Bacillus cereus* and other species of *Bacillus* isolated from Soumbala and Bikalga, African alkaline fermented food condiments. *Int J Food Microbiol* **124**:224–230.

138. Demain AL, Solomon NA. 1981. Industrial microbiology. *Sci Am* **245**:67–75.

139. Gupta R, Gigras P, Mohapatra H, Goswami VK, Chauhan B. 2003. Microbial alpha-amylases: a biotechnological perspective. *Process Biochem* **38**:1599–1616.

140. Gupta R, Beg QK, Lorenz P. 2002. Bacterial alkaline proteases: molecular approaches and industrial applications. *Appl Microbiol Biotechnol* **59**:15–32.

141. Abbas CA, Sibirny AA. 2011. Genetic control of biosynthesis and transport of riboflavin and flavin nucleotides and construction of robust biotechnological producers. *Microbiol Mol Biol Rev* **75**:321–360.

142. Glick RG, Pasternak JJ. 2003. *Molecular Biotechnology: Principles and Applications of Recombinant DNA*, 3rd ed. ASM Press, Washington, DC.

143. Harwood CR. 1992. *Bacillus subtilis* and its relatives: molecular biological and industrial workhorses. *Trends Biotechnol* **10**:247–256.

144. Sun HY, Zhao PJ, Ge XY, Xia YJ, Hao ZK, Liu JW, Peng M. 2010. Recent advances in microbial raw starch degrading enzymes. *Appl Biochem Biotechnol* **160**:988–1003.

145. Yoneda Y. 1980. Increased production of extracellular enzymes by the synergistic effect of genes introduced into *Bacillus subtilis* by stepwise transformation. *Appl Environ Microbiol* **39**:274–276.

146. Wohlgemuth R. 2009. The locks and keys to industrial biotechnology. *N Biotechnol* **25**:204–213.

147. Liu YH, Lu FP, Li Y, Yin XB, Wang Y, Gao C. 2008. Characterisation of mutagenised acid-resistant alpha-amylase expressed in *Bacillus subtilis* WB600. *Appl Microbiol Biotechnol* **78**:85–94.

148. Santhanam N, Vivanco JM, Decker SR, Reardon KF. 2011. Expression of industrially relevant laccases: prokaryotic style. *Trends Biotechnol* 29:480–489.

149. Gupta N, Farinas ET. 2010. Directed evolution of CotA laccase for increased substrate specificity using *Bacillus subtilis* spores. *Protein Eng Des Sel* 23:679–682.

150. Sharma A, Kawarabayasi Y, Satyanarayana T. 2012. Acidophilic bacteria and archaea: acid stable biocatalysts and their potential applications. *Extremophiles* 16:1–19.

151. Tu R, Martinez R, Prodanovic R, Klein M, Schwaneberg U. 2011. A flow cytometry-based screening system for directed evolution of proteases. *J Biomol Screen* 16:285–294.

152. Yao Y, Li Y, Hou S, Li C, Chen H, Liao Y. 2011. Directed evolution of neutral endoglucanase gene by error-prone PCR. *J Agric Biotechnol* 19:1136–1143.

153. Qin J, Gao W, Li Q, Li Y, Zheng F, Liu C, Gu G. 2010. Improvement of thermostability of beta-1,3-1,4-glucanase from *Bacillus amyloliquefaciens* BS5582 through in vitro evolution. *Sheng Wu Gong Cheng Xue Bao* 26:1293–1301.

154. Joo JC, Pack SP, Kim YH, Yoo YJ. 2011. Thermostabilization of *Bacillus circulans* xylanase: computational optimization of unstable residues based on thermal fluctuation analysis. *J Biotechnol* 151:56–65.

155. Van Arsdell SW, Perkins JB, Yocum RR, Luan L, Howitt CL, Chatterjee NP, Pero JG. 2005. Removing a bottleneck in the *Bacillus subtilis* biotin pathway: BioA utilizes lysine rather than S-adenosylmethionine as the amino donor in the KAPA-to-DAPA reaction. *Biotechnol Bioeng* 91:75–83.

156. Perkins JB, Sloma A, Hermann T, Theriault K, Zachgo E, Erdenberger T, Hannett N, Chatterjee NP, Williams V, Rufo GA, Hatch R, Pero J. 1999. Genetic engineering of *Bacillus subtilis* for the commercial production of riboflavin. *J Indust Microbiol Biotechnol* 22:8–18.

157. DeWulf P, Vandamme EJ. 1997. Production of D-ribose by fermentation. *Appl Microbiol Biotechnol* 48:141–148.

158. Koglin A, Doetsch V, Bernhard F. 2010. Molecular engineering aspects for the production of new and modified biosurfactants, p 158–169. *In* Sen R (ed), *Biosurfactants*, **vol 672**. Springer-Verlag Berlin, Berlin, Germany.

159. Shoda M. 2000. Bacterial control of plant diseases. *J Biosci Bioeng* 89:515–521.

160. Singh M, Patel SKS, Kalia VC. 2009. *Bacillus subtilis* as potential producer for polyhydroxyalkanoates. *Microb Cell Fact* 8:11. doi:10.1186/1475-2859-8-38.

161. Zhang XZ, Sathitsuksanoh N, Zhu Z, Zhang YH. 2011. One-step production of lactate from cellulose as the sole carbon source without any other organic nutrient by recombinant cellulolytic *Bacillus subtilis*. *Metab Eng* 13:364–372.

162. Klein HP. 1991. The Viking biology experiments: epilogue and prologue. *Orig Life Evol Biosph* 21:255–261.

163. Summons RE, Amend JP, Bish D, Buick R, Cody GD, Des Marais DJ, Dromart G, Eigenbrode JL, Knoll AH, Sumner DY. 2011. Preservation of martian organic and environmental records: final report of the Mars Biosignature Working Group. *Astrobiology* 11:157–181.

164. Ronto G, Berces A, Lammer H, Cockell CS, Molina-Cuberos GJ, Patel MR, Selsis F. 2003. Solar UV irradiation conditions on the surface of Mars. *Photochem Photobiol* 77:34–40.

165. Wassmann M, Moeller R, Reitz G, Rettberg P. 2010. Adaptation of *Bacillus subtilis* cells to Archean-like UV climate: relevant hints of microbial evolution to remarkably increased radiation resistance. *Astrobiology* 10:605–615.

166. Wassmann M, Moeller R, Reitz G, Rettberg P. 2011. Growth phase-dependent UV-C resistance of *Bacillus subtilis*: data from a short-term evolution experiment. *Arch Microbiol* 193:823–832.

167. Schuerger AC. 2004. Microbial ecology of the surface exploration of Mars with human-operated vehicles, p 363–386. *In* Cockell CS (ed), *Martian Expedition Planning*. Univelt Publishers, Santa Barbra, CA.

168. Schuerger AC, Nicholson WL. 2006. Interactive effects of hypobaria, low temperature, and CO_2 atmospheres inhibit the growth of mesophilic *Bacillus* spp. under simulated martian conditions. *Icarus* 185:143–152.

169. Nicholson WL, Fajardo-Cavazos P, Fedenko J, Ortiz-Lugo JL, Rivas-Castillo A, Waters SM, Schuerger AC. 2010. Exploring the low-pressure growth limit: evolution of *Bacillus subtilis* in the laboratory to enhanced growth at 5 kilopascals. *Appl Environ Microbiol* 76:7559–7565.

170. Waters SM, Robles-Martínez JA, Nicholson WL. 2013. *Bacillus subtilis* strains grown at 1013 vs 5 kPa pressure including controls. Gene Expression Omnibus Accession GSE 50653. http://www.ncbi.nlm.nih.gov/geo/query/acc.cgi?acc=GSE50653

171. Mansilla MC, de Mendoza D. 2005. The *Bacillus subtilis* desaturase: a model to understand phospholipid modification and temperature sensing. *Arch Microbiol* 183:229–235.

172. Fajardo-Cavazos P, Waters SM, Schuerger AC, George S, Marois JJ, Nicholson WL. 2012. Evolution of *Bacillus subtilis* to enhanced growth at low pressure: up-regulated transcription of *des-desKR*, encoding the fatty acid desaturase system. *Astrobiology* 12:258–270.

173. Fang J, Bazylinski DA. 2008. Deep sea geomicrobiology, p 237–264. *In* Michiels D, Bartlett D, Aertsen A (ed), *High-Pressure Microbiology*. ASM Press, Washington, DC.

174. Horneck G, Moeller R, Cadet J, Douki T, Mancinelli RL, Nicholson WL, Panitz C, Rabbow E, Rettberg P, Spry A, Stackebrandt E, Vaishampayan P, Venkateswaran KJ. 2012. Resistance of bacterial endospores to outer space for planetary protection purposes–experiment PROTECT of the EXPOSE-E mission. *Astrobiology* 12:445–456.

175. Horneck G, Facius R, Enge W, Beaujean R, Bartholoma KP. 1974. Microbial studies in the Biostack experiment of the Apollo 16 mission: germination and outgrowth of single Bacillus subtilis spores hit by cosmic HZE particles. *Life Sci Space Res* 12:75–83.

176. Nicholson W, Schuerger A, Race M. 2009. Migrating microbes and planetary protection. *Trends Microbiol* 17:389–392.

177. Cole JR, Wang Q, Cardenas E, Fish J, Chai B, Farris RJ, Kulam-Syed-Mohideen AS, McGarrell DM, Marsh T, Garrity GM, Tiedje JM. 2009. The Ribosomal Database Project: improved alignments and new tools for rRNA analysis. *Nucleic Acids Res* 37:D141–D145.

178. Peak KK, Duncan KE, Veguilla W, Luna VA, King DS, Heller L, Heberlein-Larson L, Reeves F, Cannons AC, Amuso P, Cattani J. 2007. *Bacillus acidiceler* sp nov., isolated from a forensic specimen, containing *Bacillus anthracis* pX02 genes. *Int J Syst Evol Microbiol* 57: 2031–2036.

179. Albert RA, Archambault J, Rossello-Mora R, Tindall BJ, Matheny M. 2005. *Bacillus acidicola* sp. nov., a novel mesophilic, acidophilic species isolated from acidic sphagnum peat bogs in Wisconsin. *Int J Syst Evol Microbiol* 55:2125–2130.

180. Jung MY, Kim JS, Chang YH. 2009. *Bacillus acidiproducens* sp. nov., vineyard soil isolates that produce lactic acid. *Int J Syst Evol Microbiol* 59:2226–2231.

181. Gugliandolo C, Maugeri TL, Caccamo D, Stackebrandt E. 2003. *Bacillus aeolius* sp. nov. a novel thermophilic, halophilic marine *Bacillus* species from Eolian Islands (Italy). *Syst Appl Microbiol* 26:172–176.

182. Shivaji S, Chaturvedi P, Suresh K, Reddy GS, Dutt CB, Wainwright M, Narlikar JV, Bhargava PM. 2006. *Bacillus aerius* sp. nov., *Bacillus aerophilus* sp. nov., *Bacillus stratosphericus* sp. nov. and *Bacillus altitudinis* sp. nov., isolated from cryogenic tubes used for collecting air samples from high altitudes. *Int J Syst Evol Microbiol* 56:1465–1473.

183. Nielsen P, Fritze D, Priest FG. 1995. Phenetic diversity of alkaliphilic *Bacillus* strains: proposal for 9 new species. *Microbiology* 141:1745–1761.

184. Xue Y, Ventosa A, Wang X, Ren P, Zhou P, Ma Y. 2008. *Bacillus aidingensis* sp. nov., a moderately halophilic bacterium isolated from Ai-Ding salt lake in China. *Int J Syst Evol Microbiol* 58:2828–2832.

185. Nogi Y, Takami H, Horikoshi K. 2005. Characterization of alkaliphilic *Bacillus* strains used in industry: proposal of five novel species. *Int J Syst Evol Microbiol* 55:2309–2315.

186. Ivanova EP, Alexeeva YA, Zhukova NV, Gorshkova NM, Buljan V, Nicolau DV, Mikhailov VV, Christen R. 2004. *Bacillus algicola* sp. nov., a novel filamentous organism isolated from brown alga *Fucus evanescens*. *Syst Appl Microbiol* 27:301–307.

187. Sorokin DY, van Pelt S, Tourova TP. 2008. Utilization of aliphatic nitriles under haloalkaline conditions by *Bacillus alkalinitrilicus* sp. nov. isolated from soda solonchak soil. *FEMS Microbiol Lett* 288:235–240.

188. Borsodi AK, Pollak B, Keki Z, Rusznyak A, Kovacs AL, Sproeer C, Schumann P, Marialigeti K, Toth EM. 2011. *Bacillus alkalisediminis* sp. nov., an alkaliphilic and moderately halophilic bacterium isolated from sediment of extremely shallow soda ponds. *Int J Syst Evol Microbiol* 61:1880–1886.

189. Lee JC, Lee GS, Park DJ, Kim CJ. 2008. *Bacillus alkalitelluris* sp. nov., an alkaliphilic bacterium isolated from sandy soil. *Int J Syst Evol Microbiol* 58:2629–2634.

190. Bae SS, Lee JH, Kim SJ. 2005. *Bacillus alveayuensis* sp. nov., a thermophilic bacterium isolated from deep-sea sediments of the Ayu Trough. *Int J Syst Evol Microbiol* 55:1211–1215.

191. Priest FG, Goodfellow M, Shute LA, Berkeley RCW. 1987. *Bacillus amyloliquefaciens* sp. nov., nom. rev. *Int J Syst Bacteriol* 37:69–71.

192. Yoon JH, Kim IG, Kang KH, Oh TK, Park YH. 2003. *Bacillus marisflavi* sp. nov. and *Bacillus aquimaris* sp. nov., isolated from sea water of a tidal flat of the Yellow Sea in Korea. *Int J Syst Evol Microbiol* 53:1297–1303.

193. Shivaji S, Suresh K, Chaturvedi P, Dube S, Sengupta S. 2005. *Bacillus arsenicus* sp. nov., an arsenic-resistant bacterium isolated from a siderite concretion in West Bengal, India. *Int J Syst Evol Microbiol* 55:1123–1127.

194. Shivaji S, Chaturvedi P, Begum Z, Pindi PK, Manorama R, Padmanaban DA, Shouche YS, Pawar S, Vaishampayan P, Dutt CBS, Datta GN, Manchanda RK, Rao UR, Bhargava PM, Narlikar JV. 2009. *Janibacter hoylei* sp. nov., *Bacillus isronensis* sp. nov. and *Bacillus aryabhattai* sp. nov., isolated from cryotubes used for collecting air from the upper atmosphere. *Int J Syst Evol Microbiol* 59:2977–2986.

195. Yumoto I, Hirota K, Yamaga S, Nodasaka Y, Kawasaki T, Matsuyama H, Nakajima K. 2004. *Bacillus asahii* sp. nov., a novel bacterium isolated from soil with the ability to deodorize the bad smell generated from short-chain fatty acids. *Int J Syst Evol Microbiol* 54:1997–2001.

196. Nakamura LK. 1989. Taxonomic relationship of black-pigmented *Bacillus subtilis* strains and a proposal for *Bacillus atrophaeus* sp. nov. *Int J Syst Bacteriol* 39:295–300.

197. Borsodi AK, Marialigeti K, Szabo G, Palatinszky M, Pollak B, Keki Z, Kovacs AL, Schumann P, Toth EM. 2008. *Bacillus aurantiacus* sp. nov., an alkaliphilic and moderately halophilic bacterium isolated from Hungarian soda lakes. *Int J Syst Evol Microbiol* 58:845–851.

198. Pichinoty F, Debarjac H, Mandel M, Asselineau J. 1983. Description of *Bacillus azotoformans* sp. nov. *Int J Syst Bacteriol* 33:660–662.

199. Heyrman J, Vanparys B, Logan NA, Balcaen A, Rodiguez-Diaz M, Felske A, De Vos P. 2004. *Bacillus novalis* sp. nov., *Bacillus vireti* sp. nov., *Bacillus soli* sp. nov., *Bacillus bataviensis* sp. nov. and *Bacillus drentensis* sp. nov., from the Drentse A grasslands. *Int J Syst Evol Microbiol* 54:47–57.

200. Qiu F, Zhang X, Liu L, Sun L, Schumann P, Song W. 2009. *Bacillus beijingensis* sp. nov. and *Bacillus ginsengi* sp. nov., isolated from ginseng root. *Int J Syst Evol Microbiol* 59:729–734.

201. Pichinoty F, Asselineau J, Mandel M. 1984. Biochemical characterization of Bacillus *benzoevorans* sp. nov., a new filamentous, sheathed mesophilic species which degrades various aromatic acids and phenols. *Ann Microbiol* (Paris) 135B:209–217.

202. Vargas VA, Delgado OD, Hatti-Kaul R, Mattiasson B. 2005. *Bacillus bogoriensis* sp. nov., a novel alkallphilic, halotolerant bacterium isolated from a Kenyan soda lake. *Int J Syst Evol Microbiol* **55**:899–902.

203. Ahmed I, Yokota A, Fujiwara T. 2007. A novel highly boron tolerant bacterium, *Bacillus boroniphilus* sp. nov., isolated from soil, that requires boron for its growth. *Extremophiles* **11**:217–224.

204. Kuisiene N, Raugalas J, Sproeer C, Kroppenstedt RM, Chitavichius D. 2008. *Bacillus butanolivorans* sp. nov., a species with industrial application for the remediation of n-butanol. *Int J Syst Evol Microbiol* **58**:505–509.

205. Newcombe D, Dekas A, Mayilraj S, Venkateswaran K. 2009. *Bacillus canaveralius* sp. nov., an alkali-tolerant bacterium isolated from a spacecraft assembly facility. *Int J Syst Evol Microbiol* **59**:2015–2019.

206. Fujita T, Shida O, Takagi H, Kunugita K, Pankrushina AN, Matsuhashi M. 1996. Description of *Bacillus carboniphilus* sp nov. *Int J Syst Bacteriol* **46**:116–118.

207. Reddy GS, Uttam A, Shivaji S. 2008. *Bacillus cecembensis* sp. nov., isolated from the Pindari glacier of the Indian Himalayas. *Int J Syst Evol Microbiol* **58**: 2330–2335.

208. Nakamura LK. 1993. DNA relatedness of *Bacillus brevis* Migula 1900 strains and proposal of *Bacillus agri* sp. nov., nom. rev., and *Bacillus centrosporus* sp. nov., nom. rev. *Int J Syst Bacteriol* **43**:20–25.

209. Frankland GC, Frankland PF. 1887. XI. Studies on some new microorganisms obtained from air. *Phil Trans Royal Soc London B* **178**:257–287.

210. Carrasco IJ, Marquez MC, Xue Y, Ma Y, Cowan DA, Jones EE, Grant WD, Ventosa A. 2007. *Bacillus chagannorensis* sp. nov., a moderate halophile from a soda lake in Inner Mongolia, China. *Int J Syst Evol Microbiol* **57**:2084–2088.

211. Cho S, Jung MY, Park M, Kim W. 2010. *Bacillus chungangensis* sp. nov., a halophilic species isolated from sea sand. *Int J Syst Evol Microbiol* **60**:1349–1352.

212. Yoon JH, Lee CH, Oh TK. 2005. *Bacillus cibi* sp. nov., isolated from Jeotgal, a traditional Korean fermented seafood. *Int J Syst Evol Microbiol* **55**:733–736.

213. Jordan EO. 1890. A report on certain species of bacteria observed in sewage. *In* Sedgwick WT (ed), *A report of the biological work of the Lawrence Experiment Station, including an account of methods employed and results obtained in the microscopical and bacteriological investigation of sewage and water. Report on water supply and sewerage, part 2.* Massachusetts State Board of Health, Boston, MA.

214. Hammer BW. 1915. Bacteriological studies on the coagulation of evaporated milk. *Iowa Agric Exp Stn Res Bull* **19**:119–131.

215. Cerritos R, Vinuesa P, Eguiarte LE, Herrera-Estrella L, Alcaraz-Peraza LD, Arvizu-Gomez J, Olmedo G, Ramirez E, Siefert JL, Souza V. 2008. *Bacillus coahuilensis* sp. nov., a moderately halophilic species from a desiccation lagoon in the Cuatro Cienegas Valley in Coahuila, Mexico. *Int J Syst Evol Microbiol* **58**: 919–923.

216. Spanka R, Fritze D. 1993. *Bacillus cohnii* sp. nov., a new, obligately alkaliphilic, oval-spore-forming *Bacillus* species with ornithine and aspartic acid instead of diaminopimelic acid in the cell wall. *Int J Syst Bacteriol* **43**: 150–156.

217. Guinebretiere MH, Auger S, Galleron N, Contzen M, De Sarrau B, De Buyser ML, Lamberet G, Fagerlund A, Granum PE, Lereclus D, De Vos P, Nguyen-The C, Sorokin A. 2012. *Bacillus cytotoxicus* sp. nov. is a new thermotolerant species of the *Bacillus cereus* group occasionally associated with food poisoning. *Int J Syst Evol Microbiol* **63**:31–40.

218. Zhang L, Xu Z, Patel BKC. 2007. *Bacillus decisifrondis* sp. nov., isolated from soil underlying decaying leaf foliage. *Int J Syst Evol Microbiol* **57**:974–978.

219. Heyrman J, Balcaen A, Rodriguez-Diaz M, Logan NA, Swings J, De Vos P. 2003. *Bacillus decolorationis* sp. nov., isolated from biodeteriorated parts of the mural paintings at the Servilia tomb (Roman necropolis of Carmona, Spain) and the Saint-Catherine chapel (Castle Herberstein, Austria). *Int J Syst Evol Microbiol* **53**: 459–463.

220. Reva ON, Smirnov VV, Pettersson B, Priest FG. 2002. *Bacillus endophyticus* sp. nov., isolated from the inner tissues of cotton plants (*Gossypium* sp.). *Int J Syst Evol Microbiol* **52**:101–107.

221. Zhang YZ, Chen WF, Li M, Sui XH, Liu H-C, Zhang XX, Chen WX. 2012. *Bacillus endoradicis* sp. nov., an endophytic bacterium isolated from soybean root. *Int J Syst Evol Microbiol* **62**:359–363.

222. Scheldeman P, Rodriguez-Diaz M, Goris J, Pil A, De Clerck E, Herman L, De Vos P, Logan NA, Heyndrickx M. 2004. *Bacillus farraginis* sp. nov., *Bacillus fortis* sp. nov. and *Bacillus fordii* sp. nov., isolated at dairy farms. *Int J Syst Evol Microbiol* **54**:1355–1364.

223. den Dooren de Jong LE. 1929. Über *Bacillus fastidiosus*. *Zentralbl Bakteriol Parasitenkd Infektionskr Hyg Abt II* **79**:344–353.

224. Werner W. 1933. Botanische Beschreibung häufiger am Buttersäureabbau beteiligter sporenbildender Bakterienspezies. *Zentralbl Bakteriol Parasitenkd Abt II* **87**:446–475.

225. Tiago I, Pires C, Mendes V, Morais PV, da Costa MS, Verissimo A. 2006. *Bacillus foraminis* sp. nov., isolated from a non-saline alkaline groundwater. *Int J Syst Evol Microbiol* **56**:2571–2574.

226. Logan NA, Lebbe L, Hoste B, Goris J, Forsyth G, Heyndrickx M, Murray BL, Syme N, Wynn-Williams DD, De Vos P. 2000. Aerobic endospore-forming bacteria from geothermal environments in northern Victoria Land, Antarctica, and Candlemas Island, South Sandwich archipelago, with the proposal of *Bacillus fumarioli* sp. nov. *Int J Syst Evol Microbiol* **50**:1741–1753.

227. Ajithkumar VP, Ajithkumar B, Iriye R, Sakai T. 2002. *Bacillus funiculus* sp. nov., novel filamentous isolates from activated sludge. *Int J Syst Evol Microbiol* **52**: 1141–1144.

228. Heyndrickx M, Logan NA, Lebbe L, Rodriguez-Diaz M, Forsyth G, Goris J, Scheldeman P, De Vos P. 2004.

Bacillus galactosidilyticus sp. nov., an alkali-tolerant beta-galactosidase producer. *Int J Syst Evol Microbiol* 54:617–621.

229. Balcazar JL, Pintado J, Planas M. 2010. *Bacillus galliciensis* sp. nov., isolated from faeces of wild seahorses (*Hippocampus guttulatus*). *Int J Syst Evol Microbiol* 60:892–895.

230. De Clerck E, Rodriguez-Diaz M, Vanhoutte T, Heyrman J, Logan NA, De Vos P. 2004. *Anoxybacillus contaminans* sp. nov. and *Bacillus gelatini* sp. nov., isolated from contaminated gelatin batches. *Int J Syst Evol Microbiol* 54:941–946.

231. Ten LN, Im WT, Baek SH, Lee JS, Oh HM, Lee ST. 2006. *Bacillus ginsengihumi* sp. nov., a novel species isolated from soil of a ginseng field in Pocheon Province, South Korea. *J Microbiol Biotechnol* 16:1554–1560.

232. Bibi F, Chung EJ, Jeon CO, Chung YR. 2011. *Bacillus graminis* sp. nov., an endophyte isolated from a coastal dune plant. *Int J Syst Evol Microbiol* 61:1567–1571.

233. Pappa A, Sanchez-Porro C, Lazoura P, Kallimanis A, Perisynakis A, Ventosa A, Drainas C, Koukkou AI. 2010. *Bacillus halochares* sp. nov., a halophilic bacterium isolated from a solar saltern. *Int J Syst Evol Microbiol* 60:1432–1436.

234. Chen YG, Zhang YQ, He JW, Klenk HP, Xiao JQ, Zhu HY, Tang SK, Li WJ. 2011. *Bacillus hemicentroti* sp. nov., a moderate halophile isolated from a sea urchin. *Int J Syst Evol Microbiol* 61:2950–2955.

235. Wieser M, Worliczek H, Kämpfer P, Busse HJ. 2005. *Bacillus herbersteinensis* sp. nov. *Int J Syst Evol Microbiol* 55:2119–2123.

236. Vaishampayan P, Probst A, Krishnamurthi S, Ghosh S, Osman S, McDowall A, Ruckmani A, Mayilraj S, Venkateswaran K. 2010. *Bacillus horneckiae* sp. nov., isolated from a spacecraft-assembly clean room. *Int J Syst Evol Microbiol* 60:1031–1037.

237. Yumoto I, Yamazaki K, Sawabe T, Nakano K, Kawasaki K, Ezura Y, Shinano H. 1998. *Bacillus horti* sp. nov., a new Gram-negative alkaliphilic *Bacillus*. *Int J Syst Bacteriol* 48:565–571.

238. Heyrman J, Rodriguez-Diaz M, Devos J, Felske A, Logan NA, De Vos P. 2005. *Bacillus arenosi* sp. nov., *Bacillus arvi* sp. nov. and *Bacillus humi* sp. nov., isolated from soil. *Int J Syst Evol Microbiol* 55:111–117.

239. Yoon JH, Kim IG, Kang KH, Oh TK, Park YH. 2004. Bacillus *hwajinpoensis* sp. nov. and an unnamed *Bacillus genomo* species, novel members of Bacillus *rRNA* group 6 isolated from sea water of the East Sea and the Yellow Sea in Korea. *Int J Syst Evol Microbiol* 54:803–808.

240. Ko KS, Oh WS, Lee MY, Lee JH, Lee H, Peck KR, Lee NY, Song JH. 2006. *Bacillus infantis* sp. nov. and *Bacillus idriensis* sp. nov., isolated from a patient with neonatal sepsis. *Int J Syst Evol Microbiol* 56:2541–2544.

241. Suresh K, Prabagaran SR, Sengupta S, Shivaji S. 2004. *Bacillus indicus* sp. nov., an arsenic-resistant bacterium isolated from an aquifer in West Bengal, India. *Int J Syst Evol Microbiol* 54:1369–1375.

242. Boone DR, Liu Y, Zhao ZJ, Balkwill DL, Drake GR, Stevens TO, Aldrich HC. 1995. *Bacillus infernus* sp. nov., an Fe(III)- and Mn(IV)-reducing anaerobe from the deep terrestrial subsurface. *Int J Syst Bacteriol* 45:441–448.

243. Albuquerque L, Tiago I, Taborda M, Nobre MF, Veríssimo A, da Costa MS. 2008. Bacillus isabeliae sp. nov., a halophilic bacterium isolated from a sea salt evaporation pond. *Int J Syst Evol Microbiol* 58:226–230.

244. Yoon JH, Kang SS, Lee KC, Kho YH, Choi SH, Kang KH, Park YH. 2001. *Bacillus jeotgali* sp. nov., isolated from jeotgal, Korean traditional fermented seafood. *Int J Syst Evol Microbiol* 51:1087–1092.

245. Lim JM, Jeon CO, Lee JC, Ju YJ, Park DJ, Kim CJ. 2006. *Bacillus koreensis* sp. nov., a spore-forming bacterium, isolated from the rhizosphere of willow roots in Korea. *Int J Syst Evol Microbiol* 56:59–63.

246. Zhang L, Wang Y, Dai J, Tang YL, Yang Q, Luo XS, Fang CX. 2009. *Bacillus korlensis* sp. nov., a moderately halotolerant bacterium isolated from a sand soil sample in China. *Int J Syst Evol Microbiol* 59:1787–1792.

247. Lim JM, Jeon CO, Lee JR, Park DJ, Kim CJ. 2007. *Bacillus kribbensis* sp. nov., isolated from a soil sample in Jeju, Korea. *Int J Syst Evol Microbiol* 57:2912–2916.

248. Yumoto I, Yamaga S, Sogabe Y, Nodasaka Y, Matsuyama H, Nakajima K, Suemori A. 2003. *Bacillus krulwichiae* sp. nov., a halotolerant obligate alkaliphile that utilizes benzoate and m-hydroxybenzoate. *Int J Syst Evol Microbiol* 53:1531–1536.

249. Ghosh A, Bhardwaj M, Satyanarayana T, Khurana M, Mayilraj S, Jain RK. 2007. *Bacillus lehensis* sp. nov., an alkalitolerant bacterium isolated from soil. *Int J Syst Evol Microbiol* 57:238–242.

250. Gibson T. 1935. The urea-decomposing microflora of soils. I. Description and classification of the organisms. *Zentbl Bakteriol Parasitenkd Infektkrankh Hyg Abt II* 92:364–380.

251. Weigmann H. 1898. Über zwei an der Käsereifung beteiligte Bakterien. *Zentralbl Bakteriol, Parasitenkd, Infektkrankh Hyg Abt II* 4:820–834.

252. Yoon JH, Oh TK. 2005. *Bacillus litoralis* sp. nov., isolated from a tidal flat of the Yellow Sea in Korea. *Int J Syst Evol Microbiol* 55:19451948.

253. Marquez MC, Carrasco IJ, de la Haba RR, Jones BE, Grant WD, Ventosa A. 2011. *Bacillus locisalis* sp. nov., a new haloalkaliphilic species from hypersaline and alkaline lakes of China, Kenya and Tanzania. *Syst Appl Microbiol* 34:424–428.

254. Logan NA, Lebbe L, Verhelst A, Goris J, Forsyth G, Rodriguez-Diaz M, Heyndrickx M, De Vos P. 2002. *Bacillus luciferensis* sp. nov., from volcanic soil on Candlemas Island, South Sandwich archipelago. *Int J Syst Evol Microbiol* 52:1985–1989.

255. Shi R, Yin M, Tang S-K, Lee J-C, Park D-J, Zhang Y-J, Kim C-J, Li W-J. 2011. *Bacillus luteolus* sp. nov., a halotolerant bacterium isolated from a salt field. *Int J Syst Evol Microbiol* 61:1344–1349.

256. Zhang T, Fan XJ, Hanada S, Kamagata Y, Fang HHP. 2006. *Bacillus macauensis* sp. nov, a long-chain bacterium isolated from a drinking water supply. *Int J Syst Evol Microbiol* 56:349–353.

257. Denizci AA, Kazan D, Erarslan A. 2010. *Bacillus marmarensis* sp. nov., an alkaliphilic, protease-producing bacterium isolated from mushroom compost. *Int J Syst Evol Microbiol* 60:1590–1594.

258. Glazunova OO, Raoult D, Roux V. 2006. *Bacillus massilensis* sp. nov., isolated from cerebrospinal fluid. *Int J Syst Evol Microbiol* 56:1485–1488.

259. de Bary HA. 1884. *Vergleichende Morphologie und Biologie der Pilze, Mycetozoen und Bakterien (Comparative Morphology and Biology of the Fungi, Slime Molds, and Bacteria)*, 2nd ed. Wilhelm Engelman, Leipzig, Germany.

260. Arfman N, Dijkhuizen L, Kirchhof G, Ludwig W, Schleifer KH, Bulygina ES, Chumakov KM, Govorukhina NI, Trotsenko YA, White D, Sharp RJ. 1992. *Bacillus methanolicus* sp. nov., a new species of thermotolerant, methano-utilizing, endospore-forming bacteria. *Int J Syst Bacteriol* 42:439–445.

261. Madhaiyan M, Poonguzhali S, Kwon SW, Sa TM. 2010. *Bacillus methylotrophicus* sp. nov., a methanol-utilizing, plant-growth-promoting bacterium isolated from rice rhizosphere soil. *Int J Syst Bacteriol* 60:2490–2495.

262. Roberts MS, Nakamura LK, Cohan FM. 1994. *Bacillus mojavensis* sp. nov., distinguishable from *Bacillus subtilis* by sexual isolation, divergence in DNA sequence, and differences in fatty acid composition. *Int J Syst Bacteriol* 44:256–264.

263. Wang L-T, Lee F-L, Tai C-J, Yokota A, Kuo H-P. 2007. Reclassification of *Bacillus axarquiensis* Ruiz-Garcia et al. 2005 and *Bacillus malacitensis* Ruiz-Garcia et al. 2005 as later heterotypic synonyms of *Bacillus mojavensis* Roberts et al. 1994. *Int J Syst Evol Microbiol* 57:1663–1667.

264. Heyrman J, Logan NA, Rodriguez-Diaz M, Scheldeman P, Lebbe L, Swings J, Heyndrickx M, De Vos P. 2005. Study of mural painting isolates, leading to the transfer of "*Bacillus maroccanus*" and "*Bacillus carotarum*" to *Bacillus simplex*, emended description of *Bacillus simplex*, re-examination of the strains previously attributed to "*Bacillus macroides*" and description of *Bacillus muralis* sp. nov. *Int J Syst Evol Microbiol* 55:119–131.

265. Borchert MS, Nielsen P, Graeber I, Kaesler I, Szewzyk U, Pape T, Antranikian G, Schafer T. 2007. *Bacillus plakortidis* sp. nov. and *Bacillus murimartini* sp. nov., novel alkalitolerant members of rRNA group 6. *Int J Syst Evol Microbiol* 57:2888–2893.

266. Flügge C. 1886. *Die Mikroorganismen: Mit Besonderer Berucksichtigung der Aetiologie der Infectionskrankheiten.* Vogel, Leipzig, Germany.

267. Chen YG, Zhang L, Zhang YQ, He JW, Klenk HP, Tang SK, Zhang YX, Li WJ. 2011. *Bacillus nanhaiensis* sp. nov., isolated from an oyster. *Int J Syst Evol Microbiol* 61:888–893.

268. Zhang JL, Wang JW, Song F, Fang CY, Xin YH, Zhang YB. 2011. *Bacillus nanhaiisediminis* sp. nov., an alkalitolerant member of *Bacillus* rRNA group 6. *Int J Syst Evol Microbiol* 61:1078–1083.

269. Venkateswaran K, Kempf M, Chen F, Satomi M, Nicholson W, Kern R. 2003. *Bacillus nealsonii* sp. nov., isolated from a spacecraft-assembly facility, whose spores are gamma-radiation resistant. *Int J Syst Evol Microbiol* 53:165–172.

270. Chen YG, Zhang YQ, Wang YX, Liu ZX, Klenk HP, Xiao HD, Tang SK, Cui XL, Li WJ. 2009. *Bacillus neizhouensis* sp. nov., a halophilic marine bacterium isolated from a sea anemone. *Int J Syst Evol Microbiol* 59:3035–3039.

271. Kwon SW, Lee SY, Kim BY, Weon HY, Kim JB, Go SJ, Lee GB. 2007. *Bacillus niabensis* sp. nov., isolated from cotton-waste composts for mushroom cultivation. *Int J Syst Evol Microbiol* 57:1909–1913.

272. Nagel M, Andreesen JR. 1991. *Bacillus niacini* sp. nov., a nicotinate-metabolizing mesophile isolated from soil. *Int J Syst Bacteriol* 41:134–139.

273. Zhang JL, Wang JW, Fang CY, Song F, Xin YH, Qu L, Ding K. 2010. *Bacillus oceanisediminis* sp. nov., isolated from marine sediment. *Int J Syst Evol Microbiol* 60:2924–2929.

274. La Duc MT, Satomi M, Venkateswaran K. 2004. *Bacillus odysseyi* sp. nov., a round-spore-forming bacillus isolated from the Mars Odyssey spacecraft. *Int J Syst Evol Microbiol* 54:195–201.

275. Nowlan B, Dodia MS, Singh SP, Patel BKC. 2006. *Bacillus okhensis* sp. nov., a halotolerant and alkalitolerant bacterium from an Indian saltpan. *Int J Syst Evol Microbiol* 56:1073–1077.

276. Li ZY, Kawamura Y, Shida O, Yamagata S, Deguchi T, Ezaki T. 2002. *Bacillus okuhidensis* sp. nov., isolated from the Okuhida spa area of Japan. *Int J Syst Evol Microbiol* 52:1205–1209.

277. Kuhnigk T, Borst EM, Breunig A, Konig H, Collins MD, Hutson RA, Kampfer P. 1995. *Bacillus oleronius* sp. nov., a member of the hindgut flora of the termite *Reticulitermes santonensis* (Feytaud). *Can J Microbiol* 41:699–706.

278. Yumoto I, Hirota K, Goto T, Nodasaka Y, Nakajima K. 2005. *Bacillus oshimensis* sp. nov., a moderately halophilic, non-motile alkaliphile. *Int J Syst Evol Microbiol* 55:907–911.

279. Ten LN, Baek SH, Im WT, Liu QM, Aslam Z, Lee ST. 2006. *Bacillus panaciterrae* sp. nov., isolated from soil of a ginseng field. *Int J Syst Evol Microbiol* 56:2861–2866.

280. Olivera N, Siñeriz F, Breccia JD. 2005. *Bacillus patagoniensis* sp. nov., a novel alkalitolerant bacterium from the rhizosphere of *Atriplex lampa* in Patagonia, Argentina. *Int J Syst Evol Microbiol* 55:443–447.

281. Amoozegar MA, Sanchez-Porro C, Rohban R, Hajighasemi M, Ventosa A. 2009. *Bacillus persepolensis* sp. nov., a moderately halophilic bacterium from a hypersaline lake. *Int J Syst Evol Microbiol* 59:2352–2358.

282. Ten LN, Baek S-H, Im W-T, Larina LL, Lee J-S, Oh H-M, Lee S-T. 2007. *Bacillus pocheonensis* sp. nov., a moderately halotolerant, aerobic bacterium isolated

from soil of a ginseng field. *Int J Syst Evol Microbiol* 57:2532–2537.

283. **Aino K, Hirota K, Matsuno T, Morita N, Nodasaka Y, Fujiwara T, Matsuyama H, Yoshimune K, Yumoto I.** 2008. *Bacillus polygoni* sp. nov., a moderately halophilic, non-motile obligate alkaliphile isolated from indigo balls. *Int J Syst Evol Microbiol* 58:120–124.

284. **Nakamura LK.** 1998. *Bacillus pseudomycoides* sp. nov. *Int J Syst Bacteriol* 48:1031–1035.

285. **Larkin JM, Stokes JL.** 1967. Taxonomy of psychrophilic strains of *Bacillus. J Bacteriol* 94:889–895.

286. **Meyer A, Gottheil O.** 1901. Botanische Beschreibung einiger Bodenbakterien. *Zentralbl Bakteriol Parasitenkd Infektkrankh Hyg Abt II* 7:680–691.

287. **Vaz-Moreira I, Figueira V, Lopes AR, Lobo-da-Cunha A, Sproeer C, Schumann P, Nunes OC, Manaia CM.** 2012. *Bacillus purgationiresistans* sp. nov., isolated from a drinking-water treatment plant. *Int J Syst Evol Microbiol* 62:71–77.

288. **Wang O-F, Li W, Liu Y-L, Cao H-H, Li Z, Guo G-Q.** 2007. *Bacillus qingdaonensis* sp. nov., a moderately haloalkaliphilic bacterium isolated from a crude sea-salt sample collected near Qingdao in eastern China. *Int J Syst Evol Microbiol* 57:1143–1147.

289. **Baik KS, Lim CH, Park SC, Kim EM, Rhee MS, Seong CN.** 2010. *Bacillus rigui* sp. nov., isolated from wetland fresh water. *Int J Syst Evol Microbiol* 60:2204–2209.

290. **Heyndrickx M, Scheldeman P, Forsyth G, Lebbe L, Rodriguez-Diaz M, Logan NA, De Vos P.** 2005. *Bacillus ruris* sp. nov., from dairy farms. *Int J Syst Evol Microbiol* 55:2551–2554.

291. **Satomi M, La Duc MT, Venkateswaran K.** 2006. *Bacillus safensis* sp. nov., isolated from spacecraft and assembly-facility surfaces. *Int J Syst Evol Microbiol* 56:1735–1740.

292. **Lim JM, Jeon CO, Lee SM, Lee JC, Xu LH, Jiang CL, Kim CJ.** 2006. *Bacillus salarius* sp. nov., a halophilic, spore-forming bacterium isolated from a salt lake in China. *Int J Syst Evol Microbiol* 56:373–377.

293. **Romano I, Lama L, Nicolaus B, Gambacorta A, Giordano A.** 2005. *Bacillus saliphilus* sp. nov., isolated from a mineral pool in Campania, Italy. *Int J Syst Evol Microbiol* 55:159–163.

294. **Schenk A, Aragno M.** 1979. *Bacillus schlegelii*, a new species of thermophilic, facultatively chemolithoautotrophic bacterium oxidizing molecular hydrogen. *J Gen Microbiol* 115:333–341.

295. **Yamamura S, Yamashita M, Fujimoto N, Kuroda M, Kashiwa M, Sei K, Fujita M, Ike M.** 2007. *Bacillus selenatarsenatis* sp. nov., a selenate- and arsenate-reducing bacterium isolated from the effluent drain of a glass-manufacturing plant. *Int J Syst Evol Microbiol* 57:1060–1064.

296. **Switzer Blum J, Bindi AB, Buzzelli J, Stolz JF, Oremland RS.** 1998. *Bacillus arsenicoselenatis*, sp. nov. and *Bacillus selenitireducens*, sp nov: two haloalkaliphiles from Mono Lake, California that respire oxyanions of selenium and arsenic. *Arch Microbiol* 171:19–30.

297. **Lee J-C, Lim J-M, Park D-J, Jeon CO, Li W-J, Kim C-J.** 2006. *Bacillus seohaeanensis* sp. nov., a halotolerant bacterium that contains L-lysine in its cell wall. *Int J Syst Evol Microbiol* 56:1893–1898.

298. **Logan NA, Lebbe L, Verhelst A, Goris J, Forsyth G, Rodriguez-Diaz M, Heyndrickx M, De Vos P.** 2004. *Bacillus shackletonii* sp. nov., from volcanic soil on Candlemas Island, South Sandwich archipelago. *Int J Syst Evol Microbiol* 54:373–376.

299. **Sumpavapol P, Tongyonk L, Tanasupawat S, Chokesajjawatee N, Luxanani P, Visessanguan W.** 2010. *Bacillus siamensis* sp. nov., isolated from salted crab (poo-khem) in Thailand. *Int J Syst Evol Microbiol* 60:2364–2370.

300. **Pettersson B, de Silva SK, Uhlen M, Priest FG.** 2000. *Bacillus siralis* sp. nov., a novel species from silage with a higher order structural attribute in the 16S rRNA genes. *Int J Syst Evol Microbiol* 50:2181–2187.

301. **Nakamura LK, Blumenstock I, Claus D.** 1988. Taxonomic study of *Bacillus coagulans* Hammer 1915 with a proposal for *Bacillus smithii* sp. nov. *Int J Syst Bacteriol* 38:63–73.

302. **Liu H, Zhou Y, Liu R, Zhang K-Y, Lai R.** 2009. *Bacillus solisalsi* sp. nov., a halotolerant, alkaliphilic bacterium isolated from soil around a salt lake. *Int J Syst Evol Microbiol* 59:1460–1464.

303. **Palmisano MM, Nakamura LK, Duncan KE, Istock CA, Cohan FM.** 2001. *Bacillus sonorensis* sp. nov., a close relative of *Bacillus licheniformis*, isolated from soil in the Sonoran Desert, Arizona. *Int J Syst Evol Microbiol* 51:1671–1679.

304. **Pettersson B, Lembke F, Hammer P, Stackebrandt E, Priest FG.** 1996. *Bacillus sporothermodurans*, a new species producing highly heat-resistant endospores. *Int J Syst Bacteriol* 46:759–764.

305. **Kanso S, Greene AC, Patel BKC.** 2002. *Bacillus subterraneus* sp. nov., an iron- and manganese-reducing bacterium from a deep subsurface Australian thermal aquifer. *Int J Syst Evol Microbiol* 52:869–874.

306. **Conn HJ.** 1930. The identity of *Bacillus subtilis. J Infect Dis* 46:341–350.

307. **Lim J-M, Jeon CO, Kim C-J.** 2006. *Bacillus taeanensis* sp. nov., a halophilic grampositive bacterium from a solar saltern in Korea. *Int J Syst Evol Microbiol* 56:2903–2908.

308. **Gatson JW, Benz BF, Chandrasekaran C, Satomi M, Venkateswaran K, Hart ME.** 2006. *Bacillus tequilensis* sp. nov., isolated from a 2000-year-old Mexican shaft-tomb, is closely related to *Bacillus subtilis. Int J Syst Evol Microbiol* 56:1475–1484.

309. **Combet-Blanc Y, Ollivier B, Streicher C, Patel BKC, Dwivedi PP, Pot B, Prensier G, Garcia J-L.** 1995. *Bacillus thermoamylovorans* sp. nov., a moderately thermophilic and amylolytic bacterium. *Int J Syst Bacteriol* 45:9–16.

310. **Demharter W, Hensel R.** 1989. *Bacillus thermocloaceae* sp. nov., a new thermophilic species from sewage sludge. *Syst Appl Microbiol* 11:272–276.

311. Coorevits A, Logan NA, Dinsdale AE, Halket G, Scheldeman P, Heyndrickx M, Schumann P, Van Landschoot A, De Vos P. 2011. *Bacillus thermolactis* sp. nov., isolated from dairy farms, and emended description of *Bacillus thermoamylovorans*. *Int J Syst Evol Microbiol* **61**:1954–1961.

312. Pérez-Ibarra BM, Flores ME, García-Varela M. 2007. Isolation and characterization of *Bacillus thioparus* sp. nov., chemolithoautotrophic, thiosulfate-oxidizing bacterium. *FEMS Microbiol Lett* **271**:289–296.

313. Berliner E. 1915. Über die Schlaffsucht der Mehlmottenraupe (Ephestia kühniella Zell) und ihren Erreger *Bacillus thuringiensis* n. sp. *Z Angew Entomol* **2**:29–56.

314. Aizawa T, Urai M, Iwabuchi N, Nakajima M, Sunairi M. 2010. *Bacillus trypoxylicola* sp. nov., xylanase-producing alkaliphilic bacteria isolated from the guts of Japanese horned beetle larvae (*Trypoxylus dichotomus septentrionalis*). *Int J Syst Evol Microbiol* **60**:61–66.

315. Bonjour F, Aragno M. 1984. *Bacillus tusciae*, a new species of thermoacidophilic, facultatively chemolithoautotrophic, hydrogen oxidizing sporeformer from a geothermal area. *Arch Microbiol* **139**:397–401.

316. Roberts MS, Nakamura LK, Cohan FM. 1996. *Bacillus vallismortis* sp. nov., a close relative of *Bacillus subtilis*, isolated from soil in Death Valley, California. *Int J Syst Bacteriol* **46**:470–475.

317. Agnew DM, Koval SF, Jarrell KF. 1995. Isolation and characterization of novel alkaliphiles from bauxite-processing waste and description of *Bacillus vedderi* sp. nov., a new obligate alkaliphile. *Syst Appl Microbiol* **18**:221–230.

318. Noguchi H, Uchino M, Shida O, Takano K, Nakamura LK, Komagata K. 2004. *Bacillus vietnamensis* sp. nov., a moderately halotolerant, aerobic, endospore-forming bacterium isolated from Vietnamese fish sauce. *Int J Syst Evol Microbiol* **54**:2117–2120.

319. Lechner S, Mayr R, Francis KP, Prüss BM, Kaplan T, Wiessner-Gunkel E, Stewartz GS, Scherer S. 1998. *Bacillus weihenstephanensis* sp. nov. is a new psychrotolerant species of the *Bacillus cereus* group. *Int J Syst Bacteriol* **48**:1373–1382.

320. Chen Y-G, Zhang Y-Q, Chen Q-H, Klenk H-P, He J-W, Tang S-K, Cui X-L, Li W-J. 2011. *Bacillus xiaoxiensis* sp. nov., a slightly halophilic bacterium isolated from non-saline forest soil. *Int J Syst Evol Microbiol* **61**:2095–2100.

321. Minana-Galbis D, Pinzon DL, Lorén JG, Manresa A, Oliart-Ros RM. 2010. Reclassification of *Geobacillus pallidus* (Scholz et al. 1988) Banat et al. 2004 as *Aeribacillus pallidus* gen. nov., comb. nov. *Int J Syst Evol Microbiol* **60**:1600–1604.

322. Romano I, Lama L, Nicolaus B, Gambacorta A, Giordano A. 2005. *Alkalibacillus filiformis* sp. nov., isolated from a mineral pool in Campania, Italy. *Int J Syst Evol Microbiol* **55**:2395–2399.

323. Yoon J-H, Kang S-J, Jung Y-T, Lee M-H, Oh T-K. 2010. *Alkalibacillus flavidus* sp. nov., isolated from a marine solar saltern. *Int J Syst Evol Microbiol* **60**:434–438.

324. Jeon CO, Lim JM, Lee JM, Xu LH, Jiang CL, Kim CJ. 2005. Reclassification of *Bacillus haloalkaliphilus* Fritze 1996 as *Alkalibacillus haloalkaliphilus* gen. nov., comb. nov and the description of *Alkalibacillus salilacus* sp. nov., a novel halophilic bacterium isolated from a salt lake in China. *Int J Syst Evol Microbiol* **55**:1891–1896.

325. Tian X-P, Dastager SG, Lee J-C, Tang S-K, Zhang Y-Q, Park D-J, Kim C-J, Li W-J. 2007. *Alkalibacillus halophilus* sp. nov., a new halophilic species isolated from hypersaline soil in Xin-Jiang province, China. *Syst Appl Microbiol* **30**:268–272.

326. Usami R, Echigo A, Fukushima T, Mizuki T, Yoshida Y, Kamekura M. 2007. *Alkalibacillus silvisoli* sp. nov., an alkaliphilic moderate halophile isolated from non-saline forest soil in Japan. *Int J Syst Evol Microbiol* **57**:770–774.

327. Sheu S-Y, Arun AB, Jiang S-R, Young C-C, Chen W-M. 2011. *Allobacillus halotolerans* gen. nov., sp. nov. isolated from shrimp paste. *Int J Syst Evol Microbiol* **61**:1023–1027.

328. Zhilina TN, Garnova ES, Tourova TP, Kostrikina NA, Zavarzin GA. 2001. *Amphibacillus fermentum* sp. nov. and *Amphibacillus tropicus* sp. nov., new alkaliphilic, facultatively anaerobic, saccharolytic bacilli from Lake Magadi. *Microbiology* **70**:711–722.

329. Wu XY, Zheng G, Zhang WW, Xu XW, Wu M, Zhu XF. 2010. *Amphibacillus jilinensis* sp. nov., a facultatively anaerobic, alkaliphilic bacillus from a soda lake. *Int J Syst Evol Microbiol* **60**:2540–2543.

330. An S-Y, Ishikawa S, Kasai H, Goto K, Yokota A. 2007. *Amphibacillus sediminis* sp. nov., an endospore-forming bacterium isolated from lake sediment in Japan. *Int J Syst Evol Microbiol* **57**:2489–2492.

331. Niimura Y, Koh E, Yanagida F, Suzuki K-I, Komagata K, Kozaki M. 1990. *Amphibacillus xylanus* gen. nov., sp. nov., a facultatively anaerobic spore-forming xylan-digesting bacterium which lacks cytochrome, quinone, and catalase. *Int J Syst Bacteriol* **40**:297–301.

332. Sorokin ID, Kravchenko IK, Tourova TP, Kolganova TV, Boulygina ES, Sorokin DY. 2008. *Bacillus alkalidiazotrophicus* sp. nov., a diazotrophic, low salt-tolerant alkaliphile isolated from Mongolian soda soil. *Int J Syst Evol Microbiol* **58**:2459–2464.

333. Zavarzina DG, Tourova TP, Kolganova TV, Boulygina ES, Zhilina TN. 2009. Description of *Anaerobacillus alkalilacustre* gen. nov., sp. nov.—Strictly anaerobic diazotrophic bacillus isolated from soda lake and transfer of *Bacillus arseniciselenatis*, *Bacillus macyae*, and *Bacillus alkalidiazotrophicus* to *Anaerobacillus* as the new combinations *A. arseniciselenatis* comb. nov., *A. macyae* comb. nov., and *A. alkalidiazotrophicus* comb. nov. *Microbiology* **78**:723–731.

334. Poli A, Esposito E, Lama L, Orlando P, Nicolaus G, de Appolonia F, Gambacorta A, Nicolaus B. 2006. *Anoxybacillus amylolyticus* sp. nov., a thermophilic amylase producing bacterium isolated from Mount Rittmann (Antarctica). *Syst Appl Microbiol* **29**:300–307.

335. Dulger S, Demirbag Z, Belduz AO. 2004. *Anoxybacillus ayderensis* sp. nov. and *Anoxybacillus kestanbolensis* sp. nov. *Int J Syst Evol Microbiol* **54**:1499–1503.

336. Zhang C-M, Huang X-W, Pan W-Z, Zhang J, Wei K-B, Klenk H-P, Tang S-K, Li W-, Zhang K-Q. 2011. *Anoxybacillus tengchongensis* sp. nov. and *Anoxybacillus eryuanensis* sp. nov., facultatively anaerobic, alkalitolerant bacteria from hot springs. *Int J Syst Evol Microbiol* 61:118–122.

337. Pikuta E, Lysenko A, Chuvilskaya N, Mendrock U, Hippe H, Suzina N, Nikitin D, Osipov G, Laurinavichius K. 2000. *Anoxybacillus pushchinensis* gen. nov., sp. nov., a novel anaerobic, alkaliphilic, moderately thermophilic bacterium from manure, and description of *Anoxybacillus falvithermus* comb. nov. *Int J Syst Evol Microbiol* 50:2109–2117.

338. Belduz AO, Dulger S, Demirbag Z. 2003. *Anoxybacillus gonensis* sp. nov., a moderately thermophilic, xylose-utilizing, endospore-forming bacterium. *Int J Syst Evol Microbiol* 53:1315–1320.

339. Kevbrin VV, Zengler K, Lysenko AM, Wiegel J. 2005. *Anoxybacillus kamchatkensis* sp. nov., a novel thermophilic facultative aerobic bacterium with a broad pH optimum from the Geyser valley, Kamchatka. *Extremophiles* 9:391–398.

340. Namsaraev ZB, Babasanova OB, Dunaevsky YE, Akimov VN, Barkhutova DD, Gorlenko VM, Namsaraev BB. 2010. *Anoxybacillus mongoliensis* sp. nov., a novel thermophilic proteinase producing bacterium isolated from alkaline hot spring, central Mongolia. *Mikrobiologiia* 79:516–523.

341. Derekova A, Sjoholm C, Mandeva R, Kambourova M. 2007. *Anoxybacillus rupiensis* sp. nov., a novel thermophilic bacterium isolated from Rupi basin (Bulgaria). *Extremophiles* 11:577–583.

342. Cihan AC, Ozcan B, Cokmus C. 2011. *Anoxybacillus salavatliensis* sp. nov., an alpha-glucosidase producing, thermophilic bacterium isolated from Salavatli, Turkey. *J Basic Microbiol* 51:136–146.

343. Poli A, Romano I, Cordella P, Orlando P, Nicolaus B, Berrini CC. 2009. *Anoxybacillus thermarum* sp. nov., a novel thermophilic bacterium isolated from thermal mud in Euganean hot springs, Abano Terme, Italy. *Extremophiles* 13:867–874.

344. Yumoto I, Hirota K, Kawahara T, Nodasaka Y, Okuyama H, Matsuyama H, Yokota Y, Nakajima K, Hoshino T. 2004. *Anoxybacillus voinovskiensis* sp. nov., a moderately thermophilic bacterium from a hot spring in Kamchatka. *Int J Syst Evol Microbiol* 54:1239–1242.

345. Marquez MC, Carrasco IJ, Xue Y, Ma Y, Cowan DA, Jones BE, Grant WD, Ventosa A. 2008. *Aquisalibacillus elongatus* gen. nov., sp. nov., a moderately halophilic bacterium of the family Bacillaceae isolated from a saline lake. *Int J Syst Evol Microbiol* 58:1922–1926.

346. Zhou Y, Wei W, Che Q, Xu Y, Wang X, Huang X, Lai R. 2008. *Bacillus pallidus* sp. nov., isolated from forest soil. *Int J Syst Evol Microbiol* 58:2850–2854.

347. Zhou Y, Xu J, Xu L, Tindall BJ. 2009. *Falsibacillus pallidus* to replace the homonym *Bacillus pallidus* Zhou et al. 2008. *Int J Syst Evol Microbiol* 59:3176–3180.

348. Schlesner H, Lawson PA, Collins MD, Weiss N, Wehmeyer U, Volker H, Thomm M. 2001. *Filobacillus milensis* gen. nov., sp nov., a new halophilic spore-forming bacterium with Orn-D-Glu-type peptidoglycan. *Int J Syst Evol Microbiol* 51:425–431.

349. Ahmad S, Scopes RK, Rees GN, Patel BKC. 2000. *Saccharococcus caldoxylosilyticus* sp. nov., an obligately thermophilic, xylose-utilizing, endospore-forming bacterium. *Int J Syst Evol Microbiol* 50:517–523.

350. Fortina MG, Mora D, Schumann P, Parini C, Manachini PL, Stackebrandt E. 2001. Reclassification of *Saccharococcus caldoxylosilyticus* as *Geobacillus caldoxylosilyticus* (Ahmad et al. 2000) comb. nov. *Int J Syst Evol Microbiol* 51:2063–2071.

351. Banat IM, Marchant R, Rahman TJ. 2004. *Geobacillus debilis* sp. nov., a novel obligately thermophilic bacterium isolated from a cool soil environment, and reassignment of *Bacillus pallidus* to *Geobacillus pallidus* comb. nov. *Int. J Syst Evol Microbiol* 54:2197–2201.

352. Nazina TN, Sokolova DS, Grigoryan AA, Shestakova NM, Mikhailova EM, Poltaraus AB, Tourova TP, Lysenko AM, Osipov GA, Belyaev SS. 2005. *Geobacillus jurassicus* sp. nov., a new thermophilic bacterium isolated from a high-temperature petroleum *Geobacillus* species reservoir, and the validation of the *Geobacillus* species. *Syst Appl Microbiol* 28:43–53.

353. Nazina TN, Tourova TP, Poltaraus AB, Novikova EV, Grigoryan AA, Ivanova AE, Lysenko AM, Petrunyaka VV, Osipov GA, Belyaev SS, Ivanov MV. 2001. Taxonomic study of aerobic thermophilic bacilli: descriptions of *Geobacillus subterraneus* gen. nov., sp. nov and *Geobacillus uzenensis* sp. nov. from petroleum reservoirs and transfer of *Bacillus stearothermophilus*, *Bacillus thermocatenulatus*, *Bacillus thermoleovorans*, *Bacillus kaustophilus*, *Bacillus thermoglucosidasius* and *Bacillus thermodenitrificans* to *Geobacillus* as the new combinations *G. stearothermophilus*, *G. thermocatenulatus*, *G. thermoleovorans*, *G. kaustophilus*, *G. thermoglucosidasius* and *G. thermodenitrificans*. *Int J Syst Evol Microbiol* 51:433–446.

354. Kuisiene N, Raugalas J, Chitavichius D. 2004. *Geobacillus lituanicus* sp. nov. *Int J Syst Evol Microbiol* 54:1991–1995.

355. Donk PA. 1920. A highly resistant thermophilic organism. *J Bacteriol* 5:373–374.

356. Schaffer C, Franck WL, Scheberl A, Kosma P, McDermott TR, Messner P. 2004. Classification of isolates from locations in Austria and Yellowstone National Park as *Geobacillus tepidamans* sp. nov. *Int J Syst Evol Microbiol* 54:2361–2368.

357. Nicolaus B, Lama L, Esposito E, Manca MC, diPrisco G, Gambacorta A. 1996. "*Bacillus thermoantarcticus*" sp. nov., from Mount Melbourne, Antarctica: a novel thermophilic species. *Polar Biol* 16:101–104.

358. Coorevits A, Dinsdale AE, Halket G, Lebbe L, de Vos P, Van Landschoot A, Logan NA. 2011. Taxonomic revision of the genus *Geobacillus*: emendation of *Geobacillus*, *G. stearothermophilus*, *G. jurassicus*, *G. toebii*, *G. thermodenitrificans* and *G. thermoglucosidans* (nom. corrig., formerly "*thermoglucosidasius*"); transfer of *Bacillus thermantarcticus* to the genus as *G. thermantarcticus*; proposal of *Caldibacillus debilis*

gen. nov., comb. nov.; transfer of *G. tepidamans* to *Anoxybacillus* as *A. tepidamans* and proposal of *Anoxybacillus caldiproteolyticus* sp. nov. *Int J Syst Evol Microbiol* **62**:1470–1485.

359. Dinsdale AE, Halket G, Coorevits A, Van Landschoot A, Busse HJ, De Vos P, Logan NA. 2011. Emended descriptions of *Geobacillus thermoleovorans* and *Geobacillus thermocatenulatus*. *Int J Syst Evol Microbiol* **61**:1802–1810.

360. Golovacheva RS, Loginova LG, Salikhov TA, Kolesnikov AA, Zaitseva GN. 1975. New thermophilic species, *Bacillus thermocatenulatus* nov. sp. *Microbiology* **44**:230–233.

361. Cihan AC, Ozcan B, Tekin N, Cokmus C. 2011. *Geobacillus thermodenitrificans* subsp. *calidus*, subsp. nov., a thermophilic and alpha-glucosidase producing bacterium isolated from Kizilcahamam, Turkey. *J Gen Appl Microbiol* **57**:83–92.

362. Manachini PL, Mora D, Nicastro G, Parini C, Stackebrandt E, Pukall R, Fortina MG. 2000. *Bacillus thermoaenitrificans* sp. nov., nom. rev. *Int J Syst Evol Microbiol* **50**:1331–1337.

363. Suzuki Y, Kishigami T, Inoue K, Mizoguchi Y, Eto N, Takagi M, Abe S. 1983. *Bacillus thermoglucosidasius* sp. nov., a new species of obligately thermophilic bacilli. *Syst Appl Microbiol* **4**:487–495.

364. Zarilla KA, Perry JJ. 1987. Bacillus thermoleovorans, sp. nov., a species of obligately thermophilic hydrocarbon utilizing endospore-forming bacteria. *Syst Appl Microbiol* **9**:258–264.

365. Sung MH, Kim H, Bae JW, Rhee SK, Jeon CO, Kim K, Kim JJ, Hong SP, Lee SG, Yoon JH, Park YH, Baek DH. 2002. *Geobacillus toebii* sp. nov., a novel thermophilic bacterium isolated from hay compost. *Int J Syst Evol Microbiol* **52**:2251–2255.

366. Caccamo D, Gugliandolo C, Stackebrandt E, Maugeri TL. 2000. *Bacillus vulcani* sp. nov., a novel thermophilic species isolated from a shallow marine hydrothermal vent. *Int J Syst Evol Microbiol* **50**:2009–2012.

367. Nazina TN, Lebedeva EV, Poltaraus AB, Tourova TP, Grigoryan AA, Sokolova DS, Lysenko AM, Osipov GA. 2004. *Geobacillus gargensis* sp. nov., a novel thermophile from a hot spring, and the reclassification of *Bacillus vulcani* as *Geobacillus vulcani* comb. nov. *Int J Syst Evol Microbiol* **54**:2019–2024.

368. Echigo A, Minegishi H, Mizuki T, Kamekura M, Usami R. 2010. *Geomicrobium halophilum* gen. nov., sp. nov., a moderately halophilic and alkaliphilic bacterium isolated from soil. *Int J Syst Evol Microbiol* **60**:990–995.

369. Ahmed I, Yokota A, Fujiwara T. 2007. *Gracilibacillus boraciitolerans* sp. nov., a highly boron-tolerant and moderately halotolerant bacterium isolated from soil. *Int J Syst Evol Microbiol* **57**:796–802.

370. Lawson PA, Deutch CE, Collins MD. 1996. Phylogenetic characterization of a novel salt-tolerant *Bacillus* species: description of *Bacillus dipsosauri* sp. nov. *J Appl Bacteriol* **81**:109–112.

371. Wainø M, Tindall BJ, Schumann P, Ingvorsen K. 1999. *Gracilibacillus* gen. nov., with description of *Gracilibacillus halotolerans* gen. nov., sp. nov.; transfer of *Bacillus dipsosauri* to *Gracilibacillus dipsosauri* comb. nov., and *Bacillus salexigens* to the genus *Salibacillus* gen. nov., as *Salibacillus salexigens* comb. nov. *Int J Syst Bacteriol* **49**:821–831.

372. Chen Y-G, Cui X-L, Zhang Y-Q, Li W-J, Wang Y-X, Xu L-H, Peng Q, Wen M-L, Jiang C-L. 2008. *Gracilibacillus halophilus* sp. nov., a moderately halophilic bacterium isolated from saline soil. *Int J Syst Evol Microbiol* **58**:2403–2408.

373. Jeon CO, Lim J-M, Jang HH, Park D-J, Xu L-H, Jiang C-L, Kim C-J. 2008. *Gracilibacillus lacisalsi* sp. nov., a halophilic Gram-positive bacterium from a salt lake in China. *Int J Syst Evol Microbiol* **58**:2282–2286.

374. Carrasco IJ, Márquez MC, Yanfen X, Ma Y, Cowan DA, Jones BE, Grant WD, Ventosa A. 2006. *Gracilibacillus orientalis* sp. nov., a novel moderately halophilic bacterium isolated from a salt lake in Inner Mongolia, China. *Int J Syst Evol Microbiol* **56**:599–604.

375. Tang S-K, Wang Y, Lou K, Mao P-H, Jin X, Jiang C-L, Xu L-H, Li W-J. 2009. *Gracilibacillus saliphilus* sp. nov., a moderately halophilic bacterium isolated from a salt lake. *Int J Syst Evol Microbiol* **59**:1620–1624.

376. Chamroensaksri N, Tanasupawat S, Akaracharanya A, Visessanguan W, Kudo T, Itoh T. 2010. *Gracilibacillus thailandensis* sp. nov., from fermented fish (pla-ra). *Int J Syst Evol Microbiol* **60**:944–948.

377. Huo Y-Y, Xu X-W, Cui H-L, Wu M. 2010. *Gracilibacillus ureilyticus* sp. nov., a halotolerant bacterium from a saline-alkaline soil. *Int J Syst Evol Microbiol* **60**:1383–1386.

378. Echigo A, Fukushima T, Mizuki T, Kamekura M, Usami R. 2007. *Halalkalibacillus halophilus* gen. nov., sp nov., a novel moderately halophilic and alkallphilic bacterium isolated from a non-saline soil sample in Japan. *Int J Syst Evol Microbiol* **57**:1081–1085.

379. Liu WY, Zeng J, Wang L, Dou YT, Yang SS. 2005. *Halobacillus dabanensis* sp. nov. and *Halobacillus aidingensis* sp. nov., isolated from salt lakes in Xinjiang, China. *Int J Syst Evol Microbiol* **55**:1991–1996.

380. Romano I, Finore I, Nicolaus G, Huertas FJ, Lama L, Nicolaus B, Poli A. 2008. *Halobacillus alkaliphilus* sp. nov., a halophilic bacterium isolated from a salt lake in Fuente de Piedra, southern Spain. *Int J Syst Evol Microbiol* **58**:886–890.

381. Yoon J-H, Kang S-J, Jung Y-T, Oh T-K. 2007. *Halobacillus campisalis* sp. nov., containing diaminopimelic acid in the cell-wall peptidoglycan, and emended description of the genus *Halobacillus*. *Int J Syst Evol Microbiol* **57**:2021–2025.

382. An S-Y, Kanoh K, Kasai H, Goto K, Yokota A. 2007. *Halobacillus faecis* sp. nov., a spore-forming bacterium isolated from a mangrove area on Ishigaki Island, Japan. *Int J Syst Evol Microbiol* **57**:2476–2479.

383. Spring S, Ludwig W, Marquez MC, Ventosa A, Schleifer KH. 1996. *Halobacillus* gen. nov., with descriptions of *Halobacillus litoralis* sp. nov. and *Halobacillus trueperi* sp. nov., and transfer of *Sporosarcina halophila* to

Halobacillus halophilus comb. nov. *Int J Syst Bacteriol* **46**:492–496.

384. Amoozegar MA, Malekzadeh F, Malik KA, Schumann P, Sproer C. 2003. *Halobacillus karajensis* sp. nov., a novel moderate halophile. *Int J Syst Evol Microbiol* **53**: 1059–1063.

385. Hua N-P, Kanekiyo A, Fujikura K, Yasuda H, Naganuma T. 2007. *Halobacillus profundi* sp. nov. and *Halobacillus kuroshimensis* sp. nov., moderately halophilic bacteria isolated from a deep-sea methane cold seep. *Int J Syst Evol Microbiol* **57**:1243–1249.

386. Yoon JH, Kang KH, Oh TK, Park YH. 2004. *Halobacillus locisalis* sp. nov., a halophilic bacterium isolated from a marine solar saltern of the Yellow Sea in Korea. *Extremophiles* **8**:23–28.

387. Soto-Ramírez N, Sanchez-Porro C, Rosas-Padilla S, Almodovar K, Jimenez G, Machado-Rodriguez M, Zapata M, Ventosa A, Montalvo-Rodriguez R. 2008. *Halobacillus mangrovi* sp. nov., a moderately halophilic bacterium isolated from the black mangrove Avicennia germinans. *Int J Syst Evol Microbiol* **58**:125–130.

388. Yoon JH, Kook HKG, Park YH. 2003. *Halobacillus salinus* sp. nov., isolated from a salt lake on the coast of the East Sea in Korea. *Int J Syst Evol Microbiol* **53**: 687–693.

389. Yoon J-H, Kang S-J, Oh T-K. 2008. *Halobacillus seohaensis* sp. nov., isolated from a marine solar saltern in Korea. *Int J Syst Evol Microbiol* **58**:622–627.

390. Yoon JH, Kang SJ, Lee CH, Oh HW, Oh TK. 2005. *Halobacillus yeomjeoni* sp. nov., isolated from a marine solar saltern in Korea. *Int J Syst Evol Microbiol* **55**: 2413–2417.

391. Cao S-J, Qu J-H, Yang J-S, Sun Q, Yuan H-L. 2008. *Halolactibacillus alkaliphilus* sp. nov., a moderately alkaliphilic and halophilic bacterium isolated from a soda lake in Inner Mongolia, China, and emended description of the genus *Halolactibacillus*. *Int J Syst Evol Microbiol* **58**:2169–2173.

392. Ishikawa M, Nakajima K, Itamiya Y, Furukawa S, Yamamoto Y, Yamasato K. 2005. *Halolactibacillus halophilus* gen. nov., sp. nov. and *Halolactibacillus miurensis* sp. nov., halophilic and alkaliphilic marine lactic acid bacteria constituting a phylogenetic lineage in *Bacillus* rRNA group 1. *Int J Syst Evol Microbiol* **55**: 2427–2439.

393. Yuan S, Ren P, Liu J, Xue Y, Ma Y, Zhou P. 2007. *Lentibacillus halodurans* sp. nov., a moderately halophilic bacterium isolated from a salt lake in Xin-Jiang, China. *Int J Syst Evol Microbiol* **57**:485–488.

394. Tanasupawat S, Pakdeeto A, Namwong S, Thawai C, Kudo T, Itoh T. 2006. *Lentibacillus halophilus* sp. nov., from fish sauce in Thailand. *Int J Syst Evol Microbiol* **56**:1859–1863.

395. Jung M-J, Roh SW, Kim M-S, Bae J-W. 2010. *Lentibacillus jeotgali* sp. nov., a halophilic bacterium isolated from traditional Korean fermented seafood. *Int J Syst Evol Microbiol* **60**:1017–1022.

396. Namwong S, Tanasupawat S, Smitinont T, Visessanguan W, Kudo T, Itoh T. 2005. Isolation of *Lentibacillus salicampi* strains and *Lentibacillus juripiscarius* sp. nov. from fish sauce in Thailand. *Int J Syst Evol Microbiol* **55**:315–320.

397. Pakdeeto A, Tanasupawat S, Thawai C, Moonmangmee S, Kudo T, Itoh T. 2007. *Lentibacillus kapialis* sp. nov., from fermented shrimp paste in Thailand. *Int J Syst Evol Microbiol* **57**:364–369.

398. Lim JM, Jeon CO, Song SM, Lee JC, Ju YJ, Xu LH, Jiang CL, Kim CJ. 2005. *Lentbacillus lacisalsi* sp. nov., a moderately halophilic bacterium isolated from a Saline Lake in China. *Int J Syst Evol Microbiol* **55**: 1805–1809.

399. Sanchez-Porro C, Amoozegar MA, Fernandez AB, Fard HB, Ramezani M, Ventosa A. 2010. *Lentibacillus persicus* sp. nov., a moderately halophilic species isolated from a saline lake. *Int J Syst Evol Microbiol* **60**: 1407–1412.

400. Jeon CO, Lim JM, Lee JC, Lee GS, Lee JM, Xu LH, Jiang CL, Kim CJ. 2005. *Lentibacillus salarius* sp. nov., isolated from saline sediment in China, and emended description of the genus *Lentibacillus*. *Int J Syst Evol Microbiol* **55**:1339–1343.

401. Yoon JH, Kang KH, Park YH. 2002. *Lentibacillus salicampi* gen. nov., sp. nov., a moderately halophilic bacterium isolated from a salt field in Korea. *Int J Syst Evol Microbiol* **52**:2043–2048.

402. Lee S-Y, Choi W-Y, Oh T-K, Yoon J-H. 2008. *Lentibacillus salinarum* sp. nov., isolated from a marine solar saltern in Korea. *Int J Syst Evol Microbiol* **58**:45–49.

403. Lee J-C, Li W-J, Xu L-H, Jiang C-L, Kim C-J. 2008. *Lentibacillus salis* sp. nov., a moderately halophilic bacterium isolated from a salt lake. *Int J Syst Evol Microbiol* **58**:1838–1843.

404. Hao MV, Kocur M, Komagata K. 1984. *Marinococcus* gen. nov., a new genus for motile cocci with meso-diaminopimelc acid in the cell wall and *Marinococcus albus* sp. nov. and *Marinococcus halophilus* (Novitsky and Kushner) comb. nov. *J Gen Appl Microbiol* **30**: 449–459.

405. Li WJ, Schumann P, Zhang YQ, Chen GZ, Tian XP, Xu LH, Stackebrandt E, Jiang CL. 2005. *Marinococcus halotolerans* sp. nov., isolated from Qinghai, North-West China. *Int J Syst Evol Microbiol* **55**:1801–1804.

406. Wang Y, Cao L-L, Tang S-K, Lou K, Mao P-H, Jin X, Jiang C-L, Xu L-H, Li W-J. 2009. *Marinococcus luteus* sp. nov., a halotolerant bacterium isolated from a salt lake, and emended description of the genus *Marinococcus*. *Int J Syst Evol Microbiol* **59**:2875–2879.

407. Sorokin ID, Zadorina EV, Kravchenko IK, Boulygina ES, Tourova TP, Sorokin DY. 2008. *Natronobacillus azotifigens* gen. nov., sp nov., an anaerobic diazotrophic haloalkaliphile from soda-rich habitats. *Extremophiles* **12**:819–827.

408. Nam J-H, Bae W, Lee D-H. 2008. *Oceanobacillus caeni* sp. nov., isolated from a *Bacillus*-dominated wastewater treatment system in Korea. *Int J Syst Evol Microbiol* **58**:1109–1113.

409. Raats D, Halpern M. 2007. *Oceanobacillus chironomi* sp. nov., a halotolerant and facultatively alkaliphilic

species isolated from a chironomid egg mass. *Int J Syst Evol Microbiol* 57:255–259.

410. Lu J, Nogi Y, Takami H. 2001. *Oceanobacillus iheyensis* gen. nov., sp. nov., a deep-sea extremely halotolerant and alkaliphilic species isolated from a depth of 1050 m on the Iheya Ridge. *FEMS Microbiol Lett* 205:291–297.

411. Namworg S, Tanasupawat S, Lee KC, Lee J-S. 2009. *Oceanobacillus kapialis* sp. nov., from fermented shrimp paste in Thailand. *Int J Syst Evol Microbiol* 59:2254–2259.

412. Whon TW, Jung M-J, Roh SW, Nam Y-D, Park E-J, Shin K-S, Bae J-W. 2010. *Oceanobacillus kimchii* sp. nov. isolated from a traditional Korean fermented food. *J Microbiol* 48:862–866.

413. Lee S-Y, Oh T-K, Kim W, Yoonl J-H. 2010. *Oceanobacillus locisalsi* sp. nov., isolated from a marine solar saltern. *Int J Syst Evol Microbiol* 60:2758–2762.

414. Yang J-Y, Huo Y-Y, Xu X-W, Meng F-X, Wu M, Wang C-S. 2010. *Oceanobacillus neutriphilus* sp. nov., isolated from activated sludge in a bioreactor. *Int J Syst Evol Microbiol* 60:2409–2414.

415. Yumoto I, Hirota K, Nodasaka Y, Nakajima K. 2005. *Oceanobacillus oncorhynchi* sp. nov., a halotolerant obligate alkaliphile isolated from the skin of a rainbow trout (*Oncorhynchus mykiss*), and emended description of the genus *Oceanobacillus*. *Int J Syst Evol Microbiol* 55:1521–1524.

416. Heyrman J, Logan NA, Busse HJ, Balcaen A, Lebbe L, Rodriguez-Diaz M, Swings J, De Vos P. 2003. *Virgibacillus carmonensis* sp. nov., *Virgibacillus necropolis* sp. nov. and *Virgibacillus picturae* sp. nov., three novel species isolated from deteriorated mural paintings, transfer of the species of the genus *Salibacillus* to *Virgibacillus*, as *Virgibacillus marismortui* comb. nov. and *Virgibacillus salexigens* comb. nov., and emended description of the genus *Virgibacillus*. *Int J Syst Evol Microbiol* 53:501–511.

417. Lee JS, Lim JM, Lee KC, Lee JC, Park YH, Kim CJ. 2006. *Virgibacillus koreensis* sp. nov., a novel bacterium from a salt field, and transfer of *Virgibacillus picturae* to the genus *Oceanobacillus* as *Oceanobacillus picturae* comb. nov with emended descriptions. *Int J Syst Evol Microbiol* 56:251–257.

418. Kim Y-G, Choi DH, Hyun S, Cho BC. 2007. *Oceanobacillus profundus* sp. nov., isolated from a deep-sea sediment core. *Int J Syst Evol Microbiol* 57:409–413.

419. Tominaga T, An S-Y, Oyaizu H, Yokota A. 2009. *Oceanobacillus soja* sp. nov. isolated from soy sauce production equipment in Japan. *J Gen Appl Microbiol* 55:225–232.

420. Mayr R, Busse HJ, Worliczek HL, Ehling-Schulz M, Scherer S. 2006. *Ornithinibacillus* gen. nov., with the species *Ornithinibacillus bavariensis* sp. nov. and *Ornithinibacillus californiensis* sp. nov. *Int J Syst Evol Microbiol* 56:1383–1389.

421. Kämpfer P, Falsen E, Lodders N, Langer S, Busse HJ, Schumann P. 2010. *Ornithinibacillus contaminans* sp. nov., an endospore-forming species. *Int J Syst Evol Microbiol* 60:2930–2934.

422. Shin NR, Whon TW, Kim MS, Roh SW, Jung MJ, Kim YO, Bae JW. 2012. *Ornithinibacillus scapharcae* sp. nov., isolated from a dead ark clam. *Antonie Leeuwenhoek* 101:147–154.

423. Chen Y-G, Cui X-L, Zhang Y-Q, Li W-J, Wang Y-X, Xu L-H, Wen M-L, Peng Q, Jiang C-L. 2009. *Paraliobacillus quinghaiensis* sp. nov., isolated from salt-lake sediment in China. *Int J Syst Evol Microbiol* 59:28–33.

424. Ishikawa M, Ishizaki S, Yamamoto Y, Yamasato K. 2002. *Paraliobacillus ryukyuensis* gen. nov., sp. nov., a new Gram-positive, slightly halophilic, extremely halotolerant, facultative anaerobe isolated from a decomposing marine alga. *J Gen Appl Microbiol* 48:269–279.

425. Nunes I, Tiago I, Pires AL, da Costa MS, Veríssimo A. 2006. *Paucisalibacillus globulus* gen. nov., sp. nov., a Gram-positive bacterium isolated from potting soil. *Int J Syst Evol Microbiol* 56:1841–1845.

426. Amoozegar MA, Sanchez-Porro C, Rohban R, Hajighasemi M, Ventosa A. 2009. *Piscibacillus halophilus* sp. nov., a moderately halophilic bacterium from a hypersaline Iranian lake. *Int J Syst Evol Microbiol* 59:3095–3099.

427. Tanasupawat S, Namwong S, Kudo T, Itoh T. 2007. *Piscibacillus salipliscarius* gen. nov., sp. nov., a moderately halophilic bacterium from fermented fish (pla-ra) in Thailand. *Int J Syst Evol Microbiol* 57:1413–1417.

428. Lim JM, Jeon CO, Song SM, Kim CJ. 2005. *Pontibacillus chungwhensis* gen. nov., sp. nov., a moderately halophilic Gram-positive bacterium from a solar saltern in Korea. *Int J Syst Evol Microbiol* 55:165–170.

429. Chen Y-G, Zhang Y-Q, Xiao H-D, Liu Z-X, Yi L-B, Shi J-X, Zhi -Y, Cui X-L, Li W-J. 2009. *Pontibacillus halophilus* sp. nov., a moderately halophilic bacterium isolated from a sea urchin. *Int J Syst Evol Microbiol* 59:1635–1639.

430. Chen Yi-G, Zhang Y-Q, Yi L-B, Li Z-Y, Wang Y-Xo, Xiao H-D, Chen Q-H, Cui X-L, Li W-J. 2010. *Pontibacillus litoralis* sp. nov., a facultatively anaerobic bacterium isolated from a sea anemone, and emended description of the genus *Pontibacillus*. *Int J Syst Evol Microbiol* 60:560–565.

431. Lim JM, Jeon CO, Park DJ, Kim HR, Yoon BJ, Kim CJ. 2005. *Pontibacillus marinus* sp. nov., a moderately halophilic bacterium from a solar saltern, and emended description of the genus *Pontibacillus*. *Int J Syst Evol Microbiol* 55:1027–1031.

432. Yang Y, Zou Z, He M, Wang G. 2011. *Pontibacillus yanchengensis* sp. nov., a moderately halophilic bacterium isolated from salt field soil. *Int J Syst Evol Microbiol* 61:1906–1911.

433. Hatayama K, Shoun H, Ueda Y, Nakamura A. 2006. *Tuberibacillus calidus* gen. nov., sp. nov., isolated from a compost pile and reclassification of *Bacillus naganoensis* Tomimura et al. 1990 as *Pullulanibacillus naganoensis* gen. nov., comb. nov. and *Bacillus laevolacticus* Andersch et al. 1994 *as Sporolactobacillus laevolacticus* comb. nov. *Int J Syst Evol Microbiol* 56:2545–2551.

434. Tomimura E, Zeman NW, Frankiewicz JR, Teague WM. 1990. Description of *Bacillus naganoensis* sp. nov. *Int J Syst Bacteriol* 40:123–125.

435. Nystrand R. 1984. *Saccharococcus thermophilus* gen. nov., sp. nov. isolated from beet sugar extraction. *Syst Appl Microbiol* 5:204–219.

436. Yoon J-H, Kang S-J, Oh T-K. 2007. Reclassification of *Marinococcus albus* Hao et al. 1985 as *Salimicrobium album* gen. nov., comb. nov. and *Bacillus halophilus* Ventosa et al. 1990 as *Salimicrobium halophilum* comb. nov., and description of *Salimicrobium luteum* sp. nov. *Int J Syst Evol Microbiol* 57:2406–2411.

437. Yoon J-H, Kang S-J, Oh K-H, Oh T-K. 2009. *Salimicrobium flavidum* sp. nov., isolated from a marine solar saltern. *Int J Syst Evol Microbiol* 59:2839–2842.

438. de la Haba RR, Yilmaz P, Sanchez-Porro C, Birbir M, Ventosa A. 2011. *Salimicrobium salexigens* sp. nov., a moderately halophilic bacterium from salted hides. *Syst Appl Microbiol* 34:435–439.

439. Ren PG, Zhou PJ. 2005. *Salinibacillus aidingensis* gen. nov., sp. nov. and *Salinibacillus kushneri* sp. nov., moderately halophilic bacteria isolated from a neutral saline lake in Xin-Jiang, China. *Int J Syst Evol Microbiol* 55:949–953.

440. Albuquerque L, Tiago I, Rainey FA, Taborda M, Nobre MF, Verissimo A, da Costa MS. 2007. *Salirhabdus euzebyi* gen. nov., sp. nov., a Gram-positive, halotolerant bacterium isolated from a sea salt evaporation pond. *Int J Syst Evol Microbiol* 57:1566–1571.

441. Cao S-J, Ou J-H, Yuan H-L, Li B-Z. 2010. *Salsuginibacillus halophilus* sp. nov., a halophilic bacterium isolated from a soda lake. *Int J Syst Evol Microbiol* 60:1339–1343.

442. Carrasco IJ, Marquez MC, Xue Y, Ma Y, Cowan DA, Jones BE, Grant WD, Ventosa A. 2007. *Salsuginibacillus kocurii* gen. nov., sp. nov., a moderately halophilic bacterium from soda-lake sediment. *Int J Syst Evol Microbiol* 57:2381–2386.

443. Wang X, Xue Y, Ma Y. 2009. *Sediminibacillus albus* sp. nov., a moderately halophilic, Gram-positive bacterium isolated from a hypersaline lake, and emended description of the genus *Sediminibacillus* Carrasco et al. 2008. *Int J Syst Evol Microbiol* 59:1640–1644.

444. Carrasco IJ, Marquez MC, Xue Y, Ma Y, Cowan DA, Jones BE, Ventosa A. 2008. *Sediminibacillus halophilus* gen. nov., sp. nov., a moderately halophilic, Gram-positive bacterium from a hypersaline lake. *Int J Syst Evol Microbiol* 58:1961–1967.

445. Li WJ, Zhang YQ, Schumann P, Tian XP, Xu LH, Jiang CL. 2006. Sinococcus qinghaiensis gen. nov., sp nov., a novel member of the order Bacillales from a saline soil in China. *Int J Syst Evol Microbiol* 56:1189–1192.

446. Li W-J, Zhi X-Y, Euzeby JP. 2008. Proposal of *Yaniellaceae* fam. nov., *Yaniella* gen. nov and *Sinobaca* gen. nov. as replacements for the illegitimate prokaryotic names Yaniaceae Li et al. 2005, Yania Li et al. 2004, emend Li et al. 2005, and Sinococcus Li et al. 2006, respectively. *Int J Syst Evol Microbiol* 58:525–527.

447. Wang X, Xue Y, Ma Y. 2011. *Streptohalobacillus salinus* gen. nov., sp. nov., a moderately halophilic, Gram-positive, facultative anaerobe isolated from subsurface saline soil. *Int J Syst Evol Microbiol* 61:1127–1132.

448. Ren PG, Zhou PJ. 2005. Tenuibacillus multivorans gen. nov., sp nov., a moderately halophilic bacterium isolated from saline soil in Xin-Jiang, China. *Int J Syst Evol Microbiol* 55:95–99.

449. Liu W, Jiang L, Guo C, Yang SS. 2010. *Terribacillus aidingensis* sp. nov., a moderately halophilic bacterium. *Int J Syst Evol Microbiol* 60:2940–2945.

450. Kim Y-G, Hwang CY, Yoo KW, Moon HT, Yoon J-H, Cho BC. 2007. *Pelagibacillus goriensis* gen. nov., sp. nov., a moderately halotolerant bacterium isolated from coastal water off the east coast of Korea. *Int J Syst Evol Microbiol* 57:1554–1560.

451. Krishnamurthi S, Chakrabarti T. 2008. Proposal for transfer of *Pelagibacillus goriensis* Kim et al. 2007 to the genus *Terribacillus* as *Terribacillus goriensis* comb. nov. *Int J Syst Evol Microbiol* 58:2287–2291.

452. An S-Y, Asahara M, Goto K, Kasai H, Yokota A. 2007. *Terribacillus saccharophilus* gen. nov., sp. nov. and *Terribacillus halophilus* sp. nov., spore-forming bacteria isolated from field soil in Japan. *Int J Syst Evol Microbiol* 57:51–55.

453. Sanchez-Porro C, Amoozegar MA, Rohban R, Hajighasemi M, Ventosa A. 2009. *Thalassobacillus cyri* sp. nov., a moderately halophilic Gram-positive bacterium from a hypersaline lake. *Int J Syst Evol Microbiol* 59:2565–2570.

454. Garcia MT, Gallego V, Ventosa A, Mellado E. 2005. *Thalassobacillus devorans* gen. nov., sp. nov., a moderately halophilic, phenol-degrading, Gram-positive bacterium. *Int J Syst Evol Microbiol* 55:1789–1795.

455. Lee S-Y, Oh T-K, Yoon J-H. 2010. *Thalassobacillus hwangdonensis* sp. nov., isolated from a tidal flat sediment. *Int J Syst Evol Microbiol* 60:2108–2112.

456. Sánchez-Porro C, Yilmaz P, de la Haba RR, Birbir M, Ventosa A. 2011. *Thalassobacillus pellis* sp. nov., a moderately halophilic, Gram-positive bacterium isolated from salted hides. *Int J Syst Evol Microbiol* 61:1206–1210.

457. Kim J, Jung M-J, Roh SW, Nam Y-D, Shin K-S, Bae J-W. 2011. *Virgibacillus alimentarius* sp. nov., isolated from a traditional Korean food. *Int J Syst Evol. Microbiol* 61:2851–2855.

458. Niederberger TD, Steven B, Charvet S, Barbier B, Whyte LG. 2009. *Virgibacillus arcticus* sp. nov., a moderately halophilic, endospore-forming bacterium from permafrost in the Canadian high Arctic. *Int J Syst Evol Microbiol* 59:2219–2225.

459. Yoon J-H, Kang S-J, Jung Y-T, Lee KC, Oh HW, Oh T-K. 2010. *Virgibacillus byunsanensis* sp. nov., isolated from a marine solar saltern. *Int J Syst Evol Microbiol* 60:291–295.

460. Lee S-Y, Kang C-H, Oh T-K, Yoon J-H. 2012. *Virgibacillus campisalis* sp. nov., from a marine solar saltern. *Int J Syst Evol Microbiol* 62:347–351.

461. Wang C-Y, Chang C-C, Ng CC, Chen T-W, Shyu Y-T. 2008. *Virgibacillus chiguensis* sp. nov., a novel halophilic bacterium isolated from Chigu, a previously commercial saltern located in southern Taiwan. *Int J Syst Evol Microbiol* 58:341–345.

462. Yoon JH, Kang SJ, Lee SY, Lee MH, Oh TK. 2005. *Virgibacillus dokdonensis* sp. nov., isolated from a Korean island, Dokdo, located at the edge of the East Sea in Korea. *Int J Syst Evol Microbiol* 55:1833–1837.

463. Denariaz G, Payne WJ, Legall J. 1989. A halophilic denitrifier, *Bacillus halodenitrificans* sp. nov. *Int J Syst Bacteriol* 39:145–151.

464. Yoon JH, Oh TK, Park YH. 2004. Transfer of *Bacillus halodenitrificans* Denariaz et al. 1989 to the genus *Virgibacillus* as *Virgibacillus halodenitrificans* comb. nov. *Int J Syst Evol Microbiol* 54:2163–2167.

465. An S-Y, Asahara M, Goto K, Kasai H, Yokota A. 2007. *Virgibacillus halophilus* sp. nov., spore-forming bacteria isolated from soil in Japan. *Int J Syst Evol Microbiol* 57:1607–1611.

466. Chen Y-G, Cui X-L, Fritze D, Chai L-H, Schumann P, Wen M-L, Wang Y-X, Xu L-H, Jiang C-L. 2008. *Virgibacillus kekensis* sp. nov., a moderately halophilic bacterium isolated from a salt lake in China. *Int J Syst Evol Microbiol* 58:647–653.

467. Arahal DR, Marquez MC, Volcani BE, Schleifer KH, Ventosa A. 1999. *Bacillus marismortui* sp. nov., a new moderately halophilic species from the Dead Sea. *Int J Syst Bacteriol* 49:521–530.

468. Quesada T, Aguilera M, Morillo JA, Ramos-Cormenzana A, Monteoliva-Sánchez M. 2007. *Virgibacillus olivae* sp. nov., isolated from waste wash-water from processing of Spanish-style green olives. *Int J Syst Evol Microbiol* 57:906–910.

469. Heyndrickx M, Lebbe L, Kersters K, Hoste B, De Wachter R, De Vos P, Forsyth G, Logan NA. 1999. Proposal of *Virgibacillus proomii* sp. nov. and emended description of *Virgibacillus pantothenticus* (*Proom and Knight 1950*) Heyndrickx et al., 1998. *Int J Syst Bacteriol* 49:1083–1090.

470. Proom H, Knight BCJ. 1950. *Bacillus pantothenticus* n. sp. *J Gen Microbiol* 4:539–541.

471. Hua N-P, Hamza-Chaffai A, Vreeland RH, Isoda H, Naganuma T. 2008. *Virgibacillus salarius* sp. nov., a halophilic bacterium isolated from a Saharan salt lake. *Int J Syst Evol Microbiol* 58:2409–2414.

472. Garabito MJ, Arahal DR, Mellado E, Marquez MC, Ventosa A. 1997. *Bacillus salexigens* sp. nov, a new moderately halophilic *Bacillus* species. *Int J Syst Bacteriol* 47:735–741.

473. Carrasco IJ, Marquez MC, Ventosa A. 2009. *Virgibacillus salinus* sp. nov., a moderately halophilic bacterium from sediment of a saline lake. *Int J Syst Evol Microbiol* 59:3068–3073.

474. Chen Y, Cui X, Wang Y, Zhang Y, Tang S, Li W, Liu Z, Wen M, Peng Q, Chen YG, Cui XL, Wang Y, Zhang YQ, Tang SK, Li WJ, Liu ZX, Wen ML, Peng Q. 2009. *Virgibacillus sediminis* sp. nov., a moderately halophilic bacterium isolated from a salt lake in China. *Int J Syst Evol Microbiol* 59:2058–2063.

475. Tanasupawat S, Chamroensaksri N, Kudo T, Itoh T. 2010. Identification of moderately halophilic bacteria from Thai fermented fish (pla-ra) and proposal of *Virgibacillus siamensis* sp. nov. *J Gen Appl Microbiol* 56:369–379.

476. Kämpfer P, Arun AB, Busse HJ, Langer S, Young CC, Chen WM, Syed AA, Rekha PD. 2011. *Virgibacillus soli* sp. nov., isolated from mountain soil. *Int J Syst Evol Microbiol* 61:275–280.

477. Wang X, Xuel Y, Mal Y. 2010. *Virgibacillus subterraneus* sp. nov., a moderately halophilic Gram-positive bacterium isolated from subsurface saline soil. *Int J Syst Evol Microbiol* 60:2763–2767.

478. Jeon CO, Kim JM, Park D-J, Xu L-H, Jiang C-L, Kim C-J. 2009. *Virgibacillus xinjiangensis* sp. nov., isolated from a salt lake of Xin-jiang province in China. *J Microbiol* 47:705–709.

479. L'Haridon S, Miroshnichenko ML, Kostrikina NA, Tindall BJ, Spring S, Schumann P, Stackebrandt E, Bonch-Osmolovskaya EA, Jeanthon C. 2006. *Vulcanibacillus modesticaldus* gen. nov., sp. nov., a strictly anaerobic, nitrate-reducing bacterium from deep-sea hydrothermal vents. *Int J Syst Evol Microbiol* 56:1047–1053.

480. Galperin MY. 2013. Genomic diversity of spore-forming *Firmicutes*. *Microbiol Spectrum* 1(2):TBS-0015-2012.

481. Mandic-Mulec I, Stefanic P, van Elsas JD. 2015. Ecology of *Bacillaceae*. *Microbiol Spectrum* 3(2):TBS-0017-2013.

482. Corel E, Lopez P, Méheust R, Bapteste E. 2016. Network-thinking: graphs to analyze microbial complexity and evolution. *Trends Microbiol* 24:224–237.

483. Puigbò P, Lobkovsky AE, Kristensen DM, Wolf YI, Koonin EV. 2014. Genomes in turmoil: quantification of genome dynamics in prokaryote supergenomes. *BMC Biol* 12:66. http://www.biomedcentral.com/1741-7007/12/66

484. Studholme DJ. 2015. Some (bacilli) like it hot: genomics of *Geobacillus* species. *Microb Biotechnol* 8:40–48.

485. Eppinger M, Bunk B, Johns MA, Edirisinghe JN, Kutumbaka KK, Koenig SS, Creasy HH, Rosovitz MJ, Riley DR, Daugherty S, Martin M, Elbourne LD, Paulsen I, Biedendieck R, Braun C, Grayburn S, Dhingra S, Lukyanchuk V, Ball B, Ul-Qamar R, Seibel J, Bremer E, Jahn D, Ravel J, Vary PS. 2011. Genome sequences of the biotechnologically important *Bacillus megaterium* strains QM B1551 and DSM319. *J Bacteriol* 193:4199–4213.

486. Vernikos G, Medini D, Riley DR, Tettelin H. 2015. Ten years of pan-genomic analysis. *Curr Opin Microbiol* 23:148–154.

487. Collins RE, Higgs PG. 2012. Testing the Infinitely Many Genes model for the evolution of the bacterial core genome and pangenome. *Mol Biol Evol* 29:3413–3425.

488. Kopac S, Wang Z, Wiedenbeck J, Sherry J, Wu M, Cohan FM. 2014. Genomic heterogeneity and ecological speciation within one subspecies of *Bacillus subtilis*. *Appl Environ Microbiol* 80:4842–4853.

489. Waters SM, Zeigler DR, Nicholson WL. 2015. Experimental evolution of enhanced growth by *Bacillus subtilis* at low atmospheric pressure: genomic changes revealed by whole-genome sequencing. *Appl Environ Microbiol* 81:7525–7532.

The Bacterial Spore: From Molecules to Systems
Edited by P. Eichenberger and A. Driks
© 2016 American Society for Microbiology, Washington, DC
doi:10.1128/microbiolspec.TBS-0017-2013

Ines Mandic-Mulec[1]
Polonca Stefanic[1]
Jan Dirk van Elsas[2]

Ecology of *Bacillaceae*

3

The most distinguishing feature of most members of the family *Bacillaceae* (phylum *Firmicutes*) is their ability to form endospores that provide high resistance to heat, radiation, chemicals, and drought, allowing these bacteria to survive adverse conditions for a prolonged period of time. *Bacillaceae* are widely distributed in natural environments, and their habitats are as varied as the niches humans have thought to sample. Over the years of microbiological research, members of this family have been found in soil, sediment, and air, as well as in unconventional environments such as clean rooms in the Kennedy Space Center, a vaccine-producing company, and even human blood (1–3). Moreover, members of the *Bacillaceae* have been detected in freshwater and marine ecosystems, in activated sludge, in human and animal systems, and in various foods (including fermented foods), but recently also in extreme environments such as hot solid and liquid systems (compost and hot springs, respectively), salt lakes, and salterns (4–6). Thus, thermophilic genera of the family *Bacillaceae* dominate the high-temperature stages of composting and have also been found in hot springs and hydrothermal vents, while representatives of halophilic genera have mostly been isolated from aquatic habitats such as salt lakes and salterns, but less often from

saline soils (7, 8). The isolates that have been obtained, in particular, from the varied extreme habitats, produce a wide range of commercially valuable extracellular enzymes, including those that are thermostable (9, 10).

Here, we focus on the ecology of selected members of the *Bacillaceae*. We first briefly address the recently revised taxonomy of this family, which encompasses strictly aerobic to facultatively and strictly anaerobic endospore-forming bacteria. Then we will focus on the ecological behavior and roles of selected organisms with special reference to members of the genus *Bacillus* that inhabit soil and the rhizosphere. These will be referred to by the name *Bacillus* except where studies focusing on particular species or different genera of the family *Bacillaceae* are discussed. We will also examine the methodology used to study the diversity, abundance, and distribution of *Bacillus* in the environment, in particular, with respect to the benefits and limitations. Finally, we examine the ecological drivers that shape the diversity and evolution of selected members of this genus and address the future goals and needs of the research aimed at furthering our understanding of how *Bacillus* communities are shaped in natural habitats.

[1]University of Ljubljana, Biotechnical Faculty, Department of Food Science and Technology, Vecna pot 111, 1000 Ljubljana, Slovenia; [2]Department of Microbial Ecology, Centre for Ecological and Evolutionary Studies, University of Groningen, Linneausborg, Nijenborgh 7, 9747AG Groningen, Netherlands.

BRIEF OVERVIEW OF THE FAMILY *BACILLACEAE*

The family *Bacillaceae* comprises mostly aerobic or facultatively anaerobic chemoorganotrophic rods with a typical Gram-positive cell wall. The majority of the taxa within the family form endospores, although exceptions are found. The aerobic or facultatively anaerobic members of the *Bacillaceae* were, until the early 1990s, positioned within the genus *Bacillus*, which stood next to the strictly anaerobic clostridia. Since then, major taxonomic changes have taken place, and consequently the family now accommodates representatives of the genus *Bacillus* and other newly formed genera with related nomenclature, examples being *Paenibacillus* ("almost" *Bacillus*), *Geobacillus*, and *Halobacillus* (see Galperin [232] for additional details of *Firmicutes* taxonomy). Currently (i.e., September 2014), the family *Bacillaceae* encompasses 62 genera (Table 1) composed of at least 457 species. New genera are continuously being described, as a result of a thorough description of a plethora of new, divergent environmental isolates. In 2009, 31 genera belonging to the *Bacillaceae* were listed in *Bergey's Manual of Systematic Bacteriology* (11), while only 2 years later Logan and Halket (12) indicated the existence of 36 genera in this family. Altogether, 25 new genera have been classified in the past 2 years, for a grand total of a staggering 62 genera (listed in Table 1). The taxonomy of the *Bacillaceae* may be rather confusing for the nonspecialist. For example, *B. pallidus* was reclassified in 2004 as *Geobacillus pallidus* (13), then later (2010) to a new genus, *Aeribacillus pallidus* (14). However, *B. pallidus* was also reclassified in 2009 to a new genus *Falsibacillus* (15). With the exception of *Anoxybacillus*, *Bacillus*, *Halobacillus*, *Geobacillus*, *Gracilibacillus*, *Lentibacillus*, *Lysinibacillus*, *Oceanobacillus*, and *Virgibacillus*, the new genera often include only one or a few species (4). See Table 1. Therefore, and very unfortunately, the taxonomy of the novel genera and species is currently often based on only one isolate per genus or species. Given this low robustness of the novel genera and the lack of sound ecological data, the ecology of these groups will not be further discussed in this review. Thus, the focus of this review will be on those representatives of the genus *Bacillus* (such as members of the *B. cereus sensu lato* and *B. subtilis*/*B. licheniformis* clades), which have gained the most scientific attention because they encompass industrially, agriculturally, or medically important species. Moreover, *B. subtilis* has been a long-standing model or reference organism for the study of gene regulation in Gram-positive bacteria, especially in the context of spore development. It is of utmost scientific interest to link the knowledge of the genetics and biochemistry of this well-studied species to that of its ecology.

The Genus *Bacillus*

The genus *Bacillus* was proposed in 1872 by Cohn, who classified its type species, *Bacillus subtilis*, as an organoheterotrophic aerobic spore-forming rod (16). Since its first description, the genus *Bacillus* has undergone many transformations due to the difficult classification of its member species (17). It is currently the largest genus within the *Bacillaceae*, presently consisting of at least 226 species (September 2014). New strains are constantly added as new species as well as being reclassified into new genera. For example, in the past 3 years, 10 existing species were reclassified into other genera and 39 new species were added to the genus only within the past year (September 2014). The inferred phylogeny of *Bacillus* species is often based on the 16S rRNA gene sequence, but this does not always distinguish species. Therefore, usage of DNA-DNA hybridization or sequencing of core genes is recommended for a better classification. This finer approach is even more important if one studies the *Bacillus* strains at the subspecies level (18, 19). In general, the different species within the genus *Bacillus* show only meager divergence of their 16S rRNA genes and this divergence is poorly correlated with phenotypic characteristics of these bacteria. Recently, Maughan and Van der Auwera (20) calculated the evolutionary relationships among 56 *Bacillus* species on the basis of their 16S rRNA gene sequences and compared these with the phenotypic traits of corresponding isolates. They found that 16S rRNA and phenotypic clustering are not congruent and that *Bacillus* species that form tight phylogenetic clusters are dispersed over the whole phenotypic tree. For example, representatives of the *B. subtilis* cluster comprise 2 subspecies (*B. subtilis* subsp. *subtilis* and *B. subtilis* subsp. *spizizenii*) and 12 species (*B. mojavensis*, *B. valismortis*, *B. amyloliquefaciens*, *B. atrophaeus*, *B. pumilus*, *B. licheniformis*, *B. sonorensis*, *B. aquimaris*, *B. oleronius*, *B. sporothermodurans*, *B. carboniphilus*, and *B. endophyticus*). In the light of this discrepancy, which is indicative of the occurrence of fluid genomes across these species, molecular data need to be used to address the phenetic and ecological relationships between closely related taxa.

As with members of the *B. subtilis* cluster, members of the *B. cereus sensu lato* group (*sensu lato* meaning "in the widest sense"), including *B. anthracis*, *B. thuringiensis*, *B. cereus*, *B. mycoides*, *B. pseudo-*

mycoides, and *B. weihenstephanensis* (21), are highly related at the genome level. Sequencing of the 16S rRNA gene (22) and even multilocus sequence typing (MLST) indicated high relatedness among the isolates of this group (21, 23, 24). This high relatedness was later confirmed at the level of gene content and synteny of their genomes (25). Traditionally, species of the *B. cereus sensu lato* group have been defined at the phenotypic level. Importantly, a relatively small number of genes, located on plasmids that are shared among *B. cereus*, *B. thuringiensis*, and *B. anthracis* strains, have a disproportionate effect on the ecological behavior (phenotype) of the three species. For example, pXO1 (181 kb) and pXO2 (96 kb) are typically found in *B. anthracis* as they carry major virulence factors associated with anthrax, but recent findings revealed pOX-1- and pOX-2-like plasmids also in *B. cereus* and *B. thuringiensis* (26–28). Interestingly, *B. anthracis* and the other strains carrying pXO1 and causing anthrax-like disease did not undergo major changes of their core genomes after plasmid acquisition, suggesting that there is no subgroup genetically predisposed to anthrax pathogenesis; instead, any number of *B. cereus sensu lato* or possibly even other *Bacillus* may be capable of gaining the ability to produce the lethal toxin (29). Thus, horizontal gene transfer of such plasmids may have a very drastic impact on the ecology of these organisms, potentially resulting in strong shifts in ecological behaviors, from the ability to infect a particular insect species to the ability to cause disease in sheep. Overall, the *B. cereus* group (including especially *B. cereus*, *B. thuringiensis*, and *B. anthracis*) contains strains of key medical and economic importance. Therefore, we advocate that species names are preserved for what we argue are very practical reasons.

LIFESTYLES AND ECOSYSTEM FUNCTION OF MEMBERS OF THE *BACILLACEAE*

Lifestyles (and thus ecological behavior) of all members of the *Bacillaceae* are tightly defined by their ability to form endospores. Spores allow survival under adverse conditions for shorter or longer time periods, even up to thousands of years (30) (see Fajardo-Cavazos et al. [233] for additional details on the isolation of ancient spores). The formation of endospores in some *Bacillus* species can be induced by nutrient deprivation and high cell population densities, and is also affected by environmental factors such as changes in pH, water content, or temperature (31). Spore survival is also influenced by specific soil parameters, including pH, organic matter content and calcium levels (32, 33).

The remarkable ability of *Bacillaceae* members to form spores and survive presumably allows these bacteria to travel large distances as living entities, even as far as between continents, using airstreams (34, 35). Recently Smith et al. (35) showed that microorganisms are abundant in the upper atmosphere and that trans-pacific plums can deliver 16S rRNA of *Bacillaceae* from Asia across the Pacific Ocean to North America (35) as well as live *Bacillus* cells (36). It has even been shown that spores survive on spacecraft exteriors especially if protected from UV radiation by soil particles (37) and speculated that spores can travel from Earth to Mars or even beyond (38). Therefore members of *Bacillaceae* are widely distributed across environmental habitats (Fig. 1), presumably also due to the longevity of their endospores. The ubiquity of *Bacillus* spp. is exemplified by the case of the insect-pathogenic *B. thuringiensis*. This organism has been isolated from all continents. Remarkably, an attempt to isolate this bacterium from 1,115 different soil samples from all over the world (using acetate selection) was successful in roughly 70% of the samples (39). Usually, isolation is performed so that the size of the sample is not taken into consideration, and only one or a few isolates are obtained from one sample and then studied further (40). Recently, however, the question about the diversity of the quorum-sensing system encoded by the *comQXPA* locus was addressed by isolation of the multiple strains of *B. subtilis* from two 1-cm^3 soil samples. This approach yielded a unique collection of strains that enabled us to address the ecology and diversity of bacilli at the micrometer distances beyond the quorum-sensing genes and are referred to below as microscale strains (41). In soil, even at micrometer distances, *Bacillus* strains showed differences in various traits and functions, e.g., colony morphology, competence for transformation, sensitivity to prophage induction by mitomycin, swarming, and metabolism (P. Stefanic, M. Črnigoj, and I. Mandic-Mulec, unpublished data).

Concerning ecosystem function, one usually thinks of biogeochemical cycling processes. Indeed, many members of the *Bacillaceae* are saprophytes that participate in the carbon, nitrogen, sulfur, and phosphorous cycles in natural habitats, e.g., in soil. Some species like *Bacillus schlegelii* isolated from geothermal soil and capable of growth from 59 to 72°C are capable of autotrophic growth by using hydrogen or thiosulfate as an energy source and carbon dioxide as a source of carbon (42). Recently, a proposal was made to transfer *B. schlegelii* to a novel genus with a novel species name *Hydrogenibacillus schlegelii* (43). In addition to

Table 1   Genera in the family *Bacillaceae*

Genus	No. of species	Isolated from	Comments	Reference(s)	Proposal of genus
Bacillus	226	Soils, animals, inner plant tissue, humans, food, domestic, industrial, hospital, marine environment, air, sediment,	38 new species since August 2013	174	1872
Aeribacillus	1	Hot springs, crude oil-contaminated soil	Reclassified from *Geobacillus pallidus*	14	2010
Alkalibacillus	6	Brine, camel dung, loam, mud, salt lake, soil, marine solar saltern, hypersaline soil	Was previously known as *Bacillus haloalkaliphilus* (Fritze 1996)	175	2005
Alkalilactibacillus	1	Cool and alkaline soil		176	2012
Allobacillus	1	Shrimp paste	1 strain	177	2011
Alteribacillus	2	Hypersaline lake		178	2012
Amphibacillus	7	Alkaline compost, lagoon and lake sediment	3 novel species have been proposed since 2011	179	1990
Anaerobacillus	3	Soda lake		180	2009
Anoxybacillus	18	Hot springs, geothermal soil, contaminant of gelatin production, cow and pig manure	6 new species since 2011	181	2000
Aquisalibacillus	1	Hypersaline lake		182	2008
Aquibacillus	3	Hypersaline lake	2 species were reclassified from *Virgibacillus* to *Aquibacillus*	183	2014
Caldalkalibacillus	2	Hot spring		184	2006
Caldibacillus	1	Soil	Reclassified from *Geobacillus debilis*	185	2012
Calditerricola	2	High-temperature compost		186	2011
Cerasibacillus	1	Kitchen refuse		187	2004
Domibacillus	1	Clean room of a vaccine-producing company		188	2012
Filobacillus	1	Sediment of marine hydrothermal vent	Species based on 1 strain	189	2001
Fictibacillus	6	Oyster, bioreactor, wall paintings, arsenic ore	Reclassification of *B. nanhaiensis*, *B. barbaricus*, *B. arsenicus*, *B. rigui*, *B. macauensis*, and *B. gelatini* as *Fictibacillus*	190	2013
Edaphobacillus	1	Hexachlorocyclohexane (HCH) contaminated soil		191	2013
Geobacillus	15	Hot springs, oilfields, spoiled canned food, milk, geothermal soil, desert sand, composts, water, ocean sediments, sugar beet juice, mud, activated sludge	Obligately thermophilic, should be reclassified as *Saccharococcus* due to morphology (Bergey's)	192	2001
Gracilibacillus	11	Salt lakes, desert soil, solar saltern, saline soil, fermented fish		193	1999
Halalkalibacillus	1	Nonsaline soil	Not included in Bergey's *Manual* (2009), only 1 strain	194	2007
Halobacillus	18	Hypersaline environments, salt marsh soils, fermented food, mural paintings		195	1996
Halolactibacillus	3	Decaying marine algae, living sponge	Possesses all the essential characteristics of lactic acid bacteria	196	2005
Jilinibacillus	1	Saline and alkali soil samples		197	2014
Lentibacillus	11	Fish sauce, salt lake, saline soil, solar saltern, saline sediment, salt field		198	2002
Hydrogenibacillus	1	Lake mud		43	2012
Lysinibacillus	10	Forest humus, soil		199	—
Marinococcus	4	Solar salterns, seawater, saline soil		200	1984

(Continued)

Table 1 *(Continued)*

Genus	No. of species	Isolated from	Comments	Reference(s)	Proposal of genus
Microaerobacter	1	Terrestrial hot spring		201	2010
Natribacillus	1	Soil		202	2012
Natronobacillus	1	Soda-rich habitats (lake sediment)		203	2008
Oceanobacillus	11	Mural paintings, algal mat from sulfurous spring, deep marine sediments, shrimp paste, fermented food, activated sludge, marine animals		204	2001
Ornithinibacillus	5	Hypersaline lake, human blood, pasteurized milk		205	2006
Paraliobacillus	2	Decomposing marine algae, salt lake sediment		206	2002
Paucisalibacillus	1	Potting soil		207	2006
Piscibacillus	2	Fermented fish, hypersaline lake		208	—
Pontibacillus	5	Solar saltern, marine animals, salt field		209	2005
Pseudogracilibacillus	1	Rhizosphere soil		210	2014
Psychrobacillus	3	Soil, mud water	Reclassified from *B. insolitus*, *B. psychrodurans*, *B. psychrotolerans*	211	2010
Saccharococcus	1	Beet sugar extraction	1 species reclassified as *Geobacillus*	212	1984
Rummeliibacillus	2	Clean room of Kennedy space center, field scale composter		213	—
Salibacillus	2	Salterns, hypersaline soils	Was reclassified from *B. salexigens*	193	1999
Salimicrobium	5	Salted hides, solar saltern	Reclassified from *Marinococcus albus* and *B. halophilus*	214	2007
Salinibacillus	2	Saline lake		215	2005
Salirhabdus	1	Sea salt evaporation pond		216	2007
Salisediminibacterium	1	Soda lake sediment		217	2012
Saliterribacillus	1	Hypersaline lake		218	2013
Salsuginibacillus	2	Lake sediment, soda lake		219	2007
Sediminibacillus	2	Hyper saline lake		220	2008
Sinibacillus	1	Tropical forest soil and a hot spring sediment		221	2014
Streptohalobacillus	1	Subsurface saline soil		222	2011
Tenuibacillus	2	Hypersaline lake		223	2005
Terribacillus	4	Soil		224	2007
Texcoconibacillus	1	Soil of the former lake Texcoco		225	2013
Thalassobacillus	4	Saline soil, hypersaline lake, tidal flat sediment		226	2005
Thermolongibacillus	2	Hot springs soil and sediment		227	2014
Tumebacillus	2	Permafrost, ginseng field		228	—
Virgibacillus	24	Soils, hypersaline soil and salterns, seawater, salt field, saline soil, mountain soil, salt lake, food, permafrost	6 new species since 2011	229	1999
Viridibacillus	3	Soil	Reclassified from *B. arvi*, *B. arenosi*, and *B. neidei*	230	2007
Vulcanibacillus	1	Deep-sea hydrothermal vents		231	2006

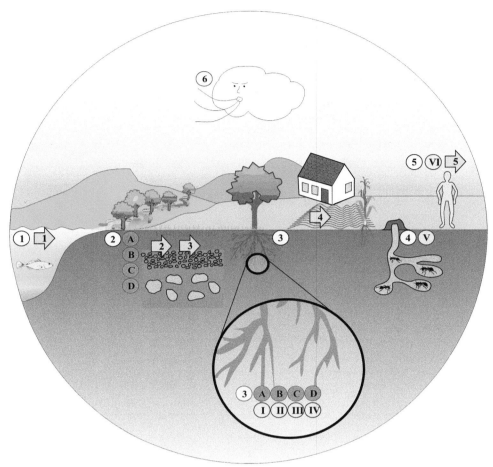

Occurrence of Bacillaceae in the environment

(1) Marine ecosystems, fresh waters, salt lakes, hot springs

(2) Soil

(3) Rhizosphere, rhizoplane

(4) Soil invertebrates

(5) Mammal pathogens

(6) Air

Biotic interactions

(I) Plant defense, microbial pesticide synthesis

(II) Symbiosis stimulation

(III) Fixation of nitrogen, phosporus, Zinc solubilization, iron aquisition

(IV) Phytohormone production

(V) Mutualism (protection from fungus)

(VI) Pathogenicity (toxin production)

Ecosystem functions

(A) Carbon cycle, degradation of soil organic matter

(B) Nitrogen cycle, nitrification, denitrification, nitrogen fixation (biofertilization)

(C) Phosporus solubilization (biofertilization)

(D) Ecoremediation of pollutants

Applications

 Hg Volatilization

 Phytoremediation of metals from polluted soil (Pb, Mn, Zn, Cr, Ni)

 Degradation of petrochemicals

 Biofertilization/Biopesticide

 Medicine - probiotics

Bacillus-mediated biogeochemical cycling, other eco-logically important functions may also be driven by members of *Bacillaceae*. As indicated in the foregoing, some representatives of the *Bacillaceae* (in particular, *B. anthracis* and *B. cereus*) are important pathogens of mammals (44), whereas *B. thuringiensis* is an insect pathogen. The pathogenicity of the latter organism is linked to plasmid-encoded *cry* genes, which are responsible for the synthesis of an insecticidal protein that interacts with receptors in the gut system of insect larvae (45). In addition, bacilli such as *B. subtilis*, *B. cereus, and B. mycoides* are known for their roles as beneficial rhizobacteria that promote plant growth (act as biofertilizers) or protect plants from plant pathogens (function as biopesticides, e.g., *B. subtilis*) (46) as illustrated in Fig. 1.

Involvement of *Bacillaceae* in the Degradation of Soil Organic Matter and Plant Litter

It is generally believed that the primary habitat of many *Bacillaceae* (e.g., *B. subtilis* and *B. cereus*) is the soil. However, almost nothing is known about the intricate mechanisms modulating the *in situ* physiology, germination, growth, and sporulation of these bacteria. Most members of the genus *Bacillus* are aerobic heterotrophic saprophytes (47) that are capable of degrading a range of polymeric carbonaceous substances. They also grow on a variety of simple compounds. Hence, it is generally assumed that actively growing bacteria in soil are associated with soil organic matter or the plant rhizosphere (Fig. 1). Thus, these organisms may prosper in microhabitats where carbon and nitrogen are not strongly limited. Several members of the genus *Bacillus* are known to be typical inhabitants of so-called hot spots for microbial activity in terrestrial habitats, e.g., where organic matter is plentiful. For example, Siala et al. (48) found that vegetative bacilli predominate in the soil A1 horizon, where organic carbon is provided by plant litter or root exudates, while spores predominate in the deeper C horizon of the soil. They also detected colonization by bacilli of organic matter aggregates interconnected by fungal hyphae. In later work in the Jansson laboratory (Uppsala), *B. cereus* was found to interact with soil mycorrhizal fungi (49). Moreover,

plant roots (rhizosphere and rhizoplane) are habitats where members of *Bacillaceae* can thrive on plant exudates. *Bacillus* strains were also isolated from animal guts and feces, suggesting that these energy-rich environments may represent suitable habitats for spore-formers from this genus. Indeed, it has been shown that spores of *B. subtilis* can germinate, grow, and go through another cycle of sporulation/germination inside the gut of a mouse (50).

Many *Bacillus* isolates have the ability to break down cellulose, hemicellulose, and pectin (51–54), which suggests their involvement in the degradation and mineralization of plant and humic materials in soil. Also, Maki et al. (55) found that a new *Bacillus* sp. strain was able to modify lignocellulose, and, owing to a variety of cellulase and xylanase activities, it displayed a high potential for lignocellulosic decomposition. In addition, chitinase activity, which facilitates the degradation of fungal cell walls and insect exoskeletons, is also common among many members of soil *Bacillaceae* (56–59). Thus, chitinolytic activity contributes to the role of *Bacillaceae* in the mineralization of soil organic matter and to their ability to protect plants from pathogens (46, 60).

Many members of *Bacillaceae* are also able to degrade proteins in soil, and some proteases produced by these bacteria have gained great scientific attention because of their industrial value. However, only a few studies have addressed proteolytic activity, including that of *Bacillaceae*, in soils. Sakurai et al. (61) showed that proteolytic activity in soil was greater following the addition of organic, rather than inorganic, fertilizer and was also greater in the rhizosphere than in bulk soil. Chu et al. (62) also found evidence for the selection of a *Bacillus*-related organism following treatment of soil with organic manure (but not inorganic fertilizer). This suggested that members of *Bacillaceae* respond to the addition of organic C and N sources to soil and may be important in the degradation of fresh organic matter in the soil.

Studies addressing the growth of *Bacillaceae* in soil or the rhizosphere are scarce. Recently, Vilain et al. (63) assessed the growth and behavior of *B. cereus* in soil extracts and demonstrated that this organism goes through a complete life cycle (germination, growth, and sporulation) in soil-mimicking conditions. Interestingly,

Figure 1 Occurrence of *Bacillaceae*, their ecosystem function, biotic interactions, and applications. The illustration shows different environments from which *Bacillaceae* have been isolated highlights their main ecosystem functions and biotic interactions, and illustrates selected existing and possible applications.

B. cereus and other related *Bacillus* isolates formed multicellular structures encased in polymeric matrices in soil extracts and soil microcosms, while in LB medium clumping of cells was never observed. This suggested that soil induces a specific physiological adaptation in cells that involves the formation of bacterial aggregates, which may be important for biofilm formation (63). *B. subtilis* isolates also show growth in autoclaved high organic matter soils with rates that are comparable to those in rich nutrient medium (Fig. 2). However, growth of these isolates in unsterilized soil, in which a microbiostatic microbiota was present, was almost undetectable.

Involvement of *Bacillaceae* in the Nitrogen Cycle

Some members of the genus *Bacillus* play key roles in particular steps of soil nitrogen-cycling processes (Fig. 1), such as denitrification and nitrogen fixation, next to their involvement in the mineralization of various nitrogen sources (64–66). Denitrification is an anaerobic process, in which nitrate serves as the terminal electron

acceptor during the oxidation of organic matter and is converted to gaseous products such as N_2O, NO, and N_2. This process is important in the removal of nitrogen in biological wastewater treatment and in degradation of organic pollutants (67) and is detrimental for soils, where it depletes nitrogen (68). Soil bacteria capable of denitrification were also found among *Bacillus* isolates (68) that were identified by sequencing their 16S rRNA genes. In addition, gene sequence analysis of *nirS* and *nirK*, both encoding nitrite reductases, was performed on more than 200 cultivated denitrifiers, but only *nirK* was detected in the *Firmicutes* (69). Although the denitrification potential of several *Bacillus* species is well known, many *Bacillus* isolates that are able to use nitrate as an electron acceptor have been misclassified as denitrifiers (68). This is because these isolates more often reduce nitrate to ammonium and not to gaseous products and therefore should be classified as microbes that perform DNRA (dissimilatory nitrate reduction to ammonium), a process that is common in environments that are rich in organic carbon (68). In luvisol soil, *Bacillus* spp. were among the most abundant members retrieved among cultured denitrifiers (66). This study

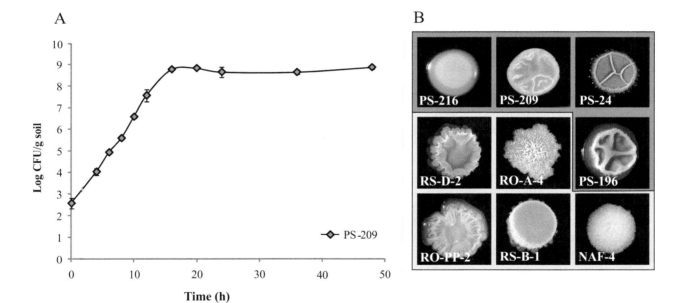

Figure 2 Growth of *B. subtilis* in soil and morphology on LB agar media. (A) Growth of riverbank isolate *B. subtilis* PS-209 (41) was grown in an autoclaved soil microcosm at 28°C, and CFU counts were performed at indicated times on Luria-Bertani (LB) medium. The experiment was performed in three replicates. Error bars represent 95% confidence intervals of means calculated from log10 transformed CFU counts (L. Pal, S. Vatovec, P. Stefanic, T. Danevčič, and I. Mandic-Mulec, unpublished data). (B) Colony morphotypes. Colony morphology was visually examined and photographed after incubation at 37°C for 48 h on LB agar medium. Riverbank microscale and desert macroscale *B. subtilis* strains are marked with green and yellow, respectively (Courtesy of P. Stefanic).

tested both, nitrate and nitrite, as electrons that were crucial to evaluate the denitrification potential among soil isolates (66). In general, however, the distribution of denitrifying bacilli in soil and their importance is poorly understood because the existing primer sets that target denitrification genes and allow monitoring of their activity in natural environments through quantitative PCR-based methods are based on sequences from Gram-negative bacteria and, therefore, may not recognize these genes in the Gram-positive *Bacillaceae*.

In addition to the denitrifier and DNRA representatives of the genus *Bacillus*, some members of *Bacillaceae* have been classified as nitrogen fixers. These organisms reduce atmospheric nitrogen to ammonia with the help of the enzyme nitrogenase. Early studies pinpointed the then-called *Bacillus polymyxa*, *Bacillus azotofixans*, and other members of *Bacillaceae* as nitrogen fixers. However, many of these have later been reclassified as *Paenibacillus* species, among which the prominent nitrogen fixers are *P. polymyxa*, *P. macerans*, *P. azotofixans*, and *P. durus* (70–72). One study showed that particular strains of *B. cereus*, *B. megaterium*, and *B. licheniformis* are able to fix nitrogen (73). In contrast, others suggested, on the basis of screenings for *nifH* genes, that nitrogen fixation within the *Bacillaceae* is limited to *Paenibacillus* species (74).

Involvement of *Bacillaceae* in the Phosphate Cycle

Phosphorus is an essential mineral for plant growth. Many of the phosphate anions in soil are in complexes with cations (Ca^{2+}, Mg^{2+}, Fe^{3+}, Al^{3+}), hindering the availability of phosphorus to plants. Many bacteria and fungi in soil possess phosphatase or phosphate-solubilizing activity, resulting in the release of phosphate from insoluble polyphosphate and consequently in the improvement of plant nutrition and growth (75–79) (Fig. 1). These phosphate solubilizers include a variety of bacilli (80–82). These naturally occurring phosphorus-solubilizing soil bacilli may serve as potential biofertilizers, either by their introduction in soils that are currently deprived of them or when they are already present by enhancing their prevalence through agronomic measures. Active phosphate solubilization will reduce the environmental phosphate load due to fertilization in phosphate-saturated regions and, in contrast, help to sustain agriculture in regions with severe phosphorus deficiencies across the globe. An example of the latter is found in Ethiopia (78), but several other regions of the world (Pakistan, India, Brazil) are in need of better phosphorus mobilization in their agricultural soils.

Bacillaceae as Plant-Beneficial Rhizobacteria

Plant growth-promoting rhizobacteria (PGPR) influence plant health, growth, and development either directly or indirectly. Direct modes of action include nitrogen fixation, phytohormone production, phosphorus solubilization, and lowering of the ethylene concentrations at the plant root to promote lengthening. In addition, indirect effects include an influence on symbiotic relationships between bacteria and plants and the repression of soil-borne plant pathogens (83, 84). Collectively, the PGPR of the genus *Bacillus* participate in nitrogen fixation (85), production of phytohormones, e.g., gibberelins (86), phosphorus solubilization (80–82), and zinc solubilization (87). Members of the *Bacillaceae* family further promote the growth of plant roots by lowering the local ethylene concentrations (88) and by increasing the assimilation of metal ions such as iron through the activation of the plant's own iron acquisition machinery (89). *Bacillus* strains can also stimulate *Rhizobium*-legume symbioses (90) and they possess a range of other plant-beneficial properties that protect plants against pathogens. This ecosystem function may depend on biofilm formation on plant roots and the production of hydrolytic enzymes, various antibiotics, and small molecules such as lipopeptides (surfactins, iturins, fengycins). Some rhizosphere *Bacillus* strains may have the ability to induce systemic resistance in plants, allowing enhanced resistance against phytopathogens. Studies reviewed by Choudhary and Johri (46) and Kloepper et al. (91) revealed that several species of the genus *Bacillus* (*B. amyloliquefaciens*, *B. subtilis*, *B. pasteurii*, *B. cereus*, *B. pumilus*, *B. mycoides*, *Lysinibacillus sphaericus*) significantly reduced the incidence or severity of various diseases on tomato, bell pepper, tobacco, *Arabidopsis* sp., and cucumber due to induced systemic resistance. Defense by *Bacillus* spp. has been reported against fungal and bacterial pathogens, systemic viruses, and root-knot nematodes (46, 91). Ryu et al. (92) showed that *Arabidopsis thaliana*, following treatment with plant-growth-promoting *Bacillus pumilus* significantly reduced symptom severity resulting from *Cucumber mosaic virus* (CMV) infections (92).

Many *Bacillus* strains have found commercial applications, mainly as biopesticides (fungicides, bactericides, viricides, actinomyceticides) as well as insecticides (93, 94). For example, tomato plants, inoculated with plant-growth-promoting *B. subtilis* isolated from the tomato rhizosphere, revealed lowered susceptibility to whitefly *Bemisia tabaci* (95). Recently, wild strains of *B. subtilis* and related organisms were shown to have biocontrol efficacy against the potato and tomato wilting agent

Ralstonia solanacearum. These strains formed robust biofilms both in defined medium and on tomato roots (95). The plant protection phenotype was found to be dependent on the genes required for biofilm formation and matrix production, which was critical for bacterial colonization of plant root surfaces (96, 97) and was induced by plant polysaccharides (98). Similarly, Bais et al. (99) suggested that biofilm formation and the production of the lipopeptide compound surfactin by *B. subtilis* on *Arabidopsis* roots is essential for the plant protection activity of this organism against *Pseudomonas syringae*. Recently, surfactin production by *Bacillus* biofilms was directly imaged on plant roots by matrix-assisted laser desorption/ionization mass spectrometry imaging (MALDI MSI) (100) and showed that *B. subtilis* strains that originate from the same plant root can dramatically differ in surfactin production, but less so in their ability to form biofilms (101). However, biofilm strength and thickness depends also on the composition of extracellular matrix components and these change in relation to available nutrients (102). It will be interesting to explore whether a good biofilm former on one plant performs similarly on other plants and whether this ability is plant specific, because this may have implications for the development of novel biopesticides. Specific *B. subtilis* strains can also trigger induced systemic resistance (ISR) in plants (103) and change transcriptional response in plants (104), which can have a significant influence on the plant: pathogen interaction and the outcome of infection.

Role of *Bacillaceae* in Soil-Dwelling Invertebrates

Some members of the genus *Bacillus* are key constituents of the bacterial communities in the intestinal systems of soil-dwelling invertebrates that degrade organic polymers (105–108). For instance, cellulose, hemicellulose, and lignin are major constituents of plant material that are broken down in the guts of termites. In the degradative process, three steps ([1] hydrolysis, [2] oxidation, and [3] methane formation) are distinguished. There is evidence that members of *Bacillaceae* may be mainly involved in the initial phases of the degradative process, i.e., steps 1 and 2. For instance, in the termite hindgut, aerobic rod-shaped spore formers have been found in high numbers, occurring in population sizes of up to 10^7 per ml in gut fluid (109). *In vitro*, the isolates obtained could produce a range of hydrolytic or other enzymatic activities that are involved in the degradation of biopolymeric compounds. Although this

suggests that such bacilli are involved in degradative processes in the termite intestine, a clear functional link or proof of their involvement is still lacking.

DISTRIBUTION AND DIVERSITY OF *BACILLACEAE* ACROSS DIFFERENT HABITATS

Habitat distribution (biogeography) and diversity of members of *Bacillaceae* in natural and man-made habitats is a fundamental ecological question. They have been investigated in this context since the first isolation of bacteria from soil at the end of the 19th century. Methods used to study diversity have followed those used for all soil bacteria and traditionally included phenotypic characterization, the application of dichotomous and numerical taxonomy, and, later on, molecular phylogenetic approaches (initiated more than 20 years ago) which rely on direct extraction of nucleic acids from the natural habitat (e.g., soil), rather than from isolates. In this review, the benefits and limitations of molecular and cultivation-based methods are critically evaluated.

Cultivation-Based Methods To Study the Distribution and Diversity of *Bacillaceae*

Bacillus strains are commonly obtained from environmental habitats after samples are treated with mild heat (e.g., 80°C for 10 to 20 min), leaving only endospores to form colonies on solid nutrient media. This is a highly selective approach, which has the great advantage of excluding all bacterial cells that are not in the spore form. Aerobic incubation of the resulting plates provides mainly colonies of the genus *Bacillus*, which can be used for further characterization and identification. Traditionally, biochemical tests are applied for the initial characterization of isolates, as described in the classical handbook by Gordon et al. (110). Also, simple miniaturized methods, such as API strips for metabolic differentiation between isolates, are in use (111). In addition, molecular methods that target specific genes (e.g., 16S rRNA) are employed for phylogenetic characterizations, with the caveat that *Bacillus* species are only slightly divergent in this respect. Thus, as mentioned, core genes like *gyrA* may be more useful for phylogenetic identifications (20). Targeting core genes, and using multilocus sequence typing (MLST) and analysis (MLSA) (both methods involving sequencing short regions of several [typically seven] housekeeping genes), may result in a better assessment of the relatedness of strains (112, 113).

Two additional rapid classification tools for *Bacillus* strains at the subspecies level are repetitive extragenic palindromic-PCR (REP-PCR) and BOX-PCR, which both target interspersed repeated sequences within the genome (114). Also, fatty acid methyl ester (FAME) analysis has been successfully used for identification of *Bacillaceae* (115). Recently, rapid bacterial identification by matrix-assisted laser desorption ionization time-of-flight mass spectrometry (MALDI-TOF MS) targeting ribosomal proteins (S10-GERMS biomarkers) was applied for identification of *Bacillus* strains. The method enables the differentiation of strains at the subspecies level (116). In addition, an alternative procedure to the standard sample preparation protocol was developed that includes microwave-accelerated tryptic digestion of the cellular material. This approach improved the discriminating power of MALDI-TOF MS by increasing the number of strain-specific peaks, in comparison with the standard method (117).

Cultivation-Independent Methods To Study Distribution and Diversity of *Bacillaceae*

We are presently able to culture in the laboratory only 1 to 10% of microbial diversity. Therefore, many as-yet-undiscovered *Bacillus* species may still be hidden within the uncultured majority. Culture-independent molecular methods are thus essential to address the distribution and diversity of *Bacillus* species in a more complete manner. The striking diversity of microbial communities, which has been unraveled in the past two decades by PCR-based molecular methods, has stimulated an unprecedented increase of investigations of various natural environments (terrestrial, aquatic, anthropogenic), including those within humans and animals (118). Representatives of the *Bacillaceae* have also often been detected in these habitats, at varying levels of abundance and richness, besides many other families and fila.

Molecular methods currently enable a direct determination of the diversity and relative abundance of representatives of *Bacillus* in total bacterial communities. Following the isolation of DNA from a particular environment, genes targeted by specific (*Bacillus*–oriented) primer sets are amplified by PCR. Amplification products obtained by PCR can then be cloned and sequenced for phylogenetic analyses, and the results compared with database sequences obtained from cultivated organisms and/or other environmental studies. Sequences can also be used to design probes for *in situ* detection using fluorescence *in situ* hybridization (FISH) or for probing of nucleic acids. Increasingly,

high-throughput sequencing techniques are being applied for the direct analysis of extracted soil DNA, which avoids the PCR-dependent cloning step (119). If high numbers of *Bacillus* sequences are obtained with these approaches, within-group diversity can also be studied. Alternatively, primers targeting specific groups can be used, e.g., for a wide suite of species within the genus *Bacillus* (120), although few such primers have been designed. In addition, the population structure and relative abundance of *Bacillaceae* can be addressed by DNA fingerprinting methods. While these cultivation-independent methods have been used only rarely to address the ecology of *Bacillaceae* that have been successfully grown in the laboratory, they are useful when many environmental samples need to be compared in a study, when we aim to decipher the relationships of *Bacillaceae* with other microbial groups *in situ*, and when we target the yet uncultured populations of these families. Fingerprinting methods most often include either denaturing gradient gel electrophoresis (DGGE), temperature gradient gel electrophoresis (TGGE), or terminal restriction fragment length polymorphism (T-RFLP). DGGE and TGGE separate amplicons (PCR products) on the basis of heterogeneities in GC content and sequence. When run on a gel containing a gradient of a denaturant or temperature, these PCR products, which are of the same size, separate into bands, visible on a gel after staining, due to differences in denaturing or melting properties. For *Bacillus* spp., this method has been applied successfully by Garbeva et al. (120). For T-RFLP, amplified DNA is digested with restriction enzymes, distinguishing amplicons with sequence polymorphisms. All these fingerprinting methods are less expensive than sequencing methods, and they allow a rapid analysis of many samples and the assessment of relative abundances of different phylotypes, but they provide less information on the identity of the organisms present. Identity is only provided by the primer set, which can be either broad (detecting all bacteria) or narrow (specific for genus *Bacillus*).

In addition to 16S rRNA genes, functional genes (e.g., those for the degradation of aromatic compounds, cellulose, proteins, and nitrate reduction) could be used to estimate the potential contribution of *Bacillaceae* members to the specific relevant ecosystem functions, although primers that target *Bacillus*-specific functional genes have not been developed yet. Moreover, by targeting RNA, it is possible to monitor the active populations in soil and the rhizosphere. The extracted RNA is reverse-transcribed into complementary DNA (cDNA), which is then amplified as described above. Targeting 16S rRNA sequences in this manner is sensitive, because cells contain more ribosomes than rRNA genes. This approach has been used to target the active soil bacterial community in acidic peat grassland soils (121). On the other hand, activity can also be assessed by techniques such as stable isotope probing (SIP) to determine which organisms are utilizing specific ^{13}C- or ^{15}N-labeled compounds, or bromodeoxyuridine (BrdU) capture, which separates organisms incorporating the thymidine analogue bromodeoxyuridine (122). Again, SIP has not been applied to *Bacillus*-specific functions yet.

Benefits and Limitations of Molecular Methods Used To Investigate *Bacillaceae* in Natural Settings

The major advantage of the nucleic acid-based techniques is their lack of dependence on laboratory cultivation of soil bacilli. Because growth media and cultivation conditions are inherently selective, it is inevitable that some strains will not be successfully cultured in the laboratory. However, molecular methods also have potential biases. Prosser et al. (122) discuss the benefits and limitations of molecular techniques and developing methods for the assessment of bacterial community diversity and activity. Nevertheless, there is now a suite of quantitative PCR methods available to identify representatives of soil *Bacillaceae*, but there is a need for better primer sets to be used to measure changes in their diversity and community structures and determine their abundance.

The presence of spores in *Bacillus* populations can introduce biases that can significantly influence specific types of studies. For example, molecular techniques are based on the extraction of nucleic acids, which requires lysis of vegetative cells. Lysis of spores, however, requires very severe conditions, and most studies achieve this through physical disruption by bead-beating. The methods for cell lysis and nucleic acid extraction require a balance between conditions and lengths of treatment that are sufficiently rigorous to optimize the lysis of cells and spores and minimize DNA degradation (which will be increased by the length of treatment). This balance is particularly difficult to achieve for *Bacillaceae* because of significant differences in the conditions required to lyse vegetative cells and spores.

Diversity of *Bacillaceae* in Natural Settings Such as the Soil and Rhizosphere

Members of the family *Bacillaceae* are often detected in soil habitats. Studies based on clone libraries of 16S rRNA genes from various soils revealed only a few *Bacillus* sequences (below 3%), indicating the generally low abundance of *Bacillaceae* (123, 124). However, in desert soils, endospore formers are usually highly abundant and diverse (124) in comparison with many other soils.

16S RNA gene-based libraries invariably contain sequences closely related to *Bacillaceae*, but the proportion varies between studies. This is not only because of the differences in environmental conditions, but it is also a result of the differences in nucleic acid isolation techniques, primers, and analytic methods. Our sequence databases have developed considerably over time, as molecular methods have continuously generated new sequence data over the past 15 to 20 years. In general, sequences matching to *Bacillus* represent 1 to 15% of total clone sequences in traditional clone libraries from soil samples (125). For example, a meta-analysis of 32 soil clone libraries (125) indicated that members of *Bacillus* encompassed less than 1 to 2% of the soil bacterial 16S rRNA gene sequences. In contrast, representatives of *Bacillus* comprised 5 to 45% of isolates from traditional cultivation-based studies. As mentioned above, DNA extraction procedures may have a very strong influence on the recovery of *Bacillus* sequences. Kraigher et al. (123) extracted DNA without bead beating and found that only 2 of 114 partial 16S rRNA sequences from a high organic grassland fen soil belonged to *Bacillus*. While high-throughput sequencing methods greatly increase the depth of coverage of soil bacterial diversity, easily providing information on more than 20,000 sequences, representation of *Bacillaceae* remains low. Roesch et al. (119) used pyrosequencing of up to 50,000 16S rRNA fragments from soils, and *Firmicutes* still comprised only 2 to 5% of the sequences. In contrast, denaturing gradient gel electrophoresis (DGGE) analysis with four primers targeting the 16S rRNA V6 region indicated that *Firmicutes* constituted between 19% and 32% of sequences in a grassland soil, and the majority of these (76 to 86%) were members of genus *Bacillus* (126). This suggests that many factors may affect measurements of *Bacillus* diversity in soils. Most diversity studies have targeted DNA, providing information on "total" communities, i.e., assessing active and dormant cells, including spores. Felske et al. (127) characterized the active soil bacterial community in acidic peat grassland soils by targeting RNA, rather than DNA. Sequencing of these clones indicated that the active community was dominated by bacilli and around 65% of all bacterial ribosomes originated from *Firmicutes*. Among these, *Bacillus* species were the most active and represented half of the detected sequences (127). Recent analyses of 16S rRNA gene sequences in the RDP database revealed many uncultivated representatives, even within the *B. cereus* or *B. subtilis* groups. In addition, 16S rRNA genes obtained from cultured representatives showed less diversity than environmental sequences classified as *Bacillus* (20). This suggests that there is still undiscovered diversity in this genus and that novel approaches are needed to study its ecology.

Diversity of *Bacillaceae* in the Rhizosphere

The rhizosphere of plants is a special habitat where microbial communities are under the influence of the plant root exudates, which serve as chemoattractants and nutrient sources for bacteria. Approximately 10^7 to 10^9 CFU of culturable rhizosphere bacteria per gram of soil have been reported (128). The rhizoplane bacteria colonize the root surface with 10^5 to 10^7 bacteria per gram of fresh root (128) and are attracted by plant exudates to the rhizosphere and rhizoplane. Specifically, malic acid has been reported to induce colonization and biofilm formation by *B. subtilis* (129). Several studies have compared the bacterial communities in bulk soil and in the rhizospheres of different plants, to assess the influence of root exudates and plant species on bacterial diversities (130–135). The prevalence of *Bacillus* representatives in the rhizosphere varied significantly between studies. Smalla et al. (136) compared bulk and rhizosphere soil microbial communities of field-grown strawberry (*Fragaria ananassa* Duch.), oilseed rape (*Brassica napus* L.), and potato (*Solanum tuberosum* L.) by DNA-based DGGE profiling. *B. megaterium* dominated the bulk soil and potato rhizosphere communities, whereas it was only detectable in strawberry and oilseed rape rhizospheres. Garbeva et al. (137) also found differences in the composition of *Bacillus* in soils under different plant species: *Bacillus* spp., *B. thuringiensis* and *B. cereus* were in particular associated with maize, *B. benzoevorans* and *B. pumilus* with pasture grass, and *Bacillus* sp. and *B. fumarioli* with oats and barley.

Differences in soil microbial communities were also associated with different plant cultivars as well as with the age of plants – young plants versus flowering plants and plants in senescence stages (138–140). Interestingly, root colonization by *Bacillus* was found to be not only plant specific but also root area specific. For example, different types of *B. amyloliquefaciens* FZB42 preferred to colonize root hairs of maize, while primary root tips and lateral roots were favored on *Arabidopsis* roots. On *Lemna*, *Bacillus* accumulated preferably along the grooves between epidermal cells of the roots

(141). Interestingly, successful colonization of lettuce rhizosphere by *B. amyloliquefaciens* FZB42 did not show durable effect on the rhizosphere community, while at the same time it did decrease disease severity caused by the pathogen *Rhizoctonia solani* (142).

Mechanisms Driving the Diversity of *Bacillaceae* in Natural Settings

Diversity in microbial communities is theoretically driven by immigration, speciation, horizontal gene transfer (HGT) and extinction. However, the shaping of these communities is also tightly linked to interactions between their members (118, 143, 144). The role of diversity in ecosystem function has attracted much scientific attention over the past 2 decades. This has been driven, in part, by the application of molecular approaches (recently including high-throughput sequencing methods) that allow us to study microbial communities without cultivation. However, molecular methods have only rarely been applied to the analysis of *Bacillaceae* population structure and diversity; most studies rely on cultured isolates. These isolates have been used as models to study the mechanisms of diversification. Recently, within defined species of the *Bacillaceae*, ecologically distinct groups, termed putative ecotypes, were identified (145). An ecotype is defined as an ecologically distinct group of organisms that fall into distinct sequence clusters (lineages), sharing a common evolutionary path. Diversity is then periodically purged by selection. Sequence clusters are initially viewed as putative ecotypes and have to be evaluated further at a more specific genetic or phenetic level or at the level of geographical distribution to be considered real ecotypes (146). Ecotype formation is thought to be countered by frequent recombination (145), which is known to occur between closely related members of *Bacillus*. To determine the actual rates of recombination, Roberts and Cohan (40) analyzed the restriction patterns of three housekeeping genes among closely related *Bacillus* isolates. Recombination within *B. subtilis* or *B. mojavensis* was too low, however, to prevent adaptive divergence between ecotypes (147, 148). Diversification of ecologically distinct populations will also increase due to neutral sequence divergence and differences in restriction/modification systems between the two populations (111, 149).

Environmental Factors Driving Diversification in Soil *Bacillaceae*

We still understand very little about the influence of environmental factors on the diversity and community

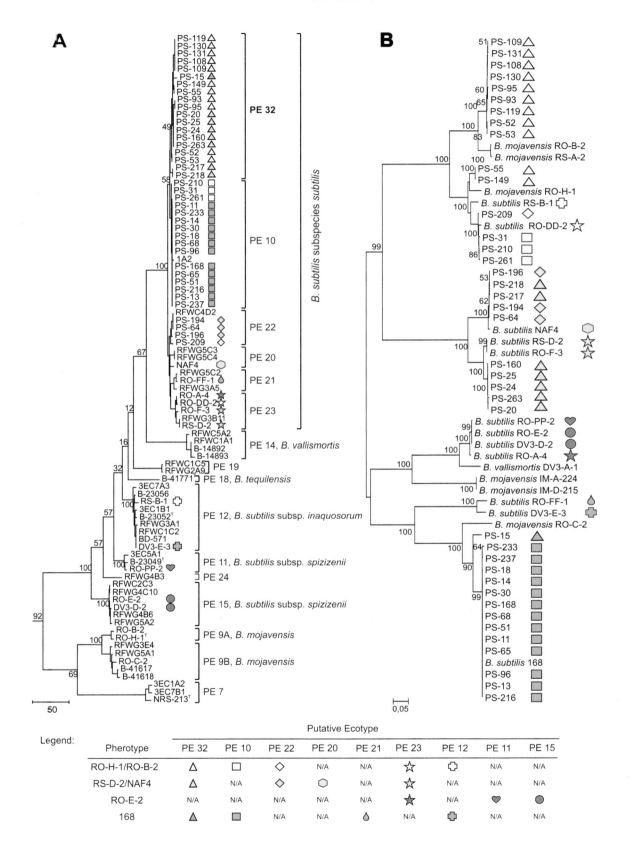

composition of soil *Bacillaceae*. Recently, an ecotype simulation algorithm (150) has been used to model the evolutionary dynamics of bacterial populations. *B. simplex* isolates obtained from the "Evolutionary Canyons" in Israel, with three major habitats: north-facing (European) slopes, south-facing (African) slopes, and the canyon bottom with largest access to water (151) were used to test the ecotype model (150). Ecotypes demarcated within the *B. simplex* collection showed a strong association with habitats of isolation (150). In agreement with this, isolates taxonomically classified as members of the *Bacillus subtilis–Bacillus licheniformis* clade sampled in the south and north slopes of Death Valley also diversified into several ecotypes, showing adaptation to solar exposure and soil texture (152). This suggests that temperature and soil type may be important environmental parameters that shape the ecological divergence of *Bacillus* ecotypes. These studies also provided evidence for the contention that ecotype clustering observed at the level of housekeeping genes and ecological distinctness may correlate.

Temperature also determines the distribution of the psychrotolerant *B. weihenstephanensis* and other *B. cereus sensu lato* representatives, which grow in the range 7 to 46°C. Von Stetten et al. (153) studied the distribution of about 1,000 mesophilic and psychrotolerant isolates obtained from tropical or temperate soil or two alpine habitats, with average annual temperatures of 28, 7, 4, and 1°C, respectively. Isolates were characterized phenotypically, in terms of their growth responses to temperature and psychrotolerance, as well as genotypically, at the level of 16S rRNA and *cspA* gene sequences. The proportions of psychrotolerant isolates in these four habitats were 0, 45, 86, and 98%, respectively, indicating strong temperature selection. Moreover, psychrotolerant and mesophilic strains exhibited growth at low or moderate temperature, respectively, and possessed psychrotolerant or mesophilic *cspA* genotypes.

Studies addressing ecotype diversity within *Bacillaceae* thus indicate a significant impact of environment (including solar exposure and soil type) on their formation, but information is still required to determine whether the variation between ecotypes is correlated with ecosystem function such as nutrient cycling or other functions such as plant protection, virulence, sporulation, development of competence, biofilm formation, and many other functions that promote survival or may involve biotic interactions.

Recently Stefanic et al. (19) tested whether quorum-sensing types of *B. subtilis* encoded by the *comQXPA* locus differ ecologically and whether they correlate with ecotypes defined by the ES model (19) (Fig. 3). *B. subtilis* strains and relatives encode a polymorphic quorum-sensing system involving the signal-processing enzyme ComQ, the signal ComX, the ComP receptor, and the response regulator ComA (19, 41, 154–159). Previous studies indicated that the *comQXP* loci show high intraspecies (within species) diversity (154). Similarity analysis of *comQXP* loci in desert and other *B. subtilis* strains indicated that they cluster into four distinct groups (19, 41, 156). Strains within the group produce similar peptide pheromones and have similar ComP receptors and are therefore able to induce quorum-sensing responses in each other. We can classify the strains that productively exchange signals as belonging to the same pherotype or communication group. In contrast, strains with divergent *comQXPA* loci (from different groups) cannot induce each other's QS response and are thus of different pherotypes (41, 154, 156, 160, 161). This exemplifies functional diversification with a potential ecological role. The study by Stefanic et al. (19) addresses the ecotype: pherotype correspondence using a collection of highly related *B. subtilis* strains that were isolated from two 1-cm^3 soil samples. Because of the relatively small size of these samples, it was assumed that *B. subtilis* isolates had been exposed to the same environmental conditions and therefore had had, at least theoretically,

Figure 3 Phylogenetic and ecotype simulation analyses of the *B. subtilis-B. mojavensis* subclade and minimum evolution tree of *com* sequences. (A) The phylogeny of *B. subtilis* isolates from riverbank microscale and desert soils is based on a maximum parsimony analysis of the recombination-free concatenation of *dnaJ*, *gyrA*, and *rpoB*, rooted by strain C-125 of *B. halodurans* (19). (B) Minimum evolution tree of *com* sequences (*comQ*, *comX*, and partial *comP* sequences, 1,402 bp) depicts four sequence clusters that correspond to previously identified pherotypes or communication groups within *B. subtilis-B. mojavensis* clade. Strains are marked with a shape representing their putative ecotype (PE) and by color representing pherotype (yellow, pherotype ROH1/RO-B-2; green, pherotype RS-D-2 /NAF4; orange, pherotype RO-E-2; and blue, pherotype 168). Unmarked strains were used as additional reference strains for tree construction (19).

a history of interactions. We refer to this collection (39 *B. subtilis* strains) as microscale strains. They diverged into three different phenotypes and three distinct ecotypes (19). The majority of strains that share the ecotype also have the same phenotype. This distribution suggested that ComX-mediated communication within ecotypes is preferred as opposed to communication between ecotypes and even other *Bacillus* species (like *B. licheniformis*). However, each ecotype also harbored a few strains representing other phenotypes (19). Based on these observations, it was hypothesized that phenotype diversity within ecotypes is driven by social interactions among strains within the ecotype. Why would this be the case? At high cell density, cells induce quorum sensing responses, which enable them to adapt to adverse environmental conditions by secreting public goods (extracellular enzymes, surfactins, antibiotics, surfactants) and develop genetic competence for transformation. These adaptive responses are costly but promote survival. If a member of this cooperative group obtained a different set of phenotype genes through horizontal gene transfer (HGT), this may confer adaptive value to the recipient organism in the next growth cycle. Then the recipient of another phenotype locus would be at low frequency and would therefore fail to induce the quorum-sensing response that requires high cell density and high concentration of the signaling molecule. However, it would still be capable of feasting on the public goods (e.g., extracellular enzymes, surfactants) produced by its ancestors. In the context of social evolution (162) the recipient of another phenotype will be a cheater and in social conflict with its ancestors. However, the competitive advantage will last only as long as its frequency increases and it is also forced to produce public goods. This hypothesis based on conflict also proposes the cycling of phenotypes within the ecotype (19) and is in agreement with the social conflict model proposed recently by Eldar (163). The model suggests that social conflict arises when the signal blind ComP mutant arises in the population of the QS wild-type cells. According to social evolution theory (162) the QS receptor mutant is a cheat (164), which, by remaining unresponsive to the signal, takes advantage of the QS wild-type cells that provide the QS-regulated public goods (extracellular enzymes, antibiotics, surfactants). By cheating, the signal blind mutant gains fitness advantage, rises in frequency, and may purge its cooperative ancestors from the population. However, the advantage of the signal blind mutants is presumably short lived because it may soon face conditions where QS is essential for survival. In this case, it is adaptive for the ComP receptor mutant

to obtain suppressive mutations in any one of the *comQXP* genes that can restore the QS response (163, 165). Therefore, this evolutionary game between cheating, frequency-dependent selection, and adaptation could be responsible for phenotype diversification under conditions that favor quorum sensing. Future experiments will show whether a different phenotype of the same ecotype is able to rise in frequency in the population dominated by another phenotype or whether the *comP* mutant has an advantage when surrounded by the wild-type cooperative cells. Our preliminary results suggest that the *comP* null mutant in competition with the wild type has an advantage (P. Stefanic and I. Mandic-Mulec, unpublished data) but it will be important to show that this holds also in conditions where QS is beneficial. It will also be interesting to test the coexistence and competition of these microscale *Bacillus* strains in soil microcosms under conditions where growth depends on quorum sensing. This would ultimately test the hypothesis of social conflict (19, 163) driving phenotype diversification. However, the observation that different *Bacillus* species such as *B. subtilis*, *B. mojavensis*, and *B. amyloliquefaciens* can share phenotypes (19, 41, 156) strongly supports the notion that HGT is also important in the distribution and evolution of phenotypes. Recently we showed that the *comQXPA* quorum-sensing genes are widespread within *Bacillaceae* and that polymorphism of this locus typical for *B. subtilis* clade is evident also in the non-*B. subtilis* clade, suggesting grossly similar evolutionary constraints in the underlying quorum-sensing systems (159, 166). Interestingly, the ComXPA QS system has built in a molecular mechanism, which by acting as a private link between signal and response in *B. subtilis* quorum sensing, may stabilize cooperative communication (167). A signal-deficient mutant showed lowered fitness in the presence of signal-producing wild-type ancestors because it overinvested into production of public goods (e.g., into production of lipopeptide antibiotic and surfactant surfactin) and became surfactin sensitive. While this punishment mechanism probably originally evolved to fine-tune the QS response, it also has implications for the stability of cooperation within a phenotype (167).

B. subtilis is one of the best-studied bacteria with respect to its molecular make-up (168). However, surprisingly little is known of its ecology and diversity in soil. Strains within the *B. subtilis* clade form two subclusters that represent two subspecies: *B. subtilis* subsp. *subtilis* and *B. subtilis* subsp. *spizizenii*. They show DNA relatedness of 58 to 69%, and differ in cell wall composition (169). Among these two subspecies, strains 168 and W23 have been most often used in research addressing various aspects of *B. subtilis* physiology (40, 169). The level of genetic diversity based on multilocus sequence typing (MLST) within subcluster W23 was found to be higher than that within 168, or in the closely related *B. mojavensis* cluster. The ecological significance of this diversity is not understood. However, microarray-based comparative genomic hybridization (M-CGH) (168) confirmed closer genomic tightness within the subclusters than between them. The level of gene sequence divergence within the species was 30%. Sequence diversity was highest for genes involved in the synthesis of secondary metabolites and of teichoic acids and for genes involved in the adaptive response to alkylating DNA damage. Recently, Earl et al. (170) published genome sequences of several closely related *Bacillus* strains. This study indicated strain-specific regions that were spread throughout the core genomic backbone. A majority of these variable sequences were smaller than 5 kb but some were also up to 100 kb in size. Many of them encoded genes involved in adaptive functions (e.g., antibiotic synthesis, competence, sporulation), which is in agreement with the high adaptive plasticity of these bacteria to diverse ecological niches.

The locus encoding the first three genes of the *comQXPA* quorum sensing system is one of the regions that is highly diverse in closely related strains of the *B. subtilis – licheniformis* group (154, 161). This diversity is present within organisms isolated from a small soil sample (1 cm^3 or even 0.25 cm^3) (41) as well as those that were isolated from soils separated by large geographical distances (156). The high heterogeneity of isolates living in close proximity may be due to a very high spatial heterogeneity of soil. Soil has a high surface-to-volume ratio and therefore provides a tremendous contact surface area for microorganisms to grow on. For example, 1 g of clay minerals, which are the smallest solid-phase component of soil and are classified as a size fraction of <2 μm in diameter, provide a surface of 93 to 800 m^2. Besides clay particles and organic matter, soils also contain the silt fraction (2 to 50 μm) and the sand fraction (50 to 2000 μm). The proportion of each fraction defines the soil texture, which influences the other two soil phases: the gaseous (soil-air) and the aqueous ones (soil-water or soil solution). The size of the soil pore increases with increased proportion of sand in soil and decreases with the increased content of clay. The soil texture affects water percolation and evaporation and influences the composition of microbial communities (171). It is believed that the soil matrix, due to its huge reactive surface, is

a habitat where geographic isolation and niche differentiation of microorganisms is already possible at small distances. Therefore, it is not that surprising to find a high diversity of *B. subtilis* in soil even at distances of 250 m (19) or below (172). The ecological and evolutionary principles, including the competition between bacteria for natural resources, acting at this scale are poorly understood and studies addressing this field have recently been reviewed by Vos et al. (173).

CHALLENGES AND GOALS

We may safely state that we currently have gathered only a rough understanding of the lifestyle of key members of the genus *Bacillus* in their natural habitat, e.g., the soil and the plant rhizosphere. Given the fact that members of *Bacillaceae* are mostly aerobic or facultatively anaerobic heterotrophic organisms with the capacity to show a rapid growth response to available organic carbon, one would expect a role of such bacteria in carbon-rich sites in nature, such as in the rhizospheres of actively exuding plants or in soil sites where organic matter is being degraded. In such sites, bacteria may become active and play roles in local processes. However, we know very little of *Bacillaceae* life dynamics in soil, because most of them are able to form spores and studies monitoring the activity of vegetative cells in soil are lacking. For example, we have no knowledge of how quorum-sensing processes operate in soil; how stable the signaling peptides are; whether the adaptive processes, such as competence development, sporulation, extracellular enzyme production, known to be controlled by quorum sensing at the laboratory conditions, are regulated in the same manner in soils. Even less is known about the temporal dynamics of these adaptive processes in soil and other natural environments. Are members of *Bacillaceae* mostly represented in soils by spores? Is vegetative growth a rare event? Do representatives of *Bacillaceae* actively compete and affect other members of the soil and rhizosphere community? How important are adaptive responses such as competence development or antibiotic production for the success of the *Bacillus* in soils and/or the rhizosphere? What is needed in future studies are direct observations, on the basis of sensitive tools from the "omics" area, of the differentiated cells and their constituents. On the methodological level, the focus should thus certainly be on the development of additional molecular tools to target such facets of *Bacillus* species in their natural habitat (e.g., soil). This may, for instance, boil down to an improvement of the methods for isolation of DNA/RNA from soil, followed by an investment in the development of sensitive tools for the detection of cell types as well as particular cellular constituents (e.g., mRNAs of different types, target proteins) that depict the make-up of the *Bacillus* population in its natural setting.

Independent studies reveal that members of the genus *Bacillus* are not usually the most abundant bacterial species in soil. Nonetheless, some *Bacillaceae* and, often, *Bacillus* species, are almost invariably present at levels of 10^6 to 10^8 per g of dry soil, which is between a thousandfold to a hundredfold less than the total bacterial density. Is the predicted role of *Bacillus* as a driver of soil organic matter mineralization in agreement with its abundance? Answers to these question call for development of *Bacillus*-specific molecular tools which will enable us to quantify the active members of the soil *Bacillus* community and also follow *in situ* their adaptive response, which have been extensively studied *in vitro*.

We here postulate that, in the light of the great diversity in ecological roles found across members of the *B. cereus* cluster, a similar great diversity of roles may be present across soil saprotrophic bacilli. This may extend into the plant rhizosphere, where such bacilli may have beneficial roles as a result of their activities as saprotrophs in this habitat. It is in this key environment, where interactions with the plant root take place that improving our understanding of *Bacillus* ecology will have the most impact. How can we promote the plant-beneficial functions exerted by such rhizospheric bacilli? A key need is to further our understanding of the plant root colonization and cell differentiation processes that presumably direct which *Bacillus* strain will be favored in the rhizosphere. It will be important to decipher cell-cell interactions within and among *Bacillus* populations *in vitro* and *in situ*. Placing such observations in the context of local conditions in the rhizosphere will ultimately pave the way to an ecology-guided strategy for the application of *Bacillus* biocontrol agents.

Citation. Mandic-Mulec I, Stefanic P, van Elsas JD. 2015. Ecology of *Bacillaceae*. Microbiol Spectrum 3(2):TBS-0017-2013.

References

1. **Vaishampayan P, Probst A, Krishnamurthi S, Ghosh S, Osman S, McDowall A, Ruckmani A, Mayilraj S, Venkateswaran K.** 2010. *Bacillus horneckiae* sp. nov., isolated from a spacecraft-assembly clean room. *Int J Syst Evol Microbiol* **60:**1031–1037.

2. **Seiler H, Wenning M, Scherer S.** 2013. *Domibacillus robiginosus* gen. nov., sp. nov., isolated from a pharmaceutical clean room. *Int J Syst Evol Microbiol* **63**(Pt 6): 2054–2061.

3. **Bottone EJ.** 2010. *Bacillus cereus*, a volatile human pathogen. *Clin Microbiol Rev* **23:**382–398.

4. **Marquez MC, Sanchez-Porro C, Ventosa A.** 2011. Hallophilic and haloakalophilic, aerobic endospore-forming bacteria in soil, p 309–339. *In* Logan NA, De Vos P (ed), *Endospore-Forming Soil Bacteria*. Springer, New York, NY.

5. **Schmidt TR, Scott EJ 2nd, Dyer DW.** 2011. Whole-genome phylogenies of the family *Bacillaceae* and expansion of the sigma factor gene family in the *Bacillus cereus* species-group. *BMC Genomics* **12:**430. doi:10.1186/1471-2164-12-430.

6. **Hoyles L, Honda H, Logan NA, Halket G, La Ragione RM, McCartney AL.** 2012. Recognition of greater diversity of *Bacillus* species and related bacteria in human faeces. *Res Microbiol* **163:**3–13.

7. **Carrasco IJ, Marquez MC, Yanfen X, Ma Y, Cowan DA, Jones BE, Grant WD, Ventosa A.** 2006. *Gracilibacillus orientalis* sp. nov., a novel moderately halophilic bacterium isolated from a salt lake in Inner Mongolia, China. *Int J Syst Evol Microbiol* **56:**599–604.

8. **Chen YG, Cui XL, Zhang YQ, Li WJ, Wang YX, Xu LH, Peng Q, Wen ML, Jiang CL.** 2008. *Gracilibacillus halophilus* sp. nov., a moderately halophilic bacterium isolated from saline soil. *Int J Syst Evol Microbiol* **58:** 2403–2408.

9. **Bischoff KM, Rooney AP, Li XL, Liu S, Hughes SR.** 2006. Purification and characterization of a family 5 endoglucanase from a moderately thermophilic strain of *Bacillus licheniformis*. *Biotechnol Lett* **28:**1761–1765.

10. **Margesin R, Schinner F.** 2001. Potential of halotolerant and halophilic microorganisms for biotechnology. *Extremophiles* **5:**73–83.

11. **Logan NA, Vos PD.** 2009. Family I. *Bacillaceae*. *In* Vos P, Garrity G, Jones D, Krieg NR, Ludwig W, Rainey FA, Schleifer K-H, Whitman W (ed), *Bergey's Manual of Systematic Bacteriology: Firmicutes*, 2nd ed, **vol 3.** Springer, New York, NY.

12. **Logan NA, Halket G.** 2011. Developments in the taxonomy of aerobic, endospore-forming bacteria, p 1–30. *In* Logan NA, De Vos P (ed), *Endospore-Forming Soil Bacteria*. Springer, New York, NY.

13. **Banat IM, Marchant R, Rahman TJ.** 2004. *Geobacillus debilis* sp. nov., a novel obligately thermophilic bacterium isolated from a cool soil environment, and reassignment of *Bacillus pallidus* to *Geobacillus pallidus* comb. nov. *Int J Syst Evol Microbiol* **54:**2197–2201.

14. **Minana-Galbis D, Pinzon DL, Loren JG, Manresa A, Oliart-Ros RM.** 2010. Reclassification of *Geobacillus pallidus* (Scholz et al. 1988) Banat et al. 2004 as *Aeribacillus pallidus* gen. nov., comb. nov. *Int J Syst Evol Microbiol* **60:**1600–1604.

15. **Zhou Y, Xu J, Xu L, Tindall BJ.** 2009. *Falsibacillus pallidus* to replace the homonym *Bacillus pallidus* Zhou et al. 2008. *Int J Syst Evol Microbiol* **59:**3176–3180.

16. **Cohn F.** 1962. Studies on the biology of the bacilli., p 49–56. *In* Dale T (ed), *Brock Milestones in Microbiology*. Prentice-Hall, Inc., Englewood Cliffs, NJ.

17. **Logan NA, Vos PD.** 2009. Genus *Bacillus* Cohn 1872, p 21–128. *In* Vos P, Garrity G, Jones D, Krieg NR, Ludwig W, Rainey FA, Schleifer K-H, Whitman W (ed), *Bergey's Manual of Systematic Bacteriology: Firmicutes*, 2nd ed, **vol 3**. Springer, New York, NY.

18. **Wang LT, Lee FL, Tai CJ, Kasai H.** 2007. Comparison of *gyrB* gene sequences, 16S rRNA gene sequences and DNA-DNA hybridization in the *Bacillus subtilis* group. *Int J Syst Evol Microbiol* **57:**1846–1850.

19. **Stefanic P, Decorosi F, Viti C, Petito J, Cohan FM, Mandic-Mulec I.** 2012. The quorum sensing diversity within and between ecotypes of *Bacillus subtilis*. *Environ Microbiol* **14:**1378–1389.

20. **Maughan H, Van der Auwera G.** 2011. *Bacillus* taxonomy in the genomic era finds phenotypes to be essential though often misleading. *Infect Genet Evol* **11:**789–797.

21. **Priest FG, Barker M, Baillie LW, Holmes EC, Maiden MC.** 2004. Population structure and evolution of the *Bacillus cereus* group. *J Bacteriol* **186:**7959–7970.

22. **Daffonchio D, Cherif A, Brusetti L, Rizzi A, Mora D, Boudabous A, Borin S.** 2003. Nature of polymorphisms in 16S-23S rRNA gene intergenic transcribed spacer fingerprinting of *Bacillus* and related genera. *Appl Environ Microbiol* **69:**5128–5137.

23. **Helgason E, Okstad OA, Caugant DA, Johansen HA, Fouet A, Mock M, Hegna I, Kolsto AB.** 2000. *Bacillus anthracis*, *Bacillus cereus*, and *Bacillus thuringiensis*–one species on the basis of genetic evidence. *Appl Environ Microbiol* **66:**2627–2630.

24. **Tourasse NJ, Kolsto AB.** 2008. SuperCAT: a supertree database for combined and integrative multilocus sequence typing analysis of the *Bacillus cereus* group of bacteria (including *B. cereus*, *B. anthracis* and *B. thuringiensis*). *Nucleic Acids Res* **36:**D461–D468.

25. **Rasko DA, Altherr MR, Han CS, Ravel J.** 2005. Genomics of the *Bacillus cereus* group of organisms. *FEMS Microbiol Rev* **29:**303–329.

26. **Rasko DA, Rosovitz MJ, Okstad OA, Fouts DE, Jiang L, Cer RZ, Kolsto AB, Gill SR, Ravel J.** 2007. Complete sequence analysis of novel plasmids from emetic and periodontal *Bacillus cereus* isolates reveals a common evolutionary history among the *B. cereus*-group plasmids, including *Bacillus anthracis* pXO1. *J Bacteriol* **189:**52–64.

27. **Ehling-Schulz M, Fricker M, Grallert H, Rieck P, Wagner M, Scherer S.** 2006. Cereulide synthetase gene cluster from emetic *Bacillus cereus*: structure and location on a mega virulence plasmid related to *Bacillus anthracis* toxin plasmid pXO1. *BMC Microbiol* **6:**20. doi:10.1186/1471-2180-6-20.

28. **Van der Auwera GA, Andrup L, Mahillon J.** 2005. Conjugative plasmid pAW63 brings new insights into the genesis of the *Bacillus anthracis* virulence plasmid pXO2 and of the *Bacillus thuringiensis* plasmid pBT9727. *BMC Genomics* **6:**103. doi:10.1186/1471-2164-6-103.

29. **Zwick ME, Joseph SJ, Didelot X, Chen PE, Bishop-Lilly KA, Stewart AC, Willner K, Nolan N, Lentz S, Thomason MK, Sozhamannan S, Mateczun AJ, Du L,** Read TD. 2012. Genomic characterization of the *Bacillus cereus* sensu lato species: backdrop to the evolution of *Bacillus anthracis*. *Genome Res* **22:**1512–1524.

30. **Setlow P.** 2006. Spores of *Bacillus subtilis*: their resistance to and killing by radiation, heat and chemicals. *J Appl Microbiol* **101:**514–525.

31. **Baril E, Coroller L, Couvert O, El Jabri M, Leguerinel I, Postollec F, Boulais C, Carlin F, Mafart P.** 2012. Sporulation boundaries and spore formation kinetics of *Bacillus* spp. as a function of temperature, pH and a(w). *Food Microbiol* **32:**79–86.

32. **Van Ness GB.** 1971. Ecology of anthrax. *Science* **172:** 1303–1307.

33. **Hugh-Jones M, Blackburn J.** 2009. The ecology of *Bacillus anthracis*. *Mol Aspects Med* **30:**356–367.

34. **Kellogg CA, Griffin DW.** 2006. Aerobiology and the global transport of desert dust. *Trends Ecol Evol* **21:** 638–644.

35. **Smith DJ, Timonen HJ, Jaffe DA, Griffin DW, Birmele MN, Perry KD, Ward PD, Roberts MS.** 2013. Intercontinental dispersal of bacteria and archaea by transpacific winds. *Appl Environ Microbiol* **79:**1134–1139.

36. **Smith DJ, Jaffe DA, Birmele MN, Griffin DW, Schuerger AC, Hee J, Roberts MS.** 2012. Free tropospheric transport of microorganisms from Asia to North America. *Microb Ecol* **64:**973–985.

37. **Horneck G, Rettberg P, Reitz G, Wehner J, Eschweiler U, Strauch K, Panitz C, Starke V, Baumstark-Khan C.** 2001. Protection of bacterial spores in space, a contribution to the discussion on Panspermia. *Orig Life Evol Biosph* **31:**527–547.

38. **Fajardo-Cavazos P, Schuerger AC, Nicholson WL.** 2007. Testing interplanetary transfer of bacteria between Earth and Mars as a result of natural impact phenomena and human spaceflight activities. *Acta Astronautica* **60:**534–540.

39. **Martin PA, Travers RS.** 1989. Worldwide abundance and distribution of *Bacillus thuringiensis* isolates. *Appl Environ Microbiol* **55:**2437–2442.

40. **Roberts MS, Cohan FM.** 1995. Recombination and migration rates in natural populations of *Bacillus subtilis* and *Bacillus mojavensis*. *Evolution* **49:**1081–1094.

41. **Stefanic P, Mandic-Mulec I.** 2009. Social interactions and distribution of *Bacillus subtilis* phenotypes at microscale. *J Bacteriol* **191:**1756–1764.

42. **Hudson AJ, Daniel RM, Morgan HW.** 1988. Isolation of a strain of *Bacillus schlegelii* from geothermally heated antarctic soil. *FEMS Microbiol Lett* **51:**57–60.

43. **Kampfer P, Glaeser SP, Busse HJ.** 2013. Transfer of *Bacillus schlegelii* to a novel genus and proposal of *Hydrogenibacillus schlegelii* gen. nov. comb. nov. *Int J Syst Evol Microbiol* **63**(Pt 5):1723–1727.

44. **Ivanova N, Sorokin A, Anderson I, Galleron N, Candelon B, Kapatral V, Bhattacharyya A, Reznik G, Mikhailova N, Lapidus A, Chu L, Mazur M, Goltsman E, Larsen N, D'Souza M, Walunas T, Grechkin Y, Pusch G, Haselkorn R, Fonstein M, Ehrlich SD, Overbeek R, Kyrpides N.** 2003. Genome sequence of *Bacillus cereus* and comparative analysis with *Bacillus anthracis*. *Nature* **423:**87–91.

45. Chattopadhyay A, Bhatnagar NB, Bhatnagar R. 2004. Bacterial insecticidal toxins. *Crit Rev Microbiol* **30**: 33–54.

46. Choudhary DK, Johri BN. 2009. Interactions of *Bacillus* spp. and plants–with special reference to induced systemic resistance (ISR). *Microbiol Res* **164**:493–513.

47. Waksman SA. 1932. *Principles of Soil Microbiology*, 2nd ed. The Williams & Wilkins Co., Baltimore, MD.

48. Siala A, Hill IR, Gray TRG. 1974. Populations of spore-forming bacteria in an acid forest soil, with special reference to *Bacillus subtilis*. *J Gen Microbiol* **81**:183–190.

49. Toljander JF, Artursson V, Paul LR, Jansson JK, Finlay RD. 2006. Attachment of different soil bacteria to arbuscular mycorrhizal fungal extraradical hyphae is determined by hyphal vitality and fungal species. *FEMS Microbiol Lett* **254**:34–40.

50. Tam NK, Uyen NQ, Hong HA, Duc le H, Hoa TT, Serra CR, Henriques AO, Cutting SM. 2006. The intestinal life cycle of *Bacillus subtilis* and close relatives. *J Bacteriol* **188**:2692–2700.

51. Soriano M, Diaz P, Pastor FI. 2005. Pectinolytic systems of two aerobic sporogenous bacterial strains with high activity on pectin. *Curr Microbiol* **50**:114–118.

52. Ouattara HG, Reverchon S, Niamke SL, Nasser W. 2011. Molecular identification and pectate lyase production by *Bacillus strains* involved in cocoa fermentation. *Food Microbiol* **28**:1–8.

53. Soares FL Jr, Melo IS, Dias AC, Andreote FD. 2012. Cellulolytic bacteria from soils in harsh environments. *World J Microbiol Biotechnol* **28**:2195–2203.

54. Okeke EC, Lu J. 2011. Characterization of a defined cellulolytic and xylanolytic bacterial consortium for bioprocessing of cellulose and hemicelluloses. *Appl Biochem Biotechnol* **163**:869–881.

55. Maki ML, Idrees A, Leung KT, Qin W. 2012. Newly isolated and characterized bacteria with great application potential for decomposition of lignocellulosic biomass. *J Mol Microbiol Biotechnol* **22**:156–166.

56. Solanki MK, Robert AS, Singh RK, Kumar S, Pandey AK, Srivastava AK, Arora DK. 2012. Characterization of mycolytic enzymes of *Bacillus strains* and their bioprotection role against *Rhizoctonia solani* in tomato. *Curr Microbiol* **65**:330–336.

57. Gomaa EZ. 2012. Chitinase production by *Bacillus thuringiensis* and *Bacillus licheniformis*: their potential in antifungal biocontrol. *J Microbiol* **50**:103–111.

58. Xiao L, Xie CC, Cai J, Lin ZJ, Chen YH. 2009. Identification and characterization of a chitinase-produced bacillus showing significant antifungal activity. *Curr Microbiol* **58**:528–533.

59. Pleban S, Chernin L, Chet I. 1997. Chitinolytic activity of an endophytic strain of *Bacillus cereus*. *Lett Appl Microbiol* **25**:284–288.

60. Hallmann J, Rodriguez-Kábana R, Kloepper JW. 1999. Chitin-mediated changes in bacterial communities of the soil, rhizosphere and within roots of cotton in relation to nematode control. *Soil Biol Biochem* **31**: 551–560.

61. Sakurai M, Suzuki K, Onodera M, Shinano T, Osaki M. 2007. Analysis of bacterial communities in soil by PCR–DGGE targeting protease genes. *Soil Biol Biochem* **39**:2777–2784.

62. Chu H, Lin X, Fujii T, Morimoto S, Yagi K, Hu J, Zhang J. 2007. Soil microbial biomass, dehydrogenase activity, bacterial community structure in response to long-term fertilizer management. *Soil Biol Biochem* **39**: 2971–2976.

63. Vilain S, Luo Y, Hildreth MB, Brozel VS. 2006. Analysis of the life cycle of the soil saprophyte *Bacillus cereus* in liquid soil extract and in soil. *Appl Environ Microbiol* **72**:4970–4977.

64. von der Weid I, Duarte GF, van Elsas JD, Seldin L. 2002. *Paenibacillus brasilensis* sp. nov., a novel nitrogen-fixing species isolated from the maize rhizosphere in Brazil. *Int J Syst Evol Microbiol* **52**:2147–2153.

65. Verbaendert I, Boon N, De Vos P, Heylen K. 2011. Denitrification is a common feature among members of the genus *Bacillus*. *Syst Appl Microbiol* **34**:385–391.

66. Ding Y, Wang J, Liu Y, Chen S. 2005. Isolation and identification of nitrogen-fixing bacilli from plant rhizospheres in Beijing region. *J Appl Microbiol* **99**:1271–1281.

67. Park SJ, Yoon JC, Shin KS, Kim EH, Yim S, Cho YJ, Sung GM, Lee DG, Kim SB, Lee DU, Woo SH, Koopman B. 2007. Dominance of endospore-forming bacteria on a Rotating Activated *Bacillus* Contactor biofilm for advanced wastewater treatment. *J Microbiol* **45**:113–121.

68. Philippot L, Hallin S, Schloter M. 2007. Ecology of denitrifying prokaryotes in agricultural soil, p 249–305. *In* Donald LS (ed), *Advances in Agronomy*, **vol 96**. Elsevier, New York, NY.

69. Heylen K, Gevers D, Vanparys B, Wittebolle L, Geets J, Boon N, De Vos P. 2006. The incidence of *nirS* and *nirK* and their genetic heterogeneity in cultivated denitrifiers. *Environ Microbiol* **8**:2012–2021.

70. Albino U, Saridakis DP, Ferreira MC, Hungria M, Vinuesa P, Andrade G. 2006. High diversity of diazotrophic bacteria associated with the carnivorous plant *Drosera villosa* var. *villosa* growing in oligotrophic habitats in Brazil. *Plant Soil* **287**:199–207.

71. Beneduzi A, Peres D, da Costa PB, Bodanese Zanettini MH, Passaglia LM. 2008. Genetic and phenotypic diversity of plant-growth-promoting bacilli isolated from wheat fields in southern Brazil. *Res Microbiol* **159**: 244–250.

72. Beneduzi A, Peres D, Vargas LK, Bodanese-Zanettini MH, Passaglia LMP. 2008. Evaluation of genetic diversity and plant growth promoting activities of nitrogen-fixing bacilli isolated from rice fields in South Brazil. *Appl Soil Ecol* **39**:311–320.

73. Różycki H, Dahm H, Strzelczyk E, Li CY. 1999. Diazotrophic bacteria in root-free soil and in the root zone of pine (*Pinus sylvestris* L.) and oak (*Quercus robur* L.). *Appl Soil Ecol* **12**:239–250.

74. Achouak W, Normand P, Heulin T. 1999. Comparative phylogeny of *rrs* and *nifH* genes in the *Bacillaceae*. *Int J Syst Bacteriol* **49**(Pt 3):961–967.

75. **Rodríguez H, Fraga R.** 1999. Phosphate solubilizing bacteria and their role in plant growth promotion. *Biotechnol Adv* **17:**319–339.

76. **Mamta, Rahi P, Pathania V, Gulati A, Singh B, Bhanwra RK, Tewari R.** 2010. Stimulatory effect of phosphate-solubilizing bacteria on plant growth, stevioside and rebaudioside-A contents of Stevia rebaudiana Bertoni. *Appl Soil Ecol* **46:**222–229.

77. **Whitelaw MA.** 1999. Growth promotion of plants inoculated with phosphate-solubilizing fungi, p 99–151. *In* Donald LS (ed), *Advances in Agronomy*, **vol 69.** Academic Press, San Diego, CA.

78. **Keneni A, Assefa F, Prabu PC.** 2010. Isolation of phosphate solubilizing bacteria from the rhizosphere of faba bean of Ethiopia and their abilities on solubilizing insoluble phosphates. *J Agric Sci Technol* **12:**79–89.

79. **Oliveira CA, Alves VMC, Marriel IE, Gomes EA, Scotti MR, Carneiro NP, Guimarães CT, Schaffert RE, Sá NMH.** 2009. Phosphate solubilizing microorganisms isolated from rhizosphere of maize cultivated in an oxisol of the Brazilian Cerrado Biome. *Soil Biol Biochem* **41:**1782–1787.

80. **Chen YP, Rekha PD, Arun AB, Shen FT, Lai WA, Young CC.** 2006. Phosphate solubilizing bacteria from subtropical soil and their tricalcium phosphate solubilizing abilities. *Appl Soil Ecol* **34:**33–41.

81. **Sandeep C, Venkat Raman R, Radhika M, Thejas MS, Patra S, Gowda T, Suresh CK, Mulla SR.** 2011. Effect of inoculation of *Bacillus megaterium* isolates on growth, biomass and nutrient content of Peppermint. *J Phytol* **3:**19–24.

82. **Pal SS.** 1998. Interactions of an acid tolerant strain of phosphate solubilizing bacteria with a few acid tolerant crops. *Plant Soil* **198:**169–177.

83. **Kloepper JW, Lifshitz R, Zablotowicz RM.** 1989. Free-living bacterial inocula for enhancing crop productivity. *Trends Biotechnol* **7:**39–44.

84. **Bhattacharyya PN, Jha DK.** 2012. Plant growth-promoting rhizobacteria (PGPR): emergence in agriculture. *World J Microbiol Biotechnol* **28:**1327–1350.

85. **Hernandez J-P, de-Bashan LE, Rodriguez DJ, Rodriguez Y, Bashan Y.** 2009. Growth promotion of the freshwater microalga *Chlorella vulgaris* by the nitrogen-fixing, plant growth-promoting bacterium *Bacillus pumilus* from arid zone soils. *Eur J Soil Biol* **45:**88–93.

86. **Gutiérrez-Mañero FJ, Ramos-Solano B, Probanza A, Mehouachi J, Tadeo FR, Talon M.** 2001. The plant-growth-promoting rhizobacteria *Bacillus pumilus* and *Bacillus licheniformis* produce high amounts of physiologically active gibberellins. *Physiol Plant* **111:**206–211.

87. **Sharma SK, Sharma MP, Ramesh A, Joshi OP.** 2012. Characterization of zinc-solubilizing *Bacillus* isolates and their potential to influence zinc assimilation in soybean seeds. *J Microbiol Biotechnol* **22:**352–359.

88. **Belimov AA, Safronova VI, Sergeyeva TA, Egorova TN, Matveyeva VA, Tsyganov VE, Borisov AY, Tikhonovich IA, Kluge C, Preisfeld A, Dietz KJ, Stepanok VV.** 2001. Characterization of plant growth promoting rhizobacteria isolated from polluted soils and containing 1-aminocyclopropane-1-carboxylate deaminase. *Can J Microbiol* **47:**642–652.

89. **Zhang H, Sun Y, Xie X, Kim MS, Dowd SE, Pare PW.** 2009. A soil bacterium regulates plant acquisition of iron via deficiency-inducible mechanisms. *Plant J* **58:**568–577.

90. **Li D-M, Alexander M.** 1988. Co-inoculation with antibiotic-producing bacteria to increase colonization and nodulation by rhizobia. *Plant Soil* **108:**211–219.

91. **Kloepper JW, Ryu CM, Zhang S.** 2004. Induced systemic resistance and promotion of plant growth by *Bacillus* spp. *Phytopathology* **94:**1259–1266.

92. **Ryu CM, Murphy JF, Mysore KS, Kloepper JW.** 2004. Plant growth-promoting rhizobacteria systemically protect *Arabidopsis thaliana* against Cucumber mosaic virus by a salicylic acid and NPR1-independent and jasmonic acid-dependent signaling pathway. *Plant J* **39:**381–392.

93. **Banerjee MR, Yesmin L, Vessey JK.** 2006. Plant-growth-promoting Rhizobacteria as biofertilizers and biopesticides, p 137–233. *In* Rai MK (ed), *Handbook of Microbial Biofertilizers*. The Haworth Press, Binghampton, NY.

94. **Ongena M, Jacques P.** 2008. *Bacillus* lipopeptides: versatile weapons for plant disease biocontrol. *Trends Microbiol* **16:**115–125.

95. **Valenzuela-Soto JH, Estrada-Hernandez MG, Ibarra-Laclette E, Delano-Frier JP.** 2010. Inoculation of tomato plants (*Solanum lycopersicum*) with growth-promoting *Bacillus subtilis* retards whitefly *Bemisia tabaci* development. *Planta* **231:**397–410.

96. **Vlamakis H, Chai Y, Beauregard P, Losick R, Kolter R.** 2013. Sticking together: building a biofilm the *Bacillus subtilis* way. *Nat Rev Microbiol* **11:**157–168.

97. **Chen Y, Yan F, Chai Y, Liu H, Kolter R, Losick R, Guo JH.** 2013. Biocontrol of tomato wilt disease by *Bacillus subtilis* isolates from natural environments depends on conserved genes mediating biofilm formation. *Environ Microbiol* **15:**848–864.

98. **Beauregard PB, Chai Y, Vlamakis H, Losick R, Kolter R.** 2013. *Bacillus subtilis* biofilm induction by plant polysaccharides. *Proc Natl Acad Sci USA* **110:**E1621–E1630.

99. **Bais HP, Fall R, Vivanco JM.** 2004. Biocontrol of *Bacillus subtilis* against infection of *Arabidopsis* roots by *Pseudomonas syringae* is facilitated by biofilm formation and surfactin production. *Plant Physiol* **134:**307–319.

100. **Debois D, Jourdan E, Smargiasso N, Thonart P, De Pauw E, Ongena M.** 2014. Spatiotemporal monitoring of the antibiome secreted by *Bacillus biofilms* on plant roots using MALDI mass spectrometry imaging. *Anal Chem* **86:**4431–4438.

101. **Oslizlo A, Stefanic P, Vatovec S, Beigot Glaser S, Rupnik M, Mandic-Mulec I.** Exploring ComQXPA quorum sensing diversity and biocontrol potential of *Bacillus* spp. isolates from tomato rhizoplane. *Microb Biotechnol*, in press.

102. **Dogsa I, Brloznik M, Stopar D, Mandic-Mulec I.** 2013. Exopolymer diversity and the role of levan in *Bacillus subtilis* biofilms. *PLoS One* **8:**e62044. doi:10.1371/journal.pone.0062044.

103. Ongena M, Jourdan E, Adam A, Paquot M, Brans A, Joris B, Arpigny JL, Thonart P. 2007. Surfactin and fengycin lipopeptides of *Bacillus subtilis* as elicitors of induced systemic resistance in plants. *Environ Microbiol* 9:1084–1090.

104. Lakshmanan V, Bais HP. 2013. Factors other than root secreted malic acid that contributes toward *Bacillus subtilis* FB17 colonization on *Arabidopsis* roots. *Plant Signal Behav* 8:e27277. doi:10.4161/psb.27277.

105. Mathew GM, Ju YM, Lai CY, Mathew DC, Huang CC. 2012. Microbial community analysis in the termite gut and fungus comb of *Odontotermes formosanus*: the implication of *Bacillus* as mutualists. *FEMS Microbiol Ecol* 79:504–517.

106. Anand AA, Vennison SJ, Sankar SG, Prabhu DI, Vasan PT, Raghuraman T, Geoffrey CJ, Vendan SE. 2010. Isolation and characterization of bacteria from the gut of *Bombyx mori* that degrade cellulose, xylan, pectin and starch and their impact on digestion. *J Insect Sci* 10:107. doi:10.1673/031.010.10701.

107. Visotto LE, Oliveira MG, Ribon AO, Mares-Guia TR, Guedes RN. 2009. Characterization and identification of proteolytic bacteria from the gut of the velvetbean caterpillar (Lepidoptera: Noctuidae). *Environ Entomol* 38:1078–1085.

108. Konig H. 2006. *Bacillus* species in the intestine of termites and other soil invertebrates. *J Appl Microbiol* 101:620–627.

109. Wenzel M, Schonig I, Berchtold M, Kampfer P, Konig H. 2002. Aerobic and facultatively anaerobic cellulolytic bacteria from the gut of the termite *Zootermopsis angusticollis*. *J Appl Microbiol* 92:32–40.

110. Gordon RE, Haynes WC, Pang H-NC. 1973. *The Genus* Bacillus. U.S. Department of Agriculture, Washington DC.

111. Roberts MS, Nakamura LK, Cohan FM. 1994. *Bacillus mojavensis* sp. nov., distinguishable from *Bacillus subtilis* by sexual isolation, divergence in DNA sequence, and differences in fatty acid composition. *Int J Syst Bacteriol* 44:256–264.

112. Sorokin A, Candelon B, Guilloux K, Galleron N, Wackerow-Kouzova N, Ehrlich SD, Bourguet D, Sanchis V. 2006. Multiple-locus sequence typing analysis of *Bacillus cereus* and *Bacillus thuringiensis* reveals separate clustering and a distinct population structure of psychrotrophic strains. *Appl Environ Microbiol* 72:1569–1578.

113. Bizzarri MF, Prabhakar A, Bishop AH. 2008. Multiple-locus sequence typing analysis of *Bacillus thuringiensis* recovered from the phylloplane of clover (*Trifolium hybridum*) in vegetative form. *Microb Ecol* 55:619–625.

114. Meintanis C, Chalkou KI, Kormas KA, Lymperopoulou DS, Katsifas EA, Hatzinikolaou DG, Karagouni AD. 2008. Application of *rpoB* sequence similarity analysis, REP-PCR and BOX-PCR for the differentiation of species within the genus *Geobacillus*. *Lett Appl Microbiol* 46:395–401.

115. Heyrman J, Mergaert J, Denys R, Swings J. 1999. The use of fatty acid methyl ester analysis (FAME) for the identification of heterotrophic bacteria present on three mural paintings showing severe damage by microorganisms. *FEMS Microbiol Lett* 181:55–62.

116. Hotta Y, Sato J, Sato H, Hosoda A, Tamura H. 2011. Classification of the genus *Bacillus* based on MALDI-TOF MS analysis of ribosomal proteins coded in S10 and spc operons. *J Agric Food Chem* 59:5222–5230.

117. Balazova T, Sedo O, Stefanic P, Mandic-Mulec I, Vos M, Zdrahal Z. 2014. Improvement in *Staphylococcus* and *Bacillus* strain differentiation by matrix-assisted laser desorption/ionization time-of-flight mass spectrometry profiling by using microwave-assisted enzymatic digestion. *Rapid Commun Mass Spectrom* 28:185–1861.

118. Prosser JI. 2012. Ecosystem processes and interactions in a morass of diversity. *FEMS Microbiol Ecol* 81:507–519.

119. Roesch LF, Fulthorpe RR, Riva A, Casella G, Hadwin AK, Kent AD, Daroub SH, Camargo FA, Farmerie WG, Triplett EW. 2007. Pyrosequencing enumerates and contrasts soil microbial diversity. *ISME J* 1:283–290.

120. Garbeva P, van Veen JA, van Elsas JD. 2003. Predominant *Bacillus* spp. in Agricultural soil under different management regimes detected via PCR-DGGE. *Microb Ecol* 45:302–316.

121. Felske A, Wolterink A, Van Lis R, De Vos WM, Akkermans AD. 2000. Response of a soil bacterial community to grassland succession as monitored by 16S rRNA levels of the predominant ribotypes. *Appl Environ Microbiol* 66:3998–4003.

122. Prosser JI, Jansson JK, Liu W-T. 2010. Nucleic-acid-based characterization of community structure and function, p 65–88. *In* Jansson JK, Liu W-T (ed), *Environmental Molecular Microbiology*. Horizon Scientific, Norfolk, UK.

123. Kraigher B, Stres B, Hacin J, Ausec L, Mahne I, van Elsas JD, Mandic-Mulec I. 2006. Microbial activity and community structure in two drained fen soils in the Ljubljana Marsh. *Soil Biol Biochem* 38:2762–2771.

124. Felske AD, Heyrman J, Balcaen A, de Vos P. 2003. Multiplex PCR screening of soil isolates for novel *Bacillus*-related lineages. *J Microbiol Methods* 55:447–458.

125. Janssen PH. 2006. Identifying the dominant soil bacterial taxa in libraries of 16S rRNA and 16S rRNA genes. *Appl Environ Microbiol* 72:1719–1728.

126. Brons JK, van Elsas JD. 2008. Analysis of bacterial communities in soil by use of denaturing gradient gel electrophoresis and clone libraries, as influenced by different reverse primers. *Appl Environ Microbiol* 74:2717–2727.

127. Felske A, Wolterink A, Van Lis R, Akkermans AD. 1998. Phylogeny of the main bacterial 16S rRNA sequences in Drentse A grassland soils (The Netherlands). *Appl Environ Microbiol* 64:871–879.

128. Benizri E, Baudoin E, Guckert A. 2001. Root colonization by inoculated plant growth-promoting rhizobacteria. *Biocontrol Sci Technol* 11:557–574.

129. Rudrappa T, Czymmek KJ, Pare PW, Bais HP. 2008. Root-secreted malic acid recruits beneficial soil bacteria. *Plant Physiol* 148:1547–1556.

130. Zeller SL, Brandl H, Schmid B. 2007. Host-plant selectivity of rhizobacteria in a crop/weed model system. *PLoS One* 2:e846. doi:10.1371/journal.pone.0000846.

131. Kremer RJ, Souissi T. 2001. Cyanide production by rhizobacteria and potential for suppression of weed seedling growth. *Curr Microbiol* 43:182–186.

132. Åström B, Gerhardson B. 1988. Differential reactions of wheat and pea genotypes to root inoculation with growth-affecting rhizosphere bacteria. *Plant Soil* 109: 263–269.

133. Long HH, Schmidt DD, Baldwin IT. 2008. Native bacterial endophytes promote host growth in a species-specific manner; phytohormone manipulations do not result in common growth responses. *PLoS One* 3: e2702. doi:10.1371/journal.pone.0002702.

134. Berg G, Roskot N, Steidle A, Eberl L, Zock A, Smalla K. 2002. Plant-dependent genotypic and phenotypic diversity of antagonistic rhizobacteria isolated from different *Verticillium* host plants. *Appl Environ Microbiol* 68:3328–3338.

135. da Silva KR, Salles JF, Seldin L, van Elsas JD. 2003. Application of a novel *Paenibacillus*-specific PCR-DGGE method and sequence analysis to assess the diversity of *Paenibacillus* spp. in the maize rhizosphere. *J Microbiol Methods* 54:213–231.

136. Smalla K, Wieland G, Buchner A, Zock A, Parzy J, Kaiser S, Roskot N, Heuer H, Berg G. 2001. Bulk and rhizosphere soil bacterial communities studied by denaturing gradient gel electrophoresis: plant-dependent enrichment and seasonal shifts revealed. *Appl Environ Microbiol* 67:4742–4751.

137. Garbeva P, van Elsas J, van Veen J. 2008. Rhizosphere microbial community and its response to plant species and soil history. *Plant Soil* 302:19–32.

138. Inceoglu O, Al-Soud WA, Salles JF, Semenov AV, van Elsas JD. 2011. Comparative analysis of bacterial communities in a potato field as determined by pyrosequencing. *PLoS One* 6:e23321. doi:10.1371/journal.pone.0023321.

139. Inceoglu O, Falcao Salles J, van Elsas JD. 2012. Soil and cultivar type shape the bacterial community in the potato rhizosphere. *Microb Ecol* 63:460–470.

140. Inceoglu O, Salles JF, van Overbeek L, van Elsas JD. 2010. Effects of plant genotype and growth stage on the betaproteobacterial communities associated with different potato cultivars in two fields. *Appl Environ Microbiol* 76:3675–3684.

141. Fan B, Borriss R, Bleiss W, Wu X. 2012. Gram-positive rhizobacterium *Bacillus amyloliquefaciens* FZB42 colonizes three types of plants in different patterns. *J Microbiol* 50:38–44.

142. Chowdhury SP, Dietel K, Randler M, Schmid M, Junge H, Borriss R, Hartmann A, Grosch R. 2013. Effects of *Bacillus amyloliquefaciens* FZB42 on lettuce growth and health under pathogen pressure and its impact on the rhizosphere bacterial community. *PLoS One* 8: e68818. doi:10.1371/journal.pone.0068818.

143. Horner-Devine MC, Carney KM, Bohannan BJ. 2004. An ecological perspective on bacterial biodiversity. *Proc Biol Sci* 271:113–122.

144. Ramette A, Tiedje JM. 2007. Biogeography: an emerging cornerstone for understanding prokaryotic diversity, ecology, and evolution. *Microb Ecol* 53:197–207.

145. Cohan FM, Perry EB. 2007. A systematics for discovering the fundamental units of bacterial diversity. *Curr Biol* 17:R373–R386.

146. Cohan FM. 2006. Towards a conceptual and operational union of bacterial systematics, ecology, and evolution. *Philos Trans R Soc Lond B Biol Sci* 361: 1985–1996.

147. Cohan FM. 2002. What are bacterial species? *Annu Rev Microbiol* 56:457–487.

148. Smith JM, Szathmáry E. 1993. The origin of chromosomes I. selection for linkage. *J Theor Biol* 164: 437–446.

149. Dubnau D, Smith I, Morell P, Marmur J. 1965. Gene conservation in *Bacillus* species. I. Conserved genetic and nucleic acid base sequence homologies. *Proc Natl Acad Sci USA* 54:491–498.

150. Koeppel A, Perry EB, Sikorski J, Krizanc D, Warner A, Ward DM, Rooney AP, Brambilla E, Connor N, Ratcliff RM, Nevo E, Cohan FM. 2008. Identifying the fundamental units of bacterial diversity: a paradigm shift to incorporate ecology into bacterial systematics. *Proc Natl Acad Sci USA* 105:2504–2509.

151. Nevo E. 1995. Asian, African and European biota meet at 'Evolution Canyon' Israel: local tests of global biodiversity and genetic diversity patterns. *Proc Biol Sci* 262: 149–155.

152. Connor N, Sikorski J, Rooney AP, Kopac S, Koeppel AF, Burger A, Cole SG, Perry EB, Krizanc D, Field NC, Slaton M, Cohan FM. 2010. Ecology of speciation in the genus *Bacillus*. *Appl Environ Microbiol* 76:1349–1358.

153. von Stetten F, Mayr R, Scherer S. 1999. Climatic influence on mesophilic *Bacillus cereus* and psychrotolerant *Bacillus weihenstephanensis* populations in tropical, temperate and alpine soil. *Environ Microbiol* 1:503–515.

154. Tortosa P, Logsdon L, Kraigher B, Itoh Y, Mandic Mulec I, Dubnau D. 2001. Specificity and genetic polymorphism of the *Bacillus* competence quorum sensing system. *J Bacteriol* 183:451–460.

155. Ansaldi M, Dubnau D. 2004. Diversifying selection at the *Bacillus* Quorum sensing locus and determinants of modification specificity during synthesis of the ComX pheromone. *J Bacteriol* 186:15–21.

156. Ansaldi M, Marolt D, Stebe T, Mandic Mulec I, Dubnau D. 2002. Specific activation of the *Bacillus* quorum-sensing systems by isoprenylated pheromone variants. *Mol Microbiol* 44:1561–1573.

157. Magnuson R, Solomon J, Grossman AD. 1994. Biochemical and genetic characterization of a competence pheromone from *B. subtilis*. *Cell* 77:207–216.

158. Schneider KB, Palmer TM, Grossman AD. 2002. Characterization of *comQ* and *comX*, two genes required for production of ComX pheromone in *Bacillus subtilis*. *J Bacteriol* 184:410–419.

159. Dogsa I, Oslizlo A, Stefanic P, Mandic-Mulec I. 2014. Social interactions and biofilm formation in *Bacillus subtilis*. *Food Technol Biotechnol* 52:149–157.

160. Mandic Mulec I, Kraigher B, Cepon U, Mahne I. 2003. Variability of the quorum sensing system in natural isolates of *Bacillus* sp. *Food Technol Biotechnol* **41**: 23–28.

161. Tran PL-S, Nagai T, Itoh Y. 2000. Divergent structure of the ComQXPA quorum-sensing components: molecular basis of strain-specific communication mechanism in *Bacillus subtilis*. *Mol Microbiol* **37**:1159–1171.

162. West SA, Griffin AS, Gardner A, Diggle SP. 2006. Social evolution theory for microorganisms. *Nat Rev Microbiol* **4**:597–607.

163. Eldar A. 2011. Social conflict drives the evolutionary divergence of quorum sensing. *Proc Natl Acad Sci USA* **108**:13635–13640.

164. Diggle SP, Griffin AS, Campbell GS, West SA. 2007. Cooperation and conflict in quorum-sensing bacterial populations. *Nature* **450**:411–414.

165. Tortosa P, Dubnau D. 1999. Competence for transformation: a matter of taste. *Curr Opin Microbiol* **2**: 588–592.

166. Dogsa I, Choudhary KS, Marsetic Z, Hudaiberdiev S, Vera R, Pongor S, Mandic-Mulec I. 2014. ComQXPA quorum sensing systems may not be unique to *Bacillus subtilis*: a census in prokaryotic genomes. *PLoS One* **9**: e96122. doi:10.1371/journal.pone.0096122.

167. Oslizlo A, Stefanic P, Dogsa I, Mandic-Mulec I. 2014. Private link between signal and response in *Bacillus subtilis* quorum sensing. *Proc Natl Acad Sci USA* **111**: 1586–1591.

168. Earl AM, Losick R, Kolter R. 2008. Ecology and genomics of *Bacillus subtilis*. *Trends Microbiol* **16**: 269–275.

169. Nakamura LK, Roberts MS, Cohan FM. 1999. Relationship of *Bacillus subtilis* clades associated with strains 168 and W23: a proposal for *Bacillus subtilis* subsp. *subtilis* subsp. nov. and *Bacillus subtilis* subsp. *spizizenii* subsp. nov. *Int J Syst Bacteriol* **49**(Pt 3): 1211–1215.

170. Earl AM, Eppinger M, Fricke WF, Rosovitz MJ, Rasko DA, Daugherty S, Losick R, Kolter R, Ravel J. 2012. Whole-genome sequences of *Bacillus subtilis* and close relatives. *J Bacteriol* **194**:2378–2379.

171. Buscot F, Varma A. 2005. Microorganisms in soils: roles in genesis and functions, p 3–16. *In* Buscot F, Varma A (ed), *Soil Biology*, **vol 3**. Springer-Verlag, Berlin.

172. Grundmann GL. 2004. Spatial scales of soil bacterial diversity – the size of a clone. *FEMS Microbiol Ecol* **48**: 119–127.

173. Vos M, Wolf AB, Jennings SJ, Kowalchuk GA. 2013. Micro-scale determinants of bacterial diversity in soil. *FEMS Microbiol Rev* **37**:936–954.

174. Cohn F. 1872. Untersuchungen uber Bakterien. *Beitrage zur Biologie der Pflanzen 1. Heft II.*

175. Jeon CO, Lim JM, Lee JM, Xu LH, Jiang CL, Kim CJ. 2005. Reclassification of *Bacillus haloalkaliphilus* Fritze 1996 as *Alkalibacillus haloalkaliphilus* gen. nov., comb. nov. and the description of *Alkalibacillus salilacus* sp. nov., a novel halophilic bacterium isolated from a salt lake in China. *Int J Syst Evol Microbiol* **55**:1891–1896.

176. Schmidt M, Prieme A, Johansen A, Stougaard P. 2012. *Alkalilactibacillus ikkensis*, gen. nov., sp. nov., a novel enzyme-producing bacterium from a cold and alkaline environment in Greenland. *Extremophiles* **16**:297–305.

177. Sheu SY, Arun AB, Jiang SR, Young CC, Chen WM. 2011. *Allobacillus halotolerans* gen. nov., sp. nov. isolated from shrimp paste. *Int J Syst Evol Microbiol* **61**: 1023–1027.

178. Didari M, Amoozegar MA, Bagheri M, Schumann P, Sproer C, Sanchez-Porro C, Ventosa A. 2012. *Alteribacillus bidgolensis* gen. nov., sp. nov., a moderately halophilic bacterium from a hypersaline lake, and reclassification of *Bacillus persepolensis* as *Alteribacillus persepolensis* comb. nov. *Int J Syst Evol Microbiol* **62**: 2691–2697.

179. Niimura Y, Koh E, Yanagida F, Suzuki K-I, Komagata K, Kozaki M. 1990. *Amphibacillus xylanus* gen. nov., sp. nov., a facultatively anaerobic sporeforming xylan-digesting bacterium which lacks cytochrome, quinone, and catalase. *Int J Syst Bacteriol* **40**:297–301.

180. Zavarzina DG, Tourova TP, Kolganova TV, Boulygina ES, Zhilina TN. 2009. Description of *Anaerobacillus alkalilacustre* gen. nov., sp. nov.—Strictly anaerobic diazotrophic bacillus isolated from soda lake and transfer of *Bacillus arseniciselenatis*, *Bacillus macyae*, and *Bacillus alkalidiazotrophicus* to *Anaerobacillus* as the new combinations *A. arseniciselenatis* comb. nov., *A. macyae* comb. nov., and *A. alkalidiazotrophicus* comb. nov. *Microbiology* **78**:723–731.

181. Pikuta E, Lysenko A, Chuvilskaya N, Mendrock U, Hippe H, Suzina N, Nikitin D, Osipov G, Laurinavichius K. 2000. *Anoxybacillus pushchinensis* gen. nov., sp. nov., a novel anaerobic, alkaliphilic, moderately thermophilic bacterium from manure, and description of *Anoxybacillus flavitherms* comb. nov. *Int J Syst Evol Microbiol* **50**(Pt 6):2109–2117.

182. Marquez MC, Carrasco IJ, Xue Y, Ma Y, Cowan DA, Jones BE, Grant WD, Ventosa A. 2008. *Aquisalibacillus elongatus* gen. nov., sp. nov., a moderately halophilic bacterium of the family *Bacillaceae* isolated from a saline lake. *Int J Syst Evol Microbiol* **58**:1922–1926.

183. Amoozegar MA, Bagheri M, Didari M, Mehrshad M, Schumann P, Sproer C, Sanchez-Porro C, Ventosa A. 2014. *Aquibacillus halophilus* gen. nov., sp. nov., a moderately halophilic bacterium from a hypersaline lake, and reclassification of *Virgibacillus koreensis* as *Aquibacillus koreensis* comb. nov. and *Virgibacillus albus* as *Aquibacillus albus* comb. nov. *Int J Syst Evol Microbiol* **64**(Pt 11):3616–3623.

184. Xue Y, Zhang X, Zhou C, Zhao Y, Cowan DA, Heaphy S, Grant WD, Jones BE, Ventosa A, Ma Y. 2006. *Caldalkalibacillus thermarum* gen. nov., sp. nov., a novel alkalithermophilic bacterium from a hot spring in China. *Int J Syst Evol Microbiol* **56**:1217–1221.

185. Coorevits A, Dinsdale AE, Halket G, Lebbe L, De Vos P, Van Landschoot A, Logan NA. 2012. Taxonomic revision of the genus *Geobacillus*: emendation of *Geobacillus*, *G. stearothermophilus*, *G. jurassicus*, *G. toebii*, *G. thermodenitrificans* and *G. thermoglucosidans* (nom. corrig., formerly 'thermoglucosidasius'); transfer

of *Bacillus thermantarcticus* to the genus as *G. thermantarcticus* comb. nov.; proposal of *Caldibacillus debilis* gen. nov., comb. nov.; transfer of *G. tepidamans* to *Anoxybacillus* as *A. tepidamans* comb. nov.; and proposal of *Anoxybacillus caldiproteolyticus* sp. nov. *Int J Syst Evol Microbiol* 62:1470–1485.

186. Moriya T, Hikota T, Yumoto I, Ito T, Terui Y, Yamagishi A, Oshima T. 2011. *Calditerricola satsumensis* gen. nov., sp. nov. and *Calditerricola yamamurae* sp. nov., extreme thermophiles isolated from a high-temperature compost. *Int J Syst Evol Microbiol* 61:631–636.

187. Nakamura K, Haruta S, Ueno S, Ishii M, Yokota A, Igarashi Y. 2004. *Cerasibacillus quisquiliarum* gen. nov., sp. nov., isolated from a semi-continuous decomposing system of kitchen refuse. *Int J Syst Evol Microbiol* 54:1063–1069.

188. Seiler H, Wenning M, Scherer S. 2012. *Domibacillus robiginosus* gen. nov., sp. nov., isolated from a pharmaceutical clean room. *Int J Syst Evol Microbiol* 63(Pt 6): 2054–2061.

189. Schlesner H, Lawson PA, Collins MD, Weiss N, Wehmeyer U, Volker H, Thomm M. 2001. *Filobacillus milensis* gen. nov., sp. nov., a new halophilic spore-forming bacterium with Orn-D-Glu-type peptidoglycan. *Int J Syst Evol Microbiol* 51:425–431.

190. Glaeser SP, Dott W, Busse HJ, Kampfer P. 2013. *Fictibacillus phosphorivorans* gen. nov. sp. nov. and proposal to reclassify *Bacillus arsenicus*, *Bacillus barbaricus*, *Bacillus macauensis*, *Bacillus nanhaiensis*, *Bacillus rigui*, *Bacillus solisalsi* and *B. gelatini* into the genus *Fictibacillus*. *Int J Syst Evol Microbiol* 63(Pt 8):2934–2944.

191. Lal D, Khan F, Gupta SK, Schumann P, Lal R. 2013. *Edaphobacillus lindanitolerans* gen. nov., sp. nov., isolated from hexachlorocyclohexane (HCH) contaminated soil. *J Basic Microbiol* 53:758–765.

192. Nazina TN, Tourova TP, Poltaraus AB, Novikova EV, Grigoryan AA, Ivanova AE, Lysenko AM, Petrunyaka VV, Osipov GA, Belyaev SS, Ivanov MV. 2001. Taxonomic study of aerobic thermophilic bacilli: descriptions of *Geobacillus subterraneus* gen. nov., sp. nov. and *Geobacillus uzenensis* sp. nov. from petroleum reservoirs and transfer of *Bacillus stearothermophilus*, *Bacillus thermocatenulatus*, *Bacillus thermoleovorans*, *Bacillus kaustophilus*, *Bacillus thermodenitrificans* to *Geobacillus* as the new combinations *G. stearothermophilus*, *G. th. Int J Syst Evol Microbiol* 51:433–446.

193. Waino M, Tindall BJ, Schumann P, Ingvorsen K. 1999. *Gracilibacillus* gen. nov., with description of *Gracilibacillus halotolerans* gen. nov., sp. nov.; transfer of *Bacillus dipsosauri* to *Gracilibacillus dipsosauri* comb. nov., and *Bacillus salexigens* to the genus *Salibacillus* gen. nov., as *Salibacillus salexigens* comb. nov. *Int J Syst Bacteriol* 49(Pt 2):821–831.

194. Echigo A, Fukushima T, Mizuki T, Kamekura M, Usami R. 2007. *Halalkalibacillus halophilus* gen. nov., sp. nov., a novel moderately halophilic and alkaliphilic bacterium isolated from a non-saline soil sample in Japan. *Int J Syst Evol Microbiol* 57:1081–1085.

195. Spring S, Ludwig W, Marquez MC, Ventosa A, Schleifer K-H. 1996. *Halobacillus* gen. nov., with descriptions of *Halobacillus litoralis* sp. nov. and *Halobacillus trueperi* sp. nov., and transfer of *Sporosarcina halophila* to *Halobacillus halophilus* comb. nov. *Int J Syst Bacteriol* 46:492–496.

196. Ishikawa M, Nakajima K, Itamiya Y, Furukawa S, Yamamoto Y, Yamasato K. 2005. *Halolactibacillus halophilus* gen. nov., sp. nov. and *Halolactibacillus miurensis* sp. nov., halophilic and alkaliphilic marine lactic acid bacteria constituting a phylogenetic lineage in Bacillus rRNA group 1. *Int J Syst Evol Microbiol* 55: 2427–2439.

197. Liu J, Wang X, Li M, Du Q, Li Q, Ma P. 2015. *Jilinibacillus soli* gen. nov., sp. nov., a novel member of the family *Bacillaceae*. *Arch Microbiol* 197:11–16.

198. Yoon JH, Kang KH, Park YH. 2002. *Lentibacillus salicampi* gen. nov., sp. nov., a moderately halophilic bacterium isolated from a salt field in Korea. *Int J Syst Evol Microbiol* 52:2043–2048.

199. Ahmed I, Yokota A, Yamazoe A, Fujiwara T. 2007. Proposal of *Lysinibacillus boronitolerans* gen. nov. sp. nov., and transfer of *Bacillus fusiformis* to *Lysinibacillus fusiformis* comb. nov. and *Bacillus sphaericus* to *Lysinibacillus sphaericus* comb. nov. *Int J Syst Evol Microbiol* 57:1117–1125.

200. Hao MV, Kocur M, Komagata K. 1984. *Marinococcus* gen. nov., a new genus for motile cocci with meso-diaminopimelic acid in the cell wall; and *Marinococcus albus* sp. nov. and *Marinococcus halophilus* (Novitsky and Kushner) comb. nov. *J Gen Appl Microbiol* 30: 449–459.

201. Khelifi N, Ben Romdhane E, Hedi A, Postec A, Fardeau ML, Hamdi M, Tholozan JL, Ollivier B, Hirschler-Rea A. 2010. Characterization of *Microaerobacter geothermalis* gen. nov., sp. nov., a novel microaerophilic, nitrate- and nitrite-reducing thermophilic bacterium isolated from a terrestrial hot spring in Tunisia. *Extremophiles* 14:297–304.

202. Echigo A, Minegishi H, Shimane Y, Kamekura M, Usami R. 2012. *Natribacillus halophilus* gen. nov., sp. nov., a moderately halophilic and alkalitolerant bacterium isolated from soil. *Int J Syst Evol Microbiol* 62: 289–294.

203. Sorokin ID, Zadorina EV, Kravchenko IK, Boulygina ES, Tourova TP, Sorokin DY. 2008. *Natronobacillus azotifigens* gen. nov., sp. nov., an anaerobic diazotrophic haloalkaliphile from soda-rich habitats. *Extremophiles* 12:819–827.

204. Lu J, Nogi Y, Takami H. 2001. *Oceanobacillus iheyensis* gen. nov., sp. nov., a deep-sea extremely halotolerant and alkaliphilic species isolated from a depth of 1050 m on the Iheya Ridge. *FEMS Microbiol Lett* 205:291–297.

205. Mayr R, Busse HJ, Worliczek HL, Ehling-Schulz M, Scherer S. 2006. *Ornithinibacillus* gen. nov., with the species *Ornithinibacillus bavariensis* sp. nov. and *Ornithinibacillus californiensis* sp. nov. *Int J Syst Evol Microbiol* 56:1383–1389.

206. Ishikawa M, Ishizaki S, Yamamoto Y, Yamasato K. 2002. *Paraliobacillus ryukyuensis* gen. nov., sp. nov., a new Gram-positive, slightly halophilic, extremely halotolerant, facultative anaerobe isolated from a

decomposing marine alga. *J Gen Appl Microbiol* **48:** 269–279

207. Nunes I, Tiago I, Pires AL, da Costa MS, Verissimo A. 2006. *Paucisalibacillus* globulus gen. nov., sp. nov., a Gram-positive bacterium isolated from potting soil. *Int J Syst Evol Microbiol* **56:**1841–1845.

208. Tanasupawat S, Namwong S, Kudo T, Itoh T. 2007. *Piscibacillus salipiscarius* gen. nov., sp. nov., a moderately halophilic bacterium from fermented fish (pla-ra) in Thailand. *Int J Syst Evol Microbiol* **57:**1413–1417.

209. Lim JM, Jeon CO, Song SM, Kim CJ. 2005. *Pontibacillus chungwhensis* gen. nov., sp. nov., a moderately halophilic Gram-positive bacterium from a solar saltern in Korea. *Int J Syst Evol Microbiol* **55:**165–170.

210. Glaeser SP, McInroy JA, Busse HJ, Kampfer P. 2014. *Pseudogracilibacillus auburnensis* gen. nov., sp. nov., isolated from the rhizosphere of *Zea mays*. *Int J Syst Evol Microbiol* **64:**2442–2448.

211. Krishnamurthi S, Ruckmani A, Pukall R, Chakrabarti T. 2010 *Psychrobacillus* gen. nov. and proposal for reclassification of *Bacillus insolitus* Larkin & Stokes, 1967, *B. psychrotolerans* Abd-El Rahman et al., 2002 and *B. psychrodurans* Abd-El Rahman et al., 2002 as *Psychrobacillus insolitus* comb. nov., *Psychrobacillus psychrotolerans* comb. nov. and *Psychrobacillus psychrodurans* comb. nov. *Syst Appl Microbiol* **33:**367–373.

212. Nystrand R. 1984. *Saccharococcus thermophilus* gen. nov., sp. nov. Isolated from Beet Sugar Extraction. *Syst Appl Microbiol* **5:**204–219.

213. Her J, Kim J. 2013. *Rummeliibacillus suwonensis* sp. nov., isolated from soil collected in a mountain area of South Korea. *J Microbiol* **51:**268–272.

214. Yoon JH, Kang SJ, Oh TK. 2007. Reclassification of *Marinococcus albus* Hao et al. 1985 as *Salimicrobium album* gen. nov., comb. nov. and *Bacillus halophilus* Ventosa et al. 1990 as *Salimicrobium halophilum* comb. nov., and description of *Salimicrobium luteum* sp. nov. *Int J Syst Evol Microbiol* **57:**2406–2411.

215. Ren PG, Zhou PJ. 2005. *Salinibacillus aidingensis* gen. nov., sp. nov. and *Salinibacillus kushneri* sp. nov., moderately halophilic bacteria isolated from a neutral saline lake in Xin-Jiang, China. *Int J Syst Evol Microbiol* **55:** 949–953.

216. Albuquerque L, Tiago I, Rainey FA, Taborda M, Nobre MF, Verissimo A, da Costa MS. 2007. *Salirhabdus euzebyi* gen. nov., sp. nov., a Gram-positive, halotolerant bacterium isolated from a sea salt evaporation pond. *Int J Syst Evol Microbiol* **57:**1566–1571.

217. Jiang F, Cao SJ, Li ZH, Fan H, Li HF, Liu WJ, Yuan HL. 2012. *Salisediminibacterium halotolerans* gen. nov., sp. nov., a halophilic bacterium from soda lake sediment. *Int J Syst Evol Microbiol* **62:**2127–2132.

218. Amoozegar MA, Bagheri M, Didari M, Shahzedeh Fazeli SA, Schumann P, Sanchez-Porro C, Ventosa A. 2013. *Saliterribacillus persicus* gen. nov., sp. nov., a moderately halophilic bacterium isolated from a hypersaline lake. *Int J Syst Evol Microbiol* **63:**345–351.

219. Carrasco IJ, Marquez MC, Xue Y, Ma Y, Cowan DA, Jones BE, Grant WD, Ventosa A. 2007. *Salsugini-*

bacillus kocurii gen. nov., sp. nov., a moderately halophilic bacterium from soda-lake sediment. *Int J Syst Evol Microbiol* **57:**2381–2386.

220. Carrasco IJ, Marquez MC, Xue Y, Ma Y, Cowan DA, Jones BE, Grant WD, Ventosa A. 2008. *Sediminibacillus halophilus* gen. nov., sp. nov., a moderately halophilic, Gram-positive bacterium from a hypersaline lake. *Int J Syst Evol Microbiol* **58:**1961–1967.

221. Yang G, Zhou S. 2014. *Sinibacillus* soli gen. nov., sp. nov., a moderately thermotolerant member of the family *Bacillaceae*. *Int J Syst Evol Microbiol* **64:**1647–1653.

222. Wang X, Xue Y, Ma Y. 2011. *Streptohalobacillus salinus* gen. nov., sp. nov., a moderately halophilic, Gram-positive, facultative anaerobe isolated from subsurface saline soil. *Int J Syst Evol Microbiol* **61:** 1127–1132.

223. Ren PG, Zhou PJ. 2005. *Tenuibacillus multivorans* gen. nov., sp. nov., a moderately halophilic bacterium isolated from saline soil in Xin-Jiang, China. *Int J Syst Evol Microbiol* **55:**95–99.

224. An SY, Asahara M, Goto K, Kasai H, Yokota A. 2007. *Terribacillus saccharophilus* gen. nov., sp. nov. and *Terribacillus halophilus* sp. nov., spore-forming bacteria isolated from field soil in Japan. *Int J Syst Evol Microbiol* **57:**51–55.

225. Ruiz-Romero E, Coutino-Coutino Mde L, Valenzuela-Encinas C, Lopez-Ramirez MP, Marsch R, Dendooven L. 2013. *Texcoconibacillus texcoconensis* gen. nov., sp. nov., alkalophilic and halotolerant bacteria isolated from soil of the former lake Texcoco (Mexico). *Int J Syst Evol Microbiol* **63:**3336–3341.

226. Garcia MT, Gallego V, Ventosa A, Mellado E. 2005. *Thalassobacillus devorans* gen. nov., sp. nov., a moderately halophilic, phenol-degrading, Gram-positive bacterium. *Int J Syst Evol Microbiol* **55:**1789–1795.

227. Cihan AC, Koc M, Ozcan B, Tekin N, Cokmus C. 2014. *Thermolongibacillus altinsuensis* gen. nov., sp. nov. and *Thermolongibacillus kozakliensis* sp. nov., aerobic, thermophilic, long bacilli isolated from hot springs. *Int J Syst Evol Microbiol* **64:**187–197.

228. Steven B, Chen MQ, Greer CW, Whyte LG, Niederberger TD. 2008. *Tumebacillus permanentifrigoris* gen. nov., sp. nov., an aerobic, spore-forming bacterium isolated from Canadian high Arctic permafrost. *Int J Syst Evol Microbiol* **58:**1497–1501.

229. Heyndrickx M, Lebbe L, Kersters K, De Vos P, Forsyth G, Logan NA. 1998. *Virgibacillus*: a new genus to accommodate *Bacillus pantothenticus* (Proom and Knight 1950). Emended description of *Virgibacillus pantothenticus*. *Int J Syst Bacteriol* **48:**99–106.

230. Albert RA, Archambault J, Lempa M, Hurst B, Richardson C, Gruenloh S, Duran M, Worliczek HL, Huber BE, Rossello-Mora R, Schumann P, Busse HJ. 2007. Proposal of *Viridibacillus* gen. nov. and reclassification of *Bacillus arvi*, *Bacillus arenosi* and *Bacillus neidei* as *Viridibacillus arvi* gen. nov., comb. nov., *Viridibacillus arenosi* comb. nov. and *Viridibacillus neidei* comb. nov. *Int J Syst Evol Microbiol* **57:**2729–2737.

231. L'Haridon S, Miroshnichenko ML, Kostrikina NA, Tindall BJ, Spring S, Schumann P, Stackebrandt E, Bonch-Osmolovskaya EA, Jeanthon C. 2006. *Vulcanibacillus modesticaldus* gen. nov., sp. nov., a strictly anaerobic, nitrate-reducing bacterium from deep-sea hydrothermal vents. *Int J Syst Evol Microbiol* **56:** 1047–1053.

232. Galperin MY. 2013. Genomic diversity of spore-forming *Firmicutes*. *Microbiol Spectrum* **1**(2):TBS-0015-2012.

233. Fajardo-Cavazos P, Nicholson WL, Maughan H. 2014. Evolution in the *Bacillaceae*. *Microbiol Spectrum* **2**(5):TBS-0020-2014. doi:10.1128/microbiolspec.TBS-0020-2014.

The Bacterial Spore: From Molecules to Systems
Edited by P. Eichenberger and A. Driks
© 2016 American Society for Microbiology, Washington, DC
doi:10.1128/microbiolspec.TBS-0013-2012

Elizabeth A. Hutchison[1]
David A. Miller[2]
Esther R. Angert[3]

Sporulation in Bacteria: Beyond the Standard Model

4

AN INTRODUCTION TO ENDOSPORE FORMATION

Bacteria thrive in amazingly diverse ecosystems and often tolerate large fluctuations within a particular environment. One highly successful strategy that allows a cell or population to escape life-threatening conditions is the production of spores. Bacterial endospores, for example, have been described as the most durable cells in nature (1). These highly resistant, dormant cells can withstand a variety of stresses, including exposure to temperature extremes, DNA-damaging agents, and hydrolytic enzymes (2). The ability to form endospores appears restricted to the *Firmicutes* (3), one of the earliest branching bacterial phyla (4). Endospore formation is broadly distributed within the phylum. Spore-forming species are represented in most classes, including the *Bacilli*, the *Clostridia*, the *Erysipelotrichi*, and the *Negativicutes* (although compelling evidence to demote this class has been presented [5]). To the best of our knowledge endospores have not been observed in members of the *Thermolithobacteria*, a class that contains only a few species that have been isolated and studied. Thus, sporulation is likely an ancient trait, established early in evolution but later lost in many lineages within the *Firmicutes* (4, 6).

Endospores occur most commonly in rod-shaped bacteria (Fig. 1), but also appear in filamentous cells and in cocci (7–11). Many endospores have been observed only in samples from nature. For instance, large, morphologically diverse helical bacteria (40 to 100 m long), named *Sporospirillum* spp., produce one or two endospore-like structures (12, 13). These bacteria have been found in the gut of batrachian tadpoles, although their affiliation within the *Firmicutes* has not been established. The diversity of endospore-producing bacteria and their varied lifestyles suggest that the sporulation pathway is finely tuned to life in a particular environment, and is an advantageous means of cellular survival, dispersal, and, in some cases, reproduction.

The basic and most familiar mode of sporulation (Fig. 2A) involves an asymmetrical cell division that leads to the formation of a mother cell and a smaller forespore (14, 15). Unique transcriptional programs within these cells result in distinct fates for the forespore and the mother cell. The initiation of sporulation in *Bacillus subtilis* is triggered by a lack of nutrients

[1]Department of Biology, SUNY Geneseo, Geneseo, NY 14454; [2]Department of Microbiology, Medical Instill Development, New Milford, CT 06776; [3]Department of Microbiology, Cornell University, Ithaca, NY 14853.

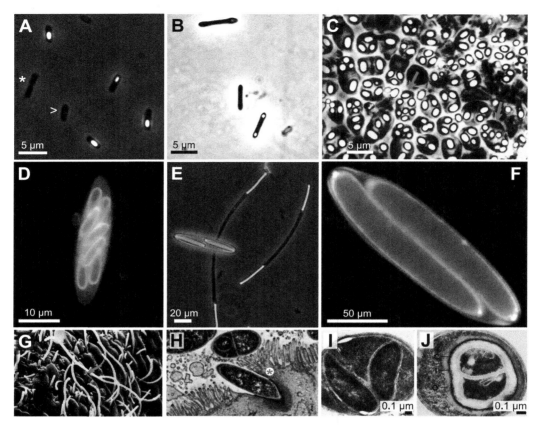

Figure 1   Bacteria that produce endospores or intracellular offspring exhibit a wide variety of morphological phenotypes. (**A**) Phase-contrast microscopy is often used to identify mature endospores (A to C and E) as these highly mineralized cells appear phase-bright. In this image of *B. subtilis*, the caret (>) indicates a cell that is not dividing or sporulating and the asterisk (*) indicates a cell undergoing binary fission. All other cells in the image contain a phase-bright endospore. (**B**) *Clostridium oceanicum* frequently produces phase-bright endospores at both ends of the cell. Image courtesy of Avigdor Eldar and Michael Elowitz, California Institute of Technology. (**C**) In this image of *Anaerobacter polyendosporus*, the arrows indicate cells with seven endospores. (**D**) The fluorescence micrograph of *Metabacterium polyspora* outlines cell membranes and spore coats stained with FM1-43. (**E**) *Epulopiscium*-like type C (cigar-shaped cell) and type J (elongated cells), each containing two phase-bright endospores. (**F**)*Epulopiscium* sp. type B with two internal daughter cells, stained with DAPI. Cellular DNA is located at the periphery of the cytoplasm in the mother cell and each offspring. (**G**) Scanning electron micrograph (SEM) of the ileum lining from a rat reveals the epithelial surface densely populated with SFB. Arrow indicates a hold-fast cell that has not yet elongated into a filament. (**H**) Transmission electron micrograph (TEM) of a thin section through the gut wall reveals the structure of the SFB holdfast cell (indicated by an asterisk). (**I to J**) TEMs illustrate the two possible fates for developing intracellular SFB: (I) two holdfast cells or (J) two endospores that are encased in a common coat (C), inner (I) and outer (O) cortex. Panel C reproduced from Siunov et al. (47) with permission from Society for General Microbiology. Panel E reproduced from Flint et al. (33) with permission from ASM Press. Panel F reproduced from Mendell et al. (93) with permission from the National Academy of Sciences, USA. Panels G and H reproduced from Erlandsen and Chase (69) with permission from the American Society for Nutrition. Panels I and J reproduced from Ferguson and Birch-Andersen (74) with permission from John Wiley and Sons.

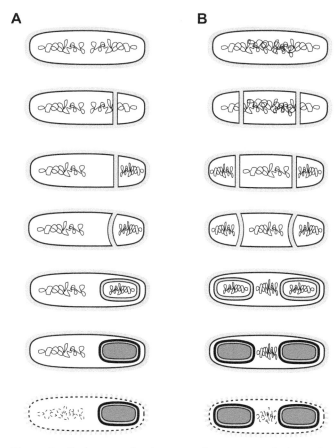

Figure 2 Endospore development. In monosporic bacteria, complete division occurs at only one end of the developing sporangium (**A**), while bacteria that produce two endospores generally divide at both poles (**B**). In some lineages, such as the SFB and *M. polyspora*, engulfed forespores undergo division (not shown). Note that at least three chromosome copies are required to produce two viable endospores. Following endospore engulfment, cortex and coat layers develop, and upon endospore maturation, the mother cell lyses, releasing one (A) or two (B) endospores.

and by high cell density (2, 15). The decision to sporulate is tightly regulated, because this energy-intensive process serves as a last resort for these starving cells. In the early stages of sporulation, gene regulation mainly depends on the stationary-phase sigma factor σ^H and the master transcriptional regulator Spo0A (16, 17). Activation of Spo0A in *B. subtilis* is governed by a phosphorelay system involving several kinases, each of which transmits information about cell condition and environmental stimuli to determine the phosphorylation state of the intracellular pool of Spo0A (18). Prior to asymmetric cell division, the chromosome replicates, and each replication origin rapidly migrates to a different pole of the cell (19). Subsequently, the origin-

proximal regions become tethered to opposite poles and the chromosomal DNA stretches from one pole to the other to form an axial filament (20, 21). During division, only ~30% of the origin-proximal portion of one chromosome is trapped within the forespore, and the rest is translocated into the forespore by SpoIIIE, a DNA transporter protein (17, 22). The other chromosome copy remains in the mother cell.

Differential activation of sporulation-specific sigma factors in the mother cell and forespore manages the fate of each cell (14). First, σ^F is activated exclusively in the forespore (17). Shortly thereafter, a signal is sent to the mother cell to process and hence activate σ^E. Both early sigma factors promote the expression of genes necessary for forespore engulfment, as well as genes needed for the production and activation of the late sporulation sigma factors (17, 23). Remodeling of septal peptidoglycan allows migration of the mother-cell membrane around the forespore (2, 17, 24, 25). Eventually, the leading edge of the migrating mother-cell membrane meets, and fission establishes the double-membrane-bound forespore within the mother cell. Completion of forespore engulfment, combined with further intercellular signaling, allows activation of σ^G in the forespore and the subsequent activation of σ^K in the mother cell. These sigma factors regulate the genes necessary for spore maturation and germination (2, 17). Ultimately, the mother cell undergoes programmed cell death and lysis, which releases the mature endospore (26, 27).

TWIN ENDOSPORE FORMATION IN *B. SUBTILIS* AND TWINS PRODUCED IN NATURE

Although sporogenesis in *B. subtilis* typically culminates in the production of a single endospore, simple mutations can vary the outcome of the program and lead to the production of two viable, mature spores (28). This is due in part to the normal assembly of functional division apparati at both ends of the cell even if only one is used. Null mutations that block activation or expression of σ^E will arrest sporulation after asymmetric division. These mutants produce abortive disporics where the developing sporangium divides sequentially at both poles and chromosome copies are transferred into each of the polar forespores, leaving the mother cell devoid of a chromosome. Forespore-specific expression of *spoIIR* is necessary for intercellular signaling to activate σ^E in the mother cell. Eldar et al. found that, by manipulating the expression of *spoIIR*, a small percentage of cells "escape" sporulation, resume chromosome replication, and then undergo division at both poles to produce viable and UV-resistant "twin endospores." When combined with mutations that increase chromosome copy number, such as those that prevent expression of the replication inhibitor *yabA*, the frequency of twins in the population elevates, provided that the mother cell retains a copy of the chromosome (28). Data from Eldar et al. and studies of other sporulation systems (discussed later) suggest that natural mutations that increase ploidy and promote bipolar division could gradually increase the occurrence of this alternative developmental outcome, thus leading to twin endospore formation as a means of reproduction (28, 29).

Several bacterial lineages naturally produce these "fraternal" twin endospores. The marine anaerobe *Clostridium oceanicum* is a rod-shaped bacterium that typically produces two endospores (Fig. 1B), depending on the temperature or medium composition (30). The DNA replication and septation events (Fig. 2B) leading to twin endospore formation in *C. oceanicum* closely resemble those of twin endospore formation in *B. subtilis* in the mutant strains described above (28). Although rare, twin endospores naturally occur in *Bacillus thuringiensis* as well (31). Other twin endosporeformers, such as the large, rod-shaped spore-forming bacteria from the intestinal tract of batrachian tadpoles and rodents (12, 13, 32, 33), have been observed, but many of these have not been phylogenetically characterized. Finally, the regular production of twin endospores has been described in *Epulopiscium*-like cells (34). Twin endospore-forming bacteria are frequently observed in the gastrointestinal tract, which suggests that these nutrient-rich ecosystems may better support increased ploidy (35), a requisite for the production of more than one endospore.

Epulopiscium spp. and their close relatives, known as "epulos," are intestinal inhabitants of certain species of surgeonfish (36). All morphotypes characterized to date are exceptionally large, with some reaching 600 m (37–39). Due to their large size, *Epulopiscium* spp. were originally classified as protists (36, 39–41), but further ultrastructural and molecular phylogenetic analyses proved that these symbionts are bacteria (37, 38). Phylogenetically, epulos group within the clostridial cluster XIVb in the *Lachnospiraceae* (34, 37, 42, 43). A survey of surgeonfish intestinal communities provided a first assessment of the distribution of epulos among host species and classified these diverse symbionts into ten morphotypes (A to J) based on their cellular and reproductive characteristics (36).

These surgeonfish symbionts exhibit a variety of novel reproductive patterns (44). Only the two largest morphotypes, A and B, are referred to as *Epulopiscium* spp., and these lineages appear to reproduce solely by the formation of multiple, nondormant intracellular offspring (39, 40, 45), which will be described below. Some of the smaller epulo morphotypes undergo binary fission, and many have the ability to produce phase-bright endospores (34, 36, 39, 40, 45). The generation of intracellular offspring in *Epulopiscium* spp. or of endospores in smaller epulo morphotypes is similar to endosporulation in other *Firmicutes*, and developmental progression can be highly synchronized in naturally occurring populations (34, 39, 45, 46).

The phase-bright endospores of the epulo C and J morphotypes from the surgeonfish *Naso lituratus* have been described in the most detail (34). Type C cells are typically cigar shaped, 40 to 130 m long, and do not undergo binary fission, while type J cells are thin filaments, 40 to 400 m long, and capable of binary fission (Fig. 1E). Endospore maturation in type C and type J epulos occurs nocturnally in a highly synchronized manner, with 95 to 100% of the cells in these populations producing spores. Endospores are not seen in fish during daylight hours, suggesting that epulos have evolved a mechanism to regulate endospore development and germination in a diurnal fashion (34). Formation of endospores may promote offspring survival by entering a period of dormancy when nutrients in the gut become depleted, as the host fish sleeps. These endospores may be more resilient than an actively growing epulo, and could aid in transfer to a new host, although the importance of spores in transmission has yet to be fully evaluated (34). Thus, type C and J epulos have modified their sporulation program to produce "fraternal" twin endospores and coordinate this developmental program with regular fluctuations in their environment.

SPORULATION PROGRAMS THAT CAN PRODUCE MORE THAN TWO ENDOSPORES

While the bacteria discussed above have the ability to form twin endospores, others have evolved the means of producing more than two endospores per cell. Endospore formation is generally considered a survival strategy, but the study of these multiple endospore-forming bacteria could provide insight into its use as a reproductive strategy, which may be better suited to some bacterial lifestyles than binary fission alone.

Anaerobacter polyendosporus was first isolated in 1985 from rice paddy soil (47). Depending on growth conditions, *A. polyendosporus* can produce up to seven endospores per cell (Fig. 1C). Cultures of

A. polyendosporus are pleomorphic (47, 48), and the varying cell types appear related to the metabolic transitions that lead cells to sporulate, although this has not been the subject of targeted studies. Thick rods with rounded ends predominate cultures in exponential growth. Cells become wider as the culture ages and eventually thick, phase-bright rods and football-shaped cells appear. All of these forms undergo binary fission (A. M. Johnson and E. R. Angert, unpublished data). Each football-shaped sporangium generally produces one or two endospores. Under certain conditions, such as growth on potato agar or in a liquid medium containing galactose, cells with more than two endospores are observed (47). Twin endospores are produced by division at both cell poles (Johnson and Angert, unpublished). Since *A. polyendosporus* is a member of the cluster I clostridia (43), some of the cell forms observed in sporulating cultures may be homologous to the phase-bright, spindle-shaped clostridial form observed in others of this group such as *Clostridium acetobutylicum* and *Clostridium perfringens* (49, 50). Little research has been conducted on *A. polyendosporus*, and many questions remain regarding its ability to produce multiple endospores, including the role of morphological transitions in sporulation and the factors that lead to the production of more than two endospores.

Metabacterium polyspora (Fig. 1D), an inhabitant of the intestinal tract of guinea pigs, has been studied in some detail, revealing insights into how and why these cells produce multiple endospores (29). Cells of *M. polyspora* pass through the digestive system and rely on the coprophagous character of guinea pigs for cycling back into its original host and for transmission to new hosts (51). The *M. polyspora* cell would not last long outside the host, and only mature endospores appear to survive transit through the mouth and stomach. Germination occurs after the passage of spores into the small intestine. While these cells have the ability to reproduce by binary fission, not all cells use this process. In those that do, binary fission occurs during a short period of time following germination (Fig. 3). After germination, the cells quickly transition to sporulation. In fact, cells with polar septa are often observed emerging from the soon-to-be discarded spore coat. The production of multiple endospores, up to nine per cell (52), allows *M. polyspora* to produce offspring that are prepared for conditions outside the host (29, 51). Considering the rapid passage of material through the gut and possibly the limited time *M. polyspora* spends inside the host, we speculate that reproduction by the instant formation of multiple endospores is advantageous to the symbiotic lifestyle of *M. polyspora* and has allowed it to move away from a reliance on binary fission.

To produce multiple endospores, cells of *M. polyspora*, like the twin endosporeformers described above, divide at both poles (Fig. 3) (17, 51). Each of the forespores receives at least one copy of the chromosome and another copy (or copies) is retained in the mother cell (53). Normally, in *B. subtilis*, DNA replication occurs once, early in sporulation, and any additional rounds of initiation are inhibited during sporulation (54, 55). In contrast, DNA replication in *M. polyspora* occurs throughout development, even after forespore engulfment (53). To form more than two spores, the fully engulfed forespore(s) divide (51). Additionally, DNA replication within forespores loads the endospores with multiple chromosomes, allowing cells to enter sporulation immediately after germination without the requisite

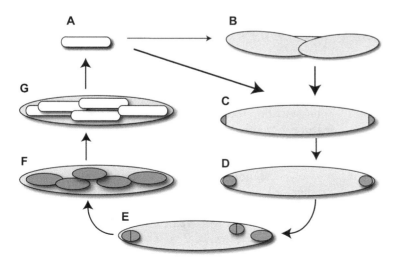

Figure 3 Life cycle of *Metabacterium polyspora*. Endospores germinate (**A**) and, during outgrowth, a cell may undergo binary fission (**B**) or immediately begin to sporulate by dividing at the poles (**C**). The forespores are engulfed (**D**), and the forespores may undergo binary fission to produce additional forespores (**E**). Forespores then elongate (**F**) and develop into mature endospores (**G**). Figure reproduced from Ward and Angert (52) with permission from John Wiley and Sons.

of binary fission or chromosome replication seen in *B. subtilis*. Bacteria that have the ability to produce two or more endospores, like *M. polyspora*, have been reported in other coprophagous rodents (33). For intestinal symbionts, it appears that spore formation not only provides protection from the harsh external environment and to the host's natural barriers to infection, but also the process may be modified to provide a consistent means of cellular propagation.

MODIFICATION OF THE SPORULATION PROGRAM FOR PRODUCTION OF NONDORMANT INTERNAL OFFSPRING

In some groups of bacteria, the sporulation program has evolved to produce multiple intracellular offspring, some of which no longer go through a dormancy period. Notably, members of the genus *Candidatus* Arthromitus, as well as members of the segmented filamentous bacteria (SFB) (also known as *Candidatus* Savagella), reproduce via filament segmentation and internal daughter cell production in addition to forming endospores (56). *Candidatus* Arthromitus and the SFB are Gram-positive, sometimes motile, endospore-forming bacteria found in the intestinal tract of a diverse array of organisms, ranging from mammals, to birds, to fish, to arthropods (56–60). The genus "Arthromitus" was first described and characterized by Joseph Leidy in the mid-1800s from his observations of filamentous bacteria in arthropods and other animals (61, 62). Phylogenetic analyses revealed that SFB (from rats, mice, chickens, and fish) form a distinct clade within the group I clostridia, while spore-forming filaments from arthropods constitute a distinct group within the *Lachnospiraceae* (56–59, 63–66). Isolates from different host species are distinct (56) and exhibit host specificity (67, 68).

As yet, none of the SFB from vertebrate hosts are available in pure culture, although the development of gnotobiotic mammalian hosts mono-associated with SFB has been successful for some lineages (69). Genome sequences derived from populations established in rodents revealed that these bacteria lack almost all biosynthetic pathways for amino acids, vitamins, cofactors, and nucleotides (63–65). The SFB likely live off simple sugars and other essential nutrients gleaned from the host and surrounding environment.

SFB can be abundant in the mammalian host (Fig. 1G) but are restricted to the distal ileum (70, 71). SFB filaments are predominantly attached to the ileal wall and localized to the Peyer's patches, specialized lymphoid follicles that function in antigen sampling and surveillance in the small intestine (72, 73). Close examination of the gut environment revealed that SFB are simultaneously present in various stages of their life cycle, including unattached teardrop-shaped cells in the intervillar spaces, and long or short filaments attached to the ileal epithelium (71). The conical tip of the teardrop-shaped cell is referred to as the holdfast, which anchors the cell to the epithelium. Upon attachment of a holdfast cell, distinct morphological changes occur. The conical tip of the holdfast protrudes into, but does not penetrate, the membrane of the host epithelial cell (Fig. 1H) (70, 71, 74–76). In the host cell cytoplasm, the area adjacent to the SFB attachment site forms an electron-dense layer that comprises predominantly actin filaments (70, 77). Although some holdfast cells appear to be phagocytosed by the host, inflammation of the epithelial tissue at the attachment site does not occur (78).

Once attached, the holdfast cell begins to elongate and septate (Fig. 4A). SFB filaments are typically 50 to 80 m, but can reach lengths up to 1,000 m (70, 73). As a filament transitions into its developmental cycle, starting at the free end of the filament, the so-called primary cells of the filament divide symmetrically, producing two equivalent secondary segments (Fig. 4A, iii) (71, 75). These divisions establish an alternating orientation of cells in the filament, with respect to new and old cell poles, which in turn appears to dictate the pattern of asymmetric division of the secondary segments. After secondary segment division, the larger cell engulfs the smaller cell, which eventually forms a spherical body within the larger mother cell (Fig. 4A, iv). These events closely resemble the early stages of endospore formation (71, 75, 76). Within each SFB mother cell, the engulfed spherical body divides by first becoming crescent-shaped (Fig. 4A, v to vi) and then constricting at the midcell, leaving a pair of cells in each mother cell (71, 73, 75). These "identical" twin offspring cells then differentiate into holdfasts. At this point, the two cells can follow one of two developmental pathways (Fig. 1I to J). In one, the cells can progress through sporulation, producing a cortex and two distinct coat layers. The emergent spores are encased in a common spore coat, a feature that appears to be unique to the SFB sporogenesis pathway. Eventually the mother cell deteriorates, releasing the spore carrying these two offspring. Alternatively, the holdfast cells are simply released upon mother cell lysis (71, 75). A free holdfast cell may establish a new filament within the host, while the spore is an effective dispersal vehicle capable of airborne infection of a naive host (73). Thus, SFB have modified their developmental program such that they can either

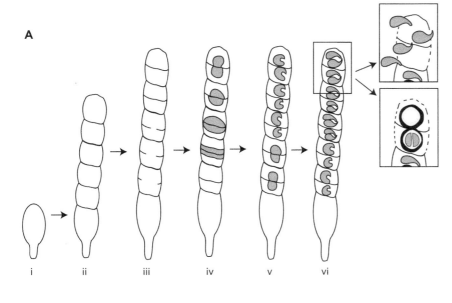

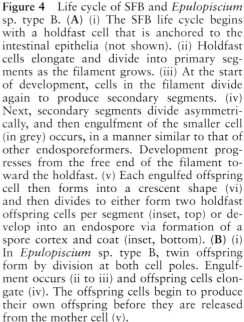

A

B

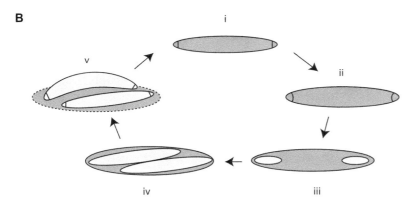

Figure 4 Life cycle of SFB and *Epulopiscium* sp. type B. (**A**) (i) The SFB life cycle begins with a holdfast cell that is anchored to the intestinal epithelia (not shown). (ii) Holdfast cells elongate and divide into primary segments as the filament grows. (iii) At the start of development, cells in the filament divide again to produce secondary segments. (iv) Next, secondary segments divide asymmetrically, and then engulfment of the smaller cell (in grey) occurs, in a manner similar to that of other endosporeformers. Development progresses from the free end of the filament toward the holdfast. (v) Each engulfed offspring cell then forms into a crescent shape (vi) and then divides to either form two holdfast offspring cells per segment (inset, top) or develop into an endospore via formation of a spore cortex and coat (inset, bottom). (**B**) (i) In *Epulopiscium* sp. type B, twin offspring form by division at both cell poles. Engulfment occurs (ii to iii) and offspring cells elongate (iv). The offspring cells begin to produce their own offspring before they are released from the mother cell (v).

produce two daughter holdfast cells or an endospore that contains two cells, likely conferring an advantage to this organism in the dynamic environment of the gut and outside the host. It is unclear how these alternative developmental processes are instigated in a given filament or how different proportions of active or dormant cells impact population dynamics.

The genetics of sporulation have not yet been characterized in detail for SFB, but genome sequence data from these organisms suggest that many components of the sporulation pathway from *B. subtilis* and clostridial genomes are conserved. Approximately 60 to 70 putative sporulation genes have been identified in SFB genomes, including those coding for sporulation sigma factors, stage-specific transcriptional regulators, and spore germination proteins (63–65). Characterization of the kinases that influence phosphorylation of Spo0A could provide insight into factors that

control the decision to sporulate or produce daughter cells, but, like other clostridia, genes encoding the phosphorelay proteins in *B. subtilis* are absent in SFB (63, 64).

SFB have been adopted as a model for examining the effect of commensals on host immune system development and homeostasis. SFB have a broad range of immunostimulatory effects (79–82), and it has been suggested that SFB affect pathogen resistance and autoimmune disease susceptibility of their host (83–89). SFB are generally considered harmless to a healthy host, and may provide critical signals for immune development (90). However, because of their intimate association with host cells and potential to trigger an inflammatory response, the SFB may contribute to disease susceptibility, depending on the genetic background of the host and composition of its resident gut microbiota (83, 91).

As an aside, sporulation in a group of unattached, multicellular filamentous gut symbionts has been described in some morphological detail. *Oscillospira guilliermondi*, later called *Oscillospira guilliermondii*, is a Gram-positive gastrointestinal bacterium found in the cecum of guinea pigs and in the rumen of cattle, sheep, and reindeer (92, 93). These filaments or ovals (5 to 100 m long) are composed of a stack of disc-shaped cells that in some ways resemble *Beggiatoa* spp., but are members of the *Ruminococcaceae* or clostridial cluster IV (92–94). Within a filament of *Oscillospira*, one or more sections may produce an endospore. While nothing is known about the genetics of sporulation in *Oscillospira*, ultrastructural images of filaments undergoing development have been published, and the process appears to have many of the hallmarks of endosporulation, including forespore engulfment and the production of a spore with a multi-layered envelope of cortex and coat. Genome sequence from a non-spore-forming, closely related bacterium, *Oscillibacter valericigenes*, isolated from the gut of a clam, revealed some conservation of sporulation genes, particularly those involved in regulating early events (95, 96).

As with the SFB described above, *Epulopiscium* spp. type A and type B live successfully in the gut of a vertebrate host and exhibit an intracellular offspring developmental program. This process has been best studied in type B cells from the host fish *Naso tonganus* (36, 45). *Epulopiscium* sp. type B cells are very large, usually 100 to 300 m long (Fig. 1F). Although some epulos, such as types C and J, reproduce via endospore production and/or binary fission, type B cells have never been observed to form endospores or undergo binary fission. Instead, *Epulopiscium* sp. type B typically produces 2 to 3 nondormant, intracellular offspring per mother cell; however, as many as 12 have been observed (29, 36).

To form offspring (Fig. 4B), *Epulopiscium* sp. type B cells undergo asymmetric cell division, much like that observed in classical endospore formation, but division occurs at both cell poles (45). A given type B cell contains tens of thousands of copies of its genome to accommodate its large size, and polar division traps only a small amount (<1%) of this DNA (45, 97). Next, the insipient offspring are engulfed and grow within the mother cell. Unlike endospore formation in *B. subtilis*, DNA replication continues in both the mother cell and offspring as the offspring grow (53). Upon completion of offspring growth, the mother cell undergoes a form of programmed cell death (45, 98). The entire developmental process occurs synchronously

within a population. Given the close phylogenetic relationship of *Epulopiscium* sp. type B and other epulos to endospore-forming bacteria, as well as the morphological similarities in the early stages of daughter cell development to that of the early stages of endospore formation, it is likely that the ancestor of all epulos produced endospores, and, with time, the program was modified to function in intracellular offspring production in these viviparous *Firmicutes* (42).

The *Epulopiscium* sp. type B genome has homologs of the *B. subtilis spoIIE* gene and the *spoIIA* operon, which contains genes coding for σ^F and its regulators SpoIIAA and SpoIIAB (46). During sporulation in *B. subtilis*, SpoIIE has dual roles: the promotion of asymmetric cell division and the activation of σ^F. The pattern of *spoIIE* expression with respect to asymmetric division and offspring development in *Epulopiscium* sp. type B populations is similar to that of *B. subtilis*, except that *spoIIE* expression peaks slightly later in *B. subtilis* and stays elevated for a longer developmental interval. Differences in expression of *spoIIE* could be a consequence of differences in the role of SpoIIE in each organism. Also, it may reflect differences in population heterogeneity because endospore formation is a last resort in *B. subtilis* and cells delay entry into sporulation as long as possible (99, 100), while development in *Epulopiscium* is essential for reproduction.

Epulopiscium sp. type B has become a model for studies of cytoarchitecture and evolutionary potential. These massive microbes are extremely polyploid and maintain tens of thousands of genome copies throughout their life cycle (97). This adaptation appears essential for maintaining an active metabolism to support such a large cytoplasmic volume (35). Likewise, polyploidy naturally provides one of the prerequisites of multiple internal offspring production. Studies of the *Epulopiscium* genome have revealed a tolerance for unstable genetic elements, which appears to be a feature shared with other polyploid symbionts (101). For *Epulopiscium* specifically, extreme polyploidy and the use of an endosporulation-derived reproduction have led to the establishment of a cell with chromosomes of differing fates (98). A small subset of chromosomes is inherited by offspring directly, and we consider these "germ line" chromosomes. Most chromosomal copies remain in the mother cell after offspring are formed, and, surprisingly, these chromosomes continue to replicate, despite the fact that they cannot be directly passed on to offspring. This suggests that replication of "somatic" chromosomes is necessary to support the metabolic needs of the mother cell and its growing offspring (98). Studies of this unconventional bacterium are providing fundamental insights into cellular biology and maintenance of genomic resources.

INSIGHTS FROM OTHER UNUSUAL NONMODEL ENDOSPOREFORMERS

Thus far, we have focused on modifications of the basic sporulation program to allow for the formation of multiple endospores or multiple nondormant, intracellular offspring. Here, we describe two other noteworthy and fruitful experimental systems that produce a single endospore per mother cell.

Pasteuria spp., parasites of nematodes and *Daphnia*, constitute another diverse group within the *Firmicutes* that forms endospores that function in a remarkable manner. Endospores of *Pasteuria* spp. consist of a spherical, opaque structure with several spore coat layers, and an additional exosporial fibrillar matrix layer that skirts the spore (102–105). This fibrillar matrix serves in host-specific attachment. The attached *Pasteuria* spore germinates and produces a germ tube that enters the host, where this obligate parasite grows and proliferates (102, 105). Sporogenesis of *Pasteuria* spp. has been characterized in microscopic detail, and a phylogenetic assessment of these members of the *Bacilli* has been carried out for the *spo0A* gene (106), yet the biology behind these unique spore structures and factors that regulate germination and host specificity have yet to be characterized fully. With the recent development of *in vitro* culturing methods by Syngenta and Pasteuria Bioscience, Inc., the structure-function relationship of this unusual spore-delivery system may soon be uncovered.

Although the term *Firmicutes* is thought of as synonymous with "low G+C Gram-positive bacteria," some members of the family *Veillonellaceae* have a Gram-negative cell envelope and can form endospores. Recently, the process of sporulation was characterized in stunning ultrastructural detail in one of these Gram-negative sporeformers, *Acetonema longum* (107). Using 3D electron cryotomographic imaging and immunodetection methods, Tocheva and colleagues show that, through engulfment, the inner and outer membranes of the *A. longum* mother cell become inverted. During outgrowth, the membrane that was previously part of the cytoplasmic membrane transforms, as outer membrane components such as lipopolysaccharide and porins assemble in this now-exposed surface of the cell envelope. The authors suggest that *A. longum* may provide insight into the mechanisms by which an outer membrane could evolve, thus providing a plausible link between early Gram-positive cell forms and the appearance of the Gram-negative envelope (107). Further, this analysis provided evidence to support a hypothesis concerning peptidoglycan dynamics in all endosporeformers. When the state of peptidoglycan of the developing spore was investigated, the authors found that, during engulfment, a thin layer of peptidoglycan is formed and this eventually becomes part of the Gram-negative periplasm (107). While analyses of this unusual Gram-negative endospore-forming bacterium aimed at elucidating unique features of this cell, its study provided additional evidence supporting a novel model of peptidoglycan remodeling in driving a key forespore developmental process, which was later confirmed in *B. subtilis* (107).

EVOLUTION OF SPORULATION FROM A COMPARATIVE GENOMICS PERSPECTIVE

Morphological comparisons between different species and early genetic work on sporulation suggested that this developmental pathway evolved only once in bacteria (6, 108–110). As complete genome sequences became available, comparative studies to look for conserved sporulation genes became feasible (108, 111–114). In one of the first extensive published surveys, Onyenwoke et al. queried a set of 52 bacterial and archaeal genomes using BLAST for 65 select *B. subtilis* sporulation genes covering all stages of sporulation (108). Genes were deemed part of the "core" sporulation pathway if they were absent in non-spore-forming lineages but present only in sporeformers or asporogenous strains (which have conserved sporulation genes but do not produce spores). With this approach, Onyenwoke et al. identified a set of 45 sporulation-specific genes (108). In addition, they noted differences between sporulation gene content in *Clostridia* versus *Bacilli* genomes, and difficulties in accurately identifying clostridial sporulation genes using sequences from *B. subtilis*.

More recently, de Hoon et al. assessed the distribution of 307 *B. subtilis* genes that are directly regulated by the sigma factors σ^H, σ^F, σ^E, σ^G, and σ^K in 24 different species of spore-forming bacteria, using BLAST (6). The authors confirmed that genes coding for the master regulator of sporulation, *spo0A*, and the main sporulation sigma factors are conserved in all sporeformers examined. Genes involved in signaling between sporulation sigma factors are also well conserved, but those genes downstream in the signaling pathway (those that function in a nonregulatory capacity) are not as conserved among sporeformers.

In an effort to improve the annotation of sporulation genes and the ability to predict sporeformers from genomic data, Galperin et al. used a clusters of orthologous genes (COG)-based approach to identify a core set of sporulation genes (96). The authors analyzed almost 400 *Firmicutes* genomes and sorted them into spore-forming and non-spore-forming based on the presence of *spoOA, sspA,* and *dpaAB* genes, which were previously known to be fairly accurate predictors of sporulation (108). The authors then compiled a list of 651 known sporulation genes and compared their distribution in sporeformers versus asporogenous strains versus nonsporeformers. The authors presented a set of approximately 60 genes conserved in members of the *Bacilli* and *Clostridia*. Consistent with the idea that these 60 genes represent the minimum gene content for spore formation, the sporulation gene complement in SFB genomes (which were published after the comparative analysis by Galperin et al.) matches the predicted core set almost exactly. SFB genomes are quite small (1.5 to 1.6 Mb) and appear streamlined (63–65, 115); therefore, the SFB may represent a minimal, yet fully functional, sporulation program. Abecasis et al. used a bidirectional BLAST approach to identify 111 genes conserved in 90% of known sporeformers (116). The authors refined this further to a sporulation signature comprising 48 genes that they used to predict sporulation competency. With comparative genomics, the authors were able to distinguish bacteria that appeared to have recently lost the ability to sporulate. In addition, they identified 22 species that have not been observed to sporulate in culture, but yet appear to have the ability to sporulate based on the presence of complete sporulation signatures.

Another general finding of these studies is that some members of the *Firmicutes* have retained many sporulation genes despite their apparent inability to form an endospore. As discussed previously in this review, *Epulopiscium* sp. type B forms multiple intracellular offspring cells using a process that is morphologically similar to sporulation. A recent study by Miller et al. used a BLAST-based approach to define and then compare the distribution of 147 highly conserved core sporulation genes in *Epulopiscium* sp. type B as well as the genome of its closest endospore-forming relative, *Cellulosilyticum lentocellum* (117). While the *C. lentocellum* genome contains 87 of the core genes, the *Epulopiscium* sp. type B genome contains 57. The conserved genes include homologs of *spoOA*, all sporulation sigma factors, and the central regulatory network that governs cell-specific transcriptional programs, as well as genes required for engulfment. Late-stage

sporulation genes that confer resistance properties, such as the synthesis and forespore transport of dipicolinic acid and germinant receptors located in the *C. lentocellum* genome, were not found in *Epulopiscium* sp. type B. Surprisingly, genes that code for small acid-soluble proteins (SASPs) and their degradation, as well as cortex biosynthesis and cortex/coat scaffolds, were conserved in both *C. lentocellum* and *Epulopiscium* sp. type B. It appears that some of these late-stage functions may still be important for *Epulopiscium*. Since endospores have never been observed in *Epulopiscium* sp. type B, it is possible that the conserved cortex-associated genes may provide a specialized envelope to support the development and rapid growth of daughter cells. SASPs may be important for DNA protection or chromosome organization in developing offspring.

In general, comparative studies have confirmed that the regulatory kinase cascade upstream of SpoOA is not conserved (108), particularly not between *Bacilli* and *Clostridia*. However, SpoOA and the sporulation sigma factors (σ^H, σ^F, σ^E, σ^G, and σ^K) are universally conserved in sporeformers. In addition, regulators of these sigma factors, for example, *spoIIAA, spoIIAB,* and the *spoIIIA* operon, are conserved. This suggests that, despite the ways in which the sporulation pathway has diverged among different *Firmicutes* lineages, these core regulatory components are ancient and essential for development. Previous morphological observations suggested that engulfment, whether it is of a developing forespore or a nondormant offspring cell, proceeds in a very similar manner to that of *B. subtilis*, and indeed genes involved in engulfment, such as *spoIID, spoIIP, spoIIM,* and *spoIIIE*, are highly conserved among sporeformers. Finally, genes involved in spore coat production and germination are not well conserved among endospore-forming bacteria, but this is not surprising given the size of some of these proteins and the wide range of environments in which sporeformers grow, sporulate, and germinate.

An additional outcome of these comparative genomics studies is the finding that asporogenous and non-sporeformers retain homologs to sporulation genes. As more of these strains are characterized with respect to sporulation, it will be interesting to see if these genes have retained functions similar to that of their sporulation homologs or if they have become functionally divergent. Among the nonmodel sporeformers, there are several species that can form more than two spores. Since much of the engulfment machinery is conserved, it is likely that these bacteria have found ways to either engulf forespores that then divide to produce multiple endospores (like *M. polyspora*), or to engulf at cellular locations other than at the poles, as sometimes occurs in *Epulopiscium* sp. type B cells (118). In the latter case, it is currently unknown how these cells regulate where, and how many, additional engulfment sites will occur. Comparative genomics approaches have provided a valuable framework with which to assess the potential to form a spore, and future work on nonmodel spore-forming organisms will provide insight into how sporulation genes evolve to function in diverse forms of bacterial reproduction and development.

THE VALUE OF COMPARATIVE APPROACHES

The sporulation pathway, as it has been classically characterized, results in a single, stress-resistant spore that allows a bacterium to survive unfavorable or even potentially lethal environmental conditions. However, bacteria have evolved and co-opted this pathway to produce a wide range of endospore phenotypes, including multiple endospores and nondormant intracellular offspring. Although it is clear that forming an endospore is advantageous for the survival of organisms in harsh environments, the environmental or developmental triggers that control endospore production in these more highly derived systems remain to be characterized fully. Of particular interest is how the production of more than two endospores in some bacteria, such as *M. polyspora* and *A. polyendosporus*, is regulated, especially since the number of spores produced varies within populations of cells. Furthermore, the nuances of why and how some bacteria alternate between multiple endospores or nondormant offspring have yet to be fully elucidated.

A common theme presented here is that many of these unusual developmental systems have been identified in anaerobic, gastrointestinal symbionts. Our work, for example, uses a comparative approach with closely related symbionts, and we have found that these systems provide informative contrasts when considering the impact of host-symbiont relations on the evolution of novel reproductive strategies (29, 34, 42). All of these intestinal symbionts are rather distant relatives of the *B. subtilis* model, and we know that *Clostridium* spp. use very different signals to trigger the onset of sporulation (109). Recent work on members of the *Clostridia* has reinforced previous observations that, while sporulation genes are conserved between *Clostridia* and *Bacilli*, frequently the regulation of these genes (including key sigma factors and their regulons) is different between these two groups of sporeformers

(119–121). For example, in *B. subtilis*, σ^K functions exclusively late in the sporulation pathway; however, in *C. botulinum* (122) and *C. perfringens* (123), σ^K is required early in sporulation. In *C. acetobutylicum*, σ^K is active both early and late in development (124). In *C. difficile*, σ^K only has a late role in sporulation, and a *sigK* mutant in *C. difficile* can be oligosporogenous (119, 121). Together, these observations illustrate that the clean, sequential model of sigma factor activation described for *B. subtilis* does not fully represent patterns seen in the *Clostridia* (119–121). We would suggest that the deep analysis of additional spore-forming anaerobes, including genomic and transcriptomic data, would provide a more robust comparative system for generating hypotheses on triggers and modifications of the basic sporulation program.

The advent of high-throughput sequencing methods has greatly expanded the ability to characterize uncultured bacteria, novel isolates with no established system for genetic dissection, and mutations that affect development. Efforts to sequence diverse bacterial genomes are providing key insights into the conservation of genes involved in sporogenesis (6, 108). In addition, the application of high-resolution microscopy, including fluorescence and cryotomographic imaging, is providing unprecedented access to the cellular structures and processes associated with developmental progression. The application of transcriptomics, proteomics, and comparative genomics to these unconventional systems will provide insight into the initiation process and potentially identify triggers that determine alternative cell fates. Together, these efforts will provide a better understanding of the conditions that repurpose sporulation, as well as the potential diversity of form and function accommodated by this complex and ancient developmental program.

Acknowledgments. We thank Avigdor Eldar and Michael Elowitz from California Institute of Technology for providing the image of C. oceanicum *and David Sannino, Jen Fownes, and Francine Arroyo for their comments on this manuscript. We are also grateful to colleagues who work with these and other unconventional model systems for their insight. Research in the Angert laboratory is supported by National Science Foundation grants 0721583 and 1244378.*

Citation. Hutchison EA, Miller DA, Angert ER. 2014. Sporulation in bacteria: beyond the standard model. Microbiol Spectrum 2(5):TBS-0013-2012.

References

1. **Nicholson WL, Munakata N, Horneck G, Melosh HJ, Setlow P.** 2000. Resistance of *Bacillus* endospores to extreme terrestrial and extraterrestrial environments. *Microbiol Mol Biol Rev* **64:**548–572.

2. Errington J. 2003. Regulation of endospore formation in *Bacillus subtilis*. *Nat Rev Microbiol* **1**:117–126.

3. Traag BA, Driks A, Stragier P, Bitter W, Broussard G, Hatfull G, Chu F, Adams KN, Ramakrishnan L, Losick R. 2010. Do mycobacteria produce endospores? *Proc Natl Acad Sci USA* **107**:878–881.

4. Ciccarelli FD, Doerks T, von Mering C, Creevey CJ, Snel B, Bork P. 2006. Toward automatic reconstruction of a highly resolved tree of life. *Science* **311**:1283–1287.

5. Yutin N, Galperin MY. 2013. A genomic update on clostridial phylogeny: Gram-negative spore formers and other misplaced clostridia. *Environ Microbiol* **15**:2631–2641.

6. de Hoon MJ, Eichenberger P, Vitkup D. 2010. Hierarchical evolution of the bacterial sporulation network. *Curr Biol* **20**:R735–R745.

7. Mazanec K, Kocur M, Martinec T. 1965. Electron microscopy of ultrathin sections of *Sporosarcina ureae*. *J Bacteriol* **90**:808–816.

8. Robinow CF. 1960. Morphology of bacterial spores, their development and germination, p 207–248. *In* Gunsalus IC, Stanier RY (ed), *The Bacteria*. Academic Press, New York, NY.

9. Zhang L, Higgins ML, Piggot PJ. 1997. The division during bacterial sporulation is symmetrically located in *Sporosarcina ureae*. *Mol Microbiol* **25**:1091–1098.

10. Chary VK, Hilbert DW, Higgins ML, Piggot PJ. 2000. The putative DNA translocase SpoIIIE is required for sporulation of the symmetrically dividing coccal species *Sporosarcina ureae*. *Mol Microbiol* **35**:612–622.

11. Chary VK, Piggot PJ. 2003. Postdivisional synthesis of the *Sporosarcina ureae* DNA translocase SpoIIIE either in the mother cell or in the prespore enables *Bacillus subtilis* to translocate DNA from the mother cell to the prespore. *J Bacteriol* **185**:879–886.

12. Delaporte B. 1964. Etude descriptive de bacteries de tres grandes dimensions. *Ann Inst Pasteur* **107**:845–862.

13. Delaporte B. 1964. Etude comparee de grande spirilles formant des spores: *Sporospirillum* (*Spirillum*) *praeclarum* (Collin) n. g., *Sporospirillum gyrini* n. sp. et *Sporospirillum bisporum* n. sp. *Ann Inst Pasteur* **107**:246–252.

14. Yudkin MD, Clarkson J. 2005. Differential gene expression in genetically identical sister cells: the initiation of sporulation in *Bacillus subtilis*. *Mol Microbiol* **56**:578–589.

15. Piggot PJ, Hilbert DW. 2004. Sporulation of *Bacillus subtilis*. *Curr Opin Microbiol* **7**:579–586.

16. Britton RA, Eichenberger P, Gonzalez-Pastor JE, Fawcett P, Monson R, Losick R, Grossman AD. 2002. Genome-wide analysis of the stationary-phase sigma factor (sigma-H) regulon of *Bacillus subtilis*. *J Bacteriol* **184**:4881–4890.

17. Hilbert DW, Piggot PJ. 2004. Compartmentalization of gene expression during *Bacillus subtilis* spore formation. *Microbiol Mol Biol Rev* **68**:234–262.

18. Stephens C. 1998. Bacterial sporulation: a question of commitment? *Curr Biol* **8**:R45–R48.

19. Webb CD, Teleman A, Gordon S, Straight A, Belmont A, Lin DC, Grossman AD, Wright A, Losick R. 1997. Bipolar localization of the replication origin regions of chromosomes in vegetative and sporulating cells of *B. subtilis*. *Cell* **88**:667–674.

20. Kay D, Warren SC. 1968. Sporulation in *Bacillus subtilis*: morphological changes. *Biochem J* **109**:819–824.

21. Ryter A, Schaeffe P, Ionesco H. 1966. Classification cytologique par leur stade de blocage des mutants de sporulation de *Bacillus subtilis Marburg*. *Ann Inst Pasteur* (Paris) **110**:305–315.

22. Burton B, Dubnau D. 2010. Membrane-associated DNA transport machines. *Cold Spring Harb Perspect Biol* **2**:a000406. doi:10.1101/cshperspect.a000406.

23. Pogliano K, Hofmeister AE, Losick R. 1997. Disappearance of the sigma E transcription factor from the forespore and the SpoIIE phosphatase from the mother cell contributes to establishment of cell-specific gene expression during sporulation in *Bacillus subtilis*. *J Bacteriol* **179**:3331–3341.

24. Gutierrez J, Smith R, Pogliano K. 2010. SpoIID-mediated peptidoglycan degradation is required throughout engulfment during *Bacillus subtilis* sporulation. *J Bacteriol* **192**:3174–3186.

25. Meyer P, Gutierrez J, Pogliano K, Dworkin J. 2010. Cell wall synthesis is necessary for membrane dynamics during sporulation of *Bacillus subtilis*. *Mol Microbiol* **76**:956–970.

26. Hosoya S, Lu Z, Ozaki Y, Takeuchi M, Sato T. 2007. Cytological analysis of the mother cell death process during sporulation in *Bacillus subtilis*. *J Bacteriol* **189**:2561–2565.

27. Nugroho FA, Yamamoto H, Kobayashi Y, Sekiguchi J. 1999. Characterization of a new sigma-K-dependent peptidoglycan hydrolase gene that plays a role in *Bacillus subtilis* mother cell lysis. *J Bacteriol* **181**:6230–6237.

28. Eldar A, Chary VK, Xenopoulos P, Fontes ME, Loson OC, Dworkin J, Piggot PJ, Elowitz MB. 2009. Partial penetrance facilitates developmental evolution in bacteria. *Nature* **460**:510–514.

29. Angert ER. 2005. Alternatives to binary fission in bacteria. *Nat Rev Microbiol* **3**:214–224.

30. Smith LD. 1970. *Clostridium oceanicum*, sp. n., a sporeforming anaerobe isolated from marine sediments. *J Bacteriol* **103**:811–813.

31. Chapman GB, Slob-van Herk A, Eguia JM. 1992. The occurrence of disporous *Bacillus thuringiensis* cells. *Antonie Van Leeuwenhoek* **61**:265–268.

32. Abadie M, Bury E. 1976. Observations sur la structure fine de la spore d'une bacterie geante parasite: *Bacillus camptospora*. *Ann Sci Nat Bot* **17**:277–286.

33. Kunstyr I, Schiel R, Kaup FJ, Uhr G, Kirchhoff H. 1988. Giant gram-negative noncultivable endospore-forming bacteria in rodent intestines. *Naturwissenschaften* **75**:525–527.

34. Flint JF, Drzymalski D, Montgomery WL, Southam G, Angert ER. 2005. Nocturnal production of endospores in natural populations of *Epulopiscium*-like surgeonfish symbionts. *J Bacteriol* **187**:7460–7470.

35. **Angert A.** 2012. DNA replication and genomic architecture of very large bacteria. *Annu Rev Microbiol* **66:** 197–212.

36. **Clements KD, Sutton DC, Choat JH.** 1989. Occurrence and characteristics of unusual protistan symbionts from surgeonfishes (Acanthuridae) of the Great Barrier Reef, Australia. *Mar Biol* **102:**403–412.

37. **Angert ER, Clements KD, Pace NR.** 1993. The largest bacterium. *Nature* **362:**239–241.

38. **Clements KD, Bullivant S.** 1991. An unusual symbiont from the gut of surgeonfishes may be the largest known prokaryote. *J Bacteriol* **173:**5359–5362.

39. **Montgomery WL, Pollak PE.** 1988. *Epulopiscium fishelsoni* n.g., n.sp., a protist of uncertain taxonomic affinities from the gut of an herbivorous reef fish. *J Protozool* **35:**565–569.

40. **Fishelson L, Montgomery WL, Myrberg AA.** 1985. A unique symbiosis in the gut of tropical herbivorous surgeonfish (Acanthuridae: Teleostei) from the red sea. *Science* **229:**49–51.

41. **Montgomery WL, Pollak PE.** 1988. Gut anatomy and pH in a Red Sea surgeonfish, *Acanthurus nigrofuscus. Mar Ecol Prog Ser* **44:**7–13.

42. **Angert ER, Brooks AE, Pace NR.** 1996. Phylogenetic analysis of *Metabacterium polyspora:* clues to the evolutionary origin of daughter cell production in *Epulopiscium* species, the largest bacteria. *J Bacteriol* **178:**1451–1456.

43. **Collins MD, Lawson PA, Willems A, Cordoba JJ, Fernandez-Garayzabal J, Garcia P, Cai J, Hippe H, Farrow JA.** 1994. The phylogeny of the genus *Clostridium:* proposal of five new genera and eleven new species combinations. *Int J Syst Bacteriol* **44:**812–826.

44. **Angert ER.** 2006. The enigmatic cytoarchitecture of *Epulopiscium* spp. *In* Shively JM (ed), *Complex Intracellular Structures in Prokaryotes.* Springer-Verlag, Berlin, Germany.

45. **Angert ER, Clements KD.** 2004. Initiation of intracellular offspring in *Epulopiscium. Mol Microbiol* **51:** 827–835.

46. **Miller DA, Choat JH, Clements KD, Angert ER.** 2011. The *spoIIE* homolog of *Epulopiscium* sp. type B is expressed early in intracellular offspring development. *J Bacteriol* **193:**2642–2646.

47. **Duda VI, Lebedinsky AV, Mushegjan MS, Mitjushina LL.** 1987. A new anaerobic bacterium, forming up to five endospores per cell - *Anaerobacter polyendosporus* gen. et spec. nov. *Arch Microbiol* **148:**121–127.

48. **Siunov AV, Nikitin DV, Suzina NE, Dmitriev VV, Kuzmin NP, Duda VI.** 1999. Phylogenetic status of *Anaerobacter polyendosporus,* an anaerobic, polysporogenic bacterium. *Int J Syst Bacteriol* **49**(pt 3)**:**1119–1124.

49. **Jones DT, Vanderwesthuizen A, Long S, Allcock ER, Reid SJ, Woods DR.** 1982. Solvent production and morphological changes in *Clostridium acetobutylicum. Appl Environ Microbiol* **43:**1434–1439.

50. **Smith AG, Ellner PD.** 1957. Cytological observations on the sporulation process of *Clostridium perfringens. J Bacteriol* **73:**1–7.

51. **Angert ER, Losick RM.** 1998. Propagation by sporulation in the guinea pig symbiont *Metabacterium polyspora. Proc Natl Acad Sci USA* **95:**10218–10223.

52. **Robinow CF.** 1957. [Short note on *Metabacterium polyspora*]. *Z Tropenmed Parasitol* **8:**225–227.

53. **Ward RJ, Angert ER.** 2008. DNA replication during endospore development in *Metabacterium polyspora. Mol Microbiol* **67:**1360–1370.

54. **Castilla-Llorente V, Munoz-Espin D, Villar L, Salas M, Meijer WJ.** 2006. Spo0A, the key transcriptional regulator for entrance into sporulation, is an inhibitor of DNA replication. *EMBO J* **25:**3890–3899.

55. **Fujita M, Losick R.** 2005. Evidence that entry into sporulation in *Bacillus subtilis* is governed by a gradual increase in the level and activity of the master regulator Spo0A. *Genes Dev* **19:**2236–2244.

56. **Snel J, Heinen PP, Blok HJ, Carman RJ, Duncan AJ, Allen PC, Collins MD.** 1995. Comparison of 16S rRNA sequences of segmented filamentous bacteria isolated from mice, rats, and chickens and proposal of "*Candidatus* Arthromitus." *Int J Syst Bacteriol* **45:** 780–782.

57. **Margulis L, Jorgensen JZ, Dolan S, Kolchinsky R, Rainey FA, Lo SC.** 1998. The *Arthromitus* stage of *Bacillus cereus:* intestinal symbionts of animals. *Proc Natl Acad Sci USA* **95:**1236–1241.

58. **Urdaci MC, Regnault B, Grimont PA.** 2001. Identification by *in situ* hybridization of segmented filamentous bacteria in the intestine of diarrheic rainbow trout (*Oncorhynchus mykiss*). *Res Microbiol* **152:**67–73.

59. **Margulis L, Olendzenski L, Afzelius BA.** 1990. Endospore-forming filamentous bacteria symbiotic in termites: ultrastructure and growth in culture of *Arthromitus. Symbiosis* **8:**95–116.

60. **Klaasen HL, Koopman JP, Van den Brink ME, Bakker MH, Poelma FG, Beynen AC.** 1993. Intestinal, segmented, filamentous bacteria in a wide range of vertebrate species. *Lab Anim* **27:**141–150.

61. **Leidy J.** 1849. On the existence of endophyta in healthy animals, as a natural condition. *Proc Natl Acad Sci Phila* **4:**225–233.

62. **Leidy J.** 1881. The parasites of termites. *J Natl Acad Sci Phila* **8:**425–447.

63. **Kuwahara T, Ogura Y, Oshima K, Kurokawa K, Ooka T, Hirakawa H, Itoh T, Nakayama-Imaohji H, Ichimura M, Itoh K, Ishifune C, Maekawa Y, Yasutomo K, Hattori M, Hayashi T.** 2011. The lifestyle of the segmented filamentous bacterium: a non-culturable gut-associated immunostimulating microbe inferred by whole-genome sequencing. *DNA Res* **18:**291–303.

64. **Prakash T, Oshima K, Morita H, Fukuda S, Imaoka A, Kumar N, Sharma VK, Kim SW, Takahashi M, Saitou N, Taylor TD, Ohno H, Umesaki Y, Hattori M.** 2011. Complete genome sequences of rat and mouse segmented filamentous bacteria, a potent inducer of Th17 cell differentiation. *Cell Host Microbe* **10:**273–284.

65. **Sczesnak A, Segata N, Qin X, Gevers D, Petrosino JF, Huttenhower C, Littman DR, Ivanov I.** 2011. The genome of Th17 cell-inducing segmented filamentous

bacteria reveals extensive auxotrophy and adaptations to the intestinal environment. *Cell Host Microbe* **10**:260–272.

66. **Thompson CL, Vier R, Mikaelyan A, Wienemann T, Brune A.** 2012. 'Candidatus Arthromitus' revised: segmented filamentous bacteria in arthropod guts are members of *Lachnospiraceae. Environ Microbiol* **14**:1454–1465.

67. **Tannock GW, Miller JR, Savage DC.** 1984. Host specificity of filamentous, segmented microorganisms adherent to the small bowel epithelium in mice and rats. *Appl Environ Microbiol* **47**:441–442.

68. **Allen PC.** 1992. Comparative study of long, segmented, filamentous organisms in chickens and mice. *Lab Anim Sci* **42**:542–547.

69. **Klaasen HL, Koopman JP, Van den Brink ME, Van Wezel HP, Beynen AC.** 1991. Mono-association of mice with non-cultivable, intestinal, segmented, filamentous bacteria. *Arch Microbiol* **156**:148–151.

70. **Erlandsen SL, Chase DG.** 1974. Morphological alterations in the microvillous border of villous epithelial cells produced by intestinal microorganisms. *Am J Clin Nutr* **27**:1277–1286.

71. **Chase DG, Erlandsen SL.** 1976. Evidence for a complex life cycle and endospore formation in the attached, filamentous, segmented bacterium from murine ileum. *J Bacteriol* **127**:572–583.

72. **Eberl G, Boneca IG.** 2010. Bacteria and MAMP-induced morphogenesis of the immune system. *Curr Opin Immunol* **22**:448–454.

73. **Klaasen HL, Koopman JP, Poelma FG, Beynen AC.** 1992. Intestinal, segmented, filamentous bacteria. *FEMS Microbiol Rev* **8**:165–180.

74. **Davis CP, Savage DC.** 1974. Habitat, succession, attachment, and morphology of segmented, filamentous microbes indigenous to the murine gastrointestinal tract. *Infect Immun* **10**:948–956.

75. **Ferguson DJ, Birch-Andersen A.** 1979. Electron microscopy of a filamentous, segmented bacterium attached to the small intestine of mice from a laboratory animal colony in Denmark. *Acta Pathol Microbiol Scand B* **87**:247–252.

76. **Snellen JE, Savage DC.** 1978. Freeze-fracture study of the filamentous, segmented microorganism attached to the murine small bowel. *J Bacteriol* **134**:1099–1107.

77. **Jepson MA, Clark MA, Simmons NL, Hirst BH.** 1993. Actin accumulation at sites of attachment of indigenous apathogenic segmented filamentous bacteria to mouse ileal epithelial cells. *Infect Immun* **61**:4001–4004.

78. **Yamauchi KE, Snel J.** 2000. Transmission electron microscopic demonstration of phagocytosis and intracellular processing of segmented filamentous bacteria by intestinal epithelial cells of the chick ileum. *Infect Immun* **68**:6496–6504.

79. **Klaasen HL, Van der Heijden PJ, Stok W, Poelma FGJ, Koopman JP, Van den Brink ME, Bakker MH, Eling WMC, Beynen AC.** 1993. Apathogenic, intestinal, segmented, filamentous bacteria stimulate the mucosal immune system of mice. *Infect Immun* **61**:303–306.

80. **Talham GL, Jiang HQ, Bos NA, Cebra JJ.** 1999. Segmented filamentous bacteria are potent stimuli of a physiologically normal state of the murine gut mucosal immune system. *Infect Immun* **67**:1992–2000.

81. **Umesaki Y, Okada Y, Matsumoto S, Imaoka A, Setoyama H.** 1995. Segmented filamentous bacteria are indigenous intestinal bacteria that activate intraepithelial lymphocytes and induce MHC class II molecules and fucosyl asialo GM1 glycolipids on the small intestinal epithelial cells in the ex-germ-free mouse. *Microbiol Immunol* **39**:555–562.

82. **Umesaki Y, Setoyama H.** 2000. Structure of the intestinal flora responsible for development of the gut immune system in a rodent model. *Microbes Infect* **2**:1343–1351.

83. **Gaboriau-Routhiau V, Rakotobe S, Lecuyer E, Mulder I, Lan A, Bridonneau C, Rochet V, Pisi A, De Paepe M, Brandi G, Eberl G, Snel J, Kelly D, Cerf-Bensussan N.** 2009. The key role of segmented filamentous bacteria in the coordinated maturation of gut helper T cell responses. *Immunity* **31**:677–689.

84. **Ivanov II, Atarashi K, Manel N, Brodie EL, Shima T, Karaoz U, Wei D, Goldfarb KC, Santee CA, Lynch SV, Tanoue T, Imaoka A, Itoh K, Takeda K, Umesaki Y, Honda K, Littman DR.** 2009. Induction of intestinal Th17 cells by segmented filamentous bacteria. *Cell* **139**:485–498.

85. **Kriegel MA, Sefik E, Hill JA, Wu HJ, Benoist C, Mathis D.** 2011. Naturally transmitted segmented filamentous bacteria segregate with diabetes protection in nonobese diabetic mice. *Proc Natl Acad Sci USA* **108**:11548–11553.

86. **Lee YK, Menezes JS, Umesaki Y, Mazmanian SK.** 2011. Proinflammatory T-cell responses to gut microbiota promote experimental autoimmune encephalomyelitis. *Proc Natl Acad Sci USA* **108**:4615–4622.

87. **Salzman NH, Hung K, Haribhai D, Chu H, Karlsson-Sjoberg J, Amir E, Teggatz P, Barman M, Hayward M, Eastwood D, Stoel M, Zhou Y, Sodergren E, Weinstock GM, Bevins CL, Williams CB, Bos NA.** 2010. Enteric defensins are essential regulators of intestinal microbial ecology. *Nat Immunol* **11**:76–83.

88. **Wu HJ, Ivanov II, Darce J, Hattori K, Shima T, Umesaki Y, Littman DR, Benoist C, Mathis D.** 2010. Gut-residing segmented filamentous bacteria drive autoimmune arthritis via T helper 17 cells. *Immunity* **32**:815–827.

89. **Chung H, Pamp SJ, Hill JA, Surana NK, Edelman SM, Troy EB, Reading NC, Villablanca EJ, Wang S, Mora JR, Umesaki Y, Mathis D, Benoist C, Relman DA, Kasper DL.** 2012. Gut immune maturation depends on colonization with a host-specific microbiota. *Cell* **149**:1578–1593.

90. **Schnupf P, Gaboriau-Routhiau V, Cerf-Bensussan N.** 2013. Host interactions with Segmented Filamentous Bacteria: an unusual trade-off that drives the post-natal maturation of the gut immune system. *Semin Immunol* **25**:342–351.

91. **Lee YK, Mazmanian SK.** 2010. Has the microbiota played a critical role in the evolution of the adaptive immune system? *Science* **330**:1768–1773.

92. Chatton E, Perard C. 1913. Schizophytes du caecum du cobaye. I. *Oscillospira guilliermondi* n. g., n. s. *C R Seances Soc Biol Paris* **74**:1159–1162.

93. Grain J, Senaud J. 1976. *Oscillospira guillermondii*, bacterie du rumen: etude ultrastructurale du trichome et de la sporulation. *J Ultrastruct Res* **55**:228–244.

94. Mackie RI, Aminov RI, Hu W, Klieve AV, Ouwerkerk D, Sundset MA, Kamagata Y. 2003. Ecology of uncultivated *Oscillospira* species in the rumen of cattle, sheep, and reindeer as assessed by microscopy and molecular approaches. *Appl Environ Microbiol* **69**:6808–6815.

95. Katano Y, Fujinami S, Kawakoshi A, Nakazawa H, Oji S, Iino T, Oguchi A, Ankai A, Fukui S, Terui Y, Kamata S, Harada T, Tanikawa S, Suzuki K, Fujita N. 2012. Complete genome sequence of *Oscillibacter valericigenes* Sjm18-20(T) (=NBRC 101213(T)). *Stand Genomic Sci* **6**:406–414.

96. Galperin MY, Mekhedov SL, Puigbo P, Smirnov S, Wolf YI, Rigden DJ. 2012. Genomic determinants of sporulation in *Bacilli* and *Clostridia*: towards the minimal set of sporulation-specific genes. *Environ Microbiol* **14**:2870–2890.

97. Mendell JE, Clements KD, Choat JH, Angert ER. 2008. Extreme polyploidy in a large bacterium. *Proc Natl Acad Sci USA* **105**:6730–6734.

98. Ward RJ, Clements KD, Choat JH, Angert ER. 2009. Cytology of terminally differentiated *Epulopiscium* mother cells. *DNA Cell Biol* **28**:57–64.

99. Chung JD, Stephanopoulos G, Ireton K, Grossman AD. 1994. Gene expression in single cells of *Bacillus subtilis*: evidence that a threshold mechanism controls the initiation of sporulation. *J Bacteriol* **176**:1977–1984.

100. Veening JW, Hamoen LW, Kuipers OP. 2005. Phosphatases modulate the bistable sporulation gene expression pattern in *Bacillus subtilis*. *Mol Microbiol* **56**:1481–1494.

101. Tamas I, Wernegreen JJ, Nystedt B, Kauppinen SN, Darby AC, Gomez-Valero L, Lundin D, Poole AM, Andersson SG. 2008. Endosymbiont gene functions impaired and rescued by polymerase infidelity at poly(A) tracts. *Proc Natl Acad Sci USA* **105**:14934–14939.

102. Davies KG, Rowe J, Manzanilla-Lopez R, Opperman CH. 2011. Re-evaluation of the life-cycle of the nematode-parasitic bacterium *Pasteuria penetrans* in root-knot nematodes, *Meloidogyne* spp. *Nematology* **13**:825–835.

103. Ebert D, Rainey P, Embley TM, Scholz D. 1996. Development, life cycle, ultrastructure and phylogenetic position of *Pasteuria ramosa* Metchnikoff 1888: rediscovery of an obligate endoparasite of *Daphnia magna* Straus. *Phil Trans R Soc Lond B* **351**:1689–1701.

104. Imbriani JL, Mankau R. 1977. Ultrastructure of the nematode pathogen, *Bacillus penetrans*. *J Invertebr Pathol* **30**:337–347.

105. Sayre RM, Wergin WP. 1977. Bacterial parasite of a plant nematode: morphology and ultrastructure. *J Bacteriol* **129**:1091–1101.

106. Trotter JR, Bishop AH. 2003. Phylogenetic analysis and confirmation of the endospore-forming nature of *Pasteuria penetrans* based on the *spo0A* gene. *FEMS Microbiol Lett* **225**:249–256.

107. Tocheva EI, Matson EG, Morris DM, Moussavi F, Leadbetter JR, Jensen GJ. 2011. Peptidoglycan remodeling and conversion of an inner membrane into an outer membrane during sporulation. *Cell* **146**:799–812.

108. Onyenwoke RU, Brill JA, Farahi K, Wiegel J. 2004. Sporulation genes in members of the low G+C Gram-type-positive phylogenetic branch (*Firmicutes*). *Arch Microbiol* **182**:182–192.

109. Paredes CJ, Alsaker KV, Papoutsakis ET. 2005. A comparative genomic view of clostridial sporulation and physiology. *Nat Rev Microbiol* **3**:969–978.

110. Sauer U, Treuner A, Buchholz M, Santangelo JD, Durre P. 1994. Sporulation and primary sigma factor homologous genes in *Clostridium acetobutylicum*. *J Bacteriol* **176**:6572–6582.

111. Brill JA, Wiegel J. 1997. Differentiation between spore forming and asporogenic bacteria using a PCR and Southern hybridization based method. *J Microbiol Methods* **31**:29–36.

112. Stragier P. 2002. A gene odyssey: exploring the genomes of endospore forming bacteria, p 519–526. *In* Soneshein AL, Hoch JA, Losick R (ed), Bacillus subtilis *and Its Closest Relatives: From Genes to Cells*. ASM Press, Washington, DC.

113. Paredes-Sabja D, Setlow P, Sarker MR. 2011. Germination of spores of *Bacillales* and *Clostridiales* species: mechanisms and proteins involved. *Trends Microbiol* **19**:85–94.

114. Xiao Y, Francke C, Abee T, Wells-Bennik MH. 2011. Clostridial spore germination versus bacilli: genome mining and current insights. *Food Microbiol* **28**:266–274.

115. Pamp SJ, Harrington ED, Quake SR, Relman DA, Blainey PC. 2012. Single-cell sequencing provides clues about the host interactions of segmented filamentous bacteria (SFB). *Genome Res* **22**:1107–1119.

116. Abecasis AB, Serrano M, Alves R, Quintais L, Pereira-Leal JB, Henriques AO. 2013. A genomic signature and the identification of new sporulation genes. *J Bacteriol* **195**:2101–2115.

117. Miller DA, Suen G, Clements KD, Angert ER. 2012. The genomic basis for the evolution of a novel form of cellular reproduction in the bacterium *Epulopiscium*. *BMC Genomics* **13**:265. doi:10.1186/1471-2164-13-265.

118. Robinow C, Angert ER. 1998. Nucleoids and coated vesicles of "Epulopiscium" spp. *Arch Microbiol* **170**:227–235.

119. Saujet L, Pereira FC, Serrano M, Soutourina O, Monot M, Shelyakin PV, Gelfand MS, Dupuy B, Henriques AO, Martin-Verstraete I. 2013. Genome-wide analysis of cell type-specific gene transcription during spore formation in *Clostridium difficile*. *PLoS Genet* **9**:e1003756. doi:10.1371/journal.pgen.1003756.

120. Pereira FC, Saujet L, Tome AR, Serrano M, Monot M, Couture-Tosi E, Martin-Verstraete I, Dupuy B, Henriques AO. 2013. The spore differentiation pathway in the enteric pathogen *Clostridium difficile*.

PLoS Genet 9:e1003782. doi:10.1371/journal.pgen.1003782

121. Fimlaid KA, Bond JP, Schutz KC, Putnam EE, Leung JM, Lawley TD, Shen A. 2013. Global analysis of the sporulation pathway of *Clostridium difficile. PLoS Genet* 9:e1003660. doi:10.1371/journal.pgen.1003660.

122. Kirk DG, Dahlsten E, Zhang Z, Korkeala H, Lindstrom M. 2012. Involvement of *Clostridium botulinum* ATCC 3502 sigma factor K in early-stage sporulation. *Appl Environ Microbiol* 78:4590–4596.

123. Harry KH, Zhou R, Kroos L, Melville SB. 2009. Sporulation and enterotoxin (CPE) synthesis are controlled by the sporulation-specific sigma factors SigE and SigK in *Clostridium perfringens. J Bacteriol* 191:2728–2742.

124. Al-Hinai MA, Jones SW, Papoutsakis ET. 2014. sigmaK of *Clostridium acetobutylicum* is the first known sporulation-specific sigma factor with two developmentally separated roles, one early and one late in sporulation. *J Bacteriol* 196:287–299.

The *Bacillaceae*: *Bacillus subtilis*

II

The Bacterial Spore: From Molecules to Systems
Edited by P. Eichenberger and A. Driks
© 2016 American Society for Microbiology, Washington, DC
doi:10.1128/microbiolspectrum.TBS-0004-2012

Nicolas Mirouze[1]
David Dubnau[2]

Chance and Necessity in *Bacillus subtilis* Development

5

Everything existing in the universe is the fruit of chance and necessity.

Democritus (as quoted by J. Monod)

Biologists often think of regulation as deterministic; causes have invariant consequences, and changes predictably beget further changes. However, with the advent of techniques for the study of individual cell phenotypes, it has become evident that random cell-to-cell variation in the amounts of mRNA and protein is prevalent and often entails significant consequences, particularly for developmental processes (1, 2). This review focuses on the developmental pathways of *Bacillus subtilis*, with attention to the roles of this variation and of stochastic reactions in competence, motility, sporulation, and biofilm formation. We do not comprehensively review the regulatory mechanisms that control these pathways but present only what is needed for our purposes. Figure 1 identifies the major players that are discussed below and illustrates the competence, motility, cannibalism, biofilm, and sporulation modules that represent the major known developmental pathways of *B. subtilis*.

When a cell population bifurcates into two subpopulations that differ in their patterns of transcription, we use the term bimodal to describe the distribution of gene expression among the cells. The term bistable is reserved for a bimodal population in which the different cell types can be epigenetically inherited. Noise, or cell-to-cell variation in the abundance of gene products, results from the stochastic nature of chemical reactions and becomes important when small numbers of reactants are involved (3, 4). Noise in gene expression results in population heterogeneity, which may or may not be bimodal in nature. Transcription is the major contributor to noise because only a few copies of each promoter are present in a cell and because the engagement of RNA polymerase with a given promoter occurs randomly, with unpredictable delays between transcription events or between clusters of such events. Other processes, including mRNA decay, translation, and protein degradation, may also contribute to noise.

Distinction is often made between intrinsic and extrinsic noise (5, 6). The first is due to the inherent properties of a promoter and other gene expression sequences of a gene. Extrinsic noise results from cell-to-cell variation in transcription factors and other molecules that determine the rates of gene expression. Intuitively it would seem that as the average number of gene

[1]UMR1319 Micalis, Bat. Biotechnologie (440), INRA, Domaine de Vilvert, 78352 Jouy-en-Josas Cedex, France; [2]Public Health Research Institute, New Jersey Medical School, Rutgers University, Newark, NJ 07103.

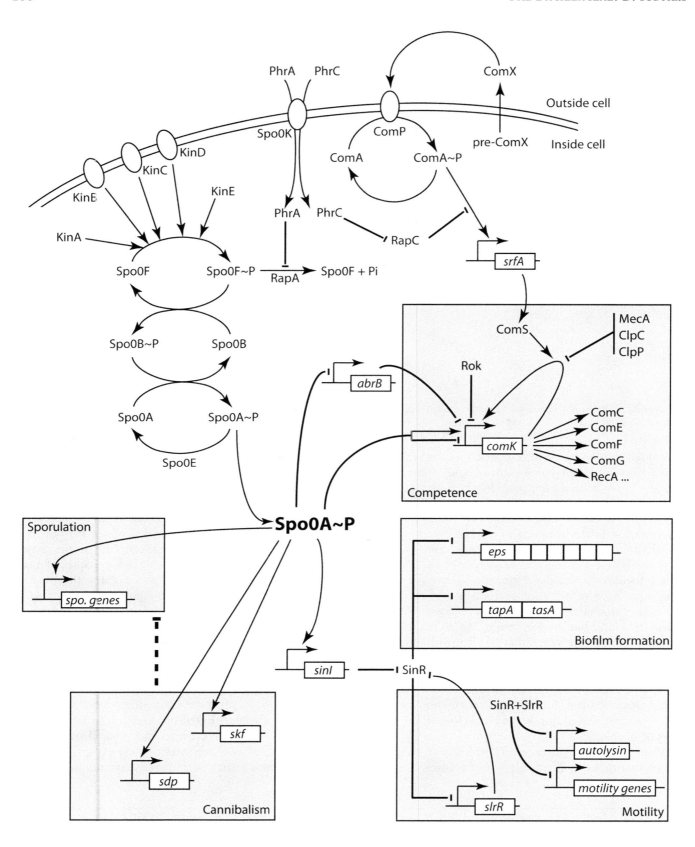

products increases, the importance of intrinsic noise will decrease and extrinsic noise will become dominant. This prediction has been confirmed, with the additional remarkable conclusion that at a given time, there is little correlation in a single cell between the number of transcripts from a gene and the number of its cognate protein molecules (7). This general result stems from the difference in mRNA and protein stability, so the number of mRNA molecules present at a given time does not predict the amount of accumulated protein.

B. subtilis provides a rich field for the investigator because of the wonderful variety of its adaptations. For example, cells can become motile and swim toward nutrients or away from repellents or may adopt a sessile lifestyle. As cultures approach stationary phase and the growth rate decreases while cell density increases, some cells express a large number of genes (in excess of 100) under the control of the transcription factor ComK (8–10). Among these so-called K-state genes are those that mediate transformation, the uptake and integration of environmental DNA. Under certain conditions, some cells elaborate products that form an extracellular matrix in which a multicellular community becomes embedded to form a biofilm (11). Within the biofilm, these matrix-producing cells produce toxic substances that cause the demise of certain of their sisters. This provides food for the aggressors, delaying their irreversible entry into the final, energy-expensive sporulation pathway. An important discovery is that the domesticated laboratory strains, derived from strain 168, have lost the ability to form robust biofilms and are modified in their motile-sessile switch behavior compared to natural isolates (12). The full display of the biofilm-associated phenotypes is only revealed in natural isolates, notably in strain 3610, which has become the industry standard. It is worth mentioning that by adopting a standard, which valuably permits results in different laboratories to be compared, we also run the risk of ignoring interesting phenotypic diversity. For example, 3610 is much less transformable than other natural isolates.

Spo0A AND THE PHOSPHORELAY AS A TEMPORAL GATEKEEPER

This review will repeatedly return to Spo0A (OA) as a temporal gatekeeper. Although originally identified as the master regulator of spore formation, 0A in its phosphorylated form governs all of the developmental pathways mentioned above. 0A is phosphorylated via the

Figure 1 Developmental modules in *B. subtilis* and their major components. All of the indicated forms of development depend on 0A and on the phosphorelay that governs the phosphorylation of this transcription factor. This figure is intended to summarize many of the major interactions mentioned in the text that govern the developmental processes. It is not exhaustive, although it may be exhausting. Lines ending in perpendiculars and arrows denote negative and positive effects, respectively. Arrows associated with right-angled lines denote transcription initiation. The dotted line from the cannibalism module indicates that the release of nutrients from dead cells delays sporulation. Several kinases deliver phosphoryl groups to the phosphorelay, which results in the formation of 0A~P. Under some conditions one or more kinase can dephosphorylate 0F~P, draining phosphate from 0A~P. RapA is one of several related proteins that can also dephosphorylate 0F~P. RapC acts by preventing ComA~P from interacting with its DNA target. These Rap proteins are inhibited by cognate secreted peptides (e.g., PhrA and PhrC), which are internalized by the oligopeptide permease Spo0K. ComX is a modified and secreted peptide which activates the autophosphorylation of ComP. ComP~P donates a phosphate to ComA, and ComA~P then activates the transcription of *srfA*. Embedded in the *srfA* operon is the gene for ComS. This small protein binds to the protease complex of MecA plus ComP plus ClpC, preventing the degradation of the transcription factor ComK. ComK is then free to activate its own expression by antagonizing the repressor Rok, activating a positive autoregulatory loop. When ComK accumulates, it, in turn, activates the transcription of many downstream genes, resulting in the induction of competence (the K-state). A low level of 0A~P is also essential for competence due to its direct interaction with the *comK* promoter and its repression of *abrB*. A low to intermediate concentration of 0A~P also activates the *sinI* promoter. SinI antagonizes SinR, lifting the repression of several transcription units that are essential for biofilm formation, as well as the repression of *slrR*. SlrR binds to SinR, further derepressing the biofilm operons. The SinR-SlrR heterocomplex represses the genes for motility as well as those that encode the autolysins that separate daughter cells following division. This results in the formation of chains of sessile cells. Low concentrations of 0A~P also activate genes that encode toxins. Toxin-producing cells (cannibals) benefit by killing other cells, thus deriving nutrients. Finally, high concentrations of 0A~P activate the sporulation genes.

famous phosphorelay (Fig. 1) (13). In the phosphorelay, one or more of five histidine kinases phosphorylates the response regulator protein SpoOF (OF), which donates its phosphoryl moiety to the phosphotransferase SpoOB (OB), which then passes the phosphoryl group to OA. The concentration of 0A~P increases as cells approach and enter stationary phase, and the consequences of this increase are multiple and profound. Programmed changes in the average concentration of 0A~P determine the frequencies at which cells enter the developmental pathways, while the choice of which cells do so is random. As described below, the phosphorelay, which is modulated by a large number of accessory proteins, has been called a "noise generator," causing the level of 0A~P to vary from cell to cell (14). Development in *B. subtilis* is thus both stochastic (chance) and deterministic (necessity). The mechanisms that regulate these changes in the concentration of 0A~P are not completely understood, involve many genes, and exert their influence on the levels of transcription (including promoter switching), phosphorylation, and dephosphorylation. Only recently has light been shed on some of the upstream signal inputs that regulate this complex system, particularly with regard to biofilm formation.

Three important generalizations are basic to the various developmental processes. First, the average concentration of 0A~P increases as cell division slows and cultures approach stationary phase. Second, promoters respond differently to 0A~P depending on their affinities for this response regulator protein, and this has consequences for development (15, 16). Third, there are marked differences in the rates at which 0A~P increases in the different cells of a single population (14, 17). This temporal heterogeneity has been assigned the useful term "heterochronicity" (17). In a number of cases, as described below, 0A~P activates and then represses a given gene, allowing transient expression. In this sense, this master regulator acts as a temporal gatekeeper. We begin our review of developmental processes with competence, for which the role of noise has received considerable attention.

COMPETENCE: THE K-STATE

An early response of *B. subtilis* to high cell density, as well as to other poorly understood signals, is competence for DNA uptake. The genes required for competence are transcribed only in the presence of the transcriptional regulator ComK (18). Although the incorporation of new genetic material potentially allows the bacteria to increase their fitness under challenging conditions, only some of the 100 or so genes under ComK control are devoted to transformation, and it is probably a mistake to discuss the evolution of the K-state only in terms of the fitness benefits that may be conferred by DNA uptake. (Despite this warning, here we use the terms competence and K-state interchangeably.) ComK not only activates transcription of the downstream K-state genes but also positively regulates its own promoter. In laboratory strains, this induction takes place in 10 to 20% of the cells. Induction in a given cell can be regarded as an all-or-none event that results in a bimodal distribution of competence gene expression. Natural isolates of *B. subtilis* vary widely in the fraction of cells that achieve the K-state, ranging from very few such cells to a frequency approaching that of the domesticated strain (10 to 20%), which has been selected in the laboratory for high transformation rates (unpublished data).

Temporal Control of Competence

During exponential growth, *B. subtilis* uses several mechanisms to prevent the development of competence (Fig. 1). First, the ComK concentration is kept low because it is targeted to the ClpC-ClpP protease complex by the MecA adaptor protein (19). In addition, *comK* transcription is inhibited by three repressors: Rok, AbrB, and CodY (20–22). This redundant inhibition of competence is necessary because K-state cells do not divide (23). As a culture approaches stationary phase and the cell density rises, a modified extracellular peptide (ComX) interacts with the ComP histidine kinase, which then donates a phosphoryl group to its cognate response regulator, ComA (24). As a result of this quorum-sensing signal transduction pathway, *comS* transcription is activated. The anti-adaptor protein ComS competes for the binding of ComK to MecA, thereby lowering the rate of ComK degradation (25). The stabilization of ComK caused by ComS thus issues a license for transition to the K-state. Since ComS is produced in all the cells, and because the addition of ComX to cultures does not increase the percentage of the cells that become competent, the cellular decision point for competence must be sought outside the ComK stabilization pathway.

Two independent studies established that the positive autoregulation of *comK* is central to bimodal expression (26, 27). When *comK* was expressed from an inducible promoter as the only source of ComK protein in the cells, the response to increasing concentrations of inducer of a ComK-dependent reporter gene was unimodal and the expression of the reporter in the entire population increased continuously to a maximum value. In contrast, a bimodal response was observed when the same inducible construct was included in cells that retained the positive-feedback loop because they carried a wild-type copy of *comK*, which was still autoregulated. Based on these studies, it was suggested that noise in the basal expression of *comK* caused only a fraction of cells to exceed a threshold concentration of ComK needed to trigger the autoregulation of *comK* and drive these cells into the K-state. Since then, several studies have verified this hypothesis.

Two systems have been used to study the stochastic process that produces competent cells. In one, dispersed cultures are allowed to reach stationary phase, at which time about 15% of the cells convert to the competence ON state (28, 29). In the other, cells are deposited on agarose pads in nutrient-poor medium in which individual cells become competent, emerge from competence over an extended period, and eventually sporulate (30–32). We begin our discussion by addressing findings from the first system, in which programmed regulatory interactions temporally adjust the chances that a cell will become competent.

Changes in the Basal Expression of *comK* Establish a Window of Opportunity for Competence

The rate of transition to competence is high around the time of entry to stationary phase and then decreases, approaching zero 1.5 to 2 h later. This "window of opportunity" limits the bimodal expression of the K-state so that only 10 to 20% of the cells become competent. Leisner et al. (28) described this in detail and showed that the average basal level of *comK* transcription increases and then decreases, proposing that these changes establish the window of opportunity. Only those cells with a ComK level above a threshold become competent, and such cells exist transiently. Maamar et al. (29) also addressed the temporal regulation of the transition rate and used fluorescence in situ hybridization to count individual *comK* transcripts, confirming a transient rise in the basal average number of *comK* mRNA transcripts per cell (the "uptick") that coincided with the timing of the window of opportunity described above. As the average mRNA content of the population increased, more cells crossed a threshold and became competent. Because this average number subsequently decreased, cells in which competence had not been induced during this period remained in the OFF state. Mathematical modeling and simulations confirmed the plausibility of this model. Following a method devised by Elowitz et al. (5), it was concluded that intrinsic noise in *comK* expression selects cells for

competence. In agreement with this conclusion, when the promoter and start codon of *comK* were manipulated to decrease this noise without reducing the mean amount of ComK protein per cell, the proportion of K-state cells was predictably decreased. Using a different stratagem to reduce noise, Süel et al. (31) also demonstrated that noise in *comK* was responsible for the competence decision.

The Uptick Is Controlled by 0A~P and by an Intrinsic Increase in Promoter Activity

The mechanism responsible for the increase and decrease in the basal rate of expression of *comK* (the uptick) is important to understand because these changes set the window of opportunity for the K-state. ComK and the positive autoregulatory loop play no role in the uptick, because the increase and decrease in the rate of transcription from the *comK* promoter (P*comK*) are unaffected in the absence of a functional *comK* gene (29).

All of the developmental adaptations of *B. subtilis* (e.g., biofilm formation, cannibalism, and sporulation) are tightly controlled by 0A~P (Fig. 1), and competence is no exception, because a *spo0A* knockout does not express competence genes (33). It has been shown that as the concentration of 0A~P increases during the approach to stationary phase, it first induces *comK* promoter activity and then represses it by direct binding (34). To accomplish the activation of basal expression, 0A~P binds to three sites and represses by binding to two downstream sites, which likely have lower affinity. Interestingly, the activation is accomplished by anti-repression with respect to the Rok repressor (34), and this appears to be accomplished without the displacement of Rok from the DNA (unpublished data). Thus, 0A~P regulates both the increasing and decreasing segments of the uptick. This is one more example of an emerging theme concerning 0A~P; this molecule can exert both positive and negative effects on a given developmental gene as its concentration increases, establishing a temporal gate for gene expression (15, 35, 36). Interestingly, the details of how this so-called "bandpass" regulation (36) is achieved differ with each target gene, suggesting that evolution has repeatedly rediscovered the same use for 0A~P. The mechanism for control of the *comK* uptick is illustrated in Fig. 2, in which additional details are provided.

Surprisingly, in the absence of 0A~P, an identical increase in *comK* basal activity takes place, but with a 10-fold-reduced amplitude (34). Thus, the roles of 0A~P in regulation of the uptick are to amplify an inherent increase in transcription and then to repress

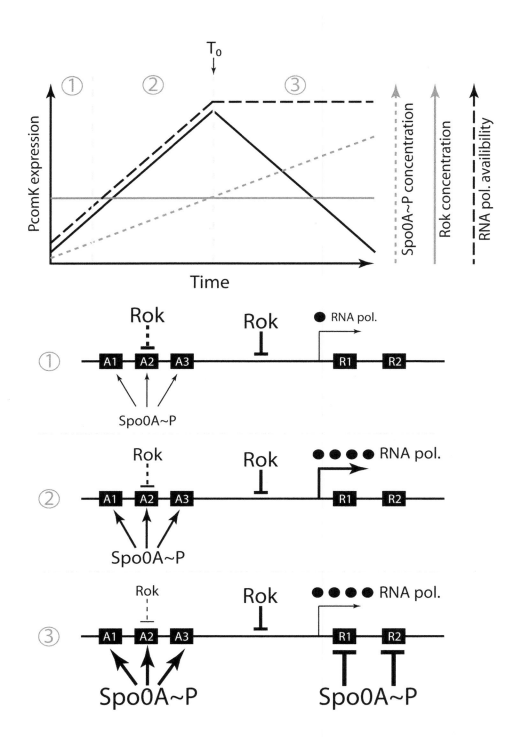

transcription from P*comK*. This inherent increase in the basal *comK* expression takes place during the approach to stationary phase and remarkably is also evident when several other promoters are studied, including completely synthetic σ^A- and σ^H-dependent promoters that consist of random sequences in which the canonical sequence motifs for these two promoter types are embedded. This inherent increase may be due to a passive mechanism in which RNA polymerase (RNApol) is released from stable RNA promoters as growth slows, thus activating repressed genes by competition with repressors as well as genes with promoters that have low RNApol affinities (Fig. 2) (37–39). Regardless of the mechanism, the inherent increase takes place as the culture approaches stationary phase and the growth rate drops, and so the competence transitions are likely to be influenced by some aspect of the metabolic state of the cells, perhaps indirectly by the same mechanisms as control transcription from stable RNA promoters as well as by the complex mechanisms that govern the formation of 0A~P. Modeling of these interactions, including the roles of 0A~P, of Rok as a repressor and of the inherent increase in *comK* activity reproduced the main features of the uptick (34).

If the amount of 0A~P determines the time course of the window of opportunity for the K-state, competent cells examined at a fixed time should be restricted to those within a limited range of 0A~P concentrations. This prediction was confirmed using spectrally distinct reporters fused to 0A~P- and ComK-dependent promoters (P*sdp* and P*spo0A* for 0A~P and P*comK* for ComK). Outside of this limited range the probability of competence decreased dramatically. However, within this range, there was no strong correlation between competence and the 0A~P concentration. This result is consistent with the conclusion that the fate-determining noise is intrinsic to the *comK* promoter (29) and does not reflect variation in extrinsic factors such as the concentration of 0A~P. Only the time course of competence development is dependent on 0A~P (34).

Competence as an Excitable System

Süel et al. (31, 32) have exploited a different experimental system in which cells growing slowly in a nutrient-poor medium randomly enter and exit the competent state. Several observations made under these conditions strongly support the stochastic nature of the K-state; sister cells become competent with independent probabilities, and within a single cell lineage, cells enter and exit competence more than once, with no apparent "memory" effect. These observations led to the notion that competence is an "excitable" system. Older observations had shown that ComK negatively regulates the transcription of *comS* (40). Although the mechanism of this negative feedback is unknown, its existence suggested an explanation for the excitable behavior of the K-state, and indeed, a negative correlation between the activity of a ComK-dependent reporter and the *comS* promoter was documented by fluorescence microscopy (31). As a test of this idea, the feedback was bypassed by expressing *comS* from a ComK-dependent promoter, so that ComS remained abundant in the competent cells. As predicted, cells entered the K-state but exited about fivefold less frequently. These experiments suggest that ComS not only contributes to the ON state by preventing the degradation of ComK but also decays when a ComK-mediated negative-feedback loop prevents *comS* transcription, thus allowing cells to enter the OFF state.

This feedback model was further supported by expressing *comK* under the control of an inducible promoter in addition to the wild-type copy of *comK* (32). Within a certain inducer concentration range, the previously excitable behavior of the system was switched to oscillatory behavior, in which cells repeatedly entered

Figure 2 Diagram of the uptick mechanism (34). The top portion shows a graphical representation of the Rok and 0A~P concentrations, as well as the availability of RNApol and the rate of *comK* basal transcription (solid black line) during the transition to stationary phase. (RNApol availability is used as a plausible stand-in for the cause of the global increase in transcription that was observed.) The peak rate of transcription coincides with T_0, the time of departure from exponential growth. When the concentrations of available RNApol and of 0A~P are low (1), Rok is dominant and the rate of *comK* transcription is also low. As the concentration of 0A~P increases further, Rok is antagonized at sites A1, A2, and A3 and at the same time RNApol becomes more available. As a result, the rate of *comK* transcription increases (2). Finally, the 0A~P concentration reaches a level that is able to repress at R1 and R2 and *comK* transcription slows (3). In reality, of course, three demarcated periods of time do not exist. Note that the concentration of Rok remains constant throughout and both RNApol and 0A~P work to counteract its effects. Rok works at an unidentified site in addition to A1 to A3, shown here between A3 and R1. For simplicity, the availability of RNApol is shown as constant after T_0, although the data would suggest that it varies somewhat (34).

and departed from the ON state. This would be expected (41) if induction raised the probability that the amount of ComK rapidly exceeded a threshold and then, with a delay, destabilized ComK by repressing *comS*. Expression of *comS* from an additional inducible promoter had little effect on the probability of competence initiation, suggesting that noise in *comK* expression remained the limiting factor. But increased expression of *comS* did have a positive effect on the average duration of competence expression. The model predicts this behavior, because extra ComS increases the time required for its decay to a level that permits the degradation of ComK. Taken together, the data from the two papers of Süel et al. strongly favor a model in which positive- and negative-feedback loops can facilitate the entry into and departure from the competence ON state and in which the noise in *comK* expression determines the probability of transitions.

To further understand the dynamics of this system, the ComS-mediated feedback loop was replaced by one in which MecA was placed under the control of a *comK*-dependent promoter (30). In this engineered feedback loop, a negative regulator (MecA) was regulated positively by ComK, whereas with the native circuitry, a positive regulator (ComS) is regulated negatively by ComK. The native and synthetic circuits differed in a number of interesting ways. For example, the time of exit from the ON state was less precisely defined with the native circuit and the mean duration of the ON state was longer. This difference can be appreciated intuitively. Since normally *comS* is repressed as ComK accumulates, causing a decrease in the number of ComS molecules per cell, noise in ComS increases in importance, thereby increasing the variability in the time of escape from competence. The opposite situation applies to the synthetic circuit. When ComK is high, so is MecA, and its noise decreases in importance. When the average amount of MecA per cell is high, ComK is sequestered and degraded and cells enter the OFF state. These experiments have at least two important and related biological implications. First, it appears that the circuit architecture can influence the dynamic properties of the system and may well have evolved to select these features. Second, the less precise timing of emergence from the K-state exhibited by the native circuit is suggested to increase fitness in the face of changing environments because for an extended time, some cells will be transformable, a bet-hedging strategy. This is certainly plausible, although it is important to recognize that the K-state involves more than transformability and the selective pressures that modulate its dynamics are therefore difficult to identify with certainty.

Two Experimental Systems for Studying Competence

Each of the two experimental systems used to analyze the regulation of competence has advantages. In the first and more classical approach, cells reach stationary phase, nutrients are depleted, and growth slows. This system seems to result in a relatively synchronous and global increase and decrease in the probability of switching to the K-state. Aside from its experimental simplicity, it displays competence as a programmed developmental system and has been used to uncover the role of the *comK* uptick as well as the importance of positive feedback and of stochastic decision making. The second system emphasizes the probabilistic nature of the transitions to both the ON and OFF states, facilitating the study of competence as an excitable system with a negative-feedback loop that governs the escape from the K-state. It is likely that both systems mimic situations that exist in nature: periods of near starvation and of transient abundance. These dual aspects, namely, programmed variation and stochastic transitions, are also evident for the sessile-motile switch, discussed below.

LIVING TOGETHER OR SWIMMING SEPARATELY

Growing cultures of *B. subtilis* contain cells in two states with respect to motility (42). In swimmers, autolysin and motility genes are ON so that daughter cells detach from one another and can swim. In the sessile state, these genes are OFF and the cells remain in nonmotile chains. Because the cells are clonal, with identical genotypes, and are growing in a uniform environment, the choice between the ON and OFF states must be random. What is more, the two states can coexist and persist for several cell cycles, providing an example of epigenetic inheritance and of bistability in gene expression (43). Motile cells can exhibit chemotaxis, swimming toward attractants or away from repellents. Sessile cells can remain in place to exploit local riches, avoiding the risk of dispersal due to the random walk associated with motility even in the absence of attractants or repellents (44). They are also primed for biofilm formation. Two mechanisms have been proposed to explain the choice between the sessile and motile states: the epigenetic and σ^D switch models.

Epigenetic Switch Model

The elegant epigenetic switch model (43) begins with the production of SinI in response to low levels of

0A~P (Fig. 3A). In turn, SinI antagonizes the activity of the SinR repressor by direct protein-protein interaction (45). When SinI is absent, SinR homodimers repress the gene that encodes SlrR (46). When SinI is present, SlrR is produced and can form a heterocomplex with SinR, which directly represses the autolysin and motility genes, while SlrR antagonizes the activity of SinR homodimers (43). SinI and SlrR are both paralogs of SinR and bind to the latter, antagonizing its activity as a repressor of *slrR* and of the matrix genes (Fig. 3A). This unique situation creates a self-reinforcing, double negative-feedback loop, in which high SinR represses *slrR* and high SlrR inactivates SinR. As a result, two states can coexist in different cells within a population; high SinR corresponds to the motility ON state and high SlrR to the OFF state (Fig. 3A). Interestingly, a similar arrangement of mutually repressing proteins had been anticipated and analyzed in a synthetic-biology experiment in *Escherichia coli* (47), and this arrangement also resembles the Cro/CI double-negative lambda phage switch (48). SinR thus has two roles: as a partner with SlrR for repression of motility and autolysin genes and, as we shall see, as a direct repressor of genes for the expression of the biofilm matrix. The source of noise that triggers the switch is not known. Chai et al. (43) suggested that noise in *sinI* transcription is the primary stochastic event. This is plausible; *sinI* knockouts are permanently ON and overproducers are OFF. Of course the real "decider" may be variation in the level of 0A~P. It has been shown recently that *sinI* transcription occurs within a window of 0A~P concentration (35). Fluctuations in the level of 0A~P during growth (39, 49) might thus cause transient spikes in the level of SinI, and decision making may be based on small, stochastic variations in the concentration of SinI or of 0A~P.

σ^D as the Switch Protein

A different mechanism has been suggested by Cozy and Kearns (50) (Fig. 3B). The motility and autolysin genes require an alternative sigma factor (σ^D) for their transcription, and σ^D is encoded near the end of the 27-kb *fla-che* operon. For unknown reasons, transcription of the 31 genes in this operon falls off with distance from a common promoter, implying that the amount of σ^D will vary from cell to cell if the transcription rate of this operon is low. Indeed, ON cells were shown to have about 10-fold more *sigD* transcript than OFF cells, and OFF cells had less σ^D protein. In this model, noise in the degree of processivity of transcription or in the degradation of *fla-che* mRNA from the 3′ end may determine whether a cell becomes motile. Similarly to *comK*

expression, positive feedback may lock cells in the ON state, because an internal σ^D-dependent promoter is present within the *fla-che* operon and another is located upstream from the major SigA-dependent $P_{fla-che}$ promoter (50).

What determines the fall-off in transcription of the *fla-che* operon? Cozy et al. (51) have recently shown that when a second copy of a gene known as *slrA* was introduced at a chromosomal ectopic site, σ^D-dependent gene expression was inhibited and the cells were completely locked into the OFF state. *slrA* had been identified previously as a direct antagonist of SinR, acting like SinI in this respect (52, 53). As noted above (Fig. 3A), SinR is a repressor of *slrR* and an SlrR-SinR heterocomplex downregulates motility genes. Thus, it is reasonable to suggest that excess SlrA will titrate SinR, resulting in derepression of *slrR* (53). This will increase the amount of the SlrR-SinR complex and trigger the OFF state with respect to motility. Up to this point, the new finding with respect to SlrA is in accord with the epigenetic switch model and its circuitry. However, Cozy et al. (51) went on to show that the second *slrA* gene does not have its effect by repressing the *fla-che* promoter, although it does reduce the amount of σ^D protein and of σ^D-dependent gene expression. Instead, it dramatically decreases the amount of *fla-che* operon transcript distal to the promoter, apparently potentiating the phenomenon previously reported, in which transcript abundance falls off with distance from the initiation site (Fig. 3B). In other words, this evidence suggests that the SlrR-SinR heterocomplex acts postinitiation on the transcription of the *fla-che* operon. As predicted by this model, artificial expression of *sigD* bypassed the effect of extra SlrA, confirming that the amount of σ^D was limiting in the strain with two copies of *slrA*. These data do not agree with the epigenetic switch model, which is based partly on gel shifts and footprinting experiments demonstrating binding of SinR plus SlrR to the promoters of autolysin genes, implying direct repression by SinR plus SlrR. Cozy et al. (51) also used their system to demonstrate that the transition from the OFF to the ON state exhibits hysteresis and hypersensitivity with respect to the amount of σ^D, both hallmarks of a bistable system.

The epigenetic and σ^D switch models differ sharply in whether the autolysin and motility genes are regulated directly by the SlrR-SinR heterocomplex or indirectly via downstream effects on *sigD*. A problem with the σ^D switch model is that the dependence of the bistable behavior on σ^D is unexplained, although as noted above, this behavior may be due to an additional σ^D-dependent promoter believed to be embedded in the

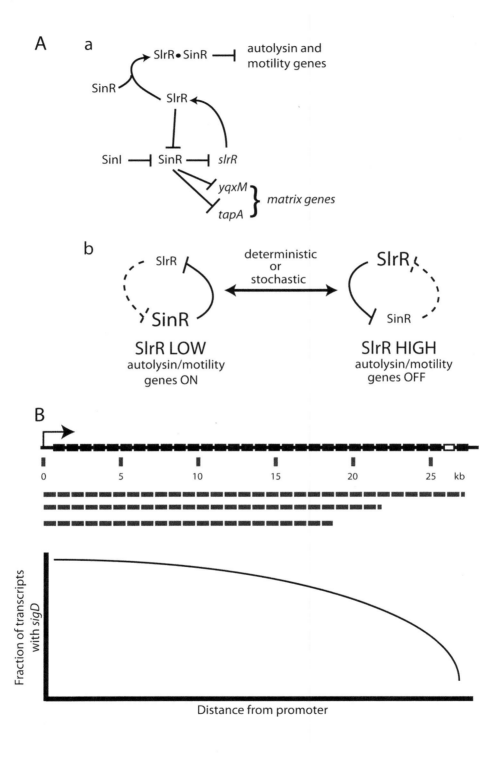

fla-che operon (50). It has been proposed that the epigenetic switch is responsible for the bistable regulation of motility but is biased by the amount of σ^D protein (43). In this view, the observation that motility ON cells exhibited enhanced *sigD* transcription, more σ^D-dependent gene expression, and more σ^D protein may reflect this bias. Nonetheless, the relationship between the two proposed mechanisms remains an important unresolved issue.

The roles of several other potentially important factors are not fully understood but hint at additional complexity (54). SwrA and SwrB bias the switch toward the ON position by increasing expression of the *fla-che* operon and by increasing the activity of σ^D, respectively (42, 55, 56). FlgM antagonizes σ^D, and DegU activates expression of *flgM* (57). Because SwrB is a membrane protein and DegU is a response regulator that is phosphorylated by DegS in response to unidentified signals, it is likely that these proteins are involved in signal transduction pathways that influence the switch bias. This would be reminiscent of the competence switch, where the probability of transition is modulated by signals that feed into the basal expression of *comK*. Also not fully understood is the role of YwcC in the control of biofilm formation (52).

SPORULATION AS A RESPONSE TO STARVATION

Spore formation, which results from the regulated expression of many genes, is the most-studied developmental adaptation of *B. subtilis* (58). The events that lead to the initiation of spore-specific gene expression are the focus of this discussion, rather than the later stages of spore formation. Classically, sporulation has been studied in poor growth media as a response to suboptimal conditions (59), although it apparently responds to different signals in the context of biofilms (60). In this section, we discuss the response to suboptimal conditions, which is typically studied in Difco Sporulation Medium (DSM) or resuspension media (61, 62).

As with competence and biofilm formation (see below), the initiation of spore formation is dependent on 0A~P (Fig. 1), and indeed, *spo0A* was first identified as a spore gene. While some promoters, like P*comK* and P*sinI*, respond to low levels of 0A~P, the spore promoters require higher levels (15). Because the average concentration of 0A~P increases during and after the transition to stationary phase, sporulation may be thought of as a later response than competence and the initiation of biofilm formation for the average cell.

Sporulation and the Phosphorelay

As noted above, the generation of 0A~P depends on the phosphorelay (13), an extraordinary pathway that consists of only four proteins at its core (a kinase, 0F, 0B, and 0A), but is modulated by a number of 0F~P phosphatases (63), several 0A~P phosphatases (64), and other proteins, such as Sda (65), YaaT (66), YlbF (67), and YmcA (68). Also, at least some of the five kinases (KinA to -E) can dephosphorylate 0F~P, and the entire pathway is reversible, so 0F can drain phosphoryl groups from 0B~P and perhaps more slowly from 0A~P. These proteins and chemical reactions represent points of information input, and the entire apparatus can be thought of as a signal integration device. In an early application of single-cell technology, the Grossman group reported that not all cells sporulate and that this difference was correlated with the low

Figure 3 Two proposed mechanisms controlling chaining and motility (A) The epigenetic switch (43). (Adapted with permission from the authors and the publisher from Fig. 1 in reference 43.) (a) SinI sequesters SinR, relieving repression of *slrR*. SlrR then binds to SinR, and the resulting complex represses the autolysin and motility genes and prevents repression of the matrix genes by SinR. (b) This circuitry allows for two metastable states. In one, when SlrR is low, the autolysin and motility genes are ON and the resulting cell is motile. In the other, when the SlrR concentration is high, these genes are OFF and the cells form chains and do not swim. The central feature of the circuitry that permits this bistable switch is the double-feedback mechanism involving repression of *slrR* by SinR and the inactivation of SinR for matrix gene repression by binding to SlrR. If SlrR is high, repression locks the cell in the motility OFF state, and vice versa. The transition between states can be stochastic, due to fluctuations in protein concentration (noise), or deterministic, in the sense that it is a programmed developmental switch. (B) Diagram of the gene position mechanism (50, 51). SigD is the penultimate gene in the 27-kb *fla-che* operon. For unknown reasons, the probability that promoter-distal genes are included in the operon transcript falls off with distance. Thus, if the mean number of transcripts per cell is low, some cells will have more SigD than others, and these cells will be motile. The distance-dependent fall-off in transcript abundance is reported to be due to the action of the SinR-SlrR heterocomplex (51).

expression of early spore genes in the nonsporulating cells (69). Presciently, it was suggested that only cells with amounts of 0A~P above a critical threshold would go on to sporulate, and this has been amply confirmed (14, 16, 17).

Because 0A~P can activate the transcription of *spo0A* via its effect on σ^H, it was suggested that this positive-feedback loop generated heterogeneity and committed cells for spore formation, by analogy with other bimodal systems such as competence (70). This hypothesis has been refuted by two publications (14, 17). Both of these studies described the kinetics of expression and the roles of phosphorelay components on the single-cell level and have further sought the source of noise that determines which cells sporulate. The first important conclusion concerned the positive-feedback loop in which 0A~P represses transcription of *abrB*, the resulting decrease in AbrB derepresses the gene encoding σ^H, and RNApol σ^H transcribes *spo0A*. This loop is not responsible for the major rise in the amount of 0A, because the rise precedes the observed decrease in the level of AbrB due to repression by 0A~P (14). The same conclusion was reached by de Jong et al. (17). Chastanet et al. (14) also examined another loop, in which 0A~P acts positively on the transcription of *spo0F*. Varying the expression of *spo0F* using an inducible promoter failed to increase the percentage of sporulating cells. Taken together, these and other results were interpreted as showing that the phosphorelay proteins are not rate limiting for spore formation. Instead, the synthesis of these proteins is adjusted to the need for 0A~P by the action of feedback loops, resulting in the coordinated synthesis of the components and of 0A~P. The feedback loops have been usefully described as comprising a "just-in-time" mechanism, which tunes the synthesis of the phosphorelay proteins (except 0B) to the needs of the system, ensuring that no component becomes rate limiting (14). An important contribution to this just-in-time mechanism is made by a complex arrangement of at least three 0A~P binding sites upstream from the two *spo0A* promoters and also translational control, which mediate promoter switching and activation of 0A synthesis (71).

Further insight was obtained by examining the single-cell expression of 0A~P activity, using a fusion of green fluorescent protein to the promoters of *spoIIA* and *spoIIE*, early spore genes that are transcribed only in the presence of 0A~P (14, 17). The expression of both promoters was unimodal and quite noisy compared to that of a control promoter, leading to the conclusion that the activity of 0A~P was likewise highly variable from cell to cell. Later in sporulation,

expression from P*spoIIA* becomes bimodal, presumably because expression continues to increase in cells committed to sporulation (17). Similar experiments with fluorescent protein fusions to other relevant promoters showed that the expression of *kinA*, *kinB*, *spo0F*, and *spo0A* was unimodal and variable (17, 72). These results suggested an important conclusion. Whereas bimodal expression and the bifurcation of the cell population into sporulating and nonsporulating components were most likely dependent on heterogeneity in the level of 0A~P, this heterogeneity was not rooted in bimodal expression of the phosphorelay components. Chastanet et al. (14) have proposed the appealing notion that the phosphorelay has evolved to be a "noise generator" and have proposed a computational model which was used to simulate the output of 0A~P. Overexpression of KinA and inactivation of *spo0E*, which encodes a 0A~P phosphatase, reduced cell-to-cell variation in 0A~P, consistent with the idea that sporulation heterogeneity is rooted in the phosphorelay.

Remarkably, during the progression to sporulation, individual cells exhibit reversible bursts of *spo0A* transcription and corresponding bursts in P*spo0F* activity, presumably due to activation of this promoter by 0A~P (73). Thus, the noise generator acts stochastically to produce pulses of 0A~P, which may be reversed by the action of Rap phosphatases acting on 0F, draining phosphate from 0A~P, or by the direct action of Spo0E or other phosphatases on 0A~P. A hallmark of spore formation is the localization of SpoIIE to the asymmetric septum (74, 75). This localization was also observed to be stochastic and reversible, so in about 2% of the cells localization of a SpoIIE fusion was seen to reverse, suggesting another layer of stochastic events on the level of protein localization (73). In contrast, the commitment to sporulation, measured by the activation of P*spoIIR*, was switch-like and not reversible; spores became apparent within a narrow window of time after P*spoIIR* was turned on. (The notion of irreversibility following activation of the σ^F-dependent *spoIIR* gene has been challenged recently [76], as explained below.) This "hybrid" model, which combines progression through reversible events with an irreversible commitment switch, was compared mathematically with two alternative models. In one, cells decide to sporulate in a single irreversible step. In the other, "reversible-only" model, even the decision to sporulate is reversible, distinguishing this model from the hybrid, or real-life, version. The three models were tested mathematically by subjecting them to stress of randomly varying durations, and the survival of the populations was determined. The hybrid model maximized survival in the face of unpredictable stress, with the reversible-only model doing well except with long stress durations. The latter result was explained by suggesting that if long periods of stress happen to be interrupted by short stress-free intervals, reversible-only cells run the risk of being caught sporeless, not able to turn on a dime and form spores. Perhaps an additional way to consider the bursts of 0A~P is that it is a bet-hedging strategy that allows cells to repeatedly enjoy the possibility of entering alternative states during a period of stress. Thus, they continue to explore "probability space" and thereby enjoy the possibility of becoming cannibals, becoming competent, entering biofilms, or sporulating. This sampling would be precluded were they to move rapidly and irreversibly to spores. As noted above, when cells encounter more sustained stress, the hybrid mechanism will enjoy an advantage over a purely reversible one (73).

A recent study (76) has addressed the issue of commitment to sporulation, by utilizing an inducible *kinA* construct in which the level of transcription can be varied by adjusting the concentration of inducer (77). Remarkably, spore formation exhibited an ultrasensitive response, so the frequency of sporulation increased about 20-fold when the added inducer concentration was increased only 2.5-fold. This nonlinear response was analyzed by a combination of modeling and *in vivo* experiments, leading to a number of conclusions. First, it was shown that the production of 0A~P exhibits a graded rather than an ultrasensitive response to *kinA* induction. Because an ultrasensitive response would be expected to generate a bimodal distribution, this is in accord with studies mentioned above, revealing that reporter gene expression for 0A~P in a sporulating population is noisy but unimodal (14, 17). The source of ultrasensitivity was then sought in two downstream events, in which the alternative sigma factors σ^F and σ^E are activated in the forespore and mother cell, respectively. Both of these events exhibited an ultrasensitive response to the induction of *kinA*. In both cases, ultrasensitivity was ascribed to the existence of well-characterized regulatory pathways, in which 0A~P directly activates the transcription of the sigma factor-encoding genes while indirectly activating each of the sigma factors posttranscriptionally. This arrangement establishes a pair of AND-gated coherent feed-forward loops (FFLs), which were shown by modeling to potentially lead to ultrasensitive responses (76) and are a common regulatory motif in both *Saccharomyces cerevisiae* and *E. coli* (78). Modeling and experimental data showed that the dynamic properties of the σ^F FFL cannot explain the ultrasensitive response of sporulation to *kinA* induction. In particular, σ^F is activated at a relatively low concentration of 0A~P, not within the range of *kinA* inducer concentrations that elicits the sporulation response. In contrast, the properties of σ^E induction were found to be impressively in accord with the observed sporulation response to *kinA*. These results lead to an elegant picture of sporulation. The phosphorelay generates the unimodal but very noisy production of 0A~P. At relatively low levels of 0A~P, σ^F is activated. This is a necessary but insufficient condition for sporulation, and cells with activated σ^F in which σ^E is not yet activated can fail to sporulate (76). Cells with a higher concentration of 0A~P cross a threshold that leads to a rapid (ultrasensitive) rise in activated σ^E, representing the point of irreversible commitment to sporulate (73, 79) and leading to the engulfment stage of spore formation. Coherent FFLs with AND-logic, operating exclusively on the level of transcription, have been described as "persistence detectors" (78). As pointed out previously (76), the analogous property in the present case may insulate cells from brief environmental fluctuations that might otherwise lead to a costly, unnecessary decision to sporulate.

The complexity of the phosphorelay, particularly with the intervention of numerous phosphatases and inputs from the cell cycle and cell-cell communication devices, offers many possibilities for generating variability while presenting a formidable challenge to deeper analysis. This complexity is underscored by the recent observation that 0A~P formation varies during the cell cycle, mediated in part by a burst of Sda synthesis when DNA replication is initiated (80). Sda inhibits the activity of kinases that feed phosphate to the phosphorelay and thus downregulates sporulation (65, 81). Because the Sda concentration is minimal prior to the completion of a round of replication, 0A~P presumably crosses the critical threshold for sporulation just before cell division. This timing mechanism tends to ensure that the developing spore will have the correct chromosome number. Also, in asynchronous cultures, only some cells will be in the appropriate stage of the division cycle for 0A~P formation, and this may contribute to the observed heterogeneity in 0A~P, in sporulation and perhaps in other forms of development.

CROSS-REGULATION VERSUS TEMPORAL COMPETITION BETWEEN COMPETENCE AND SPORULATION

Because several regulatory proteins have opposite effects on competence and sporulation, it has often been supposed that the pathways leading to these adaptations experience cross-regulation before commitment takes place (82–84). For example, SinI, which antagonizes SinR, is required for normal levels of spore formation, but inactivation of SinI has no effect on competence. In contrast, a null mutation of SinR depresses competence (40). It was reasonable to suppose that conditions leading to competence would require a low level of SinI and that prior to sporulation the SinI level would be elevated. However, the Süel group has posed a fundamental question concerning the relationship between competence and sporulation (85): does such cross-regulation take place prior to the decision point, or is there a "molecular race" in which the critical choice is made randomly, sending a cell toward one or another fate? According to this model, cross-regulation may indeed take place, but only after the decision point. This would serve to consolidate the choice, minimizing the appearance of dual-fate cells, which would be wasteful or even lethal.

To test this model, distinct fluorescent fusions of P*spo0A* (a reporter of 0A~P activity) and of P*comG* (for competence) were coexpressed and single-cell measurements were carried out (85). It was concluded that the probability of competence initiation remained constant during the approach to sporulation. These data were consistent with the molecular race model and with the notion that the cell fate decisions take place independently of one another. As a further test, the expression of SinI and of AbrB (the latter being another potential cross-regulatory molecule) was measured using promoter fusions in cells that would eventually become competent or sporulate. No differences in these promoter activities were observed prior to the decision points, suggesting the absence of cross-regulation, at least involving these promoters, although it is still possible that the amounts or activities of the SinI and AbrB proteins differed in the cells with the two ultimate fates. Interestingly, the two promoter activities did differ after the decision point, during the execution of competence and sporulation. As might be expected, P*abrB* activity was lower in spores and P*sinI* was higher. A small number (~0.1%) of "dual activity" cells which express *comK* also went on to sporulate. This number was predictable from the independent probabilities of the two cell fates. Smits et al. (86) have described another likely cross-regulation mechanism, which appears to prevent sporulation in competent cells. ComK induces the transcription of *rapH*. The RapH protein can dephosphorylate 0F~P, thus preventing sporulation. Accordingly, inactivation of RapH led to the formation of dually expressing cells. Like the cases involving SinR and AbrB, this cross-regulation must take place after the decision

point, because it depends on ComK. Since *rapH* must act after the decision point because it is produced as a product of ComK-dependent activation, it is possible to test its action as a cross-regulator by the use of a null mutant. This is not the case with *sinR* and *abrB*, because their elimination cannot distinguish between activity before or after the point of decision. Thus, although the data are consistent with the action of these molecules in cross-regulation after a decision has been made, there is no independent evidence that they do act in this manner.

Nevertheless, these findings are consistent with the following simple model for the relationship between sporulation and competence. The probabilities of competence and sporulation vary independently, partly in response to bursts in 0A~P production (73) and to a gradual increase in the average concentration of this molecule and to excursions in the basal transcription of *comK* (34). Spore initiation requires more 0A~P than competence and does, on the average, occur later. Once a cell enters one or the other pathway, cross-regulation precludes the alternative fate. If the choices to become competent and sporulate are made during a narrow interval, before the cross-regulation responses can be mounted, dual-activity cells result. Although there may be some cost to this infrequent outcome, it is presumably less than the cost of an elaborate cross-regulation mechanism that can operate before the decisions are made (85).

BIOFILM FORMATION

Like many other microorganisms, *B. subtilis* forms complex communities known as biofilms (87). Indeed, the rich developmental potential of this model organism can be appreciated only in the context of biofilm development, because here the cell types discussed above appear with their temporal and spatial specificities revealed. Because biofilms contain motile, sessile, and sporulating cells, understanding their development depends on many of the interactions discussed above. The capacity to form robust biofilms has been lost by the descendants of laboratory strain 168 and is best studied with natural isolates, such as 3610, the probable parent of 168 (12). Not all media support the development of biofilms, and MsGG, in which glycerol and glutamate are the only organic molecules, is commonly used for biofilm investigations.

Biofilms Contain Several Cell Types

As a biofilm is initiated, cells become embedded in an insoluble matrix, consisting of protein fibers (encoded by the *tapA* [*yqxM*]-*sipW*-*tasA* operon) and an uncharacterized polysaccharide (the synthesis of which is encoded by the *eps* operon) (88). An important study used fluorescent reporters to the promoters of *hag*, *tapA*, and *sspB* as surrogates for the expression of motility, matrix, and spore genes, respectively (89). Competence is missing from this picture, only because strain 3610 is very poorly competent and other isolates have not been intensively studied. Early in biofilm formation at an agar-air interface, *hag* was expressed in a majority of cells. At later times the number of motility-expressing cells decreased, and these were located at the edges and the base of the colony. Matrix-producing cells peaked in number after about 24 h and then declined. Sporulating cells increased steadily in number and were present preferentially in aerial projections from the biofilm surface, as shown previously (90). Time-lapse microscopy revealed that *hag* and *tasA* were not expressed in the same cells at the same time. Rather, cells expressing *hag* switched to *tasA* expression with time, consistent with the temporal progression noted above. Remarkably, sporulating cells arose mostly from matrix producers, resulting in a physical separation of motile and sporulating cells in different regions of the biofilm. Clearly, the various cell types tended to locate to different regions of the structure, and they showed specific temporal patterns of expression and distinct lineages. What is more, inactivation of *tasA* or *eps* resulted in a defect not only in biofilm architecture but also in the expression of a spore gene and in the regulation of the *hag* and *tasA* promoters.

The fact that different regions of a biofilm contain cells with different expression patterns tells us that geographically differentiated environmental signals influence the choice of cell fate. When studied in suspended, dispersed cultures, matrix genes are expressed in only a low percentage of the cells, and these cells are presumably selected randomly (91). The stochastic nature of decision making found for matrix production as well as for spore formation and for the expression of motility genes when dispersed cultures are studied suggests that chance as well as necessity probably plays a role in cell fate determination within a biofilm. Work emerging mainly from collaborations between R. Kolter and R. Losick and their coworkers has shed much light on the regulatory events that enable *B. subtilis* to produce biofilms.

SinR, 0A~P, and Regulation in Biofilms

SinR regulates biofilm formation by at least two pathways and is the major proximal regulator of biofilm formation (Fig. 1). As described above, reduced SinR activity leads to a decrease in motility, driving cells into

the sessile state, in which extensive chaining takes place. Additionally, SinR is a direct repressor of the biofilm-specific *eps* and *tapA* operons (92, 93). As noted above, SinI antagonizes SinR by forming SinI-SinR heterocomplexes (45). *sinI* transcription is driven by 0A~P, and as 0A~P increases in amount, the activity of SinR decreases and the repression of the *eps* and *tapA* operons is lifted. What is more, derepression of *eps* leads to the synthesis of EpsE, 1 of the 15 proteins encoded by this operon. In addition to participating in matrix synthesis, EpsE acts as a clutch protein, disconnecting the flagellar motor from its power source (94). Thus, a posttranslational mechanism contributes to the transition from the motile ON state to the OFF state accompanying biofilm formation.

In DSM starvation medium, as the amount of 0A~P continues to increase, *sinI* transcription decreases, limiting the titration of SinR. It has been shown that this bandpass regulation of *sinI* transcription is due to the direct action of 0A~P on activating and repressing sites near P*sinI* (35). On the other hand, in MsGG, as biofilms are formed, 0A~P remains at an intermediate level and the amount of SinI continues to increase, leading to the formation of matrix-producing cells. The regulation of *sinI* by 0A~P in DSM is analogous to the uptick in *comK* transcription, which is also dependent on direct binding of an increasing amount of 0A~P (34). However, whereas the uptick in *comK* temporally adjusts the probability of competence, the uptick of *sinI* transcription in sporulation medium does not cause biofilm formation, and it appears to be an unavoidable consequence of a regulatory system that has evolved for dual use: *sinI* must be repressed by 0A~P under sporulation conditions to avoid matrix production but must be expressed at an intermediate level to antagonize SinR for biofilm formation.

However, this simple description faces a potential dilemma within biofilms (35). The double negative loop proposed to regulate motility that was described above exhibits hysteresis, an inherent feature of such regulatory networks (43). In other words, when *slrR* was transiently overexpressed from an isopropyl-β-D-thiogalactopyranoside (IPTG)-inducible promoter, the system became epigenetically fixed in the motility OFF state. This is because SlrR titrates and inactivates SinR as a repressor of *slrR*, eliminating the need for the continued presence of an inducer to achieve high levels of SlrR (Fig. 2). In this motility OFF state, SinR is titrated not only by SinI but also by SlrR, and the matrix genes are derepressed. In a biofilm, most sporulating cells are derived from matrix producers. It is expected that if a matrix-producing cell goes on to sporulate, even if its

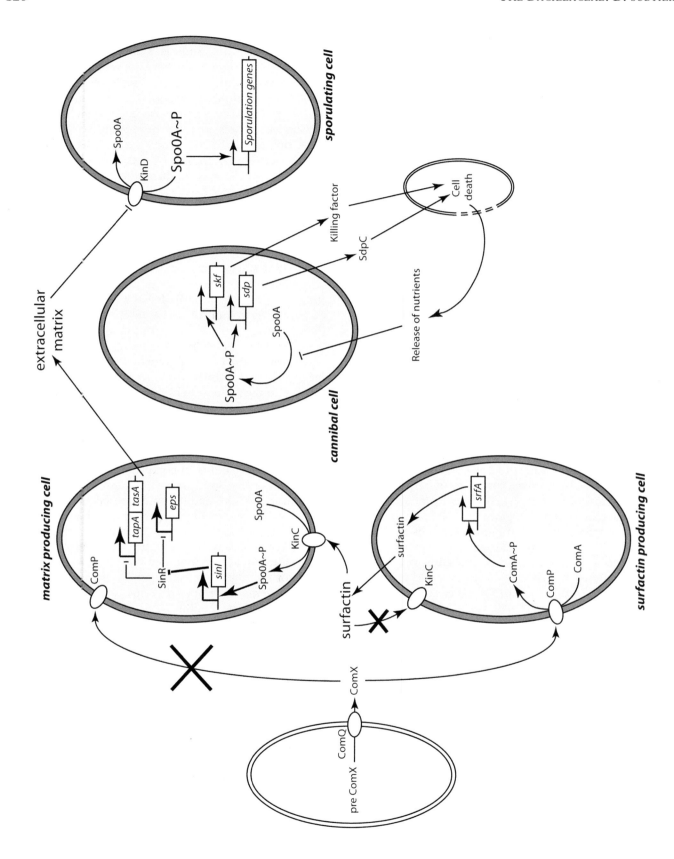

elevated level of 0A~P shuts down *sinI* expression, the previous production of SinI would have locked the cell into a matrix-producing state. But the evidence shows that under biofilm conditions, sporulating cells do not express matrix genes, although they derive from matrix producers (89). This problem is resolved by a beautifully simple proposed mechanism (35). SinR apparently binds cooperatively to its matrix gene targets but non-cooperatively to SinI. Thus, repression is hypersensitive to the concentration of SinR, but antirepression is not. Because sporulating cells contain two chromosomes, one destined for the forespore and the other for the mother cell, gene dosage of the *sinI-sinR* operon will ensure that the effect of repression by SinR will dominate over antirepression by SinI and the hysteretic effect of prior high SinI will be overcome, shutting down matrix production. In support of this model, it was shown that doubling the copy number of the *sinI-sinR* operon shuts off matrix gene expression. Thus, two effects influence the transition from matrix production to sporulation. The first depends on the presence of activation and repression sites in the *sinI* promoter for 0A~P that ensures an intermediate amount of SinI, and the second depends on gene dosage. This raises an interesting point. As noted above, in sporulating cells the activity of P*sinI* is elevated (85), likely serving to shut down competence after the decision to sporulate. For the transition from matrix production to sporulation to proceed in the biofilm, the concentration of SinI must be high enough to prevent SinR from inhibiting spore formation (95, 96), but not high enough to perpetuate matrix formation. This suggests that the reported binding of SinR to the promoters of spore genes must be relatively weak.

Why Spores in a Biofilm Come from Matrix-Producing Cells

Why are spores within a biofilm derived mainly from matrix producers, and why do mutant cells deficient

in matrix formation fail to develop spores under biofilm conditions unless they are starved for nutrients (89)? Reporter fusions to 0A~P-dependent promoters showed that in MsGG medium, matrix-deficient mutants contained decreased levels of 0A~P, consistent with their failure to sporulate (60). Remarkably, inactivation of KinD, one of the five kinases that can feed phosphoryl groups to the phosphorelay, overcame the spore deficiency of a double *eps tasA* mutant. It was proposed that KinD, like other histidine kinases (97), can act as a phosphatase as well as a kinase and in this case was preventing spore formation by limiting the accumulation of 0A~P. KinD is a membrane protein, and presumably, its phosphatase activity is down-regulated when it senses the presence of matrix. This hypothesis is consistent with the observation that co-cultivation of matrix mutants with matrix-producing sporulation mutants can complement sporulation in *trans*, demonstrating that matrix is sensed at the cell surface (60).

Quorum Sensing in Biofilm Development

It appears, then, that the decision of some cells in a biofilm to sporulate is at least in part deterministic, triggered by the production of matrix. But how are the matrix-producing cells determined? Lopez et al. (98) have discovered that a diverse set of molecules can trigger matrix production, even in a medium (LB broth) in which this does not usually occur. The common feature of these reagents is that they induce the leakage of potassium from the cell. Among the chemicals with this activity is surfactin, a natural product that is encoded by the *srfA* operon of *B. subtilis* (Fig. 4). KinC and KinD are both known to be required for biofilm formation (99), but only KinC, a membrane-localized protein, is needed for the response to surfactin (98). These and other observations strongly suggest that potassium leakage causes KinC to feed phosphate directly or via the

Figure 4 Cell type determination in biofilms (60, 98, 104, 107). Pre-ComX is processed and ComX is secreted with the aid of ComQ. ComX interacts with ComP at the cell surface, resulting in the phosphorylation of ComA and the transcriptional activation of *srfA*. The surface-active SrfA molecule induces potassium flux in a susceptible cell, activating KinC and the formation of small amounts of 0A~P. For unknown reasons, the surfactin-producing cell itself becomes refractory to activation by surfactin. In the susceptible cell, 0A~P activates the transcription of *sinI*, which interacts with SinR, relieving repression of the matrix genes. For unknown reasons, matrix producers are not activated to produce surfactin. The presence of matrix downregulates the phosphatase activity of KinD, permitting the 0A~P concentration to rise further, inducing sporulation. Matrix producers also become cannibals, because their intermediate 0A~P concentration triggers toxin production. These toxins kill nonproducers, which release nutrients, delaying sporulation. As a result, matrix producers proliferate, increasing the population of eventual sporulating cells.

phosphorelay to 0A, activating the transcription of *eps* and the *tapA* operon by the pathways described above.

These findings have several important implications. First, they establish the role of quorum-sensing systems for biofilm formation in *B. subtilis*. The two-component ComP-ComA proteins, which are activated by a quorum-sensing system, transcriptionally activate *srfA* by direct binding of ComA~P to the *srfA* promoter. Surfactin then acts as a quorum-sensing signal to modulate KinC activity. Thus, a cascade of quorum-sensing and two-component regulatory proteins are involved in biofilm formation, emphasizing the importance of two-component regulators for *Bacillus* biology and the central role of cell-cell communication. Second, because diverse natural products produced by various species of soil-dwelling bacteria, including other bacilli, cause the release of potassium, it is likely that in nature, interstrain communication can cause biofilm formation. It is noteworthy that the identification of potassium leakage as a signal for KinC activation and of matrix as a signal for KinD are the only specific input signals so far identified leading to the phosphorylation of 0A via particular kinases. Although the molecular details of the signaling mechanisms are not understood, these are important findings.

It is also worth noting that the ComX pheromone that activates ComP exhibits a fascinating diversity, in that four specificity classes have been identified in natural isolates (100–103). Activation can occur within each class, but the ComX molecules produced by members of one class cannot activate the ComP receptors of another. A sequence relatedness tree constructed for the pre-ComX proteins is congruent with that of the ComP receptors, and these sequences define the specificity classes. In contrast, the sequence relatedness of ComP and ComX is not congruent with that of housekeeping proteins. The evolutionary mechanisms that have led to this "pherotype" specificity are not understood, but somehow *Bacillus* has evolved to respond to the ComX pheromones produced by restricted classes of related bacteria, which are not necessarily the closest relatives. Whatever the selective pressures, this specificity potentially affects not only the K-state, which depends on the phosphorylation of ComA, but also the natural history of biofilms via their impact on surfactin production.

Cell Fate Determination

So far in our discussion, a linear signaling model has been described (Fig. 4). ComX activates ComP-ComA by phosphoryl transfer, and ComA~P activates *srfA*. Then surfactin, a second quorum-sensing molecule, which is the product of *srfA*, activates KinC, triggering

matrix production. Finally, matrix itself inactivates the phosphatase activity of KinD, acting extracellularly. While KinC allows sufficient 0A~P formation to produce matrix, but not enough for sporulation, the inactivation of KinD-associated phosphatase activity permits sufficient 0A~P to accumulate to trigger sporulation. How do these interactions generate cell type heterogeneity within the biofilm? By the use of promoter fusions to spectrally distinct fluorescent proteins, López et al. (104) observed that under conditions of biofilm development, only about 10% of the cells responded to the presence of the ComX pheromone (Fig. 4). This was not true for a domesticated 168 derivative, in which a more uniform response was observed. The explanation for this bimodal response is not known, but it was suggested that it may depend on a positive-feedback loop involving *phrC*, which is transcribed in the presence of ComA~P (Fig. 1) (105). Thus, cells which initially respond more to ComX would produce more PhrC, preventing RapC from inhibiting the binding of ComA~P to the *srfA* promoter, thus establishing a positive-feedback loop. Although ComA~P activates transcription of the *rapC phrC* operon, a second promoter, dependent on σ^H, drives transcription of only *phrC*, suggesting that the synthesis of PhrC may exceed that of RapC. However, this scenario seems unlikely, because PhrC is secreted and its effects would presumably be distributed to nonproducing cells. Also, if the cells that initially respond more to ComX are truly selected randomly, it is not clear why with time more and more cells do not become responders, unless some sort of additional regulation limits their number. Perhaps less-than-saturating amounts of ComX are produced by strain 3610, so only 10% of the cells cross the threshold for activation of ComP. Whatever the explanation, the stochastic selection of cells that respond to ComX provides a likely source of heterogeneity in cell fate.

Remarkably, the cells that produce surfactin in response to ComX are not the ones that produce matrix (104) (Fig. 4). This is because these responders cannot be activated by the surfactin that they produce. Also, once cells are producing matrix, they can no longer be activated by ComX to produce surfactin. Thus, the population bifurcates randomly into at least two populations with respect to matrix and surfactin production. In 168, a gene essential for competence (*comS*) is transcribed from the *srfA* promoter. Because 3610 is poorly competent, whether the surfactin producers have the potential to become competent cannot be readily studied in that strain. As noted above, matrix producers sporulate, and the bifurcation between matrix and surfactin producers has at least that additional consequence. It is interesting to note that some cells neither produce matrix nor become activated for surfactin production. Do these cells have an additional specialized role within the biofilm? Do they languish until the starvation pathway induces sporulation, do they lyse, or are they released from the biofilm to swim away (see below)?

Cannibalism in the Biofilm

In dispersed cultures grown under spore-forming conditions, some cells (cannibals) produce two toxins that kill sister cells (106). These martyrs release nutrients that are consumed by the cannibals, delaying their sporulation. The toxin-producing cells are immune to the lethal effects of the toxins. Cannibalism appears to be a "last chance" strategy that postpones the decision to sporulate. Because the toxin genes are controlled by 0A~P, and because their promoters exhibit high affinity for this activator, cannibals are among the first cells with activated 0A~P production. This is another example of the heterochronic production of 0A~P underlying heterogeneity in *B. subtilis* development.

As noted above, small to moderate amounts of 0A~P, produced in response to surfactin, trigger matrix synthesis in biofilms. It is therefore reasonable that surfactin can also activate toxin production when added under conditions (growth in LB broth) that do not ordinarily cause this to happen (107). These surfactin-responsive cells were shown to be the same ones that are triggered by surfactin to transcribe matrix genes, so the same cells in a biofilm produce toxins and matrix and only non-matrix-producing cells are killed. This has two consequences: sporulation is delayed, and the matrix toxin producers receive nutrients, which enables them to proliferate, and the proportion of matrix producers in the biofilm thus increases. It has been proposed that the production of matrix and toxins may be defensive and offensive weapons induced by other bacteria in the environment that secrete potentially damaging chemical agents causing potassium leakage.

Escape from the Biofilm

Although most cells in the biofilm eventually sporulate when nutrients are depleted, it appears that mechanisms exist that enable vegetative cells to swim away. This is perhaps one more example of bet hedging in which pioneers set out to explore the neighborhood for food or escape from toxins or predators, avoiding the costs of sporulation. Two dispersal mechanisms involve the secretion of D-amino acids, which interfere with the attachment of TasA fibers to the cell surface, facilitating breakdown of the matrix (108) and of norspermi-

dine, which apparently attacks the exopolysaccharide matrix (109). It will be interesting to discover how the synthesis of these small molecules is regulated and whether they are expressed in only a subset of cells. An additional mechanism has been proposed to contribute to a return to motility, which aids in dispersal. As described above, SlrR maintains cells in the sessile, chaining state, and it has been shown that late in the life of biofilms, SlrR becomes unstable (110). SlrR degradation appears to be due in part to autocleavage and in part to a mechanism that requires ClpC.

RARE VERSUS PROGRAMMED DEVELOPMENT: MecA AS A BUFFER

In the preceding discussion, we considered a number of developmental pathways that combine stochastic decision making and programmed regulatory events that modulate the probability that a given pathway is activated. Thus, the uptick in basal *comK* transcription or the programmed but noisy increase in 0A~P production determines the chances that random cells will embark on competence or spore formation or form biofilms. However, because the critical events in these processes respond with thresholds for transcription factor binding and because the basal expression of determining molecules (ComK and 0A~P) is nonzero and noisy, the probabilities of cells embarking on development is also not zero even during exponential growth in rich media. Thus, when reporter fusions are used, cells that express *comK*, *spoIIG* (a sporulation gene), or *eps* are readily detected in growing cells, albeit at very low frequencies (111). In fact, small numbers of mature spores and transformable cells also form in growing cultures. These considerations echo the dichotomy between excitable and programmed transitions in motility and competence.

It is plausible that the low frequency of development in exponential populations is a form of a priori bet hedging. Thus, a small investment may be made that increases the likelihood of survival in the face of an unexpected environmental catastrophe; a few cells will survive as spores, a few can receive new genes that may enhance fitness, and some may produce matrix, which primes them for biofilm formation. These stochastic switches of rare cells into developmental pathways during growth stand in contrast to the programmed transitions that occur when growth slows, and they are likely due to noise in the formation of 0A~P or of a more downstream regulator like ComK or SinI. If these transitions serve to enhance fitness, as proposed, their rates have likely been adjusted by selection.

It has been observed that the frequency of these rare developing cells increases dramatically when *mecA* is inactivated (111). As described above, MecA has been studied as an adaptor protein that targets ComK for degradation unless the targeting is relieved by the synthesis of ComS late in growth in response to quorum sensing. In loss-of-function *mecA* mutants, the autostimulation of *comK* is uncontrolled, and massive production of competent cells takes place during growth. In addition, cells that express *eps* or *spoIIG* are more frequent, as are the heat-resistant spores produced during exponential growth. This effect appears to be due to the ability of MecA to bind 0A, somehow interfering with the ability of 0A~P to activate transcription (111). This interference is not exerted via the phosphorelay, nor can MecA by itself dephosphorylate 0A~P in vitro.

MecA may have evolved as a buffer that limits the chances that cells will enter developmental pathways until programmed mechanisms come into play and the buffer is inactivated (as by ComS) or overwhelmed by the activated phosphorelay. By adjusting the affinities of MecA for its binding partners, evolution can tinker with the rates of transitions for particular pathways.

Acknowledgments. We thank Jeanie Dubnau and Dan Kearns for helpful comments on the manuscript.

Work from our lab was supported by NIH grant GM057720. The authors have no conflicts of interest regarding this manuscript.

Citation. Mirouze N, Dubnau D. 2013. Chance and necessity in *Bacillus subtilis* development. Microbiol Spectrum 1(1): TBS-0004-2012.

References

1. **Balazsi G, van Oudenaarden A, Collins JJ.** 2011. Cellular decision making and biological noise: from microbes to mammals. *Cell* **144:**910–925.

2. **Losick R, Desplan C.** 2008. Stochasticity and cell fate. *Science* **320:**65–68.

3. **Eldar A, Elowitz MB.** 2010. Functional roles for noise in genetic circuits. *Nature* **467:**167–173.

4. **Raj A, van Oudenaarden A.** 2008. Nature, nurture, or chance: stochastic gene expression and its consequences. *Cell* **135:**216–226.

5. **Elowitz MB, Levine AJ, Siggia ED, Swain PS.** 2002. Stochastic gene expression in a single cell. *Science* **297:** 1183–1186.

6. **Swain PS, Elowitz MB, Siggia ED.** 2002. Intrinsic and extrinsic contributions to stochasticity in gene expression. *Proc Natl Acad Sci USA* **99:**12795–12800.

7. **Taniguchi Y, Choi PJ, Li GW, Chen H, Babu M, Hearn J, Emili A, Xie XS.** 2010. Quantifying *E. coli* proteome and transcriptome with single-molecule sensitivity in single cells. *Science* **329:**533–538.

8. **Hamoen LW, Smits WK, de Jong A, Holsappel S, Kuipers OP.** 2002. Improving the predictive value of the competence transcription factor (ComK) binding site in *Bacillus subtilis* using a genomic approach. *Nucleic Acids Res* **30:**5517–5528.

9. **Ogura M, Yamaguchi H, Kobayashi K, Ogasawara N, Fujita Y, Tanaka T.** 2002. Whole-genome analysis of genes regulated by the *Bacillus subtilis* competence transcription factor ComK. *J Bacteriol* **184:**2344–2351.

10. **Berka RM, Hahn J, Albano M, Draskovic I, Persuh M, Cui X, Sloma A, Widner W, Dubnau D.** 2002. Microarray analysis of the *Bacillus subtilis* K-state: genome-wide expression changes dependent on ComK. *Mol Microbiol* **43:**1331–1345.

11. **Lemon KP, Earl AM, Vlamakis HC, Aguilar C, Kolter R.** 2008. Biofilm development with an emphasis on *Bacillus subtilis. Curr Top Microbiol Immunol* **322:**1–16.

12. **McLoon AL, Guttenplan SB, Kearns DB, Kolter R, Losick R.** 2011. Tracing the domestication of a biofilm-forming bacterium. *J Bacteriol* **193:**2027–2034.

13. **Burbulys D, Trach KA, Hoch JA.** 1991. Initiation of sporulation in *B. subtilis* is controlled by a multicomponent phosphorelay. *Cell* **64:**545–552.

14. **Chastanet A, Vitkup D, Yuan GC, Norman TM, Liu JS, Losick RM.** 2010. Broadly heterogeneous activation of the master regulator for sporulation in Bacillus subtilis. *Proc Natl Acad Sci USA* **107:**8486–8491.

15. **Fujita M, Gonzalez-Pastor JE, Losick R.** 2005. High- and low-threshold genes in the Spo0A regulon of *Bacillus subtilis. J Bacteriol* **187:**1357–1368.

16. **Fujita M, Losick R.** 2005. Evidence that entry into sporulation in *Bacillus subtilis* is governed by a gradual increase in the level and activity of the master regulator Spo0A. *Genes Dev* **19:**2236–2244.

17. **de Jong IG, Veening JW, Kuipers OP.** 2010. Heterochronic phosphorelay gene expression as a source of heterogeneity in *Bacillus subtilis* spore formation. *J Bacteriol* **192:**2053–2067.

18. **van Sinderen D, Luttinger A, Kong L, Dubnau D, Venema G, Hamoen L.** 1995. *comK* encodes the competence transcription factor, the key regulatory protein for competence development in *Bacillus subtilis. Mol Microbiol* **15:**455–462.

19. **Turgay K, Hahn J, Burghoorn J, Dubnau D.** 1998. Competence in *Bacillus subtilis* is controlled by regulated proteolysis of a transcription factor. *EMBO J* **17:**6730–6738.

20. **Serror P, Sonenshein AL.** 1996. CodY is required for nutritional repression of *Bacillus subtilis* genetic competence. *J Bacteriol* **178:**5910–5915.

21. **Hamoen LW, Kausche D, Marahiel MA, van Sinderen D, Venema G, Serror P.** 2003. The *Bacillus subtilis* transition state regulator AbrB binds to the −35 promoter region of comK. *FEMS Microbiol Lett* **218:**299–304.

22. **Hoa TT, Tortosa P, Albano M, Dubnau D.** 2002. Rok (YkuW) regulates genetic competence in *Bacillus subtilis* by directly repressing *comK. Mol Microbiol* **43:**15–26.

23. **Haijema BJ, Hahn J, Haynes J, Dubnau D.** 2001. A ComGA-dependent checkpoint limits growth during the escape from competence. *Mol Microbiol* **40:**52–64.

24. Magnuson R, Solomon J, Grossman AD. 1994. Biochemical and genetic characterization of a competence pheromone from *B. subtilis*. *Cell* 77:207–216.

25. Prepiak P, Dubnau D. 2007. A peptide signal for adapter protein-mediated degradation by the AAA+ protease ClpCP. *Mol Cell* 26:639–647.

26. Smits WK, Eschevins CC, Susanna KA, Bron S, Kuipers OP, Hamoen LW. 2005. Stripping *Bacillus*: ComK autostimulation is responsible for the bistable response in competence development. *Mol Microbiol* 56:604–614.

27. Maamar H, Dubnau D. 2005. Bistability in the *Bacillus subtilis* K-state (competence) system requires a positive feedback loop. *Mol Microbiol* 56:615–624.

28. Leisner M, Stingl K, Radler JO, Maier B. 2007. Basal expression rate of comK sets a 'switching-window' into the K-state of *Bacillus subtilis*. *Mol Microbiol* 63:1806–1816.

29. Maamar H, Raj A, Dubnau D. 2007. Noise in gene expression determines cell fate in *Bacillus subtilis*. *Science* 317:526–529.

30. Cagatay T, Turcotte M, Elowitz MB, Garcia-Ojalvo J, Süel GM. 2009. Architecture-dependent noise discriminates functionally analogous differentiation circuits. *Cell* 139:512–522.

31. Süel GM, Garcia-Ojalvo J, Liberman LM, Elowitz MB. 2006. An excitable gene regulatory circuit induces transient cellular differentiation. *Nature* 440:545–550.

32. Süel GM, Kulkarni RP, Dworkin J, Garcia-Ojalvo J, Elowitz MB. 2007. Tunability and noise dependence in differentiation dynamics. *Science* 315:1716–1719.

33. Albano M, Hahn J, Dubnau D. 1987. Expression of competence genes in *Bacillus subtilis*. *J Bacteriol* 169:3110–3117.

34. Mirouze N, Desai Y, Raj A, Dubnau D. 2012. Spo0A~P imposes a temporal gate for the bimodal expression of competence in *Bacillus subtilis*. *PLoS Genet* 8:e1002586.

35. Chai Y, Norman T, Kolter R, Losick R. 2011. Evidence that metabolism and chromosome copy number control mutually exclusive cell fates in *Bacillus subtilis*. *EMBO J* 30:1402–1413.

36. Sen S, Garcia-Ojalvo J, Elowitz MB. 2011. Dynamical consequences of bandpass feedback loops in a bacterial phosphorelay. *PLoS One* 6:e25102.

37. Barker MM, Gaal T, Gourse RL. 2001. Mechanism of regulation of transcription initiation by ppGpp. II. Models for positive control based on properties of RNAP mutants and competition for RNAP. *J Mol Biol* 305:689–702.

38. Zhou YN, Jin DJ. 1998. The rpoB mutants destabilizing initiation complexes at stringently controlled promoters behave like "stringent" RNA polymerases in *Escherichia coli*. *Proc Natl Acad Sci U S A* 95:2908–2913.

39. Mirouze N, Prepiak P, Dubnau D. 2011. Fluctuations in spo0A transcription control rare developmental transitions in *Bacillus subtilis*. *PLoS Genet* 7:e1002048.

40. Hahn J, Kong L, Dubnau D. 1994. The regulation of competence transcription factor synthesis constitutes a critical control point in the regulation of competence in *Bacillus subtilis*. *J Bacteriol* 176:5753–5761.

41. Barkai N, Leibler S. 2000. Circadian clocks limited by noise. *Nature* 403:267–268.

42. Kearns DB, Losick R. 2005. Cell population heterogeneity during growth of *Bacillus subtilis*. *Genes Dev* 19:3083–3094.

43. Chai Y, Norman T, Kolter R, Losick R. 2010. An epigenetic switch governing daughter cell separation in *Bacillus subtilis*. *Genes Dev* 24:754–765.

44. Berg HC. 1983. *Random Walks in Biology*. Princeton University Press, Princeton, NJ.

45. Bai U, Mandic-Mulec I, Smith I. 1993. SinI modulates the activity of SinR, a developmental switch protein of *Bacillus subtilis*, by protein-protein interaction. *Genes Dev* 7:139–148.

46. Chu F, Kearns DB, McLoon A, Chai Y, Kolter R, Losick R. 2008. A novel regulatory protein governing biofilm formation in *Bacillus subtilis*. *Mol Microbiol* 68:1117–1127.

47. Gardner TS, Cantor CR, Collins JJ. 2000. Construction of a genetic toggle switch in *Escherichia coli*. *Nature* 403:339–342.

48. Oppenheim AB, Kobiler O, Stavans J, Court DL, Adhya S. 2005. Switches in bacteriophage lambda development. *Annu Rev Genet* 39:409–429.

49. Castilla-Llorente V, Salas M, Meijer WJ. 2008. kinC/D-mediated heterogeneous expression of spo0A during logarithmical growth in *Bacillus subtilis* is responsible for partial suppression of phi 29 development. *Mol Microbiol* 68:1406–1417.

50. Cozy LM, Kearns DB. 2010. Gene position in a long operon governs motility development in *Bacillus subtilis*. *Mol Microbiol* 76:273–285.

51. Cozy LM, Phillips A, Calvo RA, Bate A, Hsueh Y-H, Bonneau R, Eichenberger P, Kearns DB. 2012. SlrA/SlrR/SinR inhibits motility gene expression upstream of a hypersensitive and hysteric switch at the level of sD in *Bacillus subtilis*. *Mol Microbiol* 83:1210–1228.

52. Kobayashi K. 2008. SlrR/SlrA controls the initiation of biofilm formation in *Bacillus subtilis*. *Mol Microbiol* 69:1399–1410.

53. Chai Y, Kolter R, Losick R. 2009. Paralogous antirepressors acting on the master regulator for biofilm formation in *Bacillus subtilis*. *Mol Microbiol* 74:876–887.

54. Patrick JE, Kearns DB. 2012. Swarming motility and the control of master regulators of flagellar biosynthesis. *Mol Microbiol* 83:14–23.

55. Werhane H, Lopez P, Mendel M, Zimmer M, Ordal GW, Marquez-Magana LM. 2004. The last gene of the *fla/che* operon in *Bacillus subtilis*, *ylxL*, is required for maximal σD function. *J Bacteriol* 186:4025–4029.

56. Calvio C, Celandroni F, Ghelardi E, Amati G, Salvetti S, Ceciliani F, Galizzi A, Senesi S. 2005. Swarming differentiation and swimming motility in *Bacillus subtilis* are controlled by *swrA*, a newly identified dicistronic operon. *J Bacteriol* 187:5356–5366.

57. Hsueh YH, Cozy LM, Sham LT, Calvo RA, Gutu AD, Winkler ME, Kearns DB. 2011. DegU-phosphate activates expression of the anti-sigma factor FlgM in *Bacillus subtilis*. *Mol Microbiol* 81:1092–1108.

58. Higgins D, Dworkin J. Recent progress in *Bacillus subtilis* sporulation. *FEMS Microbiol Rev* **36**:131–148.

59. Hoch JA. 1993. *spoO* genes, the phosphorelay, and the initiation of sporulation, p 747–755. *In* Sonenshein AL, Hoch JA, Losick R (ed.), Bacillus subtilis *and Other Gram-Positive Bacteria: Biochemistry, Physiology, and Molecular Genetics*. American Society for Microbiology, Washington, DC.

60. Aguilar C, Vlamakis H, Guzman A, Losick R, Kolter R. 2010. KinD is a checkpoint protein linking spore formation to extracellular-matrix production in Bacillus subtilis biofilms. *mBio* **1**(1):e00035-10.

61. Schaeffer P, Millet J, Aubert J-P. 1965. Catabolic repression of bacterial sporulation. *Proc Natl Acad Sci USA* **54**:704–711.

62. Sterlini JM, Mandelstam J. 1969. Commitment to sporulation in *Bacillus subtilis* and its relationship to development of actinomycin resistance. *Biochem J* **113**:29–37.

63. Perego M, Hanstein C, Welsh KM, Djavakhishvili T, Glaser P, Hoch JA. 1994. Multiple protein-aspartate phosphatases provide a mechanism for the integration of diverse signals in the control of development in *B. subtilis*. *Cell* **79**:1047–1055.

64. Perego M. 2001. A new family of aspartyl phosphate phosphatases targeting the sporulation transcription factor Spo0A of *Bacillus subtilis*. *Mol Microbiol* **42**:133–143.

65. Burkholder WF, Kurtser I, Grossman AD. 2001. Replication initiation proteins regulate a developmental checkpoint in *Bacillus subtilis*. *Cell* **104**:269–279.

66. Hosoya S, Asai K, Ogasawara N, Takeuchi M, Sato T. 2002. Mutation in *yaaT* leads to significant inhibition of phosphorelay during sporulation in *Bacillus subtilis*. *J Bacteriol* **184**:5545–5553.

67. Tortosa P, Albano M, Dubnau D. 2000. Characterization of *ylbF*, a new gene involved in competence development and sporulation in *Bacillus subtilis*. *Mol Microbiol* **35**:1110–1119.

68. Carabetta VJ, Tanner AW, Greco TM, Defrancesco M, Cristea IM, Dubnau D. 2013. A complex of YlbF, YmcA and YaaT regulates sporulation, competence and biofilm formation by accelerating the phosphorylation of Spo0A. *Mol Microbiol* **88**:283–300.

69. Chung JD, Stephanopoulos G, Ireton K, Grossman AD. 1994. Gene expression in single cells of *Bacillus subtilis*: evidence that a threshold mechanism controls the initiation of sporulation. *J Bacteriol* **176**:1977–1984.

70. Dubnau D, Losick R. 2006. Bistability in bacteria. *Mol Microbiol* **61**:564–572.

71. Chastanet A, Losick R. 2011. Just-in-time control of Spo0A synthesis in *Bacillus subtilis* by multiple regulatory mechanisms. *J Bacteriol* **193**:6366–6374.

72. Eswaramoorthy P, Dinh J, Duan D, Igoshin OA, Fujita M. Single-cell measurement of the levels and distributions of the phosphorelay components in a population of sporulating *Bacillus subtilis* cells. *Microbiology* **156**:2294–2304.

73. Kuchina A, Espinar L, Garcia-Ojalvo J, Süel G. 2011. Reversible and noisy progression towards a commitment point enables adaptable and reliable cellular decision-making. *PLoS Comp Biol* **7**:e1002273.

74. Duncan L, Alper S, Arigoni F, Losick R, Stragier P. 1995. Activation of cell-specific transcription by a serine phosphatase at the site of asymmetric division. *Science* **270**:641–644.

75. Arigoni F, Pogliano K, Webb CD, Stragier P, Losick R. 1995. Localization of protein implicated in establishment of cell type to sites of asymmetric division. *Science* **270**:637–640.

76. Narula J, Devi SN, Fujita M, Igoshin OA. 2012. Ultrasensitivity of the *Bacillus subtilis* sporulation decision. *Proc Natl Acad Sci USA* **109**:E3513–E3522.

77. Eswaramoorthy P, Duan D, Dinh J, Dravis A, Devi SN, Fujita M. 2010. The threshold level of the sensor histidine kinase KinA governs entry into sporulation in *Bacillus subtilis*. *J Bacteriol* **192**:3870–3882.

78. Alon U. 2007. *An Introduction to Systems Biology. Design Principles of Biological Circuits*. Chapman & Hall/CRC, Boca Raton, FL.

79. Dworkin J, Losick R. 2005. Developmental commitment in a bacterium. *Cell* **121**:401–409.

80. Veening JW, Murray H, Errington J. 2009. A mechanism for cell cycle regulation of sporulation initiation in *Bacillus subtilis*. *Genes Dev* **23**:1959–1970.

81. Cunningham KA, Burkholder WF. 2009. The histidine kinase inhibitor Sda binds near the site of autophosphorylation and may sterically hinder autophosphorylation and phosphotransfer to Spo0F. *Mol Microbiol* **71**:659–677.

82. Grossman AD. 1995. Genetic networks controlling the initiation of sporulation and the development of genetic competence in *Bacillus subtilis*. *Annu Rev Genet* **29**:477–508.

83. Hahn J, Roggiani M, Dubnau D. 1995. The major role of Spo0A in genetic competence is to downregulate abrB, an essential competence gene. *J Bacteriol* **177**:3601–3605.

84. Schultz D, Wolynes PG, Ben Jacob E, Onuchic JN. 2009. Deciding fate in adverse times: sporulation and competence in *Bacillus subtilis*. *Proc Natl Acad Sci USA* **106**:21027–21034.

85. Kuchina A, Espinar L, Cagatay T, Balbin AO, Zhang F, Alvarado A, Garcia-Ojalvo J, Süel GM. 2011. Temporal competition between differentiation programs determines cell fate choice. *Mol Syst Biol* **7**:557.

86. Smits WK, Bongiorni C, Veening JW, Hamoen LW, Kuipers OP, Perego M. 2007. Temporal separation of distinct differentiation pathways by a dual specificity Rap-Phr system in *Bacillus subtilis*. *Mol Microbiol* **65**:103–120.

87. Lopez D, Vlamakis H, Kolter R. 2010. Biofilms. *Cold Spring Harb Perspect Biol* **2**:a000398.

88. Branda SS, Chu F, Kearns DB, Losick R, Kolter R. 2006. A major protein component of the *Bacillus subtilis* biofilm matrix. *Mol Microbiol* **59**:1229–1238.

89. Vlamakis H, Aguilar C, Losick R, Kolter R. 2008. Control of cell fate by the formation of an architecturally complex bacterial community. *Genes Dev* **22**:945–953.

90. Branda SS, Gonzalez-Pastor JE, Ben-Yehuda S, Losick R, Kolter R. 2001. Fruiting body formation by *Bacillus subtilis*. *Proc Natl Acad Sci USA* **98**:11621–11626.

91. Chai Y, Chu F, Kolter R, Losick R. 2008. Bistability and biofilm formation in *Bacillus subtilis*. *Mol Microbiol* **67**:254–263.

92. Kearns DB, Chu F, Branda SS, Kolter R, Losick R. 2005. A master regulator for biofilm formation by *Bacillus subtilis*. *Mol Microbiol* **55**:739–749.

93. Chu F, Kearns DB, Branda SS, Kolter R, Losick R. 2006. Targets of the master regulator of biofilm formation in *Bacillus subtilis*. *Mol Microbiol* **59**:1216–1228.

94. Blair KM, Turner L, Winkelman JT, Berg HC, Kearns DB. 2008. A molecular clutch disables flagella in the *Bacillus subtilis* biofilm. *Science* **320**:1636–1638.

95. Mandic-Mulec I, Doukhan L, Smith I. 1995. The *Bacillus subtilis* SinR protein is a repressor of the key sporulation gene *spo0A*. *J Bacteriol* **177**:4619–4627.

96. Mandic-Mulec I, Gaur N, Bai U, Smith I. 1992. Sin, a stage-specific repressor of cellular differentiation. *J Bacteriol* **174**:3561–3569.

97. Huynh TN, Stewart V. 2011. Negative control in two-component signal transduction by transmitter phosphatase activity. *Mol Microbiol* **82**:275–286.

98. Lopez D, Fischbach MA, Chu F, Losick R, Kolter R. 2009. Structurally diverse natural products that cause potassium leakage trigger multicellularity in *Bacillus subtilis*. *Proc Natl Acad Sci USA* **106**:280–285.

99. Kobayashi K, Kuwana R, Takamatsu H. 2008. *kinA* mRNA is missing a stop codon in the undomesticated *Bacillus subtilis* strain ATCC 6051. *Microbiology* **154**:54–63.

100. Tran L-SP, Nagai T, Itoh Y. 2000. Divergent structure of the ComQXPA quorum sensing components: molecular basis of strain-specific communication mechanism in *Bacillus subtilis*. *Mol Microbiol* **37**:1159–1171.

101. Tortosa P, Logsdon L, Kraigher B, Itoh Y, Mandic-Mulec I, Dubnau D. 2001. Specificity and genetic polymorphism of the Bacillus competence quorum-sensing system. *J Bacteriol* **183**:451–460.

102. Ansaldi M, Marolt D, Stebe T, Mandic-Mulec I, Dubnau D. 2002. Specific activation of the *Bacillus* quorum-sensing systems by isoprenylated pheromone variants. *Mol Microbiol* **44**:1561–1573.

103. Stefanic P, Mandic-Mulec I. 2009. Social interactions and distribution of *Bacillus subtilis* phenotypes at microscale. *J Bacteriol* **191**:1756–1764.

104. López D, Vlamakis H, Losick R, Kolter R. 2009. Paracrine signaling in a bacterium. *Genes Dev* **23**:1631–1638.

105. Core L, Perego M. 2003. TPR-mediated interaction of RapC with ComA inhibits response regulator-DNA binding for competence development in *Bacillus subtilis*. *Mol Microbiol* **49**:1509–1522.

106. Gonzalez-Pastor JE, Hobbs EC, Losick R. 2003. Cannibalism by sporulating bacteria. *Science* **301**:510–513.

107. Lopez D, Vlamakis H, Losick R, Kolter R. 2009. Cannibalism enhances biofilm development in *Bacillus subtilis*. *Mol Microbiol* **74**:609–618.

108. Kolodkin-Gal I, Romero D, Cao S, Clardy J, Kolter R, Losick R. 2010. D-Amino acids trigger biofilm disassembly. *Science* **328**:627–629.

109. Kolodkin-Gal I, Cao S, Chai L, Bottcher T, Kolter R, Clardy J, Losick R. 2012. A self-produced trigger for biofilm disassembly that targets exopolysaccharide. *Cell* **149**:684–692.

110. Chai Y, Kolter R, Losick R. 2010. Reversal of an epigenetic switch governing cell chaining in *Bacillus subtilis* by protein instability. *Mol Microbiol* **78**:218–229.

111. Prepiak P, Defrancesco M, Spadavecchia S, Mirouze N, Albano M, Persuh M, Fujita M, Dubnau D. 2011. MecA dampens transitions to spore, biofilm exopolysaccharide and competence expression by two different mechanisms. *Mol Microbiol* **80**:1014–1030.

The Bacterial Spore: From Molecules to Systems
Edited by P. Eichenberger and A. Driks
© 2016 American Society for Microbiology, Washington, DC
doi:10.1128/microbiolspec.TBS-0019-2013

Ashley R. Bate[1]
Richard Bonneau[1]
Patrick Eichenberger[1]

Bacillus subtilis Systems Biology: Applications of -Omics Techniques to the Study of Endospore Formation

6

THE *BACILLUS SUBTILIS* GENOME SEQUENCE AND GENOMICS OF SPORULATION

B. subtilis Genomics

The principal *B. subtilis* laboratory strain, strain 168, is derived from a parent strain isolated in Marburg, Germany, following a mutagenesis procedure (1). The popularity of this strain arose after it was shown to be competent for genetic transformation (2, 3), which paved the way for myriad molecular genetics analyses that led to a detailed understanding of the biology of *B. subtilis* and related Gram-positive bacteria. It is therefore not surprising that strain 168 was the first Gram-positive species to have its entire genome sequenced, at a time when sequencing was a laborious and expensive process. The project to sequence the genome was set up in 1987 by a consortium of over 30 laboratories and took about 10 years to complete. Each laboratory was assigned a different region of the chromosome and used their own cloning and sequencing strategies to manage their assigned portion of the genome (4). The final genome sequence contained 4,214,810 base pairs, and the original annotation included 4,100 protein-coding genes (5). Following the development of sequencing technologies that were considerably faster and more efficient, the genome of *B. subtilis* strain 168 was resequenced and cleared of sequencing errors in 2009 (6). The most recent update of the annotation brought the total of protein-coding genes to 4,458 (7).

The improved ease and expanded access of sequencing technologies also shed light on the ancestry of several other *B. subtilis* strains. Although strain 168 is a popular model organism, it has limitations that have led numerous laboratories to study other strains. Through a process of domestication in the laboratory, strain 168 and derivatives lost swarming motility (8)

---

[1]Center for Genomics and Systems Biology, Department of Biology, New York University, New York, NY 10003.

and the ability to construct robust biofilms (9). Therefore, to study these processes, researchers have turned to the undomesticated NCIB3610 strain, which is considered to be an authentic representation of the *B. subtilis* Marburg parent strain. Furthermore, due to the presence of a *trpC2* allele, strain 168 is a tryptophan auxotroph. Thus, in order to generate Trp$^+$ derivatives of strain 168, a *trpC*$^+$ allele had to be provided by a donor strain, usually strain W23, a threonine auxotroph that appears to have arisen from an independent isolate (2). A study analyzing the genomic region surrounding *trpC* for 17 of the *B. subtilis* legacy strains has shown that a number of strain 168 Trp$^+$ derivatives have W23-parent DNA, implying that many *B. subtilis* strains used by several laboratories around the world are mosaics of strains 168 and W23 (10). The determination of strain ancestry is particularly valuable considering that some of the -omics work performed in *B. subtilis* has been done with strain PY79 (11), which has the greatest divergence from the original strain among the 168 legacy strains (10). It should be emphasized that the genome size of strain PY79 is smaller (by 180 kb) than that of strain 168, since it contains four large deletions, in particular the regions corresponding to the SPβ prophage (11) and the conjugative transposon ICEBs1 (12).

B. subtilis W23 shares a 3.6-Mb core genome with strain 168 (13). Strains W23 and 168 are fundamentally different in their cell wall composition, with teichoic acid polymers of strain W23 resembling those of *Staphylococcus aureus* (14, 15). In addition, the sporulation killing factor (SKF) and competence proteins ComQXP, two characteristic traits of strain 168, are not found in strain W23 (13). An increasing number of *B. subtilis* strains and closely related species have been sequenced recently, and analyses of these genome sequences revealed that despite a high degree of conservation, each strain harbors numerous unique regions (16–18). Most strain-specific differences appear to originate with the transfer of genetic elements including phages, plasmids, and insertion sequences. Many differences affect secondary metabolism and developmental programs (such as sporulation and competence), suggesting that these pathways are plastic and may be under greater environmental selection than other parts of the cellular machinery. This information is especially valuable to understand how *B. subtilis* can occupy the variety of environmental niches in which it has been shown to reside (see Fajardo-Cavazos et al. [105] and Mandic-Mulec et al. [106]).

Comparative Genomics of Sporulation

In addition to *B. subtilis*, the genomes of many other sporeformers have been sequenced, including many from the family *Bacillaceae* and from the order *Clostridiales*. Of particular note are several species in the *Bacillus anthracis*/*Bacillus cereus* group (19); multiple species from the family *Clostridiaceae* such as *Clostridium acetobutylicum* (20), *Clostridium perfringens* (21), and *Clostridium botulinum* (22); and the important human pathogen *Clostridium difficile* from the family *Peptostreptococcaceae* (23) (see Galperin et al. [107] for details of the phylogeny). The first phylogenomic analysis of sporulation was performed by Stragier, who analyzed the distribution of 125 sporulation genes across the five endospore-forming genome sequences (*B. subtilis*, *B. anthracis*, *Geobacillus stearothermophilus*, *C. acetobutylicum*, and *C. difficile*) available at the time (24). One of his principal observations, corroborated by Stephenson and Hoch (25), was that, although the master regulator of sporulation, SpoOA, is present in all *Clostridium* spp., the phosphorelay that initiates sporulation in *B. subtilis* by phosphorylation of SpoOA is not conserved. This suggests that the triggers for sporulation are likely to vary considerably from species to species, especially among *Clostridium* spp., where at least two different pathways of direct phosphorylation of SpoOA by histidine kinases are known to exist (26) (see also Dürre [108]).

Conversely, the sporulation sigma (σ) factors appear to always be conserved (σ is the subunit of the RNA polymerase responsible for recognizing core promoter regions on the DNA). Shortly after the initiation of sporulation (Fig. 1), the asymmetric division of the sporulating cell near one pole generates two cell types: a forespore, which will mature into a spore, and a larger mother cell, which is necessary at every stage of forespore maturation. Each cell will activate cell type-specific σ factors: σ^F and σ^G in the forespore; σ^E and σ^K in the mother cell. To a large extent, the sequential activation of the sporulation σ factors is conserved in the characteristic order: σ^F → σ^E → σ^G → σ^K (27). The spatial and temporal coordination of this regulatory cascade is achieved by several intercellular signaling mechanisms discussed by Dworkin (109). Unexpectedly, recent studies have revealed that the temporal coordination of the sporulation σ-factor cascade is less strict in *Clostridium* spp., in the sense that dependency on the preceding σ factor for activation of the subsequent σ factor is not absolute (28–30). In *C. difficile*, σ^G activity is independent of σ^E, while σ^E activity is partially independent of σ^F. Furthermore, consistent with the original observation that σ^K activity is controlled by a mother cell-specific excision of an intervening sequence element of phage origin in the *sigK* gene, σ^K activity does not require σ^G in *C. difficile* (31).

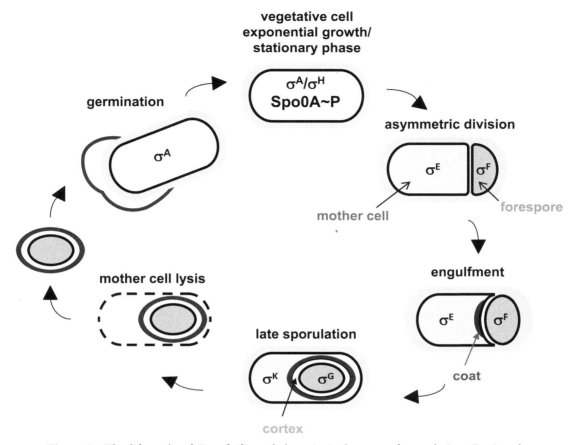

Figure 1 The life cycle of *B. subtilis* and the principal stages of sporulation. During the first stage of sporulation, the master regulator Spo0A~P is required in combination with σ^A (major σ factor) and σ^H (stationary phase σ factor) for the expression of early sporulation genes. Next, an asymmetric division of the sporulating cell creates the mother cell (light blue) and the forespore (orange). Each compartment establishes cell-specific lines of gene expression driven by σ^F in the forespore and σ^E in the mother cell. Subsequently, the mother cell engulfs the forespore. During engulfment, proteins produced in the mother cell assemble at the forespore surface to form the coat (dark red). After engulfment, σ^G substitutes for σ^F and σ^K replaces σ^E (σ^A remains active during the entire process). The cortex (yellow), made of peptidoglycan, is assembled between the inner and outer forespore membranes. Once the spore is mature, the mother cell lyses. During the germination process, the cortex is hydrolyzed and the coat is shed. Adapted from references 27, 32, and 104.

Transcriptional profiling experiments in *B. subtilis* (discussed below) vastly expanded the number of annotated sporulation genes and resulted in further phylogenetic analyses. For example, de Hoon et al. (32) traced the presence of *B. subtilis* sporulation genes in 24 spore-forming species to investigate the extent of the sporulation network conservation. The principal conclusion of this article was that, while there were differences in whether or not a gene was present in a given species, if the gene was present, its regulation was usually conserved (this was especially true of spore coat genes). Another study by Xiao et al. (33) compared

the known and putative spore germination genes of 12 *Bacillus* and 24 *Clostridium* spp. and concluded that the number of operons encoding germination receptors was lower in *Clostridium* than in *Bacillus* spp. This suggests that *Clostridium* spp. have different germination strategies that may involve yet-to-be-identified Ger proteins. For instance, in *C. difficile*, glycine and bile salts have been identified as cogerminants (34) and CspC was recently characterized as the germinant receptor involved in bile acid recognition (110).

While these studies have all depended on *B. subtilis* to provide the initial set of sporulation genes used for comparative genomics, more recent work has incorporated information from other spore-forming species. Additional sporulation genes were uncovered in phylogenetic profiling approaches. Specifically, by assessing the genomes of 46 sporeformers, eight genes previously unrecognized as sporulation genes in *B. subtilis* were found to be significantly enriched in spore-forming bacteria (35, 36). In addition, Galperin et al. (37) aimed to identify the minimal set of genes that are essential for sporulation by evaluating the distribution of 651 genes that are preferentially expressed during sporulation. This compendium of genes was compiled from phenotypic characterization of sporulation mutants, transcriptional profiling experiments in both *B. subtilis* and *C. acetobutylicum*, and proteomic analyses of spore content. *In toto*, substantial differences were noted between *Bacillaceae* and *Clostridiales* sporulation programs, particularly when it came to regulation of the onset of sporulation and the composition of the spore coat. Since numerous sporulation and germination genes characterized in *B. subtilis* do not appear in the genomes of *Clostridium* spp., sporulation and germination studies in *Clostridiales* should be expanded to better understand the differences between the two main classes of endospore-forming species.

THE *B. SUBTILIS* TRANSCRIPTOME AND REGULATORS OF GENE EXPRESSION

Transcriptomics of Sporulation

Gene arrays have existed since the 1970s in the form of dot blots, but it was the development of gridding robots in the 1980s and the automation of PCR in the 1990s that allowed for transcriptional profiling on a global scale (38). For *B. subtilis*, the first genome-wide array experiment was performed just a few years after the publication of the genome sequence, with macroarrays constructed by PCR amplifying all the identified open reading frames and spotting the cDNA onto a membrane (39).

The initial genome-wide macro- and microarray experiments in *B. subtilis* were designed to identify genes that were activated or repressed during the transition from vegetative growth to stationary phase (39–41). Fawcett et al. (39) compared wild-type cells with cells unable to produce Spo0A~P and found 586 genes with altered expression profiles during the early stages of sporulation in a *spo0A* mutant. By profiling the *sigF* mutant and using computational searches to discover conserved upstream sequence motifs for σ-factor binding, genes were identified as putatively under the

control of σ^F in the forespore or σ^E in the mother cell. Similarly, Caldwell et al. (41) compared global gene transcription from a wild-type culture to a culture with a knockout of the gene encoding the transition state regulator ScoC. In this study, it was found that ScoC directly or indirectly affected the expression of 560 genes, in particular genes involved in motility, sporulation, competence, and degradative enzyme production. The transition to stationary phase was also profiled by comparing gene transcription in wild-type cells against cells lacking the stationary phase sigma factor σ^H (40). In this experiment, Britton et al. (40) found 87 σ^H-controlled genes and discovered that, in addition to its previously characterized role in sporulation and competence, σ^H was involved in the regulation of cytochrome biogenesis, transport, cell wall metabolism, and generation of potential nutrient sources.

Comprehensive transcriptional profiling of the mother cell and forespore lines of gene transcription aimed to identify regulons for each known sporulation regulator (42–44). For the mother cell, mutants were constructed to elucidate the regulons in the mother cell hierarchical regulatory cascade of σ^E → SpoIIID/GerR → σ^K → GerE in strain PY79 (42). This also meant inactivating the downstream regulator in the cascade to limit indirect gene expression effects. Prior to this experiment, little information was available about the full set of genes governed by these mother cell regulators, with the exception of σ^E which had been previously profiled (45, 46). It was seen that approximately 9% of the genes in the *B. subtilis* genome are turned on specifically in the mother cell and that the mother cell program of gene expression consists of a series of pulses where a large number of genes are activated and then shortly after turned off by the next regulatory protein in the hierarchy. Some genes, however, show protracted expression over the course of sporulation and are under the dual control of more than one σ factor, including a few cases of genes simultaneously expressed in both the mother cell and forespore (47, 48). Subsequently, the method in reference 42 was applied to the forespore regulators to understand the forespore line of gene expression: σ^F → RsfA → σ^G → SpoVT (43).

More recently, Nicolas et al. (49) collected a large transcriptome data set for strain 168 by hybridizing 269 tiling arrays over 104 diverse experimental conditions, including aerobic and anaerobic cultures, exploiting various energy and carbon sources, in media promoting growth, motility, biofilm formation, competence, or sporulation. Transcription was detected for 85% of the annotated genes under one or more conditions. In addition, 1,583 previously unannotated RNAs were identified, and it was discovered that 13% of protein-coding sequences overlapped with antisense RNAs. Importantly, the number of *B. subtilis* promoters was found to be three times higher than previously thought, with about half of protein-coding sequences transcribed from multiple promoters. Approximately 91% of promoters had an identifiable putative σ factor-binding motif, and σ-factor activity was shown to be responsible for 66% of the variance in transcriptional activity (i.e., the condition-dependent changes in gene expression that were observed in this data set of 104 experiments). The variance was not consistent for each σ factor; sporulation σ factors had the highest values, while the major σ factor in *B. subtilis*, σ^A, had low variance. These observations suggest that σ^A-controlled promoters rely more heavily on other transcription regulators for condition-specific regulation.

Transcriptional profiling experiments have also been performed on many other spore-forming species. In fact, *B. anthracis* was the first bacterium to have a comprehensive single-nucleotide resolution view of its transcriptome using a high-throughput RNA sequencing (RNA-seq) approach (50). An extended transcriptional analysis of the *C. acetobutylicum* life cycle has been completed using DNA microarrays (51). In this study, 24 time points were sampled from exponential growth into sporulation, allowing for the investigation of sporulation dynamics. These data allowed for the first comparison of the temporal orchestration of the sporulation pathway between *Clostridium* and *Bacillus* spp. More recently, genome-wide gene expression in *C. difficile* was analyzed by transcriptional profiling during sporulation of *spo0A*, *sigF*, *sigE*, *spoIIID*, *sigG*, and *sigK* mutants. One study used RNA-seq (28), while another study relied on DNA microarrays (30). More than 300 genes were shown to be turned on during sporulation in *C. difficile* (28), with >200 genes under the direct control of a sporulation σ factor (30).

Regulators of Gene Expression in *B. subtilis*: σ Factors, Transcription Factors, and Small RNAs

Understanding the *B. subtilis* transcriptome requires determining the regulons of individual regulators of gene expression. As in other organisms, *B. subtilis* transcriptional factors (TFs) can be identified on the basis of their structural, biochemical, and genetic properties. Structural features of TFs, most commonly DNA-binding motifs, have been deposited into Pfam, a database of protein domain families (52). With the use

of the information from Pfam, the entire genome of *B. subtilis* was searched to identify all potential TFs, with the results of the analysis compiled into DBTBS, a database of *B. subtilis* promoters and transcriptional regulators (53). Currently, the repertoire of DNA-binding proteins regulating transcription in *B. subtilis* comprises 19 σ factors and 126 TFs. Importantly, this number only includes TFs with known target genes; therefore, the total number of transcriptional regulators in *B. subtilis* will increase as more TFs are characterized experimentally.

In addition to these DNA-binding proteins, noncoding RNAs regulate gene expression in *B. subtilis* and most likely other spore-forming bacteria (54). RNA regulators fall into three groups: riboswitches, which are part of the mRNA they regulate and carry out regulation by folding into alternative structures based on ligand availability (usually a specific metabolite); small RNAs (sRNAs), most of which bind to a target mRNA by base-pairing and modulate the translation and stability of the transcript; and CRISPRs, which have been shown to interfere with bacteriophage infection. The first sRNAs were identified in the 1980s; however, the prevalence of sRNAs and their contributions to numerous physiological responses was not realized until the early 2000s. Between 2001 and 2002, several studies that performed systematic computational searches for the conservation and presence of orphan promoter and terminator sequences in the intergenic regions of *Escherichia coli* resulted in the identification of many new sRNAs (55). The identification of sRNAs has further increased through the use of RNA-seq and tiled arrays with full genome coverage.

Several studies have been performed with the goal of identifying transcriptionally active regions of the *B. subtilis* genome. Using tiling arrays, Rasmussen et al. (56) profiled RNA collected from exponentially growing cells in both rich (Luria Bertani, LB) and minimal (M9) media and found 84 putative sRNAs. Using an RNA-seq approach, Irnov et al. (57) collected samples in glucose minimal medium during stationary phase, further increasing the number of sRNA candidates to over 100. The first sRNAs found to be under sporulation control were also discovered by using tiling arrays. Silvaggi et al. (58) identified three sRNAs: one under indirect Spo0A~P control, another under σ^G control, and the final under σ^K control. Using a similar approach, Schmalisch et al. (59) identified two additional sporulation sRNAs. The exact role played by these sRNAs during sporulation in *B. subtilis* and their mechanism of action remain to be determined.

GENE FUNCTION IN A SYSTEMS-WIDE CONTEXT

Essential Genes

Since the beginning of the genomics era, one specific goal of systems biology has been to identify the minimal set of genes required for growth of an organism. One of the first studies to do this was performed by using the Mollicute *Mycoplasma genitalium* (60). This bacterium was chosen because it is a self-replicating organism with a small genome containing about 500 protein-coding genes. With the use of global transposon mutagenesis, it was found that about 300 genes were essential under laboratory growth conditions. Another set of essential genes was compiled for *B. subtilis* by systematically inactivating each gene and observing if the organism would grow in rich medium (61). It was found that 271 of the 4,100 protein-coding genes annotated at the time of that study were necessary for growth; however, this number was revised recently following further analyses characterizing novel essential genes and, at the same time, showing that several genes thought to be essential are actually dispensable. The new total for *B. subtilis* stands at 254 genes (62). A modernized version of the global transposon mutagenesis approach (named *Tn*-seq) was developed to accurately map essential genes by assembling a transposon library and measuring the changes in frequency of each mutant strain and determining the sites of transposon insertion by sequencing flanking regions. When combined with measurements of fitness, quantitative genetic information can be obtained in addition to gene essentiality data (63). A *Tn*-seq screen was recently carried out in *B. subtilis* and resulted in the identification of 24 new sporulation genes (111).

Global Phenotypic Approaches: Synthetic Genetic Arrays, Chemogenomics, Proteomics, Protein-Protein Interactions, and Protein Localization on a Large Scale

While minimal genome studies are informative, essentiality is only one of the many criteria that can be used to characterize gene function. Based on the construction of gene deletion libraries, methods have been developed to further explore gene function by analyzing the phenotypes of double-gene knockouts (synthetic genetic arrays [SGAs]) and by investigating the fitness of single-gene deletions under a variety of chemical challenges (chemogenomics).

SGAs were initially developed in yeast. In this method, a query mutation is crossed to a deletion mutant library and the resulting double mutants are scored for fitness defects (64). With the use of this technique in the budding yeast *Saccharomyces cerevisiae*, multiple relationships between genes have been identified, including quantitative genetic interactions based on growth rate measurements (64–66).

To perform large-scale phenotypic profiling in *E. coli*, Nichols et al. (67) implemented a chemogenomics strategy. A single-gene deletion library, completed with partial-loss-of-function hypomorphs of essential genes and sRNA knockouts, was exposed to many chemical stresses (including challenges by antibiotics, detergents, dyes, and other molecules). The growth rate of each mutant strain was assessed to construct a data set including over 300 conditions and ~4,000 genes. By analysis of phenotypic signatures, functions were predicted for previously uncharacterized genes based on the assumption that highly correlated genes are likely to function together. By revealing which cellular function (e.g., replication or cell division) is enriched in gene clusters that correlate with application of a specific chemical challenge, a possible mode of action can be suggested for a variety of drugs. As of yet, neither SGA nor chemogenomics analyses have been performed on *B. subtilis* or any other spore-forming bacterium.

Another way to probe gene function on a global scale is to study by proteomics the occurrence of gene products in response to specific conditions. Proteomics generally refers to the analysis, by mass spectrometry, of the entire complement of proteins and protein modifications present in a sample. For example, mass spectrometry on mature *B. subtilis* spores identified the presence of 69 proteins not previously known to be present in the spore (68). Similarly, proteomic analyses of spore coat extracts revealed three new coat proteins in *B. subtilis* and six candidate spore coat proteins in *B. anthracis* (69). Nineteen additional candidates were reported in another study; however, about half of these candidates are produced from genes expressed in the *B. subtilis* forespore and are therefore unlikely to be genuine coat proteins, since coat proteins are always synthesized in the mother cell (70).

Another proteomics-based method called SILAC (stable isotope labeling with amino acids in cell culture) can provide information about relative protein concentrations and be used to compare two samples (e.g., a mutant strain and a wild type under similar conditions) (71). In this method, one sample is grown in a medium containing standard amino acids and another sample is grown with amino acids containing heavy isotopes. The samples are then pooled, trypsinized, fractionated, and analyzed with mass spectrometry. Relative protein concentrations are obtained by quantifying the difference between otherwise identical peptides that differ in mass. The SILAC strategy was applied to observe the differences in the proteome of *B. subtilis* cells grown in succinate or starved for phosphate (72); changes in levels of over 1,500 proteins were quantified in that manner. Much would be gained from expanding this work to a larger set of conditions and comparing proteomics data sets with transcriptomics data sets collected under identical conditions.

Studying the physical interactions between proteins can also aid in the elucidation of gene function, because proteins that interact with each other are likely to play a role in the same biological processes. Protein-protein interactions can be investigated on a large scale by using yeast or bacterial two-hybrid assays. In the largest yeast two-hybrid study performed in *B. subtilis*, 793 interactions were found, linking 287 proteins that encompassed many different cellular functions (73). Thirty-nine "hubs" (defined as any protein that interacted with more than 10 partners) were identified.

Similar to protein-protein interaction, protein colocalization can provide supporting evidence for functional predictions. For example, protein localization was studied in budding yeast by constructing fluorescent protein fusions for 75% of the proteome (74). Similarly, a high-throughput cloning and microscopy pipeline was developed to observe both N- and C-terminal fluorescent protein fusions in *Caulobactercrescentus*, resulting in the identification of nearly 300 proteins with specific subcellular localization in this organism (75). While no global analysis of protein localization has been performed in a spore-forming bacterium, a protein localization analysis of the spore coat proteome has been done in *B. subtilis* (76, 77).

Metabolomics (nanoDESI-MS, Annotation of Metabolic Networks)

Another type of global phenotypic approach is metabolomics, where mass spectrometry is used both to determine the spectrum of chemical compounds produced by bacterial cultures and to analyze how this chemical signature is influenced by changing environmental conditions or by mutation of specific genes. Watrous et al. (78) aimed to study the metabolic exchanges that microbes participate in to communicate with their local environments, including neighboring microbes. To do this they developed a method called nanospray desorption electrospray ionization (nanoDESI)-mass spectrometry (MS) as a way to study the compounds secreted from bacterial colonies. This process reveals many classes of compounds in a single data set; however, because no

database was available for molecules involved in metabolic exchange, a novel procedure had to be designed to simplify and explore the nanoDESI-MS output. A network-based workflow was developed in which the data were first simplified by merging identical spectra (i.e., spectra exhibiting identical mass-to-charge [m/z] ratios and fragmentation patterns). Next, the similarity for each possible pair of spectra was calculated based on the precursor m/z ratio difference and the relative intensities of the fragment ions. This information was then integrated into a network viewable in Cytoscape (an open source platform for visualizing networks, discussed below), where each node represents a spectrum and spectra are connected to similar nodes by edges.

Experiments performed with nanoDESI-MS show that the method is able to recover known molecular signals and can be used to analyze microbial communication. Using *B. subtilis* strain NCIB3610 to test the procedure, molecules known to be secreted during biofilm formation (such as surfactin) and variants of these compounds were observed to cluster together in the network. A 60-hour time series of a NCIB3610 colony was done to observe whether the metabolic output changed over time. Increases in surfactin, plipastatin, and subtilosin production were noted, as well as a major lipid production shift. This study also analyzed the interface of colonies of different species. While the interface of *Streptomyces coelicolor* and *B. subtilis* strain PY79 colonies is characterized by a known increase in SapB in *S. coelicolor* and SKF in *B. subtilis* (79), many previously unknown signals were also detected. Overall, this method will aid in furthering our understanding of how microbes communicate with other species by providing a way to study compounds that act as mediators of cell-cell signaling. Importantly, this approach can also be used to link the production of signaling compounds to the presence of a particular gene or group of genes, and thus contribute to the discovery of gene functions.

The extensive gene function information gathered from the application of genome-scale and more traditional methods can then be used to identify orthologous genes in other organisms, where they are likely to perform similar functions. The Kyoto Encyclopedia of Genes and Genomes (KEGG) is a large database that performs gene annotation using a cross-species annotation method (80), where the function of a gene experimentally characterized in one species is putatively applied to homologous genes found in other species by searching all available genomes.

For the annotation of metabolic pathways, Plata et al. (81) have designed a novel method that considers context associations to improve upon homology-based annotations. Specifically, because genes involved in the same metabolic pathway are likely to display similar patterns of expression and conservation across species, this probabilistic approach takes into account coexpression, chromosomal location, and phylogenetic distribution during the annotation procedure. This method has better precision and recall than more classic procedures, especially in cases of low sequence homology. Several predictions were experimentally validated, including those for the *sps* (spore polysaccharide synthesis) operon, whose products are involved in rhamnose synthesis, a known component of the spore surface in *B. subtilis* and *B. anthracis* (82).

MODELING REGULATORY NETWORKS

A key objective of systems biology is to take advantage of transcriptome data sets similar to those described above to construct whole-cell dynamical models of transcriptional regulation. Such models can potentially be built for any microbial organism of interest and can be thought of in two ways: as a global network, where every regulator (TF or sRNA) is connected to the gene(s) it regulates, or as a system of equations that, given a starting condition, can predict transcriptome-wide responses to defined perturbations. The global network is a comprehensive map of gene regulation hypotheses that can be used to identify new targets of known regulators, while the system of equations can reveal how the expression of each gene in the model will be affected following a specific environmental or genetic modification. When possible, models should be supplemented with other data types (e.g., proteomics) to help resolve instances where regulatory mechanisms rely on posttranscriptional events. In addition, chromatin-immunoprecipitation data (from ChIP-seq or ChIP-on-chip experiments that can reveal where a given TF binds in the chromosome) are especially helpful to distinguish between direct and indirect effects of specific TFs.

Models of Central Metabolism

Initial efforts in network construction and analysis of system-wide responses to perturbations focused on the central metabolism in both *E. coli* and *B. subtilis*. Ishii et al. (83) measured gene expression levels in various mutant strains of *E. coli* with disruptions in genes encoding enzymes involved in central carbon metabolism (i.e., the glycolysis, tricarboxylic acid cycle, and pentose phosphate pathways). The cultures were grown in a chemostat over five different dilution rates, from nearly glucose-starved to almost unlimited glucose

concentrations. Comparisons of mRNA, protein, and metabolite levels showed that the metabolic network of *E. coli* is highly robust. While it was seen that an increase in growth rate resulted in a global increase in mRNA and protein levels (likely to meet the increased metabolic need), single-gene disruptions had little effect on the regulation of mRNA and protein levels. The overall stability of the network implies that rerouting of metabolic fluxes is favored over large changes in enzyme levels as a response to a genetic perturbation.

In *B. subtilis*, Goelzer et al. (84) constructed a manually curated model of the central metabolism derived from public data and expert knowledge. All the biochemical reactions in the metabolic network were included, along with all the known levels of regulation involved in metabolic pathways. The final model included 563 reactions, 534 genes, and 456 metabolites. Another model by Oh et al. (85) reconstructed a larger network, supported by growth phenotypes of knockout strains, composed of 1,020 reactions, 844 genes, and 988 metabolites. With the use of these models to examine the organization of metabolic regulation, a clear hierarchy emerges, where local and global regulators can be readily distinguished. Typically, local regulators control one specific pathway and have potential metabolite effectors that belong to the controlled pathway (either as the end product or as an intermediate metabolite of the pathway). For instance, the tryptophan synthesis pathway is dependent on several local regulators and effectors, including tryptophan itself (86). Global regulators, such as CcpA (the main regulator for carbon catabolite repression), TnrA (the main regulator for nitrogen metabolism), and CodY (a regulator of many pathways including branched amino acid synthesis), are defined as nonlocal regulators in the sense that they are involved in the regulation of many pathways. Thus, in response to environmental changes, global regulators ensure coordination within the system, while local regulators provide a second level of regulation.

Inferring Regulatory Relationships from Large-Scale Gene Expression Data Sets

In 2007, Bonneau et al. (87) constructed a predictive model of the global transcriptional regulatory network in the archeon *Halobacterium salinarum*. This was achieved through the integration of transcriptomics and proteomics data sets, collected under a variety of genetic and environmental perturbations. New TFs were identified with protein structure predictions, whereas new operons and coregulated groups were inferred by computational analysis of genome structure. The model covered approximately 80% of the *H. salinarum* transcriptome and accurately predicted many novel transcriptional regulatory relationships.

To implement an integrative biclustering strategy, the algorithm *cMonkey* was developed. Biclustering is the grouping of genes coregulated under subsets of experimental conditions, whereas integrative biclustering incorporates additional data sources such as conserved upstream sequence motifs, genomic position, and previous information gathered from manually curated databases, e.g., KEGG. As a result, coregulated—and by extension cofunctional—groups of genes (biclusters) can be obtained via integration of expression data, *cis*-regulatory motifs, evolutionarily conserved associations among proteins, and metabolic pathways. Next, the algorithm *Inferelator* was used to predict which regulators affect the expression of these biclusters, as well as regulatory relationships between TFs and genes not included in biclusters. By the use of the model as a driver for hypothesis generation and subsequent testing, experimental validations have confirmed a number of these regulatory hypotheses, including a hierarchy of regulation for copper efflux, and assigned functions to proteins with no characterized orthologs (87).

To learn networks of conserved coregulated gene groups using multiple-species data sets, Waltman et al. (88) present an algorithm for multispecies integrative biclustering based on the *cMonkey* biclustering algorithm. Multispecies *cMonkey* (MSCM) adds another layer to traditional biclustering approaches through the consideration of evolutionary conservation, the concept being that modules composed of conserved genes are more likely to be biologically significant. MSCM biclustering was performed on all the orthologous gene pairings between *B. subtilis*, *B. anthracis*, and *Listeria monocytogenes*. Analysis of the biclusters led to the unexpected discovery that *B. anthracis*, although nonflagellated and nonmotile, expressed many flagellar motility-chemotaxis genes. Further investigation revealed that the loss of motility was likely due to the accumulation of frameshift mutations in a few of the key genes required for flagellar assembly. Another interesting result was the species-specific timing of expression of a subset of sporulation genes. Genes expressed early in the mother cell in *B. subtilis* (under the control of the sporulation sigma factor σ^E) were divided into three biclusters. The splitting of the σ^E-dependent genes into three biclusters was due in part to functional categorization, but primarily to the three biclusters having different expression profiles

in *B. anthracis*. In particular, one of these biclusters, composed of genes with related metabolic functions, is expressed earlier than the other two biclusters, and even before expression of *sigE*. This observation suggests a partial genetic rewiring in the sporulation program, implying that transcription of this bicluster in *B. anthracis* is controlled by regulators other than σ^E.

Steps Toward a Global Gene Regulatory Network Model for *B. subtilis*

To organize *B. subtilis* genes into putatively coregulated modules, Fadda et al. (89) used the integrative biclustering algorithm DISTILLER. Coexpression data from 213 publicly available microarray data sets from five studies, performed on 10 platforms and covering six different experimental conditions (DNA damage, heat challenge, peroxide stress, phosphate starvation, quorum sensing, and sporulation), were combined with sequence motif information to obtain 142 modules containing 1,153 of the genes in the *B. subtilis* genome (approximately a quarter of the protein-coding genes). The regulators for each module were first determined based on the presence of upstream motifs known to be recognized by specific regulators. When this option was unavailable, assignment of a regulator to a specific module relied on the observation that many regulators autoregulate, or regulate adjacent genes. Therefore, putative regulators were assigned based on the genomic distance between a gene encoding a regulator and a gene containing the motif of interest. This also means that any regulator present in a module would automatically be considered a putative regulator of that module. In this model, 44% of the known regulatory interactions were recovered and 417 new targets were predicted. Although several new regulators were proposed, no experimental validation was performed to follow up on those predictions.

To search for regulatory relationships between σ factors and groups of cotranscribed genes in *B. subtilis*, Nicolas et al. (49) estimated pairwise correlations between promoter activities and transcriptionally active regions (TARs) in their transcriptomics data set (see above). TARs were distributed in coregulated modules by hierarchical clustering of the transcriptional profiles. To systematically identify σ-factor regulons, an unsupervised algorithm was developed to combine DNA sequence information upstream of the transcribed region (i.e., the putative promoter region) with the clustering of transcriptional profiles to determine the conserved bipartite motif recognized by each σ factor. By this method, ~91% of the TARs had an identifiable putative σ-factor-binding site in their upstream regulatory region. In order to determine the participation of σ factors in transcriptome plasticity, the average transcription signal for all the promoters of a cluster was estimated and a linear relationship was assembled between the cluster activity and individual promoters. It was found that 66% of the changes in gene expression over the set of conditions present in the data set were due to variations in σ-factor activity. The role of other TFs (that may act in combination with the major sigma factor, σ^A) was not touched upon.

A model of the global transcriptional regulatory network of B. subtilis was published recently (112). This project was initially inspired by the strategy discussed above to obtain the H. salinarum network (87). By using two large B. subtilis gene transcription data sets (one collected specifically for the project, the other from Nicolas et al. [49] and Buescher et al. [90]) and an updated version of the Inferelator that considers known network edges and transcription factor activities to aid in making predictions (91), a global transcriptional regulatory network was constructed with all known transcriptional regulators (σ factors and TFs). This new model also includes putative regulators, such as proteins with predicted DNA-binding domains. This model recalls 74% of the previously known interactions (a substantial improvement when compared to other network inference approaches) and predicts 2,258 novel interactions. A key difference between this work and the work on H. salinarum is that it takes advantage of >2,500 previously known regulatory relationships archived in SubtiWiki (see below) (92, 93). Incorporation of this prior information dramatically improves the accuracy of the resulting network as suggested in previous work (91, 94) and confirmed experimentally for 391 (out of 635 evaluated) of the predicted novel interactions.

Network Visualization

The multiplication of -omics data sets has necessitated the development of tools to store, process, and visualize the information available. The SubtiList database was assembled as a companion to the B. subtilis genome sequencing project discussed above (5, 95). This database includes the annotated genomic sequence and several sequence analysis and visualization tools. Newly found annotations and functional assignments were added in subsequent versions, but the last update was made in 2002 (96). SubtiList was then integrated into GenoList, a database dedicated to querying and analyzing genome data from many bacterial species (97). A new wiki-

based resource called SubtiWiki has been created to store comprehensive, up-to-date information on B. subtilis genes, proteins, regulatory interactions, protein-protein interactions, pathways, and gene transcription profiles (92, 93). Because the site is a wiki, qualified scientists can contribute to the resource, allowing the site to stay up to date and allowing the collective knowledge of the B. subtilis community to be made accessible to everyone.

Cytoscape is a software package used for network visualization, data integration, and analysis, designed with systems biology research in mind (98, 99). Data can be displayed with the visual aspects of the network decided by the user. For example, a regulatory network can be generated, where genes are circular nodes and transcription factors are triangular nodes. Directional edges would connect transcription factor nodes and gene nodes based on gene expression data from microarray experiments, with the color of the edge indicating activation or repression and the width of the line reflecting the confidence score of the regulatory interaction. Gene or protein networks can also be designed with edges showing that the genes are in an operon, or that the proteins are known to physically interact. Importantly, Cytoscape affords connections to large numbers of tools (called plugins) for annotation, clustering, and visualizing networks (e.g., one plugin, described below, allows users to connect the network view presented by Cytoscape to several other tools using the Gaggle framework).

As systems biology data sets have increased in complexity, it has become necessary to use more than one visualization tool to include different types of data. Gaggle is a Java program that broadcasts gene, network, and data selection between various tools, allowing users to coordinate visualization of different aspects of gene groups, networks, and data (100). The various tools the Gaggle communicates with are referred to as geese; geese made compatible with the Gaggle include Cytoscape, Cytoscape plugins, a Global Synonym/ Ortholog Translator, and several other tools (such as a data matrix and annotation viewers) that facilitate the exploration of -omics data. Using the Gaggle, a researcher is able to take a list of genes and retrieve the gene annotations, expression in a set of experiments, and a Cytoscape network showing known associations between the genes.

We previously developed the Comparative Microbial Modules Resource (CMMR) to integrate the above tools into a single interface to the analysis, data, and meta-data for B. subtilis genomics. This resource is presented to the user as a website specific for the analysis of multispecies biclusters (101). It provides a way to search for multispecies biclusters that contain a gene of interest or are enriched for a specific annotation. Bicluster information includes the list of genes in that bicluster and their annotations, the corresponding gene expression profiles, enriched sequence motifs, and a functional enrichment summary. Furthermore, the resource utilizes the Gaggle to explore biclusters of interest and facilitate navigation between tools.

Combination of Large-Scale Approaches to Obtain Whole-Cell Models

A study from the BaSysBio consortium combined the use of multiple -omics platforms to observe changes during metabolic shifts from glucose to malate and vice versa (90). Both short-term and long-term time series data were collected for transcripts, protein levels, metabolite abundance, and promoter activity. The integration of these data types led to a detailed understanding of the metabolic shift beyond the sole measurement of transcript levels. This study was also able to uncover new transcriptionally active regions (including sRNAs), provide putative functional assignments for genes of previously unknown function, and identify posttranscriptional regulatory events to achieve a more comprehensive description of central metabolism in *B. subtilis*.

By observing 300 genes for which transcript, protein, and promoter activity data were available, 110 posttranscriptional regulation events were identified. More evidence for posttranscriptional events was gathered by performing a network component analysis for 154 transcription factors and their 1,754 known targets and observing the deviations between predicted transcription factor activity and measured mRNA abundance. It was inferred that 51 transcription factors had activity that was regulated posttranscriptionally; however, for 39 of these 51 transcription factors, the effector of posttranscriptional regulation remains unknown. Finally, it was seen that, while shifting to glucose metabolism involved more transcriptional regulation, posttranscriptional regulation played an important role in shifting to malate. This was determined by correlating the time course of the metabolic flux with corresponding enzyme abundance; a positive correlation was indicative of transcriptional regulation influencing the reaction rate. Modeling substrate usage and enzyme maintenance indicated that a delay in glucose uptake (shown to be dependent on two operons) could provide an evolutionary advantage when malate was the primary substrate. This last observation makes sense

considering that plant roots, which constitute a preferred habitat for *B. subtilis* (102), often secrete carboxylic acids such as malate (see also Mandic-Mulec et al. [107]).

Karr et al. (103) present the first whole-cell computational model of an organism using transcriptome, proteome, and metabolome data from *M. genitalium*. To construct the model, the authors first divided the cell into 28 functional modules (including replication, transcription, translation, metabolism, and cytokinesis). Each module was then modeled using the mathematical representation deemed most appropriate, and was built, parameterized, and tested independently. The resulting submodels were integrated into a complete whole-cell model by performing simulations that considered time and cell variables (such as DNA, RNA, protein, and metabolite levels and other variables including geometry and mass). The simulations were executed by running through a loop, the start of which was the beginning of the cell cycle. In these simulations, the submodels ran autonomously at each time step, but dependently on the cell variables determined by the various submodels at earlier time steps. If the simulation did not end in cell division, the loop began again and the simulation that led to full cell division was the one that resulted in the integration of the submodels into the full cell model. The model was validated by verifying that it recovered known information about *M. genitalium* and also by comparing it against many independent data sets not used in the construction of the model. It was found that the model indeed recapitulated many aspects of both the training data and independent experimental data across multiple biological functions.

An important feature of any model is not only to reiterate known information, but also to provide novel insights that can be experimentally tested. The *M. genitalium* model was able to predict features related to chromosomal occupancy by DNA-binding proteins and protein-protein collisions on the chromosome. The model also predicted that progress of DNA replication is controlled in part by the content of dNTPs at the start of replication, and therefore that availability of dNTPs is an emergent control of cell cycle duration, independent of genetic regulation. Simulations were performed in which each of the 525 genes in the *M. genitalium* genome was disrupted one by one, and 284 genes were predicted to be essential. Thus, the model agrees with the observed gene essentiality at 79% accuracy (see above); cases in which the model disagrees with the experimental results were explored further.

CONCLUSIONS

In this article, we have reviewed the impact that -omics techniques have had on the study of *B. subtilis* and other bacteria and have introduced new methods that can be used to expand the study of spore-forming organisms. While genomics and transcriptomics have been widely used to examine sporeformers, other global measurements to elucidate gene functions (such as proteomics, metabolomics, synthetic genetic arrays, and chemical genomics) have been less utilized. Because these methods are shown to expand our knowledge of other microbes, it is likely that they will soon be applied to the study of *B. subtilis* and other spore-forming bacteria.

Acknowledgments. We thank Jonathan Dworkin for critical reading of the manuscript. Work in P.E.'s laboratory is supported by grant GM081571 from the U.S. National Institutes of Health.

Citation. Bate AR, Bonneau R, Eichenberger P. 2014. *Bacillus subtilis* systems biology: applications of -omics techniques to the study of endospore formation. Microbiol Spectrum 2(2): TBS-0019-2013.

References

1. Burkholder PR, Giles NH Jr. 1947. Induced biochemical mutations in *Bacillus subtilis*. *Am J Bot* **34:**345–348.

2. Spizizen J. 1958. Transformation of biochemically deficient strains of *Bacillus subtilis* by deoxyribonucleate. *Proc Natl Acad Sci USA* **44:**1072–1078.

3. Anagnostopoulos C, Spizizen J. 1961. Requirements for transformation in *Bacillus subtilis*. *J Bacteriol* **81:** 741–746.

4. Kunst F, Vassarotti A, Danchin A. 1995. Organization of the European *Bacillus subtilis* genome sequencing project. *Microbiology* **141:**249–255.

5. Kunst F, Ogasawara N, Moszer I, Albertini AM, Alloni G, Azevedo V, Bertero MG, Bessieres P, Bolotin A, Borchert S, Borriss R, Boursier L, Brans A, Braun M, Brignell SC, Bron S, Brouillet S, Bruschi CV, Caldwell B, Capuano V, Carter NM, Choi SK, Codani JJ, Connerton IF, Danchin A, et al. 1997. The complete genome sequence of the gram-positive bacterium *Bacillus subtilis*. *Nature* **390:**249–256.

6. Barbe V, Cruveiller S, Kunst F, Lenoble P, Meurice G, Sekowska A, Vallenet D, Wang T, Moszer I, Medigue C, Danchin A. 2009. From a consortium sequence to a unified sequence: the *Bacillus subtilis* 168 reference genome a decade later. *Microbiology* **155:**1758–1775.

7. Belda E, Sekowska A, Le Fèvre F, Morgat A, Mornico D, Ouzounis C, Vallenet D, Médigue C, Danchin A. 2013. An updated metabolic view of the *Bacillus subtilis* 168 genome. *Microbiology* **159:**757–770.

8. Kearns DB, Chu F, Rudner R, Losick R. 2004. Genes governing swarming in *Bacillus subtilis* and evidence for a phase variation mechanism controlling surface motility. *Mol Microbiol* **52:**357–369.

9. McLoon AL, Guttenplan SB, Kearns DB, Kolter R, Losick R. 2011. Tracing the domestication of a biofilm-forming bacterium. *J Bacteriol* **193:**2027–2034.

10. Zeigler DR, Pragai Z, Rodriguez S, Chevreux B, Muffler A, Albert T, Bai R, Wyss M, Perkins JB. 2008. The origins of 168, W23, and other *Bacillus subtilis* legacy strains. *J Bacteriol* **190:**6983–6995.

11. Youngman P, Perkins JB, Losick R. 1984. Construction of a cloning site near one end of Tn*917* into which foreign DNA may be inserted without affecting transposition in *Bacillus subtilis* or expression of the transposon-borne *erm* gene. *Plasmid* **12:**1–9.

12. Auchtung JM, Lee CA, Monson RE, Lehman AP, Grossman AD. 2005. Regulation of a *Bacillus subtilis* mobile genetic element by intercellular signaling and the global DNA damage response. *Proc Natl Acad Sci USA* **102:**12554–12559.

13. Zeigler DR. 2011. The genome sequence of *Bacillus subtilis* subsp. *spizizenii* W23: insights into speciation within the *B. subtilis* complex and into the history of *B. subtilis* genetics. *Microbiology* **157:**2033–2041.

14. Lazarevic V, Abellan FX, Moller SB, Karamata D, Mauel C. 2002. Comparison of ribitol and glycerol teichoic acid genes in *Bacillus subtilis* W23 and 168: identical function, similar divergent organization, but different regulation. *Microbiology* **148:**815–824.

15. Qian Z, Yin Y, Zhang Y, Lu L, Li Y, Jiang Y. 2006. Genomic characterization of ribitol teichoic acid synthesis in *Staphylococcus aureus*: genes, genomic organization and gene duplication. *BMC Genomics* **7:**74. doi: 10.1186/1471-2164-7-74

16. Earl AM, Eppinger M, Fricke WF, Rosovitz MJ, Rasko DA, Daugherty S, Losick R, Kolter R, Ravel J. 2012. Whole-genome sequences of *Bacillus subtilis* and close relatives. *J Bacteriol* **194:**2378–2379.

17. Schyns G, Serra CR, Lapointe T, Pereira-Leal JB, Potot S, Fickers P, Perkins JB, Wyss M, Henriques AO. 14 February 2013. Genome of a gut strain of *Bacillus subtilis*. *Genome Announc* doi:10.1128/genomeA.00184-12.

18. Durrett R, Miras M, Mirouze N, Narechania A, Mandic-Mulec I, Dubnau D. 20 June 2013. Genome sequence of the *Bacillus subtilis* biofilm-forming transformable strain PS216. *Genome Announc* doi:10.1128/ genomeA.00288-13

19. Rasko DA, Altherr MR, Han CS, Ravel J. 2005. Genomics of the *Bacillus cereus* group of organisms. *FEMS Microbiol Rev* **29:**303–329.

20. Nolling J, Breton G, Omelchenko MV, Makarova KS, Zeng Q, Gibson R, Lee HM, Dubois J, Qiu D, Hitti J, Wolf YI, Tatusov RL, Sabathe F, Doucette-Stamm L, Soucaille P, Daly MJ, Bennett GN, Koonin EV, Smith DR. 2001. Genome sequence and comparative analysis of the solvent-producing bacterium *Clostridium acetobutylicum*. *J Bacteriol* **183:**4823–4838.

21. Shimizu T, Ohtani K, Hirakawa H, Ohshima K, Yamashita A, Shiba T, Ogasawara N, Hattori M, Kuhara S, Hayashi H. 2002. Complete genome sequence of *Clostridium perfringens*, an anaerobic flesh-eater. *Proc Natl Acad Sci USA* **99:**996–1001.

22. Sebaihia M, Peck MW, Minton NP, Thomson NR, Holden MT, Mitchell WJ, Carter AT, Bentley SD, Mason DR, Crossman L, Paul CJ, Ivens A, Wells-Bennik MH, Davis IJ, Cerdeno-Tarraga AM, Churcher C, Quail MA, Chillingworth T, Feltwell T, Fraser A, Goodhead I, Hance Z, Jagels K, Larke N, Maddison M, Moule S, Mungall K, Norbertczak H, Rabbinowitsch E, Sanders M, Simmonds M, White B, Whitehead S, Parkhill J. 2007. Genome sequence of a proteolytic (Group I) *Clostridium botulinum* strain Hall A and comparative analysis of the clostridial genomes. *Genome Res* 17:1082–1092.

23. Sebaihia M, Wren BW, Mullany P, Fairweather NF, Minton N, Stabler R, Thomson NR, Roberts AP, Cerdeño-Tárraga AM, Wang H, Holden MTG, Wright A, Churcher C, Quail MA, Baker S, Bason N, Brooks K, Chillingworth T, Cronin A, Davis P, Dowd L, Fraser A, Feltwell T, Hance Z, Holroyd S, Jagels K, Moule S, Mungall K, Price C, Rabbinowitsch E, Sharp S, Simmonds M, Stevens K, Unwin L, Whithead S, Dupuy B, Dougan G, Barrell B, Parkhill J. 2006. The multidrug-resistant human pathogen *Clostridium difficile* has a highly mobile, mosaic genome. *Nat Genet* 38:779–786.

24. Stragier P. 2002. A gene odyssey: exploring the genomes of endospore-forming bacteria, p 519–526. *In* Sonenshein AL, Hoch JA, Losick R (ed), Bacillus subtilis *and Its Closest Relatives: From Genes to Cells*. ASM Press, Washington, DC.

25. Stephenson K, Hoch JA. 2002. Evolution of signalling in the sporulation phosphorelay. *Mol Microbiol* 46:297–304.

26. Steiner E, Dago AE, Young DI, Heap JT, Minton NP, Hoch JA, Young M. 2011. Multiple orphan histidine kinases interact directly with Spo0A to control the initiation of endospore formation in *Clostridium acetobutylicum*. *Mol Microbiol* 80:641–654.

27. Losick R, Stragier P. 1992. Crisscross regulation of cell-type-specific gene expression during development in *B. subtilis*. *Nature* 355:601–604.

28. Fimlaid KA, Bond JP, Schutz KC, Putnam EE, Leung JM, Lawley TD, Shen A. 2013. Global analysis of the sporulation pathway of *Clostridium difficile*. *PLoS Genet* 8:e1003660. doi:10.1371/journal.pgen.1003660.

29. Pereira FC, Saujet L, Tomé AR, Serrano M, Monot M, Couture-Tosi E, Martin-Verstraete I, Dupuy B, Henriques AO. 2013. The spore differentiation pathway in the enteric pathogen *Clostridium difficile*. *PLoS Genet* 9:e1003782. doi:10.1371/journal.pgen.1003782.

30. Saujet L, Pereira FC, Serrano M, Soutourina O, Monot M, Shelyakin PV, Gelfand MS, Dupuy B, Henriques AO, Martin-Verstraete I. 2013. Genome-wide analysis of cell type-specific gene transcription during spore formation in *Clostridium difficile*. *PLoS Genet* 9:e1003756. doi:10.1371/journal.pgen.1003756.

31. Haraldsen JD, Sonenshein AL. 2003. Efficient sporulation in *Clostridium difficile* requires disruption of the *sigmaK* gene. *Mol Microbiol* 48:811–821.

32. de Hoon MJ, Eichenberger P, Vitkup D. 2010. Hierarchical evolution of the bacterial sporulation network. *Curr Biol* 20:R735–R745.

33. Xiao Y, Francke C, Abee T, Wells-Bennik MH. 2011. Clostridial spore germination versus bacilli: genome mining and current insights. *Food Microbiol* 28:266–274.

34. Sorg JA, Sonenshein AL. 2008. Bile salts and glycine as co-germinants for *Clostridium difficile* spores. *J Bacteriol* 190:2505–2512.

35. Traag BA, Pugliese A, Eisen JA, Losick R. 2013. Gene conservation among endospore-forming bacteria reveals additional sporulation genes in *Bacillus subtilis*. *J Bacteriol* 195:253–260.

36. Abecasis AB, Serrano M, Alves R, Quintais L, Pereira-Leal JB, Henriques AO. 2013. A genomic signature and the identification of new sporulation genes. *J Bacteriol* 195:2101–2115.

37. Galperin MY, Mekhedov SL, Puigbo P, Smirnov S, Wolf YI, Rigden DJ. 2012. Genomic determinants of sporulation in Bacilli and Clostridia: towards the minimal set of sporulation-specific genes. *Environ Microbiol* 14:2870–2890.

38. Jordan B. 2002. Historical background and anticipated developments. *Ann N Y Acad Sci* 975:24–32.

39. Fawcett P, Eichenberger P, Losick R, Youngman P. 2000. The transcriptional profile of early to middle sporulation in *Bacillus subtilis*. *Proc Natl Acad Sci USA* 97:8063–8068.

40. Britton RA, Eichenberger P, Gonzalez-Pastor JE, Fawcett P, Monson R, Losick R, Grossman AD. 2002. Genome-wide analysis of the stationary-phase sigma factor (sigma-H) regulon of *Bacillus subtilis*. *J Bacteriol* 184:4881–4890.

41. Caldwell R, Sapolsky R, Weyler W, Maile RR, Causey SC, Ferrari E. 2001. Correlation between *Bacillus subtilis scoC* phenotype and gene expression determined using microarrays for transcriptome analysis. *J Bacteriol* 183:7329–7340.

42. Eichenberger P, Fujita M, Jensen ST, Conlon EM, Rudner DZ, Wang ST, Ferguson C, Haga K, Sato T, Liu JS, Losick R. 2004. The program of gene transcription for a single differentiating cell type during sporulation in *Bacillus subtilis*. *PLoS Biol* 2:e328. doi:10.1371/journal.pbio.0020328.

43. Wang ST, Setlow B, Conlon EM, Lyon JL, Imamura D, Sato T, Setlow P, Losick R, Eichenberger P. 2006. The forespore line of gene expression in *Bacillus subtilis*. *J Mol Biol* 358:16–37.

44. Steil L, Serrano M, Henriques AO, Volker U. 2005. Genome-wide analysis of temporally regulated and compartment-specific gene expression in sporulating cells of *Bacillus subtilis*. *Microbiology* 151:399–420.

45. Eichenberger P, Jensen ST, Conlon EM, van Ooij C, Silvaggi J, Gonzalez-Pastor JE, Fujita M, Ben-Yehuda S, Stragier P, Liu JS, Losick R. 2003. The sigmaE regulon and the identification of additional sporulation genes in *Bacillus subtilis*. *J Mol Biol* 327:945–972.

46. Feucht A, Evans L, Errington J. 2003. Identification of sporulation genes by genome-wide analysis of the sigmaE regulon of *Bacillus subtilis*. *Microbiology* 149:3023–3034.

47. Imamura D, Kobayashi K, Sekiguchi J, Ogasawara N, Takeuchi M, Sato T. 2004. *spoIVH* (*ykvV*), a requisite cortex formation gene, is expressed in both sporulating compartments of *Bacillus subtilis*. *J Bacteriol* 186:5450–5459.

48. Dworkin J, Losick R. 2005. Developmental commitment in a bacterium. *Cell* 121:401–409.

49. Nicolas P, Mader U, Dervyn E, Rochat T, Leduc A, Pigeonneau N, Bidnenko E, Marchadier E, Hoebeke M, Aymerich S, Becher D, Bisicchia P, Botella E, Delumeau O, Doherty G, Denham EL, Fogg MJ, Fromion V, Goelzer A, Hansen A, Hartig E, Harwood CR, Homuth G, Jarmer H, Jules M, Klipp E, Le Chat L, Lecointe F, Lewis P, Liebermeister W, March A, Mars RA, Nannapaneni P, Noone D, Pohl S, Rinn B, Rugheimer F, Sappa PK, Samson F, Schaffer M, Schwikowski B, Steil L, Stulke J, Wiegert T, Devine KM, Wilkinson AJ, van Dijl JM, Hecker M, Volker U, Bessieres P, Noirot P. 2012. Condition-dependent transcriptome reveals high-level regulatory architecture in *Bacillus subtilis*. *Science* 335:1103–1106.

50. Passalacqua KD, Varadarajan A, Ondov BD, Okou DT, Zwick ME, Bergman NH. 2009. Structure and complexity of a bacterial transcriptome. *J Bacteriol* 191:3203–3211.

51. Jones SW, Paredes CJ, Tracy B, Cheng N, Sillers R, Senger RS, Papoutsakis ET. 2008. The transcriptional program underlying the physiology of clostridial sporulation. *Genome Biol* 9:R114. doi:10.1186/gb-2008-9-7-r114.

52. Bateman A, Birney E, Durbin R, Eddy SR, Howe KL, Sonnhammer EL. 2000. The Pfam protein families database. *Nucleic Acids Res* 28:263–266.

53. Ishii T, Yoshida K, Terai G, Fujita Y, Nakai K. 2001. DBTBS: a database of *Bacillus subtilis* promoters and transcription factors. *Nucleic Acids Res* 29:278–280.

54. Waters LS, Storz G. 2009. Regulatory RNAs in bacteria. *Cell* 136:615–628.

55. Livny J, Waldor MK. 2007. Identification of small RNAs in diverse bacterial species. *Curr Opin Microbiol* 10:96–101.

56. Rasmussen S, Nielsen HB, Jarmer H. 2009. The transcriptionally active regions in the genome of *Bacillus subtilis*. *Mol Microbiol* 73:1043–1057.

57. Irnov I, Sharma CM, Vogel J, Winkler WC. 2010. Identification of regulatory RNAs in *Bacillus subtilis*. *Nucleic Acids Res* 38:6637–6651.

58. Silvaggi JM, Perkins JB, Losick R. 2006. Genes for small, noncoding RNAs under sporulation control in *Bacillus subtilis*. *J Bacteriol* 188:532–541.

59. Schmalisch M, Maiques E, Nikolov L, Camp AH, Chevreux B, Muffler A, Rodriguez S, Perkins J, Losick R. 2010. Small genes under sporulation control in the *Bacillus subtilis* genome. *J Bacteriol* 192:5402–5412.

60. Hutchison CA, Peterson SN, Gill SR, Cline RT, White O, Fraser CM, Smith HO, Venter JC. 1999. Global transposon mutagenesis and a minimal *Mycoplasma* genome. *Science* 286:2165–2169.

61. Kobayashi K, Ehrlich SD, Albertini A, Amati G, Andersen KK, Arnaud M, Asai K, Ashikaga S, Aymerich S, Bessieres P, Boland F, Brignell SC, Bron S, Bunai K, Chapuis J, Christiansen LC, Danchin A, Debarbouille M, Dervyn E, Deuerling E, Devine K, Devine SK, Dreesen O, Errington J, Fillinger S, Foster SJ, Fujita Y, Galizzi A, Gardan R, Eschevins C, Fukushima T, Haga K, Harwood CR, Hecker M, Hosoya D, Hullo MF, Kakeshita H, Karamata D, Kasahara Y, Kawamura F, Koga K, Koski P, Kuwana R, Imamura D, Ishimaru M, Ishikawa S, Ishio I, Le Coq D, Masson A, Mauel C, Meima R, Mellado RP, Moir A, Moriya S, Nagakawa E, Nanamiya H, Nakai S, Nygaard P, Ogura M, Ohanan T, O'Reilly M, O'Rourke M, Pragai Z, Pooley HM, Rapoport G, Rawlins JP, Rivas LA, Rivolta C, Sadaie A, Sadaie Y, Sarvas M, Sato T, Saxild HH, Scanlan E, Schumann W, Seegers JF, Sekiguchi J, Sekowska A, Seror SJ, Simon M, Stragier P, Studer R, Takamatsu H, Tanaka T, Takeuchi M, Thomaides HB, Vagner V, van Dijl JM, Watabe K, Wipat A, Yamamoto H, Yamamoto M, Yamamoto Y, Yamane K, Yata K, Yoshida K, Yoshikawa H, Zuber U, Ogasawara N. 2003. Essential *Bacillus subtilis* genes. *Proc Natl Acad Sci USA* 100:4678–4683.

62. Commichau FM, Pietack N, Stulke J. 2013. Essential genes in *Bacillus subtilis*: a re-evaluation after ten years. *Mol Biosyst* 9:1068–1075.

63. van Opijnen T, Bodi KL, Camilli A. 2009. Tn-seq: high-throughput parallel sequencing for fitness and genetic interaction studies in microorganisms. *Nat Methods* 6:767–772.

64. Tong AH, Evangelista M, Parsons AB, Xu H, Bader GD, Page N, Robinson M, Raghibizadeh S, Hogue CW, Bussey H, Andrews B, Tyers M, Boone C. 2001. Systematic genetic analysis with ordered arrays of yeast deletion mutants. *Science* 294:2364–2368.

65. Jorgensen P, Nelson B, Robinson MD, Chen Y, Andrews B, Tyers M, Boone C. 2002. High-resolution genetic mapping with ordered arrays of *Saccharomyces cerevisiae* deletion mutants. *Genetics* 162:1091–1099.

66. Costanzo M, Baryshnikova A, Bellay J, Kim Y, Spear ED, Sevier CS, Ding H, Koh JL, Toufighi K, Mostafavi S, Prinz J, St Onge RP, VanderSluis B, Makhnevych T, Vizeacoumar FJ, Alizadeh S, Bahr S, Brost RL, Chen Y, Cokol M, Deshpande R, Li Z, Lin ZY, Liang W, Marback M, Paw J, San Luis BJ, Shuteriqi E, Tong AH, van Dyk N, Wallace IM, Whitney JA, Weirauch MT, Zhong G, Zhu H, Houry WA, Brudno M, Ragibizadeh S, Papp B, Pal C, Roth FP, Giaever G, Nislow C, Troyanskaya OG, Bussey H, Bader GD, Gingras AC, Morris QD, Kim PM, Kaiser CA, Myers CL, Andrews BJ, Boone C. 2010. The genetic landscape of a cell. *Science* 327:425–431.

67. Nichols RJ, Sen S, Choo YJ, Beltrao P, Zietek M, Chaba R, Lee S, Kazmierczak KM, Lee KJ, Wong A, Shales M, Lovett S, Winkler ME, Krogan NJ, Typas A, Gross CA. 2011. Phenotypic landscape of a bacterial cell. *Cell* 144:143–156.

68. Kuwana R, Kasahara Y, Fujibayashi M, Takamatsu H, Ogasawara N, Watabe K. 2002. Proteomics character-

ization of novel spore proteins of *Bacillus subtilis*. *Microbiology* 148:3971–3982.

69. **Lai EM, Phadke ND, Kachman MT, Giorno R, Vazquez S, Vazquez JA, Maddock JR, Driks A.** 2003. Proteomic analysis of the spore coats of *Bacillus subtilis* and *Bacillus anthracis*. *J Bacteriol* 185:1443–1454.

70. **Abhyankar W, Beek AT, Dekker H, Kort R, Brul S, de Koster CG.** 2011. Gel-free proteomic identification of the *Bacillus subtilis* insoluble spore coat protein fraction. *Proteomics* 11:4541–4550.

71. **Ong SE, Blagoev B, Kratchmarova I, Kristensen DB, Steen H, Pandey A, Mann M.** 2002. Stable isotope labeling by amino acids in cell culture, SILAC, as a simple and accurate approach to expression proteomics. *Mol Cell Proteomics* 1:376–386.

72. **Soufi B, Kumar C, Gnad F, Mann M, Mijakovic I, Macek B.** 2010. Stable isotope labeling by amino acids in cell culture (SILAC) applied to quantitative proteomics of *Bacillus subtilis*. *J Proteome Res* 9:3638–3646.

73. **Marchadier E, Carballido-Lopez R, Brinster S, Fabret C, Mervelet P, Bessieres P, Noirot-Gros MF, Fromion V, Noirot P.** 2011. An expanded protein-protein interaction network in *Bacillus subtilis* reveals a group of hubs: exploration by an integrative approach. *Proteomics* 11:2981–2991.

74. **Huh WK, Falvo JV, Gerke LC, Carroll AS, Howson RW, Weissman JS, O'Shea EK.** 2003. Global analysis of protein localization in budding yeast. *Nature* 425:686–691.

75. **Werner JN, Chen EY, Guberman JM, Zippilli AR, Irgon JJ, Gitai Z.** 2009. Quantitative genome-scale analysis of protein localization in an asymmetric bacterium. *Proc Natl Acad Sci USA* 106:7858–7863.

76. **McKenney PT, Driks A, Eskandarian HA, Grabowski P, Guberman J, Wang KH, Gitai Z, Eichenberger P.** 2010. A distance-weighted interaction map reveals a previously uncharacterized layer of the *Bacillus subtilis* spore coat. *Curr Biol* 20:934–938.

77. **McKenney PT, Eichenberger P.** 2012. Dynamics of spore coat morphogenesis in *Bacillus subtilis*. *Mol Microbiol* 83:245–260.

78. **Watrous J, Roach P, Alexandrov T, Heath BS, Yang JY, Kersten RD, van der Voort M, Pogliano K, Gross H, Raaijmakers JM, Moore BS, Laskin J, Bandeira N, Dorrestein PC.** 2012. Mass spectral molecular networking of living microbial colonies. *Proc Natl Acad Sci USA* 109:E1743–E1752.

79. **Yang YL, Xu Y, Straight P, Dorrestein PC.** 2009. Translating metabolic exchange with imaging mass spectrometry. *Nat Chem Biol* 5:885–887.

80. **Kanehisa M, Goto S, Sato Y, Furumichi M, Tanabe M.** 2012. KEGG for integration and interpretation of large-scale molecular data sets. *Nucleic Acids Res* 40:D109–D114.

81. **Plata G, Fuhrer T, Hsiao TL, Sauer U, Vitkup D.** 2012. Global probabilistic annotation of metabolic networks enables enzyme discovery. *Nat Chem Biol* 8:848–854.

82. **Wunschel D, Fox KF, Black GE, Fox A.** 1994. Discrimination among the *B. cereus* group, in comparison to *B. subtilis*, by structural carbohydrate profiles and ribosomal RNA spacer region PCR. *Syst Appl Microbiol* 17:625–635.

83. **Ishii N, Nakahigashi K, Baba T, Robert M, Soga T, Kanai A, Hirasawa T, Naba M, Hirai K, Hoque A, Ho PY, Kakazu Y, Sugawara K, Igarashi S, Harada S, Masuda T, Sugiyama N, Togashi T, Hasegawa M, Takai Y, Yugi K, Arakawa K, Iwata N, Toya Y, Nakayama Y, Nishioka T, Shimizu K, Mori H, Tomita M.** 2007. Multiple high-throughput analyses monitor the response of *E. coli* to perturbations. *Science* 316:593–597.

84. **Goelzer A, Bekkal Brikci F, Martin-Verstraete I, Noirot P, Bessieres P, Aymerich S, Fromion V.** 2008. Reconstruction and analysis of the genetic and metabolic regulatory networks of the central metabolism of *Bacillus subtilis*. *BMC Syst Biol* 2:20. doi:10.1186/1752-0509-2-20.

85. **Oh YK, Palsson BO, Park SM, Schilling CH, Mahadevan R.** 2007. Genome-scale reconstruction of metabolic network in *Bacillus subtilis* based on high-throughput phenotyping and gene essentiality data. *J Biol Chem* 282:28791–28799.

86. **Gollnick P, Babitzke P, Antson A, Yanofsky C.** 2005. Complexity in regulation of tryptophan biosynthesis in *Bacillus subtilis*. *Annu Rev Genet* 39:47–68.

87. **Bonneau R, Facciotti MT, Reiss DJ, Schmid AK, Pan M, Kaur A, Thorsson V, Shannon P, Johnson MH, Bare JC, Longabaugh W, Vuthoori M, Whitehead K, Madar A, Suzuki L, Mori T, Chang DE, Diruggiero J, Johnson CH, Hood L, Baliga NS.** 2007. A predictive model for transcriptional control of physiology in a free living cell. *Cell* 131:1354–1365.

88. **Waltman P, Kacmarczyk T, Bate AR, Kearns DB, Reiss DJ, Eichenberger P, Bonneau R.** 2010. Multi-species integrative biclustering. *Genome Biol* 11:R96. doi:10.1186/gb-2010-11-9-r96.

89. **Fadda A, Fierro AC, Lemmens K, Monsieurs P, Engelen K, Marchal K.** 2009. Inferring the transcriptional network of *Bacillus subtilis*. *Mol Biosyst* 5:1840–1852.

90. **Buescher JM, Liebermeister W, Jules M, Uhr M, Muntel J, Botella E, Hessling B, Kleijn RJ, Le Chat L, Lecointe F, Mader U, Nicolas P, Piersma S, Rugheimer F, Becher D, Bessieres P, Bidnenko E, Denham EL, Dervyn E, Devine KM, Doherty G, Drulhe S, Felicori L, Fogg MJ, Goelzer A, Hansen A, Harwood CR, Hecker M, Hubner S, Hultschig C, Jarmer H, Klipp E, Leduc A, Lewis P, Molina F, Noirot P, Peres S, Pigeonneau N, Pohl S, Rasmussen S, Rinn B, Schaffer M, Schnidder J, Schwikowski B, Van Dijl JM, Veiga P, Walsh S, Wilkinson AJ, Stelling J, Aymerich S, Sauer U.** 2012. Global network reorganization during dynamic adaptations of *Bacillus subtilis* metabolism. *Science* 335:1099–1103.

91. **Greenfield A, Hafemeister C, Bonneau R.** 2013. Robust data-driven incorporation of prior knowledge into the inference of dynamic regulatory networks. *Bioinformatics* 29:1060–1067.

92. **Flórez LA, Roppel SF, Schmeisky AG, Lammers CR, Stülke J.** 2009. A community-curated consensual annotation that is continuously updated: the *Bacillus subtilis*

centered wiki SubtiWiki. *Database (Oxford)* **2009:** bap012. doi:10.1093/database/bap012

93. Mader U, Schmeisky AG, Florez LA, Stulke J. 2012. SubtiWiki–a comprehensive community resource for the model organism *Bacillus subtilis. Nucleic Acids Res* **40:** D1278–D1287.

94. Marbach D, Costello JC, Küffner R, Vega NM, Prill RJ, Camacho DM, Allison KR; DREAM5 Consortium, Kellis M, Collins JJ, Stolovitzky G. 2012. Wisdom for crowds for robust gene network inference. *Nat Methods* **9:**796–804.

95. Moszer I Glaser P, Danchin A. 1995. SubtiList: a relational database for the *Bacillus subtilis* genome. *Microbiology* **141:**261–268.

96. Moszer I, Jones LM, Moreira S, Fabry C, Danchin A. 2002. SubtiList: the reference database for the *Bacillus subtilis* genome. *Nucleic Acids Res* **30:**62–65.

97. Lechat P, Hummel L, Rousseau S, Moszer I. 2008. GenoList: an integrated environment for comparative analysis of microbial genomes. *Nucleic Acids Res* **36:** D469–D474.

98. Killcoyne S, Carter GW, Smith J, Boyle J. 2009. Cytoscape: a community-based framework for network modeling. *Methods Mol Biol* **563:**219–239.

99. Smoot ME, Ono K, Ruscheinski J, Wang PL, Ideker T. 2011. Cytoscape 2.8: new features for data integration and network visualization. *Bioinformatics* **27:**431–432.

100. Shannon PT, Reiss DJ, Bonneau R, Baliga NS. 2006. The Gaggle: an open-source software system for integrating bioinformatics software and data sources. *BMC Bioinformatics* **7:**176. doi:10.1186/1471-2105-7-176

101. Kacmarczyk T, Waltman P, Bate A, Eichenberger P, Bonneau R. 2011. Comparative microbial modules resource: generation and visualization of multi-species biclusters. *PLoS Comput Biol* **7:**e1002228. doi:10.1371/journal.pcbi.1002228

102. Beauregard PB, Chai Y, Vlamakis H, Losick R, Kolter R. 2013. *Bacillus subtilis* biofilm induction by plant polysaccharides. *Proc Natl Acad Sci USA* **110:**E1621–E1630.

103. Karr JR, Sanghvi JC, Macklin DN, Gutschow MV, Jacobs JM, Bolival B Jr, Assad-Garcia N, Glass JI, Covert MW. 2012. A whole-cell computational model predicts phenotype from genotype. *Cell* **150:**389–401.

104. McKenney PT, Driks A, Eichenberger P. 2013. The *Bacillus subtilis* endospore: assembly and functions of the multilayered coat. *Nat Rev Microbiol* **11:**33–44.

105. Fajardo-Cavazos P, Maughan H, Nicholson WL. Evolution in the *Bacillaceae*. *In* Eichenberger P, Driks A (ed), *The Bacterial Spore.* ASM Press, Washington, DC, in press.

106. Mandic-Mulec I, Stefanic P, van Elsas JD. Ecology of *Bacillaceae*. *In* Eichenberger P, Driks A (ed), *The Bacterial Spore.* ASM Press, Washington, DC, in press.

107. Galperin MY. 2013. Genomic diversity of spore-forming *Firmicutes. Microbiol Spectrum* **1:**TBS-0015-2012.

108. Dürre P. 2014. Physiology and sporulation in *Clostridium. Microbiol Spectrum* **2:**TBS-0010-2012.

109. Dworkin J. 2014. Protein targeting during *Bacillus subtilis* sporulation. *Microbiol Spectrum* **2:**TBS-0006-2013.

110. Francis MB, Allen CA, Shrestha R, Sorg JA. 2013. Bile acid recognition by the *Clostridium difficile* germinant receptor, CspC, is important for establishing infection. *PLoS Pathog* **9:**e1003356.

111. Meeske AJ, Rodrigues CD, Brady J, Lim HC, Berharnd TG, Rudner DZ. 2016. High-throughput genetic screens identify a large and diverse collection of new sporulation genes in *Bacillus subtilis. PLoS Biol* **14:** e1002341.

112. Arrieta-Ortiz ML, Hafemeister C, Bate AR, Chu T, Greenfield A, Shuster B, Barry SN, Gallitto M, Liu B, Kacmarczyk T, Santoriello F, Chen J, Rodrigues CD, Sato T, Rudner DZ, Driks A, Bonneau R, Eichenberger P. 2015. An experimentally supported model of the *Bacillus subtilis* global transcriptional regulatory network. *Mol Syst Biol* **11:**839.

The Bacterial Spore: From Molecules to Systems
Edited by P. Eichenberger and A. Driks
© 2016 American Society for Microbiology, Washington, DC
doi:10.1128/microbiolspec.TBS-0006-2012

Jonathan Dworkin[1]

Protein Targeting during *Bacillus subtilis* Sporulation

7

The first published reports of sporulation were by Ferdinand Cohn in *Bacillus subtilis* and Robert Koch in *Bacillus anthracis* in 1877 in the *Zeitschrift für Plänzenbiologie* (1, 2). In addition to being among the earliest published reports of bacteria, these articles provided the initial demonstration that bacteria have an internal cellular organization. Cohn and Koch described intracellular membrane-bound compartments of different sizes and positions within the cell. Despite these striking original observations, bacteria were considered for much of the 20th century to be "bags of enzymes" that lacked any discernible spatial organization (3). However, in the past two decades, numerous examples, including dedicated intracellular compartments such as magnetosomes, the presence of a defined orientation of the chromosome, and the polarity of the chemotactic apparatus have emphatically demonstrated that bacterial cells exhibit an intracellular organization that, while less complex than eukaryotic cells, is nonetheless a critical and significant part of their physiology.

Spores look very different from growing cells. This morphological differentiation initiates with an asymmetric division near to one pole of the cell, resulting in the formation of a smaller cell, the forespore, and a larger cell, the mother cell. The sporulation septum is similar, but not identical to the normal mid-cell division septum, and contains a thinner layer of peptidoglycan separating the two compartments. Numerous proteins are specifically localized to this septum, some of which are present on only one side of the septum.

In sporulation, the smaller of the two cells resulting from the asymmetric division becomes encased within the larger cell in a process termed "engulfment" (Fig. 1A, panels iii to v). Following completion of asymmetric septation, the mother cell membranes move and eventually entirely encircle the forespore. While there are superficial similarities to membrane movements that occur during phagocytosis in eukaryotic cells, the analogy may be of limited relevance, since in sporulating cells there is a layer of peptidoglycan that surrounds the forespore and separates the two compartments. During the later stages of spore development, the inner and outer proteinaceous layers of the spore are assembled, and the spore cortex, consisting of a thick layer of peptidoglycan contained between the inner and outer spore membranes, is synthesized.

In this review, we focus on the intracellular targeting of sporulation-specific proteins that mediate these processes that occur at specific locations in the cell. Although this localization has implications for function in many cases, we do not discuss in detail how these proteins work, and direct the interested reader to several

[1]Department of Microbiology & Immunology, College of Physicians and Surgeons, Columbia University, New York, NY 10032.

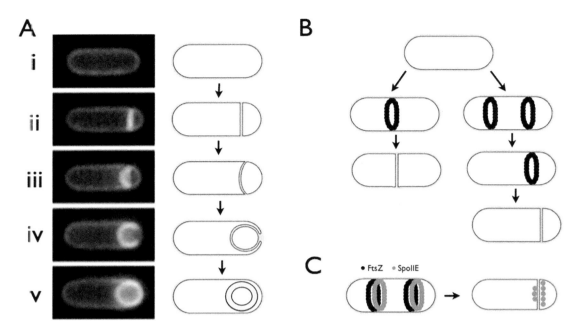

Figure 1 Morphological and protein asymmetry during early stages in sporulation. **A.** Following entry into sporulation (i), cells establish an asymmetric septum (ii), dividing the sporangium into two unequally sized compartments, the forespore and the mother cell. During engulfment, the septum begins to curve (iii) and continues to curve (iv) until it is attached to the mother cell by only a small patch. Finally, the forespore pinches off from the mother cell (v) and forms a membrane-bounded compartment containing a thick layer of peptidoglycan (gray). Shown are fluorescent microscopy images using the membrane stain FM4-64 (left) and schematic cartoons (right). **B.** FtsZ rings (Z rings; black) are located at mid-cell during growth (left), but upon entry into sporulation (right), are seen initially in a bipolar pattern and eventually in a unipolar pattern before formation of the asymmetric septum. **C.** SpoIIE (gray) initially forms "E-rings" that are seen near the bipolar Z rings (left) but following formation of the polar septum (right), SpoIIE is seen on both the mother cell and forespore faces, with apparent enrichment on the forespore face.

reviews that provide a more general perspective (4–7). Finally, as a note of caution, most of the studies described here use fluorescent protein fusions to monitor localization. While in most cases these fusions complement the endogenous proteins, very recently it has been reported that the particular intracellular distribution observed for a number of proteins such as the MreB helix (8) or the ClpX and ClpP puncta (9) is likely an artifact of the fluorescent protein fusion. Whether this bias is seen in other protein localization studies is not yet known.

ASYMMETRIC SEPTATION

Like many other bacteria, *B. subtilis* divides symmetrically during growth. However, upon entry into sporulation, an asymmetric septum forms near one pole with roughly equal probability near either the "old" or the "new" pole (10). In addition to this difference in position, the peptidoglycan layer of the sporulation

septum is noticeably thinner than vegetative septa. As with vegetative septa, the initial component of the sporulation septum is the polymeric, tubulin-like protein FtsZ. In fact, the first reported subcellular localization of a soluble protein, FtsZ, was at mid-cell (11). While the precise molecular nature of FtsZ polymer *in vivo* is not yet clear (12), it appears to form a coil-like structure that is centered at mid-cell (13). Under certain circumstances, FtsZ rings are capable of generating a contractile force (14), and, while theoretical analyses suggest that this force may be sufficient to drive cytokinesis (15, 16), whether this is true *in vivo* is not known.

When cells enter sporulation, the FtsZ ring switches its location from mid-cell to the cell poles (17) (Fig. 1B, top). While this switch maintains cellular symmetry, it should be noted that structures such as FtsZ rings that appear to be symmetric are, at higher resolution, actually asymmetric since they are composed of protein subunits with intrinsic asymmetry (18). The switch is governed by the transcription factor Spo0A, since *spo0A* mutants fail to form polar FtsZ rings (17), and requires increases in the level of FtsZ and expression of the FtsZ interacting protein SpoIIE that are both dependent on Spo0A (19). The shift in FtsZ-ring localization occurs via a spiral intermediate along the long axis of the cell in a "slinky"-like movement (19), although what causes this process to overcome the MinCD-dependent inhibition of FtsZ-ring formation near the pole is not known. Recently, a new protein, RefZ, has been described that appears to mediate the transition of FtsZ rings into spirals (20). RefZ is a DNA-binding protein that recognizes specific chromosomal sequences suggesting that the chromosome itself plays a direct role in the FtsZ ring→spiral transition (20), but the underlying mechanism has not yet been worked out.

Sporulating cells rarely make two mature polar septa (5), although this event can be observed in so-called disporic mutants that have a terminal asporogenous phenotype (21). This event is symmetry breaking since sporulating cells go from being symmetric to asymmetric with respect to septum position (22) (although the coccal bacterium *Sporosarcina ureae* undergoes symmetric division in sporulation [23]). An important clue to the mechanism underlying the transition is that the two polar septa are formed sequentially, not simultaneously. One interpretation of this observation is that cells are able to only construct a single septum at a time (24) perhaps from insufficient levels of factor(s) necessary for the simultaneous synthesis of two septa. Clearly, however, cells have sufficient levels of some key cytokinetic proteins (e.g., FtsZ) to form multiple structures. Whatever the mechanism underlying the sequential nature of septation, the cellular asymmetry is maintained by a checkpoint system whereby σ^E activation in the mother cell leads to the production of proteins (SpoIID, SpoIIM, and SpoIIP) that act together to inhibit formation of a second septum (24, 25).

A key consequence of asymmetric septation is that the septum forms over one of the two chromosomes. Thus, the origin-proximal one-third of that chromosome is located in the forespore, the smaller compartment, and genes at origin distal positions are located outside the forespore. Until the remainder of the chromosome is translocated, origin-distal genes are excluded from the forespore. This transient asymmetry has important implications for sporulation (26), since the differential activation of the compartment-specific transcription factors σ^F and σ^E is dependent on the chromosomal position of genes encoding the proteins, SpoIIAB (27) and SpoIIR (28), respectively, which are necessary for their activation. The DNA-binding protein RacA, which is under control of Spo0A, is essential for the proper attachment of this chromosome origin to a cell pole, ensuring the establishment of this polarity (29, 30). RacA localizes to the cell poles through an interaction with the protein DivIVA (29), a peripheral membrane protein that preferentially localizes to negatively curved membranes (31).

SpoIIE

FtsZ serves as a nucleating factor for proteins that mediate other cytokinetic functions (32, 33). One sporulation-specific protein that interacts directly with FtsZ is the membrane PP2C-type phosphatase SpoIIE (34). The localization of SpoIIE, at least initially, to polar positions depends on FtsZ (35, 36) (Fig. 1C), but, as is seen with FtsZ, SpoIIE becomes unipolar in its localization pattern (35, 37). Although (as discussed below), the main function of SpoIIE is in the pathway leading to forespore-specific activation of σ^F, SpoIIE may be involved also in septum formation since Δ*spoIIE* mutants have an unusually thin septum (38).

A key aspect of sporulation is the compartment-specific activation of sigma factors in the mother cell and the forespore, which follow a crisscross pattern of regulation where the activation of a sigma factor in one compartment leads to the subsequent activation in the other compartment (39). SpoIIE plays a key role in the activation of the initial forespore transcription factor σ^F by dephosphorylating SpoIIAA-P, which then relieves SpoIIAB-dependent inhibition of σ^F (40). Early models of σ^F activation hypothesized that SpoIIE was targeted to both faces of the asymmetric septum (41) and that the greater surface-to-volume ratio of the forespore in comparison with the mother cell was sufficient to account for preferential forespore activity. However, measurements of SpoIIE-green fluorescent protein (GFP) distribution in sporangia found that SpoIIE is enriched in the forespore, indicating that the relative enrichment due to surface-to-volume ratio may not be sufficient (35). In addition, detailed analysis of SpoIIE localization in single sporulating cells using PSICIC (Projected System of Internal Coordinates from Interpolated Contours) software indicates that SpoIIE is preferentially distributed on the forespore face (42) (Fig. 1C). How this asymmetry arises is unclear, but, since SpoIIE interacts with FtsZ (34), the intrinsic asymmetry of the FtsZ polymers may be responsible.

SpoIIIE

For sporulation to proceed, the distal two-thirds of one of the two replicated chromosomes must be pumped into the forespore. SpoIIIE, a large, polytopic membrane protein belonging to the FtsK family of DNA transporters, is responsible for this translocation. Members of this family have diverse physiological roles including chromosome partitioning during vegetative growth and conjugative transfer of plasmid DNA. SpoIIIE, like related proteins, is an ATPase and contains a DNA-binding domain. These proteins form hexameric aqueous channels that translocate one single double-stranded DNA molecule at a time (43). Recent evidence using superresolution microscopy indicates that DNA translocation is mediated by ~5-nm complexes each composed of ~50 SpoIIIE molecules that form a single pore (44).

SpoIIIE is observed at the asymmetric septum and is enriched toward its center (45) (Fig. 2A, panels i and ii). SpoIIIE fails to localize to the asymmetric division septum during sporulation when genetic perturbations result in the absence of DNA from the forespore, suggesting that it only assembles at the sporulation septum when DNA is trapped by the membrane (46). SpoIIIE forms a focus only in the mother cell side of the septum in wild-type cells (47). Consistent with the role of the asymmetric distribution of specific chromosomal sequences in maintaining polarity of translocation for FtsK (48), alterations in chromosome architecture switched SpoIIIE assembly to the forespore, and DNA translocation-defective SpoIIIE proteins assembled in both cells. Thus, DNA determines the "macroscale" septal localization of SpoIIIE, but it also determines the "microscale" partitioning of SpoIIIE between the forespore and mother cell faces of the septal membrane.

SpoIIIE is also observed following septation at sites of membrane fission between the mother cell and the forespore (Fig. 2A, panel iii) (49), and some *spoIIIE* mutations result in engulfment defects. SpoIIIE is observed at the leading edge of the migrating septal membrane as engulfment progresses (50), suggesting a model for passive redistribution of SpoIIIE to the poles as a consequence of this membrane movement. The basis for this movement is the subject of active study, although the synthesis (51) and degradation (52) of the peptidoglycan (as mediated by the SpoIID and SpoIIP proteins, see below) that lies between the mother cell and forespore membranes both appear to contribute. FisB is a recently identified protein that also localizes to the sites of membrane fission and is required for optimal fission (53). FisB interacts closely with cardiolipin (53), and these interactions may mediate its localization to the engulfing membrane, which has a high degree of negative curvature (54).

SpoIIGA/SpoIIR

One of the first genes to be expressed in the forespore following σ^F activation is *spoIIR*, which encodes a secreted protein that stimulates conversion of pro-σ^E to σ^E only in the mother cell. SpoIIR is likely exported into the intermembrane space of the septum where it activates the membrane-associated pro-σ^E-processing enzyme SpoIIGA. A key question has been how σ^E is selectively activated in the mother cell. Pro-σ^E is distributed uniformly along all membrane surfaces and is not confined to the mother cell face of the septum (55). Thus, the asymmetric localization of pro-σ^E is not responsible for the differential activation. In addition, SpoIIR shows a septal pattern of localization (56) that is dependent on SpoIIGA (57), indicating that the localization of SpoIIGA is central. SpoIIGA is seen very early on in polar septum formation, suggesting that it interacts with a component of the division machinery (Fig. 2B). SpoIIGA localization is not dependent on SpoIIE (58), but the role of other proteins found at the polar septum is not known. SpoIIGA may become enriched in the mother cell through a diffusion-and-capture model, but the "anchor" protein has not been identified (59). It is possible that like SpoIIE, SpoIIGA becomes enriched on the mother cell membrane through an interaction with a protein polymer like FtsZ that is intrinsically asymmetric. Thus, restriction of SpoIIR action to the mother cell face of the septum likely originates in the asymmetric distribution of SpoIIGA.

SpoIIB/SpoIID/SpoIIM/SpoIIP

The peptidoglycan located in the intermembrane space between the outer forespore and inner forespore membranes is attached to the peptidoglycan that surrounds the combined mother cell and forespore. This peptidoglycan makes a rigid connection that must be severed for engulfment to occur. Mutations in *spoIID*, *spoIIM*, or *spoIIP* result in the bulging of the forespore compartment into the mother cell and prevent engulfment from proceeding (60). SpoIID is a peptidoglycan hydrolase (61) and SpoIIP is an autolysin (62) that is both an amidase and endopeptidase that removes the stem peptides from the cell wall and cleaves their cross-links (52). Together with SpoIIM, these proteins form a so-called DMP complex that hydrolyzes the peptidoglycan between the two membranes, thereby facilitating both interactions between mother cell and forespore proteins (e.g., SpoIIQ and SpoIIIAH [63, 64]; see below) as well as the movement of the mother cell membrane around the forespore (62, 65). Localization of SpoIID depends on SpoIIP, which itself depends on SpoIIM (62). The sporulation protein SpoIIB of unknown enzymatic activity also localizes to the septal membrane (66), and SpoIIM depends on SpoIIB for its septal localization,

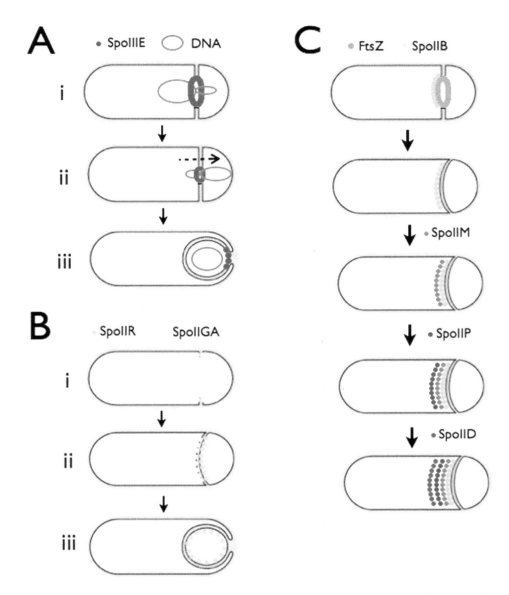

Figure 2 Localization of septal proteins early in sporulation. **A.** The SpoIIIE DNA translocase (red) localizes to the asymmetric septum (i) because of the presence of DNA (blue) and mediates DNA pumping into the forespore (ii). Following completion of septation, SpoIIIE is found in the forespore membrane at the last point of contact with the mother cell (iii). **B.** The pro-σ^E processing enzyme SpoIIGA (yellow) initially localizes to the sites of incipient septum formation (i) and then to the mother cell face of the asymmetric septum (ii). The SpoIIR signaling protein (green) is made in the forespore and crosses the forespore membrane where it presumably interacts with and activates SpoIIGA, although it is also seen in the forespore following completion of septation. **C.** SpoIIB (aqua) initially colocalizes with FtsZ (orange) during the process of Z-ring constriction and remains in the polar septum following completion of septation. The SpoIIM (light blue), SpoIIP (dark blue), and SpoIID (purple) proteins proceed to localize to the now curved asymmetric septum.

leading to a hierarchical model of localization: SpoIIB→SpoIIM→SpoIIP→SpoIID (Fig. 2C) (65). However, the DMP complex can localize in the absence of SpoIIB by apparently using the SpoIVFAB complex as a secondary mechanism (65) (see below for discussion of SpoIVFAB). Finally, in addition to their role in engulfment, SpoIID, SpoIIM, and SpoIIP play a key role in maintaining cellular asymmetry by preventing the formation of a second polar septum and the resulting formation of a terminal disporic cell (24, 25).

SpoIIQ/SpoIIIAH

When the septal peptidoglycan of sporulating cells is enzymatically removed by lysozyme, the cells engulf even in the absence of the DMP complex and instead require the forespore protein SpoIIQ and its mother cell partner SpoIIIAH (67). SpoIIQ is an integral membrane protein under the control of σ^F that is essential for the synthesis and activation of the late forespore-specific transcription factor σ^G (68). The extracellular portion of SpoIIQ contains a LytM domain found typically in endopeptidases. However, SpoIIQ is probably not an endopeptidase since its LytM domain lacks a key catalytic residue, it does not appear to interact directly with peptidoglycan *in vitro* (69), and two recent crystal structures indicate that it does not coordinate Zn2+ (70, 71). Following asymmetric septation, a SpoIIQ fluorescent protein fusion is observed at the forespore membrane, and, when engulfment initiates, it moves around the forespore. Finally, near the completion of engulfment, SpoIIQ is seen as punctate membrane-associated fluorescence (56) (Fig. 3A, purple). SpoIIQ localization is dependent on the degradation of septal peptidoglycan by SpoIID and SpoIIP, and this localization is apparently stabilized by an interaction with its partner protein SpoIIIAH (63, 64).

The polytopic membrane protein SpoIIIAH is expressed under control of σ^E in the mother cell. SpoIIIAH is part of the mother cell SpoIIIA complex, composed of eight membrane proteins, that is necessary for activation of σ^G (72). This complex forms a channel between the mother cell and the forespore (73, 74) and may allow the movement of small molecules from the mother cell into the forespore (75). SpoIIIAH is initially observed in all of the mother cell membranes, but, consistent with its role in intercompartmental signaling, SpoIIIAH becomes localized to the septum (Fig. 3A, yellow). This enrichment is dependent on SpoIIQ (76, 77) and is mediated by an interaction between the LytM domain of SpoIIQ and the extracellular YscJ domain of SpoIIIAH (69, 75) in the sporulation septum (Fig. 3B). Recent crystal structures of the SpoIIQ-SpoIIIAH complex are consistent with these biochemical studies and demonstrate that SpoIIQ and SpoIIIAH both form ring-like structures (70, 71).

This capture results in the assembly of SpoIIIAH with SpoIIQ into helical arcs and foci around the forespore (76). Photobleaching experiments indicate that the SpoIIQ multimer does not freely diffuse, so SpoIIQ and SpoIIIAH complex could function as a ratchet to irreversibly drive engulfment (67). Not much is known about how the other components of the SpoIIIA complex are targeted, but SpoIIIAH and SpoIIQ are necessary for the localization of a cyan fluorescent protein-SpoIIIAG fusion protein to the mother cell-forespore interface (78). The septal targeting of SpoIIIAH by SpoIIQ is an example of the mechanism of "diffusion and capture" (79). In other examples of this mechanism (e.g., SpoIVFA-SpoIVFB, see below) it is often unclear how the landmark anchoring protein is initially targeted. However, in the case of SpoIIIAH-SpoIIQ, one distinguishing feature of the incipient outer forespore membrane is that it is adjacent to the inner forespore membrane. Thus, specific gene expression (σ^F-dependent *spoIIQ*) in one compartment leads to asymmetrical protein distribution in another compartment. This is another example of how the intrinsic asymmetry of the chromosome is converted to a spatial asymmetry (26).

SpoIVFA/SpoIVFB/BofA

The final sigma factor to be activated is σ^K in the mother cell. σ^K is present as a pro-protein and the processing of pro-σ^K to σ^K is mediated by the membrane-embedded metalloprotease SpoIVFB, which itself is held inactive by two other membrane proteins SpoIVFA and BofA. Pro-σ^K and all three of these proteins are enriched at the outer forespore membrane as demonstrated both by immunofluorescence (pro-σ^K [80] and SpoIVFA and SpoIVFB [81]) and by complementing GFP fusions (SpoIVFA [82], SpoIVB [81], and BofA [82]). σ^K processing depends on a forespore-produced protein SpoIVB that cleaves the inhibitory protein SpoIVFA thereby freeing the SpoIVFB to process pro-σ^K (83). The signal that SpoIVB responds to is unknown, but SpoIVB fails to accumulate when engulfment is perturbed, suggesting that SpoIVB levels act as a checkpoint for the progression of engulfment (84). Given the presence of this checkpoint, it is imperative for σ^K processing to occur on the outer forespore membrane and not on the mother cell membrane. Since localization of both BofA and SpoIVFB is dependent on SpoIVFA (82), understanding the localization of SpoIVFA is key to understanding the targeting of the entire complex.

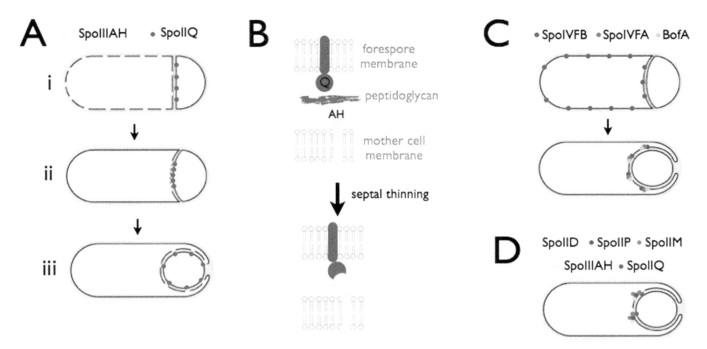

Figure 3 Localization of septal proteins later in sporulation. **A.** Expression of SpoIIIAH (lime) is under control of σ^E and it is initially found in all of the mother cell membrane (i). Expression of SpoIIQ (purple) is under control of σ^F and it is initially found in the forespore septal membrane. Interaction of SpoIIQ and SpoIIIAH in the septal intermembrane space leads to localization of SpoIIIAH to the septum (ii), and this interaction continues until late in engulfment (iii). **B.** Initially, contact between SpoIIQ in the forespore membrane and SpoIIIAH in the mother cell membrane is prevented because of the presence of peptidoglycan. However, removal of this layer allows contact between the two proteins presumably through "extracellular" domains, resulting in the enrichment of SpoIIIAH at the septum. **C.** SpoIVFB (red) is initially observed in all mother cell membranes, but it eventually is "captured" by SpoIVFA (orange) and becomes enriched at the forespore in a complex with the SpoIVFB and BofA. **D.** SpoIVFA interacts with a number of proteins in the forespore outer membrane including SpoIID (green), SpoIIP (dark blue), SpoIIM (light blue), SpoIIIAH (lime), and SpoIIQ (purple).

While SpoIVFB is initially observed on all accessible mother cell membranes, it becomes selectively enriched at the outer forespore membrane via a diffusion-and-capture mechanism (Fig. 3C). Following insertion of SpoIVFB randomly into the mother cell membrane, it undergoes free diffusion. SpoIVFB is then captured in the forespore membrane since, when the forespore membrane is not contiguous with the mother cell membrane, targeting does not occur (79). The actual anchor responsible for the "capture" is not known, but a direct biochemical interaction between the extracellular domains of SpoIIIAH and SpoIIQ and that of SpoIVFA has been demonstrated (77), and both proteins are necessary for proper SpoIVFA localization (65). While it is possible that the SpoIIQ-SpoIIIAH complex is sufficient for SpoIVFA localization, the reported dependence of this localization on SpoIID, SpoIIM, and SpoIIP (77)

suggests that a complex, multiprotein network may serve to anchor SpoIVFA to the outer forespore membrane (Fig. 3D).

SpoVM and SpoIVA

The assembling coat is synthesized in the mother cell and is targeted to the outer forespore membrane by SpoIVA (85). SpoIVA binds and hydrolyzes ATP, allowing it to self-assemble into cable-like structures (86) that form a basement layer that serves as a platform for coat assembly. A forthcoming review by A. Driks and P. Eichenberger contains a detailed discussion of protein targeting to the different layers of the coat. SpoIVA localizes to the outer forespore membrane (Fig. 4A, blue), and proper targeting of SpoIVA depends on SpoVM, a 26-amino-acid amphipathic helical peptide that is expressed under σ^E control in the mother cell

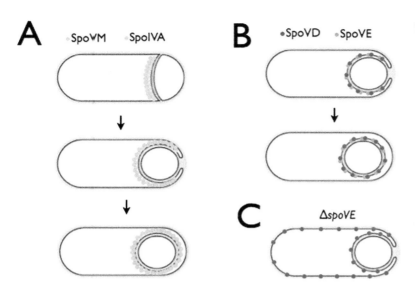

Figure 4   Localization of proteins involved in spore coat and cortex assembly. **A.** SpoVM (green) has an intrinsic affinity for the forespore. **B.** SpoVD (red) and SpoVE (orange) form a complex at the outer forespore membrane. **C.** In the absence of SpoVE, SpoVD is found throughout the mother cell membrane.

(87). A SpoVM-GFP fusion is targeted to the forespore via amino acid side chains on the hydrophobic face of the helix (88), suggesting that the SpoVM helix is oriented parallel to the membrane with the hydrophobic face buried in the lipid bilayer (89).

SpoVM-GFP produced after engulfment still was targeted to the forespore, even though this membrane was now topologically isolated (90), suggesting that the targeting was not occurring through a diffusion-and-capture mechanism. While SpoVM could be targeted by an interaction with an anchor protein present in the outer forespore membrane, its very small size poses a potential problem because of the apparent lack of a domain that might mediate this interaction. Alternatively, following engulfment, the outer forespore membrane has a high positive curvature, suggesting that SpoVM could be detecting this feature. To examine this possibility, purified SpoVM was incubated with phospholipid vesicles similar in size to the forespore (90). The binding of SpoVM was quantitatively related to the vesicle size (curvature) and was reduced by a mutation in the SpoVM amphipathic helix that also abolished localization of SpoVM to the outer forespore membrane *in vivo* (90). However, while *in vitro* the membrane curvature appears to be sufficient, the *in vivo* localization of SpoVM also depended on SpoIVA (89). The ability of a *spoIVA* suppressor mutation to allow the correct targeting of a mislocalized SpoVM mutant protein indicates that this interaction is direct. Thus, the localization of SpoVM and SpoIVA is interdependent and depends on the extreme positive curvature of the outer forespore membrane following engulfment. A key question for future work is to determine how SpoVM

senses this curvature despite its small size and its inability to multimerize *in vitro* (90), characteristics that would be expected to preclude direct measurement of curvature.

Insight into this process underlying SpoVM targeting is key not just for understanding SpoIVA targeting, but also for other proteins. For example, the small CmpA protein localizes to the outer forespore membrane in a fashion dependent on SpoVM, and CmpA is involved in spore cortex assembly (91). In addition, the proper localization of the SpoVID coat protein depends on both SpoVM and SpoIVA, although the effects of Δ*spoVM* and Δ*spoIVA* mutations on SpoVID localization differ (85). Also, whereas SpoVM localization was unaffected in cells carrying a Δ*spoVID* mutation, SpoIVA localization was incomplete (85). This complicated set of interactions is critical for the process of spore encasement where the full shell of spore coat proteins is constructed around the entire circumference of the spore (92).

SpoVD/SpoVE/SpoVB/YkvU

As a consequence of engulfment, two distinct membranes separated by a thin layer of peptidoglycan surround the developing forespore. The initial layer, called the germ cell wall, is greatly thickened following completion of engulfment by the production of the peptidoglycan cortex (see also reference 100). This results from the action of genes expressed in the mother cell compartment under control of σ^K including *spoVE* that encodes a SEDS (shape, elongation, division, and sporulation) protein. Δ*spoVE* mutants fail to form a cortex and they accumulate cytoplasmic peptidoglycan precursors (93), suggesting a defect in peptidoglycan polymerization.

Complementing SpoVE-GFP fusion proteins localize to the outer forespore membrane (Fig. 4B) and mutations in several conserved residues result in protein mislocalization (94). SpoVE targeting to this membrane is dependent on SpoIIQ, SpoIIIAH, and SpoIVFA (A. Fay and J. Dworkin, unpublished data), but it is not known whether this dependence is the result of a direct protein-protein interaction. SpoVE interacts directly both *in vivo* and *in vitro* with SpoVD, a penicillin-binding protein that is also required for spore cortex synthesis. SpoVD also localizes to the outer forespore membrane and this targeting is dependent on SpoVE (Fig. 4B and C) (95).

The SpoVB and YkvU proteins are sporulation-specific homologs of *Escherichia coli* MviN, a broadly conserved protein involved in an as yet not well-characterized fashion in peptidoglycan synthesis. SpoVB (96) and YkvU (97) proteins fused to GFP localize to the outer forespore membrane, consistent with the absence of cortex seen in Δ*spoVB* mutants (98), but nothing is known about this targeting.

FUTURE DIRECTIONS

The dramatic morphological asymmetry of sporulating *B. subtilis* has greatly facilitated the characterization of mechanisms responsible for targeting proteins to particular intracellular loci. An important question is: to what extent do these mechanisms operate in targeting proteins in other cells? Given the particularities of the morphological transformation of sporulation – specifically, the formation of a cell within a cell and the creation of a membrane-bound chromosome – some of the targeting mechanisms described above will not be generally relevant. However, since all cells localize proteins to active sites of septum formation, and positively curved membranes are seen at these positions, the mechanism that SpoVM uses to detect curvature might be generally applicable. One clear future direction is increased spatial resolution in light microscopy (e.g., references 44 and 99), which will facilitate progress on several critical issues. For example, is SpoIIR on both membranes in the space between the forespore and the mother cell? Do the FtsZ rings early in sporulation exhibit differences in structure? With increasing evidence that proteins exist in cells as part of multiprotein complexes, increased spatial resolution will also facilitate understanding of the organization of these complexes. The study of these complexes in sporulation (e.g., BofA/SpoIVFA/SpoIVFB) will likely reveal general principles underlying the nature of analogous complexes in other microbiological (and, in fact, more general cellular) pathways.

Acknowledgments. Work on sporulation in my laboratory is supported by the National Institutes of Health (R01GM081368).

Citation. Dworkin J. 2014. Protein targeting during *Bacillus subtilis* sporulation. Microbiol Spectrum 2(1):TBS-0006-2012.

References

1. **Cohn F.** 1876. Untersuchungen ueber Bakterien. IV. Beitraege zur Biologie der Bacillen. *Beitr Biol Planz* **2:**249–276.

2. **Koch R.** 1876. Die Ätiologie der Milzbrand-Krankheit, begründet auf die Entwicklungsgeschichte des *Bacillus anthracis*. *Beitr Biol Pflanz* **2:**277–231.

3. **Spitzer J.** 2011. From water and ions to crowded biomacromolecules: in vivo structuring of a prokaryotic cell. *Microbiol Mol Biol Rev* **75:**491–506.

4. **Errington J.** 2003. Regulation of endospore formation in *Bacillus subtilis*. *Nat Rev Microbiol* **1:**117–126.

5. **Hilbert DW, Piggot PJ.** 2004. Compartmentalization of gene expression during *Bacillus subtilis* spore formation. *Microbiol Mol Biol Rev* **68:**234–262.

6. **Kroos L.** 2007. The *Bacillus* and *Myxococcus* developmental networks and their transcriptional regulators. *Annu Rev Genet* **41:**13–39.

7. **Higgins D, Dworkin J.** 2012. Recent progress in *Bacillus subtilis* sporulation. *FEMS Microbiol Rev* **36:**131–148.

8. **Swulius MT, Jensen GJ.** 2012. The helical MreB cytoskeleton in *Escherichia coli* MC1000/pLE7 is an artifact of the N-terminal yellow fluorescent protein tag. *J Bacteriol* **194:** 6382–6386.

9. **Landgraf D, Okumus B, Chien P, Baker TA, Paulsson J.** 2012. Segregation of molecules at cell division reveals native protein localization. *Nat Methods* **9:**480–482.

10. **Veening JW, Stewart EJ, Berngruber TW, Taddei F, Kuipers OP, Hamoen LW.** 2008. Bet-hedging and epigenetic inheritance in bacterial cell development. *Proc Natl Acad Sci USA* **105:**4393–4398.

11. **Bi EF, Lutkenhaus J.** 1991. FtsZ ring structure associated with division in *Escherichia coli*. *Nature* **354:**161–164.

12. **Li Z, Trimble MJ, Brun YV, Jensen GJ.** 2007. The structure of FtsZ filaments in vivo suggests a force-generating role in cell division. *EMBO J* **26:**4694–4708.

13. **Michie KA, Monahan LG, Beech PL, Harry EJ.** 2006. Trapping of a spiral-like intermediate of the bacterial cytokinetic protein FtsZ. *J Bacteriol* **188:**1680–1690.

14. **Osawa M, Anderson DE, Erickson HP.** 2008. Reconstitution of contractile FtsZ rings in liposomes. *Science* **320:** 792–794.

15. **Allard JF, Cytrynbaum EN.** 2009. Force generation by a dynamic Z-ring in *Escherichia coli* cell division. *Proc Natl Acad Sci USA* **106:**145–150.

16. **Lan G, Daniels BR, Dobrowsky TM, Wirtz D, Sun SX.** 2009. Condensation of FtsZ filaments can drive bacterial cell division. *Proc Natl Acad Sci USA* **106:**121–126.

17. **Levin PA, Losick R.** 1996. Transcription factor Spo0A switches the localization of the cell division protein FtsZ from a medial to a bipolar pattern in *Bacillus subtilis*. *Genes Dev* **10:**478–488.

18. Lowe J, Amos LA. 1998. Crystal structure of the bacterial cell-division protein FtsZ. *Nature* **391**:203–206.

19. Ben-Yehuda S, Losick R. 2002. Asymmetric cell division in *B. subtilis* involves a spiral-like intermediate of the cytokinetic protein FtsZ. *Cell* **109**:257–266.

20. Wagner-Herman JK, Bernard R, Dunne R, Bisson-Filho AW, Kumar K, Nguyen T, Mulcahy L, Koullias J, Gueiros-Filho FJ, Rudner DZ. 2012. RefZ facilitates the switch from medial to polar division during spore formation in *Bacillus subtilis*. *J Bacteriol* **194**:4608–4618.

21. Young IE. 1964. Characteristics of an abortively disporic variant of *Bacillus cereus*. *J Bacteriol* **88**:242–254.

22. Dworkin J. 2009. Cellular polarity in prokaryotic organisms. *Cold Spring Harbor Perspect Biol* **1**:a003368.

23. Chary VK, Hilbert DW, Higgins ML, Piggot PJ. 2000. The putative DNA translocase SpoIIIE is required for sporulation of the symmetrically dividing coccal species *Sporosarcina ureae*. *Mol Microbiol* **35**:612–622.

24. Pogliano J, Osborne N, Sharp MD, Abanes-De Mello A, Perez A, Sun YL, Pogliano K. 1999. A vital stain for studying membrane dynamics in bacteria: a novel mechanism controlling septation during *Bacillus subtilis* sporulation. *Mol Microbiol* **31**:1149–1159.

25. Eichenberger P, Fawcett P, Losick R. 2001. A three-protein inhibitor of polar septation during sporulation in *Bacillus subtilis*. *Mol Microbiol* **42**:1147–1162.

26. Dworkin J. 2003. Transient genetic asymmetry and cell fate in a bacterium. *Trends Genet* **19**:107–112.

27. Dworkin J, Losick R. 2001. Differential gene expression governed by chromosomal spatial asymmetry. *Cell* **107**:339–346.

28. Khvorova A, Chary VK, Hilbert DW, Piggot PJ. 2000. The chromosomal location of the *Bacillus subtilis* sporulation gene *spoIIR* is important for its function. *J Bacteriol* **182**:4425–4429.

29. Ben-Yehuda S, Rudner DZ, Losick R. 2003. RacA, a bacterial protein that anchors chromosomes to the cell poles. *Science* **299**:532–536.

30. Wu LJ, Errington J. 2003. RacA and the Soj-Spo0J system combine to effect polar chromosome segregation in sporulating *Bacillus subtilis*. *Mol Microbiol* **49**:1463–1475.

31. Lenarcic R, Halbedel S, Visser L, Shaw M, Wu LJ, Errington J, Marenduzzo D, Hamoen LW. 2009. Localisation of DivIVA by targeting to negatively curved membranes. *EMBO J* **28**:2272–2282.

32. Adams DW, Errington J. 2009. Bacterial cell division: assembly, maintenance and disassembly of the Z ring. *Nat Rev Microbiol* **7**:642–653.

33. Goehring NW, Beckwith J. 2005. Diverse paths to midcell: assembly of the bacterial cell division machinery. *Curr Biol* **15**:R514–R526.

34. Lucet I, Feucht A, Yudkin MD, Errington J. 2000. Direct interaction between the cell division protein FtsZ and the cell differentiation protein SpoIIE. *EMBO J* **19**:1467–1475.

35. King N, Dreesen O, Stragier P, Pogliano K, Losick R. 1999. Septation, dephosphorylation, and the activation of sigmaF during sporulation in *Bacillus subtilis*. *Genes Dev* **13**:1156–1167.

36. Levin PA, Losick R, Stragier P, Arigoni F. 1997. Localization of the sporulation protein SpoIIE in *Bacillus subtilis* is dependent upon the cell division protein FtsZ. *Mol Microbiol* **25**:839–846.

37. Barak I, Behari J, Olmedo G, Guzman P, Brown DP, Castro E, Walker D, Westpheling J, Youngman P. 1996. Structure and function of the *Bacillus* SpoIIE protein and its localization to sites of sporulation septum assembly. *Mol Microbiol* **19**:1047–1060.

38. Barak I, Youngman P. 1996. SpoIIE mutants of *Bacillus subtilis* comprise two distinct phenotypic classes consistent with a dual functional role for the SpoIIE protein. *J Bacteriol* **178**:4984–4989.

39. Losick R, Stragier P. 1992. Crisscross regulation of cell-type-specific gene expression during development in *B. subtilis*. *Nature* **355**:601–604.

40. Duncan L, Alper S, Arigoni F, Losick R, Stragier P. 1995. Activation of cell-specific transcription by a serine phosphatase at the site of asymmetric division. *Science* **270**:641–644.

41. Arigoni F, Pogliano K, Webb CD, Stragier P, Losick R. 1995. Localization of protein implicated in establishment of cell type to sites of asymmetric division. *Science* **270**:637–640.

42. Guberman JM, Fay A, Dworkin J, Wingreen NS, Gitai Z. 2008. PSICIC: noise and asymmetry in bacterial division revealed by computational image analysis at sub-pixel resolution. *PLoS Comput Biol* **4**:e1000233. doi:10.1371/journal.pcbi.1000233.

43. Burton B, Dubnau D. 2010. Membrane-associated DNA transport machines. *Cold Spring Harbor Perspect Biol* **2**:a000406. doi:10.1101/cshperspect.a000406.

44. Fiche JB, Cattoni DI, Diekmann N, Langerak JM, Clerte C, Royer CA, Margeat E, Doan T, Nollmann M. 2013. Recruitment, assembly, and molecular architecture of the SpoIIIE DNA pump revealed by superresolution microscopy. *PLoS Biol* **11**:e1001557. doi:10.1371/journal.pbio.1001557.

45. Wu LJ, Errington J. 1997. Septal localization of the SpoIIIE chromosome partitioning protein in *Bacillus subtilis*. *EMBO J* **16**:2161–2169.

46. Ben-Yehuda S, Rudner DZ, Losick R. 2003. Assembly of the SpoIIIE DNA translocase depends on chromosome trapping in *Bacillus subtilis*. *Curr Biol* **13**:2196–2200.

47. Becker EC, Pogliano K. 2007. Cell-specific SpoIIIE assembly and DNA translocation polarity are dictated by chromosome orientation. *Mol Microbiol* **66**:1066–1079.

48. Ptacin JL, Nollmann M, Becker EC, Cozzarelli NR, Pogliano K, Bustamante C. 2008. Sequence-directed DNA export guides chromosome translocation during sporulation in *Bacillus subtilis*. *Nat Struct Mol Biol* **15**:485–493.

49. Sharp MD, Pogliano K. 1999. An in vivo membrane fusion assay implicates SpoIIIE in the final stages of engulfment during *Bacillus subtilis* sporulation. *Proc Natl Acad Sci USA* **96**:14553–14558.

50. Fleming TC, Shin JY, Lee SH, Becker E, Huang KC, Bustamante C, Pogliano K. 2010. Dynamic SpoIIIE assembly mediates septal membrane fission during *Bacillus subtilis* sporulation. *Genes Dev* **24**:1160–1172.

51. Meyer P, Gutierrez J, Pogliano K, Dworkin J. 2010. Cell wall synthesis is necessary for membrane dynamics during sporulation of *Bacillus subtilis*. *Mol Microbiol* **76:** 956–970.

52. Morlot C, Uehara T, Marquis KA, Bernhardt TG, Rudner DZ. 2010. A highly coordinated cell wall degradation machine governs spore morphogenesis in *Bacillus subtilis*. *Genes Dev* **24:**411–422.

53. Doan T, Coleman J, Marquis KA, Meeske AJ, Burton BM, Karatekin E, Rudner DZ. 2013. FisB mediates membrane fission during sporulation in *Bacillus subtilis*. *Genes Dev* **27:**322–334.

54. Ramamurthi KS, Losick R. 2009. Negative membrane curvature as a cue for subcellular localization of a bacterial protein. *Proc Natl Acad Sci USA* **106:**13541–13545.

55. Fujita M, Losick R. 2002. An investigation into the compartmentalization of the sporulation transcription factor sigmaE in *Bacillus subtilis*. *Mol Microbiol* **43:**27–38.

56. Rubio A, Pogliano K. 2004. Septal localization of forespore membrane proteins during engulfment in *Bacillus subtilis*. *EMBO J* **23:**1636–1646.

57. Diez V, Schujman GE, Gueiros-Filho FJ, de Mendoza D. 2012. Vectorial signalling mechanism required for cell-cell communication during sporulation in *Bacillus subtilis*. *Mol Microbiol* **83:**261–274.

58. Fawcett P, Melnikov A, Youngman P. 1998. The *Bacillus* SpoIIGA protein is targeted to sites of spore septum formation in a SpoIIE-independent manner. *Mol Microbiol* **28:**931–943.

59. Chary VK, Xenopoulos P, Eldar A, Piggot PJ. 2010. Loss of compartmentalization of sigma(E) activity need not prevent formation of spores by *Bacillus subtilis*. *J Bacteriol* **192:**5616–5624.

60. Frandsen N, Stragier P. 1995. Identification and characterization of the *Bacillus subtilis* spoIIP locus. *J Bacteriol* **177:**716–722.

61. Abanes-De Mello A, Sun YL, Aung S, Pogliano K. 2002. A cytoskeleton-like role for the bacterial cell wall during engulfment of the *Bacillus subtilis* forespore. *Genes Dev* **16:**3253–3264.

62. Chastanet A, Losick R. 2007. Engulfment during sporulation in *Bacillus subtilis* is governed by a multi-protein complex containing tandemly acting autolysins. *Mol Microbiol* **64:**139–152.

63. Fredlund J, Broder D, Fleming T, Claussin C, Pogliano K. 2013. The SpoIIQ landmark protein has different requirements for septal localization and immobilization. *Mol Microbiol* **89:**1053–1068.

64. Rodrigues CD, Marquis KA, Meisner J, Rudner DZ. 2013. Peptidoglycan hydrolysis is required for assembly and activity of the transenvelope secretion complex during sporulation in *Bacillus subtilis*. *Mol Microbiol* **89:** 1039–1052.

65. Aung S, Shum J, Abanes-De Mello A, Broder DH, Fredlund-Gutierrez J, Chiba S, Pogliano K. 2007. Dual localization pathways for the engulfment proteins during *Bacillus subtilis* sporulation. *Mol Microbiol* **65:**1534–1546.

66. Perez AR, Abanes-De Mello A, Pogliano K. 2000. SpoIIB localizes to active sites of septal biogenesis and spatially regulates septal thinning during engulfment in *Bacillus subtilis*. *J Bacteriol* **182:**1096–1108.

67. Broder DH, Pogliano K. 2006. Forespore engulfment mediated by a ratchet-like mechanism. *Cell* **126:**917–928.

68. Sun YL, Sharp MD, Pogliano K. 2000. A dispensable role for forespore-specific gene expression in engulfment of the forespore during sporulation of *Bacillus subtilis*. *J Bacteriol* **182:**2919–2927.

69. Meisner J, Moran CP Jr. 2011. A LytM domain dictates the localization of proteins to the mother cell-forespore interface during bacterial endospore formation. *J Bacteriol* **193:**591–598.

70. Meisner J, Maehigashi T, Andre I, Dunham CM, Moran CP Jr. 2012. Structure of the basal components of a bacterial transporter. *Proc Natl Acad Sci USA* **109:**5446–5451.

71. Levdikov VM, Blagova EV, McFeat A, Fogg MJ, Wilson KS, Wilkinson AJ. 2012. Structure of components of an intercellular channel complex in sporulating *Bacillus subtilis*. *Proc Natl Acad Sci USA* **109:**5441–5445.

72. Kellner EM, Decatur A, Moran CP Jr. 1996. Two-stage regulation of an anti-sigma factor determines developmental fate during bacterial endospore formation. *Mol Microbiol* **21:**913–924.

73. Camp AH, Losick R. 2008. A novel pathway of intercellular signalling in *Bacillus subtilis* involves a protein with similarity to a component of type III secretion channels. *Mol Microbiol* **69:**402–417.

74. Meisner J, Wang X, Serrano M, Henriques AO, Moran CP Jr. 2008. A channel connecting the mother cell and forespore during bacterial endospore formation. *Proc Natl Acad Sci USA* **105:**15100–15105.

75. Camp AH, Losick R. 2009. A feeding tube model for activation of a cell-specific transcription factor during sporulation in *Bacillus subtilis*. *Genes Dev* **23:**1014–1024.

76. Blaylock B, Jiang X, Rubio A, Moran CP Jr, Pogliano K. 2004. Zipper-like interaction between proteins in adjacent daughter cells mediates protein localization. *Genes Dev* **18:**2916–2928.

77. Doan T, Marquis KA, Rudner DZ. 2005. Subcellular localization of a sporulation membrane protein is achieved through a network of interactions along and across the septum. *Mol Microbiol* **55:**1767–1781.

78. Doan T, Morlot C, Meisner J, Serrano M, Henriques AO, Moran CP Jr, Rudner DZ. 2009. Novel secretion apparatus maintains spore integrity and developmental gene expression in *Bacillus subtilis*. *PLoS Genet* **5:** e1000566. doi:10.1371/journal.pgen.1000566.

79. Rudner DZ, Pan Q, Losick RM. 2002. Evidence that subcellular localization of a bacterial membrane protein is achieved by diffusion and capture. *Proc Natl Acad Sci USA* **99:**8701–8706.

80. Zhang B, Hofmeister A, Kroos L. 1998. The prosequence of pro-sigmaK promotes membrane association and inhibits RNA polymerase core binding. *J Bacteriol* **180:** 2434–2441.

81. Resnekov O, Alper S, Losick R. 1996. Subcellular localization of proteins governing the proteolytic activation of a developmental transcription factor in *Bacillus subtilis*. *Genes Cells* 1:529–542.

82. Rudner DZ, Losick R. 2002. A sporulation membrane protein tethers the pro-sigmaK processing enzyme to its inhibitor and dictates its subcellular localization. *Genes Dev* 16:1007–1018.

83. Campo N, Rudner DZ. 2006. A branched pathway governing the activation of a developmental transcription factor by regulated intramembrane proteolysis. *Mol Cell* 23:25–35.

84. Doan T, Rudner DZ. 2007. Perturbations to engulfment trigger a degradative response that prevents cell-cell signalling during sporulation in *Bacillus subtilis*. *Mol Microbiol* 64:500–511.

85. Wang KH, Isidro AL, Domingues L, Eskandarian HA, McKenney PT, Drew K, Grabowski P, Chua MH, Barry SN, Guan M, Bonneau R, Henriques AO, Eichenberger P. 2009. The coat morphogenetic protein SpoVID is necessary for spore encasement in *Bacillus subtilis*. *Mol Microbiol* 74:634–649.

86. Ramamurthi KS, Losick R. 2008. ATP-driven self-assembly of a morphogenetic protein in *Bacillus subtilis*. *Mol Cell* 31:406–414.

87. Price KD, Losick R. 1999. A four-dimensional view of assembly of a morphogenetic protein during sporulation in *Bacillus subtilis*. *J Bacteriol* 181:781–790.

88. van Ooij C, Losick R. 2003. Subcellular localization of a small sporulation protein in *Bacillus subtilis*. *J Bacteriol* 185:1391–1398.

89. Ramamurthi KS, Clapham KR, Losick R. 2006. Peptide anchoring spore coat assembly to the outer forespore membrane in *Bacillus subtilis*. *Mol Microbiol* 62:1547–1557.

90. Ramamurthi KS, Lecuyer S, Stone HA, Losick R. 2009. Geometric cue for protein localization in a bacterium. *Science* 323:1354–1357.

91. Ebmeier SE, Tan IS, Clapham KR, Ramamurthi KS. 2012. Small proteins link coat and cortex assembly during sporulation in *Bacillus subtilis*. *Mol Microbiol* 84:682–696.

92. McKenney PT, Eichenberger P. 2012. Dynamics of spore coat morphogenesis in *Bacillus subtilis*. *Mol Microbiol* 83:245–260.

93. Vasudevan P, Weaver A, Reichert ED, Linnstaedt SD, Popham DL. 2007. Spore cortex formation in *Bacillus subtilis* is regulated by accumulation of peptidoglycan precursors under the control of sigma K. *Mol Microbiol* 65:1582–1594.

94. Real G, Fay A, Eldar A, Pinto SM, Henriques AO, Dworkin J. 2008. Determinants for the subcellular localization and function of a nonessential SEDS protein. *J Bacteriol* 190:363–376.

95. Fay A, Meyer P, Dworkin J. 2010. Interactions between late-acting proteins required for peptidoglycan synthesis during sporulation. *J Mol Biol* 399:547–561.

96. Fay A, Dworkin J. 2009. *Bacillus subtilis* homologs of MviN (MurJ), the putative *Escherichia coli* lipid II flippase, are not essential for growth. *J Bacteriol* 191:6020–6028.

97. Eichenberger P, Jensen ST, Conlon EM, van Ooij C, Silvaggi J, Gonzalez-Pastor JE, Fujita M, Ben-Yehuda S, Stragier P, Liu JS, Losick R. 2003. The sigmaE regulon and the identification of additional sporulation genes in *Bacillus subtilis*. *J Mol Biol* 327:945–972.

98. Popham DL, Stragier P. 1991. Cloning, characterization, and expression of the *spoVB* gene of *Bacillus subtilis*. *J Bacteriol* 173:7942–7949.

99. Strauss MP, Liew AT, Turnbull L, Whitchurch CB, Monahan LG, Harry EJ. 2012. 3D-SIM super resolution microscopy reveals a bead-like arrangement for FtsZ and the division machinery: implications for triggering cytokinesis. *PLoS Biol* 10:e1001389. doi:10.1371/journal.pbio.1001389.

100. Popham DL, Bernhards CB. Spore peptidoglycan. *In* Eichenberger P, Driks A (ed), *The Bacterial Spore: From Molecules to Systems*. ASM Press, Washington, DC, in press.

The Bacterial Spore: From Molecules to Systems
Edited by P. Eichenberger and A. Driks
© 2016 American Society for Microbiology, Washington, DC
doi:10.1128/microbiolspec.TBS-0005-2012

David L. Popham[1]
Casey B. Bernhards[2]

Spore Peptidoglycan

8

The spore is simply a cell with some extremely novel properties and structural elements. The primary morphological elements shared with vegetative bacterial cells are the spore core (the cytoplasm), the inner spore membrane (the cytoplasmic membrane), and the peptidoglycan (PG) wall (Fig. 1). A spore stripped of coat protein layers outside the PG retains its dormancy and many of its resistance properties (1, 2). The primary factor contributing to spore dormancy and heat resistance, and a major factor in resistance to chemical and physical damaging agents, is the relative dehydration of the spore core (3–7). A predominant factor in maintaining this dehydration, and potentially a factor in attaining it, is the PG wall (2).

Agents that disrupt the spore PG layer lead to a loss of spore core dehydration and spore resistance properties (2). The PG is normally protected from external lytic enzymes by the proteinaceous spore coat layers (1, 2). Degradation of the spore PG is, however, an essential element of the spore germination process (Fig. 1). Spores incapable of this degradative step cannot achieve full hydration, resume metabolism, or produce colonies (8, 9). Dormant spores contain enzymes that are specifically activated during germination and that are specific for depolymerization of the spore PG (10–12).

SPORE PG STRUCTURE AND SPORE PROPERTIES

General Aspects of PG Structure

Certain aspects of PG structure are highly conserved across the domain of bacteria (Fig. 2) (reviewed in references 13 and 14). The glycan strands are composed of alternating *N*-acetylglucosamine (NAG) and *N*-acetylmuramic acid (NAM) residues. Attached to the lactyl groups of the NAM are peptide side chains. Disaccharide-pentapeptide precursors are synthesized in the cell cytoplasm and are linked to a bactoprenyl lipid carrier for movement across the membrane. The disaccharides are polymerized on the outer surface of the membrane to produce the PG glycan strands. The strands are then cross-linked via the peptide side chains. The terminal amino acid, generally D-Ala, is cleaved by a transpeptidase, and the new carboxyl terminal amino acid, generally also D-Ala, becomes linked via a peptide bond to a diamino acid in the third position of a peptide side chain on a neighboring glycan strand. In some species, an additional amino acid or short peptide cross bridge is placed between the cross-linked peptides, but this is less common in the bacilli and has never been observed in spore PG. The diamino acid in the third position in vegetative cell PG is

[1]Department of Biological Sciences, Life Sciences I, Virginia Tech, Blacksburg, VA 24061; [2]U.S. Army Edgewood Chemical Biological Center, Aberdeen Proving Ground, MD 21010.

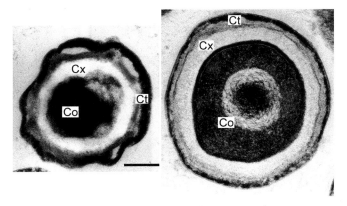

Figure 1 Structural elements of dormant and germinating *Bacillus subtilis* spores. The dormant spore (left) has a densely staining dehydrated core (Co), surrounded by the inner forespore membrane, the poorly staining cortex PG (Cx), and multiple coat layers (Ct). Within the expanded, rehydrated core of the partially germinated spore (right), the nucleoid material is visible. The cortex and coats have expanded, and the cortex now binds some stain, presumably because of reactive groups generated by cortex degradation. Both spores are photographed at the same magnification. Bar = 0.25 m.

frequently diaminopimelic acid but can be lysine or more rarely other residues.

Spore PG Layers: Germ Cell Wall versus Cortex

Spore PG is described in terms of two layers: a thin inner layer called the germ cell wall and the thicker outer cortex layer. The primary distinction between the layers is functional, although the function is likely directly related to a structural difference. The germ cell wall is defined as a PG layer that is not degraded during germination and that functions as the initial wall of the outgrowing spore, while the cortex is the PG that is depolymerized during germination (15, 16). The germ cell wall is structurally more similar to the PG of vegetative cell walls (15, 16), because it contains more tripeptide side chains and a higher level of cross-linking than the cortex PG. The cortex possesses structural modifications that may determine functions and certainly determine its susceptibility to lytic enzymes active during germination (8, 16, 17). In all species in which the spore PG structure has been directly examined, the peptide side chains contain diaminopimelic acid (18–23). There is some evidence that the germ cell wall of *Bacillus sphaericus* spores contains lysine, similar to the vegetative cell PG of this species (24), but there is no evidence that this aspect of PG structure has any effect on spore properties.

Cortex PG Structural Modifications

Cortex PG contains several structural modifications relative to vegetative cell PG. The major alteration is the removal of the peptide side chain from approximately 50% of the NAM residues and subsequent formation of muramic-δ-lactam (Fig. 2) (23). The majority of these residues are spaced quite regularly at every second NAM position along the glycan strands (23, 25). In most species, an additional 15 to 25% of the NAM residues have their side chains shortened to single L-Ala residues (Fig. 2) (18–23). In *Clostridium perfringens*, these single L-Ala side chains are not present; instead, a small percentage of the side chains are reduced to dipeptides (21). Deacetylation of a subset of sugar residues in the cortex is observed in some species, but no physiological significance has been associated with this modification (18–21). PG deacetylation contributes to lysozyme resistance of vegetative cells in several species (26, 27), but the level of deacetylation is generally higher than that observed in spore PG, and the permeability barrier of the spore coat layers is the primary mechanism of spore lysozyme resistance (1).

An outcome of modifications that remove or shorten the peptides is greatly reduced cross-linking of the spore PG. Although 33 to 45% of NAM residues in vegetative cell walls of *Bacillus subtilis* carry peptides involved in cross-links (28, 29), only ~3% of NAM residues in spore PG carry cross-linked peptides (17). While the spore PG is clearly a relatively loosely cross-linked structure, some evidence indicates that the cross-linking varies across the span of the PG layers (16, 30). The germ cell wall layer appears to have the highest cross-linking level, while the cortex immediately exterior to this is extremely loosely cross-linked. The cross-linking then increases progressively toward the outermost region of the cortex.

Relationships Between Spore PG Structure and Resistance Properties

The aspect of spore PG structure that impacts spore resistance most clearly is the amount of PG per spore. Experimental reduction of accumulation of PG during spore formation resulted in increased spore hydration and a corresponding decrease in resistance to heat and organic chemicals (31). Comparisons across strains of a species and across multiple species have revealed a correlation between the ratio of core volume to core-plus-cortex volume and spore heat resistance and dehydration (21, 32). Variation of this ratio within a species is likely a measure of cortex PG abundance (21).

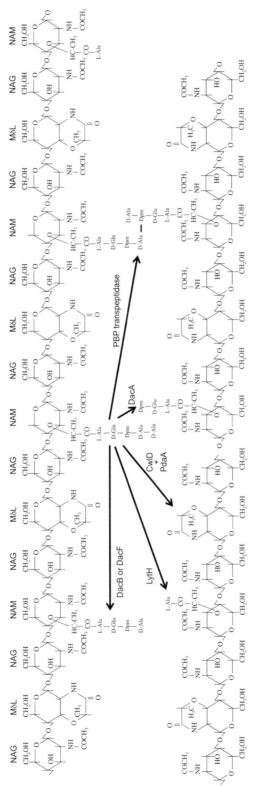

Figure 2 Spore PG structure and modification. The spore PG strands are composed of N-acetylglucosamine (NAG), muramic-δ-lactam (MδL), and N-acetylmuramic acid (NAM). Each NAM residue initially has a pentapeptide side chain composed of L-Ala, D-Glu, diaminopimelic acid (Dpm), and two D-Ala residues. The peptides can be cleaved to tetrapeptides by DacB or DacF, a reaction that regulates the degree of PG cross-linking. Many peptides in the germ cell wall are cleaved to tripeptides by DacA. LytH is an endopeptidase that produces single L-Ala side chains. The combined actions of the amidase CwlD and the deacetylase PdaA lead to the production of muramic-δ-lactam. Transpeptidase activities carried by class A and class B PBPs produce peptide cross-links between the glycan strands.

On the contrary, variation in spore PG structural parameters, such as muramic-δ-lactam abundance and peptide cross-linking, within a relatively broad range, has not revealed a clear correlation with spore dehydration and resistance properties. Spores that lack single L-Ala side chains and have a corresponding increased abundance of full-length peptide side chains are only slightly affected in their heat resistance (33). This spore PG alteration is accompanied by a very slight increase (<2-fold) in PG cross-linking (33). Other mutations that increase spore PG cross-linking 5-fold (a *dacB* mutant) (17, 34) or 9-fold (a *cwlD dacB* double mutant) (35) do not produce major alterations of spore dehydration. These alterations also remove the gradient of cross-linking across the span of the cortex, with little effect on the ability of the spores to achieve normal dehydration (16). Only when the level of cross-linking rises to more extreme levels, in a *dacB dacF* double mutant, is the relative dehydration of the spore altered (30).

The low degree of cortex PG cross-linking is striking, and the lack of an effect of this structural parameter on spore properties is surprising. It is known that the relative cross-linking of PG affects the flexibility of the structure, with more loosely cross-linked structures exhibiting greater expansion and contraction (36). It was proposed that the flexibility of loosely cross-linked spore PG could contribute to the process of spore core dehydration, either via cortex contraction (37) or anisotropic expansion (38), during spore formation. The observation that major changes in cortex PG cross-linking do not diminish dehydration (17, 34, 35) argues against this possibility. Perhaps the proper aspect of spore resistance has not been assayed relative to cortex cross-linking and flexibility. It has been demonstrated that spores expand and contract in response to changes in relative humidity of their environment (39). Presumably, a loosely cross-linked, flexible cortex might allow the spore PG layers to accommodate core volume changes. Such an ability may be very important in allowing spores to retain core dehydration, dormancy, and associated resistance properties while experiencing repeated cycles of humidity changes in their natural environment. Consistent with this idea, *B. subtilis dacB* mutant spores with increased cortex cross-linking are produced with a normal degree of core dehydration, but they are unable to maintain it and gain water content during heating (5). Interestingly, spores with increased cortex cross-linking could initiate the release of Ca^{2+}-dipicolinic acid (DPA) from the spore core more rapidly than wild-type spores following exposure to germinants (40). This affect seems to

indicate a direct effect of the cortex structure on the threshold for germination initiation within the spore.

The conversion of 50% of the NAM residues in cortex PG to muramic-δ-lactam is perhaps the most striking PG structural alteration found in the spore. Surprisingly, mutations that result in a complete lack of muramic-δ-lactam in spores do not have major effects on spore dehydration or heat resistance. Despite a >2-fold increase in peptide side chain abundance and a >2-fold increase in PG cross-linking that accompany the loss of muramic-δ-lactam, these spores are fully dehydrated and resistant (8). However, these same spores exhibit a >10^4-fold decrease in the ability to complete germination and produce colonies (8, 15, 41). Analyses of spore PG degradation and release during germination indicate that spores lacking muramic-δ-lactam are unable to depolymerize the spore PG (8, 15). This led to the proposal, since borne out experimentally (see below), that the muramic-δ-lactam serves as a specificity determination for germination-specific lytic enzymes. The abundance of muramic-δ-lactam in cortex PG and its paucity in germ cell wall PG would therefore lead to the observed specific degradation of cortex during germination, leaving the germ cell wall intact.

SPORE PEPTIDOGLYCAN SYNTHESIS AND MODIFICATION

Timing of Spore PG Synthesis

PG accumulation in *B. subtilis* cultures slows significantly during entry into stationary phase (T_0 of sporulation) and plateaus approximately 2 h later (T_2) (16, 42). PG synthesis begins again soon after T_3, coincident with the expected completion of forespore engulfment (16). An assay for the appearance of spore PG that depended on the completion of engulfment, thus protecting the spore PG from lysozyme used to produce protoplasts, allowed the detection of spore PG by $T_{3.5}$, and spore PG continued to accumulate through T_8 (16).

The involvement of proteins with lytic activities in the progression of engulfment (43–45) and inhibition of engulfment by treatments that block PG synthesis (46) have indicated that PG synthesis in the intermembrane space may play a role in the process of engulfment. Direct detection of this PG is challenging, because the PG is not visualized by transmission electron microscopy, but electron cryotomography allowed the visualization of PG between the engulfing membranes (47). The structure of this PG is unknown, because it would be difficult to physically separate from

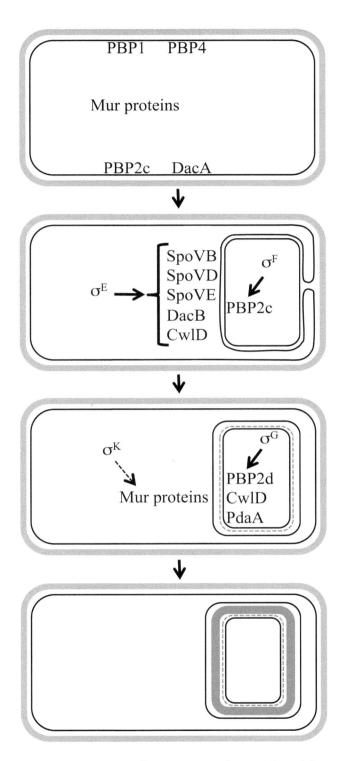

the abundant PG of the surrounding sporangium, and it would not be recovered following protoplasting of sporangia prior to the completion of engulfment. The protoplasting procedure produced no forespore-specific PG at T_3 or $T_{3.25}$, while a small amount of germ cell wall was recoverable by $T_{3.5}$ (16). This initial germ cell wall may therefore be the PG produced during engulfment.

Location of Germ Cell Wall Synthesis, Precursors, and Enzymes

There is a relatively clear division between the enzymes and substrates utilized to produce the germ cell wall and cortex PG layers. Enzymes and precursors utilized for germ cell wall synthesis are produced in the forespore, while those involved in cortex synthesis are produced in the mother cell (Fig. 3) (48). The class A penicillin-binding protein (PBP) products of *pbpF* and *pbpG*, PBP2c and PBP2d, respectively, are produced in the forespore because of transcription directed by σ^F and σ^G, respectively (49, 50). These bifunctional PBPs localize to the outer surface of the inner forespore membrane. Their linked glycosyltransferase and transpeptidase activities (13) can then polymerize the glycan strands and cross-link the peptide side chains. This general PG synthetic activity may be sufficient for germ cell wall production, which may not require any significant morphological control. Similar class A PBPs acting to produce vegetative cells of specific shapes generally require the action of one or more class B PBP transpeptidases to determine the PG morphology. While it is possible that one or more class B PBPs produced during vegetative growth remain in the inner forespore membrane, no such enzyme has been shown to play a significant role in germ cell wall synthesis.

The lipid-linked PG precursors used for germ cell wall synthesis are produced by the same *mur* gene

merization. DacA certainly is involved in shortening spore PG peptide side chains to tripeptides. Following asymmetric septation and initiation of compartmentalized gene expression, σ^E and σ^F drive expression of PG-active proteins in the two cells. Upon completion of engulfment and activation of σ^G, additional genes are expressed, and germ cell wall synthesis dependent on PBPs 2c and 2d commences adjacent to the inner forespore membrane. Mother cell-expressed proteins commence cortex synthesis adjacent to the outer forespore membrane, and a σ^K-dependent increase in Mur activity provides precursors for continued synthesis. Solid arrows indicate a direct effect on gene expression by the sigma factor. A dashed arrow indicates potential indirect effects on protein abundance or activity. Black lines represent membranes, and gray lines represent PG structures.

Figure 3 Expression of spore PG synthesis and modifying enzymes. The PBPs 1, 2c, 4, and DacA are expressed during vegetative growth and some protein persists in the sporangium. PBPs 1, 2c, and 4 may participate in cortex PG poly-products that function during vegetative growth. While several of these Mur proteins decrease in abundance during early sporulation, they do not disappear (51) and are presumably at sufficient levels within the forespore to allow germ cell wall synthesis. Assay of PG precursor synthesis activities in *B. sphaericus*, which possesses lysine in the peptides of its vegetative cell walls and diaminopimelic acid in its cortex PG, revealed that the lysine-incorporating enzyme was produced within the developing forespore while the diaminopimelic acid-incorporating enzyme was made in the mother cell compartment (24). It was thus suggested that lysine might be found in the germ cell wall, although this has never been biochemically verified.

Location of Cortex Synthesis, Precursors, and Enzymes

Substrates and enzymes required for polymerization of the cortex PG are produced in the mother cell compartment, and this polymerization takes place on the surface of the outer forespore membrane that faces the intermembrane space (Fig. 3). During early sporulation, prior to initiation of cortex synthesis, pools of nucleotide-linked PG precursors (UDP-NAM and UDP-NAM-peptide) decrease, and this correlates with decreased abundance of Mur proteins involved in their synthesis (MurAA, MurB, and MurC) (51). Only following completion of engulfment, activation of σ^G, and eventual activation of σ^K do the levels of these enzymes and their PG precursor products increase (51). Accumulation of these Mur proteins does not appear to be a result of increased transcription of their genes by σ^K-RNA polymerase, and may be a function of altered protein turnover that is affected by a σ^K-dependent process (51).

The glycosyltransferase(s) that polymerase the cortex PG strands have not been identified. Three class A PBPs that carry this activity, PBP1, PBP2c, and PBP4, are present in the vegetative cell and decline significantly in expression and abundance during early sporulation (50, 52–56). However, it is possible that enough of these proteins are available in the outer forespore membrane to perform cortex polymerization. Alternatively, a glycosyltransferase that is not associated with class A PBPs could be responsible (57). Cross-linking of the cortex PG peptide side chains could also be accomplished by these class A PBPs, but it is likely that a sporulation-specific transpeptidase plays a major role in this cross-linking. The class B PBP SpoVD is expressed in the mother cell under the control of σ^E (58) and localizes to the outer forespore membrane (59). A *spoVD* null mutant produces essentially no cortex PG

(51, 58, 60), indicating that loss of this protein blocks not only cross-linking but also disrupts polymerization of the glycan strands. Loss of either of the integral membrane proteins SpoVB or SpoVE produces a phenotype identical to that of the *spoVD* mutant (51). SpoVE is a member of the SEDS family of proteins, members of which are functionally paired with class B PBPs (59, 61, 62), and SpoVB belongs to a family of proteins also implicated in PG synthesis (63, 64). The complete block in cortex synthesis in strains lacking any of these three proteins (51) suggests the disruption of an enzyme complex. Direct physical and functional interaction between SpoVD and SpoVE has been demonstrated (59). Whether these proteins are part of a larger PG synthetic complex has not been shown.

The degree of cross-linking of the cortex PG is regulated by both the abundance of peptide side chains remaining on the glycan strands and by their modification to prevent cross-linking. The removal of 50% of the peptide side chains during muramic-δ-lactam formation, and the further cleavage of up to 25% of the peptides to single L-Ala residues, greatly reduces the potential for cross-link formation. The enzymes involved in muramic-δ-lactam formation are described below. LytH appears to function as an endopeptidase that shortens the peptide side chains to single amino acids (33). Loss of LytH results in a lack of single L-Ala side chains and a corresponding approximately 2-fold increase in tetrapeptide side chains (33). This does not result in a 2-fold increase in cross-linking, presumably because of the regulation of cross-linking by carboxypeptidases. Three low-molecular-weight PBPs, the carboxypeptidases DacA (PBP5), DacB (PBP5*), and DacF, are present during sporulation. DacA is expressed at high levels during vegetative growth and remains present at significant levels during sporulation (56, 65, 66). DacB is produced in the mother cell under control of σ^E (67), while DacF is produced in the forespore under the control of σ^F (68). Both DacB and DacF would appear to be translocated into the intermembrane space and to remain associated with a membrane. These carboxypeptidases can remove the D-Ala in the fifth position of a peptide side chain (as well as the D-Ala in the fourth position for DacA [28]), thus preventing the participation of this peptide as a donor in cross-link formation. Loss of DacA results in an increased number of longer peptide side chains, but no significant change in PG cross-linking (17, 34). DacB plays the dominant role in regulating spore PG cross-linking, because loss of DacF alone results in no significant change in spore properties (17, 30, 34, 68). Loss of DacB results in a 5-fold increase in cortex PG cross-linking (17, 30, 34) and a slight loss in heat resistance (30, 34). The absence of both DacB and DacF results in a large increase in cortex cross-linking and a failure to produce stable spores (16, 30).

The formation of muramic-δ-lactam in the cortex PG is unique, because it has not been demonstrated in any non-spore PG. Two enzymes, CwlD and PdaA, are necessary and sufficient to complete this novel process (69, 70). CwlD acts as a muramidase to remove the peptide from the NAM residue, and this is apparently followed by deacetylation and subsequent lactam ring formation catalyzed by PdaA. The order of action is suggested by the fact that CwlD is able to function in the absence of PdaA, whereas PdaA is apparently inactive in the absence of a CwlD-generated substrate (70). Both of these proteins possess signal peptides (41, 69) that direct them to the intermembrane space where the cortex is polymerized. Muramic-δ-lactam is therefore likely not produced within the PG precursors but rather during or after polymerization of the glycan strands. CwlD and PdaA have relatively unexpected expression patterns, given that their cortex PG substrate is polymerized on the surface of the outer forespore membrane. PdaA is expressed only in the forespore, under the control of σ^G (69), and must enter and traverse the intermembrane space to fully modify the cortex PG. The fact that peptide-free NAM produced by CwlD is not observed in wild-type spore PG (17, 34) indicates that PdaA is produced in excess of the amount required to convert all CwlD product to muramic-δ-lactam. CwlD is expressed in both the mother cell and the forespore, from independent σ^E- and σ^G-dependent promoters, respectively (41). CwlD expression solely from the σ^E-dependent promoter in the mother cell is sufficient to produce fully modified cortex PG, although CwlD expressed in the forespore is competent to participate in some muramic-δ-lactam formation (70).

Conservation of Spore PG Synthesis and Modification Genes

While there is significant conservation of the proteins involved in PG synthesis across species and genera (71), identification of those factors specifically involved in production of spore PG can be complicated. Most classes of enzymes involved in PG synthesis and modification are present in multiple, often functionally redundant copies within a genome. The Mur proteins that produce the nucleotide and lipid-linked PG precursors are highly conserved across all bacteria (72) and are generally present in a single copy per genome. These *mur* genes and protein products must therefore function

during both vegetative and spore wall synthesis. The genome sequences of several species representative of the breadth of the family *Bacillaceae* were searched for homologs of proteins demonstrated to be involved in polymerization and modification of spore PG in *B. subtilis*. Predictions of orthologs were made based on degree of amino acid sequence identity and shared synteny (Table 1). Across all of these genes and species, a general pattern emerged: Genes that are highly likely to be orthologs based upon shared synteny exhibit ≥45% amino acid identity, while paralogs, both within and across species, exhibit <40% amino acid identity. Identification of orthologs in the more distantly related species within the *Clostridiaceae* is more challenging, again because of the presence of multiple paralogs within each family of PG polymerization and modification genes. Good predictions can be made based on amino acid identities, but confidence in many of these predictions is not high because of greater variation in genome context.

The *Bacillaceae* species examined possess four or five class A high-molecular-weight PBPs, and the genes encoding many of these could be assigned as likely orthologs based on shared synteny. The *ponA* genes, encoding PBPs 1A and 1B in *B. subtilis*, are relatively easy to discern, and all the species possess the redundant *pbpF* and *pbpG* genes involved in germ cell wall synthesis. The additional class A PBP-encoding genes in the other species are of unknown function, but none of them appear to be orthologs of *B. subtilis pbpD*. Each species possesses genes encoding multiple low-molecular-weight PBP D,D-carboxypeptidases of the family including DacB and DacF. Likely orthologs for each of these genes can be identified in each species owing to well-conserved synteny.

Obvious orthologs of the gene encoding the sole class B high-molecular-weight PBP that is clearly involved in spore PG synthesis, SpoVD, are present in all species. These were easily differentiated from the most similar paralogs, which are orthologs of *pbpB* that is involved in cell septation. Similarly, genes encoding orthologs of the SEDS proteins that function with the class B PBPs were strongly predicted. SpoVE, which functions with SpoVD in spore PG synthesis, is encoded in a similar genome context in all species. The gene encoding FtsW, which functions with *pbpB* in septation, is slightly more variable in genome context. Surprisingly, genes with the greatest similarity to *rodA*, which functions in synthesis of the cylindrical portion of the vegetative cell wall, were found in highly variable genome contexts, making the prediction of orthologous roles more tenuous.

Table 1 Identities of orthologous genes potentially involved in synthesis of spore PG in species spanning the family Bacillaceae[a]

Protein class	B. subtilis gene name	B. subtilis gene ID	B. megaterium QM B1551	B. anthracis Ames	B. halodurans C-125	Geobacillus kaustophilus HTA426	Paenibacillus sp. JDR-2	Oceanobacillus iheyensis HTE831
Class A PBP	ponA	939044	8985840 G	1086853 G	892494 G	3186622 G	8125881 X	1018250 G
	pbpF	939766	8985035 P	1089007 P	891918 P	3183940 X	8126262 X	1017307 P
	pbpG	937200	8989631 G	1085332 P	890885 G	3185091 G	8129227 P	1016712 P
	pbpD	938851	–	–	–	–	–	
	Other	–	8987409	1084404 / 1085723	890940	3186904	8123887	1015537
Class B PBP	spoVD	936661	8988718 G	1087468 G	892777 G	3186076 G	8124796 G	1017712 G
	pbpB	939847	8988719 G	1086649 G	893157 G	3186074 G	8124797 G	1017710 G
SEDS	spoVE	936953	8988714 P	1086101 G	894451 G	3183624 P	8124791 P	1017717 P
	ftsW	935918	8985806 G	1088773 G	891080 X	3184420 G	8123874 X	1017665 G
	rodA	937294	8988461 X	1087615 X	891115 X	3185129 X	8125565 X	1016054 X
SpoVB	spoVB	938022	8989074 G	1085645 P	891687 P	3183668 P	8125539 P	1018518 P
Low-molecular-weight PBP	dacB	938953	8988813 G	1086662 P	890609 P	3186567 P	8125120 P	1018320 G
	dacF	938732	8988832 G	1087453 P	890792 P	3186697 G	8124209 G	1018334 P
PG-modifying	cwlD	938917	8984611 P	1086710 P	892258 P	3183366 P	8128773 P	1015331 P
	pdaA	936136	8984825 X*	1087807 X*	893829 X*	3186591 X*	8127140 X	1016515 X
	lytH	936510	8989424 P	1084617 P	893890 P	3185754 P	8126711 G	1015506 X*

[a]Each gene is identified by an NCBI Gene ID number. Orthologs were identified using tblastn and blastp searches (136) for similarity to the products of the indicated B. subtilis genes. In all cases, the identified ortholog exhibited the greatest amino acid sequence identity with the B. subtilis protein. Because of the presence of multiple genes within each class in each genome, the genome context of each similar gene was examined to strengthen assignments of orthologs. A letter following the Gene ID number indicates the similarity of the genome context to that found in B. subtilis: G indicates a genome context match where the adjacent genes on both sides were shared; P indicates a partial genome context match where the adjacent genes on one side were shared; X indicates that the genome context was not shared with the B. subtilis gene; X* indicates that the genome context was not shared with the B. subtilis gene but it was shared with the B. megaterium gene. Representative species were chosen to represent a broad phylogenetic spread across the family Bacillaceae (137). A dash indicates the absence of an ortholog in that organism's genome.

Proteins that are involved in modification of the spore PG structure are also well conserved. Orthologs of the genes encoding CwlD and LytH are present in similar genomic contexts. Multiple genes encoding PG deacetylases similar to PdaA are present in each genome, although none of them are in a genome context similar to *pdaA* of *B. subtilis*. However, in most species, the gene most similar to *pdaA* was found in the same context as that in *Bacillus megaterium*. If one of these is shown to perform the PdaA function in muramic-δ-lactam formation, then likely all of these will be *pdaA* orthologs.

Germ Cell Wall May Serve as Template for Cortex Deposition

Aside from the role of the germ cell wall as the initial wall of the outgrowing spore, this structure appears to play an important role in the morphogenesis of the spore, serving as a template for synthesis of the cortex. A strain lacking both PBP2c and PBP2d, which presumably does not produce a normal germ cell wall, proceeds with synthesis of a large amount of spore PG, but not in a format that allows the production of stable dormant spores (53). The cortex PG in these sporangia is not in a concentric shell around the forespore and is often in disorganized masses adjacent to the forespore (53). The conclusion drawn from this was that the germ cell wall normally serves as a template for organized synthesis of the cortex PG; synthesis that must take place in a significantly different setting than that for the vegetative cell wall.

PG synthesis in vegetative cells is normally accomplished on the convex surface of a cytoplasmic membrane, and the nascent PG is attached to the concave surface of the existing wall. This membrane is also under significant turgor, pressing out against the wall, which has been postulated to be one factor that regulates the progression of PG synthesis (73). Conversely, synthesis of the cortex PG is on the concave surface of the outer forespore membrane, and the apparent function of the germ cell wall as a template suggests that the cortex is attached to the convex outer surface of the germ cell wall. The relative turgor on the outer forespore membrane at this point of sporulation is unclear. If the intermembrane space is hypoosmotic relative to the mother cell cytoplasm, then the outer membrane may be pushing into that space, but, at the same time, the turgor of the forespore is likely increasing as dehydration of that compartment begins. These novel aspects and the unique structures found in spore PG, as well as its nonessential nature in cultivation of the organism, make further study of spore PG synthesis a fruitful area for research.

SPORE PEPTIDOGLYCAN LYSIS DURING GERMINATION

Cortex degradation is necessary for full rehydration of the core and resumption of metabolism (8, 9). Spores in which the cortex cannot be degraded successfully initiate germination and release the large depot of Ca^{2+}-DPA stored within the spore core, although this release is slower than in wild-type spores (40). The Ca^{2+}-DPA is replaced with a limited volume of water, which is accompanied by a decrease in spore refractility and a partial drop in optical density (8, 9). While these spores are able to complete stage I of germination, they cannot proceed any further. The intact cortex prevents the spore core from reaching the rehydration levels necessary for metabolism to resume, and, thus, the spores do not go through outgrowth to become vegetative cells (8, 9). Addition of lysozyme to these spores allows for the completion of germination and outgrowth, indicating the germination defect observed is due to intact cortex PG (8, 9).

Hydrolysis of the cortex during spore germination is due to the activity of germination-specific lytic enzymes (GSLEs). GSLEs can be subdivided into two categories: spore cortex-lytic enzymes (SCLEs), which preferentially degrade intact cortex PG, and cortex fragment-lytic enzymes (CFLEs), which have specificity for PG fragments that result from SCLE activity (74). Both groups require the presence of muramic-δ-lactam in the spore cortex to carry out their enzymatic activities, which ensures that the cortex is degraded while the germ cell wall remains unharmed (8, 74–76). In addition, GSLEs must be expressed during spore formation, because they must be available to degrade cortex PG before there is any metabolism in the germinating spore (9, 77).

Bacillus species possess the GSLEs SleB, CwlJ, and SleL (YaaH). Although SleL and YaaH appear to be orthologs, they are referred to by separate names as a result of their distinct enzymatic activities in different species. SleB and CwlJ must be SCLEs because either one alone is sufficient to allow completion of germination. Spores lacking both SleB and CwlJ have a severe germination defect. They release Ca^{2+}-DPA, but only lose a small percentage of their initial optical density (9, 77–81). Germination is arrested at the stage of cortex hydrolysis, resulting in no further loss of optical density, and no release of cortex muropeptides (77–80, 82). SCLE-deficient spores cannot attain the level of rehydration necessary to resume metabolism (9, 77), and thus, their colony-forming efficiency is significantly reduced by at least 10^3-fold (77–82). SleL alone cannot allow completion of germination but can accelerate the process of cortex degradation following the action of a SCLE; therefore, SleL is a CFLE.

SleB

The gene *sleB* is expressed under the control of σ^G, indicating SleB is synthesized in the developing forespore (83, 84). The arrangement of *sleB* in an operon upstream of *ypeB* is highly conserved across *Bacillus* species (12, 78, 83, 84). However, in *Bacillus anthracis* and other species of the *Bacillus cereus* group, a third gene follows *ypeB* in the operon and shows homology to the *ylaJ* and *yhcN* genes of *B. subtilis* (78). In *B. subtilis*, *ylaJ* and *yhcN* are not found in the *sleB* operon but elsewhere in the chromosome, where their transcription is also dependent on σ^G (85, 86). YlaJ is an uncharacterized sporulation-specific protein, while YhcN is predicted to be a membrane-anchored lipoprotein and appears to play a small role in spore germination or outgrowth (85, 86).

Spores lacking SleB exhibit a minor defect during germination. Studies have indicated that *sleB* mutant spores have a slight delay in loss of optical density, or they do not lose as much optical density as wild-type spores (78, 79, 82, 83, 87). In addition, spores of a *B. anthracis sleB* mutant release fewer cortex fragments and at a slower rate than wild-type spores (78). Despite their defect during germination, *sleB* mutant spores complete germination and produce viable colonies (78, 79, 82, 83). The ability of these mutant spores to form colonies indicates that SleB and CwlJ are partially redundant enzymes; the presence of one can compensate in the absence of the other.

SleB contains a classic signal sequence at its amino terminus (Fig. 4) for translocation across the inner forespore membrane, and this signal sequence is cleaved during sporulation (84, 87–91). Thus, SleB is present in a mature yet inactive form within dormant spores. The ultimate location of SleB within the dormant spore has yet to be conclusively demonstrated (Fig. 5). Immunoelectron microscopy using anti-SleB antibodies detected SleB in the outer cortex/outer

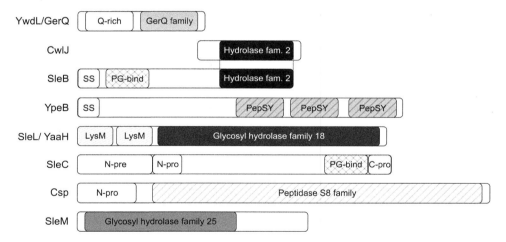

Figure 4 Domain architecture of GSLEs and interacting proteins. The proteins, along with the conserved domains and motifs shown, are drawn to scale. SleB and YpeB have signal sequences (SS) for export across the inner forespore membrane during sporulation. Both SleB and SleC have a PG-binding domain (PG-bind) (pfam01471), presumably to aid in protein localization or substrate affinity. The LysM domains (pfam01476) found in SleL/YaaH also recognize PG and are thought to play a similar role. The N-terminal pre- (N-pre) and pro- (N-pro), and C-terminal pro- (C-pro) sequences that are removed from SleC by Csp proteases during sporulation or germination are shown, as well as the N-terminal prosequence that is cleaved from Csp. YpeB contains three predicted PepSY domains (pfam03413), which play an unknown role; however, these domains have been involved in the inhibition of peptidase activity in PepSY-containing proteases. The C terminus of YwdL/GerQ is highly conserved, and a glutamine-rich (Q-rich) region is found toward the N terminus of the protein. The hydrolase family 2 (Hydrolase fam. 2) (pfam07486), glycosyl hydrolase family 18 (pfam00704), peptidase S8 family (pfam00082), and glycosyl hydrolase family 25 (pfam01183) domains contain the enzyme active sites.

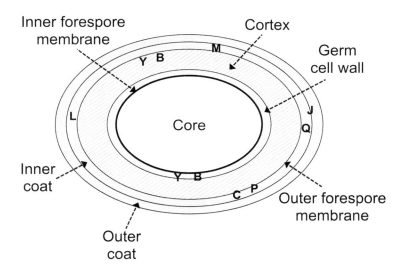

Figure 5 Localization of GSLEs and interacting proteins in dormant spores. SleB (B) and YpeB (Y) have been alternately demonstrated to be localized to the outer cortex/outer forespore membrane and the inner forespore membrane of dormant spores. The precise location of CwlJ (J) within the spore coat layers is unknown, but YwdL/GerQ (Q) is found within the inner coat. SleL/YaaH (L) has also been shown to be an inner coat protein. SleC (C), Csp proteases (P), and SleM (M) are located outside the cortex, either in the inner spore coat or outer forespore membrane. While these proteins are drawn within a single spore, in actuality, a spore only contains a subset of the proteins shown. B, Y, J, Q, and L are found in *Bacillaceae* and likely a few *Clostridiaceae*, while C, P, and M are found only in certain *Clostridiaceae*. Colocalization is shown for B-Y, J-Q, and C-P due to the requirement of YpeB and YwdL/GerQ for stable incorporation of SleB and CwlJ, respectively, into the dormant spore, and the processing of pro-SleC to active SleC by Csp proteases. However, it should be noted that there is currently no evidence that B-Y or J-Q directly interact.

forespore membrane region of spores from both *B. subtilis* and *B. cereus* (84). On the other hand, Western blots of integument (consisting of spore coats, cortex, and outer membrane), inner membrane, and soluble spore fractions showed the presence of SleB in both the integument and inner membrane fractions (75). Compared with other GSLEs, the resistance of SleB within dormant spores to detergent, heat, and alkali treatments also suggests that SleB may be localized to a more inward, protected region of the spore (75, 92).

SleB possesses two recognizable protein domains (Fig. 4). The N-terminal half of SleB has tandem sequence repeats characteristic of a PG-binding domain (pfam01471), and this motif has been found in the noncatalytic region of other PG hydrolases (87, 88, 93). Using a series of SleB-green fluorescent protein fusions, it was shown that the interaction between the PG-binding domain and muramic-δ-lactam in the spore cortex is necessary for SleB localization (93). However, this study did not utilize forms of SleB shown to retain any functional properties. To the contrary, while *in vitro* assays using SleB overexpressed and purified

from *Escherichia coli* support the dominant role of the N-terminal domain in PG binding, they also suggest that the specificity for muramic-δ-lactam is actually contained in the C-terminal domain (90).

The C-terminal half of SleB contains the PG-cleaving enzymatic domain (pfam07486) (90, 94). SleB cleaves the glycan strands of the cortex PG between the NAM and NAG residues forming anhydromuropeptide products characteristic of a lytic transglycosylase (Fig. 6) (78, 83). Lytic transglycosylase products were demonstrated to be absent in the muropeptide profiles of *sleB* mutants, supporting this enzymatic classification of SleB (78, 83, 95). Assays of purified SleB *in vitro* demonstrated that, while both protein domains are necessary for the most efficient lytic activity, the catalytic domain of SleB alone can act as a lytic transglycosylase without the PG-binding domain (90, 94).

Mutants lacking *sleB* or *ypeB* have similar germination phenotypes and cortex digestion patterns (83, 96). This is not surprising given that YpeB is required for the presence of SleB in spores, as SleB is absent in Western blots of extracts from *ypeB* mutant spores in

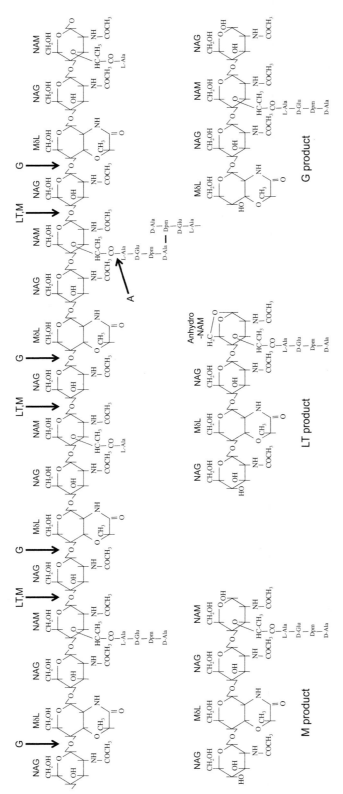

Figure 6 GSLE cleavage of cortex PG. A single strand of cortex PG is shown at the top, and the cleavage sites for the GSLE enzyme classes are indicated by arrows. A peptide cross-link to another strand is shown, but the second glycan strand is omitted. The proposed N-acetylmuramoyl-l-alanine amidase (A) activity of SleC is shown, but SleC can break peptide cross-links by cleaving a peptide from NAM. The cleavage sites and representative products (bottom) of N-acetylglucosaminidase (G, SleL), N-acetylmuramidase (M, SleM), and lytic transglycosylase (LT, SleB and SleC) are indicated.

B. subtilis and *B. anthracis* (75, 94, 96). Similarly, SleB is required for the presence of YpeB in spores (94, 96). In the absence of the partner protein, both SleB and YpeB are produced and rapidly degraded during spore formation (96). Colocalization of SleB and YpeB (Fig. 5) is likely because the spore fractionation experiments that placed SleB in the integument and inner membrane fractions detected YpeB in the same fractions (75). YpeB possesses an N-terminal signal sequence (Fig. 4) that is likely uncleaved and anchors the protein to the inner forespore membrane (75, 83). In addition, YpeB contains three putative PepSY domains (pfam03413) (Fig. 4), which in some proteins are involved in the inhibition of proteolytic activity (97). The exact relationship between SleB and YpeB, as well as the contributions of the PepSY domains, has yet to be elucidated; however, interesting possibilities include the involvement of YpeB in SleB localization, SleB stabilization against proteolysis and during prolonged spore dormancy, and holding SleB inactive prior to the triggering of spore germination. YpeB has been demonstrated to inhibit SleB activity in an *in vitro* assay, although a direct protein-protein interaction was not detected (94).

In the process of germinating, *B. cereus* spores release at least some SleB to the exudate, and this pool of SleB is active on decoated spores (98). Conversely, SleB is not released from germinating *B. subtilis* spores (75, 84, 87). SleB of a consistent size is detected in Western blots throughout germination of *B. subtilis* spores, indicating that SleB is stable and does not undergo proteolytic processing (75). While SleB is not processed during germination, there is evidence that YpeB may be proteolytically cleaved. Western blots of germinating *B. subtilis* spores show a 51-kDa band corresponding to YpeB disappear within 10 minutes, at which point a 30-kDa YpeB fragment is detected (75). This processing is also observed in *B. anthracis*, and the stable YpeB fragment was found to contain the C-terminal PepSY domains (99). The HtrC protease was shown to produce this specific cleavage of YpeB, but even in the absence of HtrC, YpeB was degraded during germination (99).

CwlJ

Unlike *sleB*, which is expressed in the forespore, *cwlJ* is a sporulation-specific gene expressed in the mother cell under the control of σ^E (82). *B. anthracis* contains two *cwlJ* homologs, *cwlJ1* and *cwlJ2* (78, 79). In many *Bacillus* species, *cwlJ* forms a bicistronic operon with *ywdL*, while in others, such as *B. subtilis*, the *cwlJ* gene is monocistronically transcribed, with *ywdL* located elsewhere in the chromosome (Table 2) (82, 100). (In *B. subtilis*, *ywdL* was renamed *gerQ* [100], but there is a different gene called *gerQ* in the *B. cereus* group [101], so the *ywdL* designation is utilized here to avoid confusion.) The gene *ywdL* is also expressed in the mother cell under the control of σ^E in *B. subtilis* (100). In *B. anthracis*, *cwlJ1* is in a bicistronic operon upstream of *ywdL*, and *cwlJ2* is monocistronic (78).

CwlJ localizes to the spore coats (Fig. 5), and some studies suggest CwlJ may oligomerize (75, 81, 100, 102, 103). CwlJ was detected in immunoblots of spore coat extracts, but not in decoated or *cotE* mutant

Table 2 Identities of orthologous genes potentially involved in degradation of spore PG in a range of species spanning the family *Bacillaceae*[a]

B. subtilis gene name	*B. subtilis* gene ID	*B. megaterium* QM B1551	*B. anthracis* Ames	*B. halodurans* C-125	*Geobacillus kaustophilus* HTA426	*Paenibacillus* sp. JDR-2	*Oceanobacillus iheyensis* HTE831
sleB	938979	8988788 P	1084998 P	891408 P	3184902 P	8126804 X 8124631 X	1018300 P
ypeB	938981	8988787 P	1086708 P	892552 P	3185430 P	8125141 X	1018299 P
cwlJ	938399	8989642 JY 8987672 X	1085358 JY 1087364 X	890546 JY	3184927 JY	8125961 X	1017903 JY
ywdL	937241	8989643 JY	1085359 JY	894423 JY	3184525 JY	8127416 X	1016708 JY
sleL/yaaH	937029	8984459 G	1089237 X	891826 X	3184040 X 3184810 X	8129825 X	1015055 P

[a]Each gene is identified by an NCBI Gene ID number. Orthologs were identified using tblastn and blastp searches (136) for similarity to the products of the indicated *B. subtilis* genes. In all cases, the identified ortholog exhibited the greatest amino acid sequence identity with the *B. subtilis* protein. Because of the presence of multiple genes within each class in some genomes, the genome context of each similar gene was examined to strengthen assignments of orthologs. A letter following the Gene ID number indicates the similarity of the genome context to that found in *B. subtilis*: G indicates a genome context match where the adjacent genes on both sides were shared; P indicates a partial genome context match where the adjacent genes on one side were shared; X indicates that the genome context was not shared with the *B. subtilis* gene; JY indicates that the *cwlJ* and *ywdL* genes were adjacent to one another, which is not the case in *B. subtilis*. Representative species were chosen to represent a broad phylogenetic spread across the family *Bacillaceae* (137).

spores in which the coat layers are disrupted or defective (102). The localization of CwlJ to the spore coats is fitting since CwlJ lacks a signal sequence necessary for transport from the mother cell across a membrane, and spore coat proteins are also synthesized in the mother cell compartment (75, 82, 104). In addition, CwlJ lacks transmembrane domains that would be characteristic of a membrane protein; thus, it is unlikely that CwlJ is associated with the outer membrane of the spore (102). Chirakkal et al. found CwlJ from *B. subtilis* to be present in Western blots of germinated spore samples with no change in band size or intensity (75). Conversely, results obtained by Bagyan and Setlow indicate CwlJ disappears during germination (102). These conflicting results may be due to the different methods of germination employed (the former study used L-alanine, while the latter used Ca^{2+}-DPA) or the CwlJ with a His-tag.

Similar to the dependence upon YpeB for the presence of SleB in the dormant, YwdL is required for the presence of CwlJ in spores (100). While CwlJ does not localize without YwdL, *cwlJ* is still expressed, suggesting YwdL plays a yet unknown role in CwlJ activity, localization, and/or stabilization (100). The relationship between the two proteins does not appear to be codependent because YwdL localizes properly in the absence of CwlJ (100). YwdL localizes to the inner spore coat layer (Fig. 5) in *B. subtilis* (100, 105) and to the inner layer of the exosporium in *B. cereus* (106, 107). In *B. subtilis*, YwdL is cross-linked into high-molecular-weight complexes, likely with other coat proteins, late in sporulation by Tgl, the spore transglutaminase (108). Three lysine residues near the N terminus of *B. subtilis* YwdL are necessary for cross-linking, but cross-linking is not required for YwdL or CwlJ localization to the coats or function during germination (100, 108, 109). There is a lack of conservation of these YwdL lysine residues across *Bacillus* species, and it is unknown if YwdL is cross-linked to the spore coats in other species (109). Fluorescence microscopy has demonstrated that YwdL does not change location during germination and remains associated with the spore coats in *B. subtilis* (100).

CwlJ is necessary for germination via Ca^{2+}-DPA, because *cwlJ* mutants fail to germinate in response to exogenous Ca^{2+}-DPA (77, 81). As expected because of its role in the incorporation of CwlJ into spores, YwdL is also necessary for germination with Ca^{2+}-DPA (106). While CwlJ from *B. anthracis* is predominately responsible for response to Ca^{2+}-DPA, CwlJ2 plays a minor part in the process (80). Germination in response to Ca^{2+}-DPA requires intact coat layers, providing further

support that CwlJ is associated with the coats and is responsible for cortex hydrolysis in response to Ca^{2+}-DPA (81). Endogenous Ca^{2+}-DPA released from the spore core during the early stages of nutrient germination is essential to activate CwlJ to degrade the cortex, although it is unclear whether CwlJ activation by Ca^{2+}-DPA is direct or indirect (81).

Interestingly, CwlJ is homologous to the C-terminal catalytic domain of SleB (pfam07486) (Fig. 4), yet CwlJ does not appear to be a lytic transglycosylase and currently has no ascribed enzymatic function (75, 78, 82). The sequence similarity between CwlJ and SleB, and the shared cleavage site of lytic transglycosylases and muramidases (Fig. 6), raise the possibility that CwlJ could be a muramidase. Numerous muramidases and lytic transglycosylases belong to a very large family of related proteins (110, 111). Current methods of PG structural analysis involve hydrolysis using a commercially available muramidase (15, 20, 80), making detection of a small amount of such an activity problematic. Attempts by several research groups to purify active CwlJ have failed, so clear demonstration of enzymatic activity may await overcoming this roadblock.

SleL (YaaH)

During sporulation, *sleL* (*yaaH*) is expressed monocistronically under the control of σ^E in the mother cell (112, 113). The amino acid sequence of SleL (YaaH) does not contain a predicted signal sequence, which is consistent with its localization to the spore coats (Fig. 5) (76, 103, 105, 112, 114). In support of the placement of SleL in the coat layers, Western blotting experiments detected SleL in the coat fraction stripped from spores and not in the remaining decoated spores (76). More specifically, fluorescent microscopy and a dependence on SafA (a major inner coat protein) for localization have demonstrated that YaaH resides in the inner spore coat (105, 114). Studies in *B. cereus* and *B. subtilis* have indicated that, after spore germination, active SleL (YaaH) is found in the exudate (75, 76).

At the N terminus of SleL are two repeated sequences that have been implicated to be involved in cell wall binding (Fig. 4) (76). These motifs are LysM domains (pfam01476), which are widely distributed in nature, bind PG and chitin, a polymer of NAG, and are most frequently encountered in PG hydrolases (115). A glycosyl hydrolase family 18 domain (pfam00704) is found at the C terminus of SleL (YaaH) (Fig. 4), but interestingly, this domain does not appear to have the same enzymatic activity across the *Bacillus* genus. High-performance liquid chromatography analyses of muropeptides collected from germinating spores indicate that YaaH has epimerase activity on cortex PG in *B. subtilis* and *B. megaterium* (75, 95). On the other hand, studies in *B. anthracis* and *B. cereus* have demonstrated SleL does not possess epimerase activity but is instead an *N*-acetylglucosaminidase that cleaves between NAG and muramic-δ-lactam (Fig. 6) (76, 113). A more recent study indicated that *B. megaterium* SleL can exhibit *N*-acetylglucosaminidase activity (116). Epimerase activity is a likely intermediate in the *N*-acetylglucosaminidase reaction (117), and the alternate production of epimerase versus *N*-acetylglucosaminidase products may be due to species and strain genetic variation (12). In *B. anthracis*, the glycosyl hydrolase domain can carry out *N*-acetylglucosaminidase activity alone; however, addition of the LysM domains results in increased affinity for its substrate resulting in greater PG degradation (118).

SleL is distinct from the other *Bacillus* GSLEs in that it is only active on cortex fragments (76). Thus, the role of SleL, where it acts as an *N*-acetylglucosaminidase, is to break down larger PG fragments created first by SleB and CwlJ (80, 113). This small contribution of SleL is not crucial for spore germination, but does explain the rapid release of muropeptides from germinating *B. anthracis* spores and the retention of the majority of cortex material in Δ*sleL* spores (80, 113). The epimerase activity of YaaH in *B. subtilis* and *B. megaterium* alters the stereochemistry of cortex fragments without hydrolyzing them, resulting in a much slower rate at which cortex material is released during germination (15, 18).

GSLEs in *Clostridiaceae*
While germination involves cortex hydrolysis in both *Bacillus* and *Clostridium* species, the mechanism through which cortex depolymerization is achieved is divergent. *Bacillus* species possess the major lytic enzymes SleB and CwlJ, either of which is sufficient to complete spore germination. Some *Clostridiaceae* contain *sleB* and *cwlJ* homologs in their genomes (Table 3), but to date, none of these homologs have been demonstrated to contribute to spore germination (12). This was found to be the case in *Clostridium difficile*, where a mutation in a *sleB*/*cwlJ* homolog had no effect on spore germination (119). Instead, many *Clostridium* species, including *C. perfringens* and *C. difficile*, contain SleC as the main GSLE (Table 3), which alone is necessary and sufficient to allow completion of germination (119, 120).

SleC acts on intact cortex PG and thus is classified as a SCLE (121, 122). Mutant strains lacking SleC have a severe germination defect. Spores without SleC experience only a small drop in optical density, release no hexosamine, and release Ca^{2+}-DPA more slowly during germination (120). In addition, these spores have a 10^3-fold decrease in colony-forming efficiency (120).

A smaller number of *Clostridiaceae*, including *C. perfringens*, also possess a second GSLE, called SleM, which plays an auxiliary role in germination (12). Akin to the role of SleL in *Bacillus*, SleM is a CFLE with specificity for cortex fragments first produced by SleC (121, 123). SleM is not required for completion of germination, but an *sleC sleM* double mutant has an even greater germination defect than a *sleC* mutant (120). Unlike the activation of CwlJ by Ca^{2+}-DPA in *Bacillus*, Ca^{2+}-DPA does not appear to trigger germination by activating SleC or SleM in *C. perfringens* (120).

SleC
In *C. perfringens*, SleC is expressed from a monocistronic gene in the mother cell compartment during

Table 3 Identities of orthologous genes potentially involved in degradation of spore PG in a range of species spanning the family *Clostridiaceae*[a]

Gene	*C. perfringens* strain 13	*C. difficile* 630	*C. lentocellum* DSM 5427	*C. leptum* DSM 753	*Thermoanaerobacter wiegelii* Rt8.B1	*Desulfotomaculum reducens* MI-1
sleC	990937	4916686 X	10331416 X	–	–	–
cspB	990939	4915699 X	10331412 X	–	–	–
sleM	989623	–	–	–	–	–
sleB	–	–	–	ZP_02079990.1 X	11084183 SY	4957310 SY
ypeB	–	–	–	ZP_02078794.1 X	11084182 SY	4958483 SY
sleL (*yaaH*)	–	–	10330877 X	ZP_02079199.1 X	11082049 X	4955511 X

[a]Each gene is identified by an NCBI Gene ID number. Orthologs were identified using tblastn and blastp searches (136) for similarity to the products of the indicated *C. perfringens* or *B. subtilis* genes. In all cases, the identified ortholog exhibited the greatest amino acid sequence identity with the query protein. Because of the presence of multiple genes within each class in some genomes, the genome context of each similar gene was examined to strengthen assignments of orthologs. A letter following the Gene ID number indicates the similarity of the genome context to that found for the query protein in its native species: SY indicates a partial genome context match where the *sleB* and *ypeB* genes are adjacent; X indicates that the genome context was not conserved. No clear *cwlJ* or *ywdL* orthologs were identified in these species. Representative species were chosen to represent a broad phylogenetic spread across the family *Clostridiaceae* (138). A dash indicates the absence of an ortholog in that organism's genome.

sporulation (124). Unlike the GSLE counterparts in *Bacillus*, SleC is synthesized as a precursor containing N-terminal pre- and prosequences and a C-terminal prosequence (Fig. 4) that are processed during stages of sporulation and germination (125, 126). During spore formation, the N-terminal presequences and the C-terminal prosequences are removed sequentially from the 50-kDa SleC precursor, and the resulting 35-kDa pro-SleC is incorporated into the dormant spore (125, 126). After cleavage, the N-terminal prepeptide remains associated with pro-SleC and is necessary for proper folding of the proenzyme (127). The inactive pro-SleC requires proteolytic cleavage of the N-terminal prosequence during germination to become the active 31-kDa form of the enzyme (125). This mature form of the enzyme contains a PG-binding domain (pfam01471) (Fig. 4) similar to that found in SleB and other PG hydrolases (125), and was proposed to possess both lytic transglycosylase and N-acetylmuramoyl-L-alanine amidase activities (Fig. 6) (121, 125). The lytic transglycosylase activity of *C. difficile* SleC has also been demonstrated (128).

In a tricistronic operon just upstream of *C. perfringens sleC*, the genes *cspA*, *cspB*, and *cspC* encode the three proteases that make up the germination-specific protease complex (GSP) needed to cleave pro-SleC to active SleC (124, 129). This is not the case for all strains of *C. perfringens* because some strains have only a single *cspB* gene upstream of *sleC* (130, 131), and this single CspB is sufficient for the processing of pro-SleC to active SleC (131). The Csp proteases are also expressed in the mother cell and contain N-terminal propeptides (Fig. 4) that likely serve as intramolecular inhibitory chaperones (124, 129, 131, 132). These proteases belong to a family of subtilisin-like serine proteases possessing a highly conserved catalytic triad and an oxyanion-binding region (129). Inactive pro-SleC and CspC localize outside the cortex, either in the inner coat or the outer membrane, within dormant spores (Fig. 5), and presumably this is also the location of the other Csp proteases (125, 129, 131, 133). After *C. perfringens* spores undergo germination, active SleC and the active Csp proteases are found in the germination exudate (125, 126, 133). The role of a CspB protease in germination-dependent processing of SleC has also recently been demonstrated in *C. difficile* (132).

SleM

Like *sleC*, *sleM* is transcribed monocistronically in the mother cell (124); however, SleM does not undergo proteolytic processing during sporulation or germination (Fig. 4) (123, 126). SleM is found as a mature enzyme within dormant spores, where it localizes outside the spore cortex to the outer forespore membrane or inner coat layers (Fig. 5) (123, 133). Release of active SleM from mechanically disrupted dormant spores provides evidence that SleM is not incorporated into the spore as a proenzyme that requires cleavage during germination (123). SleM is an *N*-acetylmuramidase (Fig. 6) and has specificity for cortex fragments rather than intact cortex PG (123). Thus, SleM is not essential for germination but likely further hydrolyzes cortex fragments first produced by SleC (120, 121). Unlike the loss of SleL in *Bacillus*, which causes spore retention of much of the muropeptides generated during germination (80, 113), muropeptides appear to be released normally by *sleM* mutant spores (120). After germination, SleM is found as an active enzyme in the spore exudate (123, 133).

INTEGRATION OF CORTEX LYSIS WITH OTHER GERMINATION EVENTS

Cortex degradation is a key step in completion of germination triggered by any stimulus, but the progression of germination events can differ based on the type of stimulus. Nutrient-stimulated germination is almost certainly the predominant process in natural environments. Interaction of nutrient germinants with *Bacillus* Ger receptors at the inner spore membrane results in the release of ions, including Ca^{2+}-DPA, and some partial hydration of the spore core (10, 134). These events trigger SleB activity via an unknown mechanism. It may be via a return of fluidity to the inner spore membrane (135), via an allosteric regulation of SleB, or via an alteration of the cortex substrate conformation. The Ca^{2+}-DPA release triggers CwlJ activity. Similar initial steps take place during nutrient-stimulated germination of *Clostridium* spores (11), resulting in activation of the SCLEs present, either SleB, CwlJ, or SleC. SleC becomes active following cleavage of its propeptide by a Csp protease, but the regulation of this processing is unknown. The combined activities of SleB and CwlJ, or SleC, result in sufficient cortex depolymerization to allow core swelling and full rehydration. If a CFLE such as SleL or SleM is present, the further breakdown of cortex fragments can speed release of muropeptides into the surrounding medium.

Germination by exogenous Ca^{2+}-DPA proceeds by a different mechanism in *Bacillus*, where activation of CwlJ is the initial step (81). Cortex depolymerization by CwlJ apparently leads to eventual destabilization of the inner spore membrane, ion release, and a return of hydration and membrane fluidity to the core. It is likely that SleB becomes activated at some point in this process, and SleL can act on cortex fragments produced by CwlJ (80). Germination of *C. perfringens* spores by exogenous Ca²⁺-DPA occurs through an unknown mechanism that does not involve direct activation of a GSLE (120).

Regardless of the stimulus that initiates the germination process, in all cases, some cortex degradation is essential for the resumption of metabolism and progression into spore outgrowth. From this discussion it is clear that the outstanding questions in this field concern the mechanisms by which the GSLEs are held inactive and stable in the dormant spore and the processes that result in their activation during germination. A full understanding of these enzymes can make them a target for efforts to stimulate highly efficient germination of spore populations followed by simplified decontamination treatments.

Citation. Popham DL, Bernhards CB. 2015. Spore peptidoglycan. Microbiol Spectrum 3(6):TBS-0005-2012.

References

1. **Driks A.** 2002. Maximum shields: the assembly and function of the bacterial spore coat. *Trends Microbiol* **10:**251–254.

2. **Koshikawa T, Beaman TC, Pankratz HS, Nakashio S, Corner TR, Gerhardt P.** 1984. Resistance, germination, and permeability correlates of *Bacillus megaterium* spores successively divested of integument layers. *J Bacteriol* **159:**624–632.

3. **Beaman TC, Gerhardt P.** 1986. Heat resistance of bacterial spores correlated with protoplast dehydration, mineralization, and thermal adaptation. *Appl Environ Microbiol* **52:**1242–1246.

4. **Nakashio S, Gerhardt P.** 1985. Protoplast dehydration correlated with heat resistance of bacterial spores. *J Bacteriol* **162:**571–578.

5. **Popham DL, Illades-Aguiar B, Setlow P.** 1995. The *Bacillus subtilis dacB* gene, encoding penicillin-binding protein 5*, is part of a three-gene operon required for proper spore cortex synthesis and spore core dehydration. *J Bacteriol* **177:**4721–4729.

6. **Setlow P.** 2006. Spores of *Bacillus subtilis*: their resistance to and killing by radiation, heat and chemicals. *J Appl Microbiol* **101:**514–525.

7. **Setlow P.** 2014. Spore resistance properties. *Microbiol Spectrum* 2(4):TBS-0003-2012. doi:10.1128/microbiolspec. TBS-0003-2012.

8. **Popham DL, Helin J, Costello CE, Setlow P.** 1996. Muramic lactam in peptidoglycan of *Bacillus subtilis* spores is required for spore outgrowth but not for spore dehydration or heat resistance. *Proc Natl Acad Sci USA* **93:**15405–15410.

9. **Setlow B, Melly E, Setlow P.** 2001. Properties of spores of *Bacillus subtilis* blocked at an intermediate stage in spore germination. *J Bacteriol* **183:**4894–4899.

10. **Moir A.** 2006. How do spores germinate? *J Appl Microbiol* **101:**526–530.

11. **Paredes-Sabja D, Setlow P, Sarker MR.** 2011. Germination of spores of *Bacillales* and *Clostridiales* species: mechanisms and proteins involved. *Trends Microbiol* **19:**85–94.

12. **Popham DL, Heffron JD, Lambert EA.** 2012. Degradation of spore peptidoglycan during germination, p 121–142. *In* Abel-Santos E (ed), *Bacterial Spores: Current Research and Applications.* Caister Academic Press, Norwich, UK.

13. **van Heijenoort J.** 2001. Formation of the glycan chains in the synthesis of bacterial peptidoglycan. *Glycobiology* **11:**25R–36R.

14. **Vollmer W.** 2008. Structural variation in the glycan strands of bacterial peptidoglycan. *FEMS Microbiol Rev* **32:**287–306.

15. **Atrih A, Zöllner P, Allmaier G, Williamson MP, Foster SJ.** 1998. Peptidoglycan structural dynamics during germination of *Bacillus subtilis* 168 endospores. *J Bacteriol* **180:**4603–4612.

16. **Meador-Parton J, Popham DL.** 2000. Structural analysis of *Bacillus subtilis* spore peptidoglycan during sporulation. *J Bacteriol* **182:**4491–4499.

17. **Atrih A, Zöllner P, Allmaier G, Foster SJ.** 1996. Structural analysis of *Bacillus subtilis* 168 endospore peptidoglycan and its role during differentiation. *J Bacteriol* **178:**6173–6183.

18. **Atrih A, Bacher G, Körner R, Allmaier G, Foster SJ.** 1999. Structural analysis of *Bacillus megaterium* KM spore peptidoglycan and its dynamics during germination. *Microbiology* **145:**1033–1041.

19. **Atrih A, Foster SJ.** 2001. Analysis of the role of bacterial endospore cortex structure in resistance properties and demonstration of its conservation amongst species. *J Appl Microbiol* **91:**364–372.

20. **Dowd MM, Orsburn B, Popham DL.** 2008. Cortex peptidoglycan lytic activity in germinating *Bacillus anthracis* spores. *J Bacteriol* **190:**4541–4548.

21. **Orsburn B, Melville SB, Popham DL.** 2008. Factors contributing to heat resistance of *Clostridium perfringens* endospores. *Appl Environ Microbiol* **74:**3328–3335.

22. **Tipper DJ, Gauthier JJ.** 1972. Structure of the bacterial endospore, p 3–12. *In* Halvorson HO, Hanson R, Campbell LL (ed), *Spores V.* American Society for Microbiology, Washington, DC.

23. **Warth AD, Strominger JL.** 1969. Structure of the peptidoglycan of bacterial spores: occurrence of the lactam of muramic acid. *Proc Natl Acad Sci USA* **64:**528–535.

24. **Tipper DJ, Linnett PE.** 1976. Distribution of peptidoglycan synthetase activities between sporangia and forespores in sporulating cells of *Bacillus sphaericus*. *J Bacteriol* **126:**213–221.

25. **Warth AD, Strominger JL.** 1972. Structure of the peptidoglycan from spores of *Bacillus subtilis*. *Biochemistry* **11:**1389–1396.

26. **Fittipaldi N, Sekizaki T, Takamatsu D, de la Cruz Domínguez-Punaro M, Harel J, Bui NK, Vollmer W, Gottschalk M.** 2008. Significant contribution of the

pgdA gene to the virulence of *Streptococcus suis*. *Mol Microbiol* **70**:1120–1135.

27. Psylinakis E, Boneca IG, Mavromatis K, Deli A, Hayhurst E, Foster SJ, Vårum KM, Bouriotis V. 2005. Peptidoglycan N-acetylglucosamine deacetylases from *Bacillus cereus*, highly conserved proteins in *Bacillus anthracis J Biol Chem* **280**:30856–30863.

28. Atrih A, Bacher G, Allmaier G, Williamson MP, Foster SJ. 1999. Analysis of peptidoglycan structure from vegetative cells of *Bacillus subtilis* 168 and role of PBP 5 in peptidoglycan maturation. *J Bacteriol* **181**:3956–3966.

29. Warth AD, Stroming_er JL. 1971. Structure of the peptidoglycan from vegetative cell walls of *Bacillus subtilis*. *Biochemistry* **10**:4349–4358.

30. Popham DL, Gilmore ME, Setlow P. 1999. Roles of low-molecular-weight penicillin-binding proteins in *Bacillus subtilis* spore peptidoglycan synthesis and spore properties. *J Bacteriol* **181**:126–132.

31. Imae Y, Stromingez JL. 1976. Relationship between cortex content and properties of *Bacillus sphaericus* spores. *J Bacteriol* **126**:907–913.

32. Beaman TC, Greenamyre JT, Corner TR, Pankratz HS, Gerhardt P. 1982. Bacterial spore heat resistance correlated with water content, wet density, and protoplast/sporoplast volume ratio. *J Bacteriol* **150**:870–877.

33. Horsburgh GJ, Atrih A, Foster SJ. 2003. Characterization of LytH, a differentiation-associated peptidoglycan hydrolase of *Bacillus subtilis* involved in endospore cortex maturation. *J Bacteriol* **185**:3813–3820.

34. Popham DL, Helin J, Costello CE, Setlow P. 1996. Analysis of the peptidoglycan structure of *Bacillus subtilis* endospores. *J Bacteriol* **178**:6451–6458.

35. Popham DL, Meador-Parton J, Costello CE, Setlow P. 1999. Spore peptidoglycan structure in a *cwlD dacB* double mutant of *Bacillus subtilis*. *J Bacteriol* **181**:6205–6209.

36. Ou L-T, Marquis RE. 1970. Electromechanical interactions in cell walls of gram-positive cocci. *J Bacteriol* **101**:92–101.

37. Lewis JC, Snell NS, Burr HK. 1960. Water permeability of bacterial spores and the concept of a contractile cortex. *Science* **132**:544–545.

38. Warth AD. 1985. Mechanisms of heat resistance, p 209–225. *In* Dring GJ, Ellar DJ, Gould GW (ed), *Fundamental and Applied Aspects of Bacterial Spores*. Academic Press, Inc, London, UK.

39. Westphal AJ, Price PB, Leighton TJ, Wheeler KE. 2003. Kinetics of size changes of individual *Bacillus thuringiensis* spores in response to changes in relative humidity. *Proc Natl Acad Sci USA* **100**:3461–3466.

40. Zhang P, Thomas S, Li YQ, Setlow P. 2012. Effects of cortex peptidoglycan structure and cortex hydrolysis on the kinetics of Ca(2+)-dipicolinic acid release during *Bacillus subtilis* spore germination. *J Bacteriol* **194**:646–652.

41. Sekiguchi J, Akeo K, Yamamoto H, Khasanov FK, Alonso JC, Kuroda A. 1995. Nucleotide sequence and regulation of a new putative cell wall hydrolase gene, *cwlD*, which affects germination in *Bacillus subtilis*. *J Bacteriol* **177**:5582–5589.

42. Wickus GG, Warth AD, Stromingeer JL. 1972. Appearance of muramic lactam during cortex synthesis in sporulating cultures of *Bacillus cereus* and *Bacillus megaterium*. *J Bacteriol* **111**:625–627.

43. Chastanet A, Losick R. 2007. Engulfment during sporulation in *Bacillus subtilis* is governed by a multi-protein complex containing tandemly acting autolysins. *Mol Microbiol* **64**:139–152.

44. Gutierrez J, Smith R, Pogliano K. 2010. SpoIID-mediated peptidoglycan degradation is required throughout engulfment during *Bacillus subtilis* sporulation. *J Bacteriol* **192**:3174–3186.

45. Morlot C, Uehara T, Marquis KA, Bernhardt TG, Rudner DZ. 2010. A highly coordinated cell wall degradation machine governs spore morphogenesis in *Bacillus subtilis*. *Genes Dev* **24**:411–422.

46. Meyer P, Gutierrez J, Pogliano K, Dworkin J. 2010. Cell wall synthesis is necessary for membrane dynamics during sporulation of *Bacillus subtilis*. *Mol Microbiol* **76**:956–970.

47. Tocheva EI, López-Garrido J, Hughes HV, Fredlund J, Kuru E, Vannieuwenhze MS, Brun YV, Pogliano K, Jensen GJ. 2013. Peptidoglycan transformations during *Bacillus subtilis* sporulation. *Mol Microbiol* **88**:673–686.

48. Dworkin J. 2014. Protein targeting during *Bacillus subtilis* sporulation. *Microbiol Spectrum* **2**(1):TBS-0006-2013. doi:10.1128/microbiolspec.TBS-0006-2013.

49. Pedersen LB, Ragkousi K, Cammett TJ, Melly E, Sekowska A, Schopick E, Murray T, Setlow P. 2000. Characterization of *ywhE*, which encodes a putative high-molecular-weight class A penicillin-binding protein in *Bacillus subtilis*. *Gene* **246**:187–196.

50. Popham DL, Setlow P. 1993. Cloning, nucleotide sequence, and regulation of the *Bacillus subtilis pbpF* gene, which codes for a putative class A high-molecular-weight penicillin-binding protein. *J Bacteriol* **175**:4870–4876.

51. Vasudevan P, Weaver A, Reichert ED, Linnstaedt SD, Popham DL. 2007. Spore cortex formation in *Bacillus subtilis* is regulated by accumulation of peptidoglycan precursors under the control of sigma K. *Mol Microbiol* **65**:1582–1594.

52. Buchanan CE, Sowell MO. 1983. Stability and synthesis of the penicillin-binding proteins during sporulation. *J Bacteriol* **156**:545–551.

53. McPherson DC, Driks A, Popham DL. 2001. Two class A high-molecular-weight penicillin-binding proteins of *Bacillus subtilis* play redundant roles in sporulation. *J Bacteriol* **183**:6046–6053.

54. Popham DL, Setlow P. 1994. Cloning, nucleotide sequence, mutagenesis, and mapping of the *Bacillus subtilis pbpD* gene, which codes for penicillin-binding protein 4. *J Bacteriol* **176**:7197–7205.

55. Popham DL, Setlow P. 1995. Cloning, nucleotide sequence, and mutagenesis of the *Bacillus subtilis ponA* operon, which codes for penicillin-binding protein (PBP) 1 and a PBP-related factor. *J Bacteriol* **177**:326–335.

56. **Sowell MO, Buchanan CE.** 1983. Changes in penicillin-binding proteins during sporulation of *Bacillus subtilis.* *J Bacteriol* **153:**1331–1337.

57. **McPherson DC, Popham DL.** 2003. Peptidoglycan synthesis in the absence of class A penicillin-binding proteins in *Bacillus subtilis.* *J Bacteriol* **185:**1423–1431.

58. **Daniel RA, Drake S, Buchanan CE, Scholle R, Errington J.** 1994. The *Bacillus subtilis spoVD* gene encodes a mother-cell-specific penicillin-binding protein required for spore morphogenesis. *J Mol Biol* **235:**209–220.

59. **Fay A, Meyer P, Dworkin J.** 2010. Interactions between late-acting proteins required for peptidoglycan synthesis during sporulation. *J Mol Biol* **399:**547–561.

60. **Bukowska-Faniband E, Hederstedt L.** 2013. Cortex synthesis during *Bacillus subtilis* sporulation depends on the transpeptidase activity of SpoVD. *FEMS Microbiol Lett* **346:**65–72.

61. **Ikeda M, Sato T, Wachi M, Jung HK, Ishino F, Kobayashi Y, Matsuhashi M.** 1989. Structural similarity among *Escherichia coli* FtsW and RodA proteins and *Bacillus subtilis* SpoVE protein, which function in cell division, cell elongation, and spore formation, respectively. *J Bacteriol* **171:**6375–6378.

62. **Joris B, Dive G, Henriques A, Piggot PJ, Ghuysen JM.** 1990. The life-cycle proteins RodA of *Escherichia coli* and SpoVE of *Bacillus subtilis* have very similar primary structures. *Mol Microbiol* **4:**513–517.

63. **Popham DL, Stragier P.** 1991. Cloning, characterization, and expression of the *spoVB* gene of *Bacillus subtilis.* *J Bacteriol* **173:**7942–7949.

64. **Vasudevan P, McElligott J, Attkisson C, Betteken M, Popham DL.** 2009. Homologues of the *Bacillus subtilis* SpoVB protein are involved in cell wall metabolism. *J Bacteriol* **191:**6012–6019.

65. **Blumberg PM, Strominger JL.** 1972. Five penicillin-binding components occur in *Bacillus subtilis* membranes. *J Biol Chem* **247:**8107–8113.

66. **Todd JA, Bone EJ, Piggot PJ, Ellar DJ.** 1983. Differential expression of penicillin-binding protein structural genes during *Bacillus subtilis* sporulation. *FEMS Microbiol Lett* **18:**197–202.

67. **Simpson EB, Hancock TW, Buchanan CE.** 1994. Transcriptional control of *dacB*, which encodes a major sporulation-specific penicillin-binding protein. *J Bacteriol* **176:**7767–7769.

68. **Wu J-J, Schuch R, Piggot PJ.** 1992. Characterization of a *Bacillus subtilis* sporulation operon that includes genes for an RNA polymerase sigma factor and for a putative DD-carboxypeptidase. *J Bacteriol* **174:**4885–4892.

69. **Fukushima T, Yamamoto H, Atrih A, Foster SJ, Sekiguchi J.** 2002. A polysaccharide deacetylase gene (*pdaA*) is required for germination and for production of muramic delta-lactam residues in the spore cortex of *Bacillus subtilis.* *J Bacteriol* **184:**6007–6015.

70. **Gilmore ME, Bandyopadhyay D, Dean AM, Linnstaedt SD, Popham DL.** 2004. Production of muramic delta-lactam in *Bacillus subtilis* spore peptidoglycan. *J Bacteriol* **186:**80–89.

71. **Galperin MY.** 2013. Genome diversity of spore-forming firmicutes. *Microbiol Spectrum* **1**(2):TBS-0015-2012. doi:10.1128/microbiolspectrum.TBS-0015-2012.

72. **van Heijenoort J.** 2007. Lipid intermediates in the biosynthesis of bacterial peptidoglycan. *Microbiol Mol Biol Rev* **71:**620–635.

73. **Koch AL.** 1983. The surface stress theory of microbial morphogenesis. *Adv Microb Physiol* **24:**301–366.

74. **Makino S, Moriyama R.** 2002. Hydrolysis of cortex peptidoglycan during bacterial spore germination. *Med Sci Monit* **8:**RA119–RA127.

75. **Chirakkal H, O'Rourke M, Atrih A, Foster SJ, Moir A.** 2002. Analysis of spore cortex lytic enzymes and related proteins in *Bacillus subtilis* endospore germination. *Microbiology* **148:**2383–2392.

76. **Chen Y, Fukuoka S, Makino S.** 2000. A novel spore peptidoglycan hydrolase of *Bacillus cereus*: biochemical characterization and nucleotide sequence of the corresponding gene, *sleL.* *J Bacteriol* **182:**1499–1506.

77. **Setlow B, Peng L, Loshon CA, Li Y-Q, Christie G, Setlow P.** 2009. Characterization of the germination of *Bacillus megaterium* spores lacking enzymes that degrade the spore cortex. *J Appl Microbiol* **107:**318–328.

78. **Heffron JD, Orsburn B, Popham DL.** 2009. Roles of germination-specific lytic enzymes CwlJ and SleB in *Bacillus anthracis.* *J Bacteriol* **191:**2237–2247.

79. **Giebel JD, Carr KA, Anderson EC, Hanna PC.** 2009. The germination-specific lytic enzymes SleB, CwlJ1, and CwlJ2 each contribute to *Bacillus anthracis* spore germination and virulence. *J Bacteriol* **191:**5569–5576.

80. **Heffron JD, Lambert EA, Sherry N, Popham DL.** 2010. Contributions of four cortex lytic enzymes to germination of *Bacillus anthracis* spores. *J Bacteriol* **192:**763–770.

81. **Paidhungat M, Ragkousi K, Setlow P.** 2001. Genetic requirements for induction of germination of spores of *Bacillus subtilis* by Ca($^{2+}$)-dipicolinate. *J Bacteriol* **183:**4886–4893.

82. **Ishikawa S, Yamane K, Sekiguchi J.** 1998. Regulation and characterization of a newly deduced cell wall hydrolase gene (*cwlJ*) which affects germination of *Bacillus subtilis* spores. *J Bacteriol* **180:**1375–1380.

83. **Boland FM, Atrih A, Chirakkal H, Foster SJ, Moir A.** 2000. Complete spore-cortex hydrolysis during germination of *Bacillus subtilis* 168 requires SleB and YpeB. *Microbiology* **146:**57–64.

84. **Moriyama R, Fukuoka H, Miyata S, Kudoh S, Hattori A, Kozuka S, Yasuda Y, Tochikubo K, Makino S.** 1999. Expression of a germination-specific amidase, SleB, of bacilli in the forespore compartment of sporulating cells and its localization on the exterior side of the cortex in dormant spores. *J Bacteriol* **181:**2373–2378.

85. **Bagyan I, Noback M, Bron S, Paidhungat M, Setlow P.** 1998. Characterization of *yhcN*, a new forespore-specific gene of *Bacillus subtilis.* *Gene* **212:**179–188.

86. **Kuwana R, Kasahara Y, Fujibayashi M, Takamatsu H, Ogasawara N, Watabe K.** 2002. Proteomics characterization of novel spore proteins of *Bacillus subtilis.* *Microbiology* **148:**3971–3982.

87. Moriyama R, Hattori A, Miyata S, Kudoh S, Makino S. 1996. A gene (sleB) encoding a spore cortex-lytic enzyme from Bacillus subtilis and response of the enzyme to L-alanine-mediated germination. J Bacteriol 178: 6059–6063.

88. Moriyama R, Kudoh S, Miyata S, Nonobe S, Hattori A, Makino S. 1996. A germination-specific spore cortex-lytic enzyme from Bacillus cereus spores: cloning and sequencing of the gene and molecular characterization of the enzyme. J Bacteriol 178:5330–5332.

89. Hu K, Yang H, Liu G, Tan H. 2007. Cloning and identification of a gene encoding spore cortex-lytic enzyme in Bacillus thuringiensis. Curr Microbiol 54:292–295.

90. Heffron JD, Sherry N, Popham DL. 2011. In vitro studies of peptidoglycan binding and hydrolysis by the Bacillus anthracis germination-specific lytic enzyme SleB. J Bacteriol 193:125–131.

91. Tjalsma H, Bolhuis A, Jongbloed JDH, Bron S, van Dijl JM. 2000. Signal peptide-dependent protein transport in Bacillus subtilis: a genome-based survey of the secretome. Microbiol Mol Biol Rev 64:515–547.

92. Atrih A, Foster SJ. 2001. In vivo roles of the germination-specific lytic enzymes of Bacillus subtilis 168. Microbiology 147:2925–2932.

93. Masayama A, Fukuoka H, Kato S, Yoshimura T, Moriyama M, Moriyama R. 2006. Subcellular localization of a germiantion-specific cortex-lytic enzyme, SleB, of bacilli during sporulation. Genes Genet Syst 81: 163–169.

94. Li Y, Butzin XY, Davis A, Setlow B, Korza G, Üstok FI, Christie G, Setlow P, Hao B. 2013. Activity and regulation of various forms of CwlJ, SleB, and YpeB proteins in degrading cortex peptidoglycan of spores of Bacillus species in vitro and during spore germination. J Bacteriol 195:2530–2540.

95. Christie G, Üstok FI, Lu Q, Packman LC, Lowe CR. 2010. Mutational analysis of Bacillus megaterium QM B1551 cortex-lytic enzymes. J Bacteriol 192:5378–5389.

96. Bernhards CB, Popham DL. 2014. Role of YpeB in cortex hydrolysis during germination of Bacillus anthracis spores. J Bacteriol 196:3399–3409.

97. Yeats C, Rawlings ND, Bateman A. 2004. The PepSY domain: a regulator of peptidase activity in the microbial environment? Trends Biochem Sci 29:169–172.

98. Makino S, Ito N, Inoue T, Miyata S, Moriyama R. 1994. A spore-lytic enzyme released from Bacillus cereus spores during germination. Microbiology 140:1403–1410.

99. Bernhards CB, Chen Y, Toutkoushian H, Popham DL. 2015. HtrC is involved in proteolysis of YpeB during germination of Bacillus anthracis and Bacillus subtilis spores. J Bacteriol 197:326–336.

100. Ragkousi K, Eichenberger P, van Ooij C, Setlow P. 2003. Identification of a new gene essential for germination of Bacillus subtilis spores with Ca2+-dipicolinate. J Bacteriol 185:2315–2329.

101. Barlass PJ, Houston CW, Clements MO, Moir A. 2002. Germination of Bacillus cereus spores in response to L-alanine and to inosine: the roles of gerL and gerQ operons. Microbiology 148:2089–2095.

102. Bagyan I, Setlow P. 2002. Localization of the cortex lytic enzyme CwlJ in spores of Bacillus subtilis. J Bacteriol 184:1219–1224.

103. McKenney PT, Eichenberger P. 2012. Dynamics of spore coat morphogenesis in Bacillus subtilis. Mol Microbiol 83:245–260.

104. Driks A. 1999. Bacillus subtilis spore coat. Microbiol Mol Biol Rev 63:1–20.

105. Imamura D, Kuwana R, Takamatsu H, Watabe K. 2010. Localization of proteins to different layers and regions of Bacillus subtilis spore coats. J Bacteriol 192: 518–524.

106. Terry C, Shepherd A, Radford DS, Moir A, Bullough PA. 2011. YwdL in Bacillus cereus: its role in germination and exosporium structure. PLoS One 6:e23801. doi:10.1371/journal.pone.0023801.

107. Liu H, Bergman NH, Thomason B, Shallom S, Hazen A, Crossno J, Rasko DA, Ravel J, Read TD, Peterson SN, Yates J III, Hanna PC. 2004. Formation and composition of the Bacillus anthracis endospore. J Bacteriol 186:164–178.

108. Ragkousi K, Setlow P. 2004. Transglutaminase-mediated cross-linking of GerQ in the coats of Bacillus subtilis spores. J Bacteriol 186:5567–5575.

109. Monroe A, Setlow P. 2006. Localization of the transglutaminase cross-linking sites in the Bacillus subtilis spore coat protein GerQ. J Bacteriol 188:7609–7616.

110. Blackburn NT, Clarke AJ. 2001. Identification of four families of peptidoglycan lytic transglycosylases. J Mol Evol 52:78–84.

111. Scheurwater E, Reid CW, Clarke AJ. 2008. Lytic transglycosylases: bacterial space-making autolysins. Int J Biochem Cell Biol 40:586–591.

112. Kodama T, Takamatsu H, Asai K, Kobayashi K, Ogasawara N, Watabe K. 1999. The Bacillus subtilis yaaH gene is transcribed by SigE RNA polymerase during sporulation, and its product is involved in germination of spores. J Bacteriol 181:4584–4591.

113. Lambert EA, Popham DL. 2008. The Bacillus anthracis SleL (YaaH) protein is an N-acetylglucosaminidase involved in spore cortex depolymerization. J Bacteriol 190:7601–7607.

114. McKenney PT, Driks A, Eskandarian HA, Grabowski P, Guberman J, Wang KH, Gitai Z, Eichenberger P. 2010. A distance-weighted interaction map reveals a previously uncharacterized layer of the Bacillus subtilis spore coat. Curr Biol 20:934–938.

115. Buist G, Steen A, Kok J, Kuipers OP. 2008. LysM, a widely distributed protein motif for binding to (peptido) glycans. Mol Microbiol 68:838–847.

116. Ustok FI, Packman LC, Lowe CR, Christie G. 2014. Spore germination mediated by Bacillus megaterium QM B1551 SleL and YpeB. J Bacteriol 196:1045–1054.

117. Papanikolau Y, Prag G, Tavlas G, Vorgias CE, Oppenheim AB, Petratos K. 2001. High resolution structural analyses of mutant chitinase A complexes with substrates provide new insight into the mechanism of catalysis. Biochemistry 40:11338–11343.

118. Lambert EA, Sherry N, Popham DL. 2012. *In vitro* and *in vivo* analyses of the *Bacillus anthracis* spore cortex lytic protein SleL. *Microbiology* **158:**1359–1368.

119. Burns DA, Heap JT, Minton NP. 2010. SleC is essential for germination of *Clostridium difficile* spores in nutrient-rich medium supplemented with the bile salt taurocholate. *J Bacteriol* **192:**657–664.

120. Paredes-Sabja D, Setlow P, Sarker MR. 2009. SleC is essential for cortex peptidoglycan hydrolysis during germination of spores of the pathogenic bacterium *Clostridium perfringens*. *J Bacteriol* **191:**2711–2720.

121. Kumazawa T, Masayama A, Fukuoka S, Makino S, Yoshimura T, Moriyama R. 2007. Mode of action of a germination-specific cortex-lytic enzyme, SleC, of *Clostridium perfringens* S40. *Biosci Biotechnol Biochem* **71:**884–892.

122. Miyata S, Moriyama R, Sugimoto K, Makino S. 1995. Purification and partial characterization of a spore cortex-lytic enzyme of *Clostridium perfringens* S40 spores. *Biosci Biotechnol Biochem* **59:**514–515.

123. Chen Y, Miyata S, Makino S, Moriyama R. 1997. Molecular characterization of a germination-specific muramidase from *Clostridium perfringens* S40 spores and nucleotide sequence of the corresponding gene. *J Bacteriol* **179:**3181–3187.

124. Masayama A, Hamasaki K, Urakami K, Shimamoto S, Kato S, Makino S, Yoshimura T, Moriyama M, Moriyama R. 2006. Expression of germination-related enzymes, CspA, CspB, CspC, SleC, and SleM, of *Clostridium perfringens* S40 in the mother cell compartment of sporulating cells. *Genes Genet Syst* **81:**227–234.

125. Miyata S, Moriyama R, Miyahara N, Makino S. 1995. A gene (*sleC*) encoding a spore-cortex-lytic enzyme from *Clostridium perfringens* S40 spores; cloning, sequence analysis and molecular characterization. *Microbiology* **141:**2643–2650.

126. Urakami K, Miyata S, Moriyama R, Sugimoto K, Makino S. 1999. Germination-specific cortex-lytic enzymes from *Clostridium perfringens* S40 spores: time of synthesis, precursor structure and regulation of enzymatic activity. *FEMS Microbiol Lett* **173:**467–473.

127. Okamura S, Urakami K, Kimata M, Aoshima T, Shimamoto S, Moriyama R, Makino S. 2000. The N-terminal prepeptide is required for the production of spore cortex-lytic enzyme from its inactive precursor during germination of *Clostridium perfringens* S40 spores. *Mol Microbiol* **37:**821–827.

128. Gutelius D, Hokeness K, Logan SM, Reid CW. 2014. Functional analysis of SleC from *Clostridium difficile*: an essential lytic transglycosylase involved in spore germination. *Microbiology* **160:**209–216.

129. Shimamoto S, Moriyama R, Sugimoto K, Miyata S, Makino S. 2001. Partial characterization of an enzyme fraction with protease activity which converts the spore peptidoglycan hydrolase (SleC) precursor to an active enzyme during germination of *Clostridium perfringens* S40 spores and analysis of a gene cluster involved in the activity. *J Bacteriol* **183:**3742–3751.

130. Myers GSA, Rasko DA, Cheung JK, Ravel J, Seshadri R, DeBoy RT, Ren Q, Varga J, Awad MM, Brinkac LM, Daugherty SC, Haft DH, Dodson RJ, Madupu R, Nelson WC, Rosovitz MJ, Sullivan SA, Khouri H, Dimitrov GI, Watkins KL, Mulligan S, Benton J, Radune D, Fisher DJ, Atkins HS, Hiscox T, Jost BH, Billington SJ, Songer JG, McClane BA, Titball RW, Rood JI, Melville SB, Paulsen IT. 2006. Skewed genomic variability in strains of the toxigenic bacterial pathogen, *Clostridium perfringens*. *Genome Res* **16:**1031–1040.

131. Paredes-Sabja D, Setlow P, Sarker MR. 2009. The protease CspB is essential for initiation of cortex hydrolysis and dipicolinic acid (DPA) release during germination of spores of *Clostridium perfringens* type A food poisoning isolates. *Microbiology* **155:**3464–3472.

132. Adams CM, Eckenroth BE, Putnam EE, Doublié S, Shen A. 2013. Structural and functional analysis of the CspB protease required for *Clostridium* spore germination. *PLoS Pathog* **9:**e1003165.

133. Miyata S, Kozuka S, Yasuda Y, Chen Y, Moriyama R, Tochikubo K, Makino S. 1997. Localization of germination-specific spore-lytic enzymes in *Clostridium perfringens* S40 spores detected by immunoelectron microscopy. *FEMS Microbiol Lett* **152:**243–247.

134. Setlow P. 2003. Spore germination. *Curr Opin Microbiol* **6:**550–556.

135. Cowan AE, Olivastro EM, Koppel DE, Loshon CA, Setlow B, Setlow P. 2004. Lipids in the inner membrane of dormant spores of *Bacillus* species are largely immobile. *Proc Natl Acad Sci USA* **101:**7733–7738.

136. Altschul SF, Gish W, Miller W, Myers EW, Lipman DJ. 1990. Basic local alignment search tool. *J Mol Biol* **215:**403–410.

137. Schmidt TR, Scott EJ II, Dyer DW. 2011. Whole-genome phylogenies of the family *Bacillaceae* and expansion of the sigma factor gene family in the *Bacillus cereus* species-group. *BMC Genomics* **12:**430.

138. Stackebrandt E, Rainey FA. 1997. Phylogenetic relationships, p 3–20. *In* Rood JI, McClane BA, Songer JG, Titball RW (ed), *The Clostridia: Molecular Biology and Pathogenesis*. Academic Press, San Diego, CA.

The Bacterial Spore: From Molecules to Systems
Edited by P. Eichenberger and A. Driks
© 2016 American Society for Microbiology, Washington, DC
doi:10.1128/microbiolspec.TBS-0023-2016

Adam Driks[1]
Patrick Eichenberger[2]

The Spore Coat

9

SPORE COAT STRUCTURE: FOLDS AND LAYERS

The coat varies considerably in width among species. In *Bacillus subtilis*, where the coat is relatively wide, it is just less than 200 nm in width, and its multilayered organization is unmistakable by transmission electron microscopy (TEM). Importantly, the number of coat layers and the presence or absence of appendages extending from the coat surface vary among species. This interspecies variation and differences in complexity drew attention as soon as spores were imaged at high resolution, and in the decades since (1–7). The coat is readily distinguished from the cortex (see reference 178) because of its higher electron density. In a large subset of species, the spore also possesses an additional layer surrounding the coat, called the exosporium (Fig. 1; see also references 8 and 9).

In *B. subtilis*, where the coat has been most deeply studied, three layers are visible by TEM (Fig. 1). A fourth layer, the basement layer, is not distinguishable by TEM, but corresponds to the innermost layer atop which the rest of the coat assembles and contains several of the coat proteins required to initiate coat assembly (i.e., SpoIVA, SpoVM, and SpoVID, see below and reference 7). Located on top of the basement layer, the inner coat is characterized by a series of fine, lightly staining lamellae. This morphology is superficially rem-iniscent of a cross section through the myelin sheath of a neuronal axon (while being entirely distinct biochemically). Not all species possess a coat layer with this appearance, but many do, including *Clostridium difficile* (10–12). Surrounding the *B. subtilis* inner coat is a coarsely layered, darkly staining layer known as the outer coat. Finally, the outermost layer of the *B. subtilis* coat is called the crust (13). The crust was detected only recently, because visualizing it requires applying the stain ruthenium red. While not typically used for imaging bacterial spores, ruthenium red was also shown to significantly enhance features in the *Bacillus anthracis* exosporium (see below) (14). It is notable that so many years after the first detailed analyses of coat structure, there is still the opportunity for new discoveries using classical methodologies.

In contrast to the morphologically complex, multilayered *B. subtilis* coat, the *B. anthracis* coat possesses a single compact layer that only occasionally resolves into two layers (4, 15) (to be explicit, here we are not considering the exosporium, which is described in dektail in reference 8). Because the inner layer stains more lightly than the outer one, these layers are also referred to as inner and outer coats (16). Other species have different coat layer arrangements (17), usually with one to three readily discerned layers (Fig. 1). Regardless of the number, all the layers usually appear to

[1]Department of Microbiology and Immunology, Stritch School of Medicine, Loyola University Chicago, Maywood, IL 60153; [2]Center for Genomics and Systems Biology, Department of Biology, New York University, New York, NY 10003.

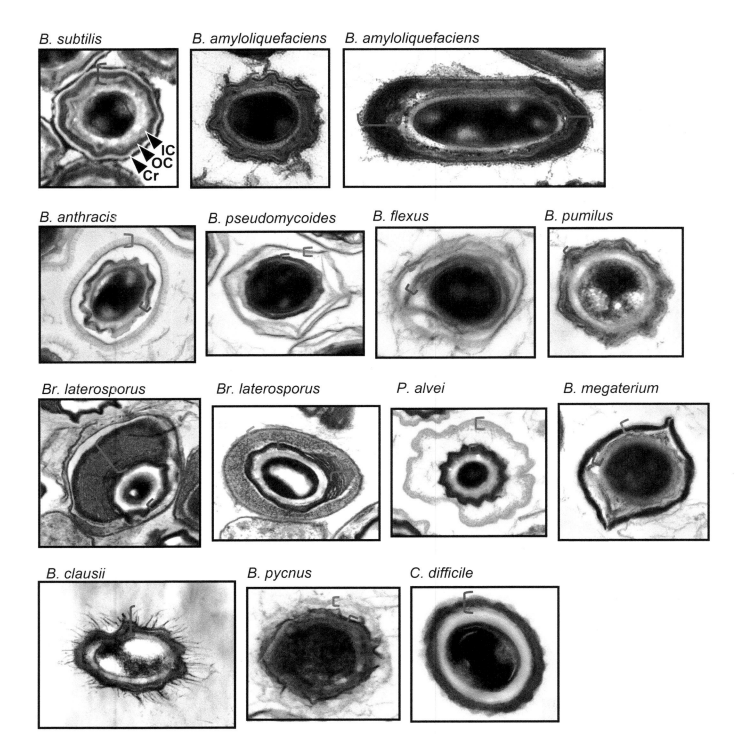

be in close contact in any given spore, suggesting that they are connected by chemical bonds (although TEM images occasionally show *B. subtilis* spores in which the inner and outer layers have partially separated, suggesting that between-layer bonds are weaker than within-layer bonds).

Taken together, these observations raise an important question: what are the benefits to spores of such extreme diversity in coat structure? It is intuitive to propose that this diversity is driven by adaptation to varying environments. While this is very likely, it does not explain the counterintuitive observation that, in many specific environments such as a single soil sample, a large number of highly diverse spore-forming species can be found (see also reference 179). Perhaps, this is due to further levels of diversity in a conventional soil sample. Alternatively (and not mutually exclusively), even in a single niche, there are adaptive benefits to a diverse community of sporeformers. This possibility, in turn, provides a motivation to better understand how the coat facilitates survival in nature. Indeed, a more sophisticated understanding of how spores interact with and survive the challenges of their niches might reveal the functions of species-specific features of the coat and, therefore, their adaptive benefits.

In most species, including *B. subtilis*, *B. anthracis*, and *C. difficile*, the coat appears to possess folds around its entire circumference (Fig. 1). The folds serve a remarkable purpose; by unfolding, they accommodate expansion in the volume of the spore core that occurs when the relative humidity changes (18, 19). It is possible that the ability to respond to changes in interior volume that take place over short timescales is important in natural environments. The dynamic properties of the coat likely have additional important and still poorly appreciated functions during dormancy and germination, as is discussed later.

COAT PROTEIN COMPOSITION

In *B. subtilis*, the coat is composed of at least 80 different proteins (Table 1) (20–22). Coat protein composition has not been characterized in as much detail in other species (even though, in the past decade, several novel coat proteins have been identified in *B. anthracis* and *C. difficile* [23–27]). It is likely, however, that most sporeformers also assemble dozens of protein species in their coat. Comparative genomic analyses indicate that coat proteins are the most species specific of all the sporulation proteins, suggesting that coat composition is highly influenced by the ecosystem in which a species resides. This is not unexpected considering that coat composition may influence the surface properties that will mediate spore dispersal in the environment or attachment to specific hosts or surfaces.

Homologs of *B. subtilis* coat proteins are found primarily in other *Bacillaceae*, more rarely in *Clostridia*, and almost never in non-spore-forming species (28) (see also reference 29). Similarly, many of the coat proteins identified in *C. difficile* (Table 2) are not conserved in *B. subtilis* or other *Bacillaceae* (11, 24–27). The coat morphogenetic proteins, SpoIVA and SpoVM (see below), however, are found in most sporeformers (28). Two proteins that are not limited to spore-forming species are CotA, a multicopper oxidase (30–32), and Tgl, a transglutaminase (33, 34). It is intriguing that so few coat proteins have discernible homologs in the protein databases. This suggests that most coat proteins have functions distinct from other proteins. But it also raises the possibility that assembly of the coat, like that of the flagellum or the pilus, is governed by a distinctive mechanism that makes the unique mechanical properties and functions of the coat (and perhaps additional as-yet-undiscovered structures) possible. If this view is correct, then the study of coat assembly could reveal fundamentally novel mechanisms of macromolecular assembly.

Figure 1 Thin-section TEM analysis of spores from diverse species. Spores were prepared as described in McKenney et al. (13). Images in the top row were fixed using ruthenium red. Other images were conventionally fixed. Images are not to scale; each image was sized to facilitate comparison. Two images of *Bacillus amyloliquefaciens* are shown (one showing a section along the long axis, the other showing a section along the short axis) to point out the thick caps of coat at the poles. The difference in thickness between the two caps is a consistent feature of this species. Two images of *Brevibacillus laterosporus* are also shown to emphasize the variation in morphology of the distinctive structure (indicated with a brown bracket) associated with the coat. The mother cell envelope, which is still present in these two spores, is indicated with a green bracket. The image of *Clostridium difficile* is taken from Semenyuk et al. (142). The crust (Cr), outer coat (OC), and inner coat (IC) are indicated in the image of *Bacillus subtilis* in the upper left. The coat and, where it is present, the exosporium are indicated with blue and red brackets, respectively. The image of *Bacillus megaterium* is courtesy of Dr. Joel Bozue at the U.S. Army Medical Research Institute of Infectious Diseases (USAMRIID).

Table 1 *B. subtilis* coat proteins (strain 168)

Name	Paralogs	Operon	Regulators (transcription)[a]	Function	Domains[b]	References
A. Morphogenetic coat proteins						
CotE (BSU17030)		cotE	σ^E/σ^K GerE(−)	Assembly of the outer coat (major)		21, 84, 110, 125, 143
CotH (BSU36060)		cotH	σ^E/σ^K SpoIIID(+) GerE(−)	Assembly of the outer coat (minor)		121, 122, 124, 127–129
CotO (YjbX) (BSU11730)		cotO	σ^E	Assembly of the outer coat (minor)		123, 126
CotX (BSU11760)		cotX	σ^E/σ^K GerE(+) GerR(+)	Assembly of the crust		13, 112
CotY (BSU11750)	CotZ	cotYZ	σ^E/σ^K GerE(+) GerR(+)	Assembly of the crust		13, 112, 113, 116, 144
CotZ (BSU11740)	CotY	cotYZ	σ^E/σ^K GerE(+) GerR(+)	Assembly of the crust		13, 21, 112, 113, 144
SafA (YrbA) (BSU27840)		safA-coxA	σ^E	Assembly of the inner coat	LysM	13, 108, 109, 119, 145, 146
SpoIVA (BSU22800)		spoIVA	σ^E SpoIIID(−)	Spore cortex formation, coat assembly and anchoring	ATPase (ATP-dependent irreversible polymerization)	94, 95, 100, 102, 103, 107, 147, 148
SpoVM (BSU15810)		spoVM	σ^E	Spore cortex formation, coat assembly, spore encasement	Atypical amphipathic α-helix (NMR structure: 2MVH and 2MVJ)	97–101, 149
SpoVID (BSU28110)		spoVID-ysxE	σ^E	Spore encasement	LysM	93, 96, 104, 107, 120
B. Other coat proteins						
B.1. Basement layer (SpoIVA-dependent, SafA-independent, CotE-independent)						
CotJA (BSU06890)		cotJ-yesJ	σ^E			150, 151
CotJB (BSU06900)		cotJ-yesJ	σ^E			150, 151
CotJC (BSU06910)	YdbD YjqC	cotJ-yesJ	σ^E	May provide protection against oxidative stress	Mn catalase (ferritin-like domain)	150, 151
CmpA (BSU09770)		cmpA	σ^E SpoIIID(+)	Adaptor for ClpXP protease, promotes degradation of SpoIVA if coat is misassembled		105, 106
LipC (YcsK) (BSU04110)		lipC	σ^K GerE(+)	Spore lipolytic enzyme	Phospholipase (GDSL lipolytic enzyme family)	59, 60
YhaX (BSU09830)		yhaX-hemZ	σ^E SpoIIID(−)		Haloacid dehydrogenase superfamily	126
YheD (BSU09770)	YheC	ybeCD	σ^E SpoIIID(+)		ATP binding	152
YjzB (BSU11320)		yjzB	σ^K GerE(+)			13, 21
YppG (BSU22250)		yppG	σ^E/σ^K GerE(+)			13, 21

B.2. Inner layer (SpoIVA- and SafA-dependent)

Protein (BSU)	Other names	Gene/operon	Regulation	Function	Domain/structure	References
CotD (BSU22200)		cotD	σ^K SpoIIID(−) GerE(+) GerR(+)			30
CotF (BSU40530)		cotF	σ^E/σ^K SpoIIID(+) GerE(−)		Ferritin-like domain	130
CotP (YdfT) (BSU05550)		ydgBA-cotP	σ^K GerE(+)			153
CotT (BSU12090)		cotT	σ^E/σ^K SpoIIID(+) GerE(−)		α-crystallin (Hsp20 family)	154
CwlJ (YcbQ) (BSU02600)	SleB YkvT	cwlJ	σ^E/σ^K SpoIIID(+)	Spore cortex lytic enzyme	Cell wall hydrolase	49–51
GerQ (YwdL) (BSU37920)		gerQ	σ^E	CwlJ inhibitor. Not to be confused with the germinant receptor GerQ in *B. cereus*		52–54
OxdD (YoaN) (BSU18670)		oxdD	σ^K GerE(−)	May provide protection against toxic compounds	Oxalate decarboxylase (Cupin superfamily)	155
Tgl (BSU31270)		tgl	σ^E/σ^K GerE(−)	Introduction of cross-links in the coat (GerQ)	Transglutaminase (X-ray diffraction structure: 4P8I and 4PA5)	33, 34, 156
YaaH (SleL) (BSU00160)	YdhD YkvQ YvbX	yaaH	σ^F SpoIIID(−)	N-Acetylglucosaminidase	LysM (2x), glycoside-hydrolase family 18 (X-ray diffraction structure: 4S3J)	13, 157, 158
YeeK (BSU06850)		yeeK	σ^K GerE(+)			159
YhjR (BSU10610)		yhjRQ	σ^E/σ^K GerE(−)		Ferritin-like domain	13, 21
YisY (BSU10900)	YdjP	yisY	σ^E/σ^K GerE(−)		α/β hydrolase superfamily	13, 21, 126
YjqC (BSU12490)	CotJC YdbD	yjqC	σ^K GerE(−)	May provide protection against oxidative stress	Mn catalase (ferritin-like domain)	13, 21
YmaG (BSU17310)		ymaG	σ^K GerE(+) GerR(+)			13, 21
YsnD (BSU28320)		ysnD	σ^E/σ^K SpoIIID(+) GerE(−)			13, 21
YsxE (BSU28100)	CotI CotS	spoVID-ysxE	σ^E	BSK (bacterial spore kinase)	Kinase-like domain	13, 21, 78
YutH (BSU32270)		yutH	σ^E SpoIIID(−)	BSK (bacterial spore kinase)	Kinase-like domain	78, 152
YuzC (BSU31730)		yuzC	σ^E			13, 21, 126
YxeE (BSU39580)		yxeE	σ^K GerE(+)			160
Yybl (BSU40630)		yybl	σ^E			13, 21, 126

B.3. Outer layer (SpoIVA- and CotE-dependent)

Protein (BSU)	Other names	Gene/operon	Regulation	Function	Domain/structure	References
CotA (BSU06300)		cotA	σ^K GerE(−)	Spore pigmentation	Multicopper oxidase (laccase); (X-ray diffraction structure: 2X87, 2BHF)	30–32, 45, 125
CotB (BSU36050)	YwrJ	cotB-ywrJ	σ^E/σ^K SpoIIID(+) GerE(+)			30, 125, 127
CotC (BSU17700)	CotU	cotC	σ^K SpoIIID(−) GerE(+)			30, 161, 162

(Continued)

Table 1 *B. subtilis* coat proteins (strain 168) *(Continued)*

Name	Paralogs	Operon	Regulators (transcription)[a]	Function	Domains[b]	References
CotG (BSU36070)		cotG	σ^E/σ^K GerE(+) GerR(+)		α-Crystallin	127, 128, 163, 164
CotM (BSU17970)		cotM	σ^E/σ^K GerE(−)		FAD-linked oxidoreductase	165
CotQ (YvdP) (BSU34520)		cotQ	σ^K GerE(+)		Kinase-like domain	13, 21, 23
CotS (BSU30900)	CotI YutH	cotSA-cotS-ytxO	σ^K GerE(+)	BSK (bacterial spore kinase)		78, 166, 167
CotU (YnzH) (BSU17670)	CotC	cotU	σ^K GerE(+) GerR(−)			23, 162, 168
GerT (YozR) (BSU19490)		gerT	σ^K GerE(−)	Not to be confused with the GerT antiporter in *B. cereus*	Hsp20 family	169
SpsB(BSU37900)		sps	σ^K			71
SpsI (BSU37840)	YfnH YtdA	sps	σ^E/σ^K	Spore polysaccharide synthesis (rhamnose synthesis)	Glucose-1-phosphate thymidylyltransferase (X-ray diffraction structure: 3HL3, 4ECM)	71–73
YknT (Cse15) (BSU14250)		yknT	σ^E SpoIIID(+)		Endonuclease-like domain (X-ray diffraction structure: 1KNV)	13, 21, 170
YlbD (BSU14970)		ylbDE	σ^K GerE(+)			13, 21
YncD (AlrB) (BSU17640)	Alr	yncD	σ^E	Conversion of L-Ala to D-Ala	Alanine racemase	13, 21
YkzQ (BSU13789)		ykvP-ykzQ-ykvQ-ykzR	σ^K GerE(+)		LysM	72
YtdA (BSU30850)	SpsI YfnH	ytdA	σ^K	Spore polysaccharide synthesis (putative)	Glucose-1-phosphate uridylyltransferase	72
YtxO (BSU30890)		cotSA-cotS-ytxO	σ^K GerE(+)			13, 21
B.4. Crust (SpoIVA, CotE-, and CotX/Y/Z-dependent)						
CgeA (BSU19780)		cgeAB	σ^K GerE(+), GerR(+)			55, 111, 171
CgeB (BSU19790)		cgeAB	σ^K GerE(+) GerR(+)			171
CotV (BSU11780)		cotVW	σ^E/σ^K GerE(+) GerR(+)			111–113, 116, 144
CotW (BSU11770)		cotVW	σ^E/σ^K GerE(+) GerR(+)			13, 111–113, 116, 144
B.5. Layer-specific localization not determined						
CotI (YtaA) (BSU30920)	CotS YutH	cotI	σ^K GerE(+)	BSK (bacterial spore kinase)	Kinase-like domain (X-ray diffraction structure: 2Q83)	23, 78
CotR (YvdO) (BSU34530)		cotR	σ^E/σ^K GerE(+)	Spore lipolytic enzyme	Phospholipase (papatin family)	125
CotSA (YtxN) (BSU30910)	YtcC	cotSA-cotS-ytxO	σ^K GerE(+)	Transfer of glycosyl groups	Glycosyltransferase 1 family	172

Protein (BSU)	Interacting proteins	Gene/operon	Regulation	Function	Domain/family (structure)	Reference [b]
GerPA (YisH) (BSU10720)	GerPF YdgA YdgB	*gerP*	σ^{K}GerE(−)			61, 62
GerPB (YisG) (BSU10710)		*gerP*	σ^{K}GerE(−)			61, 62
GerPC (YisF) (BSU10700)		*gerP*	σ^{K}GerE(−)			61, 62
GerPD (YisE) (BSU10690)		*gerP*	σ^{K}GerE(−)			61, 62
GerPE (YisD) (BSU10680)		*gerP*	σ^{K}GerE(−)			61, 62
GerPF (YisC) (BSU10670)	GerPA YdgA YdgB	*gerP*	σ^{K}GerE(−)			61, 62
SodA (BSU25020)		*sodA*	Not under sporulation control		Superoxide dismutase (X-ray diffraction structure: 2RCV)	164, 173
SscA (YhzE) (BSU09958)		*sscA*	σ^{K}GerE(+), GerR(+)			174
SpsA (BSU37910)	CgeD	*sps*	σ^K	Nucleotide-sugar-dependent glycosyltransferase	Glycosyltransferase 2 family (X-ray diffraction structure: 1QG8)	71, 76
SpsC (BSU37890)		*sps*	σ^K			71, 75
SpsK (BSU37820)		*sps*	σ^K	dTDP-4-dehydrorhamnose reductase	(X-ray diffraction structure: 3SC6)	71–73
YabG (BSU00430)		*yabG*	σ^{K}GerE(−)	Protease, substrates are in the inner coat layer (CotF, CotT, SafA, YeeK, YxeE)	Peptidase U57 family	131, 132
YdgA (BSU05560)	GerPA GerPF YdgB	*ydgBA-cotP*	σ^{K}GerE(−)			153
YdgB (BSU05570)	GerPA GerPF YdgA	*ydgBA-cotP*	σ^{K}GerE(−)			153
YdhD (BSU05710)	YaaH YkvP YvbX	*ydhD*	σ^E	Glycosylase	LysM (2x), Glycoside hydrolase family 18 (X-ray diffraction structure: 3CZ8)	175
YhbB (BSU08920)		*yhbB-cspR*	σ^E		Amidase	126
YheC (BSU09780)	YheD	*yheCD*	σ^{E}SpoIIID(+)		ATP binding	152
YkvP (BSU13780)		*ykvP-ykzQ-ykvQ-ykzR*	σ^{K}GerE(+)		LysM	175
YkvQ (BSU13790)	YaaH YdhD YvbX	*ykvP-ykzQ-ykvQ-ykzR*	σ^{K}GerE(+)	Glycosylase	Glycoside hydrolase family 18	175
YwrJ (BSU36040)	CotB	*cotB-yurJ*	σ^E/σ^{K}SpoIIID(+) GerE(+)			86

[a]Transcriptional regulation is based on information compiled in references 21 and 72.
[b]Numbers associated with protein structures are from the Protein Data Bank.

Table 2 C. *difficile* coat proteins (strain 630)

Name	Homologous coat proteins in B. subtilis	Operon	Regulators (transcription)[a]	Function	Domains	References
Alr2 (CD630_31630)	YncD	alr2	σ^E	Coat localization not confirmed	Alanine racemase	65, 66
BclA1 (CD630_03320)	No	bclA1	σ^K		Collagen-like	176
BclA2 (CD630_32300)	No	bclA2	σ^K		Collagen-like	176
BclA3 (CD630_33490)	No	bclA3	σ^K		Collagen-like	176
CD630_02130	CotF	cd0214-cd0213	σ^E			24
CD630_28640	YisY	cd2864	σ^E		α/β-Hydrolase superfamily	24
CD630_35690	YabG	cd3569	σ^K		Peptidase U57 family	24
CdeA (CD630_23750)	No	cdeA	σ^K			27
CdeB (CD630_27520)	No	cdeB				27
CdeC (CD630_10670)	No	cdeC	σ^K	Outer layer assembly		177
CdeM (CD630_15810)	No	cdeM	σ^K	Outer layer assembly		27
CotA (CD630_16130)	No	cotA	σ^K	Not to be confused with CotA in B. subtilis		25
CotB (CD630_15110)	No	cotB	σ^E	Not to be confused with CotB in B. subtilis		25
CotE (CD630_14330)	YkvQ YdhD YaaH	cotE	σ^K	Not to be confused with CotE in B. subtilis	Peroxiredoxin (glycoside hydrolase family 18)	25
CD630_05960	No	cd0596-cotJB1-cotJC1	σ^K			24
CotJB1 (CotF) (CD630_05970)	CotJB	cd0596-cotJB1-cotJC1	σ^K	Not to be confused with CotF in B. subtilis		25
CotJC1 (CotCB) (CD630_05980)	CotJC YdbD YjqC	cd0596-cotJB1-cotJC1	σ^K	Not to be confused with CotC in B. subtilis	Mn catalase (ferritin-like domain)	25
CD630_23990	No	cd2399-cotJB2-cotJC2	σ^K			24
CotJB2 (CD630_24000)	CotJB	cd2399-cotJB2-cotJC2	σ^K			24
CotJC2 (CotD) (CD630_24010)	CotJC YdbD YjqC	cd2399-cotJB2-cotJC2	σ^K	Not to be confused with CotD in B. subtilis	Mn catalase (ferritin-like domain)	25
SipL (CD630_35670)	No	sipL	σ^E	Spore encasement (functional homolog of B. subtilis SpoVID), interacts with SpoIVA	LysM	26
SleC (CD630_05510)	No	sleC	σ^K	Spore cortex lytic enzyme	Peptidoglycan binding domain	57, 58
SodA (CD630_16310)	SodA	sodA	Not under sporulation control		Superoxide dismutase	25
SpoIVA (CD630_26290)	SpoIVA	spoIVA	σ^E	Spore coat assembly, dispensable for cortex formation	ATPase (ATP-dependent irreversible polymerization)	26

[a]Transcriptional regulation is based on information from references 65 and 66.

FUNCTIONS OF THE COAT: PROTECTION, CONTROL OF GERMINATION, AND INTERACTIONS WITH ENVIRONMENT

The coat protects against a wide range of assaults, including digestion by predatory microbes and challenges by various chemicals (35–40). The mechanisms of these resistance properties remain obscure (see also reference 41), but it seems reasonable to assume that, at least to a significant degree, the coat acts as a barrier, passively excluding degradative enzymes and toxic molecules. However, the coat is not impermeable, because germinants pass through it en route to the germinant receptors in the inner membrane (see reference 180). Since the coat is porous, protection against small toxic molecules is unlikely to be due to a simple barrier function. It is possible that coat porosity increases its effective surface area (see reference 181). If the coat surface can also detoxify reactive molecules (42, 43), then the rather nonspecific protection that the coat confers against these molecules could be explained. On the other hand, more specific resistance properties, for example, against oxidative compounds, could be provided by the catalases and other enzymes that the coat possesses. Notably, catalases are found in the coat of species as distantly related as *B. subtilis* and *C. difficile* (Tables 1 and 2).

Although it is well documented that the coat protects the spore, it has been challenging to determine the role of individual coat proteins in any specific protective function (Table 1). In most cases, inactivating any coat protein gene has no measurable effect on spore resistance properties. For instance, as mentioned above, the *B. subtilis* spore coat contains the multicopper oxidase CotA, which is responsible for the characteristic pigmentation of spore-forming colonies (31, 32, 44, 45). Even though *cotA* mutant spores have altered surface topography (46), they do not seem to be severely impaired in their resistance properties. Not all coat protein gene mutations are phenotypically silent; these exceptions are in the morphogenetic coat protein genes, controlling assembly of the other coat proteins (see below). Therefore, while it can be inferred that protection is compromised when certain large subsets of coat proteins are absent (cases that always result in a major structural defect), it is unclear whether the coat's protective role depends on specific biochemical or structural properties of the individual coat proteins or requires that the coat reach a certain thickness and degree of architectural complexity.

Studies of spores lacking one or another coat layer (achieved, in initial studies, by chemical treatment [4, 47] and, later, by genetic methods [48]) revealed another function for the coat: modulating germination. CwlJ is a *B. subtilis* coat protein with a relatively well-understood role in germination (49–51). It is a cell wall hydrolase that, along with SleB, degrades the cortex peptidoglycan during stage II of germination (see reference 180). During dormancy, CwlJ is stored in the inner layer of the *B. subtilis* spore coat, where it is held inactive, most likely by forming a complex with the coat protein GerQ (not to be confused with the germination receptor GerQ in the so-called *B. cereus* group of species that includes *B. anthracis*) (52). Interestingly, GerQ is cross-linked by the transglutaminase Tgl; however, *tgl* mutant spores do not exhibit any readily detected defect in known coat functions, whereas *gerQ* mutant spores show germination defects (53, 54). Another cell wall hydrolase called YaaH in *B. subtilis* (SleL in *B. anthracis*; see reference 178) is present in the inner coat (13, 55) and contributes to peptidoglycan hydrolysis upon germination (56). Importantly, *C. difficile* spores also contain a spore-cortex lytic enzyme, SleC (CD630_05510) (57, 58). In addition to cell wall hydrolases, the *B. subtilis* spore coat contains two phospholipases (LipC and CotR) and at least one of them (LipC) is required for efficient germination. It has been hypothesized that LipC could play a role in the degradation of the spore outer membrane (59, 60).

The GerPA-GerPF group of coat proteins has also been implicated in germination, but their roles remain obscure (61). The *gerP* operon consists of six genes expressed at a late stage in sporulation. Initial studies showed that *gerP* mutant spores have a defect in receptor-mediated germination, and that this defect is overcome when a major portion of the coat is removed. These experiments are consistent with the view that GerP proteins reside in the coat and facilitate the passage of germinants. A more recent study confirms these observations (62). In particular, it strengthens the view that GerP proteins somehow mediate the flow of germinants through the coat. How this could occur remains unclear, but the *gerP* mutation reveals an intrinsic impermeability of the coat to small molecule nutrient germinants. It is possible that GerP proteins form specialized structures within the coat that act as channels for these molecules. Alternatively, GerP proteins could be distributed throughout the coat, globally altering its chemical properties such that germinant molecules pass through easily en route to the germinant receptors in the inner membrane.

An intriguing further possible role for the coat in germination is suggested by the presence of alanine racemases. Because alanine racemases can convert the germinant L-alanine to D-alanine, which does not stimulate germination in species studied so far, alanine racemases can reduce or inhibit germination when L-alanine is present. Such a role has been demonstrated in *B. anthracis* for an exosporium-associated alanine racemase (63). However, so far, no corresponding phenotype has been observed for a mutation in the *B. subtilis* coat-associated alanine racemase, YncD (64). Because there is more than one alanine racemase in *B. subtilis*, redundancy is a possible confounding factor. Interestingly, *C. difficile* also expresses a gene encoding an alanine racemase during sporulation (65, 66), even though amino acids are not sufficient to trigger germination in that species.

All spores can germinate, but the responses to specific germinants are species specific (180). For instance, in *C. difficile*, where germination must occur in the gastrointestinal tract of the host, bile is a key germinant (67–69). Furthermore, it is interesting to note that proteins involved in germination are located both in the spore interior, a structure that does not otherwise vary dramatically among species, and in the coat, which varies considerably. As already discussed, we suggest that coat variation is adaptive in survival in diverse environments. It is possible that variation in the coat also facilitates adaptive evolution of germination. In particular, it is appealing to speculate that variation in the coat provides the opportunity for optimizing the response to germinants.

There are reasons to speculate that the coat has roles beyond protection and germination. First, in some species, TEM reveals coat structures with no known functions. These include, for example, the spikelike structures that decorate the *Bacillus clausii* coat (17) (Fig. 1). Second, in species lacking the exosporium, the coat is the outermost structure and, therefore, mediates interactions with surfaces in the environment (70). Understanding these functions in any species will require better characterization of the ecologically relevant stresses that spores encounter. Therefore, future research in coat function may be driven, to a large degree, by advances in microbial ecology.

Recent work has revealed that the *B. subtilis* coat contains enzymes involved in polysaccharide synthesis (71, 72). In particular, two glucose-1P nucleotidyltransferases, SpsI and YtdA, are produced in the mother cell late in sporulation and localize to the outer layer of the spore coat. The *sps* (spore polysaccharide synthesis) operon, which includes *spsI*, is partially conserved in *B. anthracis*. In both species, it is required for production of rhamnose (73), a prominent carbohydrate of the spore surface (74). Furthermore, a specific sequence has been identified by phage display as required for docking of SpsC to the forespore surface (75), while the

crystal structure of SpsA, a nucleotide-diphospho-sugar transferase, has been reported (76). It has also been shown that spore surface properties are influenced by the presence of spore polysaccharides. For instance, disruption of *spsM* or *spsI* renders *B. subtilis* spores more hydrophobic, thus affecting spore dispersal properties (71, 72, 77). We consider spore surface properties in more detail below.

Another group of coat proteins that includes CotI (YtaA), CotS, YutH, and YsxE has been investigated recently in *B. subtilis* (78). These proteins are collectively referred to as bacterial spore kinases (BSKs), because they share a structural motif also found in eukaryotic kinases. Importantly, homologous proteins are present in other spore-forming species. It has been suggested that the majority of these BSKs are catalytically inactive because they lack some key amino acids in the active site; however, they may be able to interact with specific substrates. Interestingly, in both *B. subtilis* and *B. anthracis*, *cotI* and *cotS* are adjacent to the *ytdA* locus (which is likely involved in spore polysaccharide synthesis, see above). The simultaneous presence of BSKs and genes involved in polysaccharide synthesis or glycosyl transfer is also observed in some *Clostridia* species (78).

COAT ASSEMBLY

Coat assembly is arguably the most complex among all the assembly processes that occur during spore formation, considering that as much as 1 to 2% of the *B. subtilis* genome encodes coat proteins and at least 15% of all sporulation proteins are coat proteins. Moreover, because several hours are necessary for the synthesis, assembly, and maturation of the coat, it is also one of the most extended multiprotein assembly processes in bacteria.

Morphologically distinct stages of sporulation in *B. subtilis* (Fig. 2; see also Fig. 1 from reference 79) were identified in the 1960s by TEM analysis, and a large number of mutants blocked at these stages were isolated, and their mutations were mapped to specific loci in early studies (80, 81). About 2 h into sporulation, sporulating cells divide asymmetrically to produce a small forespore and a larger mother cell. Shortly after asymmetric division, two cell-specific lines of gene expression are established: one in the forespore, initiated under the control of the alternative sigma factor, σ^F, and the other in the mother cell, under the control of another alternative sigma factor, σ^E. About half of the known coat proteins are produced from genes that belong to the σ^E regulon (21); thus, synthesis of the first group of coat proteins commences shortly after asymmetric division, specifically in the mother cell. The next developmental stage corresponds to the engulfment of the forespore by the mother cell in a process analogous to phagocytosis. Once engulfment is complete, the forespore is physically separated from the mother cell cytoplasm by two membranes (the inner and outer forespore membranes). This morphological transition is accompanied by a transcriptional switch in both the forespore (where σ^G takes over from σ^F) and the mother cell (where σ^K replaces σ^E). Hence, synthesis of the second group of spore coat proteins is controlled by σ^K, also exclusively in the mother cell. Additional transcription factors contribute to the regulation of expression of coat genes: SpoIIID in the σ^E regulon and GerE in the σ^K regulon with minor contributions of GerR (82–89). The ultimate stage of sporulation corresponds to the lysis of the mother cell and the release of the spore in the environment. Nevertheless, coat maturation appears to continue even after mother cell lysis, in particular, because cross-linking of coat proteins may be ongoing for several days (22, 53, 90, 91).

This intricate transcriptional regulation has consequences for the production and subcellular localization dynamics of coat proteins in the mother cell (92). With the use of epifluorescence microscopy to analyze fusions of coat proteins to fluorescent proteins, six classes of coat proteins were defined based on localization kinetics (21) (Fig. 3). Most coat proteins localize to the nascent coat in two steps: initially, to the mother cell proximal (MCP) pole of the forespore (the only pole accessible before completion of engulfment) and, subsequently, to the mother cell distal (MCD) pole. The first three classes are composed of σ^E-dependent proteins, whose synthesis begins before completion of engulfment, whereas the last three correspond to σ^K-dependent proteins (i.e., produced after engulfment, when σ^K becomes active). The second factor taken into account in this classification is the stage when coat proteins complete spore encasement (in other words, the amount of time necessary to transition from a single polar cap to a complete shell surrounding the forespore [93]). Class I proteins complete encasement shortly after engulfment, as soon as the MCD pole becomes available, class II proteins when spores become visible by phase-contrast microscopy (during the main stage of cortex synthesis), and class III proteins when phase dark spores turn bright. The degree of spore brightness is correlated with the degree of dehydration of the spore core (i.e., late in sporulation, Ca^{2+}-dipicolinic acid produced in the mother cell is imported in the forespore to replace water, see references 41 and 178 and Fig. 2).

The three classes of σ^K-dependent coat proteins are distinguished based on the same morphological transitions: class IV proteins complete encasement as soon as they are produced, class V when spores are still dark, and class VI when spores become phase bright.

In addition to transcriptional regulation, coat assembly is controlled by coat morphogenetic proteins. Deposition on and around the outer forespore membrane is dependent on the coat morphogenetic protein SpoIVA, which functions as a coat anchor (94, 95). Sporulating cells harboring a mutation in *spoIVA* complete engulfment, but fail to synthesize the cortex. In addition, swirls of coat material are readily detected by TEM in the mother cell cytoplasm. Interestingly, these coat fragments retain the multilayer aspect of the mature coat, suggesting that to a significant degree, coat assembly does not depend on contact with the forespore. A related phenotype (i.e., presence of coat swirls in the cytoplasm) is observed in a *spoVID* mutant (96), albeit with an important difference, the presence of a cortex, implying that, unlike SpoIVA, SpoVID is not involved in cortex formation. In addition to SpoIVA and SpoVID, the initiation of coat assembly is further controlled by the 26-amino-acid peptide, SpoVM, which is also necessary for cortex production (97). The structure of SpoVM has recently been solved by nuclear magnetic resonance (NMR) and corresponds to an atypical amphipathic helix. This geometry is consistent with a model where SpoVM would insert deeply into membranes where it could sense variations in acyl chain packing resulting from membrane curvature (98).

It has been suggested that SpoVM is able to sense positive membrane curvature and therefore preferentially localizes to the convex outer membrane of the forespore, as opposed to the plasma membrane of the mother cell, which is negatively curved around the periphery (99). This hypothesis is well supported by *in vitro* experiments demonstrating preferential localization of SpoVM-green fluorescent protein (GFP) to membrane vesicles of similar size to the *B. subtilis* forespore and to 2-μm silica beads coated with lipid bilayers (100). In sporulating cells, however, *spoIVA* is necessary for preferential localization of SpoVM-GFP to the outer forespore membrane (93, 101). Although it is yet unclear if there is a hierarchy in the recruitment of SpoVM and SpoIVA to the outer forespore membrane, both proteins are essential for proper initiation of coat assembly.

The biochemical characterization of SpoIVA has revealed that it is an ATPase and further demonstrated that ATP hydrolysis is necessary for irreversible polymerization of SpoIVA into cablelike structures (102,

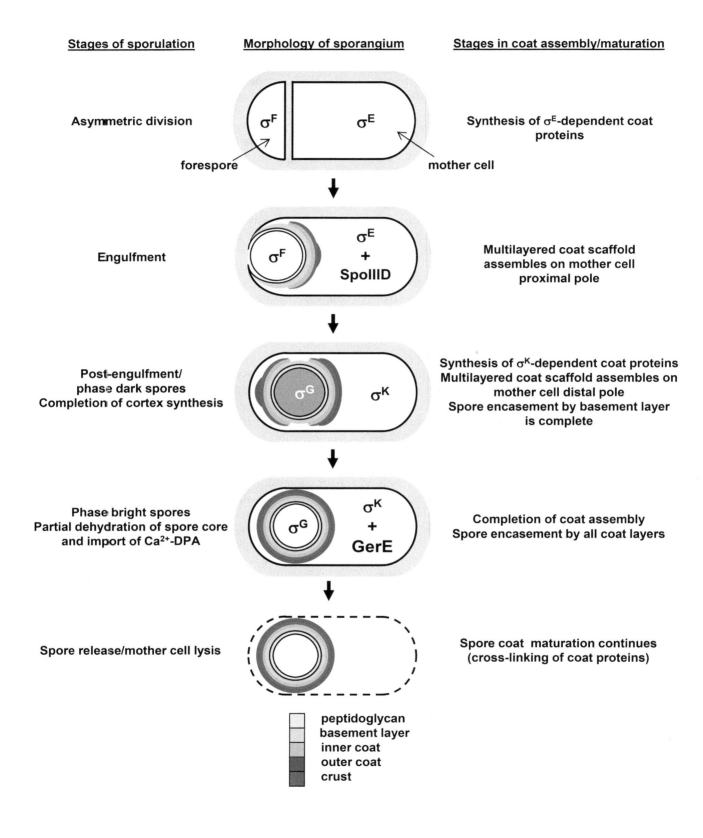

Stages of sporulation   Morphology of sporangium   Stages in coat assembly/maturation

Asymmetric division   σ^F σ^E Synthesis of σ^E-dependent coat proteins

forespore   mother cell

Engulfment   σ^F σ^E + SpoIIID   Multilayered coat scaffold assembles on mother cell proximal pole

Post-engulfment/ phase dark spores Completion of cortex synthesis   σ^G σ^K Synthesis of σ^K-dependent coat proteins Multilayered coat scaffold assembles on mother cell distal pole Spore encasement by basement layer is complete

Phase bright spores Partial dehydration of spore core and import of Ca^{2+}-DPA   σ^G σ^K + GerE   Completion of coat assembly Spore encasement by all coat layers

Spore release/mother cell lysis   σ^G Spore coat maturation continues (cross-linking of coat proteins)

peptidoglycan
basement layer
inner coat
outer coat
crust

103). Addition of SpoVM and SpoIVA to beads coated with membranes is sufficient to reconstitute the basement layer of the coat *in vitro* (100). In sporulating cells, however, SpoIVA-GFP will not complete spore encasement in the absence of *spoVID* (93). The view that SpoIVA assembly requires SpoVID is further supported by biochemical data implying that SpoIVA and SpoVID physically interact (104). Similarly, SpoIVA was shown to interact directly with SpoVM (103). Therefore, the assembly of the innermost layer of the coat (the basement layer) requires the combined contributions of three proteins: SpoIVA, SpoVM, and SpoVID. All three are dependent on σ^E for synthesis and finish spore encasement as soon as engulfment is completed (i.e., they are class I proteins) (21).

Coat assembly and cortex synthesis are largely independent events as they occur in different compartments of the sporulating cell (the intermembrane space for cortex formation versus the mother cell for coat assembly) and involve a different set of sporulation proteins (cell wall synthesis enzymes versus structural proteins). Nevertheless, as mentioned above, SpoVM and SpoIVA are necessary for both processes and provide a link between the two. In fact, a quality control mechanism has recently been characterized, suggesting that sporulating cells with defective coats are eliminated because they are unable to complete cortex synthesis (105, 106). SpoIVA resides at the core of this mechanism and is targeted for proteolysis in response to coat misassembly. The presence of spore coat defects causes the stabilization of CmpA, a small sporulation protein produced under the control of σ^E and SpoIIID (thus, shortly after the synthesis of SpoIVA, SpoVM, and SpoVID). CmpA is an adaptor protein that binds to SpoIVA and delivers it to the ClpXP protease for degradation. After elimination of SpoIVA, cortex formation is inhibited and the sporulating cell eventually lyses. Conversely, in the absence of envelope defects, CmpA is rapidly eliminated by ClpXP-mediated proteolysis.

Assembly of coat layers beyond the basement layer is dependent on additional morphogenetic proteins (107). Importantly, the formation of each layer can be associated with the presence of a specific protein: inner layer assembly is dependent on SafA (108, 109), while CotE controls morphogenesis of the outer layer (110). On top of the outer coat, a group of three proteins, CotX, CotY, and CotZ, is necessary for crust formation (13, 111). CotY and CotZ are cysteine-rich proteins that were shown to interact directly (112, 113) and may be subjected to cross-linking by formation of disulfide bonds. Purified CotY even displayed an ability to self-assemble, forming lattices reminiscent of the structure seen in the exosporium of *B. cereus* group species (114–116). Based on the mapping of genetic dependencies for more than 40 coat protein-GFP fusions, an extended network of interactions between coat proteins has been characterized (13, 117). Coat protein-GFP fusions dependent on *safA* for localization were assigned to the inner coat. Conversely, if localization required *cotE*, these proteins were thought to reside in the outer layer. An approximately equal number of coat proteins were assigned to the inner and outer layers, implying that the protein composition of both layers is of similar complexity. In all cases tested, genetic inferences were supported by image analysis at subpixel resolution (13, 55). A subset of *cotE*-dependent proteins was shown to also be dependent on the *cotX cotYZ* gene cluster and therefore categorized as crust proteins (13, 111). In addition, atomic force microscopy (AFM) was used to compare the morphology of wild-type spores and various spore coat mutants (including *spoVID*, *safA*, and *cotE* mutants [118]). Remarkably, the AFM data were consistent with the coat assembly model inferred from TEM and fluorescence microscopy studies.

Each morphogenetic protein is connected to many individual coat proteins and therefore represents a hub of the coat interaction network (13). Because recruitment of both SafA and CotE is dependent on SpoIVA, whereas recruitment of CotX, CotY, and CotZ is dependent on CotE, the organization of the coat interaction network is highly hierarchical. While SpoIVA is at the top of the hierarchy for recruitment of coat proteins, SpoVM and SpoVID are at the top of the spore encasement hierarchy. Moreover, it has been shown that encasement by CotE-GFP and SafA-GFP is dependent on *spoVM* and *spoVID*, and that SpoVID interacts directly with SafA and CotE (93, 104, 119, 120). Therefore, a model integrating the formation of

Figure 2 Model of spore coat assembly during *B. subtilis* sporulation. In the left column, we list the stages of sporulation as they appear by TEM, phase-contrast microscopy, or fluorescence microscopy in the presence of a membrane stain. The center column contains diagrams of spore coat morphogenesis. Layers of the spore coat are color coded (cyan = basement layer; yellow = inner coat; blue = outer coat; maroon = crust). In the right column, we list the stages of spore coat assembly. DPA, dipicolinic acid. Modified from McKenney et al. (21). See text for details.

multiple coat layers and the dynamics of spore encasement has been proposed (7, 21, 107). In this model (Fig. 2), a scaffold organized in multiple layers is initially assembled on one pole of the forespore. Assembly of this multilayered polar cap is dependent on morphogenetic proteins, which all localize to this structure early during coat assembly (i.e., during engulfment of the forespore by the mother cell). Subsequently, the coat encases the spore in multiple coordinated waves. The coordination of the successive waves is dependent on transcriptional regulation. Importantly, additional coat proteins can be added to this scaffold even after completion of encasement. This property implies that the coat remains permeable to proteins of a certain size

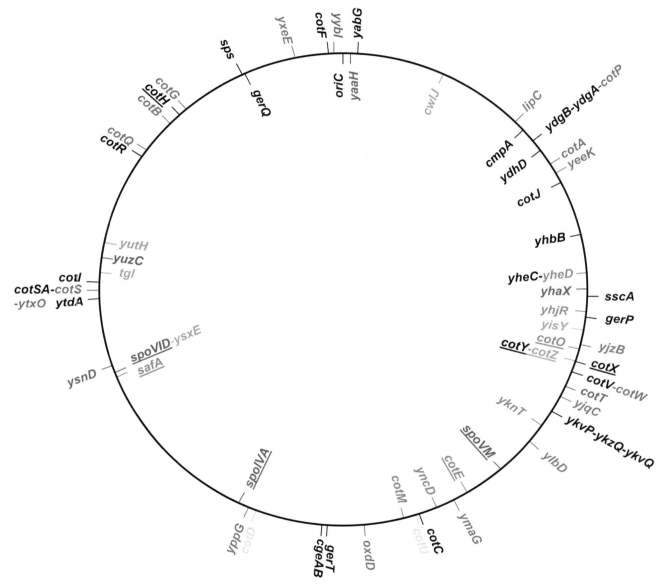

Figure 3 Classes of coat proteins based on localization kinetics. Spore coat genes are displayed according to their localization on the *B. subtilis* chromosome with the origin of replication (*oriC*) on top. Genes whose expression commences before engulfment, under the control of σ^E, are inside the circle; genes whose expression begins after engulfment, under the control of σ^K, are outside the circle. Classes are color coded (red = class 1; brown = class 2; orange = class 3; purple = class 4; blue = class 5; turquoise = class 6). Genes encoding morphogenetic proteins are underlined.

until a late stage in sporulation. For instance, the coat protein CotD is among the last to be produced (class VI), but eventually localizes to the inner coat and will have to cross the nascent outer coat layer to reach its final destination (21).

Morphogenetic proteins in addition to those described above have been documented in *B. subtilis* including, in particular, CotH and CotO (121–123). By TEM, single mutant spores of *cotH* and *cotO* have roughly similar phenotypes, characterized primarily by a disorganized outer coat structure (123). Furthermore, both proteins interact with CotE, since CotH forms a complex with CotE (124, 125), while localization of CotO-GFP (YjbX-GFP) to the coat is dependent on *cotE* (126). Nevertheless, structural differences were noted between the *cotH* and *cotO* mutants by AFM (118, 123). CotB and CotG are two outer coat proteins that have been shown to be dependent on CotH (and possibly CotO) for assembly (127). Interestingly, the *cotH* gene is surrounded by *cotB* and *cotG* on either side, and the 5′ UTR of the *cotH* mRNA overlaps with the divergently transcribed *cotG* mRNA (128). Unexpectedly, a recent study has suggested that a *cotE* mutant could be rescued, at least in part, by the overexpression of *cotH* (129).

Notably, several proteins are proteolyzed in the process of coat assembly. This was noted soon after the first coat proteins were isolated and characterized (130). It is likely that proteases, including the coat protein YabG (and possibly ClpXP, which can degrade SpoIVA), are responsible for at least some proteolytic activity in the coat and/or during coat maturation (106, 131–133). In addition to proteolysis, another important factor in spore coat maturation is protein crosslinking, which, as mentioned above, may occur even after release of the spore from the mother cell (22, 53, 90, 91).

PHYSICAL AND CHEMICAL PROPERTIES

The chemical properties of the spore surface have a major impact on the interactions that spores may establish with cells and surfaces in the environment. Therefore, it is reasonable to assume that these properties play important roles in adaptation to diverse environments. For example, the ability of pathogenic spores to adhere to host tissue, and the transport of spores through soil in the environment will be determined, to a large degree, by the chemical properties of the spore surface. These properties are still incompletely understood, in part, because of the relatively few studies that have been conducted and the complexity of this type

of analysis. In the case of *B. subtilis*, for example, the spore surface has been reported to have hydrophilic or hydrophobic characteristics, depending on the study (70, 77, 134). The differences among these studies are intriguing and underscore the need for a better understanding of which analytic methods are most useful for spore surface analysis. It is important to point out that this discussion does not address exosporium surface chemistry, which is governed by distinct proteins (70, 135–137). As alluded to above, characterizing spore surface chemical properties is important to understanding aspects of coat function in nature. Spore surface properties are also a major concern in several industries including food preparation (138), where industrial machinery can be fouled by bacteria, including spores, that adhere to equipment surfaces and interfere with the flow of materials through the equipment.

Until recently, the mechanical properties of the spore surface have received little if any study. This is somewhat surprising given that a key feature of the coat, its capacity to fold and unfold, has been evident from observations in the literature extending over decades (18, 19, 46, 139, 140). The coat's mechanical properties should be of interest to the microbiologist, as an example of a striking biological adaptation with potentially important implications for cell function, and to the biochemist, for what it might teach us about the molecular basis of flexibility at the nanoscale. In this regard, it is striking to note that electron microscopic data clearly indicate that the layers of the coat are not distorted at a fold or in its vicinity, strongly suggesting that the coat does not experience an irreversible mechanical failure when a fold appears, as might be expected from a simple model of coat structure. The molecular basis of this resiliency remains unknown.

Intriguingly, the folds are irregular; they are not uniformly spaced around the circumference of any given spore, and their sharpness and number vary from spore to spore in a population. These observations suggest that the number and heights of the folds are not dictated by specifically positioned molecular cues within the spore. More likely, the folds are an emergent property resulting from more global spore properties, and the variations in the folds' heights are the result of a variation among spores. A model at least partially explaining fold morphology and variation argues that a key event in the formation of coat folds is the decrease in spore core volume that occurs after cortex maturation (18). This event induces stress energy in the coat that will resolve by the separation of the coat from the forespore surface at several locations, dictated largely by the spore geometry, the coat stiffness (the Young's modulus), and the forces binding the coat to the fore-spore. These separation or delamination events are irreversible (because the cross-linked cortex prevents the core from expanding sufficiently for complete unfolding) and define the locations of the folds for the remainder of each spore's existence. Folds will change in height after spore release, as a function of variations in the ambient relative humidity (139, 140). This model further suggests that during germination, when the cortex is degraded and the core swells because of rehydration, the resulting complete unfolding of the coat facilitates its shedding, allowing for the further increase in cell volume that accompanies the resumption of metabolism. Further analysis of the mechanical and other physical properties of the coat will very likely reveal additional coat functions, which, in turn, should provide important future insights into the coat's roles in survival in diverse environments.

A striking consequence of the analysis just described is the insight that spores can store mechanical energy and be induced to release it very efficiently by repeated cycles of hydration and dehydration (19). This insight, in turn, raises the possibility of using spores to store the energy of evaporation in the natural environment and then releasing this energy to do practical mechanical work (141).

Acknowledgments. We thank Peter McKenney, Bentley Shuster, and Tyler Boone for critical reading of the manuscript.

Citation. Driks A, Eichenberger P. 2016. The spore coat. Microbiol Spectrum 4(3):TBS-0023-2016.

References

1. **Warth AD, Ohye DF, Murrell WG.** 1963. The composition and structure of bacterial spores. *J Cell Biol* **16:** 579–592

2. **Holt SC, Leadbetter ER.** 1969. Comparative ultrastructure of selected aerobic spore-forming bacteria: a freeze-etching study. *Bacteriol Rev* **33:**346–378.

3. **Santo LY, Doi RH.** 1974. Ultrastructural analysis during germination and outgrowth of *Bacillus subtilis* spores. *J Bacteriol* **120:**475–481.

4. **Aronson AI, Fitz-James P.** 1976. Structure and morphogenesis of the bacterial spore coat. *Bacteriol Rev* **40:** 360–402.

5. **Driks A.** 1999. *Bacillus subtilis* spore coat. *Microbiol Mol Biol Rev* **63:**1–20.

6. **Driks A.** 2007. Surface appendages of bacterial spores. *Mol Microbiol* **63:**623–625.

7. **McKenney PT, Driks A, Eichenberger P.** 2013. The *Bacillus subtilis* endospore: assembly and functions of the multilayered coat. *Nat Rev Microbiol* **11:**33–44.

8. **Bozue JA, Welkos S, Cote CK.** 2015. The *Bacillus anthracis* exosporium: what's the big "hairy" deal? *Microbiol*

Spectr 3(5):TBS-0021-2015. doi:10.1128/microbiolspec. TBS-0021-2015.

9. **Stewart GC.** 2015. The exosporium layer of bacterial spores: a connection to the environment and the infected host. *Microbiol Mol Biol Rev* **79:**437–457.

10. **Lawley TD, Croucher NJ, Yu L, Clare S, Sebaihia M, Goulding D, Pickard DJ, Parkhill J, Choudhary J, Dougan G.** 2009. Proteomic and genomic characterization of highly infectious *Clostridium difficile* 630 spores. *J Bacteriol* **191:**5377–5386.

11. **Permpoonpattana P, Tolls EH, Nadem R, Tan S, Brisson A, Cutting SM.** 2011. Surface layers of *Clostridium difficile* endospores. *J Bacteriol* **193:**6461–6470.

12. **Paredes-Sabja D, Shen A, Sorg JA.** 2014. *Clostridium difficile* spore biology: sporulation, germination, and spore structural proteins. *Trends Microbiol* **22:**406–416.

13. **McKenney PT, Driks A, Eskandarian HA, Grabowski P, Guberman J, Wang KH, Gitai Z, Eichenberger P.** 2010. A distance-weighted interaction map reveals a previously uncharacterized layer of the *Bacillus subtilis* spore coat. *Curr Biol* **20:**934–938.

14. **Waller LN, Fox N, Fox KF, Fox A, Price RL.** 2004. Ruthenium red staining for ultrastructural visualization of a glycoprotein layer surrounding the spore of *Bacillus anthracis* and *Bacillus subtilis*. *J Microbiol Methods* **58:** 23–30.

15. **Driks A.** 2009. The *Bacillus anthracis* spore. *Mol Aspects Med* **30:**368–373.

16. **Giorno R, Bozue J, Cote C, Wenzel T, Moody KS, Mallozzi M, Ryan M, Wang R, Zielke R, Maddock JR, Friedlander A, Welkos S, Driks A.** 2007. Morphogenesis of the *Bacillus anthracis* spore. *J Bacteriol* **189:**691–705.

17. **Traag BA, Driks A, Stragier P, Bitter W, Broussard G, Hatfull G, Chu F, Adams KN, Ramakrishnan L, Losick R.** 2010. Do mycobacteria produce endospores? *Proc Natl Acad Sci USA* **107:**878–881.

18. **Sahin O, Yong EH, Driks A, Mahadevan L.** 2012. Physical basis for the adaptive flexibility of *Bacillus* spore coats. *J R Soc Interface* **9:**3156–3160.

19. **Chen X, Mahadevan L, Driks A, Sahin O.** 2014. *Bacillus* spores as building blocks for stimuli-responsive materials and nanogenerators. *Nat Nanotechnol* **9:**137–141.

20. **Henriques AO, Moran CP Jr.** 2007. Structure, assembly, and function of the spore surface layers. *Annu Rev Microbiol* **61:**555–588.

21. **McKenney PT, Eichenberger P.** 2012. Dynamics of spore coat morphogenesis in *Bacillus subtilis*. *Mol Microbiol* **83:**245–260.

22. **Abhyankar W, Pandey R, Ter Beek A, Brul S, de Koning LJ, de Koster CG.** 2015. Reinforcement of *Bacillus subtilis* spores by cross-linking of outer coat proteins during maturation. *Food Microbiol* **45**(Pt A):54–62.

23. **Lai EM, Phadke ND, Kachman MT, Giorno R, Vazquez S, Vazquez JA, Maddock JR, Driks A.** 2003. Proteomic analysis of the spore coats of *Bacillus subtilis* and *Bacillus anthracis*. *J Bacteriol* **185:**1443–1454.

24. Abhyankar W, Hossain AH, Djajasaputra A, Permpoonpattana P, Ter Beek A, Dekker HL, Cutting SM, Brul S, de Koning LJ, de Koster CG. 2013. In pursuit of protein targets: proteomic characterization of bacterial spore outer layers. *J Proteome Res* **12:** 4507–4521.

25. Permpoonpattana P, Phetcharaburanin J, Mikelsone A, Dembek M, Tan S, Brisson MC, La Ragione R, Brisson AR, Fairweather N, Hong HA, Cutting SM. 2013. Functional characterization of *Clostridium difficile* spore coat proteins. *J Bacteriol* **195:**1492–1503.

26. Putnam EE, Nock AM, Lawley TD, Shen A. 2013. SpoIVA and SipL are *Clostridium difficile* spore morphogenetic proteins. *J Bacteriol* **195:**1214–1225.

27. Díaz-González F, Milano M, Olguin-Araneda V, Pizarro-Cerda J, Castro-Córdova P, Tzeng SC, Maier CS, Sarker MR, Paredes-Sabja D. 2015. Protein composition of the outermost exosporium-like layer of *Clostridium difficile* 630 spores. *J Proteomics* **123:**1–13.

28. Galperin MY, Mekhedov SL, Puigbo P, Smirnov S, Wolf YI, Rigden DJ. 2012. Genomic determinants of sporulation in *Bacilli* and *Clostridia*: towards the minimal set of sporulation-specific genes. *Environ Microbiol* **14:** 2870–2890.

29. Galperin MY. 2013. Genome diversity of spore-forming *Firmicutes*. *Microbiol Spectr* **1**(2): TBS-0015-2012. doi: 10.1128microbiolspectrum.TBS-0015-2012.

30. Donovan W, Zheng LB, Sandman K, Losick R. 1987. Genes encoding spore coat polypeptides from *Bacillus subtilis*. *J Mol Biol* **196:**1–10.

31. Hullo MF, Moszer I, Danchin A, Martin-Verstraete I. 2001. CotA of *Bacillus subtilis* is a copper-dependent laccase. *J Bacteriol* **183:**5426–5430.

32. Martins LO, Soares CM, Pereira MM, Teixeira M, Costa T, Jones GH, Henriques AO. 2002. Molecular and biochemical characterization of a highly stable bacterial laccase that occurs as a structural component of the *Bacillus subtilis* endospore coat. *J Biol Chem* **277:** 18849–18859.

33. Zilhão R, Isticato R, Martins LO, Steil L, Völker U, Ricca E, Moran CP Jr, Henriques AO. 2005. Assembly and function of a spore coat-associated transglutaminase of *Bacillus subtilis*. *J Bacteriol* **187:**7753–7764.

34. Fernandes CG, Plácido D, Lousa D, Brito JA, Isidro A, Soares CM, Pohl J, Carrondo MA, Archer M, Henriques AO. 2015. Structural and functional characterization of an ancient bacterial transglutaminase sheds light on the minimal requirements for protein cross-linking. *Biochemistry* **54:**5723–5734.

35. Milhaud P, Balassa G. 1973. Biochemical genetics of bacterial sporulation. IV. Sequential development of resistances to chemical and physical agents during sporulation of *Bacillus subtilis*. *Mol Gen Genet* **125:**241–250.

36. Nicholson WL, Munakata N, Horneck G, Melosh HJ, Setlow P. 2000. Resistance of *Bacillus* endospores to extreme terrestrial and extraterrestrial environments. *Microbiol Mol Biol Rev* **64:**548–572.

37. Klobutcher LA, Ragkousi K, Setlow P. 2006. The *Bacillus subtilis* spore coat provides "eat resistance" during phagocytic predation by the protozoan *Tetrahymena thermophila*. *Proc Natl Acad Sci USA* **103:**165–170.

38. Laaberki MH, Dworkin J. 2008. Role of spore coat proteins in the resistance of *Bacillus subtilis* spores to *Caenorhabditis elegans* predation. *J Bacteriol* **190:** 6197–6203.

39. Carroll AM, Plomp M, Malkin AJ, Setlow P. 2008. Protozoal digestion of coat-defective *Bacillus subtilis* spores produces "rinds" composed of insoluble coat protein. *Appl Environ Microbiol* **74:**5875–5881.

40. Ghosh S, Setlow B, Wahome PG, Cowan AE, Plomp M, Malkin AJ, Setlow P. 2008. Characterization of spores of *Bacillus subtilis* that lack most coat layers. *J Bacteriol* **190:**6741–6748.

41. Setlow P. 2014. Spore resistance properties. *Microbiol Spectr* **2**(5):TBS-0003-2012. doi:10.1128/microbiolspec. TBS-0003-2012.

42. Francis CA, Tebo BM. 1999. Marine *Bacillus* spores as catalysts for oxidative precipitation and sorption of metals. *J Mol Microbiol Biotechnol* **1:**71–78.

43. Francis CA, Tebo BM. 2002. Enzymatic manganese(II) oxidation by metabolically dormant spores of diverse *Bacillus* species. *Appl Environ Microbiol* **68:**874–880.

44. Sandman K, Kroos L, Cutting S, Youngman P, Losick R. 1988. Identification of the promoter for a spore coat protein gene in *Bacillus subtilis* and studies on the regulation of its induction at a late stage of sporulation. *J Mol Biol* **200:**461–473.

45. Bento I, Silva CS, Chen Z, Martins LO, Lindley PF, Soares CM. 2010. Mechanisms underlying dioxygen reduction in laccases. Structural and modelling studies focusing on proton transfer. *BMC Struct Biol* **10:**28.

46. Chada VG, Sanstad EA, Wang R, Driks A. 2003. Morphogenesis of *bacillus* spore surfaces. *J Bacteriol* **185:** 6255–6261.

47. Aronson AI, Fitz-James PC. 1971. Reconstitution of bacterial spore coat layers *in vitro*. *J Bacteriol* **108:**571–578.

48. Moir A. 1981. Germination properties of a spore coat-defective mutant of *Bacillus subtilis*. *J Bacteriol* **146:** 1106–1116.

49. Bagyan I, Setlow P. 2002. Localization of the cortex lytic enzyme CwlJ in spores of *Bacillus subtilis*. *J Bacteriol* **184:**1219–1224.

50. Paidhungat M, Ragkousi K, Setlow P. 2001. Genetic requirements for induction of germination of spores of *Bacillus subtilis* by Ca(2+)-dipicolinate. *J Bacteriol* **183:** 4886–4893.

51. Ishikawa S, Yamane K, Sekiguchi J. 1998. Regulation and characterization of a newly deduced cell wall hydrolase gene (*cwlJ*) which affects germination of *Bacillus subtilis* spores. *J Bacteriol* **180:**1375–1380.

52. Ragkousi K, Eichenberger P, van Ooij C, Setlow P. 2003. Identification of a new gene essential for germination of *Bacillus subtilis* spores with Ca2+-dipicolinate. *J Bacteriol* **185:**2315–2329.

53. Ragkousi K, Setlow P. 2004. Transglutaminase-mediated cross-linking of GerQ in the coats of *Bacillus subtilis* spores. *J Bacteriol* **186:**5567–5575.

54. Monroe A, Setlow P. 2006. Localization of the trans-glutaminase cross-linking sites in the *Bacillus subtilis* spore coat protein GerQ. *J Bacteriol* 188:7609–7616.

55. Imamura D, Kuwana R, Takamatsu H, Watabe K. 2010. Localization of proteins to different layers and regions of *Bacillus subtilis* spore coats. *J Bacteriol* 192:518–524.

56. Lambert EA, Popham DL. 2008. The *Bacillus anthracis* SleL (YaaH) protein is an N-acetylglucosaminidase involved in spore cortex depolymerization. *J Bacteriol* 190:7601–7607.

57. Burns DA, Heap JT, Minton NP. 2010. SleC is essential for germination of *Clostridium difficile* spores in nutrient-rich medium supplemented with the bile salt taurocholate. *J Bacteriol* 192:657–664.

58. Adams CM, Eckenroth BE, Putnam EE, Doublié S, Shen A. 2013. Structural and functional analysis of the CspB protease required for *Clostridium* spore germination. *PLoS Pathog* 9:e1003165. doi:10.1371/journal.ppat.1003165.

59. Masayama A, Kuwana R, Takamatsu H, Hemmi H, Yoshimura T, Watabe K, Moriyama R. 2007. A novel lipolytic enzyme, YcsK (LipC), located in the spore coat of *Bacillus subtilis*, is involved in spore germination. *J Bacteriol* 189:2369–2375.

60. Masayama A, Kato S, Terashima T, Mølgaard A, Hemmi H, Yoshimura T, Moriyama R. 2010. *Bacillus subtilis* spore coat protein LipC is a phospholipase B. *Biosci Biotechnol Biochem* 74:24–30.

61. Behravan J, Chirakkal H, Masson A, Moir A. 2000. Mutations in the *gerP* locus of *Bacillus subtilis* and *Bacillus cereus* affect access of germinants to their targets in spores. *J Bacteriol* 182:1987–1994.

62. Butzin XY, Troiano AJ, Coleman WH, Griffiths KK, Doona CJ, Feeherry FE, Wang G, Li YQ, Setlow P. 2012. Analysis of the effects of a *gerP* mutation on the germination of spores of *Bacillus subtilis*. *J Bacteriol* 194:5749–5758.

63. Chesnokova ON, McPherson SA, Steichen CT, Turnbough CL Jr. 2009. The spore-specific alanine racemase of *Bacillus anthracis* and its role in suppressing germination during spore development. *J Bacteriol* 191:1303–1310.

64. Pierce KJ, Salifu SP, Tangney M. 2008. Gene cloning and characterization of a second alanine racemase from *Bacillus subtilis* encoded by *yncD*. *FEMS Microbiol Lett* 283:69–74.

65. Fimlaid KA, Bond JP, Schutz KC, Putnam EE, Leung JM, Lawley TD, Shen A. 2013. Global analysis of the sporulation pathway of *Clostridium difficile*. *PLoS Genet* 9:e1003660. doi:10.1371/journal.pgen.1003660.

66. Saujet L, Pereira FC, Serrano M, Soutourina O, Monot M, Shelyakin PV, Gelfand MS, Dupuy B, Henriques AO, Martin-Verstraete I. 2013. Genome-wide analysis of cell type-specific gene transcription during spore formation in *Clostridium difficile*. *PLoS Genet* 9:e1003756. doi:10.1371/journal.pgen.1003756.

67. Sorg JA, Sonenshein AL. 2008. Bile salts and glycine as cogerminants for *Clostridium difficile* spores. *J Bacteriol* 190:2505–2512.

68. Wheeldon LJ, Worthington T, Hilton AC, Elliott TS, Lambert PA. 2008. Physical and chemical factors influencing the germination of *Clostridium difficile* spores. *J Appl Microbiol* 105:2223–2230.

69. Francis MB, Allen CA, Shrestha R, Sorg JA. 2013. Bile acid recognition by the *Clostridium difficile* germinant receptor, CspC, is important for establishing infection. *PLoS Pathog* 9:e1003356. doi:10.1371/journal.ppat.1003356.

70. Chen G, Driks A, Tawfiq K, Mallozzi M, Patil S. 2010. *Bacillus anthracis* and *Bacillus subtilis* spore surface properties and transport. *Colloids Surf B Biointerfaces* 76:512–518.

71. Cangiano G, Sirec T, Panarella C, Isticato R, Baccigalupi L, De Felice M, Ricca E. 2014. The *sps* gene products affect the germination, hydrophobicity, and protein adsorption of *Bacillus subtilis* spores. *Appl Environ Microbiol* 80:7293–7302.

72. Arrieta-Ortiz ML, Hafemeister C, Bate AR, Chu T, Greenfield A, Shuster B, Barry SN, Gallitto M, Liu B, Kacmarczyk T, Santoriello F, Chen J, Rodrigues CD, Sato T, Rudner DZ, Driks A, Bonneau R, Eichenberger P. 2015. An experimentally supported model of the *Bacillus subtilis* global transcriptional regulatory network. *Mol Syst Biol* 11:839. doi:10.15252/msb.20156236.

73. Plata G, Fuhrer T, Hsiao TL, Sauer U, Vitkup D. 2012. Global probabilistic annotation of metabolic networks enables enzyme discovery. *Nat Chem Biol* 8:848–854.

74. Wunschel D, Fox KF, Black GE, Fox A. 1994. Discrimination among the *B. cereus* group, in comparison to *B. subtilis*, by structural carbohydrate profiles and ribosomal RNA spacer region PCR. *Syst Appl Microbiol* 17:625–635.

75. Knurr J, Benedek O, Heslop J, Vinson RB, Boydston JA, McAndrew J, Kearney JF, Turnbough CL Jr. 2003. Peptide ligands that bind selectively to spores of *Bacillus subtilis* and closely related species. *Appl Environ Microbiol* 69:6841–6847.

76. Charnock SJ, Davies GJ. 1999. Structure of the nucleotide-diphospho-sugar transferase, SpsA from *Bacillus subtilis*, in native and nucleotide-complexed forms. *Biochemistry* 38:6380–6385.

77. Abe K, Kawano Y, Iwamoto K, Arai K, Maruyama Y, Eichenberger P, Sato T. 2014. Developmentally-regulated excision of the SPβ prophage reconstitutes a gene required for spore envelope maturation in *Bacillus subtilis*. *PLoS Genet* 10:e1004636. doi:10.1371/journal.pgen.1004636.

78. Scheeff ED, Axelrod HL, Miller MD, Chiu HJ, Deacon AM, Wilson IA, Manning G. 2010. Genomics, evolution, and crystal structure of a new family of bacterial spore kinases. *Proteins* 78:1470–1482.

79. Bate AR, Bonneau R, Eichenberger P. 2014. *Bacillus subtilis* systems biology: applications of -omics techniques to the study of endospore formation. *Microbiol Spectr* 2(2):TBS-0019-2013. doi:10.1128/microbiolspec.TBS-0019-2013.

80. Ryter A, Schaeffer P, Ionesco H. 1966. Cytologic classification, by their blockage stage, of sporulation mutants of *Bacillus subtilis* Marburg [in French]. *Ann Inst Pasteur (Paris)* **110**:305–315.

81. Piggot PJ, Coote JG. 1976. Genetic aspects of bacterial endospore formation. *Bacteriol Rev* **40**:908–962.

82. Cutting S, Panzer S, Losick R. 1989. Regulatory studies on the promoter for a gene governing synthesis and assembly of the spore coat in *Bacillus subtilis*. *J Mol Biol* **207**:393–404.

83. Kunkel B, Kroos L, Poth H, Youngman P, Losick R. 1989. Temporal and spatial control of the mother-cell regulatory gene *spoIIID* of *Bacillus subtilis*. *Genes Dev* **3**:1735–1744.

84. Zheng LB, Losick R. 1990. Cascade regulation of spore coat gene expression in *Bacillus subtilis*. *J Mol Biol* **212**:645–660.

85. Halberg R, Kroos L. 1994. Sporulation regulatory protein SpoIIID from *Bacillus subtilis* activates and represses transcription by both mother-cell-specific forms of RNA polymerase. *J Mol Biol* **243**:425–436.

86. Eichenberger P, Fujita M, Jensen ST, Conlon EM, Rudner DZ, Wang ST, Ferguson C, Haga K, Sato T, Liu JS, Losick R. 2004. The program of gene transcription for a single differentiating cell type during sporulation in *Bacillus subtilis*. *PLoS Biol* **2**:e328. doi:10.1371/journal.pbio.0020328

87. Kuwana R, Okumura T, Takamatsu H, Watabe K. 2005. The *ylbO* gene product of *Bacillus subtilis* is involved in the coat development and lysozyme resistance of spore. *FEMS Microbiol Lett* **242**:51–57.

88. Cangiano G, Mazzone A, Baccigalupi L, Isticato R, Eichenberger P, De Felice M, Ricca E. 2010. Direct and indirect control of late sporulation genes by GerR of *Bacillus subtilis*. *J Bacteriol* **192**:3406–3413.

89. de Hoon MJ, Eichenberger P, Vitkup D. 2010. Hierarchical evolution of the bacterial sporulation network. *Curr Biol* **20**:R735–R745.

90. Pandey NK, Aronson AI. 1979. Properties of the *Bacillus subtilis* spore coat. *J Bacteriol* **137**:1208–1218.

91. Sanchez-Salas JL, Setlow B, Zhang P, Li YQ, Setlow P. 2011. Maturation of released spores is necessary for acquisition of full spore heat resistance during *Bacillus subtilis* sporulation. *Appl Environ Microbiol* **77**:6746–6754.

92. Costa T, Serrano M, Steil L, Völker U, Moran CP Jr, Henriques AO. 2007. The timing of *cotE* expression affects *Bacillus subtilis* spore coat morphology but not lysozyme resistance. *J Bacteriol* **189**:2401–2410.

93. Wang KH, Isidro AL, Domingues L, Eskandarian HA, McKenney PT, Drew K, Grabowski P, Chua MH, Barry SN, Guan M, Bonneau R, Henriques AO, Eichenberger P. 2009. The coat morphogenetic protein SpoVID is necessary for spore encasement in *Bacillus subtilis*. *Mol Microbiol* **74**:634–649.

94. Roels S, Driks A, Losick R. 1992. Characterization of *spoIVA*, a sporulation gene involved in coat morphogenesis in *Bacillus subtilis*. *J Bacteriol* **174**:575–585.

95. Stevens CM, Daniel R, Illing N, Errington J. 1992. Characterization of a sporulation gene, *spoIVA*, involved in spore coat morphogenesis in *Bacillus subtilis*. *J Bacteriol* **174**:586–594.

96. Beall B, Driks A, Losick R, Moran CP Jr. 1993. Cloning and characterization of a gene required for assembly of the *Bacillus subtilis* spore coat. *J Bacteriol* **175**:1705–1716.

97. Levin PA, Fan N, Ricca E, Driks A, Losick R, Cutting S. 1993. An unusually small gene required for sporulation by *Bacillus subtilis*. *Mol Microbiol* **9**:761–771.

98. Gill RL Jr, Castaing JP, Hsin J, Tan IS, Wang X, Huang KC, Tian F, Ramamurthi KS. 2015. Structural basis for the geometry-driven localization of a small protein. *Proc Natl Acad Sci USA* **112**:E1908–E1915.

99. Ramamurthi KS, Lecuyer S, Stone HA, Losick R. 2009. Geometric cue for protein localization in a bacterium. *Science* **323**:1354–1357.

100. Wu IL, Narayan K, Castaing JP, Tian F, Subramaniam S, Ramamurthi KS. 2015. A versatile nano display platform from bacterial spore coat proteins. *Nat Commun* **6**:6777.

101. Ramamurthi KS, Clapham KR, Losick R. 2006. Peptide anchoring spore coat assembly to the outer forespore membrane in *Bacillus subtilis*. *Mol Microbiol* **62**:1547–1557.

102. Castaing JP, Nagy A, Anantharaman V, Aravind L, Ramamurthi KS. 2013. ATP hydrolysis by a domain related to translation factor GTPases drives polymerization of a static bacterial morphogenetic protein. *Proc Natl Acad Sci USA* **110**:E151–E160.

103. Ramamurthi KS, Losick R. 2008. ATP-driven self-assembly of a morphogenetic protein in *Bacillus subtilis*. *Mol Cell* **31**:406–414.

104. Müllerová D, Krajčíková D, Barák I. 2009. Interactions between *Bacillus subtilis* early spore coat morphogenetic proteins. *FEMS Microbiol Lett* **299**:74–85.

105. Ebmeier SE, Tan IS, Clapham KR, Ramamurthi KS. 2012. Small proteins link coat and cortex assembly during sporulation in *Bacillus subtilis*. *Mol Microbiol* **84**:682–696.

106. Tan IS, Weiss CA, Popham DL, Ramamurthi KS. 2015. A quality-control mechanism removes unfit cells from a population of sporulating bacteria. *Dev Cell* **34**:682–693.

107. Driks A, Roels S, Beall B, Moran CP Jr, Losick R. 1994. Subcellular localization of proteins involved in the assembly of the spore coat of *Bacillus subtilis*. *Genes Dev* **8**:234–244.

108. Takamatsu H, Kodama T, Nakayama T, Watabe K. 1999. Characterization of the *yrbA* gene of *Bacillus subtilis*, involved in resistance and germination of spores. *J Bacteriol* **181**:4986–4994.

109. Ozin AJ, Henriques AO, Yi H, Moran CP Jr. 2000. Morphogenetic proteins SpoVID and SafA form a complex during assembly of the *Bacillus subtilis* spore coat. *J Bacteriol* **182**:1828–1833.

110. Zheng LB, Donovan WP, Fitz-James PC, Losick R. 1988. Gene encoding a morphogenic protein required

in the assembly of the outer coat of the *Bacillus subtilis* endospore. *Genes Dev* 2:1047–1054.

111. **Imamura D, Kuwana R, Takamatsu H, Watabe K.** 2011. Proteins involved in formation of the outermost layer of *Bacillus subtilis* spores. *J Bacteriol* 193: 4075–4080.

112. **Zhang J, Fitz-James PC, Aronson AI.** 1993. Cloning and characterization of a cluster of genes encoding polypeptides present in the insoluble fraction of the spore coat of *Bacillus subtilis*. *J Bacteriol* 175:3757–3766.

113. **Krajcíková D, Lukácová M, Müllerová D, Cutting SM, Barák I.** 2009. Searching for protein-protein interactions within the *Bacillus subtilis* spore coat. *J Bacteriol* 191:3212–3219.

114. **Ball DA, Taylor R, Todd SJ, Redmond C, Couture-Tosi E, Sylvestre P, Moir A, Bullough PA.** 2008. Structure of the exosporium and sublayers of spores of the *Bacillus cereus* family revealed by electron crystallography. *Mol Microbiol* 68:947–958.

115. **Kailas L, Terry C, Abbott N, Taylor R, Mullin N, Tzokov SB, Todd SJ, Wallace BA, Hobbs JK, Moir A, Bullough PA.** 2011. Surface architecture of endospores of the *Bacillus cereus/anthracis/thuringiensis* family at the subnanometer scale. *Proc Natl Acad Sci USA* 108: 16014–16019.

116. **Jiang S, Wan Q, Krajcikova D, Tang J, Tzokov SB, Barak I, Bullough PA.** 2015. Diverse supramolecular structures formed by self-assembling proteins of the *Bacillus subtilis* spore coat. *Mol Microbiol* 97:347–359.

117. **Kim H, Hahn M, Grabowski P, McPherson DC, Otte MM, Wang R, Ferguson CC, Eichenberger P, Driks A.** 2006. The *Bacillus subtilis* spore coat protein interaction network. *Mol Microbiol* 59:487–502.

118. **Plomp M, Carroll AM, Setlow P, Malkin AJ.** 2014. Architecture and assembly of the *Bacillus subtilis* spore coat. *PLoS One* 9:e108560. doi:10.1371/journal.pone. 0108560.

119. **Costa T, Isidro AL, Moran CP Jr, Henriques AO.** 2006. Interaction between coat morphogenetic proteins SafA and SpoVID. *J Bacteriol* 188:7731–7741.

120. **de Francesco M, Jacobs JZ, Nunes F, Serrano M, McKenney PT, Chua MH, Henriques AO, Eichenberger P.** 2012. Physical interaction between coat morphogenetic proteins SpoVID and CotE is necessary for spore encasement in *Bacillus subtilis*. *J Bacteriol* 194:4941–4950.

121. **Zilhão R, Naclerio G, Henriques AO, Baccigalupi L, Moran CP Jr, Ricca E.** 1999. Assembly requirements and role of CotH during spore coat formation in *Bacillus subtilis*. *J Bacteriol* 181:2631–2633.

122. **Naclerio G, Baccigalupi L, Zilhao R, De Felice M, Ricca E.** 1996. *Bacillus subtilis* spore coat assembly requires *cotH* gene expression. *J Bacteriol* 178:4375–4380.

123. **McPherson DC, Kim H, Hahn M, Wang R, Grabowski P, Eichenberger P, Driks A.** 2005. Characterization of the *Bacillus subtilis* spore morphogenetic coat protein CotO. *J Bacteriol* 187:8278–8290.

124. **Isticato R, Sirec T, Vecchione S, Crispino A, Saggese A, Baccigalupi L, Notomista E, Driks A, Ricca E.** 2015. The direct interaction between two morphogenetic proteins is essential for spore coat formation in *Bacillus subtilis*. *PLoS One* 10:e0141040. doi:10.1371/journal. pone.0141040.

125. **Little S, Driks A.** 2001. Functional analysis of the *Bacillus subtilis* morphogenetic spore coat protein CotE. *Mol Microbiol* 42:1107–1120.

126. **Eichenberger P, Jensen ST, Conlon EM, van Ooij C, Silvaggi J, González-Pastor JE, Fujita M, Ben-Yehuda S, Stragier P, Liu JS, Losick R.** 2003. The sigmaE regulon and the identification of additional sporulation genes in *Bacillus subtilis*. *J Mol Biol* 327:945–972.

127. **Zilhão R, Serrano M, Isticato R, Ricca E, Moran CP Jr, Henriques AO.** 2004. Interactions among CotB, CotG, and CotH during assembly of the *Bacillus subtilis* spore coat. *J Bacteriol* 186:1110–1119.

128. **Giglio R, Fani R, Isticato R, De Felice M, Ricca E, Baccigalupi L.** 2011. Organization and evolution of the *cotG* and *cotH* genes of *Bacillus subtilis*. *J Bacteriol* 193:6664–6673.

129. **Isticato R, Sirec T, Giglio R, Baccigalupi L, Rusciano G, Pesce G, Zito G, Sasso A, De Felice M, Ricca E.** 2013. Flexibility of the programme of spore coat formation in *Bacillus subtilis*: bypass of CotE requirement by over-production of CotH. *PLoS One* 8:e74949. doi:10.1371/journal.pone.0074949.

130. **Cutting S, Zheng LB, Losick R.** 1991. Gene encoding two alkali-soluble components of the spore coat from *Bacillus subtilis*. *J Bacteriol* 173:2915–2919.

131. **Takamatsu H, Imamura A, Kodama T, Asai K, Ogasawara N, Watabe K.** 2000. The *yabG* gene of *Bacillus subtilis* encodes a sporulation specific protease which is involved in the processing of several spore coat proteins. *FEMS Microbiol Lett* 192:33–38.

132. **Takamatsu H, Kodama T, Imamura A, Asai K, Kobayashi K, Nakayama T, Ogasawara N, Watabe K.** 2000. The *Bacillus subtilis yabG* gene is transcribed by SigK RNA polymerase during sporulation, and *yabG* mutant spores have altered coat protein composition. *J Bacteriol* 182:1883–1888.

133. **Kuwana R, Okuda N, Takamatsu H, Watabe K.** 2006. Modification of GerQ reveals a functional relationship between Tgl and YabG in the coat of *Bacillus subtilis* spores. *J Biochem* 139:887–901.

134. **Faille C, Ronse A, Dewailly E, Slomianny C, Maes E, Krzewinski F, Guerardel Y.** 2014. Presence and function of a thick mucous layer rich in polysaccharides around *Bacillus subtilis* spores. *Biofouling* 30:845–858.

135. **Faille C, Lequette Y, Ronse A, Slomianny C, Garénaux E, Guerardel Y.** 2010. Morphology and physico-chemical properties of *Bacillus* spores surrounded or not with an exosporium: consequences on their ability to adhere to stainless steel. *Int J Food Microbiol* 143:125–135.

136. **Lequette Y, Garénaux E, Tauveron G, Dumez S, Perchat S, Slomianny C, Lereclus D, Guérardel Y, Faille C.** 2011. Role played by exosporium glycoproteins in the surface properties of *Bacillus cereus* spores and in

their adhesion to stainless steel. *Appl Environ Microbiol* **77**:4905–4911.

137. Lequette Y, Garénaux E, Combrouse T, Dias TL, Ronse A, Slomianny C, Trivelli X, Guerardel Y, Faille C. 2011. Domains of BclA, the major surface glycoprotein of the *B. cereus* exosporium: glycosylation patterns and role in spore surface properties. *Biofouling* **27**:751–761.

138. Eijlander RT, Abee T, Kuipers OP. 2011. Bacterial spores in food: how phenotypic variability complicates prediction of spore properties and bacterial behavior. *Curr Opin Biotechnol* **22**:180–186.

139. Driks A. 2003. The dynamic spore. *Proc Natl Acad Sci USA* **100**:3007–3009.

140. Westphal AJ, Price PB, Leighton TJ, Wheeler KE. 2003. Kinetics of size changes of individual *Bacillus thuringiensis* spores in response to changes in relative humidity. *Proc Natl Acad Sci USA* **100**:3461–3466.

141. Chen X, Goodnight D, Gao Z, Cavusoglu AH, Sabharwal N, DeLay M, Driks A, Sahin O. 2015. Scaling up nanoscale water-driven energy conversion into evaporation-driven engines and generators. *Nat Commun* **6**:7346.

142. Semenyuk EG, Laning ML, Foley J, Johnston PF, Knight KL, Gerding DN, Driks A. 2014. Spore formation and toxin production in *Clostridium difficile* biofilms. *PLoS One* **9**:e87757. doi:10.1371/journal.pone.0087757.

143. Bauer T, Little S, Stöver AG, Driks A. 1999. Functional regions of the *Bacillus subtilis* spore coat morphogenetic protein CotE. *J Bacteriol* **181**:7043–7051.

144. Zhang J, Ichikawa H, Halberg R, Kroos L, Aronson AI. 1994. Regulation of the transcription of a cluster of *Bacillus subtilis* spore coat genes. *J Mol Biol* **240**:405–415.

145. Ozin AJ, Costa T, Henriques AO, Moran CP Jr. 2001. Alternative translation initiation produces a short form of a spore coat protein in *Bacillus subtilis*. *J Bacteriol* **183**:2032–2040.

146. Ozin AJ, Samford CS, Henriques AO, Moran CP Jr. 2001. SpoVID guides SafA to the spore coat in *Bacillus subtilis*. *J Bacteriol* **183**:3041–3049.

147. Price KD, Losick R. 1999. A four-dimensional view of assembly of a morphogenetic protein during sporulation in *Bacillus subtilis*. *J Bacteriol* **181**:781–790.

148. Catalano FA, Meador-Parton J, Popham DL, Driks A. 2001. Amino acids in the *Bacillus subtilis* morphogenetic protein SpoIVA with roles in spore coat and cortex formation. *J Bacteriol* **183**:1645–1654.

149. van Ooij C, Losick R. 2003. Subcellular localization of a small sporulation protein in *Bacillus subtilis*. *J Bacteriol* **185**:1391–1398.

150. Henriques AO, Beall BW, Roland K, Moran CP Jr. 1995. Characterization of *cotJ*, a sigma E-controlled operon affecting the polypeptide composition of the coat of *Bacillus subtilis* spores. *J Bacteriol* **177**:3394–3406.

151. Seyler RW Jr, Henriques AO, Ozin AJ, Moran CP Jr. 1997. Assembly and interactions of *cotJ*-encoded proteins, constituents of the inner layers of the *Bacillus subtilis* spore coat. *Mol Microbiol* **25**:955–966.

152. van Ooij C, Eichenberger P, Losick R. 2004. Dynamic patterns of subcellular protein localization during spore coat morphogenesis in *Bacillus subtilis*. *J Bacteriol* **186**:4441–4448.

153. Reischl S, Thake S, Homuth G, Schumann W. 2001. Transcriptional analysis of three *Bacillus subtilis* genes coding for proteins with the alpha-crystallin domain characteristic of small heat shock proteins. *FEMS Microbiol Lett* **194**:99–103.

154. Bourne N, FitzJames PC, Aronson AI. 1991. Structural and germination defects of *Bacillus subtilis* spores with altered contents of a spore coat protein. *J Bacteriol* **173**:6618–6625.

155. Costa T, Steil L, Martins LO, Völker U, Henriques AO. 2004. Assembly of an oxalate decarboxylase produced under σ^K control into the *Bacillus subtilis* spore coat. *J Bacteriol* **186**:1462–1474.

156. Kobayashi K, Hashiguchi K, Yokozeki K, Yamanaka S. 1998. Molecular cloning of the transglutaminase gene from *Bacillus subtilis* and its expression in *Escherichia coli*. *Biosci Biotechnol Biochem* **62**:1109–1114.

157. Kodama T, Takamatsu H, Asai K, Kobayashi K, Ogasawara N, Watabe K. 1999. The *Bacillus subtilis yaaH* gene is transcribed by SigE RNA polymerase during sporulation, and its product is involved in germination of spores. *J Bacteriol* **181**:4584–4591.

158. Üstok FI, Chirgadze DY, Christie G. 2015. Structural and functional analysis of SleL, a peptidoglycan lysin involved in germination of *Bacillus* spores. *Proteins* **83**:1787–1799.

159. Takamatsu H, Imamura D, Kuwana R, Watabe K. 2009. Expression of *yeeK* during *Bacillus subtilis* sporulation and localization of YeeK to the inner spore coat using fluorescence microscopy. *J Bacteriol* **191**:1220–1229.

160. Kuwana R, Takamatsu H, Watabe K. 2007. Expression, localization and modification of YxeE spore coat protein in *Bacillus subtilis*. *J Biochem* **142**:681–689.

161. Isticato R, Esposito G, Zilhão R, Nolasco S, Cangiano G, De Felice M, Henriques AO, Ricca E. 2004. Assembly of multiple CotC forms into the *Bacillus subtilis* spore coat. *J Bacteriol* **186**:1129–1135.

162. Isticato R, Pelosi A, Zilhão R, Baccigalupi L, Henriques AO, De Felice M, Ricca E. 2008. CotC-CotU heterodimerization during assembly of the *Bacillus subtilis* spore coat. *J Bacteriol* **190**:1267–1275.

163. Sacco M, Ricca E, Losick R, Cutting S. 1995. An additional GerE-controlled gene encoding an abundant spore coat protein from *Bacillus subtilis*. *J Bacteriol* **177**:372–377.

164. Henriques AO, Melsen LR, Moran CP Jr. 1998. Involvement of superoxide dismutase in spore coat assembly in *Bacillus subtilis*. *J Bacteriol* **180**:2285–2291.

165. Henriques AO, Beall BW, Moran CP Jr. 1997. CotM of *Bacillus subtilis*, a member of the alpha-crystallin family of stress proteins, is induced during development and participates in spore outer coat formation. *J Bacteriol* **179**:1887–1897.

166. Abe A, Koide H, Kohno T, Watabe K. 1995. A *Bacillus subtilis* spore coat polypeptide gene, *cotS*. *Microbiology* 141:1433–1442.

167. Takamatsu H, Chikahiro Y, Kodama T, Koide H, Kozuka S, Tochikubo K, Watabe K. 1998. A spore coat protein, CotS, of *Bacillus subtilis* is synthesized under the regulation of sigmaK and GerE during development and is located in the inner coat layer of spores. *J Bacteriol* 180:2968–2974.

168. Isticato R, Pelosi A, De Felice M, Ricca E. 2010. CotE binds to CotC and CotU and mediates their interaction during spore coat formation in *Bacillus subtilis*. *J Bacteriol* 192:949–954.

169. Ferguson CC, Camp AH, Losick R. 2007. *gerT*, a newly discovered germination gene under the control of the sporulation transcription factor sigmaK in *Bacillus subtilis*. *J Bacteriol* 189:7681–7689.

170. Henriques AO, Bryan EM, Beall BW, Moran CP Jr. 1997. *cse15*, *cse60*, and *csk22* are new members of mother-cell-specific sporulation regulons in *Bacillus subtilis*. *J Bacteriol* 179:389–398.

171. Roels S, Losick R. 1995. Adjacent and divergently oriented operons under the control of the sporulation regulatory protein GerE in *Bacillus subtilis*. *J Bacteriol* 177:6263–6275.

172. Takamatsu H, Kodama T, Watabe K. 1999. Assembly of the CctSA coat protein into spores requires CotS in *Bacillus subtilis*. *FEMS Microbiol Lett* 174:201–206.

173. Liu P, Ewis HE, Huang YJ, Lu CD, Tai PC, Weber IT. 2007. Structure of *Bacillus subtilis* superoxide dismutase. *Acta Crystallogr Sect F Struct Biol Cryst Commun* 63:1003–1007.

174. Kodama T, Matsubayashi T, Yanagihara T, Komoto H, Ara K, Ozaki K, Kuwana R, Imamura D, Takamatsu H, Watabe K, Sekiguchi J. 2011. A novel small protein of *Bacillus subtilis* involved in spore germination and spore coat assembly. *Biosci Biotechnol Biochem* 75:1119–1128.

175. Kodama T, Takamatsu H, Asai K, Ogasawara N, Sadaie Y, Watabe K. 2000. Synthesis and characterization of the spore proteins of *Bacillus subtilis* YdhD, YkuD, and YkvP, which carry a motif conserved among cell wall binding proteins. *J Biochem* 128:655–663.

176. Pizarro-Guajardo M, Olguín-Araneda V, Barra-Carrasco J, Brito-Silva C, Sarker MR, Paredes-Sabja D. 2014. Characterization of the collagen-like exosporium protein, BclA1, of *Clostridium difficile* spores. *Anaerobe* 25:18–30.

177. Barra-Carrasco J, Olguín-Araneda V, Plaza-Garrido A, Miranda-Cárdenas C, Cofré-Araneda G, Pizarro-Guajardo M, Sarker MR, Paredes-Sabja D. 2013. The *Clostridium difficile* exosporium cysteine (CdeC)-rich protein is required for exosporium morphogenesis and coat assembly. *J Bacteriol* 195:3863–3875.

178. Popham DL, Bernhards CB. 2015. Spore peptidoglycan. *Microbiol Spectr* 3(6):TBS-0005-2012. doi:10.1128/microbiolspec.TBS-0005-2012.

179. Mandic-Mulec I, Stefanic P, van Elsas JD. 2015. Ecology of *Bacillaceae*. *Microbiol Spectr* 3(2):TBS-0017-2013. doi:10.1128/microbiolspec.TBS-0017-2013.

180. Moir A, Cooper G. 2014. Spore germination. *Microbiol Spectr* 3(6):TBS-0014-2012. doi:10.1128/microbiolspec.TBS-0014-2012.

181. Butterfield CN, Lee S-W, Tebo BM. 2016. The role of bacterial spores in metal cycling and their potential application in metal contaminant bioremediation. *Microbiol Spectr* 4(3):TBS-0018-2013. doi:10.1128/microbiolspec.TBS-0018-2013.

The Bacterial Spore: From Molecules to Systems
Edited by P. Eichenberger and A. Driks
© 2016 American Society for Microbiology, Washington, DC
doi:10.1128/microbiolspec.TBS-0003-2012

Peter Setlow[1]

Spore Resistance Properties

10

The extreme resistance of spores of members of the *Bacillales* and *Clostridiales* orders is probably the property most closely associated with these spores. In the past, this extreme resistance contributed to claims for spontaneous generation and, in more recent years, has contributed to the applied importance of spores in a number of different areas including the following. (i) The food industry. Given that spores of a number of species are ubiquitous in the environment, they routinely contaminate foodstuffs. Since spores of many species are vectors for food spoilage and food-borne disease, the food industry commits significant resources to eliminating spores in order to make foods sterile, in particular to eliminate extremely dangerous spores such as those of *Clostridium botulinum* (1, 2). Indeed, many of the requirements for food sterilization regimens in the United States are designed to completely inactivate *C. botulinum* spores. (ii) The medical products industry. Just as in the food industry, spores present similar concerns in the medical products industry, including the manufacture of medical devices and parenteral drugs, again because of the involvement of spores in a number of human diseases. (iii) The health care industry. There is an increasing prevalence of disease due to *Clostridium difficile* in hospital and long-term nursing care facilities, largely because of the resistance of *C. difficile* spores and thus their persistence in patient care environments unless stringent environmental decontamination regimens are followed. (iv) Vaccine development.

There is increasing interest in spores as carriers of proteins important as vaccines (3, 91), in large part because of spores' extreme stability to normal and even extreme environmental conditions. This may allow the delivery of vaccines to areas where cold storage is difficult and is facilitated by utilizing the spore coat as a means to deliver immunogens. (v) Probiotics. Since spores are dormant, as such, they will not be probiotics. However, the administration of spores with their resistance to low pH conditions in the stomach is a route to effectively deliver potentially beneficial bacteria to the lower gastrointestinal tract (4, 91). Notably, it is the *C. difficile* spore's resistance to stomach acidity that is the reason that the oral route is the major mechanism for *C. difficile* infection. (vi) Biological warfare. While the disease-causing potential of *Bacillus anthracis* is one reason that this organism has come to the fore as a biological weapon, in particular of bioterror (S. L. Welkos, unpublished data), the major reason for this organism's visibility in this area is that *B. anthracis* spores are so resistant. This makes their dispersal either in water or as an aerosol relatively simple and ensures that these spores will persist in contaminated environments and will thus require stringent decontamination methods for their elimination.

Given the applied interest in spores, in large part because of their resistance properties, it is not surprising that there has been tremendous interest in the mechanisms of spore resistance. Most of this mechanistic

[1]Department of Molecular Biology and Biophysics, University of Connecticut Health Center, Farmington, CT 06030-3305.

work has utilized *Bacillus subtilis* spores because of the large number of *B. subtilis* strains with mutations in genes that may be involved in spore resistance. However, where it has been studied (5–7), work with spores of other *Bacillus* species has generally given results similar to those with *B. subtilis*. There has, however, generally been much less work done on mechanisms of resistance of spores of *Clostridium* species. In this report, the focus will be on work with *B. subtilis* spores, unless noted otherwise. However, relevant data from spores of other species or genera will be mentioned when available. References will largely be confined to the most recent work, while references to older work can be found cited in numerous past reviews on spore resistance properties (2, 8–15).

OVERVIEW OF SPORE RESISTANCE

Spores exhibit greatly increased resistance to a large number of agents, including desiccation, freezing, thawing, elevated temperatures in either the wet or dry state, UV and γ-radiation, high pressures, and a huge number of toxic chemicals with a variety of nasty effects including oxidizing agents, alkylating agents, aldehydes, halogens, acids, and bases. Invariably, spores are much more resistant to these various agents than are growing cells of the same species (Table 1). However, spores of different strains, species, and genera can exhibit quite large differences in their resistance to various agents.

Spores are killed by damage to a number of different components, including DNA, the spore's inner membrane (IM), proteins in the spore core, and likely other components as well (Table 2). Spores also utilize a variety of

strategies to generate their extreme resistance (Table 3), including the maintenance of special outer layers to help protect sensitive spore components such as peptidoglycan (PG) from enzymatic attack and DNA from chemical attack. DNA in spores is also saturated with novel proteins that further protect the DNA against chemical attack and damage by wet heat. This novel DNA-protein complex in the spore is also important in protecting the DNA against UV and γ-radiation, as well as against dry heat and desiccation. Spores also contain a variety of novel enzymes that can rapidly repair DNA damage accumulated during spore dormancy, when spores return to life in the process of germination. Finally, the conditions inside the central spore core, where spore DNA and most spore enzymes are located, minimize damage due to agents such as wet heat and perhaps other treatments.

As will be seen when spore resistance to specific agents is discussed below, often multiple mechanisms contribute to spore resistance to one agent (Table 3); for example, both DNA protection and DNA repair contribute to spore resistance to radiation. Similarly, specific individual resistance mechanisms often contribute to resistance to more than one type of agent (Table 3). For example, a spore's outer layers contribute to resistance to both predation and many reactive chemicals. Together, all of these general mechanisms make spores one of the most resistant life forms known.

ROLE OF SPORE STRUCTURE IN SPORE RESISTANCE

Spore structure plays a major role in spore resistance, as a number of the spore layers play specific roles in

Table 1 Resistance of spores and growing cells of *B. subtilis* to various agents[a]

Treatment	Growing cells (wild type)	Spores				
		Wild type	*recA*	DPA-less	α⁻β⁻	*cotE*
None	82[b]	35[b]	35[b]	45[b]	35[b]	35[b]
Wet heat, 90°C (min)[c]	<0.05	18	23	≤1	2.5	19
Dry heat, 120°C (min)[c]	–	18	2	–	–	17
Dry heat, 105°C (min)[c]	–	95	–	16	–	–
Dry heat, 90°C (min)[c]	5[d]	–	–	–	2[d]	–
H₂O₂ (15%) (min)[c]	<0.2	50	55	40	15	45
HCHO (25 g/liter) (min)[c]	<0.1	22	10	20	5	20
HNO₂ (100 mM) (min)[c]	<0.2	100	15	95	10	11
NaOCl (50 mg/liter; pH 7)[c]	<0.1	55	55	–	58	<1
UV-254 nm (kJ/m²)[c]	36	330	185	280	18	350
Desiccation (#)[c]	<1	>20	>20	3	3	>10

[a]Data are at 23°C unless otherwise noted, and are from reference 13.
[b]Core water content as % wet wt.
[c]Time (min), radiation dose or the number of freeze-drying cycles to kill 90% of the population.
[d]Vegetative cells were dried in sucrose and spores were dried from water.

Table 2 Mechanisms of spore killing by various agents[a]

Mechanism of spore killing	Examples of agents that kill spores by this mechanism
DNA damage	EtO, nitrite, formaldehyde, dry heat, UV and γ-radiation
Inner membrane damage	Hypochlorite, ClO$_2$, ozone, some peroxides
Inactivation of core enzymes	Wet heat, perhaps H$_2$O$_2$
Germination apparatus damage	Alkali[b], dialdehydes
Unknown damage	High [HCl][c]

[a]Information is for spores of *B. subtilis* and is from references 2, 13, and 14.
[b]Alkali treatment can often generate spores that appear to be dead, but that can be revived if artificially germinated with lysozyme. However, alkali can also kill spores completely, although the mechanism of this killing is not known.
[c]High [HCl] causes spore rupture, but the primary reason for the rupture is not known.

resistance. From the outside in, the various spore layers include the exosporium, coat, outer membrane (OM), cortex, germ cell wall, IM, and core (Fig. 1). The outermost exosporium is not present in spores of all species and is absent in *B. subtilis* spores. However, in spores of *B. anthracis* the exosporium may act as a permeability barrier restricting access of antibodies to antigens present in the spore coat (16). Other than this, there is no evidence that the exosporium plays any significant role in spore resistance. The spore coat contains a large fraction of total spore protein and acts as a permeability barrier restricting access of large molecules such as enzymes to potential sensitive targets in the spore's more inner layers (A. Driks and P. Eichenberger, unpublished data). Consequently, the spore coat is responsible for protection against enzymes such as lysozyme that degrade peptidoglycan (PG), and thus for protection of spores against predation by bacteriovores (17, 18). The coats are also important in spore protection against a variety of biocidal chemicals, probably by reacting nonspecifically with and detoxifying such chemicals before they reach more essential targets further within the spore (11, 14, 16, 19). This has been shown best by using *B. subtilis cotE* spores that lack the CotE protein essential for spore coat morphogenesis (16), as *cotE* spores have greatly decreased resistance to some chemical biocides including hypochlorite and nitrous acid (Table 1). The spore coats also contain some enzymes that can detoxify potential biocidal chemicals such as peroxides, and this can further increase spore resistance to such agents. In spores of some species, the coats may contain pigments that absorb strongly in the UV region, and there is suggestive evidence that such pigments can play a significant role in spore UV resistance (11, 20).

Underlying the coat is the OM, the role of which in spore resistance is not completely clear. The OM also can contain pigments, generally carotenoids that may play a role in spore UV resistance as noted above (11, 20). However, the possible role of the OM as a permeability barrier is not clear. In general, an intact OM is not seen in electron micrographs of spores, and the disruption of the spore coat layer by mutations in key coat protein genes is sufficient to allow lysozyme to attack

Table 3 Factors important in spore resistance to various agents[a]

Type of agent	Protective factor[b]
UV radiation	DNA saturation by α/β-type SASP; DNA repair during spore outgrowth; low core water content; carotenoids in spore outer layers
γ-Radiation	DNA saturation by α/β-type SASP; DNA repair during spore outgrowth
Desiccation	DNA saturation by α/β-type SASP; DPA
Dry heat	DNA saturation by α/β-type SASP; DNA repair during spore outgrowth; DPA; perhaps divalent metal ion content
Wet heat	DNA saturation by α/β-type SASP; DPA level; low core water content; sporulation conditions including temperature; divalent metal ion content; sporulation temperature optimum
Genotoxic chemicals	Low permeability of spores' inner membrane; DNA saturation by α/β-type SASP; DNA repair during spore outgrowth; low core water content
Oxidizing agents	Spore coat protein; low permeability of spores' inner membrane; DNA saturation by α/β-type SASP; detoxifying enzymes in spore's outer layers
Dialdehydes	Spore coats
Disinfectants	Spore coats, perhaps cortex and inner membrane structure
Acids and alkali	Not understood
Plasma	Spore coat?; DNA saturation by α/β-type SASP; not yet thoroughly studied
Bacteriovores	Spore coat

[a]Information is for spores of *B. subtilis*, and is from references 2, 5, 6, 11, 17, 18, and 50.
[b]Not all protective factors are important in protecting against all chemicals of any particular type.

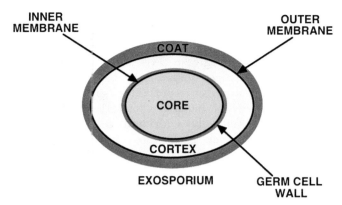

Figure 1 Schematic structure of a *Bacillus* spore. Sizes of various layers are not drawn to scale; in many species, several different coat layers can be seen; spores of some species do not have an exosporium.

spore PG layers below the OM. However, the specific role of the OM in dormant spore resistance remains an open question, although this membrane does play an essential role in spore formation. Underlying the OM are two PG layers, first the spore cortex and then the thinner germ cell wall, each with PG of slightly different structures (D. L. Popham and C. B. Bernhards, unpublished data). While both of these layers are essential for spore viability, and the cortex undoubtedly is essential for some of the novel properties of the core (see below), these two layers are not known to play any active role in spore resistance.

Under the germ cell wall is the spore's IM. While the lipid composition of the IM is not particularly unusual (21), the IM itself has some very novel properties (11). In particular, (i) lipids in the IM are largely immobile; (ii) the IM has a much higher viscosity than the germinated spore's plasma membrane; and (iii) the IM's passive permeability to small molecules is extremely low, even for molecules such as methylamine and water (11, 22–24). While the reason(s) for these novel properties of the IM is not known, these properties, especially the IM's relative impermeability, seem likely to be important in spore resistance to some biocidal chemicals by restricting these chemicals' access to targets in the spore's central core. Indeed, damage to the IM appears to be the mechanism by which a number of oxidizing agents kill spores, although the nature of this damage is unknown (11, 14). The novel properties of the spore's IM are lost when spores complete germination.

The final spore layer is the central core, which has a number of novel features that appear to play many roles in spore resistance (11, 12, 14). These include (i) the core's low water content (25 to 55% of wet

weight), important in spore wet heat resistance; (ii) the high level of pyridine-2,6-dicarboxylic acid (dipicolinic acid [DPA]) (Fig. 2) in a 1:1 complex with various divalent cations, generally mostly Ca^{2+}, and important in spore resistance to some DNA-damaging agents and in maintaining spore dormancy; and (iii) high levels of a group of novel proteins, the α/β-type small, acid-soluble spore proteins (SASPs) that saturate spore DNA and protect it from damage due to many genotoxic chemicals, desiccation, dry and wet heat, and UV and γ-radiation. The α/β-type SASPs are synthesized only late in sporulation within the developing forespore, and the genes for these 60 to 75-amino-acid (aa) proteins are transcribed by RNA polymerase with a forespore-specific sigma factor, σ^G. Spores of all *Clostridiales* and *Bacillales* species contain α/β-type SASPs, with these proteins encoded by multiple monocistronic genes. The amino acid sequence of α/β-type SASPs has been tremendously conserved throughout evolution, most likely because of the importance of these proteins' structure when bound to DNA that dramatically affects the structure of the protein-bound DNA (25). The core's low water content also ensures that α/β-type SASPs remain bound to DNA in the dormant spore even though the affinity of these proteins for DNA is not extremely high. When spores germinate and then begin to outgrow, there is significant dissociation of α/β-type SASPs from DNA and the free protein is rapidly degraded, with the degradation initiated by a SASP-specific endoprotease. Interestingly, if the affinity of α/β-type SASPs for DNA is too high, the degradation of the α/β-type SASPs during spore outgrowth is not efficient, as some of these proteins remain bound to DNA. This generally results in the death of the germinated spore. As is probably not surprising, expression of α/β-type SASPs in growing bacteria causes rapid cessation of cell growth

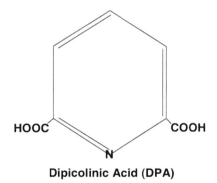

Dipicolinic Acid (DPA)

Figure 2 Structure of dipicolinic acid (DPA). Note that, at physiological pH, the two carboxyl groups will be ionized and the resultant carboxylate groups can chelate divalent cations.

as well as a loss in cell viability, and this effect is being investigated as a novel antibacterial therapy (26).

SPECIFIC SPORE RESISTANCE PROPERTIES

Radiation Resistance

UV radiation

Spores are 20- to 50-fold more resistant to UV radiation than growing cells (11) (Table 1). The magnitude of this difference depends on the species and on the UV wavelength. Not surprisingly, since DNA is the target of UV damage, 254-nm radiation is most effective in killing spores; longer and shorter wavelengths are also effective but require higher fluences than at 254 nm. There are also some differences in DNA UV photochemistry at different wavelengths (see below). Two major factors are responsible for spore UV resistance (Table 3): (i) a change in DNA's UV photochemistry due to the binding of α/β-type SASPs; and (ii) DNA repair during spore outgrowth, with this repair catalyzed in part by spore enzymes (11, 27). In addition to these major resistance factors, pigments in the outer layers of some spores may be important in shielding spores from UV damage, and spores' huge DPA depot also influences their UV resistance, as can be seen with spores that lack DPA (Table 1). Spores of several species can also be killed by high-intensity 405-nm visible light, although they are significantly more resistant than the corresponding vegetative cells (28, 29). However, nothing is yet known about factors important in spore resistance to 405-nm radiation.

UV irradiation of DNA at 254 nm *in vitro* or in growing bacteria generates a variety of photoproducts, including cyclobutane dimers (CPDs) (Fig. 3) between adjacent pyrimidines in the same DNA strand, as well as 6-4 adducts (64PP), again between adjacent pyrimidines (11). Both types of photoproducts are potentially lethal but can be repaired by a variety of mechanisms, many of which are significantly error prone. In contrast, UV irradiation of spores generates little if any CPD or 64PP, but rather a thyminyl-thymine adduct termed the spore photoproduct (SP) (Fig. 3). The SP is formed in spore DNA with a quantum efficiency not particularly different from that for CPD and 64PP formation in growing cells. However, SP formation in spores is a much less lethal photoproduct than CPD or 64PP because of its relatively error-free repair (see below). Interestingly, the photochemistry of DNA in spores changes somewhat at longer UV wavelengths, as some CPDs are formed at wavelengths >280 nm, although photoproduct formation at longer wavelengths requires

much higher fluences than at 254 nm. There have also been studies using intense white light (200 to 1,100 nm) for spore inactivation (30), but studies on the mechanisms of spore resistance to and killing by such a treatment have not yet been done.

The major reason for the formation in spores of SP rather than CPDs and 64PPs upon UV irradiation is the saturation of spore DNA by the α/β-type SASPs that changes DNA from a B-conformation to a structure between that of A and B-DNA, resulting in a change in the DNA's photochemistry (11, 25). Spores lacking the majority of their α/β-type SASPs (termed α⁻β⁻ spores) no longer have sufficient amounts of these proteins to saturate their DNA and are much less resistant than wild-type spores to 254-nm UV radiation (Table 1) as well as to other UV wavelengths including solar UV (11, 31). Irradiation of α⁻β⁻ spores by 254-nm radiation also produces large amounts of CPD and 64PP and much lower levels of SP. The spore's low water content and high level of DPA both contribute to spores' novel DNA photochemistry, and this has been duplicated *in vitro* with purified components (11). Although slight changes in core water content do not affect spore resistance to 254-nm UV, spores with elevated core water content are more sensitive to environmental UV radiation of >280 nm (31). The structure of the α/β-type SASP-DNA complex has been determined at high resolution, and analysis of this structure has indicated why UV irradiation of spores generates SP and not CPD or 64PP (25). The precise nature of the sporulation medium can also alter spores' resistance to UV radiation between 280 and 400 nm somewhat; in particular, the presence of potential radioprotective agents in the medium such as cysteine can result in spores with

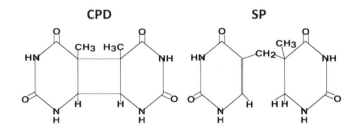

Figure 3 Structures of major photoproduct formed in growing cells (CPDs) and dormant spores (SP). The structures shown are as if these were formed between adjacent bases, with the nitrogen normally linked to the sugar in nucleosides shown with a hydrogen atom attached. The CPD shown is the major one, formed between two adjacent thymidine residues on the same DNA strand, although CPDs can also form between two adjacent cytidine residues and between adjacent cytidine and thymidine residues. SP is formed only between adjacent thymidine residues.

elevated resistance to UV as well as to γ-radiation, although these effects are eliminated if spores are first decoated (32). There is also one report that a spore's Mn level is important in its UV resistance (33), but the mechanism of this effect has not been studied in detail.

The altered UV photochemistry of spore DNA alone is just one part of spores' elevated UV resistance, as SP is potentially a lethal photoproduct. The second major factor in spore UV resistance is the repair of UV damage to DNA during spore outgrowth. This repair is catalyzed by multiple independent enzymes/enzyme systems including recombination repair, nucleotide excision repair, repair of abasic sites, and SP-specific repair (11, 12, 34, 35). The first two repair systems use enzymes that are also involved in the repair of CPDs and 64PPs in growing cell DNA. However, repair of abasic sites and SP in spore DNA can use spore-specific enzymes, with SP repaired by spore photoproduct lyase (Spl), an *S*-adenosylmethionine (SAM)-dependent enzyme that monomerizes SP back to two thymine residues by using a radical-SAM mechanism (11, 12, 36, 37). The *spl* gene is expressed only in the developing spore under control of the same RNA polymerase sigma factor, σ^G, that directs transcription of genes that encode α/β-type SASPs.

While most studies of the effects of α/β-type SASPs on spore UV resistance have been performed with spores of *B. subtilis*, the α/β-type SASPs are also a major factor in the UV resistance of *Clostridium perfringens* spores (38, 39). Indeed, a *C. perfringens* α/β-type SASP can largely restore the UV resistance of $\alpha^-\beta^-$ *B. subtilis* spores, as well as their resistance to at least one genotoxic chemical (40). Clostridial α/β-type SASPs also have the same effects on DNA photochemistry *in vitro* as do the homologous *B. subtilis* proteins (11).

γ-Radiation

Spores are also more resistant to γ-radiation than are growing cells. Again the α/β-type SASPs contribute to spore γ-radiation resistance, although precisely how is not known. Repair of γ-radiation damage during spore outgrowth is also an important factor in spore γ-radiation resistance (41–43). As with UV damage in spores, multiple enzymes are involved in the repair of γ-radiation damage to spore DNA, and at least some of these enzymes are spore-specific, including enzymes involved in repair of double-strand breaks in DNA. One area that has not been well studied in spore γ-radiation resistance is the precise spectrum of DNA damage generated by γ-radiation in spores, in comparison with γ-radiation damage generated in growing cells and in spores lacking α/β-type SASPs.

Chemical Resistance
Factors Involved in Spore Resistance to Chemicals
Spores are much more resistant than growing cells to a host of toxic chemicals, including aldehydes, oxidizing agents, alkylating agents, acids, and bases (Table 1) (2, 1, 12). A number of factors are important in spore resistance to chemicals (Table 3), including (i) detoxifying enzymes in spore coats and/or exosporia; (ii) nonspecific detoxification by spore coat components; (iii) low permeation rates of toxic chemicals up to and through the spore's IM; (iv) protection of DNA against chemical attack by α/β-type SASP binding; and (v) repair of chemically induced DNA damage during spore outgrowth.

Detoxifying enzymes in spore outer layers
Spores of several species have enzymes in their outer layers that can potentially detoxify toxic chemicals, including catalase to destroy hydrogen peroxide and superoxide dismutase (SOD) to eliminate superoxide (16, 44). There is some evidence that at least SODs in spore outer layers are important in *B. anthracis* spore pathogenicity (45), perhaps by minimizing spore killing by superoxide generated inside host cell compartments. Enzymes such as catalase and SOD and others are also important in the resistance of growing cells to oxidative stress. However, while these enzymes are present in the spore core, these core enzymes play no role in dormant spore resistance, presumably because core enzymes are generally inactive due to the low core water content and protein immobility in the spore core (11, 14, 46).

Nonspecific detoxification of chemicals by spore outer layers
The presence of an intact spore coat is a major factor in spore resistance to many toxic chemicals including halogens, larger oxidizing agents, and aldehydes, and *cotE* spores are often more sensitive to such chemical biocides (Table 1). The reasons for this are not clear, but this is either because the rate of permeation of many of these molecules through the coats to more sensitive layers further within the spore is slow, or the huge amount of coat protein reacts with and detoxifies reactive chemicals before they reach the inner membrane (11, 14, 16). At present, it is not completely possible to decide between these possibilities—indeed, both of these explanations may be correct. Spores of at least one species also contain significant levels of a polycyclic terpenoid in outer spore layers that may serve to detoxify oxidizing agents (19).

Low permeability of the spore's IM

Much work has shown that the spore's IM has extremely low permeability to small molecules, even including water (11, 24). Thus, the IM's permeability to potential DNA-damaging chemicals would be expected to be low. Indeed, there is significant evidence that changes in the permeability of the spore IM by sporulation at different temperatures result in spores with changes in sensitivity to DNA-damaging agents that parallel changes in IM permeability (11, 14, 22). For at least some chemicals, in particular hydrogen peroxide and low-molecular-weight DNA alkylating agents, the coats play a minimal role in protection against such agents (11, 22).

DNA protection by α/β-type SASP binding

In addition to protecting against UV and γ-radiation, the saturation of DNA by α/β-type SASPs also protects spore DNA against a number of genotoxic chemicals, including hydrogen peroxide, nitrous acid, and formaldehyde (Table 1) (2, 11, 14). Invariably, $\alpha^-\beta^-$ spores are more sensitive to the latter chemical agents (Table 1). Protection against genotoxic chemicals is also given by α/β-type SASPs in C. *perfringens* spores (2, 11, 14, 39). As a consequence of the protection of DNA by α/β-type SASPs against genotoxic oxidizing agents such as hydrogen peroxide (H_2O_2), such agents do not kill spores by DNA damage, and loss of much DNA repair capacity by a *recA* mutation does not decrease otherwise wild-type spores' resistance to H_2O_2 (Table 1). However, H_2O_2 does kill $\alpha^-\beta^-$ spores via DNA damage. Presumably, the protection of DNA against H_2O_2 in wild-type spores by α/β-type SASP binding is so strong that damage to some other spore component kills spores. Indeed, proteins important in repair of oxidative damage to DNA in growing cells, such as MutT and RecA, have no protective effects against H_2O_2 in spores (Table 1) (47). Formaldehyde and nitrous acid do kill wild-type spores by DNA damage, but $\alpha^-\beta^-$ spores are much more sensitive to these agents (Table 1).

It was initially surprising that α/β-type SASP binding does not protect spores against alkylating agents such as ethylene oxide (EtO) gas or ethyl methanesulfonate. However, α/β-type SASPs also do not prevent DNA alkylation *in vitro*. Ultimately, the determination of the α/β-type SASP-DNA structure at high resolution showed that DNA groups sensitive to alkylation are not at all shielded in the α/β-type SASP-DNA complex, as these sensitive groups are in the DNA's major groove and α/β-type SASPs bind in DNA's minor groove (25).

Repair of DNA damage

While in most cases chemical biocides do not kill spores by DNA damage, this is the case for some chemicals, including formaldehyde and nitrous acid as noted above. For these agents, DNA repair in spore outgrowth is also an important component of spore resistance to such agents, as shown by *recA* spores' increased sensitivity to these two chemicals (Table 1). Enzymes present in dormant spores are also important in repairing damage caused by oxidizing agents that can be generated during spore germination and outgrowth (11, 14).

Mechanisms Whereby Chemical Biocides Kill Spores

In addition to multiple mechanisms used by spores to resist chemical biocides, there are also multiple ways in which chemical biocides kill spores (Table 2), including (i) DNA damage; (ii) damage to the spore's IM; (iii) damage to one or more key spore core enzymes; (iv) damage to the spore germination apparatus; (v) breaching all spore permeability barriers; and (vi) unknown mechanisms.

DNA damage

As noted above, some chemical biocides kill spores by DNA damage, including formaldehyde and nitrous acid, as well as alkylating agents such as EtO. However, for at least some potentially genotoxic chemicals, spore DNA is so well protected that damage to other spore components is how spores are killed.

IM damage

A large number of oxidizing agents kill spores by causing some kind of damage to the spore's IM, such that this membrane readily ruptures when spores germinate, leading to rapid spore death (2, 11–14). However, the precise nature of this damage is not known, although it is not oxidation of unsaturated fatty acids. This likely IM damage can also sensitize spores to killing by other treatments such as wet or dry heat or desiccation (11, 14, 48). Indeed, mild pretreatment of spores with oxidizing agents potentiates DPA release from these spores when they are given a normally sublethal heat treatment (11–14).

Damage to key spore enzyme(s)

Several small oxidizing agents, H_2O_2 being the best studied, do not kill spores by DNA damage, yet they readily penetrate the spore core (11, 14). These agents cause significant inactivation of sensitive enzymes in the core, and in a few cases this enzyme inactivation precedes spore killing. While it is thus tempting to speculate that inactivation of one or more key spore core enzymes is the mechanism by which H_2O_2 and a few other peroxides kill spores, this has not been proven, and a key spore enzyme whose inactivation by H_2O_2 might lead to spore death has not been identified.

Damage to the spore germination apparatus

While there is no question that some chemical agents can inactivate one or more essential germination components, it is important to note that, in many cases, this type of defect can often be bypassed. In other words, spores killed by such a mechanism may be only conditionally dead and can be revived given proper treatment. The best-studied example of this phenomenon is spores treated with NaOH that appear dead, since the NaOH has inactivated the lytic enzymes needed to complete spore germination (11, 14). However, when plated with small amounts of lysozyme, these NaOH-treated spores exhibit normal viability. Other chemicals that also have effects on spore germination include some peroxides and dialdehydes. However, the specific targets of these agents have not been identified.

Breaching all spore permeability barriers

Strong acids have a remarkable sporicidal effect. When spores of many species are rapidly mixed with high concentrations of mineral acids, the spores undergo what is called "acid-popping," in which the spores appear to rupture violently, releasing all spore contents (2, 11, 14). The cause of this dramatic effect is not well understood, but may well tell us something about the forces that maintain the intact spore structure. Interestingly, the silicon layer present in the outer layers of spores of some *Bacillus* species, but not *B. subtilis*, is important in protecting spores against killing by acid (49).

Unknown factors

A huge number of chemicals have been tested and found to have at least some sporicidal efficacy (11, 14). Not surprisingly, there is minimal if any knowledge of the mechanism of spore killing by many of these chemicals.

Heat Resistance

Spores are much more resistant to high temperatures than growing cells, both in the dry and wet states (11) (Table 1). In water, spores are resistant to 40°C higher temperatures than are growing cells of the same species, and to ~30°C higher temperatures when dry. While some of the factors in spore resistance to wet

and dry heat are identical (Table 3), the mechanisms of spore killing by wet and dry heat are different (Table 2), with protein damage likely killing spores exposed to wet heat, while DNA damage is a major mechanism whereby spores are killed by dry heat (5, 6, 11, 50–52).

Dry Heat

The major known mechanism that protects spores from dry heat is the saturation of spore DNA with α/β-type SASPs, and, as expected, $\alpha^-\beta^-$ spores have greatly decreased dry heat resistance. This is actually similar to the dry heat resistance of growing cells (Table 1) (11). However, even α/β-type SASP saturation of spore DNA is not sufficient to prevent dry heat from killing spores by DNA damage, with at least some of this damage being depurination. Consequently, DNA repair during spore outgrowth, at least some of which is RecA dependent, is also an important factor in spore dry heat resistance (Table 1), and a number of proteins important in this repair have been identified (11, 12, 53). Spores' large DPA depot is also important in protecting spores against dry heat (54) (Table 1); however, the mechanism of this effect is not clear. The specific divalent ions chelated to DPA may also be important in spore resistance to dry heat, with Mn^{2+} being a cation whose level may be particularly important, at least with spores of some species (11, 32, 55). However, Mn^{2+} levels appear to play no role in spores' γ-radiation resistance. Spores can also be sensitized to dry heat treatment by pretreatment with oxidizing agents (48).

Wet Heat

Resistance to wet heat is probably the property most associated with spores, as spores of some species are resistant to ≥100°C for extended periods of time (2, 11, 12). A large number of factors are involved in spore resistance to wet heat (Table 3), including (i) DNA saturation with α/β-type SASPs; (ii) DPA; (ii) divalent metal ion content and identity; (iv) growth temperature optimum of the strain; (v) sporulation temperature; (vi) solid versus liquid sporulation medium; (vii) core water content; and (viii) additives present in the solution, in particular the pH, during wet heat treatment. There is also significant evidence that wet heat resistance varies considerably between individuals in spore populations, although the reason for this heterogeneity is not known (see below).

α/β-Type SASP

Wet heat does not kill spores by DNA damage (Table 1), which is somewhat surprising since the temperatures to which spores are resistant would be expected to lead to significant DNA damage, specifically depurination. This indicates that there must be significant protection of DNA against wet heat damage in spores, and this is due to the saturation of spore DNA with α/β-type SASPs in spores of both *Bacillus* and *Clostridium* species (Table 1) (2, 11, 39, 40). Presumably, the protection of spore DNA due to α/β-type SASP binding is so great that damage to some other spore component is what kills spores. Indeed, spores lacking α/β-type SASPs are killed at lower temperatures than wild-type spores and are killed by DNA damage (Table 1) (11). Much of the wet heat damage in $\alpha^-\beta^-$ spores is base loss likely via depurination, and α/β-type SASP binding has been shown to block depurination of DNA *in vitro* (11).

DPA

DPA may play no direct role in spore resistance to wet heat, but it certainly plays an indirect role, as when DPA is not accumulated in the spore core, the core's water content does not decrease as much as in DPA-replete spores (11, 54) (Table 1). This is also the case in DPA-less *C. perfringens* spores (56). Since core water content is a major factor in spore wet heat resistance (see below), anything that results in increased spore core water content will decrease spore wet heat resistance. Consequently, DPA-less spores of both *Bacillus* and *Clostridium* species have higher core water content and significantly reduced wet heat resistance (Table 1) (11, 56).

Divalent metal ions

The spore core's DPA depot is present primarily as a 1:1 chelate with divalent ions. The most common ion chelated with DPA in spores is Ca^{2+}, although other divalent cations can be substituted, and even monovalent cations can replace the divalent cations. The precise type of cation associated with DPA has a significant effect on spore wet heat resistance, with Ca^{2+} generally giving the highest resistance and monovalent cations the lowest (4, 11, 57). There are also reports that altering the levels of some DPA-associated divalent cations, in particular Mn^{2+}, can influence the wet heat resistance of spores in some but not all species, although the reason for this effect is not clear (2, 32, 55, 57).

Growth temperature optimum of the strain

It has long been known that spores of thermophiles have higher wet heat resistance than spores of mesophiles, with the latter generally being more wet heat resistant than the spores of psychrophiles (2, 11). The major reason for this effect is most likely that proteins in organisms that grow optimally at high temperatures are generally more thermostable than proteins in organisms that grow optimally at lower temperatures. Given that it appears most likely that it is damage to one or more key proteins that results in spore killing by wet heat (see below), it will thus generally require higher temperatures to inactivate a thermophile's protein in spores. It is also notable that enzymes in spores exhibit resistance to ~40°C higher temperatures than do the same enzymes *in vitro*.

Sporulation temperature

Within a temperature range at which sporulation remains reasonably efficient, sporulation of a variety of different species and strains at higher temperatures results in spores with higher wet heat resistance than that of spores produced at lower temperatures (2, 11, 40, 58–63). The major factor responsible for this effect appears to be spore core water content, which is lower in spores produced at higher temperatures, and there is a reasonably coherent inverse relationship between core water content and spore wet heat resistance (see below). While other spore properties may also change depending on the sporulation temperature, no changes other than to core water content have been directly correlated with changes in levels of spore wet heat resistance.

Solid versus liquid sporulation

Sporulation at the same temperature on solid or liquid media also results in spores with slightly different wet heat resistance, and spores' wet heat resistance is also altered by specific characteristics of solid sporulation media (64, 65). However, the reasons for these effects are not known.

Core water content

The water content in the spore core is probably the major factor in spore wet heat resistance, certainly for values between ~30 and 50% of core wet weight as water (11). The presumption is that a low core water content results in reduced molecular mobility of core proteins and thus elevated protein resistance to thermal inactivation. Indeed, protein mobility in the dormant spore core is extremely low, as a normally soluble core protein, green fluorescent protein, is immobile in dormant spores, while its diffusion coefficient increases >4 orders of magnitude when spores complete their germination (46). There are, however, a number of unknowns about spore core water content, including (i) how core water content is lowered during spore formation—the spore cortex structure and the SpmA/B and DacB proteins play roles in determining spore core water content (11, 66–68), but how is not known—and (ii) how much spore core water is free water and how much is bound water; available evidence suggests that there is normally very little free water in the spore core, but this is based on only a very few measurements.

Additives present during wet heat treatment

There is an extensive older literature on the effects of additives on spore killing by wet heat, with pH being one variable that has been extremely well studied (2, 69). Thus pH values above 8 and below 6 during wet heat treatment decrease spores' wet heat resistance, with larger effects as the pH diverges further from these values. There is also a recent report that several antimicrobial peptides can decrease spores' wet heat resistance, although this effect may be species specific (70).

Mechanism of Wet Heat Killing of Spores

While dry heat kills spores by DNA damage, this is not how wet heat kills spores (Tables 1 and 2). Indeed, wet heat-treated spore populations accumulate no mutations and no DNA damage, and DNA repair defects, including a *recA* mutation, do not sensitize spores to wet heat (Table 1) (2, 11). Analysis of the kinetics of spore killing by wet heat indicates that release of the spore's DPA takes place after spore killing (5, 6, 50). However, spore killing is paralleled and even preceded by damage to spore core proteins, including denaturation. These results suggest that wet heat kills spores by damage to one or more key spore proteins. However, the identity of these key proteins has not been established. Spores given a sublethal wet heat treatment also commonly exhibit slow germination, most likely because of inactivation of cortex-lytic enzymes essential for completion of spore germination (71). However, inactivation of cortex-lytic enzymes alone is not how wet heat treatment kills spores, since there is minimal if any recovery of wet heat-treated spores by subsequent lysozyme treatment.

Heterogeneity in Spore Wet Heat Resistance

As is probably not surprising, there is significant evidence that the wet heat resistance of individual spores in a population is quite heterogeneous (14, 40, 51, 52, 72–76). The most striking evidence for this is the analysis of the behavior of individual spores in water incubated at elevated temperatures when the level of spore DPA and the state of spore proteins is monitored throughout the incubation. Strikingly, DPA release from an individual spore incubated at 80 to 90°C takes only a few minutes, but begins only after an extremely

variable lag period of minutes to hours in individual spores. Unfortunately, the factors that determine the length of this lag period prior to DPA release at elevated temperatures are not known, although DPA release is preceded by changes in spores' protein spectrum that suggest there is some protein denaturation just prior to fast DPA release (5, 51, 52). Presumably, the spores with the longest lag periods prior to initiation of DPA release at elevated temperatures are the most wet-heat-resistant spores in populations. That spore populations do indeed have some spores with much higher levels of wet heat resistance has also been shown directly by isolation of superdormant spores as 1 to 2% of spore populations and demonstration that these spores have higher wet heat resistance than the general spore population (73). These latter spores also had a lower core water content than the spore population as a whole.

Miscellaneous Spore Resistance Properties

High Pressure

Spores are much more resistant than growing cells to extremely high pressures (HPs), and they are also germinated by HP (77). The germination of spores, in particular, the release of spores' DPA, is an essential step leading to spore inactivation by HP, which is most often by an elevated temperature (77–80). The reason for spores' high resistance to HP is not known, but this does not require either α/β-type SASPs or an intact spore coat. The mechanism by which HP alone kills spores is also not known; as noted above, efficient HP killing of spores generally requires high temperatures, though these temperatures are significantly lower than needed to kill spores in the absence of concomitant HP treatment.

Abrasion

Treatment of spores with abrasives in either the wet or dry state can lead to spore disruption with or without prior spore germination (81, 82). Dormant spores are significantly more resistant to killing by abrasion than are growing cells or germinated spores. An intact spore coat is not required for spore resistance to abrasion. Attempts to generate *B. subtilis* spores that completely lack an outer extremely insoluble protein layer termed the "rind" have been unsuccessful, so it remains possible that it is this structure that is responsible for spore abrasion resistance.

Predation by Bacteriovores

B. subtilis spores are readily ingested by soil- and water-dwelling protozoa and nematodes, and wild-type

B. subtilis spores are excreted by these organisms with their viability unchanged by their passage through digestive systems or phagocytic vacuoles (17, 18). In contrast, spores with coat defects that render the spores sensitive to lysozyme are rapidly digested by these bacteriovores, with the spore contents used to support the predators' growth. However, not all spore components are digested, as the highly insoluble rind structure noted above is excreted. It has been suggested that the complex spore coat structure might have evolved in part to preclude spore destruction by predators that prey on bacteria.

Freeze-Thawing

Spores are routinely resistant to multiple cycles of freezing and thawing, even in the absence of exogenous osmotic stabilizers, and even if the freezing process is extremely slow (11). In contrast, growing bacteria are often killed by multiple freeze-thaw cycles. This resistance property has not been well studied, although neither an intact spore coat nor α/β-type SASPs are involved in spore resistance to freezing and thawing.

Desiccation

Wild-type spores are resistant to multiple cycles of desiccation and rehydration, while growing bacteria are often killed by a single desiccation treatment unless specific compatible solutes are present in the solution from which the growing cells are dried (Table 1) (2, 11, 54). Two factors appear to play a role in spore desiccation resistance: (i) α/β-type SASPs and (ii) DPA (Table 1). In the absence of α/β-type SASPs, desiccation treatments do kill spores by DNA damage, and the viability of $\alpha^-\beta^-$ DPA-less spores is extremely low. Interestingly, while wild-type spores are essentially completely resistant to freeze-drying under low or high vacuums, desiccation under ultra-high vacuum ($\leq 10^{-9}$ torr) does result in spore killing (79). This latter killing appears to be due to DNA damage, something also seen with growing cells. However, factors that might be involved in spore resistance to ultra-high vacuum have not been studied.

Gas Dynamic Heating

There has been significant interest recently in the killing of spores in a high-temperature gas environment at high shock pressures, undoubtedly because of interest in destroying *B. anthracis* spores being stored or developed as a potential bioweapon. Spores are indeed killed in such environments in milliseconds at temperatures >500 K (83, 84). However, the mechanism of spore killing in this type of environment has not been studied, nor have factors that help spores resist such treatment.

Plasma

There is currently significant interest in the use of nonthermal gas discharge plasmas for spore killing, in particular in decontamination of medical devices or packaged materials (85, 86). Unfortunately, there have been very few thorough studies of factors involved in spore resistance to or killing by nonthermal plasma. While a number of studies have noted severe morphological damage to spores by plasma, most of this damage seems likely to have taken place long after spore killing. There are a few studies indicating that plasma with significant associated UV radiation kills spores by DNA damage. However, plasma without associated UV radiation also can kill spores, suggesting there is a lethal target in addition to DNA, but this target has not been identified.

Supercritical Fluids

Recent work has examined the use of supercritical fluids, in particular supercritical CO_2, for spore inactivation (14, 87, 88, 92). While the use of supercritical CO_2 alone requires relatively high temperatures for spore killing, small amounts of additives such as H_2O_2, peracetic acid, or water allow spore killing at more moderate temperatures and pressures. Spore killing by these supercritical fluids can take place with little obvious change in spore morphology or permeability. The mechanisms of spore killing by and resistance to supercritical fluids are not known. However, recent work indicates that supercritical CO_2 plus peracetic acid kills spores by some type of inner membrane damage (92).

FINAL THOUGHTS

While much is known about the mechanisms of spore resistance, there is still much to be learned. In spore resistance to wet heat, the precise mechanisms that modulate spore core water content remain unknown, as is the identity of specific proteins that are the targets of spore wet heat killing. In spore resistance to chemical biocides, the reason for the spore IM's low permeability to such chemicals is also unknown, as is the precise likely IM damage that causes spore killing by many oxidizing agents. There are also a number of general questions that invite further work. Foremost among these is the question of whether mechanisms of spore resistance established largely with *B. subtilis* spores are also the case with spores of other *Bacillales* species as well as with spores of *Clostridiales* species. While this seems likely, given the extreme applied importance of spores of a number of *Clostridium* species, this would appear to be an important question to focus on, especially now that methods are available for genetic manipulation of a number of *Clostridium* species.

Another question that has arisen recently concerns the heterogeneity in the resistance properties of individual spores in populations (8, 51, 52, 72, 74, 89). There is now significant evidence for heterogeneity in the resistance of spores to at least wet heat, and this seems likely to be true for resistance to other agents as well. What are the causes of this heterogeneity, and can the answer to this question give us further information on the mechanisms of spore resistance? There has been significant interest recently in heterogeneity in gene expression between individuals in bacterial populations, including sporulating cells, and this has been shown to play a major role in the heterogeneity in germination properties of spores in populations. While the causes of this heterogeneity in gene expression in sporulation are not completely understood, it seems likely to be in large part due to stochasticity. Might this also be the cause of heterogeneity in resistance properties of individual spores in populations?

Finally, it should be noted that there is increasing evidence that spore resistance is not completely static, but can change as spores "mature" (63, 75, 76). Thus, *B. subtilis* spore wet heat resistance appears to increase markedly even after spores are released from the sporangium, although the precise mechanism for this effect is not known. In addition, there is recent evidence that there are significant changes in RNAs present in dormant spores, again well after their release from sporangia (90). While the precise meaning of the latter changes is unknown, it certainly seems possible that these changes could also modulate spores' intrinsic resistance well after spores are released from sporangia. All in all, the resistance of spores undoubtedly still has a number of new and fascinating surprises in store for us.

Acknowledgments. Work in the author's laboratory on spore resistance has been supported by grants from the National Institutes of Health (GM19698) and the Army Research Office and by support from a Department of Defense Multidisciplinary University Research Initiative through the U.S. Army Research Laboratory and the U.S. Army Research Office under contract number W911NF-09-1-0286.

Citation. Setlow P. 2014. Spore resistance properties. Microbiol Spectrum 2(5):TBS-0003-2012.

References

1. Peck MW, Stringer SC, Carter AT. 2011. *Clostridium botulinum* in the post-genomic era. *Food Microbiol* **28**: 183–191.

2. Setlow P, Johnson EA. 2012. Spores and their significance, p 45–79. *In* Doyle MP, Bucanan R (ed), *Food Microbiology: Fundamentals and Frontiers*, 4th ed. ASM Press, Washington, DC.

3. Potot S, Serra CR, Henriques AO, Schyns G. 2010. Display of recombinant proteins on *Bacillus subtilis* spores, using a coat-associated enzyme as the carrier. *Appl Environ Microbiol* **76**:5926–5933.

4. Bader J, Albin A, Stahl U. 2012. Spore-forming bacteria and their utilization as probiotics. *Benef Microbes* **3**: 67–75.

5. Coleman WH, Chen D, Li Y-q, Cowan AE, Setlow P. 2007. How moist heat kills spores of *Bacillus subtilis*. *J Bacteriol* **189**:8458–8466.

6. Coleman WH, Zhang P, Li Yq, Setlow P. 2010. Mechanism of killing of spores of *Bacillus cereus* and *Bacillus megaterium* by wet heat. *Lett Appl Microbiol* **50**:507–514.

7. Setlow B, Parish S, Zhang P, Li YQ, Neely C, Setlow P. 2014. Mechanism of killing of spores of *Bacillus anthracis* in a high-temperature gas environment, and analysis of DNA damage generated by various decontamination treatments of spores of *Bacillus anthracis*, *Bacillus subtilis* and *Bacillus thuringiensis*. *J Appl Microbiol* **116**:805–814.

8. Brul S, van Beilen J, Caspers M, O'Brien A, de Koster C, Oomes S, Smelt J, Kort R, Ter Beek A. 2011. Challenges and advances in systems biology analysis of *Bacillus* spore physiology; molecular differences between an extreme heat resistant spore forming *Bacillus subtilis* food isolate and a laboratory strain. *Food Microbiol* **28**: 221–227.

9. Leggett MJ, McDonnell G, Denyer SP, Setlow P, Maillard J-Y. 2012. Bacterial spore structures and their protective role in biocide resistance. *J Appl Microbiol* **113**:485–499.

10. Maillard J-Y. 2011. Innate resistance to sporicides and potential failure to decontaminate. *J Hosp Infect* **77**: 204–209.

11. Setlow P. 2006. Spores of *Bacillus subtilis*: their resistance to radiation, heat and chemicals. *J Appl Microbiol* **101**:514–525.

12. Setlow P. 2007. I will survive: DNA protection in bacterial spores. *Trends Microbiol* **15**:172–180.

13. Setlow P. 2010. Resistance of bacterial spores, p 319–332. *In* Storz G, Hengge R (ed), *Bacterial Stress Response*, 2nd ed. American Society for Microbiology, Washington, DC.

14. Setlow P. 2012. Resistance of bacterial spores to chemical agents, p 121–130. *In* Maillard J-Y, Fraise A, Sattar S (ed), *Principles and Practice of Disinfection, Preservation & Sterilization*, 4th ed. Wiley-Blackwell, Oxford, United Kingdom.

15. Ter Beek A, Brul S. 2010. To kill or not to kill *Bacilli*: opportunities for food biotechnology. *Curr Opin Biotechnol* **21**:168–174.

16. Henriques AO, Moran CP Jr. 2007. Structure, assembly, and function of the spore surface layers. *Annu Rev Microbiol* **61**:555–588.

17. Klobutcher LA, Ragkousi K, Setlow P. 2006. The *Bacillus subtilis* spore coat provides "eat resistance" during phagosomal predation of the protozoan *Tetrahymena thermophila*. *Proc Natl Acad Sci USA* **103:**165–170.

18. Laaberki MH, Dworkin J. 2008. Role of spore coat proteins in the resistance of *Bacillus subtilis* spores to *Caenorhabditis elegans* predation. *J Bacteriol* **190:**197–203.

19. Bosak T, Losick RM, Pearson A. 2008. A polycyclic terpenoid that alleviates oxidative stress. *Proc Natl Acad Sci USA* **105:**6725–6729.

20. Khaneja R, Perez-Fons L, Fakhry S, Baccigalupi L, Steiger S, To E, Sandmann G, Dong TC, Ricca E, Fraser PD, Cutting SM. 2010. Carotenoids found in *Bacillus*. *J Appl Microbiol* **108:**1889–1902.

21. Griffiths KK, Setlow P. 2009. Effects of modification of membrane lipid composition on *Bacillus subtilis* sporulation and spore properties. *J Appl Microbiol* **106:**2064–2078.

22. Cortezzo DE, Setlow P. 2005. Analysis of factors that influence the sensitivity of spores of *Bacillus subtilis* to DNA damaging chemicals. *J Appl Microbiol* **98:**606–617.

23. Loisan P, Hosny NA, Gervais P, Champion D, Kuimova MK, Perrier-Cornet JM. 2013. Direct investigation of viscosity of an atypical inner membrane of *Bacillus* spores: a molecular rotor/FLIM study. *Biochim Biophys Acta* **1828:**2436–2443.

24. Sunde EP, Setlow P, Hederstedt L, Halle B. 2009. The physical state of water in bacterial spores. *Proc Natl Acad Sci USA* **106:**19334–19339.

25. Lee KS, Bumbaca D, Kosman J, Setlow P, Jedrzejas MJ. 2008. Structure of a protein-DNA complex essential for DNA protection in spores of *Bacillus* species. *Proc Natl Acad Sci USA* **105:**2806–2811.

26. Fairhead H. 2009. SASP gene delivery: a novel antibacterial approach. *Drug News Perspect* **22:**197–203.

27. Ramírez-Gaudiana FH, Barraza-Salas M, Ramírez-Ramírez N, Ortiz-Cortés M, Setlow P, Pedraza-Reyes M. 2012. Alternative excision repair of ultraviolet B- and C-induced DNA damage in dormant and developing spores of *Bacillus subtilis*. *J Bacteriol* **194:**6096–6104.

28. Maclean M, Murdoch LE, MacGregor SJ, Anderson JG. 2013. Sporicidal effects of high-intensity 405 nm visible light on endospore-forming bacteria. *Photochem Photobiol* **89:**120–126.

29. Tyler GSD, Dai T, Hamblin MR. 2013. Killing bacterial spores with blue light: when innate resistance meets the power of light. *Photochem Photobiol* **89:**2–4.

30. Levy C, Aubert X, Lacour B, Carlin F. 2012. Relevant factors affecting microbial surface decontamination by pulsed light. *Int J Food Microbiol* **152:**168–174.

31. Moeller R, Setlow P, Reitz G, Nicholson WL. 2009. Roles of small, acid-soluble spore proteins and core water content in survival of *Bacillus subtilis* spores exposed to environmental solar UV radiation. *Appl Environ Microbiol* **75:**5202–5208.

32. Moeller R, Wassmann M, Reitz G, Setlow P. 2011. Effect of radioprotective agents in sporulation medium on *Bacillus subtilis* spore resistance to hydrogen peroxide, wet heat and germicidal and environmentally relevant UV radiation. *J Appl Microbiol* **110:**1485–1494.

33. Ghosh S, Ramirez-Peralta A, Gaidamakova E, Zhang P, Li Y-Q, Daly MJ, Setlow P. 2011. Effects of Mn levels on resistance of *Bacillus megaterium* spores to heat, radiation and hydrogen peroxide. *J Appl Microbiol* **111:**663–670.

34. Ibarra JR, Orozco AD, Rojas JA, Lopez K, Setlow P, Yasbin RE, Pedraza-Reyes M. 2008. Role of the Nfo and ExoA apurinic/apyrimidinic endonucleases in repair of DNA damage during outgrowth of *Bacillus subtilis* spores. *J Bacteriol* **190:**2031–2038.

35. Moeller R, Setlow P, Pedraza-Reyes M, Okayasu R, Reitz G, Nicholson WL. 2011. Role of the Nfo and ExoA apurinic/apyrimidinic (AP) endonucleases in the radiation resistance and radiation-induced mutagenesis of *Bacillus subtilis* spores. *J Bacteriol* **193:**2875–2879.

36. Chandra T, Silver SC, Zilinskas E, Shepard EM, Broderick WE, Broderick JB. 2009. Spore photoproduct lyase catalyzes specific repair of the 5R but not the 5S spore photoproduct. *J Am Chem Soc* **131:**2420–2421.

37. Yang L, Nelson RS, Benjdia A, Lin G, Telser J, Stoll S, Schlichting I, Li L. 2013. A radical transfer pathway in spore photoproduct lyase. *Biochemistry* **52:**3041–3050.

38. Li J, Paredes-Sabja D, Sarker MR, McClane BA. 2009. Further characterization of *Clostridium perfringens* small acid soluble protein-4 (Ssp4) properties and expression. *PLoS One* **4:**e6249. doi:10.1371/journal.pone.0006249.

39. Raju D, Walters M, Setlow P, Sarker MR. 2006. Investigating the role of small, acid-soluble spore proteins (SASPs) in the resistance of *Clostridium perfringens* spores to heat. *BMC Microbiol* **6:**50. doi:10.1186/1471-2180-6-50.

40. Leyva-Illades JF, Setlow B, Sarker MR, Setlow P. 2007. Effect of a small, acid-soluble spore protein from *Clostridium perfringens* on the resistance properties of *Bacillus subtilis* spores. *J Bacteriol* **189:**7927–7931.

41. Moeller R, Raguse M, Reitz G, Okayasu R, Li Z, Klein S, Setlow P, Nicholson WL. 2014. Resistance of *Bacillis subtilis* spore DNA to lethal ionizing radiation damage relies heavily on spore core components and DNA repair, with minor effects of oxygen radical detoxification. *Appl Environ Microbiol* **80:**104–109.

42. Moeller R, Setlow P, Horneck G, Berger T, Reitz G, Rutberg P, Doherty AJ, Okayasu R, Nicholson WL. 2008. Role of major small, acid-soluble spore proteins, spore specific and universal DNA repair mechanisms in the resistance of *Bacillus subtilis* spores to ionizing radiation from X-rays and high energy charged (HZE) particle bombardment. *J Bacteriol* **190:**1134–1140.

43. Vlasic I, Mertens R, Seco EM, Carrasco B, Ayora S, Reitz G, Commichau FM, Alonso JC, Moeller R. 2014. *Bacillus subtilis* RecA and its accessory factors, RecF, RecO, RecR and RecX, are required for spore resistance to DNA double-strand breaks. *Nucleic Acids Res* **42:**2295–2307.

44. Checinska A, Burbank M, Paszczynski AJ. 2012. Protection of *Bacillus pumilus* spores by catalases. *Appl Environ Microbiol* **78:**6413–6422.

45. Cybulski RJ Jr, Sanz P, Alem F, Stibitz S, Bull RL, O'Brien AD. 2009. Four superoxide dismutases contribute to *Bacillus anthracis* virulence and provide spores with redundant protection from oxidative stress. *Infect Immun* 77:274–285.

46. Cowan AE, Koppel DE, Setlow B, Setlow P. 2003. A soluble protein is immobile in dormant spores of *Bacillus subtilis* but is mobile in germinated spores: implications for spore dormancy. *Proc Natl Acad Sci USA* 100: 4209–4214.

47. Castellanos-Juarez FX, Alvarez-Alvarez C, Yasbin RE, Setlow B, Setlow P, Pedraza-Reyes M. 2006. YtkD and MutT protect vegetative cells but not spores of *Bacillus subtilis* from oxidative stress. *J Bacteriol* 188:2285–2289.

48. Armas A de Benito, Padula NL, Setlow B, Setlow P. 2008. Sensitization of *Bacillus subtilis* spores to dry heat and desiccation by pre-treatment with oxidizing agents. *Lett Appl Microbiol* 46:492–497.

49. Hirota R, Hata Y, Ikeda T, Ishida T, Kuroda A. 2010. The silicon layer supports acid resistance of *Bacillus cereus* spores. *J Bacteriol* 192:111–116.

50. Coleman WH, Setlow P. 2009. Analysis of damage due to moist heat treatment of spores of *Bacillus subtilis*. *J Appl Microbiol* 106:1600–1607.

51. Zhang P, Kong L, Setlow P, Li YQ. 2010. Characterization of wet heat inactivation of single spores of *Bacillus* species by dual-trap Raman spectroscopy and elastic light scattering. *Appl Environ Microbiol* 76:1796–1805.

52. Zhang P, Kong L, Wang G, Setlow P, Li Y-Q. 2011. Monitoring the wet-heat inactivation dynamics of single spores of *Bacillus* species using Raman tweezers, differential interference contrast and nucleic acid dye fluorescence microscopy. *Appl Environ Microbiol* 77:4754–4769.

53. Barraza-Salas M, Ibarra-Rodriguez JR, Mellado SJ, Salas-Pacheco J-M, Setlow P, Pedraza-Reyes M. 2010. Effects of forespore-specific overexpression of apurinic/ apyrimidinic-endonuclease Nfo on the DNA-damage resistance properties of *Bacillus subtilis* spores. *FEMS Lett* 302:159–165.

54. Magge A, Granger AC, Wahome PG, Setlow B, Vepachedu VR, Loshon CA, Peng L, Chen D, Li Y-q, Setlow P. 2008. Role of dipicolinic acid in the germination, stability and viability of spores of *Bacillus subtilis*. *J Bacteriol* 190:4798–4807.

55. Granger AC, Gaidamakova EK, Matrosova VY, Daly MJ, Setlow P. 2011. Effects of levels of Mn and Fe on *Bacillus subtilis* spore resistance, and effects of Mn^{2+}, other divalent cations, orthophosphate, and dipicolinic acid on resistance of a protein to ionizing radiation. *Appl Environ Microbiol* 77:32–40.

56. Paredes-Sabja D, Setlow B, Setlow P, Sarker MR. 2008. Characterization of *Clostridium perfringens* spores that lack SpoVA proteins and dipicolinic acid. *J Bacteriol* 190: 4648–4659.

57. Mah JH, Kang DH, Tang J. 2008. Effects of minerals on sporulation and heat resistance of *Clostridium sporogenes*. *Int J Food Microbiol* 128:385–389.

58. Garcia D, van der Voort M, Abee T. 2010. Comparative analysis of *Bacillus weihenstephanensis* KBAB4 spores obtained at different temperatures. *Int J Food Microbiol* 140:146–153.

59. Huo Z, Zhang N, Raa W, Huang X, Yong X, Liu Y, Wang D, Li S, Shen Q, Zhang R. 2012. Comparison of the spores of *Paenibacillus polymyxa* prepared at different temperatures. *Biotechnol Lett* 34:925–933.

60. Mah JH, Kang DH, Tang J. 2009. Comparison of viability and heat resistance of *Clostridium sporogenes* stored at different temperatures. *J Food Sci* 74:M23–M27.

61. Planchon S, Dargaignaratz C, Levy C, Ginies C, Broussolle V, Carlin F. 2011. Spores of *Bacillus cereus* strain KBAB4 produced at 10°C and 30°C display variations in their properties. *Food Microbiol* 28:291–297.

62. Baril E, L. Coroller L, Couvert O, Leguerinel I, Postollec F, Boulais C, Carlin F, Mafart P. 2012. Modeling heat resistance of *Bacillus weihenstephanensis* and *Bacillus licheniformis* spores as function of sporulation temperature and pH. *Food Microbiol* 30:29–36.

63. Minh HNT, Durand A, Loison P, Perrier-Cornet J-M, Gervais P. 2011. Effect of sporulation conditions on the resistance of *Bacillus subtilis* spores to heat and high pressure. *Appl Micro Cell Physiol* 90:1409–1417.

64. Rose R, Setlow B, Monroe A, Malozzi M, Driks A, Setlow P. 2007. Comparison of the properties of *Bacillus subtilis* spores made in liquid or on plates. *J Appl Microbiol* 103:691–699.

65. Stecchini ML, Spaziani M, Del Torre M, Pacor S. 2009. *Bacillus cereus* cell and spore properties as influenced by the micro-structure of the medium. *J Appl Microbiol* 106:1838–1848.

66. Orsburn B, Sucre K, Popham DL, Melville SB. 2009. The SpmA/B and DacF proteins of *Clostridium perfringens* play important roles in spore heat resistance. *FEMS Microbiol Lett* 291:188–194.

67. Paredes-Sabja D, Sarker N, Setlow B, Setlow P, Sarker M. 2008. Roles of DacB and Spm proteins in *Clostridium perfringens* spore resistance to moist heat and chemicals. *Appl Environ Microbiol* 74:3730–3738.

68. Popham DL, Sengupta S, Setlow P. 1995. Heat, hydrogen peroxide and UV resistance of *Bacillus subtilis* spores with an increased core water content and with or without major DNA-binding proteins. *Appl Environ Microbiol* 61:3633–3638.

69. Derossi A, Fiore AG, De Pilli T, Severini C. 2011. A review on acidifying treatments for vegetable canned food. *Crit Rev Food Sci Nutr* 51:955–964.

70. Cabo ML, Torres B, Herrera JJ, Bernardez M, Pastoriza L. 2009. Application of nisin and pediocin against resistance and germination of *Bacillus* spores in sous vide products. *J Food Prot* 72:515–523.

71. Wang G, Zhang P, Setlow P, Li YQ. 2011. Kinetics of germination of wet heat-treated individual spores of *Bacillus* species as followed by Raman spectroscopy and differential interference contrast microscopy. *Appl Environ Microbiol* 77:3368–3379.

72. Eijlander RT, Abee T, Kuipers OP. 2011. Bacterial spores in food: how phenotypic variability complicates

prediction of spore properties and bacterial behavior. *Curr Opin Biotechnol* **22**:180–186.

73. Ghosh S, Zhang P, Li Y-q, Setlow P. 2009. Superdormant spores of *Bacillus* species have elevated wet heat resistance and temperature requirements for heat activation. *J Bacteriol* **191**:5584–5591.

74. Hornstra LM, Ter Beek A, Smelt JP, Kallemeijn WW, Brul S. 2009. On the origin of heterogeneity in (preservation) resistance of *Bacillus* spores: Input for a 'systems' analysis approach of bacterial spore outgrowth. *Int J Food Microbiol* **134**:9–15.

75. Rodriguez-Palacios A, Lejeune JT. 2011. Moist-heat resistance, spore aging, and superdormancy in *Clostridium difficile*. *Appl Environ Microbiol* **77**:3085–3091.

76. Sanchez-Salas J-L, Setlow B, Zhang P, Li YQ, Setlow P. 2011. Maturation of released spores is necessary for acquisition of full spore heat resistance during *Bacillus subtilis* sporulation. *Appl Environ Microbiol* **77**:6746–6754.

77. Setlow P. 2007. Germination of spores of *Bacillus subtilis* by high pressure, p 15–40. *In* Doona CJ, Feeherry FE (ed), *High Pressure Processing of Foods*. Wiley-Blackwell, Hoboken, NJ.

78. Georget E, Kapoor S, Winter R, Reineke K, Song Y, Callanan M, Ananta E, Heinz V, Mathys A. 2014. *In situ* investigation of *Geobacillus stearothermophilus* spore germination and inactivation mechanisms under moderate high pressure. *Food Microbiol* **41**:8–18.

79. Nicholson WL, Munakata N, Horneck G, Melosh HJ, Setlow P. 2000. Resistance of *Bacillus* endospores to extreme terrestrial and extraterrestrial environments. *Microbiol Mol Biol Rev* **64**:548–572.

80. Reineke K, Mathys A, Heinz V, Knorr D. 2013. Mechanisms of endospore inactivation under high pressure. *Trends Microbiol* **21**:296–304.

81. Reineke K, Schlumbach K, Baier D, Mathys A, Knorr D. 2013. The release of dipicolinic acid—the rate-limiting step of *Bacillus* endospore inactivation during the high pressure thermal sterilization process. *Int J Food Microbiol* **162**:55–63.

82. Jones CA, Padula NL, Setlow P. 2005. Effect of mechanical abrasion on the viability, disruption and germination of spores of *Bacillus subtilis*. *J Appl Microbiol* **99**:1484–1494.

83. Gates SD, McCartt AD, Jeffries JB, Hanson RK, Hokama LA, Mortelmans KE. 2011. Extension of *Bacillus* endospore gas dynamic heating studies to multiple species and test conditions. *J Appl Microbiol* **111**:925–931.

84. Gates SD, McCartt AD, Lappas P, Jeffries JB, Hanson RK, Hokama LA, Mortelmans KE. 2010. *Bacillus* endospore resistance to gas dynamic heating. *J Appl Microbiol* **109**:1591–1598.

85. Tseng S, Abramzon N, Jackson JO, Lin W-J. 2012. Gas discharge plasmas are effective in inactivating *Bacillus* and *Clostridium* species. *Appl Microbiol Biotechnol* **93**:2563–2570.

86. Yardimci O, Setlow P. 2010. Plasma sterilization: opportunities and microbial assessment strategies in medical device manufacturing. *IEEE Trans Plasma Sci* **38**:973–981.

87. Checinska A, Firth IA, Green TL, Crawford RL, Paszczynski AJ. 2011. Sterilization of biological pathogens using supercritical carbon dioxide containing water and hydrogen peroxide. *J Microbiol Methods* **87**:70–75.

88. Shieh E, Paszczynski A, Wai CM, Lang Q, Crawford RL. 2009. Sterilization of *Bacillus pumilus* spores using supercritical fluid carbon dioxide containing various modifier solutions. *J Microbiol Methods* **76**:247–252.

89. Setlow P, Liu J, Faeder JR. 2012. Heterogeneity in bacterial spore populations, p 201–216. *In* Abel-Santos E (ed), *Bacterial Spores: Current Research and Applications*. Horizon Scientific Press, Norwich, United Kingdom.

90. Segey E, Smith Y, Ben-Yehuda S. 2012. RNA dynamics in aging bacterial spores. *Cell* **148**:139–149.

91. Isticato R, Ricca E. 2014. Spore surface display. *Microbiol Spectrum* **2**(5):TBS-0011-2012.

92. Setlow B, Korza G, Blatt KM, Fey J, Setlow P. 2016. Mechanism of *Bacillus subtilis* spore inactivation by and resistance to supercritical CO_2 plus peracetic acid. *J Appl Microbiol* **120**:57–69.

The Bacterial Spore: From Molecules to Systems
Edited by P. Eichenberger and A. Driks
© 2016 American Society for Microbiology, Washington, DC
doi:10.1128/microbiolspec.TBS-0014-2012

Anne Moir[1]
Gareth Cooper[1]

Spore Germination

11

A PERSPECTIVE ON GERMINATION STUDIES

The specialized structure that maintains the dormancy and resistance properties of endospores provides an opportunity for wide dispersal of spores in the environment, and for survival over long periods under conditions unfavorable for growth. Although dormant and uniquely resistant to environmental insult (1, 206), a spore remains sensitive to changes in its environment. Specific germinants are detected by receptors in the spore inner membrane; this signal is then transduced by mechanisms that are not understood in detail, but result in the activation of proteins that variously allow movement of small molecules across the membrane and deconstruct protective layers, restoring normal hydration and active metabolism (2, 3).

In addition to recent reviews of germination (4, 5), the literature also provides a valuable earlier perspective. The germination behavior of a variety of species, including bacilli (6, 7) and clostridia, was reviewed by Gould (8). Biochemical and physiological changes during germination and outgrowth were examined (3, 9). In the decades following the 1970s, classical and molecular genetics, mainly in *Bacillus subtilis*, then allowed identification of many of the gene products involved in the germination process (2, 3, 10–12). More recently, genome sequencing has allowed wider comparisons across endosporeformers, and has led to direct testing of the contribution of candidate gene products to the overall germination behavior in other species in which gene knockouts are possible, as reviewed elsewhere (13–18). Multiple receptors in a species would allow effective germination in different environmental conditions, and are commonly found.

Common germinants include amino acids, sugars, purine nucleosides, inorganic salts, or combinations of these molecules (7, 19). Although there is no need for them to be metabolized, or to enter the spore core, they are often described as "nutrient" germinants to distinguish them from "nonnutrient" germinants, such as Ca^{2+}-dipicolinic acid (CaDPA) or dodecylamine. The amino acid L-alanine and its analogues are frequently effective (6) as sole germinants or in combination with others. Recently, the germinant receptors of *B. subtilis* have been shown to cluster in a "germinosome," and definitive evidence has been obtained for the role of SpoVA proteins in release of CaDPA from the spore core. Other current areas of research include detailed measurements of events at the single spore level, structure/function relationships in germination proteins, both *in vitro* and *in vivo*, and an appreciation of the properties of "superdormant" spores within a spore population.

Spore germination can also be induced by other means in a receptor-independent process. These agents

---

[1]Krebs Institute, Department of Molecular Biology & Biotechnology, University of Sheffield, Sheffield S10 2TN, United Kingdom.

include exogenous CaDPA (20), lysozyme (21), dodecylamine (22), or very high pressure (23). In addition, fragments of vegetative cell peptidoglycan (24, 25) induce spore germination in a protein kinase-dependent manner that is not at all understood.

KINETICS OF GERMINATION

In general, the response of a spore population to a germinant stimulus is heterogeneous. The timing of germination of individual spores was studied by phase-contrast microscopy (26, 27), and two parameters were defined: microlag, the time between the addition of germinant and the start of phase darkening, and microgermination, the time for germination-associated phase darkening to occur, as a result of rehydration of the spore core, which is much more rapid. The heterogeneity of germination was due to the variation in this microlag. Conveniently, the response of spores to germinant can be measured in suspension, where observations such as loss of heat resistance, release of dipicolinic acid (DPA) into the supernatant, and reduction in optical density reflect the aggregated behavior of the population. As a result, the kinetics of such overall changes will reflect the distribution of lag times in the spore population.

Dissection of germination events in large numbers of single spores of wild type and mutants, measured in real time, is now possible, using technologies such as Raman spectroscopy with laser tweezers to follow DPA release from single spores, which is concomitant with the loss of refractility, and differential interference contrast (DIC) microscopy, which can also detect the rehydration that accompanies cortex hydrolysis (28–30). Such approaches have confirmed and extended the earlier work, showing for example that the level of germinant receptors in the spore affects the lag time, but not the 0.5 to 3 min taken to release the bulk of the DPA. This bulk release is slowed in a CwlJ mutant and is faster in a spore with higher-than-normal levels of SpoVA proteins. In some spores, up to 15% of DPA may be released during the "lag" period, suggesting that limited germination-associated changes in the spore may occur well before the point when the spore is committed to germinate.

From a germinating spore population, the spores that are less responsive to germinant and have remained dormant can be separated on a density gradient, and the behavior of this "superdormant" fraction, which is of considerable applied significance, has been examined (31–34). The variation in response appears to be related to heterogeneity in individual spores within a population, in terms of the amount of receptor and water content, for example. A recent review (35) discusses this in more detail from a mathematical perspective, while another review (36) focuses on the consequences of heterogeneity from the applied perspective of food microbiology. Progress through germination and outgrowth of *Clostridium botulinum* has also been examined at the single-spore level (37, 38) to inform predictive food microbiology.

A Cautionary Reminder

In practical terms, data in the literature only provide a general guide to the behavior of the reader's favorite spore-forming species or strain: the overall germination rate of the spore suspension can also vary considerably, depending on the conditions of sporulation, such as temperature or medium used, or length of time for maturation after the completion of sporulation, whether in a psychrotrophic *Bacillus cereus* (39) or the laboratory strain of *B. subtilis* (40–42). Germination response is often measured at a single pH, temperature, and germinant concentration; this may not be optimal and may underplay the contribution of different receptors to different environmental conditions. Different wild isolates from the same species may behave very differently (43). In addition, the consequences of domestication may need to be considered. While it is important that longer-term subculture or storage of *Bacillus* strains on non-sporulating media is to be avoided (44, 45), long-term storage on sporulation agar may favor the survival of spores that do not rapidly germinate—some *B. subtilis* 168 isolates from different laboratories show differences in germination behavior, and even mutations in recognized *ger* genes (46).

STAGES OF GERMINATION

The change from dormant spore to germinated spore (Fig. 1A and B, respectively) and then to vegetative cell can be rather arbitrarily divided into four main stages: activation, stages I and II of germination, the latter dependent on cortex hydrolysis, and then the last stage, where a fully hydrated germinated spore resumes metabolism and outgrows. These are discussed in turn, based largely on experiments with the paradigm organism, *B. subtilis*.

Activation

Depending on the strain, sporulation conditions, and the extent of washing to remove surface-bound molecules that may potentiate germination, the average spore will respond more rapidly to germinants if the

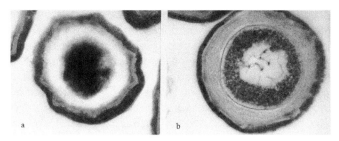

Figure 1 (a) Dormant spore, showing dehydrated core, electron transparent cortex, and extensive coat. (b) Germinated spore, showing rehydrated core and part-degraded cortex and coat layers.

population has first been activated, for example, by exposure to sublethal heating or by extended storage at 4°C (47). A classic analysis of *B. subtilis* (48) showed that temperatures in between these extremes also lead to activation. Spores of *Geobacillus stearothermophilus* may be effectively activated by incubation at 37°C (49, 50).

For laboratory-based study, a brief sublethal heating of *Bacillus* spores accelerates the activation of germination via nutrient receptors, by reducing the microlag; the degree of activation required depends on the receptor that is stimulated (51). Stimulation by multiple germinants, targeting different receptors, reduces the need for heat activation in the general spore population (33). Sublethal heating improved germination in a *gerP* mutant of *B. cereus* (52), suggesting that activation may also increase the permeability of coat layers. Germination of superdormant spores that are more resistant to wet heat and have a lower water content could also be stimulated by heat activation, but at higher temperatures than the average spore in the population

(33). In contrast, spores that lack the spore maturation *spmA* or *spmB* genes have a higher core water content and respond to germinants more quickly than wild type (53, 54). Speculatively, it may be that the tipping point for signal transduction downstream of the receptor is influenced by the local hydration state.

EVENTS IN SPORE GERMINATION

The full molecular details of signal transduction in spore germination are not yet clear, but many of the proteins involved in the process are identified, and reasonable hypotheses can be constructed with the available information (Fig. 2). Most of the evidence is derived from studies with *B. subtilis*, the genetic and biochemical paradigm sporeformer. Large-scale transport or metabolism of germinant is not required for germination (55), and neither energy metabolism nor macromolecular synthesis is required for the earliest stages. However, the assumption that the germination process is essentially a biophysical and degradative one is challenged by recent work (56) using dormant spores of *B. subtilis* permeabilized to antibiotics; late stages of germination (after loss of DPA and heat resistance, but before the complete loss of phase brightness) were then dependent on protein synthesis. How this is to be integrated with other observations of late-germination events must await a more detailed analysis of the properties of such part-germinated spores.

How Does the Germinant Access the Receptor?

First, the germinant must traverse the outer layers of the spore to access nutrient germinant receptors in the

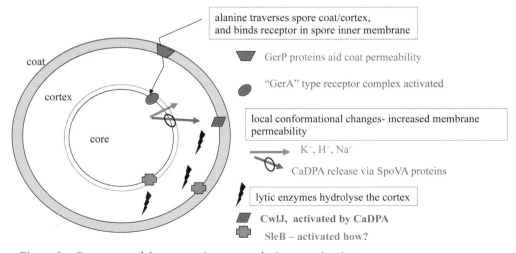

Figure 2 Cartoon model representing events during germination.

inner membrane The spore coats act as a molecular sieve, excluding enzymes, but small molecules such as germinants ought to be able to penetrate them. In the *B. cereus/B. anthracis* family, and to a lesser extent *B. subtilis*, the loss of GerP proteins from the spore coat (52, 57) reduced the effectiveness of nutrient germinants and CaDPA, suggesting that the permeability of spore coats to germinants is reduced if GerP proteins are absent. The coats may also influence germination rates because of the presence of enzymes that can modify or degrade the germinant, such as alanine racemase in spore coats (58, 59), and, along with inosine hydrolase, in the exosporium of the *B. anthracis/B. cereus/B. thuringiensis* family. As discussed later, some spore cortex lytic enzymes, such as CwlJ, are sequestered at or near the spore coat.

Stage I Germination

The germinant(s) interact with a membrane-located receptor complex, which transduces the stimulus, activating directly or indirectly the membrane-associated changes and hydrolytic reactions that occur in germination. The mechanism of signal transduction is not known, but it is likely to involve some conformational change of receptor protein(s) as a result of germinant binding. That moderate physical pressure induces germination via such receptors (60) may suggest that it also induces conformational change. Germinant receptors are discussed in considerable detail later in this review. Once the receptor has been activated, and the signal transduced, the spore is committed to germinate, and events will proceed even after removal of the germinant, or addition of D-alanine to competitively inhibit further L-alanine binding (61).

Large-scale excretion of monovalent cations, including H^+, Na^+, and K^+, occurs very early (62). The release of H^+ from the core results in an elevation of core pH, which will be required to activate phosphoglycerate mutase that metabolizes stores of 3-phosphoglycerate, generating ATP during late germination or outgrowth (63). The proteins involved in these ion fluxes are unknown, but, since the number of receptors in the spore is low (see below), it is unlikely to be mediated directly by receptor proteins, although it may be initiated by them. Later in stage I germination, the large store of CaDPA within the spore core is released and is replaced by water. The influx of water acts to partially rehydrate the spore core and, as a result, there is a marked reduction in wet-heat resistance (64). Proteins of the *spoVA* operon mediate CaDPA release during germination (65), as well as CaDPA uptake from the mother cell compartment during sporulation (66), and roles of

individual proteins are being identified (67, 68). Bacilli and clostridial genomes all encode at least a minimal set of SpoVAC, SpoAD, and SpoAE proteins (15, 207), emphasizing the central role of these proteins in endospore biology. A transient increase in spore permeability, resulting in the release of free amino acids (glutamate and arginine) and other small molecules from the core also occurs at about the same time as the release of CaDPA, but its mechanism is not known: the retention of adenine nucleotides and 3-phosphoglycerate reveals that there is selectivity in the excretion of these various molecules (69).

After the early events of germination, the extreme heat resistance of the spore has been lost, but protein mobility in the core (70) and bulk lipid mobility in the inner membrane (71) have not yet increased. Up to this point there is no detectable metabolism and the majority of common high-energy compounds are found in their lower-energy states, such as ADP and AMP, coenzyme A, and NAD (9). Enzyme activation and ATP synthesis do not occur without complete rehydration of the spore core, which requires cortex lysis in stage II.

Stage II Germination

In *B. subtilis*, cortex hydrolysis is dependent on two, partially redundant, cortex lytic enzymes, CwlJ and SleB (72, 73). In stage II of germination, cortex lytic enzymes are activated, and cleave peptidoglycan containing muramic δ-lactam, leaving the germ cell wall intact. Spores with a mutation in *cwlD*, or in *pdaA*, a gene encoding a polysaccharide deacetylase gene, have no muramic δ-lactam, and are defective in nutrient-induced germination, because their cortex is no longer recognized by the lytic enzymes (64, 74). Cortex breakdown is completely blocked in a double *sleB cwlJ* mutant, although early germination-associated changes, including some release of CaDPA, occur (75).

The SleB/CwlJ pair of enzymes are common in bacilli (15), and have been studied in *B. subtilis* (73, 76, 77); homologues play equivalent roles in *Bacillus anthracis* (78–80) and *Bacillus megaterium* (81, 82). Because CwlJ can be activated, bypassing the germinant receptors, by addition of endogenous CaDPA (20), it would be activated somehow in nutrient-stimulated germination by the CaDPA released from the spore core. The two cortex lytic enzymes are located differently in the spore. The CwlJ protein is synthesized in the mother cell and is localized in the spore by the coat protein YwdL (GerQ) (83). In contrast, SleB is synthesized in the forespore and secreted, and is present in two locations in the mature spore (73)—at the inner membrane and in the outer layers, presumably in association with the coexpressed YpeB protein, which is required for the retention of the SleB protein in the spore (73, 84). The role of YpeB may extend beyond localization (85, 86).

Spore coat depletion, which removes CwlJ and the outer component of SleB, does not completely block spore germination, and, therefore, the SleB associated with the inner membrane is sufficient for cortex hydrolysis. How the SleB protein is activated during germination is not yet clear. SleB has a lytic transglycosylase activity and is active on cortex, either intact or fragmented (87). The recently determined 3D structure of the catalytic domain of SleB protein from the *B. cereus/B. anthracis* group demonstrates the specificity for cortex peptidoglycan (88, 89) and the active site topology. In contrast, the YaaH (SleL) protein (90) is an *N*-acetylglucosaminidase that is active on cortex fragments during germination, but it is not required for the spore to germinate (91).

Cortex lysis leads to full rehydration and expansion of the core; the inner membrane and germ cell wall expand, and the germinated spore regains more normal cellular properties (Fig. 1B). With the rehydration and expansion of the spore core, proteins and lipids in the inner membrane become mobile again. This marks the end of dormancy and the spore's sensitivity to wet heat is now similar to that of vegetative cells. Small acid-soluble proteins (SASPs), including those of alpha and beta type that are DNA associated, are degraded following the activation of the specific GSP protease (92), freeing the DNA for transcription and providing a source of amino acids for biosynthesis during outgrowth. DNA repair proteins already present in the spore are now active (93, 94) to repair damage incurred during spore dormancy. ATP is generated from the spore's store of 3-phosphoglycerate, because phosphoglyceromutase is activated as the pH in the core rises.

Spore Coat Degradation

Somehow, the germination signal is transduced to activate partial spore coat breakdown. Although remnants of spore coat remain even after germination, there is significant local degradation, beginning at about the same time as cortex hydrolysis, visible in a series of beautiful electron micrographs of thin sections of germinating and outgrowing spores (95) and in atomic force microscopy images (96). The outer coat thins from the inside, and gaps appear in the inner multilamellar coats. Presumably, local machinery is present in the coat to initiate degradation. Nothing is yet known of the proteases involved, or how they are activated; some spore coat changes are seen in germinating spores of a *cwlD* mutant that cannot hydrolyze the cortex (A. Moir, unpublished data), so some signal to activate spore coat degradation must result from receptor activation or other early events, independent of cortex hydrolysis or complete core rehydration. A proteomics study of *B. anthracis* spores from dormancy through germination and outgrowth has noted the reduction in levels of spore proteins during germination, including some coat proteins and a number of uncharacterized small proteins (97).

Outgrowth

The resumption of metabolism and *de novo* synthesis of macromolecules in the cell results in outgrowth of the germinated spore into a vegetative cell. In the first few minutes, energy and intermediates must depend on catabolism of molecules carried over in the dormant spore. Sources of energy and intermediates include stores of 3-phosphoglyceric acid (3-PGA) and amino acids chiefly generated by degradation of the large stores of SASPs. Lipid synthesis resumes early, followed by bulk protein and DNA synthesis. In general, genes required during outgrowth (98, 99) are also required for vegetative growth. A detailed transcriptome analysis, following *B. subtilis* spores through germination and outgrowth (100), demonstrated a clear pattern of gene expression during outgrowth from earliest times; transcripts encoding transporters, DNA helicases, and DNA repair proteins were all synthesized early. Several late, probably long-lived, forespore sigma G-dependent transcripts were carried through in dormant spores (100). There are other reports of mRNAs in dormant spores in *Clostridium novyi* (101) and *B. subtilis* (102). The term "spore revival" has been used to encompass late germination and outgrowth (56); these authors have identified proteins synthesized during this period in *B. subtilis*, and subsequent analysis of mutants has revealed additional components important in the process. Most strikingly, the very early translated proteins, translation-associated factors Tig and RpmE, are required for the latest stages of spore germination, and malic enzymes for utilization of endogenous malate in the spore are required for the "ripening" phase, where metabolically active spores have not yet shown significant cell outgrowth (56).

GERMINANT RECEPTORS FOR NUTRIENT-INDUCED GERMINATION IN *B. SUBTILIS*

The GerA receptor of *B. subtilis*, the first to be identified (103), is the paradigm for germination receptor complexes and lends its name to this family of receptors. Expressed at a low level in the forespore during sporulation, the three protein components (GerAA, GerAB, and GerAC) are all membrane associated and function together, probably in a complex. In addition, some receptor operons also encode a small (60 to 85 amino acids) protein that influences (either up or down) the response to the cognate receptor (104).

Five germinant receptor operons are encoded in the genome of *B. subtilis* 168. *B. subtilis* spores with a deletion in either the *gerB* or *gerK* operons are defective in their response to a cogerminant mixture of an amino acid plus sugars glucose and fructose, and K$^+$ ions (AGFK), but are normal in their response to L-alanine as a single germinant, which is mediated by the GerA receptor. If all three are deleted, only one spore in 10^4 germinates to form a colony on rich medium (105). Deletion of the other two operons (*yndDEF* and *yfkQRT*) has no effect on the ability of spores to germinate in response to nutrients under standard conditions (105). The levels of GerA, GerB, and GerK receptor proteins in *B. subtilis* spores have been estimated in comparison with the more abundant GerD and SpoVAD proteins (106).

Regulation of Expression of the Receptor Proteins

Much of the germination machinery, including the germinant receptor subunits, GerD, SpoVA proteins, and SleB-YpeB proteins, is expressed in the developing forespore, directed by sigma G. The SpoVT protein, an AbrB homologue, moderates expression, reducing transcription of *gerA* and *gerB* and increasing expression of *spoVA* and *gerD* (107). Overexpression of RNA polymerase binding factor YlyA (108) results in downregulation of germinant receptor proteins, and receptor expression may also be moderated by the protein phosphatase PrpE (109), although its role is not well characterized.

Receptors Encoded in Genomes of Other Species

The sequences of hundreds of germinant receptor operons are now available; attempts have been made to classify these in terms of molecular phylogeny (17, 110). Such receptors are encoded in the genomes of almost all *Bacillales* and *Clostridiales* whose genomes have been sequenced to date, with the notable exception of *Clostridium difficile* and closely related species. The majority are organized as tricistronic operons, although there are many variations on this theme,

especially in clostridia, where individual genes are common. Several are encoded on plasmids; for example, a pBtoxis plasmid-encoded operon confers on *B. thuringiensis* var. *israelensis* the ability to germinate in the alkaline conditions of the insect midgut (111). The *B. anthracis gerX* operon is encoded in the pathogenicity island on the toxin-encoding plasmid pXO1 (112), and is associated with a transposase and a site-specific recombinase, in an apparent cointegrative transposon (113), while the *gerU* operon and *gerVB* gene of *B. megaterium* are also plasmid located (114).

The Germinosome

Providing a major advance in our conceptual image of the germination process, recent technically demanding experiments have visualized germinant receptor proteins as green fluorescent protein (GFP) fusions in spores (115), in an apparent cluster named a "germinosome." The spores used were coat defective (116) so that autofluorescing coat proteins did not interfere. Within the limits of the resolution of the microscopy, all three types of subunit of the germinant receptors are present at a single focus on the spore inner membrane, and the different GerA, GerB, and GerK receptors all colocalize in one cluster. This extends the earlier evidence, from Western blotting of proteins after spore fractionation, that germinant receptors are located in the relatively protected environment of the spore's inner membrane, for GerAA and GerAC (117) and for GerBA (118) in *B. subtilis*. The work also defines a role for the GerD lipoprotein (119–121), which colocalizes in the same cluster with the receptors; without the GerD protein, the receptor proteins are dispersed around the spore inner membrane. The GerD protein that organizes the receptor proteins in the germinosome crystallizes as an α-helical trimeric, rodlike structure (122). The cluster of germinant receptors in a germinosome would facilitate the integration of stimuli from different receptors (123, 124), and provide a means of signal transduction, either to other proteins in the locality or by creating a sufficient local change in the membrane properties or ionic environment. In contrast, the inner membrane-associated SpoVA protein complex that would represent the next known component in the germination cascade is distributed throughout the membrane (115).

How might a receptor be organized? At least in the paradigm *B. subtilis* GerA receptor, all three coevolved component proteins are required for function. Indirect but strong evidence that the subunits of an individual receptor interact comes from a study of point mutants; some single-amino-acid changes in the GerAA and GerAB subunits result in a likely failure to assemble a stable receptor, because levels of the GerAC lipoprotein are low or absent in the spore (125, 126). In general, the receptors are functionally distinct, suggesting that the subunits are not typically interchangeable, although interchangeable behavior is seen in *B. megaterium* QM B1551 for the more closely related GerUB, GerVB, and GerWB subunits. Clustering of receptors could facilitate competitive as well as cooperative interactions (127, 128).

STRUCTURE/FUNCTION RELATIONSHIPS IN GERMINANT RECEPTOR PROTEINS

The GerAA and GerAC components have no homologues outside the endosporeformers, whereas those of the GerAB family represent a distinct branch of the APC (amino acid polyamine organocation) superfamily of membrane-associated single-component membrane transporters (129). The homology between members of the component families of receptor subunits suggests that all would have an equivalent domain structure and general topology.

GerAA Subunit

The GerAA subunit, ca. 53 kDa, is an integral membrane protein composed of three domains. As yet, no protein or individual domain has been successfully overexpressed for structural studies. Hydropathy profiles have suggested that the central domain might contain five membrane spans (2), but programs that predict likely membrane-spanning helices based on amino acid sequence give very variable results for this protein family. The only experimental data so far have been for the important GerHA receptor protein of *B. anthracis*, based on an analysis of GFP fusions to internal positions (130). For GerHA, and the A subunits in general, the possible organization suggested would include an N-terminal hydrophilic domain located on the spore core side of the membrane, followed by a large hydrophobic domain containing four membrane-spanning α-helices, then a C-terminal hydrophilic domain that would also be located in the spore core (Fig. 3). Substitutions in various positions in the GerAA protein (126) may be interpreted in the light of this likely topology. For example, the apparent (if somewhat unusual) second transmembrane helix, commencing with a conserved PFPP hinge-like motif, contains several conserved acidic residues and appears to be of particular significance: substitution of a serine for the first proline increases the responsiveness of the GerA receptor to L-alanine and its analogues as germinants; this is also the case in GerBA, where the equivalent

GerBA* substitution (131) results in this receptor responding to L-alanine or L-asparagine as sole germinant (132), without the need for a contribution from the GerK receptor (132). In GerAA, changing the second proline results in the release of phase dark (i.e., germinated) spores, as though the receptor is permanently activated (126); removing conserved acidic groups from this second transmembrane helix destabilizes the receptor, resulting in little or no receptor complex. Whether the GerAA subunit binds germinant is unclear, but increases in responsiveness would suggest that this region could play a role in germinant binding and/or be subject to conformational changes during germination.

GerAB Subunit

The GerAB subunit and its homologues in sporeformers are integral membrane proteins with 10 predicted membrane spans and represent a subunit that is almost certain to bind germinants. The strongest evidence for this is the alteration of germinant specificity in the *B. megaterium* germinant receptor, depending on which of the GerUB, GerVB, or GerWB subunits are associated with the GerU receptor (114, 133, 134). Also, two mutant alleles in GerAB, with amino acid changes in likely transmembrane helices 1 and 6, reduce the responsiveness of spores to L-alanine by 8-fold and 100-fold, respectively (2, 135). It is not likely that germinant receptors mediate bulk transport of germinant (55), but there are precedents within the APC superfamily of proteins acting as sensors of extracellular substrates (136). Topology predictions and experiments with GFP fusions in the GerHB subunit of *B. anthracis* both suggest 10 membrane-spanning helices (130). Site-directed mutations in GerAB of *B. subtilis* (125) and in GerVB of *B. megaterium* (133, 134) that alter receptor function have been described. The recent determination of 3D structures of several members of the APC family, including AdiC (137) and ApcT (138), with and without bound ligand, has highlighted the importance of common structural elements and demonstrated major conformational changes. Extended regions in the otherwise helical structure of transmembrane helices 1 and 6 are conserved and of importance to structure and binding of the cognate transported substrate; such extended regions within these helices would also be predicted in GerAB family proteins. Changes in the TM6 region of GerVB, informed by the structural data for these transporters, identified effects on the concentrations of glucose, and sometimes of other germinants, required to stimulate germination (133).

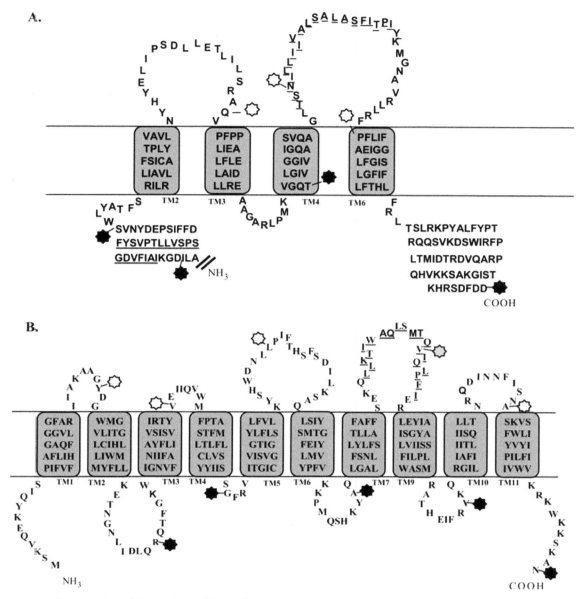

Figure 3 Possible topology of *B. anthracis* germinant receptor subunits, on the basis of GFP fusion constructs in *E. coli*. Reproduced with permission from reference 130. Open stars, surface-exposed locations; closed stars, intracellular locations. (**A**) GerHA, not showing the complete N-terminal region Underlined residues represent predicted membrane spans that were not confirmed by fusion data. (**B**) GerHB.

GerAC Protein

The GerAC protein and its homologues are predicted to be hydrophilic, with the exception of the conserved N-terminal prelipoprotein signal sequence. GerAC is synthesized in the forespore, secreted, and covalently attached to lipid, anchoring it to the outer surface of the inner spore membrane. Without this lipomodification, the protein is not retained in the spore (125).

Uniquely among these receptor proteins, a 3D structure has been reported for GerBC (139). The protein consists of three domains, each with a unique fold; there is an extended linker between domains I and II. Sequence alignments suggest that the other members of the family are likely to adopt the same overall structure. Single and multiple alanine substitutions have been introduced into conserved regions, and the extended linker deleted (140). This has defined several residues important for function in germination, where the altered protein is still present in spores at normal levels, whereas changes to other residues (including the linker) result in the reduction or loss of the protein from the spore. An understanding of these observations will have to await structural information on interactions of the C subunit within the complete receptor complex.

Biochemical evidence for the accessibility of germination proteins in decoated dormant spores of *B. subtilis* confirms the location of the majority of the proteins at the outer surface of the inner membrane (141). However, some proteins (GerAA, GerD, SpoVAD) are not fully accessible until germination has occurred, suggesting that there is some reorganization of the proteins or their environment during germination.

GERMINATION RECEPTORS AND THEIR FUNCTION IN *B. CEREUS SENSU LATO*

Apart from *B. subtilis*, the paradigm sporeformer, the next most extensively examined species are members of the *B. cereus*/*B. anthracis*/*B. thuringiensis* family, described as *B. cereus sensu lato*. Although named on the basis of their pathogenicity, *B. cereus* and *B. thuringiensis* strains are intermingled in the phylogenetic group, and *B. anthracis* represents a clonal internal lineage (142, 143). The functions of individual germinant receptors have been studied by mutation in *B. anthracis* and in two *B. cereus* strains, ATCC 10876 and ATCC 14579. Strains of the same species may share some, but not all, *ger* receptor operons, as discussed below.

B. cereus ATCC10876

The first germination mutants in *B. cereus* (defining receptor operons *gerI*, *gerQ*, and *gerL*, and the *gerP* operon) were generated by transposon mutagenesis in strain ATCC10876, before genome sequencing in any of this group. The draft genome sequence reveals the additional presence of homologues of the *gerR*, *gerG*, *gerS*, and *gerK* operons of ATCC14579, and also one (ZP_04321160) not found in ATCC14579 but present in many other *B. cereus* and *B. thuringiensis* strains.

Of the three characterized receptors, GerI is necessary for germination in response to inosine as a sole germinant and also has a role in L-alanine germination (144). GerQ, in combination with GerI, is required for germination responses in inosine as a sole germinant, but is not involved in L-alanine germination (145). GerI and GerQ receptors have different spectra of inhibition by nucleoside analogues (146), and germination by

inosine alone may also involve release of endogenous alanine (147).

Spores lacking GerI receptor function will still germinate in L-alanine, albeit at a slower rate than the wild-type parent, and another receptor, GerL, is responsible for the major L-alanine germination response in this strain (145). The GerI and GerL receptors function at distinctly different optimal temperatures, pH, and germinant concentrations, extending the range of conditions under which germination can be triggered by L-alanine.

The GerI receptor has several unusual properties. Germination in inosine (but not in L-alanine) is inhibited by K^+, whereas most germination receptors are stimulated by K^+ (7). Like its homologue in *B. anthracis*, GerH, the GerIA protein has a much longer N-terminal domain than most receptors, containing multiple glutamine-rich repeats (144) of unknown function. Next, all GerI-mediated responses in *B. cereus* ATCC 10876 depend on GerN, an electrogenic Na^+/H^+, K^+ ion antiporter, which is particularly important for inosine germination (148–150). A second GerN-like protein in *B. cereus*, GerT, will substitute for GerN in inosine germination if the stimulus is high enough, i.e., at the highest inosine concentrations, but has a more significant role in outgrowth in high salt or at alkaline pH (150). The function of other germinant receptors in *B. cereus* does not require GerN or GerT, so whatever their role, it is specific to the GerI receptor. An interesting speculation is that they may be required to restore ion or charge balance during the function of this, but not other, receptors. It is likely that other ion transport proteins are responsible for the bulk ion movements that occur early in germination in response to all nutrient germinants.

The earliest report of a GerN-like protein (GrmA) required for germination was in *B. megaterium*, although this was not confirmed for another *B. megaterium* strain (114, 151). A GerN homologue in *Clostridium perfringens*, named GerO, is required for germination (152), but is expressed in the mother cell, unlike the forespore-expressed GerN; it is therefore less likely to be appropriately located to participate in inner membrane events during germination. There is no close GerN homologue in *B. subtilis* (148).

B. cereus ATCC14579

Seven *gerA* operon orthologues were identified in the genome sequence of this type strain. Six of the operons, *gerQ*, *gerL*, *gerI*, *gerK*, *gerS*, and *gerG*, have the same organization as *gerA* of *B. subtilis* with the individual genes ordered A-B-C. The final homologue, *gerR* (not to be confused with the identically named late sporulation transcription factor), is ordered *gerRA-gerRC-gerRB*. Very differently from the ATCC10876 strain, the *gerR* operon proved to be the most important, because the first screens for germination mutants identified *gerR*; germination in L-alanine and in inosine (as individual components but not in combination) was defective in a *gerR* mutant (153). A wider study of mutants with targeted disruptions in each operon individually (154) revealed that the GerR receptor is crucial to germination in all amino acids except L-glutamine, which was mediated by the GerG receptor. GerI and GerQ contributed with GerR to germination in nucleosides, and GerI and GerG were both required for a response to glutamine/inosine combination. Mutations in *gerK*, *gerL*, and *gerS* had no effect on germination, possibly as a result of redundancy in function with other receptors such as the dominant GerR.

B. anthracis

Of a clonal lineage, *B. anthracis* has a complement of receptor genes that is common to all isolates so far described. The chromosome encodes four functional germinant receptors, GerH, GerS, GerL, and GerK, and the fifth, GerX, is encoded on the virulence plasmid pXO1; two other chromosomal operons, named *gerY* and *gerA*, harbor frameshifts that would lead to loss of function (155), and are closely related to *gerR* and *gerG* of *B. cereus* ATCC14579, respectively (17). The GerH proteins (156) are extremely similar to the GerI proteins of *B. cereus*, except that the unusual glutamine-rich repetitive sequence close to the N terminus of GerIA/HA is slightly different in sequence. The GerS and GerL receptors are closely related in sequence to those with the same names in *B. cereus*.

A detailed analysis of germination behavior *in vitro* has been undertaken in the attenuated *B. anthracis* Sterne strain, which lacks pXO2, the plasmid encoding capsule biosynthetic enzymes. Each receptor operon was knocked out individually (157). Unlike the situation in *B. cereus*, inosine is not effective as sole germinant, probably because of the absence of GerQ and the nonfunctionality of GerY/R. In addition, strong germination responses to L-alanine are only seen at high concentrations (≥100 mM). In contrast, when used as cogerminants, either together or separately with other amino acids, 1 mM concentrations of alanine and inosine are effective. In general, the stimulation of at least two receptors by distinct germinants leads to the most rapid germination in *B. anthracis*. The GerH operon is required for response to cogerminant mixes containing inosine. The GerK and GerL

receptors contribute individually to germination in high alanine, while also being individually required for the recognition of cogerminants (GerK: methionine and proline; GerL: serine and valine) in inosine-induced germination. GerS is required, in combination with another receptor, for germination in combinations of alanine or inosine with an aromatic amino acid.

No specific germinant could be identified for the pXO1-encoded GerX receptor, and a significant role in *B. anthracis* virulence is not proven, despite its location within the pathogenicity island. Germination of spores with a *gerX* defect was significantly reduced within phagocytic macrophages, suggesting that *gerX* may be important for pathogenicity (112), and reduced germination of a *gerX* mutant in murine macrophages was independently observed (noted in the discussion in reference 158). In contrast, studies of strains lacking GerX, or with GerX as sole remaining receptor, found no change in virulence in comparison with the receptor-complete parent or the complete receptor null, respectively (13). The latter study attempted to define the role of individual receptors in germination *in vitro*, and their contribution to virulence in a mouse model, by fully deleting all the functional receptors, except one. Under the conditions used, the GerH receptor was sufficient for germination in inosine combinations, and GerK and GerL receptors individually allowed germination in high alanine concentrations. Spores of a strain lacking all five functional receptors germinated very poorly indeed, with a colony-forming efficiency of 1 colony per 1,000 spores inoculated on the very rich Brain Heart Infusion agar; retention of any one of the H, K, L, or S receptors would restore this to near wild-type levels. Equally, the deletion of all of these four receptors would dramatically reduce the virulence in mice, whether inoculated intratracheally or subcutaneously. Any one of the four was sufficient for a fully virulent infection by the intratracheal route, although time to death might be extended. In contrast, GerH was the only receptor required for virulence in subcutaneous infection and for spore germination in blood (13). An independent transposon mutagenesis study to identify spores that did not germinate in macrophages gave multiple hits to the *gerH* operon (159), suggesting its particular importance.

B. megaterium

Unusually, the tricistronic *gerU* receptor operon on pMB700 in *B. megaterium* QMB1551 is followed by an adjacently encoded monocistronic *gerVB*. The A and C subunits of the receptor can interact either with the B subunit or with the chromosomally encoded

GerWB (133). The *gerU* operon restored the germination response to glucose and leucine as single germinants, and to KBr as a cogerminant, in a plasmidless strain of *B. megaterium*. Expression from *gerVB* along with *gerU* confers the germination response to proline and KBr as single germinants, while also enhancing the response to glucose and leucine as single germinants. Introducing a plasmid carrying this gene cluster into *B. megaterium* KM, a strain that germinates in alanine (65), allowed germination in all these additional germinants (114).

GERMINANT RECEPTORS IN CLOSTRIDIA

Germination in clostridia often requires multiple germinants, and there is an extensive literature over the years covering germination requirements of different strains, as, for example, in references 8, 160, 161, and 162. Homologues of many, but not all, of the germination proteins identified in *B. subtilis* are found in clostridia (15, 18). A review of germinant receptors (14) considers clostridial examples, and germinant receptors in *C. botulinum* and *C. sporogenes* have recently been characterized (163).

A detailed study of *C. perfringens* spore germination has revealed that there may be some differences in receptor function from the *Bacillus* paradigm. For example, the germinant receptor GerKC-KA can recognize germinants without the presence of the divergently transcribed GerKB component, although the latter does improve germination (19, 164, 165). There is no GerD protein in clostridia, so the question of whether receptors cluster together in germinosomes, and, if so, how this is mediated, remains an open one.

An Alternative System for Cortex Lysis

The prevalence of encoded cortex lytic enzymes in clostridial genomes has been explored (15, 18). Some *C. botulinum* strains encode SleB, YpeB, and CwlJ proteins, as do *C. novyi and C. kluyveri* (18); in many others, including *C. perfringens* and Group II *C. botulinum* strains, a different enzyme, SleC, is essential for cortex lysis. This cortex lytic enzyme was first described in *C. perfringens* S40, and is expressed with its cognate protease(s), encoded by the *cspA*, *B*, and *C* genes (166, 167). SleC is processed from a precursor during spore formation (168) and localized as a proform to the outer cortex/coat boundary along with the Csp proteins (169), where it is further activated by processing during germination; the signal stimulating such processing is not known. Homologues of SleC and its protease are found in many clostridia, and SleC

proteins have been shown to be required for germination in other *C. perfringens* strains (170, 171). An additional lytic enzyme, SleM, commonly found in *C. perfringens* (167), degrades cortical fragments but is not essential for germination or outgrowth.

C. difficile encodes no germinant receptors at all, yet responds to a combination of the bile salt taurocholate, with glycine and histidine (172–174). The clinical importance of this species has led to a concentrated effort to explore its spore biology. Here, the SleC cortex lytic enzyme is essential for germination, and the activation protease Csp proteins appear to have evolved to contribute more directly to germinant sensing, CspC providing a bile acid-specific receptor (175), although nothing is yet known of the mechanism of glycine sensing (176). Unlike in other clostridia encoding receptors, and where cortex lysis is achieved by downstream activation of SleC protein (177, 178), the direct activation of SleC in *C. difficile* results in cortex hydrolysis preceding DPA release (179).

THE LIPID ENVIRONMENT AND GERMINATION

The changes in membrane properties during spore germination, including the passage of ions and water, and the restoration of membrane fluidity, highlight the importance of the membrane environment around the germinant receptor. Mutations in the *gerF* (*lgt*) gene of *B. subtilis* greatly reduced nutrient-induced germination responses, while non-nutrient-induced germination remained like that of wild type (46, 180). The same phenotype is seen in *B. anthracis* (44). This can be explained on the basis of its importance to germinant receptors—the C subunits of germinant receptors, and the GerD protein, are all predicted lipoproteins and dependent for processing on GerF/Lgt. Mutations in the prelipoprotein processing site of GerD (121) or of GerAC (125) resulted in the absence of these proteins from the spore, although the effect on GerBC and GerKC receptors may be less marked (180).

The lipid environment of receptors or their associated proteins may influence germination efficiency. Spores of *B. subtilis* with only traces of cardiolipin in their membranes germinated poorly in L-alanine and AGFK, as measured by the loss of turbidity (181), although approximately 15 to 20% of the normal level of DPA release was seen in both germinants. In another study, an equivalent reduction in cardiolipin levels had a less dramatic effect, although still slowing germination significantly (182). Both experiments used very high saturating concentrations of germinants, although at different pH values; it might be interesting to test under conditions where the germinant stimulus is less strong, such as with valine rather than alanine as germinant. In another example where alterations in spore lipid properties may influence germination, the spore coat-associated phospholipase LipC (YcsK) influences spore fatty acid levels (183): mutant spores do not germinate efficiently in L-alanine, although there was no defect in AGFK germination. Our lack of understanding of the nonfluid lipid environment of the inner membrane of the dormant spore hampers understanding of these observations.

INFLUENCE OF THE SPORE COAT ON GERMINATION BEHAVIOR

The role of three relevant proteins has already been described: that of the GerP proteins, required for access of germinants through the spore coats, and the cortex lytic enzyme, CwlJ, localized in the spore coat by the GerQ/YwdL protein. Mutations in most of the individual *B. subtilis* coat genes do not affect germination significantly, unless they have a wider morphogenetic effect on spore coat assembly.

The *gerE* gene, encoding a transcriptional regulator, activating expression of many late-expressed spore coat protein genes, and repressing others (184), is a classic example. A *gerE* mutant is blocked after the initial, receptor-dependent, loss of heat resistance (185), and the pleiotropic coat defect could mean that there are several reasons for the poor progress through late germination. One of the genes in the GerE-σ^K regulon is *gerT* (previously named *yozR*), which encodes a spore coat protein whose expression is downregulated by GerE (186), and whose assembly around the spore depends on the morphogenetic protein CotE. Mutants of *gerT* were impaired at an early stage of germination, responding poorly to nutrient germinants and to nonnutrient CaDPA, a phenotype reminiscent of mutations in the similarly regulated *gerP* operon that appears to be required for spore coat permeability to germinants (52, 57).

In *B. subtilis*, a small (28 amino acids) spore coat protein SscA is required for correct assembly of the outer spore coat proteins CotH, G, and B, and *sscA* (previously named *yhzE*) mutants germinate more slowly, presumably because of their coat defect (187). A *cotH* mutant has already been described as defective in germination (188), especially in combination with a defect in the major morphogenetic protein CotE (189), which itself is required for normal germination (190). A *cotT* mutant, which has an altered inner coat,

germinates poorly (191). Despite all these effects, the coat is not required for the earliest receptor-dependent changes in the spore that precede cortex lysis, because spores of a double mutant of *cotE* and *gerE* contain only a small amount of residual coat material but still show a relatively normal stage I germination response (116), as do chemically coat-depleted spores. Coat permeability to germinants could influence germination, and the coat region is important to the localization of various cortex lytic enzymes that are sequestered in an inactive form until germination.

At least in a few bacilli spore coats generally contain an alanine racemase activity that is probably important in preventing germination during spore development, and would be responsible for generation of D-alanine, a competitive inhibitor of germination (192). In the *B. anthracis/B. cereus/B. thuringiensis* group, alanine racemase and inosine hydrolase are both located in the exosporium or between it and the coat (a region known as the interspace), the outermost layer around the spore (193), and may influence germination rates (58, 59, 194). By contrast, the alanine racemase YncD is located in the outer coat in *B. subtilis*.

GERMINATION IN RESPONSE TO NONNUTRIENT ENVIRONMENTAL CUES

Spores are able to germinate in response to a variety of stimuli, in addition to the common nutrient germinants. These may be chemical (CaDPA), enzymic (lysozyme), or physical (pressure); dodecylamine also causes "germination," but kills the spore.

CaDPA-Induced Germination

Activation of the cortex lytic enzyme CwlJ of *B. subtilis* by endogenous CaDPA released from the core is a later event in receptor-mediated germination (20). However, exogenous CaDPA was found to independently activate CwlJ, entirely bypassing the nutrient-induced germination via the GerA receptor family, and effective in the absence of all such receptors (83, 105). CaDPA is an effective germinant in clostridia as well as bacilli, even though the former frequently do not encode a CwlJ protein. In the case of *C. perfringens* (19), which uses a SleC cortex lytic enzyme, the response to CaDPA requires the GerK nutrient germinant receptor, but it is not yet known whether this is a general mechanism that also applies in other SleC-containing species.

Pressure-Induced Germination

Germination in response to pressure has been shown to follow two distinct pathways, differently dependent on the extent of pressure applied to spores. Relatively low pressures (100 to 300 MPa) activate a germination process that resembles that of nutrient-induced germination (195, 196) and depends on germinant receptors (60, 196, 197). In contrast, spores germinating under higher pressures of 500 to 800 MPa do not generate ATP or degrade SASPs and retain their resistances to UV light and H_2O_2 (195), and the process is receptor independent. High pressure seems to cause direct release of DPA, and at least one of the cortex lytic enzymes is required to complete germination (198).

Dodecylamine-Induced Germination

Long-chain alkyl amines act as potent germinative agents up to temperatures of 70°C (199). Spores germinating at 37°C in dodecylamine release CaDPA from their core, and their cortex is degraded normally; at 70°C, CaDPA is released, but the cortex is not degraded, presumably because cortex lytic enzymes are inactivated. Dodecylamine-germinated spores have relatively dehydrated cores, and show no metabolism; SASPs are not degraded, and the spores are rapidly killed (22). Germination is independent of germinant receptors, cortex lysis, or the presence of a spore coat; dodecylamine induces germination by a separate mechanism to other nonnutrient stimuli, perhaps via direct membrane effects (22) leading to CaDPA release via the SpoVA proteins (200).

Lysozyme-Induced Germination

The spore coat acts as a barrier to exogenous enzymes, but once this has been breached, for example, by chemical decoating procedures, the spore cortex peptidoglycan can be degraded by lysozyme, initiating the release of the core's CaDPA, and the rehydration of the spore core. In some clostridia, lysozyme-containing agar can be used to improve colony formation from spores (201, 202). In contrast, in B. subtilis, lysozyme also degrades the spore's germ cell wall, meaning that spores germinated in this way are unable to outgrow without osmotic stabilization (64).

Germination Induced by Peptidoglycan Fragments

Spores may germinate in response to growth signals from other cells. Disaccharide tri- or tetrapeptide muropeptide fragments from vegetative cells (disaccharide tri- or pentapeptides, with a meso-diaminopimelate at the third position in the peptide side chain) are able to induce germination (25). A membrane-associated serine/threonine kinase PrkC, located in the spore inner membrane and with a triple extracellular PASTA domain able to bind peptidoglycan, is essential for this response to the cognate muropeptides (25). How PrkC functions in germination is not known, but presumably it phosphorylates some important component; germination stimulated by muropeptides, or by bryostatin, which activates PrkC, does not require germinant receptors and is effective in superdormant spores (203). These are effective in B. cereus as well as in B. subtilis (204); the full range of species that will respond to muropeptides has not yet been established, although PrkC homologues are found in bacilli and clostridia. This family of eukaryotic-like Ser/Thr kinases has a variety of roles in bacteria (24). In B. subtilis vegetative cells, PrkC, activated by muropeptides, induces expression of an extracellular muralytic enzyme YocH by an unknown mechanism. This YocH enzyme would then generate muropeptides from peptidoglycan fragments, amplifying its own expression (205). How PrkC functions in B. subtilis germination is not known; germination stimulated by muropeptides, or by bryostatin, which activates PrkC, does not require germinant receptors, and is effective in superdormant spores (203).

OVERVIEW

Exciting progress has been made in the detailed description of the germination process at the single spore level, and in possible structure/function relationships in nutrient receptors, both at the level of individual protein subunits, and their potential clustering to provide a node for receptor-associated changes, such as local rehydration and increased ion permeability. There is the beginning of an appreciation of some wider membrane-associated events in germination, notably the likely role of a SpoVA channel in the bulk release of CaDPA from the core across the inner membrane. We understand none of the mechanisms at the molecular level, and there are many important questions to be addressed, including studying candidate proteins that are now clearly defined. With available genome sequences of many sporeformers, and recent technologies for making mutants in a wider variety of bacilli and clostridia, has come the appreciation that there is considerable diversity in the setting for the germination machinery, and in some of the components, that remains largely to be explored.

Citation. Moir A, Cooper G. 2014. Spore germination. Microbiol Spectrum 3(6):TBS-0014-2012.

References

1. Setlow P. 2006. Spores of *Bacillus subtilis*: their resistance to and killing by radiation, heat and chemicals. *J Appl Microbiol* **101**:514–525.

2. Moir A, Corfe BM, Behravan J. 2002. Spore germination. *Cell Mol Life Sci* **59**:403–409.

3. Paidhungat M, Setlow P. 2002. Spore germination and outgrowth, p 537–548. *In* Sonenshein AL, Hoch JA, Losick R (ed), Bacillus subtilis *and Its Closest Relatives: From Genes to Cells*. ASM Press, Washington, DC.

4. Setlow P. 2013. Summer meeting 201–when the sleepers wake: the germination of spores of *Bacillus* species. *J Appl Microbiol* **115**:1251–1268.

5. Setlow P. 2014. Germination of spores of *Bacillus* species: what we know and do not know. *J Bacteriol* **196**: 1297–1305.

6. Woese CR, Morowitz HJ, Hutchison CA III. 1958. Analysis of action of L-alanine analogues in spore germination. *J Bacteriol* **76**:578–588.

7. Foerster HF, Foster JW. 1966. Response of *Bacillus* spores to combinations of germinative compounds. *J Bacteriol* **91**:1168–1177.

8. Gould GW. 1969. Germination, p 397–444. *In* Gould GW, Hurst A (ed), *The Bacterial Spore*. Academic Press Inc, London, United Kingdom.

9. Setlow P. 1983. Germination and outgrowth, p 211–254. *In* Hurst A, Gould GW (ed), *The Bacterial Spore*, vol 2. Academic Press, London, United Kingdom.

10. Moir A, Smith DA. 1990. The genetics of bacterial spore germination. *Annu Rev Microbiol* **44**:531–553.

11. Setlow P. 2003. Spore germination. *Curr Opin Microbiol* **6**:550–556.

12. Moir A. 2006. How do spores germinate? *J Appl Microbiol* **101**:526–530.

13. Carr KA, Lybarger SR, Anderson EC, Janes BK, Hanna PC. 2010. The role of *Bacillus anthracis* germinant receptors in germination and virulence. *Mol Microbiol* **75**:365–375.

14. Ross C, Abel-Santos E. 2010. The Ger receptor family from sporulating bacteria. *Curr Issues Mol Biol* **12**: 147–158.

15. Paredes-Sabja D, Setlow P, Sarker MR. 2011. Germination of spores of *Bacillales* and *Clostridiales* species: mechanisms and proteins involved. *Trends Microbiol* **19**:85–94

16. Peck MW, Stringer SC, Carter AT. 2011. *Clostridium botulinum* in the post-genomic era. *Food Microbiol* **28**: 183–191.

17. Abee T, Groot MN, Tempelaars M, Zwietering M, Moezelaar R, van der Voort M. 2011. Germination and outgrowth of spores of *Bacillus cereus* group members: diversity and role of germinant receptors. *Food Microbiol* **28**:199–208.

18. Xiao Y, Francke C, Abee T, Wells-Bennik MH. 2011. Clostridial spore germination versus bacilli: genome mining and current insights. *Food Microbiol* **28**: 266–274.

19. Paredes-Sabja D, Torres JA, Setlow P, Sarker MR. 2008. *Clostridium perfringens* spore germination: characterization of germinants and their receptors. *J Bacteriol* **190**:1190–1201.

20. Paidhungat M, Ragkousi K, Setlow P. 2001. Genetic requirements for induction of germination of spores of *Bacillus subtilis* by Ca^{2+}-dipicolinate. *J Bacteriol* **183**: 4886–4893.

21. Popham DL, Meador-Parton J, Costello CE, Setlow P. 1999. Spore peptidoglycan structure in a cwlD dacB double mutant of *Bacillus subtilis*. *J Bacteriol* **181**: 6205–6209.

22. Setlow B, Cowan AE, Setlow P. 2003. Germination of spores of *Bacillus subtilis* with dodecylamine. *J Appl Microbiol* **95**:637–648.

23. Paidhungat M, Setlow B, Daniels WB, Hoover D, Papafragkou E, Setlow P. 2002. Mechanisms of induction of germination of *Bacillus subtilis* spores by high pressure. *Appl Environ Microbiol* **68**:3172–3175.

24. Dworkin J, Shah IM. 2010. Exit from dormancy in microbial organisms. *Nat Rev Microbiol* **8**:890–896.

25. Shah IM, Laaberki MH, Popham DL, Dworkin J. 2008. A eukaryotic-like Ser/Thr kinase signals bacteria to exit dormancy in response to peptidoglycan fragments. *Cell* **135**:486–496.

26. Vary JC, Halvorson HO. 1965. Kinetics of germination of *Bacillus* spores. *J Bacteriol* **89**:1340–1347.

27. Hashimoto T, Frieben WR, Conti SF. 1969. Microgermination of *Bacillus cereus* spores. *J Bacteriol* **100**: 1385–1392.

28. Kong L, Zhang P, Wang G, Yu J, Setlow P, Li YQ. 2011. Characterization of bacterial spore germination using phase-contrast and fluorescence microscopy, Raman spectroscopy and optical tweezers. *Nat Protoc* **6**:625–639.

29. Zhang P, Kong L, Wang G, Scotland M, Ghosh S, Setlow B, Setlow P, Li YQ. 2012. Analysis of the slow germination of multiple individual superdormant *Bacillus subtilis* spores using multifocus Raman microspectroscopy and differential interference contrast microscopy. *J Appl Microbiol* **112**:526–536.

30. Zhang P, Garner W, Yi X, Yu J, Li YQ, Setlow P. 2010. Factors affecting variability in time between addition of nutrient germinants and rapid dipicolinic acid release during germination of spores of *Bacillus* species. *J Bacteriol* **192**:3608–3619.

31. Ghosh S, Setlow P. 2009. Isolation and characterization of superdormant spores of *Bacillus* species. *J Bacteriol* **191**:1787–1797.

32. Ghosh S, Setlow P. 2010. The preparation, germination properties and stability of superdormant spores of *Bacillus cereus*. *J Appl Microbiol* **108**:582–590.

33. Ghosh S, Zhang P, Li YQ, Setlow P. 2009. Superdormant spores of *Bacillus* species have elevated wet-heat resistance and temperature requirements for heat activation. *J Bacteriol* **191**:5584–5591.

34. Ghosh S, Scotland M, Setlow P. 2012. Levels of germination proteins in dormant and superdormant spores of *Bacillus subtilis*. *J Bacteriol* **194**:2221–2227.

35. Setlow P, Liu G, Faeder JR. 2012. Heterogeneity in bacterial spore populations, p 119–214. *In* Abel-Santos E (ed), *Bacterial Spores: Current Research and Applications*. Caister Academic Press, Poole, United Kingdom.

36. Eijlander RT, Abee T, Kuipers OP. 2011. Bacterial spores in food: how phenotypic variability complicates prediction of spore properties and bacterial behavior. *Curr Opin Biotechnol* **22**:180–186.

37. Stringer SC, Webb MD, Peck MW. 2009. Contrasting effects of heat treatment and incubation temperature on germination and outgrowth of individual spores of nonproteolytic *Clostridium botulinum* bacteria. *Appl Environ Microbiol* **75**:2712–2719.

38. Stringer SC, Webb MD, Peck MW. 2011. Lag time variability in individual spores of *Clostridium botulinum*. *Food Microbiol* **28**:228–235.

39. Planchon S, Dargaignaratz C, Levy C, Ginies C, Broussolle V, Carlin F. 2011. Spores of *Bacillus cereus* strain KBAB4 produced at 10 °C and 30 °C display variations in their properties. *Food Microbiol* **28**: 291–297.

40. Melly E, Genest PC, Gilmore ME, Little S, Popham DL, Driks A, Setlow P. 2002. Analysis of the properties of spores of *Bacillus subtilis* prepared at different temperatures. *J Appl Microbiol* **92**:1105–1115.

41. Rose R, Setlow B, Monroe A, Mallozzi M, Driks A, Setlow P. 2007. Comparison of the properties of *Bacillus subtilis* spores made in liquid or on agar plates. *J Appl Microbiol* **103**:691–699.

42. Sanchez-Salas JL, Setlow B, Zhang P, Li YQ, Setlow P. 2011. Maturation of released spores is necessary for acquisition of full spore heat resistance during *Bacillus subtilis* sporulation. *Appl Environ Microbiol* **77**:6746–6754.

43. Alzahrani OM, Moir A. 2014. Spore germination and germinant receptor genes in wild strains of *Bacillus subtilis*. *J Appl Microbiol* **117**:741–749.

44. Okugawa S, Moayeri M, Pomerantsev AP, Sastalla I, Crown D, Gupta PK, Leppla SH. 2012. Lipoprotein biosynthesis by prolipoprotein diacylglyceryl transferase is required for efficient spore germination and full virulence of *Bacillus anthracis*. *Mol Microbiol* **83**:96–109.

45. Sastalla I, Rosovitz MJ, Leppla SH. 2010. Accidental selection and intentional restoration of sporulation-deficient *Bacillus anthracis* mutants. *Appl Environ Microbiol* **76**:6318–6321.

46. Moir A, Lafferty E, Smith DA. 1979. Genetics analysis of spore germination mutants of *Bacillus subtilis* 168: the correlation of phenotype with map location. *J Gen Microbiol* **111**:165–180.

47. Keynan A, Evenchick Z. 1969. Activation, p 359–396. *In* Gould GW, Hurst A (ed), *The Bacterial Spore*. Academic Press, London, United Kingdom.

48. Busta FF, Ordal ZJ. 1964. Heat-activation kinetics of endospores of *Bacillus subtilis*. *J Food Sci* **29**:345–353.

49. Foerster HF. 1983. Activation and germination characteristics observed in endospores of thermophilic strains of *Bacillus*. *Arch Microbiol* **134**:175–181.

50. Foerster HF. 1985. The effects of alterations in the suspending medium on low-temperature activation of spores of *Bacillus stearothermophilus*-NGB101. *Arch Microbiol* **142**:185–189.

51. Luu S, Cruz-Mora J, Setlow B, Feeherry FE, Doona CJ, Setlow P. 2015. The effects of heat activation on *Bacillus* spore germination, with nutrients or under high pressure, with or without various germination proteins. *Appl Environ Microbiol* **81**:2927–2938.

52. Behravan J, Chirakkal H, Masson A, Moir A. 2000. Mutations in the *gerP* locus of *Bacillus subtilis* and *Bacillus cereus* affect access of germinants to their targets in spores. *J Bacteriol* **182**:1987–1994.

53. Popham DL, Illades-Aguiar B, Setlow P. 1995. The *Bacillus subtilis dacB* gene, encoding penicillin-binding protein 5*, is part of a three-gene operon required for proper spore cortex synthesis and spore core dehydration. *J Bacteriol* **177**:4721–4729.

54. Orsburn B, Sucre K, Popham DL, Melville SB. 2009. The SpmA/B and DacF proteins of *Clostridium perfringens* play important roles in spore heat resistance. *FEMS Microbiol Lett* **291**:188–194.

55. Scott IR, Ellar DJ. 1978. Metabolism and the triggering of germination of *Bacillus megaterium*. Use of L-[3H] alanine and tritiated water to detect metabolism. *Biochem J* **174**:635–640.

56. Sinai L, Rosenberg A, Smith Y, Segev E, Ben-Yehuda S. 2015. The molecular timeline of a reviving bacterial spore. *Mol Cell* **57**:695–707.

57. Carr KA, Janes BK, Hanna PC. 2010. Role of the *gerP* operon in germination and outgrowth of *Bacillus anthracis* spores. *PLoS One* **5**:e9128. doi:10.1371/journal.pone.0009128.

58. Yasuda Y, Kanda K, Nishioka S, Tanimoto Y, Kato C, Saito A, Fukuchi S, Nakanishi Y, Tochikubo K. 1993. Regulation of L-alanine-initiated germination of *Bacillus subtilis* spores by alanine racemase. *Amino Acids* **4**: 89–99.

59. Chesnokova ON, McPherson SA, Steichen CT, Turnbough CL Jr. 2009. The spore-specific alanine racemase of *Bacillus anthracis* and its role in suppressing germination during spore development. *J Bacteriol* **191**: 1303–1310.

60. Doona CJ, Ghosh S, Feeherry FF, Ramirez-Peralta A, Huang Y, Chen H, Setlow P. 2014. High pressure germination of *Bacillus subtilis* spores with alterations in levels and types of germination proteins. *J Appl Microbiol* **117**:711–720.

61. Yi X, Setlow P. 2010. Studies of the commitment step in the germination of spores of bacillus species. *J Bacteriol* **192**:3424–3433.

62. Swerdlow BM, Setlow B, Setlow P. 1981. Levels of H+ and other monovalent cations in dormant and germinating spores of *Bacillus megaterium*. *J Bacteriol* **148**:20–29.

63. Magill NG, Cowan AE, Leyva-Vazquez MA, Brown M, Koppel DE, Setlow P. 1996. Analysis of the relationship between the decrease in pH and accumulation of 3-phosphoglyceric acid in developing forespores of *Bacillus* species. *J Bacteriol* **178**:2204–2210.

64. Popham DL, Helin J, Costello CE, Setlow P. 1996. Muramic lactam in peptidoglycan of *Bacillus subtilis* spores is required for spore outgrowth but not for spore dehydration or heat resistance. *Proc Natl Acad Sci USA* 93:15405–15410.

65. Li Y, Davis A, Korza G, Zhang P, Li YQ, Setlow B, Setlow P, Hao B. 2012. Role of a SpoVA protein in dipicolinic acid uptake into developing spores of *Bacillus subtilis*. *J Bacteriol* 194:1875–1884.

66. Tovar-Rojo F, Chander M, Setlow B, Setlow P. 2002. The products of the spoVA operon are involved in dipicolinic acid uptake into developing spores of *Bacillus subtilis*. *J Bacteriol* 184:584–587.

67. Perez-Valdespino A, Li Y, Setlow B, Ghosh S, Pan D, Korza G, Feeherry FE, Doona CJ, Li Y-Q, Hao B, Setlow P. 2014. Function of the SpoVAEa and SpoVAF proteins of *Bacillus subtilis* spores. *J Bacteriol* 196:2077–2088.

68. Velásquez J, Schuurman-Wolters G, Birkner JP, Abee T, Poolman B. 2014. *Bacillus subtilis* spore protein SpoVAC functions as a mechanosensitive channel. *Mol Microbiol* 92:813–823.

69. Setlow B, Wahome PG, Setlow P. 2008. Release of small molecules during germination of spores of *Bacillus* species *J Bacteriol* 190:4759–4763.

70. Cowan AE, Koppel DE, Setlow B, Setlow P. 2003. A soluble protein is immobile in dormant spores of *Bacillus subtilis* but is mobile in germinated spores: implications for spore dormancy. *Proc Natl Acad Sci USA* 100:4209–4214.

71. Cowan AE, Olivastro EM, Koppel DE, Loshon CA, Setlow B, Setlow P. 2004. Lipids in the inner membrane of dormant spores of *Bacillus* species are largely immobile. *Proc Natl Acad Sci USA* 101:7733–7738.

72. Boland FM, Atrih A, Chirakkal H, Foster SJ, Moir A. 2000. Complete spore-cortex hydrolysis during germination of *Bacillus subtilis* 168 requires SleB and YpeB. *Microbiology* 146:57–64.

73. Chirakkal H, O'Rourke M, Atrih A, Foster SJ, Moir A. 2002. Analysis of spore cortex lytic enzymes and related proteins in *Bacillus subtilis* endospore germination. *Microbiology* 148:2383–2392.

74. Fukushima T, Yamamoto H, Atrih A, Foster SJ, Sekiguchi J. 2002. A polysaccharide deacetylase gene (*pdaA*) is required for germination and for production of muramic delta-lactam residues in the spore cortex of *Bacillus subtilis*. *J Bacteriol* 184:6007–6015.

75. Zhang P, Thomas S, Li YQ, Setlow P. 2012. Effects of cortex peptidoglycan structure and cortex hydrolysis on the kinetics of Ca(2+)-dipicolinic acid release during *Bacillus subtilis* spore germination. *J Bacteriol* 194:646–652.

76. Moriyama R, Fukuoka H, Miyata S, Kudoh S, Hattori A, Kozuka S, Yasuda Y, Tochikubo K, Makino S. 1999. Expression of a germination-specific amidase, SleB, of bacilli in the forespore compartment of sporulating cells and its localization on the exterior side of the cortex in dormant spores. *J Bacteriol* 181:2373–2378.

77. Atrih A, Foster SJ. 2001. In vivo roles of the germination-specific lytic enzymes of *Bacillus subtilis* 168. *Microbiology* 147:2925–2932.

78. Heffron JD, Lambert EA, Sherry N, Popham DL. 2010. Contributions of four cortex lytic enzymes to germination of *Bacillus anthracis* spores. *J Bacteriol* 192:763–770.

79. Heffron JD, Orsburn B, Popham DL. 2009. Roles of germination-specific lytic enzymes CwlJ and SleB in *Bacillus anthracis*. *J Bacteriol* 191:2237–2247.

80. Giebel JD, Carr KA, Anderson EC, Hanna PC. 2009. The germination-specific lytic enzymes SleB, CwlJ1, and CwlJ2 each contribute to *Bacillus anthracis* spore germination and virulence. *J Bacteriol* 191:5569–5576.

81. Setlow B, Peng L, Loshon CA, Li YQ, Christie G, Setlow P. 2009. Characterization of the germination of *Bacillus megaterium* spores lacking enzymes that degrade the spore cortex. *J Appl Microbiol* 107:318–328.

82. Christie G, Ustok FI, Lu Q, Packman LC, Lowe CR. 2010. Mutational analysis of *Bacillus megaterium* QM B1551 cortex-lytic enzymes. *J Bacteriol* 192:5378–5389.

83. Ragkousi K, Eichenberger P, van Ooij C, Setlow P. 2003. Identification of a new gene essential for germination of *Bacillus subtilis* spores with Ca²⁺-dipicolinate. *J Bacteriol* 185:2315–2329.

84. Bernhards CB, Popham DL. 2014. Role of YpeB in cortex hydrolysis during germination of *Bacillus anthracis* spores. *J Bacteriol* 196:3399–3409.

85. Li Y, Butzin XY, Davis A, Setlow B, Korza G, Üstok FI, Christie G, Setlow P, Hao B. 2013. Activity and regulation of various forms of CwlJ, SleB, and YpeB proteins in degrading cortex peptidoglycan of spores of *Bacillus* species in vitro and during spore germination. *J Bacteriol* 195:2530–2540.

86. Üstok FI, Packman LC, Lowe CR, Christie G. 2014. Spore germination mediated by *Bacillus megaterium* QM B1551 SleL and YpeB. *J Bacteriol* 196:1045–1054.

87. Heffron JD, Sherry N, Popham DL. 2011. In vitro studies of peptidoglycan binding and hydrolysis by the *Bacillus anthracis* germination-specific lytic enzyme SleB. *J Bacteriol* 193:125–131.

88. Li Y, Jin K, Setlow B, Setlow P, Hao B. 2012. Crystal structure of the catalytic domain of the *Bacillus cereus* SleB protein, important in cortex peptidoglycan degradation during spore germination. *J Bacteriol* 194:4537–4545.

89. Jing X, Robinson HR, Heffron JD, Popham DL, Schubot FD. 2012. The catalytic domain of the germination-specific lytic transglycosylase SleB from *Bacillus anthracis* displays a unique active site topology. *Proteins* 80:2469–2475.

90. Kodama T, Takamatsu H, Asai K, Kobayashi K, Ogasawara N, Watabe K. 1999. The *Bacillus subtilis* yaaH gene is transcribed by SigE RNA polymerase during sporulation, and its product is involved in germination of spores. *J Bacteriol* 181:4584–4591.

91. Lambert EA, Popham DL. 2008. The *Bacillus anthracis* SleL (YaaH) protein is an N-acetylglucosaminidase

involved in spore cortex depolymerization. *J Bacteriol* **190**:7601–7607.

92. **Sanchez-Salas JL, Santiago-Lara ML, Setlow B, Sussman MD, Setlow P.** 1992. Properties of *Bacillus megaterium* and *Bacillus subtilis* mutants which lack the protease that degrades small, acid-soluble proteins during spore germination. *J Bacteriol* **174**:807–814.

93. **Ibarra JR, Orozco AD, Rojas JA, López K, Setlow P, Yasbin RE, Pedraza-Reyes M.** 2008. Role of the Nfo and ExoA apurinic/apyrimidinic endonucleases in repair of DNA damage during outgrowth of *Bacillus subtilis* spores. *J Bacteriol* **190**:2031–2038.

94. **Setlow P.** 2007. I will survive: DNA protection in bacterial spores. *Trends Microbiol* **15**:172–180.

95. **Santo LY, Doi RH.** 1974. Ultrastructural analysis during germination and outgrowth of *Bacillus subtilis* spores. *J Bacteriol* **120**:475–481.

96. **Plomp M, Leighton TJ, Wheeler KE, Hill HD, Malkin AJ.** 2007. In vitro high-resolution structural dynamics of single germinating bacterial spores. *Proc Natl Acad Sci USA* **104**:9644–9649.

97. **Jagtap P, Michailidis G, Zielke R, Walker AK, Patel N, Strahler JR, Driks A, Andrews PC, Maddock JR.** 2006. Early events of *Bacillus anthracis* germination identified by time-course quantitative proteomics. *Proteomics* **6**:5199–5211.

98. **Nessi C, Albertini AM, Speranza ML, Galizzi A.** 1995. The outB gene of *Bacillus subtilis* codes for NAD synthetase. *J Biol Chem* **270**:6181–6185.

99. **Murray T, Popham DL, Pearson CB, Hand AR, Setlow P.** 1998. Analysis of outgrowth of *Bacillus subtilis* spores lacking penicillin-binding protein 2a. *J Bacteriol* **180**:6493–6502.

100. **Keijser BJF, Ter Beek A, Rauwerda H, Schuren F, Montijn R, van der Spek H, Brul S.** 2007. Analysis of temporal gene expression during *Bacillus subtilis* spore germination and outgrowth. *J Bacteriol* **189**:3624–3634.

101. **Bettegowda C, Huang X, Lin J, Cheong I, Kohli M, Szabo SA, Zhang X, Diaz LA Jr, Velculescu VE, Parmigiani G, Kinzler KW, Vogelstein B, Zhou S.** 2006. The genome and transcriptomes of the anti-tumor agent *Clostridium novyi*-NT. *Nat Biotechnol* **24**:1573–1580.

102. **Segev E, Smith Y, Ben-Yehuda S.** 2012. RNA dynamics in aging bacterial spores. *Cell* **148**:139–149.

103. **Zuberi AR, Moir A, Feavers IM.** 1987. The nucleotide sequence and gene organization of the *gerA* spore germination operon of *Bacillus subtilis* 168. *Gene* **51**:1–11.

104. **Ramirez-Peralta A, Gupta S, Butzin XY, Setlow B, Korza G, Leyva-Vazquez M-A, Christie G, Setlow P.** 2013. Identification of new proteins that modulate the germination of spores of *Bacillus* species. *J Bacteriol* **195**:3009–3021.

105. **Paidhungat M, Setlow P.** 2000. Role of ger proteins in nutrient and nonnutrient triggering of spore germination in *Bacillus subtilis*. *J Bacteriol* **182**:2513–2519.

106. **Stewart K-AV, Setlow P.** 2013. Numbers of individual nutrient germinant receptors and other germination proteins in spores of *Bacillus subtilis*. *J Bacteriol* **195**:3575–3582.

107. **Bagyan I, Hobot J, Cutting S.** 1996. A compartmentalized regulator of developmental gene expression in *Bacillus subtilis*. *J Bacteriol* **178**:4500–4507.

108. **Traag BA, Ramirez-Peralta A, Wang Erickson AF, Setlow P, Losick R.** 2013. A novel RNA polymerase-binding protein controlling genes involved in spore germination in *Bacillus subtilis*. *Mol Microbiol* **89**:113–122.

109. **Hinc K, Nagórska K, Iwanicki A, Wegrzyn G, Séror SJ, Obuchowski M.** 2006. Expression of genes coding for GerA and GerK spore germination receptors is dependent on the protein phosphatase PrpE. *J Bacteriol* **188**:4373–4383.

110. **Ross CA, Abel-Santos E.** 2010. Guidelines for nomenclature assignment of Ger receptors. *Res Microbiol* **161**:830–837.

111. **Abdoarrahem MM, Gammon K, Dancer BN, Berry C.** 2009. Genetic basis for alkaline activation of germination in *Bacillus thuringiensis* subsp. *israelensis*. *Appl Environ Microbiol* **75**:6410–6413.

112. **Guidi-Rontani C, Pereira Y, Ruffie S, Sirard JC, Weber-Levy M, Mock M.** 1999. Identification and characterization of a germination operon on the virulence plasmid pXO1 of *Bacillus anthracis*. *Mol Microbiol* **33**:407–414.

113. **Van der Auwera G, Mahillon J.** 2005. TnXO1, a germination-associated class II transposon from *Bacillus anthracis*. *Plasmid* **53**:251–257.

114. **Christie G, Lowe CR.** 2007. Role of chromosomal and plasmid-borne receptor homologues in the response of *Bacillus megaterium* QM B1551 spores to germinants. *J Bacteriol* **189**:4375–4383.

115. **Griffiths KK, Zhang J, Cowan AE, Yu J, Setlow P.** 2011. Germination proteins in the inner membrane of dormant *Bacillus subtilis* spores colocalize in a discrete cluster. *Mol Microbiol* **81**:1061–1077.

116. **Ghosh S, Setlow B, Wahome PG, Cowan AE, Plomp M, Malkin AJ, Setlow P.** 2008. Characterization of spores of *Bacillus subtilis* that lack most coat layers. *J Bacteriol* **190**:6741–6748.

117. **Hudson KD, Corfe BM, Kemp EH, Feavers IM, Coote PJ, Moir A.** 2001. Localization of GerAA and GerAC germination proteins in the *Bacillus subtilis* spore. *J Bacteriol* **183**:4317–4322.

118. **Paidhungat M, Setlow P.** 2001. Localization of a germinant receptor protein (GerBA) to the inner membrane of *Bacillus subtilis* spores. *J Bacteriol* **183**:3982–3990.

119. **Pelczar PL, Igarashi T, Setlow B, Setlow P.** 2007. Role of GerD in germination of *Bacillus subtilis* spores. *J Bacteriol* **189**:1090–1098.

120. **Pelczar PL, Setlow P.** 2008. Localization of the germination protein GerD to the inner membrane in *Bacillus subtilis* spores. *J Bacteriol* **190**:5635–5641.

121. **Mongkolthanaruk W, Robinson C, Moir A.** 2009. Localization of the GerD spore germination protein in the *Bacillus subtilis* spore. *Microbiology* **155**:1146–1151.

122. **Li Y, Jin K, Ghosh S, Devarakonda P, Carlson K, Davis A, Stewart K-AV, Cammett E, Pelczar Rossi P, Setlow**

B, Lu M, Setlow P, Hao B. 2014. Structural and functional analysis of the GerD spore germination protein of *Bacillus* species. *J Mol Biol* **426**:1995–2008.

123. Yi X, Liu J, Faeder JR, Setlow P. 2011. Synergism between different germinant receptors in the germination of *Bacillus subtilis* spores. *J Bacteriol* **193**:4664–4671.

124. Luu H, Akoachere M, Patra M, Abel-Santos E. 2011. Cooperativity and interference of germination pathways in *Bacillus anthracis* spores. *J Bacteriol* **193**:4192–4198.

125. Cooper GR, Moir A. 2011. Amino acid residues in the GerAB protein important in the function and assembly of the alanine spore germination receptor of *Bacillus subtilis* 168. *J Bacteriol* **193**:2261–2267.

126. Mongkolthanaruk W, Cooper GR, Mawer JSP, Allan RN, Moir A. 2011. Effect of amino acid substitutions in the GerAA protein on the function of the alanine-responsive germinant receptor of *Bacillus subtilis* spores. *J Bacteriol* **193**:2268–2275.

127. Gupta S, Ustok FI, Johnson CL, Bailey DMD, Lowe CR, Christie G. 2013. Investigating the functional hierarchy of *Bacillus megaterium* PV361 spore germinant receptors. *J Bacteriol* **195**:3045–3053.

128. Stewart KAV, Yi X, Ghosh S, Setlow P. 2012. Germination protein levels and rates of germination of spores of *Bacillus subtilis* with overexpressed or deleted genes encoding germination proteins. *J Bacteriol* **194**:3156–3164.

129. Jack DL, Paulsen IT, Saier MH. 2000. The amino acid/polyamine/organocation (APC) superfamily of transporters specific for amino acids, polyamines and organocations. *Microbiology* **146**:1797–1814.

130. Wilson MJ, Carlson PE, Janes BK, Hanna PC. 2012. Membrane topology of the *Bacillus anthracis* GerH germinant receptor proteins. *J Bacteriol* **194**:1369–1377.

131. Paidhungat M, Setlow P. 1999. Isolation and characterization of mutations in *Bacillus subtilis* that allow spore germination in the novel germinant D-alanine. *J Bacteriol* **181**:3341–3350.

132. Atluri S, Ragkousi K, Cortezzo DE, Setlow P. 2006. Cooperativity between different nutrient receptors in germination of spores of *Bacillus subtilis* and reduction of this cooperativity by alterations in the GerB receptor. *J Bacteriol* **188**:28–36.

133. Christie G, Götzke H, Lowe CR. 2010. Identification of a receptor subunit and putative ligand-binding residues involved in the *Bacillus megaterium* QM B1551 spore germination response to glucose. *J Bacteriol* **192**:4317–4326.

134. Christie G, Lazarevska M, Lowe CR. 2008. Functional consequences of amino acid substitutions to GerVB, a component of the *Bacillus megaterium* spore germinant receptor. *J Bacteriol* **190**:2014–2022.

135. Sammons RL, Moir A, Smith DA. 1981. Isolation and properties of spore germination mutants of *Bacillus subtilis* 168 deficient in the initiation of germination. *J Gen Microbiol* **124**:229–241.

136. Taylor PM. 2009. Amino acid transporters: éminences grises of nutrient signalling mechanisms? *Biochem Soc Trans* **37**:237–241.

137. Gao X, Zhou L, Jiao X, Lu F, Yan C, Zeng X, Wang J, Shi Y. 2010. Mechanism of substrate recognition and transport by an amino acid antiporter. *Nature* **463**:828–832.

138. Shaffer PL, Goehring A, Shankaranarayanan A, Gouaux E. 2009. Structure and mechanism of a Na$^+$-independent amino acid transporter. *Science* **325**:1010–1014.

139. Li Y, Setlow B, Setlow P, Hao B. 2010. Crystal structure of the GerBC component of a *Bacillus subtilis* spore germinant receptor. *J Mol Biol* **402**:8–16.

140. Li Y, Catta P, Stewart KAV, Dufner M, Setlow P, Hao B. 2011. Structure-based functional studies of the effects of amino acid substitutions in GerBC, the C subunit of the *Bacillus subtilis* GerB spore germinant receptor. *J Bacteriol* **193**:4143–4152.

141. Korza G, Setlow P. 2013. Topology and accessibility of germination proteins in the *Bacillus subtilis* spore inner membrane. *J Bacteriol* **195**:1484–1491.

142. Priest FG, Barker M, Baillie LWJ, Holmes EC, Maiden MCJ. 2004. Population structure and evolution of the *Bacillus cereus* group. *J Bacteriol* **186**:7959–7970.

143. Tourasse NJ, Helgason E, Klevan A, Sylvestre P, Moya M, Haustant M, Økstad OA, Fouet A, Mock M, Kolstø AB. 2011. Extended and global phylogenetic view of the *Bacillus cereus* group population by combination of MLST, AFLP, and MLEE genotyping data. *Food Microbiol* **28**:236–244.

144. Clements MO, Moir A. 1998. Role of the *gerI* operon of *Bacillus cereus* 569 in the response of spores to germinants. *J Bacteriol* **180**:6729–6735.

145. Barlass PJ, Houston CW, Clements MO, Moir A. 2002. Germination of *Bacillus cereus* spores in response to L-alanine and to inosine: the roles of gerL and gerQ operons. *Microbiology* **148**:2089–2095.

146. Dodatko T, Akoachere M, Jimenez N, Alyarez Z, Abel-Santos E. 2010. Dissecting interactions between nucleosides and germination receptors in *Bacillus cereus* 569 spores. *Microbiology* **156**:1244–1255.

147. Dodatko T, Akoachere M, Muehlbauer SM, Helfrich F, Howerton A, Ross C, Wysocki V, Brojatsch J, Abel-Santos E. 2009. *Bacillus cereus* spores release alanine that synergizes with inosine to promote germination. *PLoS One* **4**:e6398. doi:10.1371/journal.pone.0006398.

148. Thackray PD, Behravan J, Southworth TW, Moir A. 2001. GerN, an antiporter homologue important in germination of *Bacillus cereus* endospores. *J Bacteriol* **183**:476–482.

149. Southworth TW, Guffanti AA, Moir A, Krulwich TA. 2001. GerN, an endospore germination protein of *Bacillus cereus*, is an Na(+)/H(+)-K(+) antiporter. *J Bacteriol* **183**:5896–5903.

150. Senior A, Moir A. 2008. The *Bacillus cereus* GerN and GerT protein homologs have distinct roles in spore germination and outgrowth, respectively. *J Bacteriol* **190**:6148–6152.

151. Tani K, Watanabe T, Matsuda H, Nasu M, Kondo M. 1996. Cloning and sequencing of the spore germination gene of *Bacillus megaterium* ATCC 12872: similarities to the NaH-antiporter gene of *Enterococcus hirae*. *Microbiol Immunol* 40:99–105.

152. Paredes-Sabja D, Setlow P, Sarker MR. 2009. GerO, a putative Na$^+$/H$^+$-K$^+$ antiporter, is essential for normal germination of spores of the pathogenic bacterium *Clostridium perfringens*. *J Bacteriol* 191:3822–3831.

153. Hornstra LM, de Vries YP, de Vos WM, Abee T, Wells-Bennik MHJ. 2005. gerR, a novel ger operon involved in L-alanine- and inosine-initiated germination of *Bacillus cereus* ATCC 14579. *Appl Environ Microbiol* 71: 774–781.

154. Hornstra LM, de Vries YP, Wells-Bennik MHJ, de Vos WM, Abee T. 2006. Characterization of germination receptors of *Bacillus cereus* ATCC 14579. *Appl Environ Microbiol* 72:44–53.

155. Read TD, Peterson SN, Tourasse N, Baillie LW, Paulsen IT, Nelson KE, Tettelin H, Fouts DE, Eisen JA, Gill SR, Holtzapple EK, Okstad OA, Helgason E, Rilstone J, Wu M, Kolonay JF, Beanan MJ, Dodson RJ, Brinkac LM, Gwinn M, DeBoy RT, Madpu R, Daugherty SC, Durkin AS, Haft DH, Nelson WC, Peterson JD, Pop M, Khour HM, Radume D, Benton JL, Mahamoud Y, Jiang L, Hance IR, Weidman JF, Berry KJ, Plaut RD, Wolf AM, Watkins KL, Nierman WC, Hazen A, Cline R, Redmond C, Thwaite JE, White O, Salzberg SL, Thomason B, Friedlander AM, Koehler TM, Hanna PC, Kolstø AB, Fraser CM. 2003. The genome sequence of *Bacillus anthracis* Ames and comparison to closely related bacteria. *Nature* 423:81–86.

156. Weiner MA, Read TD, Hanna PC. 2003. Identification and characterization of the gerH operon of *Bacillus anthracis* endospores: a differential role for purine nucleosides in germination. *J Bacteriol* 185:1462–1464.

157. Fisher N, Hanna P. 2005. Characterization of *Bacillus anthracis* germinant receptors in vitro. *J Bacteriol* 187: 8055–8062.

158. Hu H, Emerson J, Aronson AI. 2007. Factors involved in the germination and inactivation of *Bacillus anthracis* spores in murine primary macrophages. *FEMS Microbiol Lett* 272:245–250.

159. Barua S, McKevitt M, DeGiusti K, Hamm EE, Larabee J, Shakir S, Bryant K, Koehler TM, Blanke SR, Dyer D, Gillaspy A, Ballard JD. 2009. The mechanism of *Bacillus anthracis* intracellular germination requires multiple and highly diverse genetic loci. *Infect Immun* 77:23–31.

160. Ramirez N, Abel-Santos E. 2010. Requirements for germination of *Clostridium sordellii* spores in vitro. *J Bacteriol* 192:418–425.

161. Waites WM, Wyatt LR. 1974. The effect of pH, germinants and temperature on the germination of spores of *Clostridium bifermentans*. *J Gen Microbiol* 80:253–258.

162. Adam KH, Brunt J, Brightwell G, Flint SH, Peck MW. 2011. Spore germination of the psychrotolerant, red meat spoiler, *Clostridium frigidicarnis*. *Lett Appl Microbiol* 53:92–97.

163. Brunt J, Plowman J, Gaskin DJH, Itchner M, Carter AT, Peck MW. 2014. Functional characterisation of germinant receptors in *Clostridium botulinum* and *Clostridium sporogenes* presents novel insights into spore germination systems. *PLoS Pathog* 10:e1004382. doi:10.1371/journal.ppat.1004382.

164. Paredes-Sabja D, Setlow P, Sarker MR. 2009. Role of GerKB in germination and outgrowth of *Clostridium perfringens* spores. *Appl Environ Microbiol* 75:3813–3817.

165. Olguín-Araneda V, Banawas S, Sarker MR, Paredes-Sabja D. 2015. Recent advances in germination of *Clostridium* spores. *Res Microbiol* 166:236–243.

166. Urakami K, Miyata S, Moriyama R, Sugimoto K, Makino S. 1999. Germination-specific cortex-lytic enzymes from *Clostridium perfringens* S40 spores: time of synthesis, precursor structure and regulation of enzymatic activity. *FEMS Microbiol Lett* 173:467–473.

167. Makino S, Moriyama R. 2002. Hydrolysis of cortex peptidoglycan during bacterial spore germination. *Med Sci Monit* 8:RA119–RA127.

168. Okamura S, Urakami K, Kimata M, Aoshima T, Shimamoto S, Moriyama R, Makino S. 2000. The N-terminal prepeptide is required for the production of spore cortex-lytic enzyme from its inactive precursor during germination of *Clostridium perfringens* S40 spores. *Mol Microbiol* 37:821–827.

169. Banawas S, Korza G, Paredes-Sabja D, Li Y, Hao B, Setlow P, Sarker MR. 2015. Location and stoichiometry of the protease CspB and the cortex-lytic enzyme SleC in *Clostridium perfringens* spores. *Food Microbiol* 50: 83–87.

170. Paredes-Sabja D, Setlow P, Sarker MR. 2009. SleC is essential for cortex peptidoglycan hydrolysis during germination of spores of the pathogenic bacterium *Clostridium perfringens*. *J Bacteriol* 191:2711–2720.

171. Paredes-Sabja D, Setlow P, Sarker MR. 2009. The protease CspB is essential for initiation of cortex hydrolysis and dipicolinic acid (DPA) release during germination of spores of *Clostridium perfringens* type A food poisoning isolates. *Microbiology* 155:3464–3472.

172. Sorg JA, Sonenshein AL. 2008. Bile salts and glycine as cogerminants for *Clostridium difficile* spores. *J Bacteriol* 190:2505–2512.

173. Howerton A, Ramirez N, Abel-Santos E. 2011. Mapping interactions between germinants and *Clostridium difficile* spores. *J Bacteriol* 193:274–282.

174. Wheeldon LJ, Worthington T, Lambert PA. 2011. Histidine acts as a co-germinant with glycine and taurocholate for *Clostridium difficile* spores. *J Appl Microbiol* 110: 987–994.

175. Francis MB, Allen CA, Shrestha R, Sorg JA. 2013. Bile acid recognition by the *Clostridium difficile* germinant receptor, CspC, is important for establishing infection. *PLoS Pathog* 9:e1003356. doi:10.1371/journal.ppat. 1003356.

176. Paredes-Sabja D, Shen A, Sorg JA. 2014. *Clostridium difficile* spore biology: sporulation, germination, and spore structural proteins. *Trends Microbiol* 22:406–416.

177. Burns DA, Heap JT, Minton NP. 2010. SleC is essential for germination of *Clostridium difficile* spores in nutrient-rich medium supplemented with the bile salt taurocholate. *J Bacteriol* 192:657–664.

178. Burns DA, Heap JT, Minton NP. 2010. *Clostridium difficile* spore germination: an update. *Res Microbiol* 161:730–734.

179. Francis MB, Allen CA, Sorg JA. 2015. Spore cortex hydrolysis precedes dipicolinic acid release during *Clostridium difficile* spore germination. *J Bacteriol* 197:2276–2283.

180. Igarashi T, Setlow B, Paidhungat M, Setlow P. 2004. Effects of a *gerF* (*lgt*) mutation on the germination of spores of *Bacillus subtilis*. *J Bacteriol* 186:2984–2991.

181. Kawai F, Hara H, Takamatsu H, Watabe K, Matsumoto K. 2006. Cardiolipin enrichment in spore membranes and its involvement in germination of *Bacillus subtilis* Marburg. *Genes Genet Syst* 81:69–76.

182. Griffiths KK, Setlow P. 2009. Effects of modification of membrane lipid composition on *Bacillus subtilis* sporulation and spore properties. *J Appl Microbiol* 106:2064–2078.

183. Masayama A, Kato S, Terashima T, Mølgaard A, Hemmi H, Yoshimura T, Moriyama R. 2010. *Bacillus subtilis* spore coat protein LipC is a phospholipase B. *Biosci Biotechnol Biochem* 74:24–30.

184. Henriques AO, Moran CP. 2007. Structure, assembly, and function of the spore surface layers. *Annu Rev Microbiol* 61:555–588.

185. Moir A. 1981. Germination properties of a spore coat-defective mutant of *Bacillus subtilis*. *J Bacteriol* 146:1106–1116.

186. Ferguson CC, Camp AH, Losick R. 2007. gerT, a newly discovered germination gene under the control of the sporulation transcription factor sigmaK in *Bacillus subtilis*. *J Bacteriol* 189:7681–7689.

187. Kodama T, Matsubayashi T, Yanagihara T, Komoto H, Ara K, Ozaki K, Kuwana R, Imamura D, Takamatsu H, Watabe K, Sekiguchi J. 2011. A novel small protein of *Bacillus subtilis* involved in spore germination and spore coat assembly. *Biosci Biotechnol Biochem* 75:1119–1128.

188. Naclerio G, Baccigalupi L, Zilhao R, De Felice M, Ricca E. 1996. *Bacillus subtilis* spore coat assembly requires *cotH* gene expression. *J Bacteriol* 178:4375–4380.

189. Little S, Driks A. 2001. Functional analysis of the *Bacillus subtilis* morphogenetic spore coat protein CotE. *Mol Microbiol* 42:1107–1120.

190. Zheng LB, Donovan WP, Fitz-James PC, Losick R. 1988. Gene encoding a morphogenic protein required in the assembly of the outer coat of the *Bacillus subtilis* endospore. *Genes Dev* 2:1047–1054.

191. Bourne N, FitzJames PC, Aronson AI. 1991. Structural and germination defects of *Bacillus subtilis* spores with altered contents of a spore coat protein. *J Bacteriol* 173:6618–6625.

192. Stewart ET, Halvorson HO. 1953. Studies on the spores of aerobic bacteria. I. The occurrence of alanine racemase. *J Bacteriol* 65:160–166.

193. Todd SJ, Moir AJG, Johnson MJ, Moir A. 2003. Genes of *Bacillus cereus* and *Bacillus anthracis* encoding proteins of the exosporium. *J Bacteriol* 185:3373–3378.

194. Liang L, He XH, Liu G, Tan HR. 2008. The role of a purine-specific nucleoside hydrolase in spore germination of *Bacillus thuringiensis*. *Microbiology* 154:1333–1340.

195. Wuytack EY, Boven S, Michiels CW. 1998. Comparative study of pressure-induced germination of *Bacillus subtilis* spores at low and high pressures. *Appl Environ Microbiol* 64:3220–3224.

196. Wuytack EY, Soons J, Poschet F, Michiels CW. 2000. Comparative study of pressure- and nutrient-induced germination of *Bacillus subtilis* spores. *Appl Environ Microbiol* 66:257–261.

197. Black EP, Koziol-Dube K, Guan D, Wei J, Setlow B, Cortezzo DE, Hoover DG, Setlow P. 2005. Factors influencing germination of *Bacillus subtilis* spores via activation of nutrient receptors by high pressure. *Appl Environ Microbiol* 71:5879–5887.

198. Black EP, Wei J, Atluri S, Cortezzo DE, Koziol-Dube K, Hoover DG, Setlow P. 2007. Analysis of factors influencing the rate of germination of spores of *Bacillus subtilis* by very high pressure. *J Appl Microbiol* 102:65–76.

199. Rode LJ, Foster JW. 1961. Germination of bacterial spores with alkyl primary amines. *J Bacteriol* 81:768–779.

200. Vepachedu VR, Setlow P. 2007. Role of SpoVA proteins in release of dipicolinic acid during germination of *Bacillus subtilis* spores triggered by dodecylamine or lysozyme. *J Bacteriol* 189:1565–1572.

201. Peck MW, Fairbairn DA, Lund BM. 1992. Factors affecting growth from heat-treated spores of nonproteolytic *Clostridium botulinum*. *Lett Appl Microbiol* 15:152–155.

202. Peck MW, Fairbairn DA, Lund BM. 1993. Heat-resistance of spores of nonproteolytic *Clostridium botulinum* estimated on medium containing lysozyme. *Lett Appl Microbiol* 16:126–131.

203. Wei J, Shah IM, Ghosh S, Dworkin J, Hoover DG, Setlow P. 2010. Superdormant spores of bacillus species germinate normally with high pressure, peptidoglycan fragments, and bryostatin. *J Bacteriol* 192:1455–1458.

204. van Melis CCJ, Nierop Groot MN, Abee T. 2011. Impact of sorbic acid on germinant receptor-dependent and -independent germination pathways in *Bacillus cereus*. *Appl Environ Microbiol* 77:2552–2554.

205. Shah IM, Dworkin J. 2010. Induction and regulation of a secreted peptidoglycan hydrolase by a membrane Ser/Thr kinase that detects muropeptides. *Mol Microbiol* 75:1232–1243.

206. Setlow P. 2014. Spore resistance properties. *Microbiol Spectrum* 2(4):TBS-0003-2012. doi:10.1128/microbiolspec.TBS-0003-2012.

207. Galperin MY. 2013. Genomic diversity of spore-forming *Firmicutes*. *Microbiol Spectrum* 1(2):TBS-0015-2012.

The *Bacillaceae*: *Bacillus anthracis*

III

The Bacterial Spore: From Molecules to Systems
Edited by P. Eichenberger and A. Driks
© 2016 American Society for Microbiology, Washington, DC
doi:10.1128/microbiolspec.TBS-0012-2012

Richard T. Okinaka[1]
Paul Keim[1]

The Phylogeny of *Bacillus cereus sensu lato*

12

The three main species of the *Bacillus cereus sensu lato*, *B. cereus*, *B. thuringiensis*, and *B. anthracis*, were recognized and established by the early 1900s because they each exhibited distinct phenotypic traits. *B. thuringiensis* isolates and their parasporal crystal proteins have long been established as a natural pesticide and insect pathogen (1). *B. anthracis*, the etiological agent for anthrax, was used by Robert Koch in the 19th century as a model to develop the germ theory of disease (2), and *B. cereus*, a common soil organism, is also an occasional opportunistic pathogen of humans (3–5). In addition to these three historical species designations, are three less-recognized and -understood species: *B. mycoides*, *B. weihenstephanensis*, and *B. pseudomycoides*. All of these "species" combined comprise the *B. cereus sensu lato* group. Despite these apparently clear phenotypic definitions, early molecular approaches to separate the first three by various DNA hybridization and 16S/23S ribosomal sequence analyses led to some "confusion" because there were limited differences to differentiate between these species (6). These and other results have led to frequent suggestions that a taxonomic change was warranted to reclassify this group to a single species (7, 8). But the pathogenic properties of *B. anthracis* and the biopesticide applications of *B. thuringiensis* appear to "have outweighed pure taxonomic considerations" and the separate species categories are still being maintained (9). *B. cereus sensu lato* represents a classic example of a now common bacterial species taxonomic quandary where relatively new molecular data must somehow be incorporated into a traditional hierarchical classification system (10).

AFLP APPROACH LEADS TO PHYLOGENETIC RESOLUTION OF *B. ANTHRACIS* AND *B. CEREUS SENSU LATO*

In the mid-1990s an amplified fragment length polymorphism (AFLP) method was developed to examine restriction fragment length polymorphisms (RFLPs) in whole genomes using restriction enzyme digestion coupled to PCR analysis and high-resolution electrophoresis (11). The advantage of AFLP was that the ends of the restricted fragments could be linked to specific primer sequences that then served as targets of PCR amplification for internal sequences that had not previously been described. This approach immediately offered greater genome coverage of RFLP sites and proved to be useful in resolving the monomorphic

---

[1]Center for Microbial Genetics and Genomics, Northern Arizona University, Flagstaff, AZ 86011-4073.

B. anthracis lineage and its close relatives, *B. cereus*, *B. thuringiensis*, and *B. mycoides* (12). This initial study uncovered 357 AFLP characters (polymorphic fragments) that could be used in cladistic and phenetic analyses to construct a phylogeny of the *B. cereus* group in addition to several distant relatives. In this instance, *B. anthracis* could clearly be distinguished from its two closest relatives, *B. cereus/B. thuringiensis,* and another member of this group, *B. mycoides*. This initial AFLP analysis not only provided the first evidence of significant diversity and a DNA-based phylogeny of 78 *B. anthracis* isolates, but it also provided an experimental approach to examine the phylogenetic relationship between diverse isolates contained in large *B. cereus/B. thuringiensis* collections (13, 14).

In the latter analysis, fluorescent AFLP analysis was performed on the DNA from an extensive collection of 332 diverse *B. cereus, B. thuringiensis,* and *B. anthracis* isolates (14). This analysis included 34 diverse Norwegian soil isolates (13, 15); 222 *B. thuringiensis* isolates representing 36 different serovars from a U.S. Department of Agriculture collection; 24 diverse *B. anthracis* isolates; 42 *B. cereus* isolates recovered either from contaminated food products or from clinical samples by the Food Research Institute, University of Wisconsin; and finally, 8 *B. thuringiensis/B. cereus* isolates from the American Type Culture Collection. These latter samples included several type strains that were useful in the comparison of these results with nearly simultaneous studies that were using multiple locus sequence typing (MLST) analysis to examine similar collections of isolates (15, 16).

A diagrammatic representation of a basal tree from this AFLP analysis is illustrated in Fig. 1. The key features of this tree illustrate that the 332 isolates are dispersed into 3 major clusters and into 10 distinct branches (labeled A to K) within the 3 clusters. More importantly, each of the three clusters contains representatives of both *B. cereus* and *B. thuringiensis* isolates. These results portray an important and consistent general theme, i.e., the increased overall resolution offered by AFLP does not separate *B. cereus* and *B. thuringiensis* isolates into distinct clusters. Instead representatives of both species are found scattered throughout the three main clusters defined by the AFLP analysis. These findings have major implications for the evolution of all the subgroups under the nomenclature umbrella for *E. cereus sensu lato*. It supports the notion that a *B. cereus sensu lato* genomic background has evolved in a primarily clonal fashion to form the major phylogenetic branches of this group. However, the species designations for this group are based primarily on the horizontal gene transfer of various plasmids, genomic islands, etc., that contained specialized factors that helped to define specific phenotypes that became the hallmark features of, e.g., *B. anthracis* and *B. thuringiensis* (8).

Figure 1 is a representation of the entire AFLP tree without illustrating significant branch resolution, but highly resolved individual branches (Fig. 3 to 8, Hill et al. [14]) within this tree reveal a potentially wider spectrum of evolutionary developments. Branches A and C, for example, consist mostly of *B. thuringiensis* isolates with only a limited number of related serovars, suggesting that a clonal expansion and fitness of specific *B. thuringiensis* isolates may have founded these clades. This is also reflected in the *B. anthracis* lineage, which is a monophyletic, clonal expansion of a relatively young branch that has sparse diversity and exchange within all the known *B. anthracis* isolates (12, 17). Conversely, other branches in the AFLP tree, e.g., clusters 2 and 3, appear to have several branches where a specific *B. cereus sensu lato* lineage may have given rise to both *B. cereus* and *B. thuringiensis* isolates as a result of horizontal transfer events that occurred later in the evolutionary time scale. A similar phenomenon appears to have occurred more recently in the AFLP F branch, where *B. cereus* isolates, not in the monophyletic clade that is *B. anthracis*, have acquired a pXO1 plasmid (*B. cereus* G9241 [18]) or both the pXO1 and pXO2 plasmids (*B. cereus* CI, biovar *anthracis* [19]). Various typing schemes indicate that these two isolates (*B. cereus* G9241 and *B. cereus* CI) are close relatives of the *B. anthracis* lineage and that both would belong to AFLP group 1 (Fig. 1).

MLST

MLST (http://pubmlst.org/) was originally designed as a molecular typing method that could take advantage of the portability and exchange of sequence data between laboratories (20). The basic idea was to identify seven or more "housekeeping genes" from closely related populations such as the *B. cereus sensu lato* group and to generate ~500-bp sequences from each of these seven genes in each isolate of interest. The ~3,500 bp of sequence for each isolate then defined its sequence type (ST) when the sequence profiles were established by comparative analysis to all other isolates in the database. The MLST database for *B. cereus* currently (as of 29 January 2016) contains 1,185 sequence types among 1,518 isolates.

Three independent *B. cereus sensu lato* MLST studies were published in 2004 (15, 16, 21) and four other

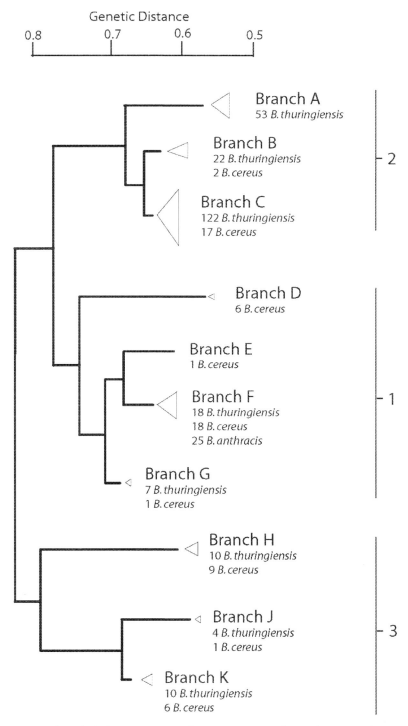

Figure 1 AFLP-based phylogenetic tree of *B. cereus sensu lato*. This is a schematic representation redrawn from Hill et al. (12) of 332 isolates. While 10 distinct branches were identified, they formed three main clusters labeled as 1, 2, and 3 to correspond to subgroups identified by Priest et al. (15) to maintain consistency between AFLP and MLST trees based on the positions of known matching isolates in both trees.

reports have appeared since (22–25). The singular consistent observation between these data sets and four independent AFLP reports (13, 14, 26, 27) is that there are three main clusters that appear to be conserved in all the studies (25). Taken together, these findings support the widely held view that the *B. cereus sensu lato* group, in general, has a basic clonal population structure (15, 16). It should be reiterated that these conclusions are based on both MLST data, which target "housekeeping" genes that are presumed to be more conserved, and AFLP data, which are often driven by small repeated elements (i.e., variable number tandem repeats [17]) that can mutate and evolve more rapidly than single-nucleotide polymorphisms (SNPs).

The most recent study to date (27, 28) has a composite analysis of 2,213 isolates in the *B. cereus sensu lato* group assembled by various combinations of MLST, AFLP, and multienzyme electrophoresis (MEE [29]). "Altogether, the global analysis confirms and extends the results underlying the opportunistic nature of *B. cereus* group organisms, and the fact that isolates responsible for disease outbreaks and contamination of foodstuffs can originate from various genetic backgrounds" (27). Again, the data sets illustrate that *B. cereus*, *B. thuringiensis*, and *B. mycoides* isolates are dispersed throughout the basal *B. cereus sensu lato* phylogenetic tree. This does not, however, preclude the existence of clonal expansions where specific lineages may have become significant factors as clinical or insecticidal or environmental clusters. Examples include the highly conserved *B. anthracis* lineage, clonal clinical complexes causing periodontal or emetic disease in humans (25), and clonal complexes that might be specific for certain *B. thuringiensis* isolates and insecticidal toxins (e.g., branch C in reference 14).

HOMOLOGOUS RECOMBINATION

Asexual reproduction in bacteria defines clonal population structure where all the progeny are derived from a single parental cell. However, recombination within these populations can scramble and redistribute the DNA polymorphisms and cause a decomposition of this clonal pattern after many generations. The early MLST studies indicated that there was a strong clonal structure to the *B. cereus sensu lato* populations defined by specific lineages that contained fixed SNPs that appeared to be unique to each lineage. These same analyses indicated that the distribution of a limited number of SNPs was not in congruence with a strict clonal structure, i.e., there were a limited number of SNPs that were not fixed to a specific branch but were

shared between different lineages (16). This would be evidence for genetic exchange by homologous or non-homologous exchange or recombination between different lineages of the *B. cereus* subgroup.

From a historical perspective, a large body of evidence has helped to define homologous recombination and DNA excision repair as part of the general strategy that cells use to repair the two main forms of damage to DNA: double-strand breaks and nucleotide damage, respectively (30). The homologous recombination repair system is a *RecA* protein-driven process where the damaged DNA (a double-strand break) initiates a cascade of events including the eventual alignment and repair synthesis of a damaged region by using an undamaged template from a sister chromatid (30). This is an important "housekeeping" function because most bacteria, e.g., *Escherichia coli* and *Bacillus subtilis*, cannot tolerate a single unrepaired double-strand break in its chromosome. In humans, the disease ataxia telangiectasia, a deficiency in double-strand break repair, is short-lived, and patients with this disease are susceptible to many ionizing radiation-induced maladies (30).

While the "housekeeping" chores of the homologous exchange system to repair damaged DNA go unnoticed in large populations of bacteria, ancillary processing by these same proteins can also cause rare but effective homologous exchange between closely related bacteria. These events are rare because they first have to involve a form of "sexual" exchange of DNA between related bacteria via a number of processes, e.g., conjugation/transformation, transduction, etc. In addition, this transferred DNA must either be incorporated into the new genome as "foreign DNA" via insertion sequences, "genomic islands," etc., or as in homologous recombination as donor DNA that can recombine with a homologous template in the recipient cell. If the donor DNA is sufficiently different from the homologous stretch in a recipient cell, i.e., it contains several SNPs that are "foreign" to the recipient DNA, then this newly incorporated stretch of DNA can be detected by programs like eBurst (31–33) and ClonalFrame (34). This is the case for the *B. cereus sensu lato* group.

The housekeeping genes used in MLST were designed to be separated by sufficient distances to lessen the probability that more than one of the seven selected loci would be involved in any single recombination event (20), and this typing scheme also led to the use of allele designations for each unique sequence at each locus. For example, in the *B. cereus* MLST site (http://pubmlst.org/bcereus/) there are 240 different sequence alleles in the *glp* fragment (the first of 7 MLST sequences) that were discovered in 1,518 isolates. These allele designations led to the development of methods to establish genetic relationships utilizing cluster analysis of the seven alleles in each sequence type, e.g., eBurst (31–33).

A more recent inference model (ClonalFrame; git clone https://github.com/xavierdidelot/ClonalFrameML) indicates that MLST data can be used to determine the clonal relationship of bacteria while also providing the chromosomal position of homologous recombination that can potentially disrupt clonal patterns (34, 35). Unlike eBurst, ClonalFrame does not treat every allele designation with equal weight, and it recognizes that recombination events occur at a constant rate of substitutions to a contiguous region of sequence. When ClonalFrame encounters two strains with six of seven alleles having only one or two differences between each of these alleles and then a seventh allele having many nucleotide differences, it does not dismiss these two strains as being unrelated. Rather, it suggests that they have a clonal relationship with a homologous recombination in the seventh allele originating from an outside source.

ClonalFrame was used to reconstruct the evolutionary history of 667 strains in the *B. cereus sensu lato* group from MLST data (9). This analysis again confirmed the presence of three major clonal clusters and also demonstrated a variety of genetic exchanges between and within these clusters including a high number of exchanges with sources external to any of the clades. As an example, Didelot et al. (9) describe a subset of 35 recombination events that were inferred by ClonalFrame involving the specific cluster 2 (16). To illustrate and simplify how ClonalFrame was able identify and define these events, we have examined a single clade in Priest's original cluster 2 in detail. A maximum likelihood MLST phylogenetic tree for a clade defined as Sotto is illustrated in Fig. 2A, and it shows the relationship between seven sequence types (STs) in this clade. Note that ST-49, ST-55, and ST-9 are positioned at points that are distal to the remaining STs that otherwise form a tight cluster.

ClonalFrame identified a single gene sequence (*ilv*) to be responsible for creating the distal relationship between ST-49 and ST-55 from the cluster containing ST-12, ST-16, ST-23, and ST-56. Figure 2B illustrates a MLST maximum parsimony tree of six of the MLST genes (minus the *ilv* allele) in two subclades (Priest's Tolworthi and Sotto). Notice how the removal of the *ilv* sequence has caused all of the ST types to form a tight Sotto cluster. In Fig. 2C, a maximum parsimony tree of the *ilv* locus by itself illustrates that the *ilv* allele in ST-55 is identical to isolates from a subclade, Tolworthi, which is located on a different branch on

the same cluster 2. This is indicative of a homologous exchange between an isolate from the Sotto clade with an isolate from the Tolworthi clade in the region containing the *ilv* locus. The *ilv* allele for ST-49, on the other hand, was identical to the *ilv* allele for ST-13, a sequence type that is found in another cluster 2 subclade named Kurstaki (16).

This single subclade (Sotto) contains six sequence types (excluding ST-9), and, in 42 isolates, this group has two presumed examples of homologous exchange between (i) isolates from the Tolworthi and Sotto subclades and (ii) isolates from the Kurstaki and Sotto subclades that involve the regions surrounding the *ilv* locus. This example exemplifies the overall state of recombination in this subgroup of bacilli. ClonalFrame has demonstrated that, despite the use of only seven MLST fragments per genome, covering ~0.05% of the average *B. cereus sensu lato* genome, there is a considerable amount of recombination in the *B. cereus sensu lato* group. Didelot et al. (9) estimated that the *B. cereus* group is significantly less clonal than its close relative, *Staphylococcus aureus*, as measured by the relative impact of recombination and mutation (r/m) values ranging between 0.69 and 2.90 in *B. cereus* versus a value of 0.1 for *S. aureus*.

Two other significant conclusions included the observation that the pathogenic *B. cereus* strains are distributed throughout the first two clades of the phylogenetic tree and that increased or decreased rates of homologous recombination were not apparent within the pathogenic lineages of *B. cereus*. These results are consistent with the notion that *B. cereus sensu lato* is an "opportunistic" pathogen without specific predispositions that appear to be associated with pathogenicity.

THE PANGENOME OF *B. CEREUS SENSU LATO*

It has become evident that genetic content in individual isolates from any given bacterial species can vary considerably (36–39). The advent of next-generation sequencing technologies has caused an escalation in attempts to define the pangenome, or the whole genome complement or gene variation in a single clade or species. This was accomplished by the process of multiple genome comparisons of isolates within a given species (38, 40, 41). The crux of the initial pangenome discovery was that any single strain of *Streptococcus agalactiae* had a significant percentage of genes that were unique to that strain, and additional strains each provided another subset of new genes (40). After eight genome comparisons, the number of new genes had

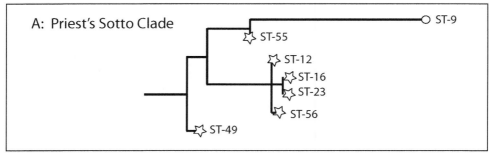

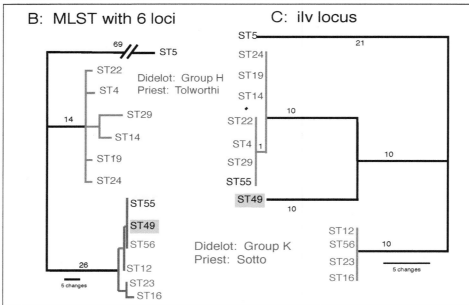

Figure 2   An example of homologous recombination identified in MLST profiles. (A) The diversity of an original MLST subclade (Sotto) based on seven MLST fragments (15). The same branch is shown, but is now dissected by ClonalFrame (33) and separated into six consistent fragments (B) and a second inconsistent fragment (*ilv*) (C) to show that the *ilv* fragment has two sequence types, 55 and 49, that had experienced recombination events with two other distinct clades.

reached an exponential decay function that had stabilized to 33 new genes per genome. This led to the eventual concept that certain species, e.g., *S. agalactiae*, *E. coli*, *Streptococcus pneumoniae*, *Prochlorococcus marinus*, and *B. cereus* have "open" genomes where new gene discovery would be unbounded. This would be in contrast to several apparently "closed" genomes, e.g., *S. aureus*, *Streptococcus pyogenes*, *Ureaplasma urealyticum*, and *B. anthracis* (41).

The initial concept of the pangenome was defined by two basic components: a "core" genome that consisted of genes shared by all strains in a species and a "dispensable" genome that consisted of genes that are found in some but not all of the strains. The core genes would conceptually provide the functions that define

the basic biology of the species and the dispensable genes would define the diversity and impact selective advantages such as adaptation and antibiotic resistance. In *B. cereus sensu lato* it is easy to visualize a "core" genome as consisting of a conserved framework that maintains functions that would allow an "opportunistic" or "fit" isolate to adapt to a new environment that is suitable for a lifestyle as a common soil organism or as a pathogen of mammals, humans, and insects. It is also appropriate to think of the "dispensable" genome as consisting, in part, of the plethora of small and large plasmids that are uniquely associated with the phenotypes that define each of these "species." What is not obvious is the role of the core genome in the overall diversity of these species.

These basic concepts have recently been examined within the *B. cereus sensu lato* subgroup by analyses of 58 diverse genomes (42). This study included the generation of high-redundancy whole-genome sequences by 454 pyrosequencing (43) of 45 *B. cereus sensu lato* strains containing an array of isolates based on geographical, phenotypic, and phylogenetic diversity. These data sets were combined with 13 previously sequenced genomes to establish the pangenome for *B. cereus sensu lato*. This included only a single *B. anthracis* genome, the Ames ancestor (44), to avoid overrepresenting this highly conserved and closed pangenome (38, 41).

Zwick et al. (42) defined the pangenome of *B. cereus sensu lato* using the expanded terminology of Lapierre and Gogarten (45). They found a typical bimodal distribution for 22,975 gene clusters in the 58 genomes (Fig. 3). They also defined gene families found in six or fewer genomes as "accessory" genes, gene families found in 49 or more genomes as "extended core," and those in between as "character genes." The core genome (genes found in every genome) consisted of 1,754 genes and the extended core took this number for *B. cereus* up to 3,904. As mentioned earlier, the overall analysis of the whole genome (accessory, character, and extended core genes) of these 58 diverse *B. cereus sensu lato* isolates defines an "open" pangenome for this group of closely related microorganisms (41, 61). This analysis included the *B. anthracis* lineage that represents a classic example of a "closed" genome and may eventually describe several other lineages that might be deemed as closed genomes as part of specific clonal human or insect pathogenic lineages (27).

A phylogenetic tree constructed by using a distance-based approach and concatenated chromosomal core proteins for the 58 genomes is illustrated in Fig. 4. This whole-genome phylogeny agreed with previous MLST and AFLP studies indicating that the *B. cereus sensu lato* group is separated into three major clusters or clades (14–16). This core gene data analysis also was consistent with the notion that, individually, the isolates defined as *B. cereus*, *B. thuringiensis*, and *B. mycoides* are not confined to discrete clades. To avoid further confusion, the authors chose the designations used by Priest et al. (16) to indicate that clade 1 contains the well-defined *B. anthracis* lineage and clade 2 has a large presence of *B. thuringiensis* isolates, and hence the subclades named after serotypes Tolworthi, Kurstaki, Sotto, and Thuringiensis. The article by Tourasse et al. (27) provides a comprehensive analysis to establish the relationships between the clade and cluster designations in AFLP, MLST, and MEE in the various studies on the *B. cereus sensu lato* group.

The analysis of the whole-genome sequences of 58 diverse *B. cereus sensu lato* isolates defines an "open" pangenome for this group of closely related microorganisms. The data garnered from the "core genomes" of these isolates confirm that the group has a clonal phylogenetic structure and that isolates designated as either *B. thuringiensis* or *B. cereus* are scattered throughout the "core genome" tree. These core data represent unparalleled DNA signatures, the whole-genome sequences from 58 diverse *B. cereus sensu lato* isolates, and they support the idea that this group consists of a relatively conserved genomic background that could be treated as a single phylogenetic entity.

Clade 3 was in a group that Priest et al. (16) called "others" and included a *B. mycoides* isolate and two *B. weihenstephanensis* isolates. The "core gene"

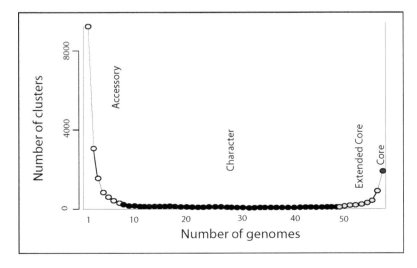

Figure 3 A graph of the distribution of gene families across *B. cereus sensu lato* genomes redrawn from Zwick et al. (41). This figure is based on the definition of the extended core as genes encoding proteins present in 49 or more genomes and accessory genes as those present in <6 genomes. The class between these extremes defined the character gene set.

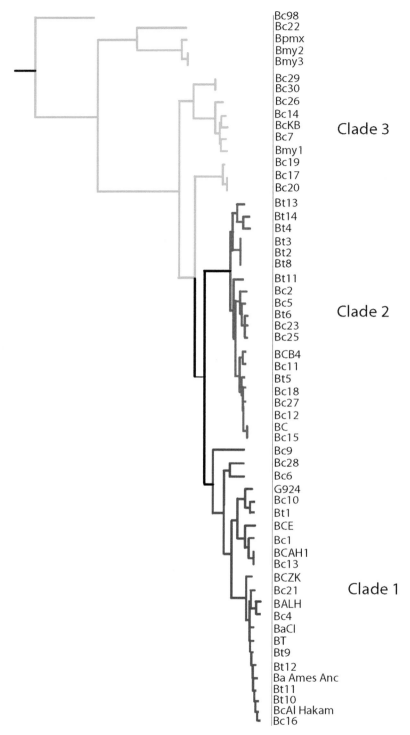

Figure 4   Whole-genome phylogeny of *B. cereus sensu lato*. This tree was redrawn based on data sets of concatenated, conserved protein sequences by using a neighbor-joining algorithm (41). Note that the relative distribution of the isolates based on a conserved whole-genome phylogeny is essentially the same as those observed in numerous MLST and AFLP studies and separated into three major clades.

whole-genome phylogeny expanded this clade to 15 isolates and demonstrated that clade 3 is in reality a polyphyletic grouping containing several new clades. Recent descriptions of psychrotolerant strains appear to fall into categories that include additional *B. mycoides* and *B. weihenstephanensis* strains and analysis of a selection of these genomes may help to define these new clades in line with this pangenome analysis (26–28).

The program ClonalFrame was used to revisit the patterns of homologous recombination within the 2.74 Mb core genomes in the 58 *B. cereus sensu lato* genomes. The ratio for the overall effects of homologous recombination and mutation (r/m) was 2.91, which is somewhat higher than the results obtained from ClonalFrame analysis of the 7 MLST fragments of 667 *B. cereus sensu lato* isolates (9). Nevertheless, these results are consistent with the idea that the r/m values for this group of bacteria are intermediate in relationship to a highly recombining population such as *Helicobacter pylori* (r/m = 13.6 [46]) and *Burkholderia* species (r/m ~ 25 [47]) versus a highly clonal population such as *S. aureus* (r/m = 0.1 [9]).

The analysis of the pangenome of *B. cereus sensu lato* confirms the relatively conserved clonal structure and a potential role for homologous recombination in the evolution of this group of bacilli. These are generalizations that had already been indicated since the "confusing" phylogenetic mixtures of *B. cereus*, *B. thuringiensis*, and *B. mycoides* were first acknowledged by DNA hybridization and 16S RNA studies (6, 48, 49) and then again supported by more recent AFLP, MEE, and MLST investigations (13–16, 21–27). However, Zwick et al. (42) have also used pangenome analyses to generate additional "global" insights concerning the evolution of this group of "opportunistic" pathogens. Pangenomes contain gene inventories that allow us to determine precise differences between pathogenic and nonpathogenic strains of a single species (39, 50). In many instances, these differences are measured by the loss and/or gain of sharply defined cassettes of genes. Two prime examples in the *B. cereus sensu lato* are (i) *B. anthracis*, which acquired the two plasmids (pXO1 and pXO2) containing the tripartite toxin gene complex and a capsule synthesis gene set, respectively, and (ii) *B. thuringiensis*, whose properties are mainly attributed to plasmids containing specific insecticidal toxin genes. From a more global perspective, gene acquisition and gene loss are now perceived as major underlying factors in the emergence and evolution of bacterial pathogens (51).

Several early whole-genome analyses of different microbes have revealed that bacteria have tendencies toward trimming and streamlining their genomes on the basis of functional needs (51–54). This "decay" of genomes appears to be fostered by selective pressures dictated along evolutionary timelines. Ochman and Davalos (51) have illustrated these evolutionary forces by the comparison of the relative genome sizes of free-living species versus the genome sizes of related facultative and obligate pathogens. These analyses suggest that free-living bacteria maintain relatively large genome sizes (5 to 10 Mb), because most of their genome functions are required for survival in diverse environments. Conversely, related facultative and obligate pathogens rely on host functions and nutrients and are maintained in smaller population sizes that eventually lead to accumulation of nonsynonymous mutations in extraneous genes and the eventual loss of unnecessary elements within these genomes. The net effect is that recent and facultative pathogens have genomes of intermediate size (2 to 5 Mb), while obligate symbiotic or pathogenic bacteria have relatively small genomes (0.5 to 1.5 Mb).

By these criteria, the *B. cereus sensu lato* subgroup, in general, classified as a soil organism, has a number of recently emerged pathogens of humans, insects, and other species. Zwick et al. indicate that there were very few character genes that could be used to describe any one clade within the *B. cereus sensu lato*. In addition, there was very little evidence that core genes were specifically being deleted from any single clade. A very specific analysis seeking evidence for gene cassettes directed toward a predisposition and adaptation toward pathogenicity included the comparison of the genomes in the clade that included three isolates that contained the pXO1 plasmid: *B. anthracis*, *B. cereus* G2941, and *B. cereus* CI (18, 19). These whole-genome comparisons to other isolates in this clade did not reveal any genes that were unique to the three isolates except for those acquired from the pXO1 and/or pXO2 plasmids. These results support the notion that these three genomes are not uniquely adapted to a lifestyle that would include the invasion and a natural fitness for growth in mammals.

Although not directly involved with phylogenetic issues, there is one emerging concept that does impact how particular isolates might readily adapt to different niches. This idea indicates that specific phenotypic differences can be caused by altered gene expression for a variety of *trans*-acting factors (55), rather that the loss or gain of specific coding regions. Prime examples are the effects attributed to the plcR and SigP-RsiP regulons in *B. anthracis* (56, 57). PlcR is a regulon whose expression regulates multiple genes related to virulence in *B. cereus* but is not expressed in *B. anthracis*. Similarly, mutations in the SigP-RsiP sigma factor regulon have recently been shown to affect the expression of the beta-lactamase genes that affect the sensitivity of *B. anthracis* to penicillin (57).

Consistent with this trend, Zwick et al. (42) did not detect significant evidence for excessive gene loss and the accumulation of nonsynonymous mutations in clades 1 and 2, which house the main human and insect pathogenic strains. These results suggest the absence of the early signs for genome decay that is frequently associated with facultative pathogens and includes phenomena such as the accumulation of pseudo-genes in superfluous genomic elements (51). *B. cereus* is often described as an "opportunistic" pathogen and, while this designation is usually reserved for infections involving nonpathogenic organisms becoming pathogenic in compromised individuals, this terminology appears appropriate when considered alongside the analysis of the pangenome of this subgroup. Zwick et al. (42) chose to describe the clade 1 and 2 isolates as "hopeful monsters" (58, 59) because these typically soil isolates appear to contain genomic complements that would allow them to exist as human or insect pathogens "in waiting" – waiting for the appropriate "toxin"-containing plasmid and/or an environmental niche that might allow a specific lineage to rise to prominence.

Curiously, genome "decay" is apparent in clade 3 of *B. cereus sensu lato* as measured by gene loss, accumulation of nonsynonymous mutations, and higher rates of homologous recombination relative to mutation. This clade was not well represented in earlier studies, and more recent data suggest that this group may contain several clades with greater phylogenetic and phenotypic diversity than observed with clades 1 and 2 (26–28, 42). An unusual isolate, NVH391-98, a severe foodborne pathogen whose genome is severely reduced in size (5.2 to 5.5 Mb in *B. cereus* versus 4.0 Mb in NVH391-98), has been used as an outgroup to root a *B. cereus sensu lato* 16S RNA tree (60) and a whole-genome core tree (42). The NVH391-98 genome does have collinear aspects with respect to the *B. cereus sensu lato* genome but this distant relative, while suitable as an outgroup, is currently being proposed as a separate species, *B. cytotoxis*. Nevertheless, this rooting does point to the diverse clade 3 (potentially clades 4, 5, 6, and 7) as an ancestral branch of the *B. cereus sensu lato* and suggests that NVH391-98 may share a common ancestor with certain isolates in clade 3 that are also exhibiting signs of genome decay in their core genome (e.g., see top of Fig. 1B, Zwick et al. [42]).

CONCLUSIONS

Despite evidence for homologous recombination, the phylogeny of *B. cereus sensu lato* as a whole remains one of clonal expansion. This has held true from conclusions based on earliest 16S RNA analysis through a more recent history that includes MEE, AFLP, MLST, and whole-genome sequence analyses. More importantly, the central debate about nomenclature and the *B. cereus sensu lato* group remains unchanged. These numerous studies have solidified the idea that, while the phenotype of each individual isolate may be determined by a variety of physiological tests and marker analyses, the phylogeny of the species as a whole has a conserved, clonal structure. This produces a structure that has mixtures of *B. cereus*, *B. thuringiensis*, *B. mycoides*, and others, scattered throughout the phylogenetic tree. Recent pangenome analyses have revealed that two of the first three "clades" of *B. cereus sensu lato* have pathogenic lineages that have NOT begun to show signs of genome "decay" that is more "typical" of facultative and obligate pathogenic lineages in other species. These findings suggest that potentially "facultative pathogens" and "free-living" *B. cereus* isolates have similar, relatively unaltered, basal genomic structure and size; and this, in turn, suggests that the movement of different kinds of pathogenic elements remains an ongoing phenomenon in this species. These notions support the idea that the *B. cereus sensu lato* group is in part "hopeful monsters" that can be transformed into new pathogenic lineages under the right set of circumstances. A potential example of this phenomenon, in progress, may be *B. cereus* isolates from the *B. anthracis*-containing clade 1 that have acquired pXO1 (BcG9241) and/or pXO2 (*B. cereus* or BA CI) and may be new, emerging lineages that can cause anthrax-like diseases.

Citation. Okinaka RT, Keim P. 2016. The phylogeny of *Bacillus cereus sensu lato*. Microbiol Spectrum 4(1):TBS-0012-2012.

References

1. **Aronson AI.** 1993. Insecticidal toxins, p 953–963. *In* Sonenshein AL, Hoch JA, Losick R (ed), *Bacillus subtilis and Other Gram-Positive Bacteria*. American Society for Microbiology, Washington, DC.

2. **Turnbull PCB.** 2002. Introduction: anthrax history, disease and ecology, p 1–19. *In* Koehler TM (ed), *Anthrax*, **vol 271**. Springer-Verlag, Berlin, Germany.

3. **Drobniewski FA.** 1993. *Bacillus cereus* and related species. *Clin Microbiol Rev* **6**:324–338.

4. **Granum PE, Lund T.** 1997. *Bacillus cereus* and its food poisoning toxins. *FEMS Microbiol Lett* **157**:223–228.

5. Helgason E, Caugant DA, Olsen I, Kolstø AB. 2000. Genetic structure of population of *Bacillus cereus* and *B. thuringiensis* isolates associated with periodontitis and other human infections. *J Clin Microbiol* **38**:1615–1622.

6. Priest FG. 1993. Systematics and ecology of bacillus, p 3–33. *In* Sonenshien A, Hoch JA, Losick R (ed), *Bacillus subtilis and Other Gram-Positive Bacteria*. American Society for Microbiology, Washington, DC.

7. Helgason E, Okstad OA, Caugant DA, Johansen HA, Fouet A, Mock M, Hegna I, Kolstø AB. 2000. *Bacillus anthracis, Bacillus cereus,* and *Bacillus thuringiensis*—one species on the basis of genetic evidence. *Appl Environ Microbiol* **66**:2627–2630.

8. Schnepf E, Crickmore N, Van Rie J, Lereclus D, Baum J, Feitelson J, Zeigler DR, Dean DH. 1998. *Bacillus thuringiensis* and its pesticidal crystal proteins. *Microbiol Mol Biol Rev* **62**:775–806.

9. Didelot X, Barker M, Falush D, Priest FG. 2009. Evolution of pathogenicity in the *Bacillus cereus* group. *Syst Appl Microbiol* **32**:81–90.

10. Staley JT. 2006. The bacterial species dilemma and the genomic-phylogenetic species concept. *Philos Trans R Soc Lond B Biol Sci* **361**:1899–1909.

11. Vos P, Hogers R, Bleeker M, Reijans M, van de Lee T, Hornes M, Friters A, Pot J, Paleman J, Kuiper M, Zabeau M. 1995. AFLP: a new technique for DNA fingerprinting. *Nucleic Acids Res* **23**:4407–4414.

12. Keim P, Kalif A, Schupp J, Hill K, Travis SE, Richmond K, Adair DM, Hugh-Jones M, Kuske CR, Jackson P. 1997. Molecular evolution and diversity in *Bacillus anthracis* as detected by amplified fragment length polymorphism markers. *J Bacteriol* **179**:818–824.

13. Ticknor LO, Kolstø AB, Hill KK, Keim P, Laker MT, Tonks M, Jackson PJ. 2001. Fluorescent amplified fragment length polymorphism analysis of Norwegian *Bacillus cereus* and *Bacillus thuringiensis* soil isolates. *Appl Environ Microbiol* **67**:4863–4873.

14. Hill KK, Ticknor LO, Okinaka RT, Asay M, Blair H, Bliss KA, Laker M, Pardington PE, Richardson AP, Tonks M, Beecher DJ, Kemp JD, Kolstø AB, Wong AC, Keim P, Jackson PJ. 2004. Fluorescent amplified fragment length polymorphism analysis of *Bacillus anthracis, Bacillus cereus,* and *Bacillus thuringiensis* isolates. *Appl Environ Microbiol* **70**:1068–1080.

15. Helgason E, Tourasse NJ, Meisal R, Caugant DA, Kolstø AB. 2004. Multilocus sequence typing scheme for bacteria of the *Bacillus cereus* group. *Appl Environ Microbiol* **70**:191–201.

16. Priest FG, Barker M, Baillie LW, Holmes EC, Maiden MC. 2004. Population structure and evolution of the *Bacillus cereus* group. *J Bacteriol* **186**:7959–7970.

17. Keim P, Price LB, Klevytska AM, Smith KL, Schupp JM, Okinaka R, Jackson PJ, Hugh-Jones ME. 2000. Multiple-locus variable-number tandem repeat analysis reveals genetic relationships within *Bacillus anthracis*. *J Bacteriol* **182**:2928–2936.

18. Hoffmaster AR, Hill KK, Gee JE, Marston CK, De BK, Popovic T, Sue D, Wilkins PP, Avashia SB, Drumgoole R, Helma CH, Ticknor LO, Okinaka RT, Jackson PJ. 2006. Characterization of *Bacillus cereus* isolates associated with fatal pneumonias: strains are closely related to *Bacillus anthracis* and harbor *B. anthracis* virulence genes. *J Clin Microbiol* **44**:3352–3360.

19. Klee SR, Brzuszkiewicz EB, Nattermann H, Brüggemann H, Dupke S, Wollherr A, Franz T, Pauli G, Appel B, Liebl W, Couacy-Hymann E, Boesch C, Meyer FD, Leendertz FH, Ellerbrok H, Gottschalk G, Grunow R, Liesegang H. 2010. The genome of a *Bacillus* isolate causing anthrax in chimpanzees combines chromosomal properties of *B. cereus* with *B. anthracis* virulence plasmids. *PLoS One* **5**:e10986. doi:10.1371/journal.pone.0010986.

20. Maiden MC, Bygraves JA, Feil E, Morelli G, Russell JE, Urwin R, Zhang Q, Zhou J, Zurth K, Caugant DA, Feavers IM, Achtman M, Spratt BG. 1998. Multilocus sequence typing: a portable approach to the identification of clones within populations of pathogenic microorganisms. *Proc Natl Acad Sci USA* **95**:3140–3145.

21. Ko KS, Kim JW, Kim JM, Kim W, Chung SI, Kim IJ, Kook YH. 2004. Population structure of the *Bacillus cereus* group as determined by sequence analysis of six housekeeping genes and the plcR gene. *Infect Immun* **72**:5253–5261.

22. Cardazzo B, Negrisolo E, Carraro L, Alberghini L, Patarnello T, Giaccone V. 2008. Multiple-locus sequence typing and analysis of toxin genes in *Bacillus cereus* foodborne isolates. *Appl Environ Microbiol* **74**:850–860.

23. Olsen JS, Skogan G, Fykse EM, Rawlinson EL, Tomaso H, Granum PE, Blatny JM. 2007. Genetic distribution of 295 *Bacillus cereus* group members based on adk-screening in combination with MLST (Multilocus Sequence Typing) used for validating a primer targeting a chromosomal locus in *B. anthracis*. *J Microbiol Methods* **71**:265–274.

24. Sorokin A, Candelon B, Guilloux K, Galleron N, Wackerow-Kouzova N, Ehrlich SD, Bourguet D, Sanchis V. 2006. Multiple-locus sequence typing analysis of *Bacillus cereus* and *Bacillus thuringiensis* reveals separate clustering and a distinct population structure of psychrotrophic strains. *Appl Environ Microbiol* **72**:1569–1578.

25. Tourasse NJ, Helgason E, Økstad OA, Hegna IK, Kolstø AB. 2006. The *Bacillus cereus* group: novel aspects of population structure and genome dynamics. *J Appl Microbiol* **101**:579–593.

26. Guinebretière MH, Thompson FL, Sorokin A, Normand P, Dawyndt P, Ehling-Schulz M, Svensson B, Sanchis V, Nguyen-The C, Heyndrickx M, De Vos P. 2008. Ecological diversification in the *Bacillus cereus* group. *Environ Microbiol* **10**:851–865.

27. Tourasse NJ, Helgason E, Klevan A, Sylvestre P, Moya M, Haustant M, Økstad OA, Fouet A, Mock M, Kolstø AB. 2011. Extended and global phylogenetic view of the *Bacillus cereus* group population by combination of MLST, AFLP, and MLEE genotyping data. *Food Microbiol* **28**:236–244.

28. Maughan H, Van der Auwera G. 2011. Bacillus taxonomy in the genomic era finds phenotypes to be essential though often misleading. *Infect Genet Evol* **11**:789–797.

29. Helgason E, Caugant DA, Lecadet MM, Chen Y, Mahillon J, Lövgren A, Hegna I, Kvaløy K, Kolstø AB. 1998. Genetic diversity of *Bacillus cereus*/*B. thuringiensis* isolates from natural sources. *Curr Microbiol* 37:80–87.

30. Friedberg EC, Walker GC, Siede W, Schultz RA, Ellenberger T. 2006. *DNA Repair and Mutagenesis*, 2nd ed. ASM Press, Washington, DC.

31. Feil EJ, Li BC, Aanensen DM, Hanage WP, Spratt BG. 2004. eBURST: inferring patterns of evolutionary descent among clusters of related bacterial genotypes from multilocus sequence typing data. *J Bacteriol* 186:1518–1530.

32. Jolley KA, Wilson DJ, Kriz P, McVean G, Maiden MC. 2005. The influence of mutation, recombination, population history, and selection on patterns of genetic diversity in *Neisseria meningitidis*. *Mol Biol Evol* 22:562–569.

33. Spratt BG, Hanage WP, Li B, Aanensen DM, Feil EJ. 2004. Displaying the relatedness among isolates of bacterial species – the eBURST approach. *FEMS Microbiol Lett* 241:129–134.

34. Didelot X, Falush D. 2007. Inference of bacterial microevolution using multilocus sequence data. *Genetics* 175: 1251–1266.

35. Didelot X, Maiden MC. 2010. Impact of recombination on bacterial evolution. *Trends Microbiol* 18:315–322.

36. Hayashi T, Makino K, Ohnishi M, Kurokawa K, Ishii K, Yokoyama K, Han CG, Ohtsubo E, Nakayama K, Murata T, Tanaka M, Tobe T, Iida T, Takami H, Honda T, Sasakawa C, Ogasawara N, Yasunaga T, Kuhara S, Shiba T, Hattori M, Shinagawa H. 2001. Complete genome sequence of enterohemorrhagic *Escherichia coli* O157:H7 and genomic comparison with a laboratory strain K-12. *DNA Res* 8:11–22.

37. Lawrence JG, Ochman H. 1998. Molecular archaeology of the *Escherichia coli* genome. *Proc Natl Acad Sci USA* 95:9413–9417.

38. Rasko DA, Rosovitz MJ, Myers GS, Mongodin EF, Fricke WF, Gajer P, Crabtree J, Sebaihia M, Thomson NR, Chaudhuri R, Henderson IR, Sperandio V, Ravel J. 2008. The pangenome structure of *Escherichia coli*: comparative genomic analysis of *E. coli* commensal and pathogenic isolates. *J Bacteriol* 190:6881–6893.

39. Welch RA, Burland V, Plunkett G III, Redford P, Roesch P, Rasko D, Buckles EL, Liou SR, Boutin A, Hackett J, Stroud D, Mayhew GF, Rose DJ, Zhou S, Schwartz DC, Perna NT, Mobley HL, Donnenberg MS, Blattner FR. 2002. Extensive mosaic structure revealed by the complete genome sequence of uropathogenic *Escherichia coli*. *Proc Natl Acad Sci USA* 99:17020–17024.

40. Tettelin H, Masignani V, Cieslewicz MJ, Donati C, Medini D, Ward NL, Angiuoli SV, Crabtree J, Jones AL, Durkin AS, Deboy RT, Davidsen TM, Mora M, Scarselli M, Margarit y Ros I, Peterson JD, Hauser CR, Sundaram JP, Nelson WC, Madupu R, Brinkac LM, Dodson RJ, Rosovitz MJ, Sullivan SA, Daugherty SC, Haft DH, Selengut J, Gwinn ML, Zhou L, Zafar N, Khouri H, Radune D, Dimitrov G, Watkins K, O'Connor KJ, Smith S, Utterback TR, White O, Rubens CE, Grandi G, Madoff LC, Kasper DL, Telford JL, Wessels MR, Rappuoli R, Fraser CM. 2005. Genome analysis of multiple pathogenic isolates of *Streptococcus agalactiae*: implications

41. Tettelin H, Riley D, Cattuto C, Medini D. 2008. Comparative genomics: the bacterial pan-genome. *Curr Opin Microbiol* 11:472–477.

for the microbial "pan-genome." *Proc Natl Acad Sci USA* 102:13950–13955.

42. Zwick ME, Joseph SJ, Didelot X, Chen PE, Bishop-Lilly KA, Stewart AC, Willner K, Nolan N, Lentz S, Thomason MK, Sozhamannan S, Mateczun AJ, Du L, Read TD. 2012. Genomic characterization of the *Bacillus cereus* sensu lato species: backdrop to the evolution of *Bacillus anthracis*. *Genome Res* 22:1512–1524.

43. Wheeler DA, Srinivasan M, Egholm M, Shen Y, Chen L, McGuire A, He W, Chen YJ, Makhijani V, Roth GT, Gomes X, Tartaro K, Niazi F, Turcotte CL, Irzyk GP, Lupski JR, Chinault C, Song XZ, Liu Y, Yuan Y, Nazareth L, Qin X, Muzny DM, Margulies M, Weinstock GM, Gibbs RA, Rothberg JM. 2008. The complete genome of an individual by massively parallel DNA sequencing. *Nature* 452:872–876.

44. Ravel J, Jiang L, Stanley ST, Wilson MR, Decker RS, Read TD, Worsham P, Keim PS, Salzberg SL, Fraser-Liggett CM, Rasko DA. 2009. The complete genome sequence of *Bacillus anthracis* Ames "Ancestor". *J Bacteriol* 191:445–446.

45. Lapierre P, Gogarten JP. 2009. Estimating the size of the bacterial pan-genome. *Trends Genet* 25:107–110.

46. Vos M, Didelot X. 2009. A comparison of homologous recombination rates in bacteria and archaea. *ISME J* 3: 199–208.

47. Pearson T, Giffard P, Beckstrom-Sternberg S, Auerbach R, Hornstra H, Tuanyok A, Price EP, Glass MB, Leadem B, Beckstrom-Sternberg JS, Allan GJ, Foster JT, Wagner DM, Okinaka RT, Sim SH, Pearson O, Wu Z, Chang J, Kaul R, Hoffmaster AR, Brettin TS, Robison RA, Mayo M, Gee JE, Tan P, Currie BJ, Keim P. 2009. Phylogeographic reconstruction of a bacterial species with high levels of lateral gene transfer. *BMC Biol* 7:78. doi: 10.1186/1741-7007-7-78.

48. Ash C, Collins MD. 1992. Comparative analysis of 23S ribosomal RNA gene sequences of *Bacillus anthracis* and emetic *Bacillus cereus* determined by PCR-direct sequencing. *FEMS Microbiol Lett* 73:75–80.

49. Ash C, Farrow JA, Dorsch M, Stackebrandt E, Collins MD. 1991. Comparative analysis of *Bacillus anthracis*, *Bacillus cereus*, and related species on the basis of reverse transcriptase sequencing of 16S rRNA. *Int J Syst Bacteriol* 41:343–346.

50. Lawrence JG. 2005. Common themes in the genome strategies of pathogens. *Curr Opin Genet Dev* 15:584–588.

51. Ochman H, Davalos LM. 2006. The nature and dynamics of bacterial genomes. *Science* 311:1730–1733.

52. Andersson SG, Zomorodipour A, Andersson JO, Sicheritz-Pontén T, Alsmark UC, Podowski RM, Näslund AK, Eriksson AS, Winkler HH, Kurland CG. 1998. The genome sequence of *Rickettsia prowazekii* and the origin of mitochondria. *Nature* 396:133–140.

53. Cole ST, Eiglmeier K, Parkhill J, James KD, Thomson NR, Wheeler PR, Honoré N, Garnier T, Churcher C, Harris D, Mungall K, Basham D, Brown D, Chillingworth

T, Connor R, Davies RM, Devlin K, Duthoy S, Feltwell T, Fraser A, Hamlin N, Holroyd S, Hornsby T, Jagels K, Lacroix C, Maclean J, Moule S, Murphy L, Oliver K, Quail MA, Rajandream MA, Rutherford KM, Rutter S, Seeger K, Simon S, Simmonds M, Skelton J, Squares R, Squares S, Stevens K, Taylor K, Whitehead S, Woodward JR, Barrell BG. 2001. Massive gene decay in the leprosy bacillus. *Nature* 409:1007–1011.

54. Parkhill J, Dougan G, James KD, Thomson NR, Pickard D, Wain J, Churcher C, Mungall KL, Bentley SD, Holden MT, Sebaihia M, Baker S, Basham D, Brooks K, Chillingworth T, Connerton P, Cronin A, Davis P, Davies RM, Dowd L, White N, Farrar J, Feltwell T, Hamlin N, Haque A, Hien TT, Holroyd S, Jagels K, Krogh A, Larsen TS, Leather S, Moule S, O'Gaora P, Parry C, Quail M, Rutherford K, Simmonds M, Skelton J, Stevens K, Whitehead S, Barrell BG. 2001. Complete genome sequence of a multiple drug resistant *Salmonella enterica* serovar Typhi CT18. *Nature* 413:848–852.

55. Toby IT, Widmer J, Dyer DW. 2014. Divergence of protein-coding capacity and regulation in the *B. cereus sensu lato* group. *BMC Bioinformatics* 15(Suppl 11):S8. doi:10.1186/1471-2105-15-S11-S8.

56. Agaisse H, Gominet M, Okstad OA, Kolstø AB, Lereclus D. 1999. PlcR is a pleiotropic regulator of extracellular virulence factor gene expression in *Bacillus thuringiensis*. *Mol Microbiol* 32:1043–1053.

57. Ross CL, Thomason KS, Koehler TM. 2009. An extracytoplasmic function sigma factor controls beta-lactamase gene expression in *Bacillus anthracis* and other *Bacillus cereus* group species. *J Bacteriol* 191:6683–6693.

58. Gould SJ. 1977. This view of life: the return of hopeful monsters. *Nat Hist* 86:22–30.

59. Keim PS, Wagner DM. 2009. Humans and evolutionary and ecological forces shaped the phylogeography of recently emerged diseases. *Nat Rev Microbiol* 7:813–821.

60. Lapidus A, Goltsman E, Auger S, Galleron N, Ségurens B, Dossat C, Land ML, Broussolle V, Brillard J, Guinebretiere MH, Sanchis V, Nguen-The C, Lereclus D, Richardson P, Wincker P, Weissenbach J, Ehrlich SD, Sorokin A. 2008. Extending the *Bacillus cereus* group genomics to putative food-borne pathogens of different toxicity. *Chem Biol Interact* 171:236–249.

61. Papazisi L, Rasko DA, Ratayake S, Bock GR, Remortel BG, Appalla L, Liu J, Dracheva T, Braisted JC, Shallome S, Jarrahi B, Snesrud E, Ahn S, Sun Q, Rilstone J, Okstad OA, Kolsto A-B, Fleischmann RD, Peterson SN. 2011. Investigating the genome diversity of *B. cereus* and evolutionary aspects of *B. anthracis* emergence. *Genomics* 98:26–39.

The Bacterial Spore: From Molecules to Systems
Edited by P. Eichenberger and A. Driks
© 2016 American Society for Microbiology, Washington, DC
doi:10.1128/microbiolspec.TBS-0021-2015

Joel A. Bozue[1]
Susan Welkos[1]
Christopher K. Cote[1]

The *Bacillus anthracis* Exosporium: What's the Big "Hairy" Deal?

13

In some *Bacillus* species, including *Bacillus subtilis*, the coat and a glycoprotein layer referred to as the spore crust are the outermost layer of the spore (Fig. 1). These spore structures are discussed in more detail in Driks and Eichenberger (108). In others, such as the *Bacillus cereus* family, there is an additional layer that envelops the coat, called the exosporium, which is distinct from the crust. In the case of *Bacillus anthracis*, one of the three pathogenic species of the *B. cereus* family, a series of fine hair-like projections, also referred to as a "hair-like" nap, extends from the exosporium basal layer (1–4) (Fig. 1). Other exosporium-producing *Bacillus* species, such as *Bacillus megaterium*, lack this "hairy" nap (Fig. 1). Separating the exosporium from the rest of the spore structure is an area referred to as the interspace (5).

The exact role of the exosporium in *B. anthracis*, or for any of the *Bacillus* species possessing this structure, remains unclear. However, it has been assumed that the exosporium would play some role in infection for *B. anthracis* because it is the outermost structure of the spore and would make initial contact with host and immune cells during infection. In addition, because the spores are present in soil, the exosporium will interact with environmental material and other organisms. Therefore, the exosporium has been a topic of great interest, and over the past decade much progress has been made to understand its composition, biosynthesis, and potential roles. However, several key aspects of this spore structure are still debated and remain undetermined. In this review, we focus primarily on the exosporium of *B. anthracis* and highlight important findings from the literature. A complete compilation and review of all the primary literature is beyond the scope of this article, and any failure to cite significant sources is unintended.

EXOSPORIUM COMPOSITION

The exosporium is the outermost integument surrounding the mature *B. anthracis* spore and is a loose-fitting, balloon-like structure. Although it appears to be loosely associated with the spore, the exosporium is not readily removed. Exosporium removal by physical means

[1]U.S. Army Medical Research Institute of Infectious Diseases, Division of Bacteriology, Fort Detrick, MD 21702.

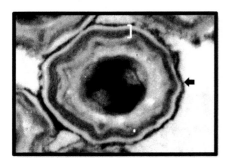

B. subtilis

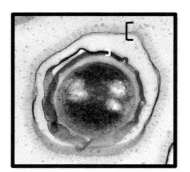

B. anthracis

Figure 1 Electron micrographs of *B. subtilis*, *B. anthracis*, and *B. megaterium* spores. The spore coat (white bar) or exosporium (black bar) as present is labeled for the various species. The arrow indicates the presence of a crust on the *B. subtilis* spore. The *B. subtilis* micrograph is courtesy of Adam Driks, Stritch School of Medicine, Loyola University Chicago, Maywood, IL.

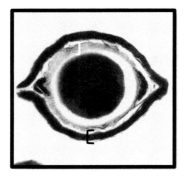

B. megaterium

requires multiple passages through a French press; however, strain differences in ease of removal have been reported (6).

The *B. anthracis* exosporium is composed of two layers: the outer-layer nap and the basal layer (Fig. 1). The inner layer consists of a paracrystalline basal layer with a hexagonal lattice structure (3). A more in-depth study of the exosporium crystal fragments has been performed from *B. cereus*, *B. thuringiensis*, and *B. anthracis* strains. From this study, three distinct crystal types were found between the various species (7). The so-called type II crystals were common to all three *Bacillus* species examined and form the outermost layer of the exosporium. Type I crystals were rare in *B. cereus* and *B. anthracis* exosporium fractions. The type III crystals were found in all three *Bacillus* species but were the least frequently observed and their presence was dependent on the strain and method of spore preparation. The arrays formed by type I and II crystals would presumably form a semipermeable barrier around the spore. The barrier would include pores approximately 23 to 34 Å in diameter that would allow passage of germinants to the endospore but inhibit access of antibodies and enzymes (7).

The exosporium of the *Bacillus* species is a biochemically complex structure that cross-reacts with lectins

(8) and can bind to polysaccharides (9). The *B. cereus* exosporium constitutes approximately 2% of the spore mass. Some, but not all, studies indicate that the exosporium is more hydrophobic than the spore coat, and the surface hydrophobicity of spores appears to depend mainly on components of the exosporium. In these analyses, spores having an exosporium are more hydrophobic than those lacking this structure (10–12). In addition, the composition of the exosporium was determined to be approximately 50% protein, 20% lipid, 20% neutral polysaccharides, and 10% other components (9).

Exosporium and spore coat proteins have the capacity to cross-link into high-molecular-weight complexes (13–19). Over the past several years, numerous laboratories have purified the exosporium from different strains of *B. anthracis* and *B. cereus*, identifying the specific proteins composing this structure. Many of the same proteins identified were common between studies, but some differences have been noted that could be intrinsic or artifact due to the methods of bacterial growth, spore harvesting, exosporium purification, or analyses employed. Details of the various identified proteins will be discussed later in this chapter.

One of the first proteomic studies of the exosporium analyzed the Sterne 34F2 strain of *B. anthracis* (17).

The spores were prepared in Difco sporulation broth, washed, and separated from vegetative debris through a density gradient. The exosporium was purified after passage through a French press. Five major protein bands were identified by amino-terminal sequencing: alanine racemase, iron/manganese superoxide dismutase, BxpA, BclA, and BxpB (also referred to as ExsFA).

In contrast, the number of identified exosporium proteins was much greater in a separate study of the Sterne 34F2 strain of *B. anthracis* (20). In this study, 137 exosporium proteins were identified. Several differences were noted between this study and the one described above. The *B. anthracis* strain was prepared and sporulated in G medium and, more importantly, spores were only water washed to better model environmental conditions in nature. The exosporium could potentially have protein and debris associated with and/or bound to it that might play some role in environmental persistence, transmission, or infectivity of the spores (21).

Other studies identifying proteins of the exosporium expanded these observations to include multiple strains: the fully virulent Ames strain, an Ames derivative lacking pXO2, and a Sterne derivative (UM23CL2) lacking both pXO1 and pXO2 (22). Spores were prepared on New Sporulation medium agar and water washed. The exosporium was purified following sonication rather than a French press procedure because of the equipment constraints and the potential for formation of aerosols within containment laboratories. Analysis of the exosporium proteins was performed on unwashed as well as salt- and detergent-washed proteins. As expected, many of the proteins identified from the unwashed exosporium contain vegetative cell proteins, such as EA1, an S-layer protein (23). The washed exosporium preparations from the three strains listed above were used to identify those exosporium proteins that are tightly bound and integral to this structure. The proteins identified included BclA, alanine racemase, inosine-preferring nucleoside hydrolase, BxpB/ExsFA, CotB, CotY, ExsY, and ExsK. From this study, no differences were identified in the exosporium profile between the three strains examined (22); however, as discussed below, strain-related differences in the sizes of some exosporium proteins have been shown (16, 17, 24).

A thorough study of the protein profile of the exosporium from the *B. cereus* strain ATCC 10876 has been reported (25). The spores were prepared in CCY (casein hydrolysate yeast-containing) medium and purified by water washes. To obtain exosporium, the spores were processed twice through a French press. Then, the

exosporial fragments were further purified and washed in salt and detergent. From the washed preparations, ten exosporium-associated proteins were identified: alanine racemase, nucleoside hydrolase, ExsB, ExsC, ExsD, ExsE, ExsF, ExsG, ExsJ, and CotE. Interestingly, in *B. subtilis*, CotE is a morphogenetic protein involved in spore coat assembly located between the inner and outer coat (26). Only the N-terminal portion of the *B. cereus* CotE protein was found associated with the exosporium profile. The presence of CotE within the exosporium suggested a link between coat and exosporium assembly that has since been demonstrated (27). In *B. anthracis*, CotE has been shown to play a role in exosporium assembly. Interestingly, in *B. anthracis*, CotE has been detected only within the spore coat (27, 28). However, this does not discount the possibility of CotE being present in multiple locations of the spore. The role of CotE in the assembly of exosporium and spore coat is discussed in more detail below and by Driks and Eichenberger (108).

EXOSPORIUM PROTEINS

Collagen-Like Proteins

BclA (*Bacillus* Collagen-Like Protein A)

One of the first exosporium proteins to be identified and the most extensively studied to date is BclA (17, 18). BclA is a complex glycoprotein and the structural constituent of the hair-like filaments on the outer layer of the exosporium (18). BclA, along with other exosporium proteins, forms a covalently linked complex that is displayed as a very large-molecular-weight band when viewed on SDS-PAGE gels (13, 17–19, 22, 29–33).

The BclA protein contains three domains: the N-terminal domain, the central region, and the carboxy-terminal domain (CTD). The N-terminal domain forms the proximal end of the BclA and is involved in targeting and anchoring the protein to the basal layer of the exosporium following a cleavage event (34, 35).

The BclA central region is composed of GXX collagen-like repeats that are directly correlated with filament length, which can vary for strains from 140 to 608 Å (24). Collagen-like sequences are rare in microorganisms but have been identified occasionally (36, 37). However, antibodies to BclA were shown not to cross-react with human collagen (38). The central region is the attachment site of two O-linked carbohydrate components: a 715-Da tetrasaccharide (three rhamnose residues, an unusual component termed anthrose, and a protein-bound N-acetylgalactosamine) and a 324-Da disaccharide (one rhamnose residue and 3-O-methyl rhamnose) (39). The exact number of each oligosaccharide attached to BclA and the precise attachment sites on the protein remain to be determined. Anthrose was originally thought to be unique to *B. anthracis*; however, after further analysis, the sugar is more widespread than originally reported, because the genes and product are present in other *Bacillus* species (40). The BclA-associated oligosaccharides account for most of the carbohydrate in the exosporium (39).

The CTD of the BclA forms the distal end of the filament and is the immunodominant protein epitope on the intact spore surface, because most of the anti-BclA monoclonal antibodies generated against spores or exosporium react to the CTD (35). However, antibody responses to anthrose-containing trisaccharide were found to be useful as a diagnostic for nonhuman primates that survived experimental pneumonic exposure to *B. anthracis* (41). In addition, recombinant CTD trimers easily formed crystalline sheets that were resistant to elevated temperatures, detergents, and collagenases. It is the CTD of BclA that provides a rugged shield surrounding the spore (35). Interestingly, it was determined that in the three-dimensional structure of BclA, the collagen-like N-terminal portion of the protein was disordered. In contrast, the CTD of BclA formed trimers as described above that would fold into a β structure with a "jelly roll" topology (42). The folds made by the BclA crystal structure are similar to the tumor necrosis factor (TNF) family of proteins, especially C1q. While BclA does not possess many of the C1q activities, C1q functions as a bridge in the adherence of the BclA fibers to the integrin of epithelial cells (43).

BclA may play a role in interacting with and recognizing hydrophobic components in the external environment or in the infected host. In fact, many studies have demonstrated that removing BclA from spores alters the overall hydrophobic, adhesive, and surface properties of the spore outer surface (11, 44–49). The details of many of these studies are discussed below.

BclB

BclB is a partially surface-exposed, collagen-like glycoprotein containing tandem repeats (50, 51). Also, similar to BclA, BclB contains spore-specific sugars (rhamnose, 3-O-methyl rhamnose, and galactosamine) (51) and an N-terminal domain that targets the protein to the exosporium (50). The role of the BclB protein continues to emerge. However, a *bclB* mutant strain has been constructed and spores of the mutant possessed a fragile exosporium. Electron microscopic observations of these spores revealed a mixed population: free exosporial basal layers, spores with damaged exosporia, as well as many spores with no visible defects to the exosporia (50). The mutation did not affect spore germination or kinetics of spore survival within macrophages (50).

BetA (*Bacillus* Exosporium-Targeted A)

The BetA protein was recently identified and found to contain a small collagen-like repeat region which exhibits strain-associated variation in the number of repeats. It is believed that BetA is located within the basal layer of the exosporium underneath the BclA protein (33). It contains a conserved amino acid sequence in the N-terminal domain needed for targeting the protein to the exosporium (33). Currently, the role of the BetA has not been defined.

ExsH and ExsJ

These collagen-like glycoproteins are present within the exosporia of *B. thuringiensis* (52) and *B. cereus* (21, 25) but have not been identified within *B. anthracis*. The proteins and their function have not been well characterized to date, but both contain a large region of GXX repeats. Recently, the impact of the *exsH* gene on the properties of the exosporium was examined in the *B. cereus* ATCC 14579 strain (48). Spores of a mutant strain deleted for *exsH* demonstrated no visible changes to the exosporium basal layer or hair-like nap.

Exosporium Enzymes

Superoxide Dismutases (SODs)

The *B. anthracis* genome encodes for four superoxide dismutases that are enzymes capable of detoxifying oxygen radicals. Of these, two of the SODs (SOD15 and SODA1) are present within the exosporium (17, 20). To determine if the exosporium SODs play a role in pathogenesis, a double mutant Δ*sod15* Δ*sodA1* was constructed in the Sterne strain of *B. anthracis* by Cybulski et al. (53). The mutant strain was slower to germinate and also was more sensitive to oxidative stress in comparison with the parental strain. However, no differences in virulence were determined between wild-type and mutant spores in the A/J mouse model of lethal Sterne infection. When a quadruple-SOD knockout strain was constructed (the two exosporium-associated and two non-exosporium-associated SODs), spores of this mutant were attenuated in mice following intranasal challenge.

Alanine Racemase (Alr)

The Alr enzyme converts L-alanine (a germinant) to the D-alanine form (a germination inhibitor) (54). The Alr protein is present in the basal layer of the exosporium but not present within the "cap-like" region that serves as the region for the germinating bacillus to emerge during germination and outgrowth (6). In addition, Alr is also present within the coat layer of the spore (29). The Alr enzyme presumably suppresses spore germination in areas that would be inhospitable for vegetative cell survival, whether in the soil environment or in a host, until there is a level of germinant high enough to overcome the Alr activity (55).

An *alr* deletion mutant was constructed in the Sterne strain of *B. anthracis*. Spores of the mutant strain were able to more efficiently germinate in low L-alanine environments in comparison with wild-type spores, since the Alr enzyme was not present to convert the germinant into the D-alanine form (56). However, no difference was noted in germination rates with other germinants. A second, related role for Alr was also identified from this experiment. Alr appears necessary for producing mature, resistant spores by converting L-alanine to D-alanine within the mother cell. During sporulation, approximately half of the *alr* mutant spores converted from a phase bright/heat-resistant form to a phase dark/heat-sensitive form.

IunH (Inosine Hydrolase)

Previous fractionation studies identified IunH in the exosporium (22, 25). However, a separate study based on a green fluorescent protein fusion construct localized the protein to the interspace region, which is between the coat and the exosporium (5). This enzyme presumably regulates the rate of inosine-induced germination and perhaps spore germination within macrophages.

Other Basal Layer Exosporium Proteins

ExsY and CotY

The ExsY and CotY proteins are homologs of the *B. subtilis* coat proteins CotY and CotZ (57, 108) and were identified as components of the *B. anthracis* exosporium within the high-molecular-weight complex (22). ExsY and CotY are paralogs and have ~90% amino acid identity to one another, except for their N-terminal domains (13). In *B. cereus* ATCC 14579, spores of the *exsY* mutant were missing the exosporium basal layer and the coat did not stain evenly by electron microscopy (EM) analysis, suggesting an effect on the spore coat. As such, the spores of *exsY* mutant were more sensitive to lysozyme than to wild type. Spores of a

cotY mutant still contained an exosporium, although it appeared to be closer to the coat, in contrast to the larger distance between the coat and exosporium surrounding wild-type spores (13).

The ExsY protein has also been investigated in the Sterne strain of *B. anthracis*. The ability of the *exsY* mutant to form an exosporium was dependent on culture conditions. When grown on solid media, spores of the mutant still formed an exosporium, based on reaction with an anti-BclA antibody. However, the labeling of the antibody was restricted to the polar cap-like region of the spore. In contrast, when grown in liquid media, spores of the mutant did not possess an exosporium (29). Interestingly, in comparison with the results with spores of the *B. cereus exsY* mutant (13), the *B. anthracis exsY* mutant was not sensitive to various environmental stresses, including exposure to lysozyme (29).

ExsA

The ExsA protein of *B. cereus* was identified in a transposon mutagenesis study by enriching for spores that were less hydrophobic following exposure to a hexadecane/aqueous-phase partitioning system (10). The ExsA protein is the equivalent of SafA (YrbA) of *B. subtilis*, which has a role in spore coat assembly (58). In *B. cereus*, the ExsA protein contains an N-terminal region that shows 86% identity to *B. subtilis* SafA over the first 49 amino acids. In contrast to SafA, the ExsA protein contains multiple tandem oligopeptide repeats. The ExsA protein was examined in detail in *B. cereus* ATCC 10876 in which the gene was inactivated. In spores of the mutant, the exosporium was thinner, looser, and irregularly shaped. Many spores were also observed lacking any associated exosporium. In addition, the spore coat was similarly affected in an *exsA* mutant. By EM observations, some spores lacked a spore coat, whereas others had a partially assembled spore coat. Not surprisingly, the spores of the *exsA* mutant were very sensitive to lysozyme and overall were less stable compared with wild-type spores. The gene encoding the ExsA protein is also present in the *B. anthracis* genome (59); however, the protein has not been examined to date. Nevertheless, it is hypothesized to play a role similar to that observed in *B. cereus*.

BxpA

The BxpA protein was originally identified by Steichen et al. (17) and Liu et al. (20) as an exosporium protein and has many interesting characteristics. Only the carboxy-terminal fragment of this protein was found within the exosporium profile, suggesting proteolysis.

The location of the amino-terminal portion of BxpA was not determined. The protein displayed a highly unusual amino acid composition with 54 glycine and 38 proline residues. In addition, the protein is composed of three pairs of identical tandem repeats. The BxpA proteins from the Ames and Sterne strains of *B. anthracis* differ by a noted 14-bp codon deletion in the Sterne strain (17). A second study utilized immunoelectron microscopy to localize the majority of the BxpA protein to the spore cortex and not the exosporium of Ames spores (16). One potential reason for these differing results between studies is that BxpA may be present both in the exosporium and associated with the cortex, but the concentration of exosporium-associated BxpA may be below the limit of detection of immune-electron microscopy. The presence of spore proteins in multiple locations is not unprecedented (25, 30, 60).

Moody et al. (16) described that the maturation of BxpA depends on the proteins YabG and Tgl. It appears that YabG proteolyzes and Tgl cross-links BxpA. Similarly, several *B. subtilis* coat proteins are cleaved during or after assembly (61–63). At least some of this proteolytic processing is under the control of YabG (58, 61, 63) Spore proteins can also be modified through cross-linking. It has been shown that at least some coat protein cross-linking is controlled by the coat-associated transglutaminase Tgl (14, 64, 65). The discovery of roles for YabG and Tgl in BxpA maturation is consistent with the possibility that, as in *B. subtilis*, YabG cleaves and Tgl cross-links *B. anthracis* spore proteins.

ExsFA/BxpB and ExsFB
The ExsFA/BxpB and ExsFB proteins are part of the high-molecular-mass spore protein complex. These proteins are required for localization of the BclA protein on the spore surface (5, 19, 31, 66) and stability of the exosporium crystalline layers (19). The ExsFA/BxpB protein has been localized to the basal layer of the exosporium (5, 31). In contrast, ExsFB is located in the interspace region of the spore between the spore coat and the exosporium (5).

ExsM
ExsM is a small basic protein that was identified from a proteomic analysis of the exosporium of *B. cereus* strain ATCC 4342 (67). When the *exsM* gene was inactivated, spores of the mutant were morphologically different. In the parental strain, the exosporium was elongated. In contrast, the *exsM* mutant spores had the exosporium more closely associated to the cell, and the

spores were significantly shorter. In addition, spores lacking the ExsM protein were encased in two distinct exosporium layers with the hair-like nap present on each of the exosporium layers facing the exterior. The double exosporium spores were more resistant to treatment with lysozyme, which is consistent with the exosporium being a semipermeable barrier to macromolecules. The double exosporium spores were also altered in their germination kinetics. Specifically, the *exsM* spores showed an initial low rate of germination in the presence of L-alanine. The authors theorized that the escape mechanism of the cell may not be forceful enough to rapidly burst from the double exosporium.

The spores of an *exsM*-deficient mutant in the *B. anthracis* ΔSterne strain were shorter and rounder than those of the wild-type parent strain. The exosporium of the mutant was partially or fully detached from the spores, and approximately 30% of the mutant spores had altered exosporia (either partial or double exosporium) (67).

ExsK
The ExsK protein was initially identified in exosporium proteomic studies (20, 22, 25) and further characterized by Severson et al. (30) in the Sterne and Ames *B. anthracis* strains. It is part of the BclA-associated high-molecular-weight exosporium protein complex. Also, by microscopy, ExsK was found in two distinct locations of the spore. ExsK is on the outside of the exosporium basal layer, but its recognition by antibody is partially blocked by the presence of BclA. In addition, when examining spores of the *cotE* mutant that lack an exosporium (27), ExsK was still detected on the surface. These data suggest that ExsK is both on the exosporium surface and within the spore, beneath the exosporium.

To determine the role of ExsK in *B. anthracis*, mutant strains were constructed. Ames *exsK* mutants were tested in two different animal modes of infection. However, no loss of virulence was detected (30). When germination was measured by a fluorescent-dye uptake assay (68), the *exsK* mutant spores germinated slightly but significantly faster than wild-type spores. The exact mechanism of this effect of ExsK on germination remains to be determined.

EXOSPORIUM BIOSYNTHESIS
Assembly of the *B. anthracis* spore coat and exosporium is still poorly understood, although this has been an active area of interest. Many coat proteins are conserved between *B. anthracis* and the more thoroughly

studied *B. subtilis*, suggesting some overlap in the mechanism controlling coat assembly (69, 70). In addition, in the *B. cereus* family, some homologs of *B. subtilis* spore proteins function in both coat assembly and formation of the exosporium (10, 13, 27, 29).

The exosporium proteins first appear during stage III of sporulation, and the exosporium is observed as a small lamellar structure in the mother cell cytoplasm that is synthesized concurrently with the spore coat (71, 72). The SpoIVA protein, which is important for directing assembly of the coat to the *B. subtilis* forespore (73), is similarly responsible for assembly of the *B. anthracis* coat and indirectly exosporium (27). Bacilli of *B. anthracis spoIVA* mutants were able to enter sporulation normally but were never able to progress through sporulation. Swirls of material were observed within the mother cell that likely contained both coat and exosporium material (27). Thus, the proteins required for spore production were present within the mother cell, but assembly was severely impacted.

In *B. subtilis*, the CotE protein is a major morphogenetic protein responsible for assembling the outer coat proteins (26). In contrast, when the *cotE* gene is mutated in *B. anthracis*, the resulting spores have a significant but limited defect in the spore coat. Only minor changes were observed by EM observations to the coat and spore protein profile. In addition, spores of the *B. anthracis cotE* mutant did not become sensitive to lysozyme. The exosporium of the *cotE* mutant is synthesized but fails to encircle and attach to the forespore during sporulation (27).

CotE may direct the synthesis of other proteins involved with exosporium biosynthesis, ExsY and CotY. These proteins are orthologs of CotE-controlled protein, CotZ, in *B. subtilis* (74). In *B. cereus*, ExsY and CotY are both necessary for proper coat and exosporium formation (13). If the *exsY* gene in *B. anthracis* is mutated, the exosporium production is affected where only a cap-like fragment assembles onto one end of the forespore closest to the mother cell midpoint (29). Perhaps ExsY then directs the remainder of the polymerization of the sheet of the exosporium shell to close upon itself.

In a *bclA* mutant strain of *B. anthracis*, the BclB protein is also localized at the exosporium cap (50). Likewise, the integrity of the exosporium of a *bclB* mutant is compromised, leading the structure to slough off the spore (50, 51). During sporulation, the BclA protein and exosporium basal layer localize to the central pole of the forespore. The N-terminal domain of BclA is required for targeting and incorporating this protein into the exosporium basal layer following cleavage and release of 19 amino acids (34).

Other proteins shown to be involved in correct assembly of BclA into the exosporium are ExsFA/BxpB and ExsFB. Mutations of these corresponding genes lead to spores that are devoid of or decreased in the amount of the hair-like nap that extends from the exosporium basal layer (5, 19, 31, 66). Likewise, Giorno et al. (5) reported that in an *exsFA/bxpB* mutant, BclA accumulates at the mother cell pole of the forespore. Further data demonstrate that BclA and ExsFA/BxpB are coexpressed, form a complex, and then localize to the mother cell-central pole of the spore. As the exosporium continues to enclose the spore, BclA is glycosylated and the N-terminal 19-amino-acid peptide is cleaved from BclA (66).

FUNCTION OF THE EXOSPORIUM IN GERMINATION

Spore germination is a somewhat paradoxical phenomenon, in that spores must remain resistant to chemical and physical insults; however, spores must also be prepared to germinate upon introduction of germinants, such as L-alanine, inosine, etc. (75). Germination is essential to the pathogenesis of *B. anthracis* and subsequent disease. Spores that have altered germination potential result in significantly different pathology and outcomes of infection (55, 76). Much work has been accomplished trying to understand germination (see Popham and Bernhards [109]); however, many facets of germination still remain uncertain. In particular, the role of the exosporium in spore germination remains rather undefined.

Germinant receptors are present in the inner spore membrane, several layers beneath the exosporium. Perhaps an additional function of the exosporium is to dictate which molecules/compounds or what concentrations of these potential germinants come into contact with the spore. The exosporium basal layer is somewhat permeable. Early studies by Gerhardt and Black indicated that the exosporium is clearly permeable to small molecules. Based on modeling criteria at that time, the pore sizes ranged from 10 to 200 Å, and these pore sizes were thought to be uniform across the exosporium surface (77). More recently, Ball and coworkers detailed the structure of the exosporium based on data generated through crystallography (7). In these studies, the exosporium was demonstrated to contain pores ranging in size from 23 to 34 Å (7). This size pore would be large enough for germinants, such as inosine and L-alanine, to permeate but would be restrictive to potentially degradative enzymes or host-generated antibodies (7). However, it has been shown that anti-whole-spore antibodies have a germination inhibitory activity (68, 78); these antibodies could possibly interact with structural components of the exosporium, block the pores, and subsequently interfere with the translocation of potential germinants across this structure. While this inhibition of germination is largely transient, these antibodies significantly alter the germination potential of *B. anthracis* spores. These antibodies have also been shown to be opsonic in *in vitro* assays (potentially altering the spore-macrophage interactions) (50, 79) and anti-spore antibodies provided limited passive protection to mice, suggesting that these antibodies are potentially protective, as described in more detail below (60, 79).

The impact of specific exosporium structural components on germination has been difficult to ascertain. For example, Bozue et al. demonstrated that there were no significant differences associated with the rate or extent of germination in Δ*bclA* versus wild-type Ames spores (44). These results were confirmed using two different assays for germination (fluorometric and loss of absorbance). However, Brahmbhatt et al. reported that Δ*bclA* spores germinated slightly (but significantly) better than Sterne wild type (11). One study used a fully virulent strain of *B. anthracis*, whereas the other used the attenuated Sterne strain of *B. anthracis*. Additionally, there were variations in procedures employed that make these data difficult to compare directly. Giorno et al. demonstrated that spores devoid of ExsFA/BxpB exhibited a modest but significant defect in a potentially early step in germination (5). However, Steichen at al. reported that their version of these mutant spores germinated better than the parental strain (31). Again, there is enough variation between these studies to make direct comparisons of subtle germination phenotypes challenging. Removal of the ExsA protein, another exosporium protein from *B. cereus* with a homologue in *B. anthracis*, appeared to slow germination in comparison with wild-type spores (10). It has also been reported that *exsK* mutant spores of *B. anthracis* germinate more rapidly than wild-type spores (30).

Other exosporium proteins may not be structural proteins, but are thought to be tightly adsorbed and accordingly integrated into the exosporium. Examples of this phenomenon include a nucleoside hydrolase and an alanine racemase (25, 80). Inosine may act as a primary germinant for *B. cereus* and *B. thuringiensis* (inosine induces significant spore germination in the absence of other germinants) and a cogerminant for *B. anthracis*

(inosine significantly lowers the required concentration of a primary germinant, such as L-alanine). The nucleoside hydrolase would degrade inosine, thus regulating environmental levels and potentially lessening the impact of inosine as a cogerminant. L-Alanine is an important germinant for both *B. anthracis* and *B. cereus*. The alanine racemase is thought to, in part, regulate germination in several instances. The enzymatic activity of alanine racemase involves converting L-alanine (germination trigger) to D-alanine (germination inhibitor). Thus, this enzyme can manipulate L-alanine concentrations to ensure germination is only initiated when conditions are desirable. When alanine racemase activity has been inhibited, *B. cereus* spores germinate more readily (81, 82). It has been proposed that this alanine racemase activity may be part of a germination-inducing quorum-sensing mechanism in *B. cereus* (81). In these studies, the authors demonstrated that *B. cereus* germination was enhanced by higher spore concentrations and conditioned medium (81). This conditioned medium contained L-alanine released by previously germinated spores. This quorum sensing has not been observed in *B. anthracis*, presumably because *B. anthracis* spores do not release amino acids, and under normal conditions, require two germinants (e.g., a germinant and a cogerminant) to germinate (81). Additionally, it has been documented that *B. anthracis* spore germination rates are negatively impacted by higher spore densities *in vitro* (83).

In the case of *B. anthracis*, Alr activity plays a role in spore-macrophage interactions (55, 82, 84). While cogerminants are necessary for optimal germination of *B. anthracis* spores, the Alr enzyme is sufficient to inhibit germination induced by L-alanine. In mouse models of anthrax, the importance of appropriately timed germination is underscored, because spores with altered germination profiles result in significantly different disease progression (55, 76). Additionally, Chesnokova and colleagues have shown that Alr activity prevents germination of late-stage spores within the sporulating mother cells, suggesting that this enzymatic function is also necessary to prevent premature germination during sporulation, thus ensuring the production of environmentally stable endospores (56, 82, 84).

The "cap-like" region of the exosporium is at one pole of the spore and serves as the exit route of the outgrowing cell during germination. The "cap-like" region corresponds to the area where the exosporium begins assembling, as described above, within the mother cell during sporulation. In a mature spore, this area of the exosporium does not bind anti-Alr monoclonal antibodies, demonstrating that exosporium assembly is not a uniform process. The germinating cells always escape through the exosporium by popping through the cap with enough force to allow the cell to rapidly and completely separate from the exosporium. Perhaps this escape from the spore is important for the vegetative cells when facing the *in vivo* environment and host's immune cells (6).

ROLE OF EXOSPORIUM IN VIRULENCE

The role of the exosporium in virulence, especially for the BclA protein itself, has been a topic of great interest. When examining *B. anthracis* spores lacking the complete exosporium, as generated from *cotE* mutant spores in the Ames strain, no difference in virulence was detected following intramuscular challenge in guinea pigs or intranasal challenge in mice (27). These challenges were performed by using low doses of the spores, and no difference was noted in comparison with the challenge with wild-type Ames spores. However, median lethal dose (LD_{50}) analyses were not performed with the exosporium minus strain. In addition, genetic mutants of the Ames strains have been constructed that lack the nap-like fibers but retain the exosporium basal layer, and no differences in virulence have been detected through multiple animal *B. anthracis* challenge models (5, 44). Similar results were first described with a *bclA* mutant in the Sterne strain of *B. anthracis* following subcutaneous challenge in mice (18). Other studies indicate that lack of the BclA fibers in the Sterne vaccine strain render the mutant more virulent than the parental strain as determined through differences in mean time to death or LD_{50} measurements (11, 49). It can be concluded that the BclA protein or the complete exosporium structure is not needed for *B. anthracis* spores to remain virulent in animal models of anthrax. While BclA is certainly immunodominant and accordingly an attractive target for vaccine and detection technologies, this protein could easily be removed from a *B. anthracis* variant engineered to subvert vaccine efficacy or evade detection systems yet retain full virulence. Thus, a multi-antigen vaccine strategy, making vaccine subversions difficult, would be ideal.

INTERACTION OF THE EXOSPORIUM WITH HOST CELLS

Several lines of evidence indicate that the BclA fibers play a critical role in targeting spores to mammalian phagocytes. Using a *bclA* mutant in the Ames strain, Bozue et al. (45) demonstrated that spores lacking the

fibers were significantly more adherent to epithelial, fibroblast, and endothelial cells than spores of the parental wild-type strain (Figs. 2 and 3). In contrast, no difference in spore binding to macrophages was detected (45). Furthermore, BclA was shown to be the ligand for the integrin Mac-1 receptor on phagocytes (49). Oliva and coworkers subsequently demonstrated that CD14 acts as coreceptor for spores in the Mac-1 pathway. It was shown that CD14 binds to rhamnose residues within the BclA protein and significantly enhances the internalization of spores via this Mac-1-

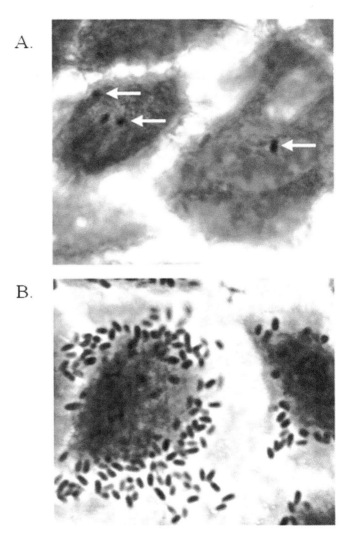

Figure 2 Association of spores to bronchial epithelial cells. Micrographs of spores associated with the bronchial epithelial cells. Samples were stained with spore stain (malachite green) and counterstained with a Wright-Giemsa stain. (**A**) Ames spores (indicated by white arrows) with the cells. (**B**) *bclA* mutant spores with the cells. Reprinted from reference 45 with permission.

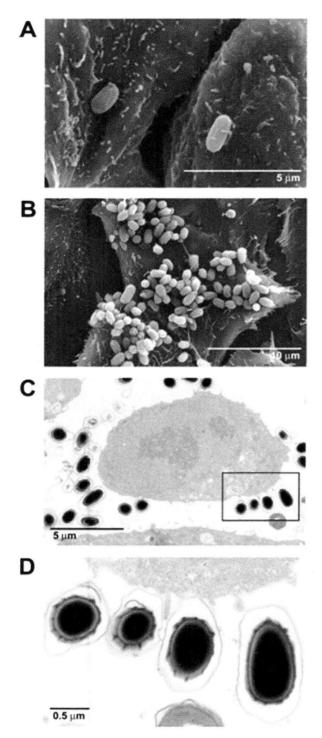

dependent mechanism (85). Interestingly, the relationship between rhamnose and spore-macrophage interactions had been shown previously; however, the reasons for these observations were unclear at that time (46). Taken together, these results demonstrate that the BclA fibers may inhibit nonspecific interactions between *B. anthracis* spores with nonprofessional phagocytic cells and direct the spores toward uptake by phagocytes during infection.

In addition to its perceived role in attachment to macrophages, the exosporium may play a role in the survival of the spore once engulfed by macrophages. In one study, murine peritoneal macrophages were infected with Sterne 34F2 spores with or without an exosporium (removed by sonication). The sonicated spores (exosporium deficient) did not survive as well intracellularly as spores retaining exosporia (86). The method by which the spores evade killing remains to be determined. However, recent evidence suggests that the exosporium may play a role in inhibiting the macrophage-oxidative burst. The presence of arginase within the exosporium has been demonstrated by several studies (22, 87–89). The arginase activity of the exosporium may affect the nitric oxide production of the macrophage (87). The efficiency of spore killing was further increased by macrophages when pretreated with interferon-γ (IFN-γ). In contrast, if the exosporium-deficient spores were opsonized with immune-IgG (collected from a serum sample from a human who had received the United Kingdom-licensed anthrax vaccine), no differences were observed in macrophage killing between treated and untreated sonicated spores. These data suggest that there is a target for IgG on the exosporium. Other investigators have observed the presence of the protective antigen (PA) associated with *B. anthracis* spores, which leads to increased uptake and killing by macrophages (68, 79, 90).

Further evidence indicates that the exosporium structure may play a role during the interaction between spores and macrophages based on the cytokine response of the macrophage. When Sterne spores lacking an exosporium (removed by sonication) were added to macrophages, the cytokine mRNA response was significantly upregulated for IFN-β, interleukin-1β (IL-1β), TNF-α, and IL-6 in comparison with spores containing an intact exosporium (91). The exact means by which the exosporium inhibits the cytokine response remains

Figure 3 Scanning and transmission electron micrographs of the interaction between *B. anthracis* spores and bronchial epithelial cells. (**A**) Adherence of Ames spores to cells 2 h postinfection. Bar corresponds to 5 m. (**B**) Adherence of *bclA* mutant spores to cells 2 h postinfection. Bar corresponds to 10 m. (**C**)*bclA* mutant spores incubated with cells. Bar corresponds to 5 m. (**D**) Close-up view from boxed area in panel C. Bar corresponds to 0.5 m. Reprinted from reference 45 with permission.

undetermined. It is possible that the exosporium cloaks structures that lead to the cytokine induction or enzymes are present within the exosporium that potentially alter the response.

A recent study has demonstrated that plasminogen was able to bind to the outer surface of *B. anthracis* spores (92). If the exosporium was removed, little plasminogen was observed to bind to the spores. Plasminogen is a central component of the fibrinolytic system and is activated to plasmin. Once activated, it can degrade extracellular and matrix and basal layer proteins. The metalloprotease InhA, located within the exosporium (93), can accelerate the activation of plasminogen.

In addition, BclA has also been shown to bind the pulmonary surfactant SP-C (42), further suggesting that BclA may be used for targeting host cells. *B. anthracis* spores may also interact with lung epithelial cells during infection (94, 95) and the receptor for entry is the integrin α2β1 (43). Recent work has demonstrated that BclA is required for spores to enter epithelial cells (43). However, other reports have shown that spores lacking BclA are readily able to gain entry within nonprofessional phagocytic cells (45, 49). BclA also appears to interact with the complement component C1q to bind to α2β1 (43) on epithelial cells (43).

EXOSPORIUM AS A VACCINE TARGET

While the current and next-generation anthrax vaccines are effective, they target primarily only a single protein, PA (96). PA is responsible for translocation of the toxin effector molecules (edema and lethal factors) into the cytosol of host cells (97). Antibody against PA interferes with this process and as a result prevents significant intoxication of the host. However, there are several drawbacks associated with this strategy. Since these vaccine strategies are antitoxin, essentially the bacteria are permitted to infect the host until the antitoxin vaccine begins to take effect. Additionally, from a biodefense perspective, it would be beneficial to have a multicomponent anthrax vaccine to combat emerging or engineered threats that may be refractory to current vaccine approaches (98, 99). Because the exosporium is the outermost surface of a *B. anthracis* spore and a source of numerous antigens of vaccine targets, the exosporium has been the focus of numerous novel vaccination approaches. Accordingly, several laboratories have investigated the concept of vaccines based on inactivated spores or spore structures, such as the exosporium.

The most promising data indicating protection attributed to an anti-spore immune response have been generated by vaccinating small animals with formaldehyde-inactivated spores (FIS). Brossier and colleagues demonstrated that vaccine formulations containing PA and FIS provided significantly greater levels of protection in mice and guinea pigs when compared with vaccination with PA alone (100). These studies involved parenteral challenge routes. These observations were confirmed by Gauthier and coworkers who utilized similar vaccine formulations; however, the mice and guinea pigs in this study were challenged with virulent spores delivered via the inhalation route (both intranasal instillation and aerosol delivery) (101). In these studies, it was demonstrated that by targeting both the earliest phase of infection (entry of ungerminated *B. anthracis* into hosts) and also the later stage of infection (intoxication), the experimental animals were more likely to survive infection with lethal doses of *B. anthracis*. These observations were consistent among studies regardless of the infection route.

Cote et al. presented data suggesting that mice vaccinated with irradiated spores were more likely to survive an infection with aerosolized *B. anthracis* spores compared with unvaccinated controls (102). It was demonstrated that irradiated Ames strain spores and irradiated Δ*bclA* spores offered similar levels of protection to the mice following a challenge with a low dose of aerosolized Ames spores. These data suggested other non-BclA epitopes associated with irradiated spores may be somewhat protective (102). Of more recent note, Vergis et al. (103) demonstrated that immunization with FIS from nontoxigenic and nonencapsulated *B. cereus* provided significant levels of protection to mice challenged with the Ames strain of *B. anthracis*. These studies demonstrated that this protection was independent of PA or BclA being present in the vaccine.

In an ideal situation, whole killed spores would be replaced in vaccine formulations with a defined individual or group of protective epitopes. Recombinant exosporium components have been shown to augment the efficacy of a PA vaccination. Brahmbhatt et al. demonstrated that the immune responses generated in mice receiving a single suboptimal injection of PA were significantly augmented by a later injection of BclA (38) and that the mice receiving both antigens were better protected against a lethal Sterne spore challenge than those receiving either PA or BclA alone. These results were similar to those published earlier by Hahn et al. (104). In the latter studies, it was demonstrated that vaccination with a DNA construct encoding PA in combination with a construct encoding BclA was superior in efficacy to vaccination with PA constructs alone (104). Hahn et al. demonstrated partial protection against challenge with the fully virulent Ames strain.

It has also been shown that other immunogens besides BclA may offer an advantage to vaccinated animals. Cybulski and colleagues observed significant protection when ExsFA/BxpB or p5303 (a hypothetical protein associated with the exosporium of *B. anthracis*) was used in vaccine formulations (105). The same recombinant exosporium components (BclA, ExsFA/BxpB, and p5303) significantly improved the outcome of infection in mice and guinea pigs (102). These resulting anti-exosporium component antibodies augmented protection of both mice and guinea pigs when challenged with ΔAmes (pXO1⁻/pXO2⁺) and fully virulent Ames, respectively (102).

Data from passive protection studies performed using antisera generated against whole spores have been less impressive. It was previously reported that passive transfer of rabbit anti-spore antibodies could significantly protect mice against challenge with a fully virulent strain of *B. anthracis* (78). The antibodies used were generated in rabbits against a strain of *B. anthracis* that was cured of both pXO1 and PXO2, so that antibodies would be elicited against spore structures (presumably the surface-exposed exosporium layer) and not against either toxin components or capsule. In contrast, Glomski et al. could not demonstrate significant levels of passive protection when mouse immune FIS sera were transferred to naïve mice (106). Cote et al. later demonstrated that slight protection was afforded by passive transfer of rabbit anti-irradiated spore IgG into naïve mice, but this protection was only statistically significant when compared with completely naïve mice (60). When the survival was compared with mice receiving normal rabbit IgG, the observed protection was statistically insignificant, suggesting that some of the protection was not spore specific. When taken together, these data indicate that there are trends toward protection afforded by passive transfer of anti-spore antibodies; however, these trends lack reliable statistical significance.

Augmented protection afforded by an active anti-spore immune response has been detailed by several laboratories; however, this protection was not reproduced in passive protection studies, suggesting a mechanism that may not be completely antibody mediated. These data encouraged Glomski et al. to further investigate the mechanisms associated with the added protection offered by FIS when added to a PA vaccine regimen (106). These authors elegantly demonstrated a cellular component to this anti-spore immunity. When splenocytes harvested from FIS immune mice were transferred to naïve mice, significant protection was achieved. Through further experimentation using transgenic animals and

in vivo lymphocyte depletion methodology, it was demonstrated that INF-γ-producing CD4 lymphocytes were, in part, responsible for the vaccine augmentation associated with FIS (106).

Regardless of the exact mechanism associated with the augmented levels of protection, this body of evidence clearly demonstrates that an anti-spore, in particular, an anti-exosporium immune response, benefits the vaccinated host. Efforts to more precisely elucidate these benefits are needed so as to further optimize current anthrax vaccine strategies as well as to increase preparedness against future emerging and/or engineered threats.

CONCLUSIONS

Over the past decade, great strides have been made in better understanding the composition and biosynthesis of the exosporium of *B. anthracis*. Although insights have been gained regarding the interaction of the exosporium with the host during infection, the exact roles and significance of this complex structure remain to be determined. Because this is the outermost structure of the spore and affects adherence to mammalian cells, the exosporium most likely plays a role for persistence in the natural soil environment, as recent evidence indicates (107). Furthermore, because the exosporium is a highly antigenic structure, future strategies for the next-generation anthrax vaccine should pursue its inclusion as a component to provide protection against the spore itself during the initial stages of anthrax.

Acknowledgments. Opinions, interpretations, conclusions, and recommendations are those of the authors and are not necessarily endorsed by the U.S. Army.

Citation. Bozue JA, Welkos S, Cote CK. 2015. The *Bacillus anthracis* exosporium: what's the big "hairy" deal? Microbiol Spectrum 3(5):TBS-0021-2015.

References

1. **Aronson AI, Fitz-James P.** 1976. Structure and morphogenesis of the bacterial spore coat. *Bacteriol Rev* **40:** 360–402.
2. **Gerhardt P.** 1967. Cytology of *Bacillus anthracis. Fed Proc* **26:**1504–1517.
3. **Gerhardt P, Ribi E.** 1964. Ultrastructure of the exosporium enveloping spores of *Bacillus cereus. J Bacteriol* **88:**1774–1789.
4. **Vary PS.** 1994. Prime time for *Bacillus megaterium. Microbiology* **140:**1001–1013.
5. **Giorno R, Mallozzi M, Bozue J, Moody KS, Slack A, Qiu D, Wang R, Friedlander A, Welkos S, Driks A.** 2009. Localization and assembly of proteins comprising the outer structures of the *Bacillus anthracis* spore. *Microbiology* **155:**1133–1145.

6. Steichen CT, Kearney JF, Turnbough CL Jr. 2007. Non-uniform assembly of the *Bacillus anthracis* exosporium and a bottle cap model for spore germination and outgrowth. *Mol Microbiol* **64:**359–367.

7. Ball DA, Taylor R, Todd SJ, Redmond C, Couture-Tosi E, Sylvestre P, Moir A, Bullough PA. 2008. Structure of the exosporium and sublayers of spores of the *Bacillus cereus* family revealed by electron crystallography. *Mol Microbiol* **68:**947–958.

8. Cole HB, Ezzell JW Jr, Keller KF, Doyle RJ. 1984. Differentiation of *Bacillus anthracis* and other *Bacillus* species by lectins. *J Clin Microbiol* **19:**48–53.

9. Matz LL, Beaman TC, Gerhardt P. 1970. Chemical composition of exosporium from spores of *Bacillus cereus*. *J Bacteriol* **101:**196–201.

10. Bailey-Smith K, Todd SJ, Southworth TW, Proctor J, Moir A. 2005. The ExsA protein of *Bacillus cereus* is required for assembly of coat and exosporium onto the spore surface. *J Bacteriol* **187:**3800–3806.

11. Brahmbhatt TN, Janes BK, Stibitz ES, Darnell SC, Sanz P, Rasmussen SB, O'Brien AD. 2007. *Bacillus anthracis* exosporium protein BclA affects spore germination, interaction with extracellular matrix proteins, and hydrophobicity. *Infect Immun* **75:**5233–5239.

12. Koshikawa T, Yamazaki M, Yoshimi M, Ogawa S, Yamada A, Watabe K, Torii M. 1989. Surface hydrophobicity of spores of *Bacillus* spp. *J Gen Microbiol* **135:**2717–2722.

13. Johnson MJ, Todd SJ, Ball DA, Shepherd AM, Sylvestre P, Moir A. 2006. ExsY and CotY are required for the correct assembly of the exosporium and spore coat of *Bacillus cereus*. *J Bacteriol* **188:**7905–7913.

14. Kobayashi K, Hashiguchi K, Yokozeki K, Yamanaka S. 1998. Molecular cloning of the transglutaminase gene from *Bacillus subtilis* and its expression in *Escherichia coli*. *Biosci Biotechnol Biochem* **62:**1109–1114.

15. Kobayashi K, Suzuki SI, Izawa Y, Miwa K, Yamanaka S. 1998. Transglutaminase in sporulating cells of *Bacillus subtilis*. *J Gen Appl Microbiol* **44:**85–91.

16. Moody KL, Driks A, Rother GL, Cote CK, Brueggemann EE, Hines HB, Friedlander AM, Bozue J. 2010. Processing, assembly and localization of a *Bacillus anthracis* spore protein. *Microbiology* **156:**174–183.

17. Steichen C, Chen P, Kearney JF, Turnbough CL Jr. 2003. Identification of the immunodominant protein and other proteins of the *Bacillus anthracis* exosporium. *J Bacteriol* **185:**1903–1910.

18. Sylvestre P, Couture-Tosi E, Mock M. 2002. A collagen-like surface glycoprotein is a structural component of the *Bacillus anthracis* exosporium. *Mol Microbiol* **45:**169–178.

19. Sylvestre P, Couture-Tosi E, Mock M. 2005. Contribution of ExsFA and ExsFB proteins to the localization of BclA on the spore surface and to the stability of the *Bacillus anthracis* exosporium. *J Bacteriol* **187:**5122–5128.

20. Liu H, Bergman NH, Thomason B, Shallom S, Hazen A, Crossno J, Rasko DA, Ravel J, Read TD, Peterson SN, Yates J III, Hanna PC. 2004. Formation and composition of the *Bacillus anthracis* endospore. *J Bacteriol* **186:**164–178.

21. Charlton S, Moir AJ, Baillie L, Moir A. 1999. Characterization of the exosporium of *Bacillus cereus*. *J Appl Microbiol* **87:**241–245.

22. Redmond C, Baillie LW, Hibbs S, Moir AJ, Moir A. 2004. Identification of proteins in the exosporium of *Bacillus anthracis*. *Microbiology* **150:**355–363.

23. Mignot T, Mesnage S, Couture-Tosi E, Mock M, Fouet A. 2002. Developmental switch of S-layer protein synthesis in *Bacillus anthracis*. *Mol Microbiol* **43:**1615–1627.

24. Sylvestre P, Couture-Tosi E, Mock M. 2003. Polymorphism in the collagen-like region of the *Bacillus anthracis* BclA protein leads to variation in exosporium filament length. *J Bacteriol* **185:**1555–1563.

25. Todd SJ, Moir AJ, Johnson MJ, Moir A. 2003. Genes of *Bacillus cereus* and *Bacillus anthracis* encoding proteins of the exosporium. *J Bacteriol* **185:**3373–3378.

26. Little S, Driks A. 2001. Functional analysis of the *Bacillus subtilis* morphogenetic spore coat protein CotE. *Mol Microbiol* **42:**1107–1120.

27. Giorno R, Bozue J, Cote C, Wenzel T, Moody KS, Mallozzi M, Ryan M, Wang R, Zielke R, Maddock JR, Friedlander A, Welkos S, Driks A. 2007. Morphogenesis of the *Bacillus anthracis* spore. *J Bacteriol* **189:**691–705.

28. Lai EM, Phadke ND, Kachman MT, Giorno R, Vazquez S, Vazquez JA, Maddock JR, Driks A. 2003. Proteomic analysis of the spore coats of *Bacillus subtilis* and *Bacillus anthracis*. *J Bacteriol* **185:**1443–1454.

29. Boydston JA, Yue L, Kearney JF, Turnbough CL Jr. 2006. The ExsY protein is required for complete formation of the exosporium of *Bacillus anthracis*. *J Bacteriol* **188:**7440–7448.

30. Severson KM, Mallozzi M, Bozue J, Welkos SL, Cote CK, Knight KL, Driks A. 2009. Roles of the *Bacillus anthracis* spore protein ExsK in exosporium maturation and germination. *J Bacteriol* **191:**7587–7596.

31. Steichen CT, Kearney JF, Turnbough CL Jr. 2005. Characterization of the exosporium basal layer protein BxpB of *Bacillus anthracis*. *J Bacteriol* **187:**5868–5876.

32. Swiecki MK, Lisanby MW, Shu F, Turnbough CL Jr, Kearney JF. 2006. Monoclonal antibodies for *Bacillus anthracis* spore detection and functional analyses of spore germination and outgrowth. *J Immunol* **176:**6076–6084.

33. Thompson BM, Hoelscher BC, Driks A, Stewart GC. 2011. Localization and assembly of the novel exosporium protein BetA of *Bacillus anthracis*. *J Bacteriol* **193:**5098–5104.

34. Thompson BM, Stewart GC. 2008. Targeting of the BclA and BclB proteins to the *Bacillus anthracis* spore surface. *Mol Microbiol* **70:**421–434.

35. Boydston JA, Chen P, Steichen CT, Turnbough CL Jr. 2005. Orientation within the exosporium and structural stability of the collagen-like glycoprotein BclA of *Bacillus anthracis*. *J Bacteriol* **187:**5310–5317.

36. Leski TA, Caswell CC, Pawlowski M, Klinke DJ, Bujnicki JM, Hart SJ, Lukomski S. 2009. Identification and classification of bcl genes and proteins of *Bacillus cereus* group organisms and their application in *Bacillus anthracis* detection and fingerprinting. *Appl Environ Microbiol* 75:7163–7172.

37. Rasmussen M, Jacobsson M, Björck L. 2003. Genome-based identification and analysis of collagen-related structural motifs in bacterial and viral proteins. *J Biol Chem* 278:32313–32316.

38. Brahmbhatt TN, Darnell SC, Carvalho HM, Sanz P, Kang TJ, Bull RL, Rasmussen SB, Cross AS, O'Brien AD. 2007. Recombinant exosporium protein BclA of *Bacillus anthracis* is effective as a booster for mice primed with suboptimal amounts of protective antigen. *Infect Immun* 75:5240–5247.

39. Daubenspeck JM, Zeng H, Chen P, Dong S, Steichen CT, Krishna NR, Pritchard DG, Turnbough CL Jr. 2004. Novel oligosaccharide side chains of the collagen-like region of BclA, the major glycoprotein of the *Bacillus anthracis* exosporium. *J Biol Chem* 279:30945–30953.

40. Dong S, McPherson SA, Tan L, Chesnokova ON, Turnbough CL Jr, Pritchard DG. 2008. Anthrose bio-synthetic operon of *Bacillus anthracis*. *J Bacteriol* 190:2350–2359.

41. Saile E, Boons GJ, Buskas T, Carlson RW, Kannenberg EL, Barr JR, Boyer AE, Gallegos-Candela M, Quinn CP. 2011. Antibody responses to a spore carbohydrate antigen as a marker of nonfatal inhalation anthrax in rhesus macaques. *Clin Vaccine Immunol* 18:743–748.

42. Réty S, Salamitou S, Garcia-Verdugo I, Hulmes DJ, Le Hégarat F, Chaby R, Lewit-Bentley A. 2005. The crystal structure of the *Bacillus anthracis* spore surface protein BclA shows remarkable similarity to mammalian proteins. *J Biol Chem* 280:43073–43078.

43. Xue Q, Gu C, Rivera J, Höök M, Chen X, Pozzi A, Xu Y. 2011. Entry of *Bacillus anthracis* spores into epithelial cells is mediated by the spore surface protein BclA, integrin $\alpha 2\beta 1$ and complement component C1q. *Cell Microbiol* 13:620–634.

44. Bozue J, Cote CK, Moody KL, Welkos SL. 2007. Fully virulent *Bacillus anthracis* does not require the immunodominant protein BclA for pathogenesis. *Infect Immun* 75:508–511.

45. Bozue J, Moody KL, Cote CK, Stiles BG, Friedlander AM, Welkos SL, Hale ML. 2007. *Bacillus anthracis* spores of the bclA mutant exhibit increased adherence to epithelial cells, fibroblasts, and endothelial cells but not to macrophages. *Infect Immun* 75:4498–4505.

46. Bozue JA, Parthasarathy N, Phillips LR, Cote CK, Fellows PF, Mendelson I, Shafferman A, Friedlander AM. 2005. Construction of a rhamnose mutation in *Bacillus anthracis* affects adherence to macrophages but not virulence in guinea pigs. *Microb Pathog* 38:1–12.

47. Chen G, Driks A, Tawfiq K, Mallozzi M, Patil S. 2010. *Bacillus anthracis* and *Bacillus subtilis* spore surface properties and transport. *Colloids Surf B Biointerfaces* 76:512–518.

48. Lequette Y, Garénaux E, Tauveron G, Dumez S, Perchat S, Slomianny C, Lereclus D, Guérardel Y, Faille C. 2011. Role played by exosporium glycoproteins in the surface properties of *Bacillus cereus* spores and in their adhesion to stainless steel. *Appl Environ Microbiol* 77:4905–4911.

49. Oliva CR, Swiecki MK, Griguer CE, Lisanby MW, Bullard DC, Turnbough CL Jr, Kearney JF. 2008. The integrin Mac-1 (CR3) mediates internalization and directs *Bacillus anthracis* spores into professional phagocytes. *Proc Natl Acad Sci USA* 105:1261–1266.

50. Thompson BM, Waller LN, Fox KF, Fox A, Stewart GC. 2007. The BclB glycoprotein of *Bacillus anthracis* is involved in exosporium integrity. *J Bacteriol* 189:6704–6713.

51. Waller LN, Stump MJ, Fox KF, Harley WM, Fox A, Stewart GC, Shahgholi M. 2005. Identification of a second collagen-like glycoprotein produced by *Bacillus anthracis* and demonstration of associated spore-specific sugars. *J Bacteriol* 187:4592–4597.

52. García-Patrone M, Tandecarz JS. 1995. A glycoprotein multimer from *Bacillus thuringiensis* sporangia: dissociation into subunits and sugar composition. *Mol Cell Biochem* 145:29–37.

53. Cybulski RJ Jr, Sanz P, Alem F, Stibitz S, Bull RL, O'Brien AD. 2009. Four superoxide dismutases contribute to *Bacillus anthracis* virulence and provide spores with redundant protection from oxidative stress. *Infect Immun* 77:274–285.

54. Yasuda Y, Kanda K, Nishioka S, Tanimoto Y, Kato C, Saito A, Fukuchi S, Nakanishi Y, Tochikubo K. 1993. Regulation of L-alanine-initiated germination of *Bacillus subtilis* spores by alanine racemase. *Amino Acids* 4:89–99.

55. McKevitt MT, Bryant KM, Shakir SM, Larabee JL, Blanke SR, Lovchik J, Lyons CR, Ballard JD. 2007. Effects of endogenous D-alanine synthesis and autoinhibition of *Bacillus anthracis* germination on *in vitro* and *in vivo* infections. *Infect Immun* 75:5726–5734.

56. Chesnokova ON, McPherson SA, Steichen CT, Turnbough CL Jr. 2009. The spore-specific alanine racemase of *Bacillus anthracis* and its role in suppressing germination during spore development. *J Bacteriol* 191:1303–1310.

57. Zhang J, Fitz-James PC, Aronson AI. 1993. Cloning and characterization of a cluster of genes encoding polypeptides present in the insoluble fraction of the spore coat of *Bacillus subtilis*. *J Bacteriol* 175:3757–3766.

58. Ozin AJ, Henriques AO, Yi H, Moran CP Jr. 2000. Morphogenetic proteins SpoVID and SafA form a complex during assembly of the *Bacillus subtilis* spore coat. *J Bacteriol* 182:1828–1833.

59. Read TD, et al. 2003. The genome sequence of *Bacillus anthracis* Ames and comparison to closely related bacteria. *Nature* 423:81–86.

60. Cote CK, Bozue J, Moody KL, DiMezzo TL, Chapman CE, Welkos SL. 2008. Analysis of a novel spore antigen in *Bacillus anthracis* that contributes to spore opsonization. *Microbiology* 154:619–632.

61. **Kuwana R, Takamatsu H, Watabe K.** 2007. Expression, localization and modification of YxeE spore coat protein in *Bacillus subtilis*. *J Biochem* **142**:681–689.

62. **Kuwana R, Okuda N, Takamatsu H, Watabe K.** 2006. Modification of GerQ reveals a functional relationship between Tgl and YabG in the coat of *Bacillus subtilis* spores. *J Biochem* **139**:887–901.

63. **Takamatsu H, Kodama T, Imamura A, Asai K, Kobayashi K, Nakayama T, Ogasawara N, Watabe K.** 2000. The *Bacillus subtilis yabG* gene is transcribed by SigK RNA polymerase during sporulation, and *yabG* mutant spores have altered coat protein composition. *J Bacteriol* **182**:1883–1888.

64. **Monroe A, Setlow P.** 2006. Localization of the transglutaminase cross-linking sites in the *Bacillus subtilis* spore coat protein GerQ. *J Bacteriol* **188**:7609–7616.

65. **Zilhão R, Isticato R, Martins LO, Steil L, Völker U, Ricca E, Moran CP Jr, Henriques AO.** 2005. Assembly and function of a spore coat-associated transglutaminase of *Bacillus subtilis*. *J Bacteriol* **187**:7753–7764.

66. **Thompson BM, Hsieh HY, Spreng KA, Stewart GC.** 2011. The co-dependence of BxpB/ExsFA and BclA for proper incorporation into the exosporium of *Bacillus anthracis*. *Mol Microbiol* **79**:799–813.

67. **Fazzini MM, Schuch R, Fischetti VA.** 2010. A novel spore protein, ExsM, regulates formation of the exosporium in *Bacillus cereus* and *Bacillus anthracis* and affects spore size and shape. *J Bacteriol* **192**:4012–4021.

68. **Welkos SL, Cote CK, Rea KM, Gibbs PH.** 2004. A microtiter fluorometric assay to detect the germination of *Bacillus anthracis* spores and the germination inhibitory effects of antibodies. *J Microbiol Methods* **56**:253–265.

69. **Driks A.** 2002. Maximum shields: the assembly and function of the bacterial spore coat. *Trends Microbiol* **10**:251–254.

70. **Henriques AO, Moran CP Jr.** 2007. Structure, assembly, and function of the spore surface layers. *Annu Rev Microbiol* **61**:555–588.

71. **DesRosier JP, Lara JC.** 1984. Synthesis of the exosporium during sporulation of *Bacillus cereus*. *J Gen Microbiol* **130**:935–940.

72. **Ohye DF, Murrell WG.** 1973. Exosporium and spore coat formation in *Bacillus cereus* T. *J Bacteriol* **115**:1179–1190.

73. **Roels S, Driks A, Losick R.** 1992. Characterization of *spoIVA*, a sporulation gene involved in coat morphogenesis in *Bacillus subtilis*. *J Bacteriol* **174**:575–585.

74. **Kim H, Hahn M, Grabowski P, McPherson DC, Otte MM, Wang R, Ferguson CC, Eichenberger P, Driks A.** 2006. The *Bacillus subtilis* spore coat protein interaction network. *Mol Microbiol* **59**:487–502.

75. **Moir A, Corfe BM, Behravan J.** 2002. Spore germination. *Cell Mol Life Sci* **59**:403–409.

76. **Cote CK, Bozue J, Twenhafel N, Welkos SL.** 2009. Effects of altering the germination potential of *Bacillus anthracis* spores by exogenous means in a mouse model. *J Med Microbiol* **58**:816–825.

77. **Gerhardt P, Black SH.** 1961. Permeability of bacterial spores. II. Molecular variables affecting solute permeation. *J Bacteriol* **82**:750–760.

78. **Enkhtuya J, Kawamoto K, Kobayashi Y, Uchida I, Rana N, Makino S.** 2006. Significant passive protective effect against anthrax by antibody to *Bacillus anthracis* inactivated spores that lack two virulence plasmids. *Microbiology* **152**:3103–3110.

79. **Welkos S, Little S, Friedlander A, Fritz D, Fellows P.** 2001. The role of antibodies to *Bacillus anthracis* and anthrax toxin components in inhibiting the early stages of infection by anthrax spores. *Microbiology* **147**:1677–1685.

80. **Liang L, He X, Liu G, Tan H.** 2008. The role of a purine-specific nucleoside hydrolase in spore germination of *Bacillus thuringiensis*. *Microbiology* **154**:1333–1340.

81. **Dodatko T, Akoachere M, Muehlbauer SM, Helfrich F, Howerton A, Ross C, Wysocki V, Brojatsch J, Abel-Santos E.** 2009. *Bacillus cereus* spores release alanine that synergizes with inosine to promote germination. *PLoS One* **4**:e6398. doi:10.1371/journal.pone.0006398.

82. **Gould GW.** 1966. Stimulation of L-alanine-induced germination of *Bacillus cereus* spores by D-cycloserine and O-carbamyl-D-serine. *J Bacteriol* **92**:1261–1262.

83. **Titball RW, Manchee RJ.** 1987. Factors affecting the germination of spores of *Bacillus anthracis*. *J Appl Bacteriol* **62**:269–273.

84. **Omotade TO, Heffron JD, Klimko CP, Marchand CL, Miller LL, Halasahoris SA, Bozue JA, Welkos SL, Cote CK.** 2013. D-Cycloserine or similar physiochemical compounds may be uniquely suited for use in *Bacillus anthracis* spore decontamination strategies. *J Appl Microbiol* **115**:1343–1356.

85. **Oliva C, Turnbough CL Jr, Kearney JF.** 2009. CD14-Mac-1 interactions in *Bacillus anthracis* spore internalization by macrophages. *Proc Natl Acad Sci USA* **106**:13957–13962.

86. **Kang TJ, Fenton MJ, Weiner MA, Hibbs S, Basu S, Baillie L, Cross AS.** 2005. Murine macrophages kill the vegetative form of *Bacillus anthracis*. *Infect Immun* **73**:7495–7501.

87. **Raines KW, Kang TJ, Hibbs S, Cao GL, Weaver J, Tsai P, Baillie L, Cross AS, Rosen GM.** 2006. Importance of nitric oxide synthase in the control of infection by *Bacillus anthracis*. *Infect Immun* **74**:2268–2276.

88. **Soru E.** 1983. Chemical and immunological properties of *B. anthracis* arginase and its metabolic involvement. *Mol Cell Biochem* **50**:173–183.

89. **Weaver J, Kang TJ, Raines KW, Cao GL, Hibbs S, Tsai P, Baillie L, Rosen GM, Cross AS.** 2007. Protective role of *Bacillus anthracis* exosporium in macrophage-mediated killing by nitric oxide. *Infect Immun* **75**:3894–3901.

90. **Welkos S, Friedlander A, Weeks S, Little S, Mendelson I.** 2002. In-vitro characterisation of the phagocytosis and fate of anthrax spores in macrophages and the effects of anti-PA antibody. *J Med Microbiol* **51**:821–831.

91. **Basu S, Kang TJ, Chen WH, Fenton MJ, Baillie L, Hibbs S, Cross AS.** 2007. Role of *Bacillus anthracis*

spore structures in macrophage cytokine responses. *Infect Immun* 75:2351–2358.

92. Chung MC, Tonry JH, Narayanan A, Manes NP, Mackie RS, Gutting B, Mukherjee DV, Popova TG, Kashanchi F, Bailey CL, Popov SG. 2011. *Bacillus anthracis* interacts with plasmin(ogen) to evade C3b-dependent innate immunity. *PLoS One* 6:e18119. doi:10.1371/journal.pone.0018119.

93. Mukhopadhyay S, Akmal A, Stewart AC, Hsia RC, Read TD. 2009. Identification of *Bacillus anthracis* spore component antigens conserved across diverse *Bacillus cereus sensu lato* strains. *Mol Cell Proteomics* 8:1174–1191.

94. Russell BH, Liu Q, Jenkins SA, Tuvim MJ, Dickey BF, Xu Y. 2008. *In vivo* demonstration and quantification of intracellular *Bacillus anthracis* in lung epithelial cells. *Infect Immun* 76:3975–3983.

95. Russell BH, Vasan R, Keene DR, Xu Y. 2007. *Bacillus anthracis* internalization by human fibroblasts and epithelial cells. *Cell Microbiol* 9:1262–1274.

96. Friedlander AM, Pittman PR, Parker GW. 1999. Anthrax vaccine: evidence for safety and efficacy against inhalational anthrax. *JAMA* 282:2104–2106.

97. Young JA, Collier RJ. 2007. Anthrax toxin: receptor binding, internalization, pore formation, and translocation. *Annu Rev Biochem* 76:243–265.

98. Wycoff KL, Belle A, Deppe D, Schaefer L, Maclean JM, Haase S, Trilling AK, Liu S, Leppla SH, Geren IN, Pawlik J, Peterson JW. 2011. Recombinant anthrax toxin receptor-Fc fusion proteins produced in plants protect rabbits against inhalational anthrax. *Antimicrob Agents Chemother* 55:132–139.

99. Sharma S, Thomas D, Marlett J, Manchester M, Young JA. 2009. Efficient neutralization of antibody-resistant forms of anthrax toxin by a soluble receptor decoy inhibitor. *Antimicrob Agents Chemother* 53:1210–1212.

100. Brossier F, Levy M, Mock M. 2002. Anthrax spores make an essential contribution to vaccine efficacy. *Infect Immun* 70:661–664.

101. Gauthier YP, Tournier JN, Paucod JC, Corre JP, Mock M, Goossens PL, Vidal DR. 2009. Efficacy of a vaccine based on protective antigen and killed spores against experimental inhalational anthrax. *Infect Immun* 77:1197–1207.

102. Cote CK, Kaatz L, Reinhardt J, Bozue J, Tobery SA, Bassett AD, Sanz P, Darnell SC, Alem F, O'Brien AD, Welkos SL. 2012. Characterization of a multi-component anthrax vaccine designed to target the initial stages of infection as well as toxaemia. *J Med Microbiol* 61:1380–1392.

103. Vergis JM, Cote CK, Bozue J, Alem F, Ventura CL, Welkos SL, O'Brien AD. 2013. Immunization of mice with formalin-inactivated spores from avirulent *Bacillus cereus* strains provides significant protection from challenge with *Bacillus anthracis* Ames. *Clin Vaccine Immunol* 20:56–65.

104. Hahn UK, Boehm R, Beyer W. 2006. DNA vaccination against anthrax in mice-combination of anti-spore and anti-toxin components. *Vaccine* 24:4569–4571.

105. Cybulski RJ Jr, Sanz P, McDaniel D, Darnell S, Bull RL, O'Brien AD. 2008. Recombinant *Bacillus anthracis* spore proteins enhance protection of mice primed with suboptimal amounts of protective antigen. *Vaccine* 26:4927–4939.

106. Glomski IJ, Corre JP, Mock M, Goossens PL. 2007. Cutting Edge: IFN-gamma-producing CD4 T lymphocytes mediate spore-induced immunity to capsulated *Bacillus anthracis*. *J Immunol* 178:2646–2650.

107. Williams G, Linley E, Nicholas R, Baillie L. 2013. The role of the exosporium in the environmental distribution of anthrax. *J Appl Microbiol* 114:396–403.

108. Driks A, Eichenberger P. The spore coat. *In* Eichenberger P, Driks A (ed), *The Bacterial Spore*. ASM Press, Washington, DC, in press.

109. Popham DB, Bernhards CB. Spore peptidoglycan. *In* Eichenberger P, Driks A (ed), *The Bacterial Spore*. ASM Press, Washington, DC, in press.

The Bacterial Spore: From Molecules to Systems
Edited by P. Eichenberger and A. Driks
© 2016 American Society for Microbiology, Washington, DC
doi:10.1128/microbiolspec.TBS-0001-2012

Susan Welkos,[1] Joel A. Bozue,[1]
Nancy Twenhafel,[2] and Christopher K. Cote[1]

Animal Models for the Pathogenesis, Treatment, and Prevention of Infection by *Bacillus anthracis*

14

Anthrax is primarily a zoonotic disease caused by the Gram-positive spore-forming bacterium *Bacillus anthracis*, which occurs in domesticated and wild animals, primarily herbivores. *B. anthracis* is found throughout the world and exists in the soil as the dormant, highly resistant spore form. Animals become infected when they ingest spores while grazing on contaminated land or ingesting spore-laden feed, although mechanical spreading by flies or vultures from one animal environmental locus to another could be possible (1–3). Humans are incidental and relatively rare hosts that, in natural settings, become infected by contact with infected animals or contaminated animal products (3–7). These infections occur most often in agricultural settings and lesser-developed countries. Before the development of effective vaccines and disinfection practices, industrial exposures were also common in European and North American countries. The latter were associated with the processing of animal materials (hides, hair, wool, and bones), as illustrated by the highly lethal illness known as woolsorter's disease. This disease occurred after inhalation of spore-laden dust and aerosols in wool and textile mills in England and the industrialized regions of the northeastern United States (4, 6, 8).

Descriptions of the course and pathology of human anthrax are available elsewhere and are summarized briefly here (3–7, 9–16). Human disease is manifested in one of three major forms, cutaneous, oropharyngeal/gastrointestinal, and inhalational; secondary meningitis can occur in all three forms of anthrax. More than 95% of human cases reported worldwide are cutaneous infections. Although all three forms are potentially fatal if untreated, cutaneous anthrax is often self-limited, with mortality rates in untreated cases of about 20% (10 to 40%) (3, 4, 7, 10). The pathognomonic characteristic of cutaneous anthrax is a painless black lesion referred to as an eschar (3, 7, 10). The oropharyngeal/gastrointestinal form of anthrax is usually caused by consuming contaminated and inadequately cooked meat. Lesions can be observed in the oral cavity, stomach, or

[1]Bacteriology Division, United States Army Medical Research Institute of Infectious Diseases, Frederick, MD 21702; [2]Pathology Division, United States Army Medical Research Institute of Infectious Diseases, Frederick, MD 21702.

intestinal tract and can result in swelling and obstruction (e.g., of the airway), perforation, or hemorrhage; untreated cases have fatality rates of 25 to 60% (3, 7, 10, 17, 18). Finally, inhalation of B. anthracis spores results in the most severe form of anthrax, which is almost 100% fatal if left untreated. Because of its nonspecific symptoms, early diagnosis can be challenging; the disease can progress to a stage that is no longer amenable to antibiotic treatment because antimicrobials do not counter the effects of the bacterial-secreted anthrax toxins. A dramatic widening of the mediastinum as observed radiographically is the result of hemorrhagic lymphadenitis attributable to bacterial replication and toxin production in the mediastinal lymph nodes. The onset of severe respiratory distress and hypotension can lead to death within 24 h (4, 7, 19).

The incidence of human anthrax in developed countries is very low, and "natural" disease is not a major health concern today (3, 4, 7). Occasional outbreaks and smaller clusters of primarily cutaneous anthrax, usually linked to handling and consuming meat from infected animals, have been reported (1, 20). Also, in recent years, unusual and often more lethal cases of anthrax associated with new sources of exposure (i.e., use of "bongo" drums made from spore-contaminated hides) or routes of infection (septicemic anthrax associated with intravenous [IV] injection of spore-contaminated heroin) have been described (21–25). However, the major impetus for research on anthrax is the threat of B. anthracis being used as a biological weapon. This scenario was given credence by several episodes of accidental or intentional release of spores that have occurred in recent times (4, 16, 26–31).

For obvious ethical reasons, research and testing of anthrax countermeasures cannot be performed directly in humans, and the low incidence of human anthrax makes field studies impractical. Thus, animals have played a crucial role in anthrax research beginning with the earliest efforts to understand anthrax pathogenesis, detection, and cure. Koch is generally considered to be the first to definitely isolate B. anthracis in culture and establish in animals the microbial etiology of an infectious disease (32). In the late 1800s, Pasteur and Greenfield were the first to develop a successful bacterial vaccine when they induced effective protection against anthrax in sheep with an attenuated derivative of B. anthracis (33–35). Others studied the pathogenesis of inhalational anthrax and showed that, although lethal disease ensued after exposing various animal species to spore-laden aerosols, no primary anthrax pneumonia was observed and the site of spore germination was unknown (36, 37). Several animal species were

subsequently employed in studies which showed that B. anthracis spore dissemination involved the lymphatic system, with no evidence of lung injury (37–39). Using guinea pigs exposed either intratracheally or by inhalation to virulent B. anthracis, Ross observed that the spores were rapidly taken up by lung phagocytes migrating toward the lymphatics. They were then detectable in phagocytes in the lymph glands from which they invaded the bloodstream after germination into the vegetative bacilli (39). The lungs did not exhibit major histopathological changes. These data overall supported the conclusion that the site of germination of inhaled spores is in the regional lymph nodes, where those spores that survive transit from the lungs germinate and spread systemically. Studies on nonhuman primates (NHPs) infected with spore aerosols (40–42), as well as gross and histopathological analyses at autopsy of human cases (19, 26, 43–46) supported this scenario.

Early research studies were being done in parallel to elucidate the mechanisms of B. anthracis virulence and pathogenesis; and animal models were again pivotal. Bail showed that organisms that had lost the ability to produce a capsule were nearly avirulent in animals (47), and Sterne demonstrated that such capsule-deficient mutants induced protective immunity, a finding that led to the use of toxigenic nonencapsulated B. anthracis spores as effective live vaccines in animals and some human populations (48, 49). A role for toxins in anthrax pathogenesis was shown in studies in which guinea pigs were injected with sterile fluids from infected animals (50, 51); others showed that filtrates of B. anthracis cultures could produce edema after being injected into the skin of rabbits (52). Interestingly, the susceptibility of numerous animal species to anthrax was shown to vary inversely with sensitivity to the injection of anthrax toxin (53). This inverse relationship holds true when comparing inbred strains of mice (54, 55). Thus, this early body of work, in which animal models had a predominant role, was pivotal in our current understanding of B. anthracis pathogenesis and subsequent disease.

ANIMAL MODELS AND HUMAN ANTHRAX

It is well known that different animal species vary significantly in their susceptibility to infection with B. anthracis and to intoxication, as highlighted many years ago by Lincoln et al. (53) and reviewed since then (3, 49, 56). For instance, parenteral LD_{50} ranging from ≤ 10 in some mice and guinea pigs to $>10^{10}$ in dogs have been reported, and inhalational values have been reported to span $<10^4$ CFUs in NHPs to $>10^7$ in dogs and swine. The relative susceptibility of humans has been modeled and estimated, as described previously (16, 45, 56–63). These estimates were based on extrapolations from animal data and analyses of rare large-scale human exposures, such as the Sverdlovsk release (30, 43). The values for pulmonary exposure have usually ranged from about 2,500 to 10,000 spores; however estimates more than fivefold higher have also been cited (3, 16, 57), while the results of some studies with NHPs have implied that much smaller doses are infectious (16, 56, 64, 65). Nevertheless, it is generally perceived that there is a threshold for resistance to inhalational anthrax (56), and that, by using values from animal data sets amenable to dose-response modeling, a pooled interspecies $LD_{50}(s)$ applicable to humans is possible (57). While the data derived from modeling are very useful, given the many biological and environmental variables inherent in these efforts (involving the host, infecting strain, conditions of exposure [Table 1]), it may be possible to only derive estimates of the "true" infectious or lethal human doses. Also, the basis of the differences in species susceptibility to infection and toxins is not known and is hampered by the lack of inbred strains and immunological reagents needed to pursue such studies in most animals except mice (55, 66–71). Thus, this review will not focus on susceptibility differences as being the primary basis of selecting a particular species of animal as an anthrax model. Rather, attributes such as disease course and pathogenesis and nature of the host responses will be considered.

The selection of a model for studies on *B. anthracis* is complex. It is first dependent on the specific aim of the research or test (72) and should then take into consideration the differing attributes of the animal species and of the manner and route of exposure. NHP and rabbit models have been developed for anthrax and are generally considered to be the best animal models for definitive evaluation of anthrax countermeasures being developed for human use, despite well-described differences in the disease exhibited in these models and anthrax in humans (5, 73). However, especially for NHPs, factors including costs, supply limitations, the housing and maintenance requirements, and animal activism concerns usually restrict their use to final evaluation of products in advanced development toward licensure for human use. Thus, small animals, primarily mice, guinea pigs, and rats, have been used extensively to investigate the pathogenesis of anthrax and the ability of novel countermeasures to prevent or modulate its course. We have attempted to review the major animal models utilized for *B. anthracis* studies and highlight their contributions to understanding anthrax. However, it is important to also be cognizant of known limitations and differences between humans and animals (anatomical, physiological, and immunological) that

Table 1 Methods for delivery of *Bacillus anthracis* by the pulmonary route[a]

Method	Advantages	Disadvantages
Exposure to aerosol	1. Mimics natural transmission of airborne infections 2. Simultaneous infection of groups of animals 3. Uniform, predictable deposition from animal to animal 4. Simulation of response to large (upper airway) and small (lower lung) particles possible 5. Symmetric bilateral infection permits efficient use of lung tissue 6. No anesthesia required	1. Costly aerosol exposure system required 2. Deposition on eyes, pelt in whole-animal exposures 3. Animals restrained for nose-only exposures 4. Some pathogens survive poorly in aerosols 5. Upper limit on achievable inoculum
Transtracheal injection	1. Mimics oropharyngeal aspiration 2. Precise dosing to lower respiratory tract 3. Can be directed unilaterally 4. Simple procedure	1. General anesthesia required 2. Surgical cut down required 3. Nonuniform, asymmetric deposition 4. Staggered start time to experiment
Peroral intubation	1. Mimics oropharyngeal aspiration 2. Precise dosing to lower respiratory tract 3. Can be directed unilaterally 4. No surgical wound	1. General anesthesia required 2. Technically difficult 3. Nonuniform, asymmetric deposition 4. Staggered start time to experiment 5. Risk of contamination with oral flora
Intranasal inoculation	1. Mimics oropharyngeal aspiration 2. Models infection of upper and lower respiratory tract 3. Simple procedure	1. General anesthesia required 2. Highly variable deposition in lungs 3. Nonuniform, asymmetric deposition 4. Risk of contamination with oral flora

[a]This table is a modification of the one in reference 95, copyright © 2008 by the American Physiological Society, and is used with permission from the publisher.

might qualify the use of a particular species. Nevertheless, the information obtained from these models has enriched the breadth and depth of knowledge on anthrax, as we hope to illustrate.

ANIMAL MODELS BY SPECIES

Models for all three forms of anthrax (inhalational, gastrointestinal, and cutaneous) have been described. For purposes of modeling human disease, the major focus of this review will be on those for inhalational anthrax. However, models that utilize nonpulmonary routes of infection for defined purposes (e.g., to screen for virulence-associated mutations or investigate the effects of the toxins on host responses) will be discussed also. Table 1 details the various means by which an inoculum can be delivered by the pulmonary route and the attributes/risks of each method. The latter will be alluded to in various places in the review.

Mice

Mice are used extensively in research on anthrax for reasons that include their size, housing requirements, and relatively modest purchase/maintenance costs, statistical considerations, and the availability of extensive murine genetic databases and mutant strains. Inbred mouse strains were shown to be susceptible to parenteral and pulmonary challenge with spores of fully virulent strains of *B. anthracis* and to develop acute disease characterized by extensive edema and large concentrations of bacilli in the blood and organs (e.g., spleen, liver, lungs) (55, 74–78). However, mice are probably most often used because of the availability of certain strains that are unusually susceptible to lethal infection with attenuated strains of *B. anthracis* that are toxin-producing (tox+) but nonencapsulated (cap–), as discussed below. Such strains are safe to handle and are used as veterinary vaccines (Sterne) and, in some countries, as human vaccines (48). These tox+ cap– strains cause a disease which resembles that due to virulent tox + cap+ *B. anthracis* strains, yet do not require biosafety level 3 containment and the huge cost and extensive regulatory requirements involved in working with virulent wild-type strains (79–85).

In many respects, mice are a reasonable reflection of human biology and responses. Nevertheless, all results using mice are tempered by certain differences in responses to infection and vaccination between mice and humans and by their phylogenetic differences. First, both fully virulent tox+ cap+ *B. anthracis* strains and some strains that are encapsulated but nontoxigenic are lethal for mice and exhibit similar LD_{50} values and

duration of morbidity (86–88). This susceptibility to tox– cap+ *B. anthracis* strains appears to be unique among animal models of anthrax, with some exceptions as will be detailed in a later section. Guinea pigs, rabbits, and NHPs are generally resistant to infection by such *B. anthracis* strains (89, 90). Second, aluminum-adjuvanted protective antigen (PA) vaccines such as anthrax vaccine adsorbed (AVA) are not protective for mice against challenge with the virulent Ames strain, unlike the full protection observed in rabbits and NHPs and partial protection in guinea pigs (90, 91). Additionally, despite the genetic conservation between mice and humans (92), significant differences exist in not only anatomy/physiology, but also in immune system development and activation, and in the innate and adaptive response to challenge. For example, differences in pulmonary anatomy and physiology result in a proportionately larger nasal surface area and fewer airway branches in mice than in humans, and the absence of bronchioles in mice; the latter are obligate nose breathers and, in comparison with humans, mice exhibit a weak or absent cough reflex (93–95). Nevertheless, mice exhibit significant bronchus-associated lymphoid tissue, whereas the latter is mostly absent in healthy people (93, 94). Immunologically, there are many differences in the functions and effectors of the murine and human immune systems, as discussed extensively elsewhere (72, 93–95). Such differences might be expected since the two species diverged millions of years ago, differ greatly in both size and lifespan, and have evolved in different ecological niches where widely different pathogenic challenges need to be met.

The Role of Mice in Advanced Development of Candidate Human Products

As discussed above, evaluation of vaccines and therapeutics for anthrax, especially inhalational, cannot be performed in humans, and thus licensure of any new anthrax vaccine or therapeutic will rely on efficacy data obtained in animal studies and on human safety/immunogenicity data, as detailed in the U.S. Food and Drug Administration (FDA) animal rule (21 CFR 601.91 Subpart H, "Approval of Biological Products When Human Efficacy Studies Are Not Ethical or Feasible"). Mice are not considered primary candidates for use in the advanced development of such new products, NHPs and rabbits being the more optimal human comparison models, as described in Table 2. However, mice have adjunctive roles in these efforts, as demonstrated with the development of a mouse inhalational model to establish the efficacy of new antibiotics and provide data supporting the data derived from a NHP model

Table 2 Animal species: comparative use as models

Animal	Model use	Advantages and use	Disadvantages and restrictions
NHP	General	• Phylogenetic closeness to humans: considered essential for extrapolations and predictive outcome to humans	• Cost, housing and animal care requirements, statistics (animal numbers) • Potential health risk to humans; especially herpes B virus
	Disease pathogenesis and host response	• Confirm and refine knowledge acquired in lower animals • Telemetry and other procedures may be performed to collect clinical data and establish basic model parameters and similarity to human anthrax	• Model usually restricted to final-stage confirmation
	Identification of virulence factors and potential mechanisms/genetic bases of host susceptibility	• Confirm and refine only	• Lack of widely available genetically defined strains to confirm genetic bases of susceptibility
	Baseline screening of countermeasure		• Not appropriate
	Evaluation of selected vaccine/therapeutic candidates for advanced development	• Generally accepted best human model; similar pharmacokinetics/ pharmacodynamics to humans with exceptions (64, 343, 393) • Estimate human administrations and dosages	
Rabbit	General	• Usage more feasible than NHPs • Larger animals can accommodate larger sample/tissue collections	• Cost, housing and animal care requirements, statistics (animal numbers) – While not as restrictive as for NHPs, these issues can prove limiting for some laboratories
	Disease pathogenesis and host response	• Confirm and refine knowledge acquired in lower animals • Can perform telemetry for collecting clinical data • Numerous similarities documented between this model and human anthrax (i.e., cardiac events)	• Presence of certain specific differences between rabbit and human anthrax (i.e., significant meningeal anthrax) • Disease more acute in rabbits than humans; attributed to greater rabbit sensitivity
	Identification of virulence factors and potential mechanisms/genetic bases of host susceptibility	• May be appropriate, but will likely serve to confirm and refine	• Rapidity of infection might hinder identification of novel virulence factors • Lack of widely available genetically defined strains for susceptibility studies
	Baseline screening of countermeasures	• Confirm/refine findings in lower species	• Usually not appropriate
	Evaluation of selected vaccine/therapeutic candidates for advanced development	• Generally accepted good vaccine model and therapeutic testing model (second only to NHP); see disadvantage • Well-established *in vitro* correlates of immunity	• Prescreening for toxicity required due to inherent sensitivities to some antimicrobials (e.g., certain antibiotics)
Guinea pig	General	• Purchase and maintenance costs • Smaller size (compared with rabbits) and ease of maintenance allowing use of statistically adequate numbers	• May have unique characteristics affecting pathogenesis and not observed in other species (259, 279, 280, 285)

(Continued)

Table 1   *(Continued)*

Animal	Model use	Advantages and use	Disadvantages and restrictions
		• Larger size (compared with mice) allows for increased sample/tissue collection as well as limited telemetry for collection of clinical data	
	Disease pathogenesis and host response	• Appropriate	• See above • Overt clinical signs in guinea pigs can be difficult to discern
	Identification of virulence factors and potential mechanisms/genetic bases of host susceptibility	• Appropriate	• See above • Absence of widely available genetically defined strains and genetic methods
	Baseline screening of countermeasures	• Generally appropriate	• See above • Toxicity prescreening of therapeutic candidates required due to inherent sensitivities (as per rabbit)
	Evaluation of selected vaccine/therapeutic candidates for advanced development	• Useful model in characterizing next-generation vaccine strategies such as PA + nontoxin (e.g., capsule or spore antigens) and potentially novel therapeutics	• Relatively more difficult to protect with PA vaccines • Toxicity prescreening of therapeutics required, as per rabbits
Mouse	General	• Purchase and maintenance costs • Smaller size (compared with guinea pigs) and ease of maintenance allowing use of statistically adequate numbers • Extensively characterized genetic polymorphisms • Extensively characterized inbred strains resulting in differences in resistance to *B. anthracis* (e.g., the attenuated Sterne strain is lethal to AJ mice)	• Extreme sensitivity (with some variation associated with mouse strains) to fully virulent *B. anthracis* strains • Nearly unique susceptibility to cap +/tox− strains of *B. anthracis*
	Disease pathogenesis and host response	• Models to study the roles of toxins (C5-deficient mice and Sterne strain of *B. anthracis*) • Models to study nontoxin antigens (normal mice, cap+/tox− strains of *B. anthracis*)	• Physiologic, anatomic, and immunological differences from humans (152, 235, 240)
	Identification of virulence factors and potential mechanisms/genetic bases of host susceptibility	• Appropriate	
	Baseline screening of countermeasures	• Appropriate	
	Evaluation of selected vaccine/therapeutic candidates for advanced development	• Supporting role; use in potency assay development	• Ancillary role only in fulfillment of FDA Animal Rule (51, 89, 154, 324)
Rat	General	• Purchase and maintenance costs (similar to that of mice) • *In vivo* detection/quantification of lethal toxin (LT)	• Usually not appropriate except for genetically defined strains in exploring basis of LT (and possibly infection) resistance (253, 256, 287)
	Disease pathogenesis and host response		• Not appropriate, with possible exceptions (see above)
		• Toxin-related only	• See above

(Continued)

Table 2 Animal species: comparative use as models *(Continued)*

Animal	Model use	Advantages and use	Disadvantages and restrictions
	Identification of virulence factors and potential mechanisms/genetic bases of host susceptibility		
	Baseline screening of countermeasures	• Sensitive test of antitoxin therapeutics and vaccines if rats are challenged with anthrax toxin	• Not appropriate for spore challenges (extremely resistant to spore challenge)
	Evaluation of selected vaccine/therapeutic candidates for advanced development	• Potential use in potency assay of toxin-based challenge materials and countermeasures (superseded by *in vitro* toxin neutralization assays)	• Not useful for direct fulfillment of FDA Animal Rule

(77, 96). In addition, some of the new improved potency assays being developed for PA-based anthrax vaccines are based on the immunogenicity of the vaccine for mice. Sera from vaccinated mouse strains such as A/J are used *in vitro* to measure their quantitative antitoxin antibody levels or their toxin-neutralizing activity (97, 98). Such assays are being developed as a replacement for the current assay that is based on the lethal challenge of guinea pigs with virulent *B. anthracis* spores. A passive transfer model employing the A/J mouse strain and the attenuated Sterne animal vaccine strain of *B. anthracis* (discussed below) has also been used in studies to develop an immune correlate of human protection (99, 100). These new potency and mouse passive protection strategies are also advantageous since they can be performed in a biosafety level 2 laboratory instead of higher containment conditions.

Five basic murine models for anthrax have been described, four infection models involving mice differing in immune competency or challenge strains differing in virulence, and one toxin challenge model. Attributes and examples of uses of these models will be discussed. Bacterial strain virulence differences are based on the ability of the *B. anthracis* strain to express the two factors that are essential for maximal virulence of *B. anthracis* in most animals, anthrax toxins and an antiphagocytic capsule. There are two toxins, each of which contains the cell-binding PA component associated with either edema factor (EF), a calcium and calmodulin-dependent adenylate cyclase, to form edema toxin (ET) or with lethal factor (LF), a zinc-dependent metalloprotease, to form lethal toxin (LT).

Immunodeficient (C5⁻) Mouse/Attenuated Challenge

Some mouse strains, e.g., A/J and DBA/2J, succumb to relatively low doses of tox+ cap– strains of *B. anthracis*

such as the Sterne strain (intraperitoneal [IP] or intranasal [IN] LD_{50} doses of 10^3 to 10^4 CFU), while other strains, e.g., BALB/c and C57BL6, are susceptible only to high doses (10^6 to 10^8 CFU) (55, 101). Deficient expression of the *Hc* locus, encoding the C5 component of complement, was genetically associated with susceptibility to Sterne, and functionally associated with deficiencies in host inflammatory responses to infection (102, 103). In addition to antibacterial activities, C5 is the major serum chemoattractant and susceptibility to Sterne was associated with the extent and rate of mobilization to the site of infection of neutrophil and macrophage phagocytes (55, 103, 104). The pathogenesis of, and host response to, infection with Sterne in the susceptible strains were similar to those produced by fully virulent tox+ cap+ strains (81, 84, 103).

Disease and host responses

The model employing a C5-deficient strain of mouse and tox+ cap– *B. anthracis* (C5⁻/tox+ cap–) has been useful primarily for studies on the role of toxins in pathogenesis and on the efficacy of toxin-based vaccine and therapeutics. For instance, several groups examined the roles of the two major anthrax toxins, LT and ET, in the dissemination and course of disease and the induction of protective adaptive immunity in pulmonary challenge models (79, 82, 85, 105–107). After exposing A/J mice to aerosolized Sterne spores, Loving et al. documented histological and bacteriological findings (lung, liver, spleen) that were similar to those observed in NHP and rabbit anthrax models (e.g., splenic lymphocytic depletion with necrosis and occasional hemorrhage, and presence of dormant spores in the lungs at early time points). LT was required for outgrowth and dissemination of infection by their cap– strains and the development of protective immunity (85). The protective immunity induced by infection with the tox+ cap– bacteria was characterized as were the cytokine responses (85, 101, 105, 106, 108, 109). However, the aerosol model utilized required exposure conditions that would likely not be encountered outside the laboratory, i.e., very high inocula delivered by aerosol continuously for a prolonged period of time (84, 110).

Using luminescent derivatives of tox+ cap– and tox– cap+ strains, Glomski and colleagues found that spores could germinate and disseminate from several different infection sites (pulmonary, skin, gastrointestinal) without requiring transport to lymph nodes. In contrast, in other animal models, spores reaching the lungs after aerosol challenge did not germinate until they were engulfed by phagocytes and transported to regional lymph nodes (39, 41, 43, 75, 84, 106, 111–116). In addition, the absence of capsule in tox+ cap– strains greatly delayed and modified the dissemination pattern. After aerosol exposure of A/J mice, tox+ cap– B. anthracis were largely confined to and germinated in the nasal-associated lymphoid tissue (NALT) (79). Systemic spread occurred late in infection and there was minimal colonization of the spleen. In contrast, tox– cap+ strains disseminated from the initial sites of infection, e.g., the NALT, to the draining lymph nodes, spleen, and then lungs and blood (79, 82, 83). Interestingly, studies with these models showed that infection patterns in the lungs after pulmonary exposure were dependent on dose size and mode of delivery. Aerosolized or moderate IN doses of spores which reached the lungs did not germinate until they were transported by phagocytes to lymphoid tissue, as expected. Higher doses led to initiation/germination in the lungs (mainly in association with macrophages) and/or NALT (79, 107). These findings overall implied that spore interactions with local entry sites are critical in the development of systemic disease and might provide new targets and approaches toward anthrax prevention, concepts recently expanded upon by Weiner and Glomski (383). They also highlight the importance of experimental design differences on the development of anthrax infections and emphasize that findings using mouse models such as the C5–/tox+ cap– model should be confirmed in models employing fully virulent B. anthracis strains and higher order animal species.

Pathogenetic mechanisms and virulence factors

The C5–/tox– cap– model has also been used to identify and investigate the mechanisms of novel virulence factors. For instance, in studies employing DBA/2 mice intratracheally infected with Sterne, Carr et al. (117) analyzed the role of the five germinant receptors in

germination and virulence by using mutants and found that a mutant lacking all five receptors was attenuated, whereas single germinant receptor-expressing mutants were not. However, the route of inoculation modified and influenced the germination requirements. The role of the bacterial inhibitor of germination, D-alanine, in facilitating infection was also studied in the C5–/tox+ cap– intratracheal (IT) challenge model. McKevitt and coworkers (118) found that the delivery of D-alanine with Sterne spores IT resulted in a reduced LD$_{50}$ dose compared with untreated spores, suggesting that the rate of germination may be impacted by this amino acid early during infection (118). However, it was demonstrated in an inhalational challenge model (using the fully virulent Ames strain) that germination-altering strategies are likely not beneficial to the infected host or useful as potential therapeutic options (119), as described below. Factors including mouse/bacterial strain and mode/route of infection should be considered in analyzing murine data (118–120).

A number of extracellular proteins and enzymes produced by B. anthracis have been characterized recently and their roles in virulence have been studied in mice. These factors include, for example, phospholipase C enzymes, secreted metalloproteases, anthrolysin O, nitric oxide synthase, and siderophores, sortases, and other proteins involved in iron acquisition (121–131). Nevertheless, the precise mechanisms and roles of many of these and other similar candidate proteins in the virulence of B. anthracis have not been definitively evaluated in higher-order animal models.

The importance of structural proteins of the spore, especially those in the exosporium, have also been studied in this and other murine models, as reported previously and described in detail by Bozue and coauthors (112, 132–135, 392).

Vaccines and therapeutics

Mice are often used as a facile and inexpensive first-stage in vivo test of a candidate product after its in vitro effectiveness is demonstrated. Such tests provide a proof of concept to support advancement to subsequent more costly higher-containment trials with fully virulent B. anthracis strains in rodents, or, depending on the maturity and intended use of the product, in rabbits and NHPs.

Vaccines. The C5–/tox+ cap– strain model has proven to be a reliable and robust model for the characterization of the immunogenicity and efficacy of PA-based vaccines, such as the current U.S. anthrax vaccine, AVA, licensed as BioThrax (Emergent

BioSolutions). PA vaccines are capable of fully protecting C5-deficient mice against challenge with tox+ cap– strains. Immune responses to PA and protective efficacy of defined protein vaccines (presented with different adjuvants and on different platforms such as a microsphere) and whole recombinant bacterial or virus-based vaccines have been demonstrated. Examples include: (i) protein vaccines consisting of PA alone (101, 102, 136–138); (ii) PA supplemented with proteins such as spore antigens (139, 140); and (iii) recombinant PA-expressing live (bacterial or virus) or replicon vaccines (102, 138, 141–148).

This model has also contributed to discerning the roles of EF and LT in acquired immunity and their contribution to PA-stimulated immunity. For example, Zeng et al. (148) evaluated a nontoxic N-terminal fragment of EF as a candidate antigen using an adenoviral vector in the Sterne lethal challenge model. The vaccine elicited anti-EF antibody, yielded significant protection, and provided evidence in support of including EF in a next-generation multicomponent subunit vaccine. The contribution of antigens other than toxin components (i.e., spore-specific antigens) to vaccine-induced immunity is also beginning to be explored using the C5⁻/tox+ cap– strain model, as shown in studies by O'Brien and coworkers (139, 140, 149).

Treatments (pre- or postexposure). As demonstrated by studies involving antitoxin monoclonal antibodies and toxin receptor decoys, mouse models have been employed in evaluating potential therapeutics. With the use of an A/J-Sterne challenge model, it has been demonstrated that one dose of a human anti-PA monoclonal antibody or an anti-LF monoclonal antibody could provide preexposure protection against parenteral spore challenge (99); similar protective efficacy in mice as been reported for other anti-PA monoclonal antibodies as well (150–152). These and other anti-PA antibodies appear to have direct antispore as well as anthrax toxin-neutralizing activities; and the toxin- and spore-directed activities of these PA-specific antibodies might contribute to the protection of a host animal against a lethal spore infection (150, 151, 153–155).

Another new category of therapeutic based on the anthrax toxin host cell receptor has also been evaluated in this mouse model. These inhibitors are generally a fusion of the cell receptor (ANTXR2) PA-binding domain (VWA) with the Fc portions of human IgG molecules. Receptor decoy-IgG2 proteins protected mice against killing after IT administration of Sterne spores

to DBA/2 mice (156). However further studies showed that the toxin-inhibitory decoy complex can disassociate and result in delayed toxicity and host fatality (M. Manchester and J. A. Young, unpublished data).

Immunoproficient Mouse/Attenuated tox+ cap– Challenge

Models involving the infection of C5-proficient mice with Sterne-like strains have also been utilized, despite the relative resistance of such animals to nonencapsulated *B. anthracis*. However the reduced sensitivity of such mice to Sterne (in comparison with the extreme sensitivity to the Ames strain, for example) has permitted the protective effects of countermeasures and the identification and potential roles in infection of virulence factors to be more readily detected. These models have also facilitated an extensive accumulation of information on host responses to infection.

Disease and host responses

The pathological events that occur during pulmonary or parenteral infection with Sterne have been reported to be similar to those produced during infection with wild-type (wt) virulent strains (80, 157–163). However, some evidence suggests that cap–, and perhaps even fully virulent *B. anthracis*, are not disseminated in the same manner in mice as in humans or rabbits. On autopsy, infected humans show evidence of vegetative bacilli throughout the body and in the blood (19, 164). In contrast, Kalns et al. (165) observed that in C57Bl/5 mice infected by subcutaneous (SC) injection with Sterne, the bacteria were only observed at injection sites, even in moribund mice, suggesting that death of animals was to due to the systemic effects of the toxin rather than bacteremia. Alterations in dissemination of tox+ cap– *B. anthracis* compared with cap+ *B. anthracis* after exposure of BALB/c (and A/J) mice were also documented by using an *in vivo* imaging system, as described in "Immunodeficient (C5⁻) mouse/attenuated challenge" above (79, 82). Further studies with attenuated and fully virulent *B. anthracis* are needed to clarify these potential differences in pathogenesis. Thus, this model offers a facile system for an initial evaluation of the effects on infection of novel countermeasures or of genetic alterations in the bacterium or host; however, significant results require confirmation in models more closely related phylogenetically to humans.

A model for cutaneous anthrax was developed in mice with the use of inoculations of Sterne spores onto intact or abraded skin (166). In contrast to most pulmonary infections, cutaneous infections in this model appear to lead to germination and vegetative growth of

B. anthracis in the absence of cellular uptake. The major host-defensive response involves a neutrophilic inflammatory cell exudation into the epicutaneous fluids of infected skin, resulting in bacterial destruction and prevention of further invasion, the success of which may depend on the extent of the cellular response and other host factors (167).

Pathogenetic mechanisms and virulence factors

The importance of proteins on the outer surface of the spore in virulence and host interactions have been studied in this murine model, as described in "Immunodeficient (C5⁻) mouse/attenuated challenge" above and by Bozue et al. (392). In addition, the roles of the anthrax toxins in virulence have been explored in these mice. Using mutants of Sterne with specific deletions in the toxin component genes, Mock and coworkers definitively established the role of the individual LF and EF components in virulence and immune protection in immunocompetent inbred mice (18, 157, 159). The specific actions of the toxins *in vivo* are currently being dissected. For example, using a mouse/Sterne model and an *in vitro* blood-brain barrier (BBB) model, van Sorge et al. (168) showed that *B. anthracis* invades human brain endothelial cells. Mice challenged with Sterne but not a tox– variant developed histologically confirmed meningitis (168). These findings agree with those reported for humans (7, 13–15, 19, 27), supporting the utility of this mouse model of anthrax meningitis. In later studies, the differential roles of ET and LT in BBB penetration and meningitis were analyzed in mice (169). Others have also demonstrated roles for nontoxin enzymes in meningeal pathogenesis, as described in "Immunodeficient (C5⁻) mouse/attenuated challenge" above. The examination of these results in a mouse model of infection with a tox+ cap+ strain is necessary since data documenting meningeal sequelae in mice infected with virulent strains are lacking or describe rare occurrences of intravascular brain damage that might be attributable to the generalized bacteremia (78).

An involvement of the host macrophage (MΦ) in the pathogenesis of anthrax has long been assumed and is being further clarified in mouse models. It was initially demonstrated *in vitro* that LT induces a rapid lytic death of murine MΦs from certain inbred mouse strains (170, 171). Furthermore, studies with several animal models of inhalational anthrax documented that alveolar MΦs from bronchoalveolar lavage fluids (BAL) phagocytose spores. However, the nature of the spore-phagocyte interactions and the identity of the lung phagocytic cells that subsequently migrated to the thoracic lymph nodes had not yet been precisely defined. Guidi-Rontani et al. observed that spores germinated within alveolar MΦs but not in the extracellular spaces in lungs; the toxin genes are expressed very early during spore germination and are required for survival within phagocytic cells (115, 116, 172, 173). Using a C57Bl/6 mouse-Sterne challenge model, Cleret and colleagues confirmed that spores are initially phagocytosed by alveolar MΦs, but lung dendritic cells subsequently engulf and transport the spores to the regional thoracic lymph nodes (80, 174–176). Numerous subsequent studies employing this and related murine models have characterized many aspects of spore-phagocytic host cell interactions, as discussed below.

Genetic mechanisms of host resistance to infection

Early workers highlighted the inverse nature of the susceptibility of animals to spores of *B. anthracis* and to the toxins, as described in the introduction. The initial observations, reviewed by Lincoln et al. (53), were later extended to inbred strains of mice in an attempt to reproduce the range of differences in animal species by using inbred strains. Similar to the earlier observations, a dichotomy was observed in mice. Strains observed to be resistant to infection by *B. anthracis* were more susceptible to challenge by its toxin and the inverse was true for infection-susceptible species (54, 55, 102, 103). As described in the section "Immunodeficient (C5⁻) mouse/attenuated challenge" above, Welkos et al. implicated the *Hc* locus, encoding C5 production in the susceptibility of C5⁻ mice to Sterne and in associated deficiencies in phagocytic cell mobilization. Harvill and coworkers supported a role for the *Hc* locus in resistance to *B. anthracis* by demonstrating the increased susceptibility of C5⁺ mice treated with complement-depleting cobra venom factor to Sterne (110).

Current studies exploiting genetically defined and mutant strains of inbred mice are focused on the role of host systems which recognize microbial components and activate innate immunity, such as the Toll-like receptors (TLRs) and the NOD-like receptor (NLR) inflammasome system, in anthrax pathogenesis. *B. anthracis* infection activates the TLR and NLR signaling systems in MΦs, resulting in the production of tumor necrosis factor alpha through a mitogen-activated protein kinase (MAPK) signaling pathway (177). The presence of LT blocks this response by cleaving MAPK kinase proteins and inducing apoptosis (177–182). The production of cAMP by the anthrax ET (183) also affects cell signaling pathways. The disrupted pathways play important roles in numerous cellular functions. However, the NLR signaling system can counteract the immunosuppressive effect of the LT. The protein Nlrp1b is the product of the major LT sensitivity locus on chromosome 11 (184). The LT-sensitive allele of *nlrp1b* confers sensitivity to rapid lysis of MΦs and activates caspase-1, which in turn induces production of the proinflammatory cytokines interleukin 1β (IL-1β) and interleukin 18 (IL-18) (185–187). These cytokines recruit neutrophil and mononuclear phagocytic cells (74, 75, 108). As shown in recent studies, mice expressing the sensitive Nlrp1b^S locus and having LT-sensitive MΦs are resistant to infection, unlike mice harboring LT-resistant MΦs. Thus, it has been proposed that the control of *B. anthracis* infection is at least partially due to the LT-mediated and Nlrp1b-dependent activation of caspase-1 activation and release of IL-1β, findings that could partially explain the sensitivity results of previous reports (165, 188).

The data thus substantiate an inverse relationship between murine MΦ sensitivity to LT and mouse susceptibility to spore infection. Clearly, the response in mice to the intact bacterium is different from the response to the purified toxin. Furthermore, it was shown that MΦ sensitivity to toxin was not the major mechanism involved in the terminal events of anthrax attributed to the anthrax toxins; sensitivity to toxin injection appears to be mediated by multiple genetic loci and cells (66, 189, 190). Using mutant C57Bl/6 mice that lacked myeloid cells (e.g., polymorphonuclear leukocyte [PMN], MΦs) that express the toxin receptor CMG2, Liu et al. showed that these mutant mice retained full sensitivity to injected LT and ET. Nevertheless, the targeting of CMG2-expressing MΦs and neutrophils by toxin has a role in resistance to infection by spores of toxigenic *B. anthracis* (66, 68, 69, 189, 191–193); mice lacking CMG2+ myeloid cells are resistant to spore challenge (66). A mouse model for anthrax toxin is evolving in which two distinct pathogenic roles are hypothesized, an immunosuppressive role early during infection and a lethal role late in disease culminating in host death.

Thus, differential sensitivity of mouse strains to parenteral infection with tox+ cap− *B. anthracis* appears to be influenced by allelic variations in at least two genes: *Nlrp1b* and *Hc* (55, 68, 69, 104, 110, 138, 190). Allelic variation of these genes does not, however, fully account for differential sensitivity to infection or to intoxication by LT and current studies are discerning the role of additional genes and signaling pathways (190).

Vaccines and therapeutics

Models employing the infection of C5-proficient mice with Sterne-like *B. anthracis* have allowed further characterization of toxin-based vaccine immunogenicity and efficacy. In previous studies, derivatives of Sterne strain 7702 were constructed for use as live vaccines; they harbored defined mutations or deletions of one or more of the three toxin components (157). The different extents of protection from a lethal Sterne challenge observed in mice vaccinated with a recombinant vaccine strain not only confirmed the role of PA as the major protective antigen in the humoral response (90) but also delineated significant contributions of immune responses to LF and EF to protection (161, 162).

The efficacy of postexposure antibiotic therapy has been demonstrated in C5+/Sterne mouse models. For example, postexposure treatment of C57Bl/6 mice with doxycycline was protective if given at the time of, or 4 h after, challenge with Sterne but it did not prevent death when given ≥24 h postchallenge (158). Similarly, antibodies or antidotes to cytokines may prevent death against an LT challenge (188) and, in some cases, against Sterne infection (165), although the involvement of cytokine overproduction in facilitation of infection is clearly a matter of ongoing debate. The high challenge doses required in C5+/Sterne model suggest that preliminary efficacy data for therapeutics and vaccines should be confirmed in other models that can model challenges with lower lethal doses or more virulent strains.

Immunoproficient Mouse/Virulent tox+ cap+ Challenge

The usefulness of mice in studies with virulent strains, particularly those involving countermeasure testing, is tempered by a few issues, as discussed at the beginning of the mouse section. One is the extreme sensitivity of different inbred mouse strains to "fully virulent" tox+, cap+ *B. anthracis* strains, with LD_{50} by parenteral routes of <1 CFU to approximately 150 (3, 55, 102, 194) or 500 CFU (109). However, some naturally isolated and/or laboratory-manipulated tox+, cap+ strains have been described that exhibit differences in virulence and reduced sensitivity of the murine response to *B. anthracis* (195, 196). Also, mice display a uniquely elevated susceptibility to challenge with some tox− cap+ strains of *B. anthracis*, as discussed below. Given the central role of the toxins in anthrax pathogenesis and of PA in protection against anthrax, the tox− cap+ sensitivity of mice leads to the question of the relevance of mice as models for studies involving fully virulent tox+ cap+ strains. Certainly this capsule-associated sensitive phenotype must be taken into account when evaluating murine anthrax data. However, because of its unique responses, the mouse may prove to be a useful tool for screening the relative virulence of different isolates of pX01+/pX02+ strains, similar to the utility of guinea pigs in distinguishing *B. anthracis* strains that were refractory to vaccine strategies (91, 197). Employing mouse susceptibility differences to differentiate strains was proposed by early investigators (54) and has been used more recently in studies on vaccine efficacy. Brossier, Gauthier, and coworkers characterized two different natural pX01+/pX02+ isolates, one of high and the other of lower virulence, which varied tenfold in LD_{50} by the SC route for mice (194, 196).

Disease and host responses

Several laboratories recently characterized mouse models of anthrax exposure by a pulmonary route by using the virulent tox+ cap+ Ames strain. Exposure routes employed were typically IN or IT (71, 76, 78, 198), although models employing aerosol challenge with Ames spores of inbred and outbred mouse strains were also established (77). In the latter, the pathogenesis and efficacy results using a BALB/c mouse model correlated well with clinical and histological observations of inhalational anthrax in humans and with earlier antibiotic studies in the rhesus NHP inhalational anthrax model. Specifically, after lethal aerosol challenge of BALB/c mice with spores of the Ames strain, ungerminated spores, only, were cultured from the lungs; and, subsequently, vegetative organisms appeared in the mediastinum and spleen. The most common histological finding was the presence of bacteria in a variety of organs. Less commonly found were focally extensive areas of mediastinitis, splenic lesions with myriad bacilli, occasional mild PMN infiltrates, and moderate lymphoid depletion. Lung pathology was minimal and likely a consequence of the secondary bacteremia. Antibiotic efficacies are currently being determined with this model (section below). Additional models involving virulent spore challenge by IN and IT routes (78, 198) have been characterized by systemic histological and bacteriological findings similar to those reported for other murine aerosol models (77, 84) and to those observed in NHPs and humans that succumbed after inhalation exposure (27, 43, 114, 199). Such models have been useful in studies characterizing host responses and genetic bases of susceptibility (71, 198) and countermeasure efficacy (200, 201). Nevertheless the initial disease process in the IT/IN models can be dissimilar to that observed in other reported animal and human studies in that significant germination of the spores in the lungs is sometimes observed (39, 79, 107). In contrast, spores were shown to remain dormant in the lung tissue of subjects exposed by aerosol (39, 41, 43, 75,

84, 106, 111–116). This discrepancy might be due to various differences between aerosol and IN/IT exposure (Table 1). Small-particle aerosols can deliver spores deeper and in larger numbers to the alveoli, while the IN and IT routes likely deposit spores primarily at the level of the bronchioles where there may be increased germination owing to the localized presence of a large spore bolus to an immunodeficient (C5⁻) host, delivery-associated tissue damage, or postcollection processing (39, 75, 78, 79, 86, 107). Inhalational models are likely more relevant to the route of spore exposure that would be expected in a bioterror scenario.

The first description of real-time tracking of anthrax in mice of a virulent tox+/cap+ strain was reported by Dumetz et al. in studies with an *in vivo* bioluminescence imaging system (IVIS) (76). BALB/c mice were challenged with the fully virulent strain or toxin gene deletion derivatives by pulmonary or cutaneous routes, with analysis focused on responses in the spleen. The results provide new insights into the complex route- and time-associated effects of the anthrax toxins in the host; support the increasing evidence that ET plays a major role in pathogenesis and may be an important diagnostic and countermeasure target; and reinforce data showing that, among the toxin components, LT is usually detected earliest in infection (116, 202–204). The findings should be confirmed in models that involve higher-order animals, evaluate host targets in addition to the spleen, and utilize methodologies capable of detecting small bacterial concentrations.

Murine models to assess innate immune cell responses to infection utilizing fully virulent *B. anthracis* have been developed. In contrast to the *in vitro* immunosuppressive response of host cells to anthrax LT, the infection of mice by *B. anthracis* induces a general proinflammatory innate immune response. A significant stimulation of cytokines were observed in mice infected with Ames spore (109) and Sterne spores (106, 205). Murine models have also been used to characterize the roles of host MΦ and neutrophil responses in combating infection with Ames as well as the early processes in spore germination *in vivo*. Treating mice with MΦ-depleting regimens was found to promote infection with Ames spores instead of preventing it (74, 75); similar strategies revealed a distinct but smaller role of neutrophils in protective cellular responses to pulmonary and parenteral Ames challenge. Thus, these results suggested MΦs have a role in host protection as well as potentially in supporting spore germination (115). This model also demonstrated bacterial-host interactions early in infection that influence the balance between spore clearance and development of a lethal infection;

MΦ and PMN appear to be early *in vivo* targets of infection-associated LT, with MΦ interactions being mediated mainly through the CMG2 toxin receptor (66, 206, 207). Finally, these models have revealed the impact of the germination state of the infecting spore on disease pathogenesis (118, 119). Results of inhalational challenges employing Ames spores demonstrated that neither stimulating nor inhibiting germination within the lungs could protect mice from the subsequent lethal infection (119). These studies again emphasize that factors such as the bacterial strain, mouse strain, and mode/route of infection should be considered in analyzing murine data.

Pathogenetic mechanisms, virulence factors, and genetic basis of host resistance

Many recent efforts have focused on identifying factors in addition to the anthrax toxins and capsule that are potentially utilized by *B. anthracis* to facilitate infection. The role of spore exosporium and coat antigens in virulence of tox+ cap+ *B. anthracis* has been studied in mice, as described previously (112, 132, 208–211) and reviewed in greater detail in reference 392.

Although studies with genetically defined strains of mice indicated that resistance to Sterne spores is influenced by allelic variations in the *hc* gene and in the *nlrp1* locus and other genes on chromosome 11, resistance to fully virulent *B. anthracis* strains appears to be multifactorial and likely even more complex genetically. IT and IN exposure models were exploited to examine the genetic basis of resistance to *B. anthracis* strain Ames (71, 78). Lyons et al. reported significant LD_{50} differences among seven strains of mice, data suggesting that multiple genetic factors play a role in establishing resistance; no gender- or age-related effects were observed. Recently, Yadov and coworkers examined genetic control of resistance in 14 inbred strains exposed IT to Ames. As observed before (55, 78), three susceptibility groups were identified based on the differences in survival time. The data provided the first evidence of a gender bias in host susceptibility, confirmed a role for multiple genes in the differential control of infection, and suggested quantitative trait loci on as many as five chromosomes were most closely linked with susceptibility.

Vaccines and therapeutics

Except as a stringent early test of vaccine efficacy, mice do not appear to be an optimal model for evaluating protective efficacy of PA-based protein or DNA vaccines against challenge with fully virulent tox+ cap+ *B. anthracis* strains. This situation is likely due to the high susceptibility of mice to most such strains and their sensitivity to nontoxin virulence factors. Despite stimulation of high titers of toxin-neutralizing anti-PA antibody, the licensed human anthrax vaccines and newer recombinant PA (rPA) vaccine (89) adsorbed to an aluminum (AL)-based adjuvant, failed to protect mice; and the passive administration of anti-PA antiserum likewise had little or no effect (102, 137, 138, 194, 200, 212–215). However, mice are often exploited to characterize the immunogenicity of new generation vaccines (212–215) or the responses to novel vaccination routes (i.e., IN or intracutaneous) designed to improve efficacy of the current human vaccines (216–220).

Nevertheless, PA has been shown to elicit significant protection in mice against virulent *B. anthracis* strain challenge under certain conditions. Protection has been shown when PA has been delivered by novel routes such as transcutaneous, combined with other adjuvants (e.g., Ribi TriMix or *Escherichia coli* enterotoxins), purified from recombinant baculovirus-infected cell cultures, conjugated to itself, or produced by live vaccines (90, 102, 220–223). Protective live vaccines have included, for instance, *B. subtilis* harboring rPA (223); aromatic-deficient *B. anthracis* mutants (223); and recombinant PA-producing virus vectors such as vaccinia (222). Since the toxins are not required for lethality in mice, the basis of the efficacy of some PA-based vaccines is unclear. In previous studies, protective immune serum from AVA-vaccinated animals and monoclonal/polyclonal anti-PA antibodies exhibited antispore properties *in vitro*; for instance, they enhanced *in vitro* phagocytosis and intracellular killing of spores (150, 151, 153–155, 224). One can conclude from these data and newer studies that actively acquired immunity in mice to fully virulent *B. anthracis* strains usually requires PA; however, complete protection may require additional antigens and/or alterations in form or delivery that potentially induce immune mechanisms other than or in addition to Th2-biased humoral immunity. Mouse models can provide a stringent *in vivo* system for identifying additional antigens other than toxin components and capsule, such as spore or vegetative cell antigens, which may augment or extend the prophylactic capacity of rPA against virulent *B. anthracis* (235, 384, 385). For instance, Uchida et al. demonstrated the value of extractable antigen 1 (EA1), a major S-layer component of vegetative *B. anthracis*, as an effective potential supplement to PA in a next generation multicomponent anthrax vaccine (385). The combination of PA and cell wall preparation of covalently linked bacterial cell wall polyglutamate capsule antigen and peptidoglycan (GluPG) completely protected mice against subcutaneous challenge with a fully virulent *B. anthracis* (384). A further discussion of the role of spore antigens as vaccine immunogens is included in the review by Bozue et al. (392). Finally, in a departure from these studies on multicomponent toxin-supplemental antigen vaccines, Vergis and colleagues demonstrated that vaccination with formaldehyde-inactivated ungerminated spores (FIS) of avirulent *Bacillus cereus* strains can protect mice against lethal *B. anthracis* challenge (149), suggesting that the presentation of an optimal arrangement of cross-reactive or homologous nontoxin spore antigens to the mouse immune systems may offer significant protection against virulent *B. anthracis*.

Despite the disadvantages detailed above of the murine tox+ cap+ *B. anthracis* models, successful treatment of mice has provided supporting preclinical data for efficacy of a candidate therapeutic, i.e., as a stringent low-cost early-phase assessment. For example, in studies on the protective role of human group IIA phospholipase A2 (PLA2-IIA) against anthrax, Piris-Gimenez et al. observed that mice pretreated with human recombinant PLA2-IIA were resistant to infection with virulent tox+ and tox– encapsulated *B. anthracis* strains; postexposure therapeutic value was also demonstrated (225, 226). Antibiotic efficacy studies have also illustrated the utility of mouse models. Antibiotics are still considered to be of major importance for treatment of individuals with inhalational anthrax, although new countermeasures such as antitoxin antibodies, toxin antidotes, and postexposure vaccines are being developed. New anthrax therapeutics will likely include combinations of these countermeasures. Although *in vitro* antibiotic sensitivities of *B. anthracis* to clinically relevant antibiotics such as ciprofloxacin and doxycycline are established, *in vivo* susceptibility data are limited. Thus, small-animal models are needed that facilitate downselection of antibiotics based on comparisons of efficacy and pharmacokinetic characteristics; subsequent evaluation in higher-order animal models could then be done on the most effective treatments. Peterson et al. tested the efficacy of two fluoroquinolones against Ames challenge by the IN route in three small-animal species. Effective treatment could be delayed up to 24 h after challenge only and protection of mice required high and prolonged doses of antibiotic (227). These results agree with those observed previously in NHPs and mice (228) as described below. An *in vitro* hollow-fiber infection model that reproduces the pharmacokinetics of antibiotics observed in humans or in animals was developed and used as the basis for

the antibiotic administration schedules evaluated in a mouse inhalational model (229). Deziel et al. (96) determined a levofloxacin treatment regimen that was effective for *B. anthracis* infection and postexposure prophylaxis; and the murine inhalational data together with the treatment results from a rhesus monkey model supported a successful application to the FDA for an indication for levofloxacin for treatment and postexposure prophylaxis of *B. anthracis*. In similar studies with a BALB/c aerosol model (77), prolonged treatment with ciprofloxacin and doxycycline resulting in eventual culture negativity for all tissues evaluated, and the findings overall were similar to those described in other small-animal and NHP therapeutic models (41, 113, 227, 228). These findings support the use of mouse models in early characterizations of potentially clinically useful antibiotics and as an adjunct model to support FDA applications under the animal rule.

Therapeutics that target the anthrax toxins or capsule are also being actively evaluated in mice. Human monoclonal anti-PA antibody was shown to be synergistic with ciprofloxacin in protecting mice, while either alone showed only minimal protection (230). Monoclonal antibodies to anthrax capsule were also passively protective of BALB/c mice when given before, or up to 20 h after, challenge with Ames by IT route (201, 231). Thus, such anticapsule antibodies might be used alone or with antitoxin monoclonal antibodies, and either could be combined with antibiotics, for effective postexposure treatment.

Immunoproficient Mouse/Virulent tox– cap+ Challenge

As noted above, mice display a unique susceptibility to lethal infection with some nontoxigenic encapsulated strains of *B. anthracis* (86–88, 194, 232). In most animals examined, toxin production is required for anthrax infection and tox– strains are highly attenuated (3, 7, 18, 90, 233). However, Brossier et al. showed that a tox– cap+ *B. anthracis* derivative of strain 9602 was lethal for guinea pigs at high doses (194); and guinea pigs were recently reported to be relatively sensitive to lethal infection with a PA-deficient nontoxigenic Ames mutant; PA vaccination did not protect animals against SC infection with this strain (234). Although this phenotype is potentially aberrant, it has proven useful in identifying virulence factors and pathogenetic mechanisms unrelated to toxins and discerning the mechanisms of effective countermeasures (antibacterial or antitoxin) and the stage of infection influenced.

Disease and host responses

Encapsulated tox– mutants of *B. anthracis* have been shown to be lethal for mice at relatively low doses by different routes to include SC, cutaneous, gastrointestinal, intravenous, and pulmonary (IN/IT/aerosol) (82, 86–88, 232). In early studies, some *B. anthracis* strains cured of the toxin-encoding pX01 plasmid but retaining the capsule-associated pX02 plasmid (such as ΔAmes) remained unexpectedly fully virulent by the SC route for mice (87, 88). Using defined mutants of *B. anthracis* and a model of IT infection (86, 232), Drysdale, Heninger, and coworkers confirmed and extended earlier findings (87, 88). The LD_{50} and mean time to death of the tox– cap+ mutants did not differ significantly from those of the tox+ cap+ parent, and capsule synthesis was required for dissemination of the infection. However, tissue histopathology data revealed differences in responses to infections by these strains (e.g., more extensive neutrophil infiltrates and greater proliferation of the tox– cap+ strain) and suggested that the toxins and especially ET suppressed the host inflammatory responses (86). In models employing bioluminescent strains of *B. anthracis*, Glomski, Dumetz, and coworkers showed that after exposure by all routes tested, all toxin and capsule variants initially infected the local site of infection (e.g., nasopharynx after aerosol) to a detectable extent. However the absence of capsule or of the toxins delayed bacterial dissemination to the major organs, or altered its pattern, respectively (76, 79). As observed in many other models, tox– cap+ *B. anthracis* did not germinate/grow in the lungs of mice.

Pathogenetic mechanisms and virulence factors

The mechanisms involved in lethal infection by tox– cap+ strains are poorly understood. In the earlier studies comparing different strains that had been differentially cured of their toxin- and/or capsule-encoding plasmids, the virulence of *B. anthracis* for mice was shown to be strain related and influenced to a large extent by factors encoded on pX02 and the chromosome (87, 88). Later studies utilizing mutants with genetic inactivations of toxin and capsule loci confirmed the major role for capsule in virulence and dissemination of the infection (79, 86, 232). However, these data do not rule out a role for ET and LT in the pathogenesis of murine anthrax. Further studies with different strains of *B. anthracis* showed that the dominant virulence factors (toxins, capsule) used to establish infection by *B. anthracis* depend on the route of inoculation and the bacterial strain (78, 86, 195). Each factor likely contributes in a specific albeit not completely determined manner to a successful progressive infection.

Vaccines and therapeutics

Models involving challenge with tox– cap+ mutants of *B. anthracis* have been exploited to determine the contribution of nontoxin antigens in adaptive immunity. Utilizing the pX01-cured ΔAmes challenge mouse model, Chabot et al. evaluated the role of capsule in protection. The anthrax capsule by itself protected outbred mice against >100 LD_{50}s of the tox– cap+ ΔAmes strain and elicited high IgM titers and opsonic activity; the addition of PA was required to elicit protection against fully virulent Ames (221). The tox– cap+ model has also revealed a role for spore antigens in protective immunity. Brossier et al. observed that FIS alone protected 50% of mice against SC challenge with a PA-deleted version of the virulent *B. anthracis* 9602 strain; in contrast, PA alone was nonprotective (194). Vaccination of mice with PA combined with purified exosporium proteins elicited greater protection than PA alone against a ΔAmes spore challenge and stimulated higher titers of anti-PA antibody than PA alone (235). Studies employing tox– cap+ *B. anthracis* spore challenges also provided the first direct evidence for the role of cell-mediated immunity (CMI) in protection. Whereas passively transferred humoral antibody did not protect against encapsulated *B. anthracis* (tox– cap+), the transfer of interferon-γ-producing CD4 T lymphocytes produced a protective cell-mediated immune response (236).

Novel candidate therapeutics have been evaluated with this model. For example, Piris-Gimenez et al. reported that transgenic mice expressing recombinant human group IIA phospholipase A2 (sPLA2-IIA) were resistant to infection with virulent cap+ *B. anthracis* (tox+ and tox–), by both SC and IN routes. Treatment of outbred mice with sPLA2-IIA protected against intradermal challenge with the tox– cap+ strain, thus demonstrating this bactericidal enzyme's therapeutic efficacy (226). Finally, spore challenges with the ΔAmes strain were employed to confirm the effectiveness of the CapD capsule depolymerase in treating infection with spores or bacilli (237).

Mouse Toxin Challenge Models

Anthrax is an infection due to a microbe that is toxigenic but also invasive. The anthrax toxins, when given intravenously, exhibit relatively low potency in most animals in comparison with others such as botulinal toxin (238). Most animals (e.g., guinea pigs, mice) are relatively resistant to the toxins, but can be killed with comparatively few spores; conversely, rats display the inverse phenotype, as will be discussed below (239). Thus, although LT and ET are essential virulence factors and clearly important in the progression of infection, challenge with LT or ET does not pose a relevant natural or bioterror-associated risk for humans. The cause of death in lethal infections is still unresolved; the irreversibility of a *B. anthracis* infection has been linked not only to toxin-mediated effects, but also to the presence of widespread, often high levels of bacteremia in the terminally infected host (4, 240, 241). Furthermore, the histopathology reported for human cases of inhalational anthrax differs from that found in LT intoxication. For instance, whereas hemorrhagic thoracic lymphadenitis and mediastinitis were uniformly found, and hemorrhagic meningitis was observed frequently postmortem in the Sverdlovsk and 2001 letters cases (13, 19, 27, 43, 241), these findings were not reported in mice that were administered toxic amounts of purified LT (189, 193, 242). Finally, as detailed above, various studies suggest that different mechanisms may be involved in animal susceptibility and host responses to spore versus toxin challenge (53, 69, 108, 190).

Pathogenesis and host responses

Models employing administration to mice of LT or ET by injection, although not clearly relevant for studies on the pathogenesis of anthrax, have offered useful insights on several questions. These include the potential mechanisms of toxin action and toxin-mediated effects on the host response and the genetic bases for host susceptibility to infection and to intoxication (55, 66, 67, 108, 179, 189, 192, 193). For instance, Moayeri et al. (243) showed that the injection of LT at the levels found in terminally ill anthrax-infected animals induces an atypical type of vascular collapse and shock that appeared to be the sequelae of vascular insufficiency. The heart was identified as a major target of the toxin, and a potential role for nitric acid in countering the LT-induced cardiac damage was shown. Firoved et al. demonstrated that IV injection of ET caused extensive multiorgan damage and rapid lethality in mice, underscoring the renewed focus on ET as an important virulence factor and potential target for countermeasure development (76, 148, 169, 244, 245).

Vaccines and therapeutics

Mice have been exploited to characterize vaccine epitopes associated with protective efficacy against challenge with toxin and to illustrate the protection elicited by novel toxin-based vaccines. For example, Crowe observed that antigenic regions of LT were targeted in the sera of individuals vaccinated with AVA; antibodies specific for these epitopes protected mice against challenge with LT (246), thus showing a possible role of not only PA but also LF in Biothrax/AVA induced immunity. Aulinger et al. (247) developed a countermeasure that could potentially serve as both a vaccine and therapeutic. A dominant-negative inhibitory variant of PA was conjugated to capsule antigen and shown to elicit significantly higher anti-PA and anticapsule antibody titers than conjugates with unmodified PA; it protected against IV toxin challenge. The mouse toxin model also has been exploited to demonstrate protection by passive anti-toxin antibody treatment against lethal parenteral challenge with toxins (246, 248–250).

Guinea Pigs

There is a long history of the use of guinea pigs as models by anthrax investigators for basic research on mechanisms and pathogenesis of *B. anthracis* infection and the toxins. This is especially true for the extensive amount of biodefense research performed in early and later efforts at the U.S. Army Medical Research Institute of Infectious Diseases (USAMRIID) and the Defence Science and Technology Laboratory (DSTL) in the United Kingdom. From the 1930s to the present, guinea pigs have proven to be a key model in understanding the basis of the pathogenesis of inhalational anthrax including spore deposition, translocation, and germination (38, 39, 51, 239, 251–256). Other studies examined the novel aspects of anthrax virulence, and the data generated from the guinea pig model facilitated the major early strides made toward an effective anthrax vaccine. In more recent years, there has been a decline in the use of guinea pigs as a model for testing "cell-free" or subunit vaccines because of the perceived deficient protection elicited. However, these animals continue to be used to characterize and evaluate *B. anthracis* strain virulence, potency of licensed vaccines, *in vivo* correlate markers of protection, efficacy of therapeutics and postexposure prophylactics, and protection afforded by whole-cell vaccines as well as other new candidate vaccines. The guinea pig model of infection offers practical benefits to researchers. Their size makes them ideal for biomedical research. Guinea pigs are large enough to collect adequate amounts of tissues (organ tissue, blood, sera, etc.) for analyses, a situation that can be problematical in studies with mice. Furthermore, guinea pigs are small enough to allow adequate yet cost-effective housing and animal husbandry, which is not always the case when working with rabbits or NHPs.

Disease and Host Responses

Guinea pigs are highly sensitive to *B. anthracis* infection and have been an important animal model for pathogenesis studies. Although these animals are susceptible to spore infections, they are relatively resistant to anthrax toxins (53). After challenge with spores by the aerosol route, the pathology of infection in this host involves widespread edema and hemorrhage to include the spleen, lungs, and lymph nodes. In addition, consistent with other animal models of infection, the inflammatory response in guinea pigs is limited, as observed with a fulminating septicemia rather than a primary pulmonary infection (39). Lincoln and others reported that, unlike other species, guinea pigs do not exhibit a fever but rather a rapid fall in body temperature during the terminal stages of anthrax. Infected guinea pigs exhibited normal body temperatures of 102°F, but, immediately before death, the body temperature was reported as approximately 86°F (257, 258). In contrast, rabbits, swine, dogs, and humans develop a fever in response to infection, while rats and chimpanzees do not display significant fluctuations in temperature during infection.

A highly useful feature of the guinea model of anthrax infection is the relatively consistent kinetics of the infection observed after intramuscular (IM) inoculation with the Ames strain, particularly in comparison with the pathogenesis of infection in mice (J. Bozue, C. Cote, and S. Welkos; unpublished data). Of course, this feature may vary depending on the route of infection or strain of *B. anthracis* used. The terminal bacterial loads in the guinea pig model are also remarkably uniform. Early studies by Keppie et al. (51) demonstrated that the concentrations of vegetative *B. anthracis* 13 h preceding death increased from 0.2×10^6 to 1×10^9 chains per ml of blood at death (51, 255). In a recent study, the terminal concentrations of *B. anthracis* bacteria in the blood and spleen were remarkably consistent (approximately 9×10^8 CFU/ml) following a relatively low IM challenge dose with Ames spores (259). Interestingly, while guinea pigs reach impressive levels of terminal bacteremia, such high levels have not been reported for mice or rats (258). These differences may be attributable to varying degrees of sensitivities to anthrax toxins observed in these different species.

Pathogenetic Mechanisms and Virulence Factors

Guinea pigs have been a standard animal model for studying anthrax infection for nearly a century. Of particular note are the seminal studies by Ross in which she elegantly tracked the dissemination of *B. anthracis* spores from the lungs to the mediastinal lymph nodes (39). From these studies, it was demonstrated that pulmonary anthrax (initiated by either IT administration

or exposure to aerosolized spores) was not a pneumonic disease and that the lungs serve only as a portal of entry under usual conditions of IT instillations and aerosol exposures. In these studies, localized spore germination in the lung was only noted after the lung had been compromised or injured by the procedure used to deliver the spores (i.e., physical tissue damage caused by IT instillation). Additional histopathologic evidence demonstrated that spores rely on phagocytes for translocation out of the alveolar spaces and subsequent infection.

Guinea pigs are routinely used to demonstrate the importance (or lack thereof) of various virulence factors. The importance of the clonality and copy numbers of the *B. anthracis* virulence plasmids pXO1 and pXO2 was demonstrated in the guinea pig model (260). Also, guinea pigs have been used to demonstrate the requirement of specific *B. anthracis* proteins for optimal virulence. Proteins associated with the vegetative form of the bacteria such as HtrA protease, PurH biosynthetic enzyme, MntA manganese transporter component, BslA S-layer adhesion, ClpA protease, and CapA capsule biosynthetic enzyme (J. Bozue, unpublished data) have been shown to be important for virulence in guinea pigs (259, 261). In contrast, it was shown that various structural proteins present within the coat and exosporium of the dormant spore (i.e., CotE, CotH, ExsFA, BclA, Rml) do not have a role in pathogenesis after spore challenge in guinea pigs (112, 208, 209, 262). Guinea pigs have been used to demonstrate the immunogenicity of novel defined *B. anthracis* antigens (263), although none of the resulting immune responses protected against challenge with virulent *B. anthracis* spores.

Competitive models of infection have been developed in guinea pigs to test the *in vivo* fitness of *B. anthracis*. These studies are helpful in determining whether a mutant strain may be partially attenuated, as shown by its reduced ability to compete with the wt strain during infection. In these studies, guinea pigs are challenged with approximately equal numbers of spores of the wt strain and a mutant (containing an antibiotic resistance gene). When the animal becomes moribund, the spleen and/or blood are harvested and the terminal numbers of wt and mutant are determined by quantitative plating (112). The ratio of wt to mutant spores that was administered at challenge is compared with the final ratio of vegetative cells harvested from the organs of moribund animals. Applying this model, it was shown that the Ames *bclA* mutant is able to compete as well in the guinea pig host as the wt strain after IM challenge, despite the altered bacterial phenotype (112). In contrast, complete loss of virulence was demonstrated for a *purH* mutant, which harbors a defective purine biosynthetic gene (259). However, the *in vivo* fitness model may be most useful for detecting a slight attenuation of a mutant strain that would not be demonstrated by comparison of survival after lethal challenge, as shown with the Ames *soaA* mutant, which is deficient in a protein (SoaA) that was shown to facilitate spore phagocytosis by macrophages (264). This assay was also employed to determine the impact of overexpressing the CapD protein on the *in vivo* fitness and overall virulence of *B. anthracis* (265).

Vaccines and Therapeutics

The performance of a potency assay is required by the FDA (21 CFR 620.23) before the release of each new lot of AVA, the human anthrax vaccine. The current potency assay employs guinea pigs as models to assess the protective efficacy of AVA. In the assay, guinea pigs are vaccinated SC with various dilutions of the vaccine lot and then challenged subcutaneously with approximately 1,000 spores of the virulent Vollum strain of *B. anthracis*.

The protective efficacy induced by anthrax vaccines in guinea pigs has varied. The protective efficacy of AVA in guinea pigs differed depending on the challenge strain of *B. anthracis* used (91, 197, 260, 266, 267). In contrast, AVA has been shown to afford protection to rabbits and NHPs against challenge with different geographical isolates of virulent *B. anthracis*. These latter studies suggest that responses to certain vaccines in guinea pigs may not be a good indicator of the immune response in other animal species. Nevertheless, they might be considered to be sensitive indicators of strains of virulent *B. anthracis* that have the potential to be more refractory to vaccine-induced immunity. In contrast to the responses to AVA, guinea pigs are fully protected by live vaccines against challenge with various strains of *B. anthracis* (197, 261, 267). Guinea pigs have also been used successfully to demonstrate passive protection with anti-PA polyclonal antisera as well as monoclonal anti-PA antibodies (268–270). Furthermore, recently, there has been renewed interest in using the guinea pig model for studies on next-generation anthrax vaccine strategies. Brossier and colleagues confirmed that guinea pigs were indeed difficult to protect with PA-based vaccines alone (194). However, when paired with formalin-inactivated spores, complete protection was achieved after a SC challenge with a virulent strain of *B. anthracis*. These results were confirmed and expanded upon by Gauthier and coworkers (196) who again showed an enhancement of PA-induced

immunity by the FIS, this time against pulmonary challenge. The results suggest a role for antigens on the surface of ungerminated spores in active immunity against anthrax. Specifically, recombinant spore proteins given with PA significantly improved survival of vaccinated guinea pigs after an intradermal challenge with the Ames strain (235). Lee et al. and Garufi and coworkers also demonstrated that complete protection could be achieved when the capsule component poly-γ-D-glutamic acid (PGA) was conjugated to PA or to the receptor binding domain of PA (234, 271); and high levels of anti-PA, anti-PGA, and toxin-neutralizing antibodies were induced. Finally, Skoble et al. (272) showed that killed but metabolically active anthrax vaccines were partially protective (comparable to current licensed vaccine) using the guinea pig model.

Guinea pigs have been used to demonstrate adjuvant efficacy in anthrax vaccine strategies. Gu et al. showed that the addition of CpG oligonucleotides (ODNs) to the human vaccine enhanced survival of BioThrax-vaccinated guinea pigs and with lower doses of vaccine (217). These CpG ODNs are short synthetic single-stranded DNA molecules. Recently, the guinea pig model has been used to evaluate the efficacy of a next-generation PA-based vaccine adjuvanated with CpG ODNs (273). Humans are known to respond strongly to CpG ODNs; however, rabbits may not (273). Therefore, in this case, guinea pigs were an appropriate alternative to rabbits to measure an immune response augmented by CpG ODNs. Finally, Bielinska et al. (212) demonstrated the efficacy of a novel nanoemulsion adjuvant using the guinea pig model. Thus, these data indicate that, while guinea pigs may not be the best model for demonstrating protection afforded by a PA-based vaccine, they appear to be a suitable model for demonstrating adjunct immunity provided by the addition of other antigens or adjuvants.

The innate sensitivity of guinea pigs to certain antibiotics limits their use in some antibiotic-based therapeutic research (274). Nonetheless, several groups have found the guinea pig model to be useful in evaluating antibiotic and/or postexposure vaccination strategies (227, 275–277). Recently, the guinea pig was used to characterize the efficacy of current anthrax countermeasures against a strain of *B. cereus* isolated from a patient with anthrax-like inhalational disease (278). Palmer and coworkers evaluated the efficacy of anthrax vaccines (both recombinant proteins and DNA-based vaccines) and monoclonal anti-PA antibodies in guinea pigs challenged with aerosolized *B. cereus* G9241 spores (278). Thus, the guinea pig model appears to play an important role in characterization of emerging *Bacillus* pathogens and efficacious novel medical countermeasures.

Considerations for Use of This Model in Anthrax Research

A recent study demonstrated nutritional differences in guinea pigs that led to discrepancies in the virulence assessment of a *B. anthracis* mutant in guinea pigs compared with rabbits and mice. An intact purine biosynthetic pathway, encoded by the *pur* operon, was shown to be required for virulence of the Ames strain in guinea pigs (259). Spores of the *purH* mutants were as virulent as the wt parent strain in mouse and rabbit infection models. In contrast, *purH* mutant spores were highly attenuated in guinea pigs regardless of administration route. To date, this is the only example of a host-specific nutrient requirement associated with a mutation in *B. anthracis*. However, other pathogens have shown host-dependent variations, especially in regard to virulence in guinea pigs (279–281).

Levy and colleagues (282) reported that the LF component of anthrax LT may not be as necessary for virulence in the guinea pig model (and rabbits) as previously suggested. In this study, no significant differences were noted between mortality rates or times to death in guinea pigs challenged with wild-type Vollum spores compared with spores from a Vollum strain lacking the gene encoding the LF. More recently, these workers demonstrated the reciprocal, showing that a construct of the Vollum strain containing a deletion of edema factor retained virulence (386). Thus, in the guinea pig, the presence of just one of the toxins, ET or LT, is sufficient for the development of lethal infection, at least in this model that employs intranasal or SC challenge with the Vollum strain of *B. anthracis*. Although the toxins were required for optimal virulence (lowest LD_{50} and shortest mean TTD), the deletion of all three toxin genes (PA, LF, and EF) attenuated the strain and resulted in a more prolonged TTD, but could produce lethal infection (387, 388). Similarly, Garufi demonstrated the PA-deficient, and thus toxin-negative, derivative of the Ames strain could produce lethal disease in guinea pigs challenged subcutaneously, a situation that partially resembles the virulence of capsule-producing nontoxigenic strains in mice (86, 88, 234). These results may indicate that *B. anthracis* harbors additional virulence factors in addition to the toxins that require the expression of the pX01-encoded AtxA virulence regulator in the presence of the pXO2 plasmid. The role of these nontoxin mechanisms in the context of exposure to spores of strains such as Ames by the aerosol route requires further study, but these mechanisms may have a role in the later stages of infection by fully virulent strains.

Rabbits

Rabbits have been used to assess the virulence associated with *B. anthracis* isolates and to characterize the immune response induced by vaccination with PA-based anthrax vaccines. The rabbit model provides some advantages over guinea pig or NHP models. As discussed earlier, guinea pigs may not be ideally suited for anthrax vaccine research (particularly for vaccines that only contain PA). However, the costs of acquisition and maintenance, safety concerns, and extensive husbandry conditions associated with the use of NHPs can be prohibitive. Additionally, FDA licensure of any new anthrax vaccine or therapeutic for human use requires fulfillment of the animal rule, i.e., the candidate product must be thoroughly evaluated in well-characterized animal models expected to respond in a manner predictive of that in humans. It appears that two models will be needed. Thus, the rabbit model has been developed in part to fill this crucial niche.

Disease and Host Responses

In studies with fully virulent strains of *B. anthracis*, rabbits are generally infected by either SC injection, IN instillation, or aerosol exposure; however, there are documented studies using intracutaneous, IP, or IV routes of infection. Most rabbit model development has been performed by using New Zealand white (NZW) rabbits (164, 283). However, one of the first reports detailing extensive telemetry data collected from infected rabbits characterized anthrax disease progression in dwarf Dutch belted rabbits after IN instillation (276). The data collected from the different rabbit strains are generally in agreement. The pathology of infection in rabbits is reported to be similar to that seen in both NHPs and humans, but the findings tend to be less severe in rabbits (73, 164). Lesions involving hemorrhage, edema, and inflammation in the lymphoid tissues that are often observed in human disease are less severe in rabbits, and anthrax meningitis is reported to be rare in these animals. (However, a recent report by Levy et al. [388] reported meningeal infection in rabbits challenged by the IV or SC route by germinated bacilli of mutant strains of Vollum. The ability to cross into the brain from a peripheral [SC or IV] route required the presence of pX02 and the pX01-encoded *atxA* virulence regulator, but not the genes encoding the lethal and edema toxins.) The less robust pathology might be attributable to the exquisite sensitivity of rabbits to *B. anthracis*. Historically, rabbits survive approximately 2 to 4 days postinfection (164, 257, 284). There have been slightly longer times to death reported (i.e., 5 and 6 days), but this finding is likely due to *B. anthracis* strain differences or sources of rabbits. For example, using a *B. anthracis* strain (A0843) isolated from a sheep that had died in an anthrax outbreak to challenge rabbits, Fasanella and colleagues observed deaths from 2 to 5 days after SC infection with approximately 20 LD_{50} equivalents (285). Levy and coworkers recently reported that deaths occurred as late as 6 days after IN administration of Vollum spores, but when the rabbits were infected by SC injection, the time to death remained within the 2- to 4-day window (282). Even with these few documented deviations, the time to death or euthanasia of moribund animals does not vary dramatically based on route of infection or dose of spores administered. Recent evidence suggested that the time to death attributed to inhalational anthrax in a rabbit model can be affected by a prior or underlying infection (286). Rabbits pretreated with bacterial lipopolysaccharide (LPS) survived inhalational anthrax significantly longer (90 h) compared with naïve rabbits (41 h) (286). Interestingly, even with the addition of LPS, the majority of the rabbits succumb to infection between days 2 and 4.

The susceptibility of rabbits to *B. anthracis* appears to be relatively unaffected by the size of the dose. The LD_{50} dose for the SC route was reported to be 1.56×10^3 with an LD_{99} of 2.83×10^6. Rabbits appear to exhibit significant variability in their individual responses to SC infection; however, rabbits that succumbed to infection with SC doses ranging from 43 to 156,000 spores all died within 2 to 4 days (164). Likewise, the time to death/morbidity of rabbits exposed to aerosolized spores is relatively unaffected by the magnitude of the dose. Although the LD_{50} dose of aerosolized spores was reported to be 1.05×10^5 CFU (LD_{99} of 1.36×10^5 CFU), rabbits exposed to aerosolized doses ranging from 8.34×10^4 to 1.03×10^7 spores all succumbed to the ensuing infection on day 2 or day 3. This lack of an obvious dose response in the course of infection in rabbits was also documented in 1967 by Walker et al. (257). In this early report, rabbits were challenged IP with Vollum 1B and there were no obvious trends suggesting a dose response in doses ranging from 1.4×10^4 to 1.4×10^{-0} (257). The rabbits succumbed in less than 3 days (20 h to 70 h) postinfection with no association between dose and time to death. Additionally, the rabbits exhibited a somewhat unusual temperature response to anthrax infection. Two temperature trends were apparent: (i) When rabbits were challenged IP

with high doses of spores, the body temperature was significantly elevated by 10 h after infection and remained elevated until the rabbit succumbed to the infection. (ii) Lower doses of spores resulted in an elevated body temperature approximately 40 h after infection, which then declined until death. Peterson et al. also confirmed a difference in fever profiles depending on the dose of spores instilled IN; higher spore doses (300 LD_{50} equivalents) resulted in prolonged fevers (276). Interestingly, it was proposed in early reports that rabbits would not be an appropriate model in which to study anthrax pathogenesis or anthrax vaccination strategies because of these observations regarding body temperature and the lack of a clear dose response (257). Nevertheless, as described below, the rabbit has been an invaluable model for characterizing vaccination strategies, particularly in research within the past decade.

Pathogenetic Mechanisms and Virulence Factors

There are several similarities between the pathogenesis of anthrax in rabbits and the pathogenesis of anthrax in NHPs and humans. Real-time telemetry data collected from rabbits infected by IN instillation indicate that similar cardiac events (i.e., decrease in mean arterial pressure), hemodynamic changes (i.e., altered serum chemistries), and respiratory symptoms (i.e., respiratory distress) occur in the experimental rabbit model and case reports of human infections (276). Lawrence et al. also recently demonstrated the direct effects of IV toxin administration on the hemodynamics and heart function in a rabbit (287).

Additional shared pathology includes lymphoid tissue destruction and, albeit rare, hemorrhaging within the meninges and brain (73, 164, 274). Lymphoid tissue destruction is typically seen in the spleen and, depending on the route of infection, in certain regional lymph nodes (i.e., mediastinal, mandibular, or axillary lymph nodes). This destruction is typically associated with hemorrhage, necrosis, inflammation, and the presence of a large number of *B. anthracis* bacilli (73, 164, 274). Meningitis has also been reported in rabbits infected with *B. anthracis*; however, the frequency and extent of this pathology may not indicate the meningitis seen in NHPs or humans. This lack of consistent meningitis described postmortem may be due to the inherent sensitivity of rabbits to anthrax and the resulting abbreviated disease time course.

In general, similar gross pathology was observed in rabbits regardless of infection route (164). The main difference noted was the location of the affected lymph nodes. In the case of SC injection, axillary lymph nodes were affected, whereas rabbits exposed to aerosolized spores exhibited the most severe pathology in mandibular or mediastinal lymph nodes (73, 164, 274). The commonalities in gross pathology and physiological events of rabbits and NHPs in response to *B. anthracis* spores, in addition to similar protective immune responses induced by PA-based anthrax vaccination strategies (described below), lend validity to the use of rabbits as suitable models of anthrax.

Obvious clinical symptoms have been reported to be difficult to discern in rabbits infected with *B. anthracis*. Several recent reports describing the natural history of the disease in rabbits have indicated that rabbits appear physically normal until shortly before succumbing to inhalational anthrax. This was the case regardless if rabbits were exposed by IN instillation (276) or to aerosolized spores (164, 283). This is a common trait of anthrax in other small-animal models such as mice and guinea pigs. While this observation does not negatively impact the data collected, it can make early endpoint euthanasia somewhat challenging. To prevent animals from suffering during terminal illness, researchers attempt to implement early endpoint euthanasia when clinical signs indicate that the animal is moribund beyond recovery. However, to compensate for the lack of overt clinical symptoms in rabbits, there have been several studies designed to identify other indicators, such as biomarkers, of an anthrax infection. An early report indicated the main clinical manifestations of anthrax in rabbits infected intravenously with virulent spores were increasing dyspnea (shortness of breath) and the appearance of extremely low concentrations of oxygen in the blood of terminal rabbits in the later stages of disease (288). The oxygen levels were noted to be abnormally low within 30 min of death, leading the authors to hypothesize that the rabbits were succumbing to asphyxia (288). The rabbit and guinea pig models were used to characterize the kinetics of PA expression during infection in relationship to the levels of bacteremia (289). In this study, it was determined that the average time to death in rabbits that had been inoculated IN with Vollum spores was approximately 16 h after the appearance of bacteremia and PA in blood. Sela-Abramovich and colleagues demonstrated that three *B. anthracis*-secreted proteins could potentially be used as biomarkers of infection in rabbits (290). These three *B. anthracis*-specific proteins (a serine protease, an endopeptidase, and a protein of unknown function) could be detected within rabbit sera at relatively early stages of bacteremia (approximately $\leq 10^5$ CFU/ml). Such studies are offering clues for developing clinical tests

that can diagnose *B. anthracis* more rapidly than currently possible, as described below.

Recently, the rabbit model was employed to examine the virulence associated with a pathogenic "anthrax-like" *B. cereus* isolate (291). This particular *B. cereus* strain (G9241) was isolated from a near-fatal anthrax-like pulmonary infection in a welder in Louisiana, USA. Plasmids encoding putative anthrax toxins and capsules were identified within this *B. cereus* isolate (292). However, testing in the rabbit model indicated that this strain was nearly avirulent by the SC route of infection, and the LD$_{50}$ dose of aerosolized spores was approximately 100-fold higher than fully virulent *B. anthracis* (291). These findings suggest that fulminant disease by such strains in humans requires the presence of an underlying predisposing and/or immunosuppressive condition(s). Thus, the rabbit model will continue to be an important tool for ascertaining virulence associated with *Bacillus* species as emerging infectious agents.

Vaccines and Therapeutics

Rabbits have been used for decades in studies characterizing vaccination strategies involving the PA component of the anthrax toxins (91, 293–296) as well as novel vaccination or therapeutic strategies. Such strategies include novel protein formulations (271, 297–300), various liposomal formulations (389), DNA-based vaccines (301–303), and a killed but metabolically active vaccine strategy (272); novel routes of vaccination such as needle-free skin delivery are also being evaluated (389). The rabbit has also served as a useful model to investigate immune therapies such as monoclonal anti-PA antibodies or other inhibitors of anthrax toxin components (230, 304–307). As demonstrated recently (53), it is sometimes more difficult to demonstrate protection with new countermeasures against lethal challenge in rabbits than in NHPs, a finding that is likely a consequence of their extreme very high sensitivity and rapid time to morbidity.

While the literature regarding antibiotic testing in rabbits is less extensive than that of vaccine strategies, there are reports describing the use of the rabbit model for assessing antibiotic efficacy (227, 283) as well as testing antibiotics as adjunct therapies (230, 283). Yee and colleagues demonstrated that IV levofloxacin is an effective treatment against inhalational anthrax in rabbits (283), and Peterson and coworkers used the rabbit model to demonstrate that the protection afforded by a human monoclonal antibody against PA is augmented with the addition of the antibiotic ciprofloxacin (230). However, rabbits have unexpected sensitivities to certain therapeutic compounds or antibiotics that might limit their usefulness in some therapeutic studies (274).

In Vitro Correlates of Immunity

Perhaps the most important aspect of the rabbit model of anthrax is its role in the development of *in vitro* correlates of immunity. In adjunctive efforts to the extensive vaccine studies done in rabbits, the rabbit sera were used to demonstrate that a robust immune response to PA is a satisfactory correlate of immunity to anthrax in vaccinated animals (277, 294, 308–310). In addition to determining quantitative anti-PA IgG titers in rabbit sera, a toxin neutralization assay (TNA) was developed to measure antibody function and has since been used extensively to evaluate sera from vaccinated rabbits (308, 310). This assay measures the ability of anti-PA antibodies to neutralize the *in vitro* intoxication of macrophage-like tissue culture cells. Thus, research performed using the rabbit model yielded well-characterized quantitative candidate assays as potential *in vitro* correlates of immunity for PA-based anthrax vaccines. Both the quantitative anti-PA IgG enzyme-linked immunosorbent assay (ELISA) and TNA are significant predictors of protection after exposure to aerosolized spores of vaccinated rabbits (309, 310), and these assays have been considered to be the "gold standard" when characterizing the human immune response to the currently licensed human anthrax vaccine as well as next generation PA-based vaccines (311, 312).

Considerations for Use of This Model in Anthrax Research

As with any animal model of infection, there have been a few anomalies associated with the rabbit model of anthrax. For instance, a recent report suggested that LF may not be required for lethality in rabbits. In this study, Levy et al. demonstrated that the IN LD_{50} for spores of the Vollum strain of *B. anthracis* was 1.6×10^4 CFU and the LD_{50} for the SC route was 20 CFU. Although the *lef* Vollum mutant (which does not produce the LF component of anthrax LT) was partially attenuated, the LD_{50} doses were increased to only 1.5×10^6 CFU and 500 CFU, respectively. This residual virulence indicated that LF may not be required for fatal anthrax disease in rabbits (282). More recently, evidence that lethal disease can be produced by nontoxigenic strains was provided by a study of rabbits infected by the IV route with bacilli of an LT- and ET-negative derivative of the Vollum strain, in a model of the septicemic phase of infection (387). This activity required the presence of an unidentified factor encoded on the pX02 plasmid and the pX01-encoded

atxA virulence regulator gene (but not the *tox* genes on pX01). As described above, similar results were reported by these authors for guinea pig models of infection (282, 387, 388). These data bring into question the pathogenesis associated with the rabbit model and the presumed major role for LT and highlight the potential importance of the anthrax ET in pathogenesis in the rabbit, as has recently been demonstrated in other animal models (76, 169, 245). However, Lovchik and coworkers recently showed that production of either LT or ET by mutant derivatives of Ames is sufficient for *B. anthracis* virulence by the intratracheal route (390). The *B. anthracis* strain and the route/manner of infection may be important in determining the exact roles of the anthrax toxin in anthrax in rabbits. Another possible potential rabbit-associated anomaly is exemplified by the current evidence suggesting that an epitope of PA (loop-neutralization determinant [LND]), which appears to be immunogenic in humans, is cryptic or unrecognized by the immune response in rabbits (313–315). In these studies, Oscherwitz and colleagues achieved a strong humoral response in rabbits against multiple antigenic peptides that targeted a LND of PA, but when rabbits were inoculated with full-length PA, this region was not antigenic (315). The finding that the human immune system recognizes this region of PA (even when presented as full-length PA-83), but rabbits do not, emphasizes that there are indeed differences between the human immune response and those of animals used as models of infection (313). Nonetheless, Oscherwitz et al. (391) demonstrated recently that immunization of rabbits with peptide vaccines directed specifically against a region in PA domain 2 containing the LND and synthesized with a T-helper peptide epitope elicited toxin-neutralizing antibodies and protected rabbits against aerosolized spores of *B. anthracis* strain Ames. The immunogenicity and efficacy of these peptides in NHPs or humans have not been determined.

Nonhuman Primates

Aerosolized *B. anthracis* studies in NHPs have been conducted since the 1950s and 1960s. Although several species of NHPs were utilized, until the turn of the century, the majority of those studies were performed in rhesus macaques (*Macaca mulatta*) and that particular species was considered the "gold standard" inhalational anthrax animal model (111, 114, 316, 317). Since then, detailed pathogenesis studies have been conducted in cynomolgus macaques (*Macaca fascicularis*) and African green monkeys (*Chlorocebus aethiops*) showing their value as animal models of inhalational anthrax (65, 199). Most recently, initial studies have been performed in marmosets (*Callithrix jacchus*) to begin characterizing this species as a potential inhalational anthrax animal model (64, 318).

Modern day aerosol challenges of NHPs are primarily conducted in a head-only device exposing anesthetized animals to a steady-state aerosol that is delivered at a known rate in a dynamic inhalation chamber (319, 320). It is worth noting that these devices expose the entire head to the challenge atmosphere, and therefore spores may be deposited in the eyes, lips, ears, hair, and skin of the head. Sampling of the atmosphere from the exposure chamber enables an investigator to calculate spore concentrations. This, coupled with measured physiologic parameters of the animal including body weight, respiratory rate, and tidal volume provide data for calculating an estimated inhaled dose. The inhaled dose is not necessarily equivalent to the dose an animal actually receives because the fate of inhaled particles is highly variable. The size of particles within the aerosol can be regulated and range from small to large, typically 0.1 to 10 μm (319, 320). Each aerosol particle may contain single or multiple spores that are deposited anywhere within the oropharyngeal cavity, along the upper and lower respiratory tree, or simply exhaled. NHPs are unique among the animal models because they are truly oronasal breathers, similar to humans. The physiologic mechanism of how air is inhaled and enters the respiratory system plays a role in determining the location and quantity of particle deposition, factors that may ultimately impact spore uptake and disease pathogenesis. In general, larger particles settle in the upper respiratory tree (nasopharyngeal area, trachea, and bronchi), whereas the smaller particles are inhaled into the lower respiratory tree (bronchioles and alveoli). Settled particles may be swallowed, spit out in saliva, blown out in nasal mucus, and coughed out in tracheopharyngeal and bronchial mucus. Through respiratory secretions and the mucociliary apparatus, even particles deposited deep in the respiratory tract can be efficiently shuttled out of the animal's body. Once deposited in the respiratory system, spores may be retained or taken up by macrophages, dendritic cells or their transepithelial extensions, other phagocytic cells, and infrequently or rarely by epithelial cells or direct entry into blood and lymphatic fluid (74, 75, 80, 174, 175, 321). Retained spores may remain dormant in the respiratory system for extended periods (41, 74, 75, 86, 171, 321).

Disease and Host Responses

NHPs have shown consistency across studies in developing the full range of classic lesions of human inhala-

tional anthrax making them valuable as animal models. Important features of human disease include a widened mediastinum seen radiographically, fluid in the chest cavity, lymphadenopathy of pleural and mediastinal lymph nodes, pneumonia, splenic lymphoid depletion, meningitis, hepatic, gastrointestinal and urogenital hemorrhage and/or inflammation, and *B. anthracis* bacteremia and toxemia. Death occurs between 4 and 6 days and the mortality rate of untreated patients is nearly 100% (12, 13, 27, 28, 43, 45, 89, 322–328). The critical aspects of human disease that are also found in infected NHPs include bacteremia, increased white blood cell counts, meningitis, lymphadenitis, splenitis, mediastinitis, pneumonia, pleural effusion, hepatitis, adenitis, vasculitis; and hemorrhage, congestion, necrosis, and edema in multiple tissues. When challenged with calculated aerosol doses between 2×10^4 and 5×10^{10} CFU of virulent aerosolized *B. anthracis*, the majority of animals succumb within 7 days with a reported range of death between 2 and 25 days of exposure (40, 44, 65, 111, 114, 199, 274, 316, 317).

NHPs are the only animal model of inhalational anthrax that consistently develop anthrax meningitis. Statistical information gathered by the Centers for Disease Control and Prevention (CDC) on human inhalational anthrax cases indicates that meningitis rarely occurs in the nonpulmonary forms of anthrax. Specifically, less than 5% of humans with cutaneous or gastrointestinal anthrax develop meningitis; however, meningitis is believed to occur in at least 50% of the patients with inhalational anthrax (4, 329). Additionally, long-term and devastating sequelae to anthrax meningitis can occur despite successful elimination of the bacterial infection (330). Little is known about the bacteria's predilection for the meninges after aerosol infection. In NHPs, meningitis occurs 3 to 5 days after aerosol exposure in up to 77% of experimental NHP cases (329). Inflammation, bacteria, and hemorrhage may be found in one or more layers of the meninges. More studies designed to elucidate the pathogenesis of this unique aspect of inhalational anthrax are needed to begin developing more effective medical treatments and improve long-term prognosis of human patients (169).

Pathogenesis and Detection of Infection

Early diagnosis and medical intervention are the keys to successfully treating a patient with inhalational anthrax. In humans and animals, blood cultures become positive within 2 to 3 days of exposure. These cultures may require 24 h or longer to yield results and may be falsely negative owing to varying bacilli concentrations in the bloodstream at the time the blood is taken (289, 290, 331–334). For these and other reasons, efforts to discover early diagnostic biomarkers and develop rapid diagnostic tests have accelerated. Diagnostic markers for anthrax have mainly included blood levels of the *B. anthracis* capsule and toxins and of the LF, EF, and PA components of these toxins, as illustrated in the animal models discussed in this review and more recently using NHPs; the assays employed in the latter studies have utilized technologies such as PCR, ELISA, electrochemiluminescence (ECL), and others (203, 331–333, 335). Serum levels of LF and PA have been detected concurrently with or before detectable bacteremia, as have been other unique diagnostic markers (203, 331, 332, 335). For example, Rossi and coworkers studied anthrax diagnostics in the African green monkey model (65) and confirmed that these NHPs exhibit a disease course similar to that observed in rhesus macaques and in humans (301). They demonstrated that detection of the biomarkers PA and capsule by PCR and an ECL immunoassay were associated with the blood levels of *B. anthracis* in African green monkeys. The results of the ECL immunoassays, and, in particular, the PA-ECL, correlated with culture results, were rapid and simple to perform as well as being highly sensitive and specific, and were available in <1 to 2 h; thus the usefulness of ECL in real-time diagnosis of anthrax and as an indicator for treatment was substantiated (331). More recently, Boyer et al. characterized the blood levels of bacteria, the toxin components (LF, PA, and EF), LT, and capsule; and detected the PA gene (*pagA*) by PCR during the course of inhalation anthrax in the rhesus model (203, 332). With the use of sensitive mass spectrometry assays for LF and EF and capture ELISA for toxin, toxin components, and capsule, LF and capsule were shown to exhibit a triphasic profile of positivity, similar to those associated with bacteremia and neutrophil levels. These data suggested that these biomarkers can be used to identify a time during the infection after which treatment is ineffective and a lethal outcome inevitably ensues. Specific "point-of-no-return" concentrations of LF, LT, and EF were suggested; and LF was espoused as an especially important marker for early diagnosis of inhalational anthrax before the onset of a fulminant systemic infection. Unlike previous results with PA (289, 334, 336) in which PA was detected after culture positivity and sometimes only later in infection, LF was detected by mass spectrometry as early as 24 h postexposure and over the course of a 4- to 5-day infection period.

Additionally, anti-PA IgG is an important confirmatory diagnostic marker of anthrax, a predictor of serum antitoxin activity, and a marker of immunological memory against anthrax (113, 337). Antibody responses in NHPs to other antigens such as anthrose, the terminal carbohydrate of the outermost spore exosporium protein BclA, also appear to be of diagnostic value in confirming an aerosol exposure to *B. anthracis*, even in animals treated successfully with antibiotics (335).

Vaccines and Therapeutics

Immunogenicity, toxicity, and safety testing of medical products and pharmaceuticals being developed for human use are often performed in NHPs because these animals most closely mimic human physiological and immune responses. These types of studies are conducted to safeguard human test subjects and provide results that improve the overall health and welfare of people (338). Their similarities to humans in pathogenesis of inhalational anthrax also make the NHP an ideal animal model for conducting studies that evaluate vaccines and therapeutics for the treatment and prevention of anthrax. However, despite the obvious value of this animal model, multiple factors may limit their use, as described in Table 1. To date, many studies have been conducted successfully in NHPs to evaluate new and established vaccines and adjuvants, postexposure antibiotic treatments, combination vaccination with postexposure antibiotics, postexposure treatment with antibodies such as anti-PA monoclonal antibodies, and others (41, 53, 64, 73, 113, 304, 338–348).

The efficacy of a licensed human anthrax vaccine (AVA/BioThrax, formerly Michigan Department of Public Health [MDPH]) against lethal challenge by virulent *B. anthracis* was evaluated for the first time in NHPs by Ivins et al. (347). Rhesus macaques were exposed by aerosol to spores of the Ames strain. All vaccinated animals survived challenge with doses from 200 to >400 LD$_{50}$s at ≤9 months and 88% survived when challenged 2 years after vaccination; a uniform serum antibody response to PA was shown (347). The authors further characterized the efficacy of purified PA combined with different candidate adjuvants in the same NHP model and documented the humoral and cellular responses to the vaccines and the occurrence of post-challenge bacteremia. One dose of each PA vaccine protected >90% of vaccinees challenged at 6 weeks (348, 349). These early studies confirmed the substantial efficacy afforded by PA vaccines in NHPs under standard but rigorous challenge conditions. These results were extended in a study that evaluated the efficacy of AVA in several models to include rhesus macaques against aerosol challenge by *B. anthracis* isolates of diverse geographical origin. Strains differing genetically and in virulence for vaccinated guinea pigs were also included (91, 197). Although AVA only partially protected guinea pigs against Ames strain spores and eight other isolates (13 to 57% survival), it provided 80 to 100% protection against representative virulent strains in rabbits and NHPs. Also, although transient bacteremia was detected in the vaccinated rabbits and NHPs, the number of rhesus monkeys that became bacteremic was less than for the vaccinated rabbits. These data confirmed previous reports showing that AVA is highly protective in the rhesus NHP model (73).

Although licensed human vaccines are available that were shown in animals to protect against inhalational anthrax, improvements to include reductions in the number of doses and the time required to achieve full protection are needed. One example of a strategy proposed to address these deficiencies was described by Galen et al. (339). In their study, a recombinant *Salmonella* vaccine strain that expressed PA fused with the export protein ClyA was evaluated for vaccine potential in a prime-boost strategy employing rhesus and cynomolgus macaques. The live vaccine was delivered mucosally by the IN route, followed by a parenteral booster with purified PA plus alum or BioThrax. Monkeys primed with the live vaccine developed high levels of serum toxin-neutralizing antibodies just 7 days after the booster dose, whereas unprimed controls lacked serum toxin-neutralizing activity. The data supported the possibility of a preexposure vaccination strategy in which an attenuated live vector anthrax vaccine is administered to prime individuals immunologically so that they will produce a rapid, protective antibody response within days of receiving a dose of PA-based vaccine. This approach might reduce the total number of vaccine doses needed to produce solid protection.

In addition to improvements to optimize serum responses to the current vaccine, data are needed to characterize more fully the humoral and CMI responses that accompany vaccination of NHPs with AVA or other anthrax vaccines. This information is critical for defining robust correlate markers of protection that can be used to predict the efficacy of anthrax vaccines in humans under FDA guidelines. It was shown that, for the recombinant rPA anthrax vaccine, the titer of toxin-neutralizing serum IgG was predictive of protection. IgG derived from the serum of rPA-vaccinated rhesus macaques and passively transferred into mice protected the animals against subsequent challenge with *B. anthracis* (100, 312). The protection conferred by passive transfer of the IgG correlated significantly

with the *in vitro* TNA titer of the donor macaque sera (312). Nevertheless, while the TNA titer induced to rPA is a significant immune correlate for this vaccine, the ability of anti-PA to prevent toxin uptake and killing of host cells does not encompass all of the protective immune mechanisms that are elicited in an infection by *B. anthracis*, and it is apparent that other responses such as CMI and LF- and EF-specific immune responses are also important (76, 100, 236, 312, 350–352). In addition, other antigens that might enhance the efficacy of PA-based anthrax vaccines, such as spore and capsule antigens, are being evaluated in animals, although with a few exceptions, studies have yet to be performed in NHPs. For instance, the protective efficacy of a *B. anthracis* capsule fusion construct was shown in NHPs by Chabot, Friedlander, and coworkers (353). The inclusion of multiple antigens of *B. anthracis* in an anthrax vaccine might extend its protection to strains resistant to vaccination with one component, or could afford protection to recipients that are nonresponsive to the single-component vaccine.

NHPs have also been useful in assessing the value of therapeutics and prophylactic treatments. For purposes of discussion, treatments will be regarded as compounds administered to prevent or ameliorate the effects of an infection and, unlike vaccines, do not induce active immunity. These treatments can be administered after exposure, but before the development of any signs or symptoms of infection and are thus presymptomatic prophylactics. Alternately, treatments administered after infection is established are considered to be therapeutics (postsymptomatic).

Antibiotics are still considered to be the primary treatment of individuals with inhalational anthrax, although new countermeasures are being explored, as illustrated in other sections of this review.

NHPs have been invaluable in evaluating the efficacy of specific antibiotics in treating anthrax, establishing dose and pharmacokinetic parameters, and estimating the latest point in the course of disease when treatment can be initiated and still influence outcome (41, 64, 113, 340, 342, 345). Early studies demonstrated the efficacy of antibiotics such as penicillin in treating anthrax and also showed that antibiotics are active only after spores have germinated (38, 340, 354). Later groups demonstrated efficacy attributable to antibiotics such as the tetracyclines and fluoroquinolones (113, 342, 343, 345). However, it was apparent that residual spores can persist after antibiotics are discontinued and may subsequently germinate and cause disease. Animal experiments to include those with NHPs have confirmed the prolonged persistence of spores after aerosol exposure and a potential extended period before germination (41, 113). In one study, rhesus macaques were protected during a 30-day course of antibiotics after aerosol challenge. However, some animals developed fatal infection after the antibiotics were discontinued (113). Thus, complete protection against high doses of aerosolized spores required a prolonged period of treatment (41, 77, 113, 227, 228).

Since prolonged antibiotic treatment can lead to increases in adverse side effects and a decline in adherence rates among humans, modifications of this treatment scheme, or alternate nonantibiotic treatment strategies, have been investigated in NHPs. For instance, Vietri et al. determined that a short course of ciprofloxacin treatment alone, when given after development of symptoms (i.e., therapeutically), instead of prophylactically, could effectively treat inhalational anthrax and prevent disease caused by the germination of spores after the discontinuation of antibiotics (345). In a symptomatic individual, bacterial replication will probably have occurred before the beginning of antibiotic treatment, and the immune response generated should protect against disease resulting from subsequent spore germination (345). However, this approach can be risky since it depends on the ability to detect and treat an active infection in time to ensure patient recovery. Several groups are evaluating the protective efficacy of monoclonal antibodies specific for the anthrax toxin components given pre- or postexposure. For example, Vitale et al. reported that a fully human monoclonal anti-PA antibody was effective prophylactically, and partially protective when given postsymptomatically ($\leq$48 h), to rabbits exposed to aerosolized *B. anthracis* Ames spores. One dose given 1 h after exposure also protected cynomolgus monkeys challenged with aerosolized spores (346). Additional studies are required to determine the extent and kinetics of postsymptomatic protection and the duration of protection with this antibody. Migone and coworkers evaluated the efficacy of a human anti-PA monoclonal as a prophylactic agent and therapeutic in monkeys. A single dose of the anti-PA increased the survival compared to untreated controls when given either prophylactically or postsymptomatically (i.e., after PA and bacteria were detected in blood) to cynomolgus monkeys exposed by aerosol to Ames strain spores; survival beyond the 28-day period of the study was not determined (304); similar protective results were reported by Pitt et al. and Henning et al., using different human anti-PA monoclonal antibodies and the African green monkey or cynomolgus macaque models, respectively (355, 357).

Finally, combination regimens given pre- or postexposure are being explored as viable treatment strategies. Pre- or postexposure vaccination when combined with antibiotics improved efficacy (41, 113), and, as shown in later studies, potentially enabled a shorter course of antibiotic treatment (344). Using the rhesus macaque model of inhalational anthrax, Vietri et al. demonstrated that NHPs administered ciprofloxacin postexposure for 14 days had a high mortality rate after antibiotics were discontinued. Adding the AVA vaccine to postexposure antibiotic prophylaxis enhanced the protection afforded by just 14 days of ciprofloxacin prophylaxis alone and completely protected animals against inhalational anthrax (344). An additional approach to postexposure prophylaxis involves combining antitoxin antibodies with antibiotics to target both the toxin and the organism, as illustrated with different animal models and described elsewhere in this review. Similar studies evaluating such combination strategies in NHPs are currently being evaluated. Dyer et al. developed a therapeutic African green monkey model for inhalational anthrax for use in demonstrating specifically the value of combination therapies. In this model, African green monkeys were treated with IV ciprofloxacin at various times after PA was detected in the blood; a delay in treatment resulted in reduced survival and thus permitted the demonstration of added benefit of adjunctive therapies such as antibodies (356). Using a similar scheme, Henning et al. assessed the efficacy of a monoclonal anti-PA given in combination with ciprofloxacin in a cynomolgus macaque therapeutic model (357). Antibiotic alone or in combination with the monoclonal anti-PA was given 24 h after confirmed bacteremia and toxemia and, whereas 62% of the animals given the combination therapy survived, only 15% of those given antibiotic alone lived. However, not withstanding these positive results, the addition of antitoxin antibodies to an antibiotic regimen might not be optimal for preventing anthrax resulting from the residual spores that may germinate after discontinuation of the antibiotic, and repeated doses of antibodies may be required.

In 2002, the FDA developed regulations allowing approval of human drugs and biological products based solely on animal studies when human efficacy studies would be unethical, as discussed above and elsewhere (358). The similarities between anthrax in NHPs and humans combined with the FDA regulatory requirements highlight the value of this model for research aimed at the development of improved medical countermeasures for inhalational anthrax.

Rats

Although relatively resistant to parenteral challenge with spores, rats such as the Fisher 344 strain are extremely sensitive to the lethal effects of injected anthrax toxin (359–363). Death can occur in as little as 37 min, and this precise lethal phenotype was utilized by earlier workers as a biofunctional assay for anthrax toxin measurement (364). The role of LT in this lethal phenotype was confirmed more recently in studies employing purified toxin components. More detailed descriptions of the rat toxin model are available in a recent review (179).

The rat was recently exploited as another animal model for dissecting the genetic basis of host susceptibility to LT (365, 366). A recombinant inbred panel of 19 rat strains generated from LT-sensitive and LT-resistant progenitors was used to map LT sensitivity in rats to a locus on chromosome 10 that includes the inflammasome NLR sensor, Nlrp1 (rat homolog of mouse Nlrp1b) (365–367). In mice, Nlrp1b controls MΦ sensitivity to LT *in vitro*, but mouse susceptibility to LT is influenced by additional factors. In contrast, an absolute correlation between *in vitro* MΦ sensitivity to LT-induced lysis and animal susceptibility to the toxin was found in rats. These data suggest that toxin-mediated lethality in rats as well as MΦ sensitivity in this animal model are controlled by a single locus on chromosome 10 that appears to be the inflammasome NLR sensor, Nlrp1.

The rat toxin model with its well-defined endpoint provides a very sensitive test of the efficacy of the various antitoxin countermeasures being developed, such as monoclonal antibodies, small-molecule inhibitors or inactive substrate analogs, toxin receptors, and dominant-negative toxins (368–372). For instance, Scobie et al. reported the first use of a soluble toxin receptor decoy inhibitor, based on the von Willebrand factor A (VWA) domain of the toxin receptor ANTXR2, as a candidate anthrax therapeutic (371). This molecule protected rats against killing by LT (PA + LF) and also neutralized altered forms of PA that were resistant to anti-PA monoclonal antibody neutralization (373). Thus, such countermeasures might be especially useful to treat disease caused by either vaccine- or antibiotic-resistant bacterial strains. More recent efforts have employed modified forms generated by fusing the receptor decoy to human IgG to optimize serum half life (Manchester and Young, unpublished) (156). In addition, the therapeutic efficacy of viral-like particles (VLPs) complexed with toxin receptor was demonstrated in rats, and the vaccine potential of VLPs complexed with PA was shown; the latter elicited high anti-PA titers and

protected rats from anthrax LT challenge after a single immunization (374). Finally, Little et al. characterized the toxin-neutralizing or toxin-enhancing activities of monoclonal anti-PA antibodies in a rat model, and demonstrated for the first time that certain anti-PA monoclonal antibodies can enhance LT intoxication *in vivo* (375). Such studies could help guide the development of protective human monoclonal antitoxin antibodies to ensure their specificity for LT-neutralizing and not LT-enhancing epitopes.

Other Models

Other live model systems have been described and generally employed for specific purposes in anthrax research. The reader is referred to other sources for more information on these models (3, 53, 376). Such models span a wide range of living systems and applications; for example, the zebra fish, a cold-blooded animal used for transparent studies on the anthrax toxins and their interactions with host cells (377); highly resistant species, such as dogs and swine, for studies on anthrax pathogenesis (37, 53, 376); Syrian hamsters and their responses to live and defined chemical vaccines (378–380); wild or domesticated animals and their possible roles as intermediate hosts allowing genetic exchange between *B. cereus* group members (381); and invertebrates and plants and their possible roles in exploring the hypothesized existence of potential cycles of germination and sporulation of *B. anthracis* outside mammalian hosts (382).

CONCLUSIONS AND COMMENTS ON SELECTION OF AN ANIMAL MODEL

There has been a long history of the use in anthrax research of animals, from species in lower phylogenetic orders to higher mammals, and each has provided valuable contributions. As clearly evident from this review, the selection of a model for these studies is complex, as befitting such a toxigenic and invasive pathogen, and it is dependent on the specific aim of the research or test (72, 93, 376). It should take into consideration the attributes (advantages and disadvantages) both of the different animal species and of the manner and route of exposure, as well as other factors such as economic and ethical considerations. Table 2 provides a summary of some of these considerations. NHPs are generally considered the animals most amenable for extrapolation to humans. However, no one animal species is equivalent to a human, just as no one individual or group of people can represent responses of all humans, which are influenced by factors associated with biology (e.g., immunodeficiencies), genetics, and socioeconomic conditions (72, 94, 95). In addition, the ongoing debate on anthrax pathogenesis and the ultimate cause(s) of an irreversibly terminal infection argue that information and insights from research involving various animal models continue to be leveraged. Therefore, as reiterated in other recent reviews (72, 376), by assembling the data appropriately contributed by each animal model, a more complete view of anthrax will evolve that can then be more effectively extrapolated to the disease in humans.

Note in Proof. Recently an article was published (reference 394) describing how the positive results obtained using anthrax animal models (rabbit and nonhuman primate) were used to drive FDA licensure of a monoclonal antibody for prevention and treatment of anthrax in people. This antibody, Raxibacumab, was the first biologic product to be approved for licensure through the FDA animal efficacy rule ("Animal Rule"). This product was approved for the prevention (prophylaxis) of inhalational anthrax when necessary, and for the treatment of the disease when administered in combination with an appropriate antibiotic.

Acknowledgments. The views expressed in this report are those of the authors and do not reflect official policy or position of the Department of the Army, Department of Defense, or the U.S. Government.

Citation. Welkos S, Bozue J, Twenhafel N, Cote C. 2015. Animal models for the pathogenesis, treatment, and prevention of infection by Bacillus anthracis. Microbiol Spectrum 3(1): TBS-0001-2012.

References

1. Davies JC. 1983. A major epidemic of anthrax in Zimbabwe. Part II. *Cent Afr J Med* 29:8–12.
2. De Vos V. 1990. The ecology of anthrax in the Kruger National Park, South Africa. *Salisbury Med Bull* 87: 26–30.
3. Turnbull P. 2008. *Anthrax in Humans and Animals*, 4th ed. WHO Press, Geneva, Switzerland.
4. Brachman P, Friedlander A, Grabenstein J. 2008. Anthrax vaccine, p 111–126. *In* Plotkin S, Orenstein W, Offit P (ed), *Vaccines*, 5th ed. Saunders Elsevier Press, Philadelphia, PA.
5. Friedlander A, Little S, Gurwith M. 2009. Anthrax vaccines, p 851–863. *In* Levine MM, Dougan G, Good MF, Liu MA (ed), *New Generation Vaccines*, 4th ed. Informa Healthcare, New York, NY.
6. Laforce FM. 1978. Woolsorters' disease in England. *Bull N Y Acad Med* 54:956–963.
7. Purcell BK, Worsham PL, Friedlander AM. 2007. Anthrax, p 69–90. *In* Dembek ZF (ed), *Medical Aspects of Biological Warfare*. Borden Institute, Walter Reed Army Medical Center, Washington, DC.
8. Plotkin SA, Brachman PS, Utell M, Bumford FH, Atchison MM. 1960. An epidemic of inhalation anthrax, the first in the twentieth century. I. Clinical features. *Am J Med* 29:992–1001.
9. Cieslak TJ, Henretig FM. 2001. Medical consequences of biological warfare: the Ten Commandments of management. *Mil Med* 166:11–12.
10. Bell DM, Kozarsky PE, Stephens DS. 2002. Clinical issues in the prophylaxis, diagnosis, and treatment of anthrax. *Emerg Infect Dis* 8:222–225.
11. Bell JH, Fee E, Brown TM. 2002. Anthrax and the wool trade. 1902. *Am J Public Health* 92:754–757.
12. Dixon TC, Meselson M, Guillemin J, Hanna PC. 1999. Anthrax. *N Engl J Med* 341:815–826.
13. Guarner J, Del Rio C. 2011. Pathology, diagnosis, and treatment of anthrax in humans, p 251–268. *In* Bergman NH (ed), *Bacillus anthracis and anthrax*. John Wiley & Sons, Inc., Hoboken, NJ.
14. Holty JE, Bravata DM, Liu H, Olshen RA, McDonald KM, Owens DK. 2006. Systematic review: a century of inhalational anthrax cases from 1900 to 2005. *Ann Intern Med* 144:270–280.
15. Holty JE, Kim RY, Bravata DM. 2006. Anthrax: a systematic review of atypical presentations. *Ann Emerg Med* 48:200–211.
16. Inglesby TV, O'Toole T, Henderson DA, Bartlett JG, Ascher MS, Eitzen E, Friedlander AM, Gerberding J, Hauer J, Hughes J, McDade J, Osterholm MT, Parker G, Perl TM, Russell PK, Tonat K. 2002. Anthrax as a biological weapon, 2002: updated recommendations for management. *JAMA* 287:2236–2252.
17. Beatty ME, Ashford DA, Griffin PM, Tauxe RV, Sobel J. 2003. Gastrointestinal anthrax: review of the literature. *Arch Intern Med* 163:2527–2531.
18. Mock M, Fouet A. 2001. Anthrax. *Annu Rev Microbiol* 55:647–671.
19. Grinberg LM, Abramova FA, Yampolskaya OV, Walker DH, Smith JH. 2001. Quantitative pathology of inhalational anthrax I: quantitative microscopic findings. *Mod Pathol* 14:482–495.
20. ProMED-mail. 2005. *Anthrax, human-Russia (Northern Osetia)*. http://www.promedmail.org. Accessed 14 September 2005.
21. Mayo L, Dionne-Odom J, Talbot EA, Adamski C, Bean C, Daly ER. 2009. Gastrointestinal anthrax after an animal-hide drumming event – New Hampshire and Massachusetts, 2009. *MMWR Morb Mortal Wkly Rep* 59:872–877.
22. ProMED-mail. 2009. *Anthrax, human, 2006-UK*. http://www.promedmail.org. Accessed 21 April 2009.
23. ProMED-mail. 2010. *Anthrax-United Kingdom: (Scotland)*. http://www.promedmail.org. Accessed 29 January 2010.
24. Ringertz SH, Hoiby EA, Jensenius M, Maehlen J, Caugant DA, Myklebust A, Fossum K. 2000. Injectional anthrax in a heroin skin-popper. *Lancet* 356:1574–1575.
25. Walsh J, Fraser G, Hunt B, Husband B, Nalluswami K, Pollard K. 2006. Inhalation anthrax associated with dried animal hides – Pennsylvania and New York City, 2006. *MMWR Morb Mortal Wkly Rep* 55:280–282.

26. Abramova AA, Grinberg LM. 1993. Pathology of anthrax sepsis according to materials of the infectious outbreak in 1979 in Sverdlovsk (microscopic changes). *Arkh Patol* 55:18–23. (In Russian.)

27. Guarner J, Jernigan JA, Shieh WJ, Tatti K, Flannagan LM, Stephens DS, Popovic T, Ashford DA, Perkins BA, Zaki SR. 2003. Pathology and pathogenesis of bioterrorism-related inhalational anthrax. *Am J Pathol* 163:701–709.

28. Jernigan JA, Stephens DS, Ashford DA, Omenaca C, Topiel MS, Galbraith M, Tapper M, Fisk TL, Zaki S, Popovic T, Meyer RF, Quinn CP, Harper SA, Fridkin SK, Sejvar JJ, Shepard CW, McConnell M, Guarner J, Shieh WJ, Malecki JM, Gerberding JL, Hughes JM, Perkins BA. 2001. Bioterrorism-related inhalational anthrax: the first 10 cases reported in the United States. *Emerg Infect Dis* 7:933–944.

29. Keim P, Smith KL, Keys C, Takahashi H, Kurata T, Kaufmann A. 2001. Molecular investigation of the Aum Shinrikyo anthrax release in Kameido, Japan. *J Clin Microbiol* 39:4566–4567.

30. Meselson M, Guillemin J, Hugh-Jones M, Langmuir A, Popova I, Shelokov A, Yampolskaya O. 1994. The Sverdlovsk anthrax outbreak of 1979. *Science* 266:1202–1208.

31. Zilinskas RA. 1997. Iraq's biological weapons. The past as future? *JAMA* 278:418–424.

32. Koch R. 1938. Beitrage zur Biologie der Pflanzen. *Med Classics* 2:787–820.

33. Pasteur L. 1881. De l'attenuation des virus et de leur retour à la virulence. *C R Acad Sci Agric Bulg* 1881, 92:429–435.

34. Pasteur L, Chamberland, Roux. 2002. Summary report of the experiments conducted at Pouilly-le-Fort, near Melun, on the anthrax vaccination, 1881. *Yale J Biol Med* 75:59–62.

35. Turnbull PCB, Shadomy SV. 2011. Anthrax from 5000 BC to AD2010, p 1–16. *In* Bergman NH (ed), *Bacillus anthracis and anthrax*. John Wiley & Sons, Inc., Hoboken, NJ.

36. Buchner H, Merkel F, Enderlen E. 1888. *Arch Hyg* (Berlin) 8:145.

37. Young GA, Zelle MR, Lincoln RE. 1946. Respiratory pathogenecity of *Bacillus anthracis* spores. I. Methods of study and observation on pathogenesis. *J Infect Dis* 79:233–246.

38. Barnes JM. 1947. The development of anthrax following the administration of spores by inhalation. *Br J Exp Pathol* 28:385–393.

39. Ross JM. 1957. The pathogenesis of anthrax following the administration of spores by the respiratory route. *J Pathol Bacteriol* 73:485–494.

40. Albrink WS, Goodlow RJ. 1959. Experimental inhalation anthrax in the chimpanzee. *Am J Pathol* 35:1055–1065.

41. Henderson DW, Peacock S, Belton FC. 1956. Observations on the prophylaxis of experimental pulmonary anthrax in the monkey. *J Hyg* (Lond) 54:28–36.

42. Lincoln RE, Hodges DR, Klein F, Mahlandt BG, Jones WI Jr, Haines BW, Rhian MA, Walker JS. 1965. Role of the lymphatics in the pathogenesis of anthrax. *J Infect Dis* 115:481–494.

43. Abramova FA, Grinberg LM, Yampolskaya OV, Walker DH. 1993. Pathology of inhalational anthrax in 42 cases from the Sverdlovsk outbreak of 1979. *Proc Natl Acad Sci USA* 90:2291–2294.

44. Albrink WS, Brooks SM, Biron RE, Kopel M. 1960. Human inhalation anthrax. A report of three fatal cases. *Am J Pathol* 36:457–471.

45. Brachman PS, Plotkin SA, Bumford FH, Atchison MM. 1960. An epidemic of inhalation anthrax: the first in the twentieth century. II. Epidemiology. *Am J Hyg* 72:6–23.

46. Grinberg LM, Abramova AA. 1993. Pathology of anthrax sepsis according to materials of the infectious outbreak in 1979 in Sverdlovsk (various aspects of morpho-, patho- and thanatogenesis). *Arkh Patol* 55:23–26. (In Russian.)

47. Bail O. 1914. Changes to bacteria inside the animal body. IX. Regarding the correlation between capsule formation, spore formation and infectiousness of the anthrax bacillus. *Zentb Bakteriol* 75:159–173.

48. Shlyakhov EN, Rubinstein E. 1994. Human live anthrax vaccine in the former USSR. *Vaccine* 12:727–730.

49. Turnbull PC. 1991. Anthrax vaccines: past, present and future. *Vaccine* 9:533–539.

50. Keppie J, Smith H, Harris-Smith PW. 1953. The chemical basis of the virulence of bacillus anthracis. II. Some biological properties of bacterial products. *Br J Exp Pathol* 34:486–496.

51. Keppie J, Smith H, Harris-Smith PW. 1955. The chemical basis of the virulence of Bacillus anthracis. III. The role of the terminal bacteraemia in death of guinea-pigs from anthrax. *Br J Exp Pathol* 36:315–322.

52. Evans DG, Shoesmith JG. 1954. Production of toxin by Bacillus anthracis. *Lancet* 263:136. doi:10.1016/S0140-6736(54)90984-8.

53. Lincoln RE, Walker JS, Klein F, Rosenwald AJ, Jones WI Jr. 1967. Value of field data for extrapolation in anthrax. *Fed Proc* 26:1558–1562.

54. Abalakin VA, Cherkasskii BL. 1978. Use of inbred mice as a model for the indication and differentiation of Bacillus-anthracis strains. *Zh Mikrobiol Epidemiol Immunobiol*146–147. (In Russian.)

55. Welkos SL, Keener TJ, Gibbs PH. 1986. Differences in susceptibility of inbred mice to *Bacillus anthracis*. *Infect Immun* 51:795–800.

56. Coleman ME, Thran B, Morse SS, Hugh-Jones M, Massulik S. 2008. Inhalation anthrax: dose response and risk analysis. *Biosecur Bioterror* 6:147–160.

57. Bartrand TA, Weir MH, Haas CN. 2008. Dose-response models for inhalation of *Bacillus anthracis* spores: interspecies comparisons. *Risk Anal* 28:1115–1124.

58. Brookmeyer R, Johnson E, Barry S. 2005. Modelling the incubation period of anthrax. *Stat Med* 24:531–542.

59. Cohen M, Whalen T. 2007. Implications of low level human exposure to respirable *B. anthracis*. *Appl Biosafety* 12:109–115.

60. Glassman HN. 1966. Industrail inhalational anthrax. Discussion. *Bacteriol Rev* 30:657–659.

61. Webb GF, Blaser MJ. 2002. Mailborne transmission of anthrax: Modeling and implications. *Proc Natl Acad Sci USA* 99:7027–7032.

62. Wilkening DA. 2006. Sverdlovsk revisited: modeling human inhalation anthrax. *Proc Natl Acad Sci USA* 103:7589–7594.

63. Wilkening DA. 2008. Modeling the incubation period of inhalational anthrax. *Med Decis Making* 28:593–605.

64. Nelson M, Stagg AJ, Stevens DJ, Brown MA, Pearce PC, Simpson AJ, Lever MS. 2011. Post-exposure therapy of inhalational anthrax in the common marmoset. *Int J Antimicrob Agents* 38:60–64.

65. Twenhafel NA, Leffel E, Pitt ML. 2007. Pathology of inhalational anthrax infection in the african green monkey. *Vet Pathol* 44:716–721.

66. Liu S, Miller-Randolph S, Crown D, Moayeri M, Sastalla I, Okugawa S, Leppla SH. 2010. Anthrax toxin targeting of myeloid cells through the CMG2 receptor is essential for establishment of *Bacillus anthracis* infections in mice. *Cell Host Microbe* 8:455–462.

67. McAllister RD, Singh Y, du Bois WD, Potter M, Boehm T, Meeker ND, Fillmore PD, Anderson LM, Poynter ME, Teuscher C. 2003. Susceptibility to anthrax lethal toxin is controlled by three linked quantitative trait loci. *Am J Pathol* 163:1735–1741.

68. Moayeri M, Crown D, Newman ZL, Okugawa S, Eckhaus M, Cataisson C, Liu S, Sastalla I, Leppla SH. 2010. Inflammasome sensor Nlrp1b-dependent resistance to anthrax is mediated by caspase-1, IL-1 signaling and neutrophil recruitment. *PLoS Pathog* 6: e1001222. doi:10.1371/journal.ppat.1001222.

69. Terra JK, Cote CK, France B, Jenkins AL, Bozue JA, Welkos SL, LeVine SM, Bradley KA. 2010. Cutting edge: resistance to Bacillus anthracis infection mediated by a lethal toxin sensitive allele of Nalp1b/Nlrp1b. *J Immunol* 184:17–20.

70. Watters JW, Dietrich WF. 2001. Genetic, physical, and transcript map of the Ltxs1 region of mouse chromosome 11. *Genomics* 73:223–231.

71. Yadav JS, Pradhan S, Kapoor R, Bangar H, Burzynski BB, Prows DR, Levin L. 2011. Multigenic control and sex bias in host susceptibility to spore-induced pulmonary anthrax in mice. *Infect Immun* 79:3204–3215.

72. Hau J. 2008. Animal models for human diseases, p 3–8. *In* Conn PM (ed), *Sourcebook of Models for Biomedical Research*. Humana Press, Inc., Towota, NJ.

73. Phipps AJ, Premanandan C, Barnewall RE, Lairmore MD. 2004. Rabbit and nonhuman primate models of toxin-targeting human anthrax vaccines. *Microbiol Mol Biol Rev* 68:617–629.

74. Cote CK, Rea KM, Norris SL, van Rooijen N, Welkos SL. 2004. The use of a model of in vivo macrophage depletion to study the role of macrophages during

infection with Bacillus anthracis spores. *Microb Pathog* 37:169–175.

75. Cote CK, van Rooijen N, Welkos SL. 2006. The roles of macrophages and neutrophils in the early host response to *Bacillus anthracis* spores using a mouse model of infection. *Infect Immun* 74:469–480.

76. Dumetz F, Jouvion G, Khun H, Glomski IJ, Corre JP, Rougeaux C, Tang WJ, Mock M, Huerre M, Goossens PL. 2011. Noninvasive imaging technologies reveal edema toxin as a key virulence factor in anthrax. *Am J Pathol* 178:2523–2535.

77. Heine HS, Bassett J, Miller L, Hartings JM, Ivins BE, Pitt ML, Fritz D, Norris SL, Byrne WR. 2007. Determination of antibiotic efficacy against *Bacillus anthracis* in a mouse aerosol challenge model. *Antimicrob Agents Chemother* 51:1373–1379.

78. Lyons CR, Lovchik J, Hutt J, Lipscomb MF, Wang E, Heninger S, Berliba L, Garrison K. 2004. Murine model of pulmonary anthrax: kinetics of dissemination, histopathology, and mouse strain susceptibility. *Infect Immun* 72:4801–4809.

79. Glomski IJ, Dumetz F, Jouvion G, Huerre MR, Mock M, Goossens PL. 2008. Inhaled non-capsulated *Bacillus anthracis* in A/J mice: nasopharynx and alveolar space as dual portals of entry, delayed dissemination, and specific organ targeting. *Microbes Infect* 10:1398–1404.

80. Cleret A, Quesnel-Hellmann A, Vallon-Eberhard A, Verrier B, Jung S, Vidal D, Mathieu J, Tournier JN. 2007. Lung dendritic cells rapidly mediate anthrax spore entry through the pulmonary route. *J Immunol* 178:7994–8001.

81. Duong S, Chiaraviglio L, Kirby JE. 2006. Histopathology in a murine model of anthrax. *Int J Exp Pathol* 87:131–137.

82. Glomski IJ, Corre JP, Mock M, Goossens PL. 2007. Noncapsulated toxinogenic Bacillus anthracis presents a specific growth and dissemination pattern in naive and protective antigen-immune mice. *Infect Immun* 75:4754–4761.

83. Glomski IJ, Piris-Gimenez A, Huerre M, Mock M, Goossens PL. 2007. Primary involvement of pharynx and Peyer's patch in inhalational and intestinal anthrax. *PLoS Pathog* 3:e76. doi:10.1371/journal.ppat.0030076.

84. Loving CL, Kennett M, Lee GM, Grippe VK, Merkel TJ. 2007. Murine aerosol challenge model of anthrax. *Infect Immun* 75:2689–2698.

85. Loving CL, Khurana T, Osorio M, Lee GM, Kelly VK, Stibitz S, Merkel TJ. 2009. Role of anthrax toxins in dissemination, disease progression, and induction of protective adaptive immunity in the mouse aerosol challenge model. *Infect Immun* 77:255–265.

86. Heninger S, Drysdale M, Lovchik J, Hutt J, Lipscomb MF, Koehler TM, Lyons CR. 2006. Toxin-deficient mutants of *Bacillus anthracis* are lethal in a murine model for pulmonary anthrax. *Infect Immun* 74:6067–6074.

87. Welkos SL. 1991. Plasmid-associated virulence factors of non-toxigenic (pX01-) *Bacillus anthracis*. *Microb Pathog* 10:183–198.

88. Welkos SL, Vietri NJ, Gibbs PH. 1993. Non-toxigenic derivatives of the Ames strain of Bacillus anthracis are fully virulent for mice: role of plasmid pX02 and chromosome in strain-dependent virulence. *Microb Pathog* **14**:381–388.

89. Friedlander AM, Welkos SL, Ivins BE. 2002. Anthrax vaccines. *Curr Top Microbiol Immunol* **271**:33–60.

90. Ivins BE, Ezzell JW Jr, Jemski J, Hedlund KW, Ristroph JD, Leppla SH. 1986. Immunization studies with attenuated strains of Bacillus anthracis. *Infect Immun* **52**: 454–458.

91. Fellows PF, Linscott MK, Ivins BE, Pitt ML, Rossi CA, Gibbs PH, Friedlander AM. 2001. Efficacy of a human anthrax vaccine in guinea pigs, rabbits, and rhesus macaques against challenge by Bacillus anthracis isolates of diverse geographical origin. *Vaccine* **19**:3241–3247.

92. Waterston RH, Lindblad-Toh K, Birney E, Rogers J, Abril JF, Agarwal P, Agarwala R, Ainscough R, Alexandersson M, An P, Antonarakis SE, Attwood J, Baertsch R, Bailey J, Barlow K, Beck S, Berry E, Birren B, Bloom T, Bork P, Botcherby M, Bray N, Brent MR, Brown DG, Brown SD, Bult C, Burton J, Butler J, Campbell RD, Carninci P, Cawley S, Chiaromonte F, Chinwalla AT, Church DM, Clamp M, Clee C, Collins FS, Cook LL, Copley RR, Coulson A, Couronne O, Cuff J, Curwen V, Cutts T, Daly M, David R, Davies J, Delehaunty KD, Deri J, Dermitzakis ET, Dewey C, Dickens NJ, Diekhans M, Dodge S, Dubchak I, Dunn DM, Eddy SR, Elnitski L, Emes RD, Eswara P, Eyras E, Felsenfeld A, Fewell GA, Flicek P, Foley K, Frankel WN, Fulton LA, Fulton RS, Furey TS, Gage D, Gibbs RA, Glusman G, Gnerre S, Goldman N, Goodstadt L, Grafham D, Graves TA, Green ED, Gregory S, Guigó R, Guyer M, Hardison RC, Haussler D, Hayashizaki Y, Hillier LW, Hinrichs A, Hlavina W, Holzer T, Hsu F, Hua A, Hubbard T, Hunt A, Jackson I, Jaffe DB, Johnson LS, Jones M, Jones TA, Joy A, Kamal M, Karlsson EK, Karolchik D, Kasprzyk A, Kawai J, Keibler E, Kells C, Kent WJ, Kirby A, Kolbe DL, Korf I, Kucherlapati RS, Kulbokas EJ, Kulp D, Landers T, Leger JP, Leonard S, Letunic I, Levine R, Li J, Li M, Lloyd C, Lucas S, Ma B, Maglott DR, Mardis ER, Matthews L, Mauceli E, Mayer JH, McCarthy M, McCombie WR, McLaren S, McLay K, McPherson JD, Meldrim J, Meredith B, Mesirov JP, Miller W, Miner TL, Mongin E, Montgomery KT, Morgan M, Mott R, Mullikin JC, Muzny DM, Nash WE, Nelson JO, Nhan MN, Nicol R, Ning Z, Nusbaum C, O'Connor MJ, Okazaki Y, Oliver K, Overton-Larty E, Pachter L, Parra G, Pepin KH, Peterson J, Pevzner P, Plumb R, Pohl CS, Poliakov A, Ponce TC, Ponting CP, Potter S, Quail M, Reymond A, Roe BA, Roskin KM, Rubin EM, Rust AG, Santos R, Sapojnikov V, Schultz B, Schultz J, Schwartz MS, Schwartz S, Scott C, Seaman S, Searle S, Sharpe T, Sheridan A, Shownkeen R, Sims S, Singer JB, Slater G, Smit A, Smith DR, Spencer B, Stabenau A, Stange-Thomann N, Sugnet C, Suyama M, Tesler G, Thompson J, Torrents D, Trevaskis E, Tromp J, Ucla C, Ureta-Vidal A, Vinson JP, Von Niederhausern AC, Wade CM, Wall M, Weber RJ, Weiss RB, Wendl MC, West AP, Wetterstrand K, Wheeler R, Whelan S,

Wierzbowski J, Willey D, Williams S, Wilson RK, Winter E, Worley KC, Wyman D, Yang S, Yang SP, Zdobnov EM, Zody MC, Lander ES. 2002. Initial sequencing and comparative analysis of the mouse genome. *Nature* **420**:520–562.

93. Haley PJ. 2003. Species differences in the structure and function of the immune system. *Toxicology* **188**:49–71.

94. Mestas J, Hughes CC. 2004. Of mice and not men: differences between mouse and human immunology. *J Immunol* **172**:2731–2738.

95. Mizgerd JP, Skerrett SJ. 2008. Animal models of human pneumonia. *Am J Physiol Lung Cell Mol Physiol* **294**: L387–L398.

96. Deziel MR, Heine H, Louie A, Kao M, Byrne WR, Basset J, Miller L, Bush K, Kelly M, Drusano GL. 2005. Effective antimicrobial regimens for use in humans for therapy of Bacillus anthracis infections and postexposure prophylaxis. *Antimicrob Agents Chemother* **49**: 5099–5106.

97. Castelan-Vega J, Corvette L, Sirota L, Arciniega J. 2011. Reduction of immunogenicity of anthrax vaccines subjected to thermal stress, as measured by a toxin neutralization assay. *Clin Vaccine Immunol* **18**:349–351.

98. Little SF, Webster WM, Ivins BE, Fellows PF, Norris SL, Andrews GP. 2004. Development of an in vitro-based potency assay for anthrax vaccine. *Vaccine* **22**: 2843–2852.

99. Albrecht MT, Li H, Williamson ED, LeButt CS, Flick-Smith HC, Quinn CP, Westra H, Galloway D, Mateczun A, Goldman S, Groen H, Baillie LW. 2007. Human monoclonal antibodies against anthrax lethal factor and protective antigen act independently to protect against *Bacillus anthracis* infection and enhance endogenous immunity to anthrax. *Infect Immun* **75**: 5425–5433.

100. Williamson ED, Duchars MG, Kohberger R. 2010. Predictive models and correlates of protection for testing biodefence vaccines. *Expert Rev Vaccines* **9**:527–537.

101. Beedham RJ, Turnbull PC, Williamson ED. 2001. Passive transfer of protection against Bacillus anthracis infection in a murine model. *Vaccine* **19**:4409–4416.

102. Welkos S, Becker D, Friedlander A, Trotter R. 1989. Pathogenesis and host response to *Bacillus anthracis*: a mouse model. *Salisbury Med Bull* **87**:49–52.

103. Welkos SL, Friedlander AM. 1988. Pathogenesis and genetic control of resistance to the Sterne strain of *Bacillus anthracis*. *Microb Pathog* **4**:53–69.

104. Welkos SL, Trotter RW, Becker DM, Nelson GO. 1989. Resistance to the Sterne strain of *B. anthracis*: phagocytic cell responses of resistant and susceptible mice. *Microb Pathog* **7**:15–35.

105. Harvill ET, Osorio M, Loving CL, Lee GM, Kelly VK, Merkel TJ. 2008. Anamnestic protective immunity to *Bacillus anthracis* is antibody mediated but independent of complement and Fc receptors. *Infect Immun* **76**: 2177–2182.

106. Pickering AK, Osorio M, Lee GM, Grippe VK, Bray M, Merkel TJ. 2004. Cytokine response to infection with *Bacillus anthracis* spores. *Infect Immun* **72**:6382–6389.

107. Sanz P, Teel LD, Alem F, Carvalho HM, Darnell SC, O'Brien AD. 2008. Detection of *Bacillus anthracis* spore germination in vivo by bioluminescence imaging. *Infect Immun* **76**:1036–1047.

108. Kang TJ, Basu S, Zhang L, Thomas KE, Vogel SN, Baillie L, Cross AS. 2008. *Bacillus anthracis* spores and lethal toxin induce IL-1beta via functionally distinct signaling pathways. *Eur J Immunol* **38**:1574–1584.

109. Popov SG, Popova TG, Grene E, Klotz F, Cardwell J, Bradburne C, Jama Y, Maland M, Wells J, Nalca A, Voss T, Bailey C, Alibek K. 2004. Systemic cytokine response in murine anthrax. *Cell Microbiol* **6**:225–233.

110. Harvill ET, Lee G, Grippe VK, Merkel TJ. 2005. Complement depletion renders C57BL/6 mice sensitive to the *Bacillus anthracis* Sterne strain. *Infect Immun* **73**:4420–4422.

111. Gleiser CA, Berdjis CC, Hartman HA, Gochenour WS. 1963. Pathology of experimental respiratory anthrax in *Macaca mulatta*. *Br J Exp Pathol* **44**:416–426.

112. Bozue J, Cote CK, Moody KL, Welkos SL. 2007. Fully virulent *Bacillus anthracis* does not require the immunodominant protein, BclA, for pathogenesis. *Infect Immun* **75**:508–511.

113. Friedlander AM, Welkos SL, Pitt ML, Ezzell JW, Worsham PL, Rose KJ, Ivins BE, Lowe JR, Howe GB, Mikesell P. 1993. Postexposure prophylaxis against experimental inhalation anthrax. *J Infect Dis* **167**:1239–1243.

114. Fritz DL, Jaax NK, Lawrence WB, Davis KJ, Pitt ML, Ezzell JW, Friedlander AM. 1995. Pathology of experimental inhalation anthrax in the rhesus monkey. *Lab Invest* **73**:691–702.

115. Guidi-Rontani C. 2002. The alveolar macrophage: the Trojan horse of *Bacillus anthracis*. *Trends Microbiol* **10**:405–409.

116. Guidi-Rontani C, Weber-Levy M, Labruyere E, Mock M. 1999. Germination of Bacillus anthracis spores within alveolar macrophages. *Mol Microbiol* **31**:9–17.

117. Carr KA, Lybarger SR, Anderson EC, Janes BK, Hanna PC. 2010. The role of *Bacillus anthracis* germinant receptors in germination and virulence. *Mol Microbiol* **75**:365–375.

118. McKevitt MT, Bryant KM, Shakir SM, Larabee JL, Blanke SR, Lovchik J, Lyons CR, Ballard JD. 2007. Effects of endogenous D-alanine synthesis and autoinhibition of *Bacillus anthracis* germination on in vitro and in vivo infections. *Infect Immun* **75**:5726–5734.

119. Cote CK, Bozue J, Twenhafel N, Welkos SL. 2009. Effects of altering the germination potential of *Bacillus anthracis* spores by exogenous means in a mouse model. *J Med Microbiol* **58**:816–825.

120. Fisher N, Shetron-Rama L, Herring-Palmer A, Heffernan B, Bergman N, Hanna P. 2006. The dltABCD operon of Bacillus anthracis sterne is required for virulence and resistance to peptide, enzymatic, and cellular mediators of innate immunity. *J Bacteriol* **188**:1301–1309.

121. Cendrowski S, MacArthur W, Hanna P. 2004. Bacillus anthracis requires siderophore biosynthesis for growth in macrophages and mouse virulence. *Mol Microbiol* **51**:407–417.

122. Chung MC, Popova TG, Jorgensen SC, Dong L, Chandhoke V, Bailey CL, Popov SG. 2008. Degradation of circulating von Willebrand factor and its regulator ADAMTS13 implicates secreted *Bacillus anthracis* metalloproteases in anthrax consumptive coagulopathy. *J Biol Chem* **283**:9531–9542.

123. Chung MC, Popova TG, Millis BA, Mukherjee DV, Zhou W, Liotta LA, Petricoin EF, Chandhoke V, Bailey C, Popov SG. 2006. Secreted neutral metalloproteases of Bacillus anthracis as candidate pathogenic factors. *J Biol Chem* **281**:31408–31418.

124. Cowan GJ, Atkins HS, Johnson LK, Titball RW, Mitchell TJ. 2007. Immunisation with anthrolysin O or a genetic toxoid protects against challenge with the toxin but not against *Bacillus anthracis*. *Vaccine* **25**:7197–7205.

125. Heffernan BJ, Thomason B, Herring-Palmer A, Shaughnessy L, McDonald R, Fisher N, Huffnagle GB, Hanna P. 2006. *Bacillus anthracis* phospholipases C facilitate macrophage-associated growth and contribute to virulence in a murine model of inhalation anthrax. *Infect Immun* **74**:3756–3764.

126. Honsa ES, Fabian M, Cardenas AM, Olson JS, Maresso AW. 2011. The five near-iron transporter (NEAT) domain anthrax hemophore, IsdX2, scavenges heme from hemoglobin and transfers heme to the surface protein IsdC. *J Biol Chem* **286**:33652–33660.

127. Honsa ES, Maresso AW. 2011. Mechanisms of iron import in anthrax. *Biometals* **24**:533–545.

128. Mosser EM, Rest RF. 2006. The *Bacillus anthracis* cholesterol-dependent cytolysin, Anthrolysin O, kills human neutrophils, monocytes and macrophages. *BMC Microbiol* **6**:56. doi:10.1186/1471-2180-6-56.

129. Mukherjee DV, Tonry JH, Kim KS, Ramarao N, Popova TG, Bailey C, Popov S, Chung MC. 2011. *Bacillus anthracis* protease InhA increases blood-brain barrier permeability and contributes to cerebral hemorrhages. *PLoS One* **6**:e17921. doi:10.1371/journal.pone.0017921.

130. Shannon JG, Ross CL, Koehler TM, Rest RF. 2003. Characterization of anthrolysin O, the *Bacillus anthracis* cholesterol-dependent cytolysin. *Infect Immun* **71**:3183–3189.

131. Shatalin K, Gusarov I, Avetissova E, Shatalina Y, McQuade LE, Lippard SJ, Nudler E. 2008. *Bacillus anthracis*-derived nitric oxide is essential for pathogen virulence and survival in macrophages. *Proc Natl Acad Sci USA* **105**:1009–1013.

132. Bozue J, Moody KL, Cote CK, Stiles BG, Friedlander AM, Welkos SL, Hale ML. 2007. Bacillus anthracis spores of the bclA mutant exhibit increased adherence to epithelial cells, fibroblasts, and endothelial cells but not to macrophages. *Infect Immun* **75**:4498–4505.

133. Brahmbhatt TN, Janes BK, Stibitz ES, Darnell SC, Sanz P, Rasmussen SB, O'Brien AD. 2007. *Bacillus anthracis* exosporium protein BclA affects spore germination, interaction with extracellular matrix proteins, and hydrophobicity. *Infect Immun* **75**:5233–5239.

134. Oliva C, Turnbough CL Jr, Kearney JF. 2009. CD14-Mac-1 interactions in *Bacillus anthracis* spore

internalization by macrophages. *Proc Natl Acad Sci USA* 106:13957–13962.

135. Sylvestre P, Couture-Tosi E, Mock M. 2002. A collagen-like surface glycoprotein is a structural component of the *Bacillus anthracis* exosporium. *Mol Microbiol* 45:169–178.

136. Flick-Smith HC, Walker NJ, Gibson P, Bullifent H, Hayward S, Miller J, Titball RW, Williamson ED. 2002. A recombinant carboxy-terminal domain of the protective antigen of *Bacillus anthracis* protects mice against anthrax infection. *Infect Immun* 70:1653–1656.

137. Flick-Smith HC, Waters EL, Walker NJ, Miller J, Stagg AJ, Green M, Williamson ED. 2005. Mouse model characterisation for anthrax vaccine development: comparison of one inbred and one outbred mouse strain. *Microb Pathog* 38:33–40.

138. Welkos SL, Friedlander AM. 1988. Comparative safety and efficacy against Bacillus anthracis of protective antigen and live vaccines in mice. *Microb Pathog* 5:127–139.

139. Brahmbhatt TN, Darnell SC, Carvalho HM, Sanz P, Kang TJ, Bull RL, Rasmussen SB, Cross AS, O'Brien AD. 2007. Recombinant exosporium protein BclA of Bacillus anthracis is effective as a booster for mice primed with suboptimal amounts of protective antigen. *Infect Immun* 75:5240–5247.

140. Cybulski RJ Jr, Sanz P, McDaniel D, Darnell S, Bull RL, O'Brien AD. 2008. Recombinant *Bacillus anthracis* spore proteins enhance protection of mice primed with suboptimal amounts of protective antigen. *Vaccine* 26:4927–4939.

141. Duc le H, Hong HA, Atkins HS, Flick-Smith HC, Durrani Z, Rijpkema S, Titball RW, Cutting SM. 2007. Immunization against anthrax using *Bacillus subtilis* spores expressing the anthrax protective antigen. *Vaccine* 25:346–355.

142. Galen JE, Zhao L, Chinchilla M, Wang JY, Pasetti MF, Green J, Levine MM. 2004. Adaptation of the endogenous *Salmonella enterica* serovar Typhi clyA-encoded hemolysin for antigen export enhances the immunogenicity of anthrax protective antigen domain 4 expressed by the attenuated live-vector vaccine strain CVD 908-htrA. *Infect Immun* 72:7096–7106.

143. Garmory HS, Titball RW, Griffin KF, Hahn U, Bohm R, Beyer W. 2003. *Salmonella enterica* serovar Typhimurium expressing a chromosomally integrated copy of the *Bacillus anthracis* protective antigen gene protects mice against an anthrax spore challenge. *Infect Immun* 71:3831–3836.

144. Lee JS, Groebner JL, Hadjipanayis AG, Negley DL, Schmaljohn AL, Welkos SL, Smith LA, Smith JF. 2006. Multiagent vaccines vectored by Venezuelan equine encephalitis virus replicon elicits immune responses to Marburg virus and protection against anthrax and botulinum neurotoxin in mice. *Vaccine* 24:6886–6892.

145. Lee JS, Hadjipanayis AG, Welkos SL. 2003. Venezuelan equine encephalitis virus-vectored vaccines protect mice against anthrax spore challenge. *Infect Immun* 71:1491–1496.

146. Stokes MG, Titball RW, Neeson BN, Galen JE, Walker NJ, Stagg AJ, Jenner DC, Thwaite JE, Nataro JP, Baillie LW, Atkins HS. 2007. Oral administration of a Salmonella enterica-based vaccine expressing *Bacillus anthracis* protective antigen confers protection against aerosolized B. anthracis. *Infect Immun* 75:1827–1834.

147. Tan Y, Hackett NR, Boyer JL, Crystal RG. 2003. Protective immunity evoked against anthrax lethal toxin after a single intramuscular administration of an adenovirus-based vaccine encoding humanized protective antigen. *Hum Gene Ther* 14:1673–1682.

148. Zeng M, Xu Q, Hesek ED, Pichichero ME. 2006. N-fragment of edema factor as a candidate antigen for immunization against anthrax. *Vaccine* 24:662–670.

149. Vergis JM, Cote CK, Bozue J, Alem F, Ventura CL, Welkos SL, O'Brien AD. 2013. Immunization of mice with formalin-inactivated spores from avirulent *Bacillus cereus* strains provides significant protection from challenge with *Bacillus anthracis* Ames. *Clin Vacc Immunol* 20:56–65.

150. Basu S, Kang T, Whitford M, Meldorf M, Lowry I, Cross A. 2009. Mechanisms of an anti-PA antibody (Valortim) that mediate protection against B. *anthracis* infection, abstr *Bacillus*-ACT: The International Conference on *Bacillus anthracis, B. cereus, and B. thuringiensis*, Santa Fe, NM, 30 August to 3 September, 2009.

151. Basu S, Kang TJ, Langermann S, Riddle V, Lowy I, Cross AS. 2007. A monoclonal antibody that neutralizes lethal toxin mediates the killing of B. anthracis by murine macrophages, abstr *Bacillus*-ACT: The International Conference on *Bacillus anthracis, B. cereus, and B. thuringiensis*, Oslo, Norway, 17 to 21 June 2007.

152. Mett V, Chichester JA, Stewart ML, Musiychuk K, Bi H, Reifsnyder CJ, Hull AK, Albrecht MT, Goldman S, Baillie LW, Yusibov V. 2011. A non-glycosylated, plant-produced human monoclonal antibody against anthrax protective antigen protects mice and non-human primates from B. *anthracis* spore challenge. *Hum Vaccin* 7(Suppl):183–190.

153. Cote CK, Rossi CA, Kang AS, Morrow PR, Lee JS, Welkos SL. 2005. The detection of protective antigen (PA) associated with spores of *Bacillus anthracis* and the effects of anti-PA antibodies on spore germination and macrophage interactions. *Microb Pathog* 38:209–225.

154. Welkos S, Friedlander A, Weeks S, Little S, Mendelson I. 2002. In-vitro characterisation of the phagocytosis and fate of anthrax spores in macrophages and the effects of anti-PA antibody. *J Med Microbiol* 51:821–831.

155. Welkos S, Little S, Friedlander A, Fritz D, Fellows P. 2001. The role of antibodies to Bacillus anthracis and anthrax toxin components in inhibiting the early stages of infection by anthrax spores. *Microbiology* 147:1677–1685.

156. Vuyisich M, Gnanakaran S, Lovchik JA, Lyons CR, Gupta G. 2008. A dual-purpose protein ligand for effective therapy and sensitive diagnosis of anthrax. *Protein J* 27:292–302.

157. Brossier F, Weber-Levy M, Mock M, Sirard JC. 2000. Role of toxin functional domains in anthrax pathogenesis. *Infect Immun* **68**:1781–1786.

158. Kalns J, Morris J, Eggers J, Kiel J. 2002. Delayed treatment with doxycycline has limited effect on anthrax infection in BLK57/B6 mice. *Biochem Biophys Res Commun* **297**:506–509.

159. Pezard C, Berche P, Mock M. 1991. Contribution of individual toxin components to virulence of *Bacillus anthracis*. *Infect Immun* **59**:3472–3477.

160. Pezard C, Duflot E, Mock M. 1993. Construction of *Bacillus anthracis* mutant strains producing a single toxin component. *J Gen Microbiol* **139**(Pt 10):2459–2463.

161. Pezard C, Sirard JC, Mock M. 1996. Protective immunity induced by *Bacillus anthracis* toxin mutant strains. *Adv Exp Med Biol* **397**:69–72.

162. Pezard C, Weber M, Sirard JC, Berche P, Mock M. 1995. Protective immunity induced by *Bacillus anthracis* toxin-deficient strains. *Infect Immun* **63**:1369–1372.

163. Russell BH, Liu Q, Jenkins SA, Tuvim MJ, Xu Y. 2008. In vivo demonstration and quantification of intracellular *Bacillus anthracis* in lung epithelial cells. *Infect Immun* **76**:3975–3983.

164. Zaucha GM, Pitt LM, Estep J, Ivins BE, Friedlander AM. 1998. The pathology of experimental anthrax in rabbits exposed by inhalation and subcutaneous inoculation. *Arch Pathol Lab Med* **122**:982–992.

165. Kalns J, Scruggs J, Millenbaugh N, Vivekananda J, Shealy D, Eggers J, Kiel J. 2002. TNF receptor 1, IL-1 receptor, and iNOS genetic knockout mice are not protected from anthrax infection. *Biochem Biophys Res Commun* **292**:41–44.

166. Hahn BL, Sharma S, Sohnle PG. 2005. Analysis of epidermal entry in experimental cutaneous *Bacillus anthracis* infections in mice. *J Lab Clin Med* **146**:95–102.

167. Hahn BL, Bischof TS, Sohnle PG. 2008. Superficial exudates of neutrophils prevent invasion of *Bacillus anthracis* bacilli into abraded skin of resistant mice. *Int J Exp Pathol* **89**:180–187.

168. van Sorge NM, Ebrahimi CM, McGillivray SM, Quach D, Sabet M, Guiney DG, Doran KS. 2008. Anthrax toxins inhibit neutrophil signaling pathways in brain endothelium and contribute to the pathogenesis of meningitis. *PLoS One* **3**:e2964. doi:10.1371/journal.pone.0002964.

169. Ebrahimi CM, Sheen TR, Renken CW, Gottlieb RA, Doran KS. 2011. Contribution of lethal toxin and edema toxin to the pathogenesis of anthrax meningitis. *Infect Immun* **79**:2510–2518.

170. Friedlander AM. 1986. Macrophages are sensitive to anthrax lethal toxin through an acid-dependent process. *J Biol Chem* **261**:7123–7126.

171. Friedlander AM, Bhatnagar R, Leppla SH, Johnson L, Singh Y. 1993. Characterization of macrophage sensitivity and resistance to anthrax lethal toxin. *Infect Immun* **61**:245–252.

172. Guidi-Rontani C, Levy M, Ohayon H, Mock M. 2001. Fate of *germinated Bacillus anthracis* spores in primary murine macrophages. *Mol Microbiol* **42**:931–938.

173. Guidi-Rontani C, Pereira Y, Ruffie S, Sirard JC, Weber-Levy M, Mock M. 1999. Identification and characterization of a germination operon on the virulence plasmid pXO1 of *Bacillus anthracis*. *Mol Microbiol* **33**:407–414.

174. Brittingham KC, Ruthel G, Panchal RG, Fuller CL, Ribot WJ, Hoover TA, Young HA, Anderson AO, Bavari S. 2005. Dendritic cells endocytose *Bacillus anthracis* spores: implications for anthrax pathogenesis. *J Immunol* **174**:5545–5552.

175. Cleret A, Quesnel-Hellmann A, Mathieu J, Vidal D, Tournier JN. 2006. Resident CD11c+ lung cells are impaired by anthrax toxins after spore infection. *J Infect Dis* **194**:86–94.

176. Tournier JN, Quesnel-Hellmann A, Mathieu J, Montecucco C, Tang WJ, Mock M, Vidal DR, Goossens PL. 2005. Anthrax edema toxin cooperates with lethal toxin to impair cytokine secretion during infection of dendritic cells. *J Immunol* **174**:4934–4941.

177. Hsu LC, Ali SR, McGillivray S, Tseng PH, Mariathasan S, Humke EW, Eckmann L, Powell JJ, Nizet V, Dixit VM, Karin M. 2008. A NOD2-NALP1 complex mediates caspase-1-dependent IL-1beta secretion in response to *Bacillus anthracis* infection and muramyl dipeptide. *Proc Natl Acad Sci USA* **105**:7803–7808.

178. Duesbery NS, Webb CP, Leppla SH, Gordon VM, Klimpel KR, Copeland TD, Ahn NG, Oskarsson MK, Fukasawa K, Paull KD, Vande Woude GF. 1998. Proteolytic inactivation of MAP-kinase-kinase by anthrax lethal factor. *Science* **280**:734–737.

179. Moayeri M, Leppla SH. 2009. Cellular and systemic effects of anthrax lethal toxin and edema toxin. *Mol Aspects Med* **30**:439–455.

180. Park JM, Greten FR, Li ZW, Karin M. 2002. Macrophage apoptosis by anthrax lethal factor through p38 MAP kinase inhibition. *Science* **297**:2048–2051.

181. Pellizzari R, Guidi-Rontani C, Vitale G, Mock M, Montecucco C. 1999. Anthrax lethal factor cleaves MKK3 in macrophages and inhibits the LPS/IFNgamma-induced release of NO and TNFalpha. *FEBS Lett* **462**:199–204.

182. Vitale G, Pellizzari R, Recchi C, Napolitani G, Mock M, Montecucco C. 1998. Anthrax lethal factor cleaves the N-terminus of MAPKKs and induces tyrosine/threonine phosphorylation of MAPKs in cultured macrophages. *Biochem Biophys Res Commun* **248**:706–711.

183. Leppla SH. 1982. Anthrax toxin edema factor: a bacterial adenylate cyclase that increases cyclic AMP concentrations of eukaryotic cells. *Proc Natl Acad Sci USA* **79**:3162–3166.

184. Roberts JE, Watters JW, Ballard JD, Dietrich WF. 1998. Ltx1, a mouse locus that influences the susceptibility of macrophages to cytolysis caused by intoxication with Bacillus anthracis lethal factor, maps to chromosome 11. *Mol Microbiol* **29**:581–591.

185. Boyden ED, Dietrich WF. 2006. Nalp1b controls mouse macrophage susceptibility to anthrax lethal toxin. *Nat Genet* 38:240–244.

186. Martinon F. 2007. Orchestration of pathogen recognition by inflammasome diversity: variations on a common theme. *Eur J Immunol* 37:3003–3006.

187. Tschopp J. 1989. NALPs: resistance to Sterne strain of *Bacillus anthracis*: phagocytic cell response of resistant and susceptible mice. *Microb Pathog* 7:15–35.

188. Hanna PC, Acosta D, Collier RJ. 1993. On the role of macrophages in anthrax. *Proc Natl Acad Sci USA* 90:10198–10201.

189. Moayeri M, Martinez NW, Wiggins J, Young HA, Leppla SH. 2004. Mouse susceptibility to anthrax lethal toxin is influenced by genetic factors in addition to those controlling macrophage sensitivity. *Infect Immun* 72:4439–4447.

190. Terra JK, France B, Cote CK, Jenkins A, Bozue J, Welkos SL, Lusis AJ, Davis R, LeVine SM, Bradley KA. 2011. Identification of a novel quantitative trait locus controlling inflammatory response to anthrax lethal toxin and resistance to *Bacillus anthracis*. *PLoS Pathog* 7(12):e1002469.

191. Banks DJ, Barnajian M, Maldonado-Arocho FJ, Sanchez AM, Bradley KA. 2005. Anthrax toxin receptor 2 mediates *Bacillus anthracis* killing of macrophages following spore challenge. *Cell Microbiol* 7:1173–1185.

192. Liu S, Crown D, Miller-Randolph S, Moayeri M, Wang H, Hu H, Morley T, Leppla SH. 2009. Capillary morphogenesis protein-2 is the major receptor mediating lethality of anthrax toxin in vivo. *Proc Natl Acad Sci USA* 106:12424–12429.

193. Moayeri M, Haines D, Young HA, Leppla SH. 2003. *Bacillus anthracis* lethal toxin induces TNF-alpha-independent hypoxia-mediated toxicity in mice. *J Clin Invest* 112:670–682.

194. Brossier F, Levy M, Mock M. 2002. Anthrax spores make an essential contribution to vaccine efficacy. *Infect Immun* 70:661–664.

195. Chand HS, Drysdale M, Lovchik J, Koehler TM, Lipscomb MF, Lyons CR. 2009. Discriminating virulence mechanisms among *Bacillus anthracis* strains by using a murine subcutaneous infection model. *Infect Immun* 77:429–435.

196. Gauthier YP, Tournier JN, Paucod JC, Corre JP, Mock M, Goossens PL, Vidal DR. 2009. Efficacy of a vaccine based on protective antigen and killed spores against experimental inhalational anthrax. *Infect Immun* 77:1197–1207.

197. Little SF, Knudson GB. 1986. Comparative efficacy of *Bacillus anthracis* live spore vaccine and protective antigen vaccine against anthrax in the guinea pig. *Infect Immun* 52:509–512.

198. Moen ST, Yeager LA, Lawrence WS, Ponce C, Galindo CL, Garner HR, Baze WB, Suarez G, Peterson JW, Chopra AK. 2008. Transcriptional profiling of murine organ genes in response to infection with *Bacillus anthracis* Ames spores. *Microb Pathog* 44:293–310.

199. Vasconcelos D, Barnewall R, Babin M, Hunt R, Estep J, Nielsen C, Carnes R, Carney J. 2003. Pathology of inhalation anthrax in cynomolgus monkeys (*Macaca fascicularis*). *Lab Invest* 83:1201–1209.

200. Hahn UK, Boehm R, Beyer W. 2006. DNA vaccination against anthrax in mice-combination of anti-spore and anti-toxin components. *Vaccine* 24:4569–4571.

201. Kozel TR, Murphy WJ, Brandt S, Blazar BR, Lovchik JA, Thorkildson P, Percival A, Lyons CR. 2004. mAbs to Bacillus anthracis capsular antigen for immunoprotection in anthrax and detection of antigenemia. *Proc Natl Acad Sci USA* 101:5042–5047.

202. Barr JR, Boyer AE, Hoffmaster AR, Quinn CP, Gallegos-Candela M, Lins RC, Shadomy SV, Lehman M, Pesik N, Jahangir H, Chakraborty A, Ramsay CN, Smith TL, Pirkle JL. 2011. Detection and quantification of anthrax toxins by mass spectrometry-application to clinical anthrax, abstr *Bacillus*-ACT: The International Conference on *Bacillus anthracis, B. cereus, and B. thuringiensis*, Bruges, Belgium, 7 to 11 August 2011.

203. Boyer AE, Lins RC, Gallegos-Candella M, Kuklenyik Z, Quinn CP, Hoffmaster A, Marston CK, Beesley CA, Barr JR. 2011. Kinetics of anthrax toxins, lethal factor, edmea factor, and lethal toxin during inhalational anthrax in rhesus macaques, abstr *Bacillus*-ACT: The International Conference on *Bacillus anthracis, B. cereus, and B. thuringiensis*, Bruges, Belgium, 7 to 11 August 2011.

204. Meister G, Blosser E, Herr-Calomeni P, Padgett N, Mott J. 2007. Development of an inhalational Bacillus anthracis exposure therapeutic model in New Zealand White rabbits, abstr *Bacillus*-ACT: The International Conference on *Bacillus anthracis, B. cereus, and B. thuringiensis*, Oslo, Norway, 17 to 21 June 2007.

205. Bergman NH, Passalacqua KD, Gaspard R, Shetron-Rama LM, Quackenbush J, Hanna PC. 2005. Murine macrophage transcriptional responses to *Bacillus anthracis* infection and intoxication. *Infect Immun* 73:1069–1080.

206. Bradley KA, LeVine SM. 2010. Anthrax toxin delivers a one-two punch. *Cell Host Microbe* 8:394–395.

207. Cote CK, Dimezzo TL, Banks DJ, France B, Bradley KA, Welkos SL. 2008. Early interactions between fully virulent *Bacillus anthracis* and macrophages that influence the balance between spore clearance and development of a lethal infection. *Microbes Infect.* 10:613–619.

208. Giorno R, Bozue J, Cote C, Wenzel T, Moody KS, Mallozzi M, Ryan M, Wang R, Zielke R, Maddock JR, Friedlander A, Welkos S, Driks A. 2007. Morphogenesis of the *Bacillus anthracis* spore. *J Bacteriol* 189:691–705.

209. Giorno R, Mallozzi M, Bozue J, Moody KS, Slack A, Qiu D, Wang R, Friedlander A, Welkos S, Driks A. 2009. Localization and assembly of proteins comprising the outer structures of the *Bacillus anthracis* spore. *Microbiology* 155:1133–1145.

210. Mallozzi M, Bozue J, Giorno R, Moody KS, Slack A, Cote C, Qiu D, Wang R, McKenney P, Lai EM, Maddock JR, Friedlander A, Welkos S, Eichenberger P, Driks A. 2008. Characterization of a *Bacillus anthracis*

spore coat-surface protein that influences coat-surface morphology. *FEMS Microbiol Lett* **289**:110–117.

211. Severson KM, Mallozzi M, Bozue J, Welkos SL, Cote CK, Knight KL, Driks A. 2009. Roles of the *Bacillus anthracis* spore protein ExsK in exosporium maturation and germination. *J Bacteriol* **191**:7587–7596.

212. Bielinska AU, Janczak KW, Landers JJ, Makidon P, Sower LE, Peterson JW, Baker JR Jr. 2007. Mucosal immunization with a novel nanoemulsion-based recombinant anthrax protective antigen vaccine protects against *Bacillus anthracis* spore challenge. *Infect Immun* **75**:4020–4029.

213. Boyaka PN, Tafaro A, Fischer R, Leppla SH, Fujihashi K, McGhee JR. 2003. Effective mucosal immunity to anthrax: neutralizing antibodies and Th cell responses following nasal immunization with protective antigen. *J Immunol* **170**:5636–5643.

214. Shivachandra SB, Li Q, Peachman KK, Matyas GR, Leppla SH, Alving CR, Rao M, Rao VB. 2007. Multicomponent anthrax toxin display and delivery using bacteriophage T4. *Vaccine* **25**:1225–1235.

215. Williamson ED, Beedham RJ, Bennett AM, Perkins SD, Miller J, Baillie LW. 1999. Presentation of protective antigen to the mouse immune system: immune sequelae. *J Appl Microbiol* **87**:315–317.

216. Gaur R, Gupta PK, Banerjea AC, Singh Y. 2002. Effect of nasal immunization with protective antigen of *Bacillus anthracis* on protective immune response against anthrax toxin. *Vaccine* **20**:2836–2839.

217. Gu M, Hine PM, James Jackson W, Giri L, Nabors GS. 2007. Increased potency of BioThrax anthrax vaccine with the addition of the C-class CpG oligonucleotide adjuvant CPG 10109. *Vaccine* **25**:526–534.

218. Matyas GR, Friedlander AM, Glenn GM, Little S, Yu J, Alving CR. 2004. Needle-free skin patch vaccination method for anthrax. *Infect Immun* **72**:1181–1183.

219. Mikszta JA, Sullivan VJ, Dean C, Waterston AM, Alarcon JB, Dekker JP 3rd, Brittingham JM, Huang J, Hwang CR, Ferriter M, Jiang G, Mar K, Saikh KU, Stiles BG, Roy CJ, Ulrich RG, Harvey NG. 2005. Protective immunization against inhalational anthrax: a comparison of minimally invasive delivery platforms. *J Infect Dis* **191**:278–288.

220. Peachman KK, Rao M, Alving CR, Burge R, Leppla SH, Rao VB, Matyas GR. 2006. Correlation between lethal toxin-neutralizing antibody titers and protection from intranasal challenge with *Bacillus anthracis* Ames strain spores in mice after transcutaneous immunization with recombinant anthrax protective antigen. *Infect Immun* **74**:794–797.

221. Chabot DJ, Scorpio A, Tobery SA, Little SF, Norris SL, Friedlander AM. 2004. Anthrax capsule vaccine protects against experimental infection. *Vaccine* **23**:43–47.

222. Iacono-Connors LC, Welkos SL, Ivins BE, Dalrymple JM. 1991. Protection against anthrax with recombinant virus-expressed protective antigen in experimental animals. *Infect Immun* **59**:1961–1965.

223. Ivins BE, Welkos SL, Knudson GB, Little SF. 1990. Immunization against anthrax with aromatic compound-dependent (Aro-) mutants of *Bacillus anthracis* and with recombinant strains of *Bacillus subtilis* that produce anthrax protective antigen. *Infect Immun* **58**:303–308.

224. Welkos SL, Cote CK, Rea KM, Gibbs PH. 2004. A microtiter fluorometric assay to detect the germination of *Bacillus anthracis* spores and the germination inhibitory effects of antibodies. *J Microbiol Methods* **56**:253–265.

225. Piris Gimenez A, Wu Y-Z, Paya M, Delclaux C, Touqui L, Goosens PL. 2004. High bactericidal efficiency of type IIA phospholipase A2 against *Bacillus anthracis* and inhibition of its secretion by the lethal toxin. *J Immunol* **173**:521–530.

226. Piris-Gimenez A, Paya M, Lambeau G, Chignard M, Mock M, Touqui L, Goossens PL. 2005. In vivo protective role of human group IIa phospholipase A2 against experimental anthrax. *J Immunol* **175**:6786–6791.

227. Peterson JW, Moen ST, Healy D, Pawlik JE, Taormina J, Hardcastle J, Thomas JM, Lawrence WS, Ponce C, Chatuev BM, Gnade BT, Foltz SM, Agar SL, Sha J, Klimpel GR, Kirtley ML, Eaves-Pyles T, Chopra AK. 2010. Protection afforded by fluoroquinolones in animal models of respiratory infections with *Bacillus anthracis*, *Yersinia pestis*, and *Francisella tularensis*. *Open Microbiol J* **4**:34–46.

228. Steward J, Lever MS, Simpson AJ, Sefton AM, Brooks TJ. 2004. Post-exposure prophylaxis of systemic anthrax in mice and treatment with fluoroquinolones. *J Antimicrob Chemother* **54**:95–99.

229. Louie A, Heine HS, Kim K, Brown DL, VanScoy B, Liu W, Kinzig-Schippers M, Sorgel F, Drusano GL. 2008. Use of an in vitro pharmacodynamic model to derive a linezolid regimen that optimizes bacterial kill and prevents emergence of resistance in *Bacillus anthracis*. *Antimicrob Agents Chemother* **52**:2486–2496.

230. Peterson JW, Comer JE, Noffsinger DM, Wenglikowski A, Walberg KG, Chatuev BM, Chopra AK, Stanberry LR, Kang AS, Scholz WW, Sircar J. 2006. Human monoclonal anti-protective antigen antibody completely protects rabbits and is synergistic with ciprofloxacin in protecting mice and guinea pigs against inhalation anthrax. *Infect Immun* **74**:1016–1024.

231. Chen Z, Schneerson R, Lovchik J, Lyons CR, Zhao H, Dai Z, Kubler-Kielb J, Leppla SH, Purcell RH. 2011. Pre- and postexposure protection against virulent anthrax infection in mice by humanized monoclonal antibodies to *Bacillus anthracis* capsule. *Proc Natl Acad Sci USA* **108**:739–744.

232. Drysdale M, Heninger S, Hutt J, Chen Y, Lyons CR, Koehler TM. 2005. Capsule synthesis by *Bacillus anthracis* is required for dissemination in murine inhalation anthrax. *EMBO J* **24**:221–227.

233. Ivins BE, Welkos SL. 1988. Recent advances in the development of an improved, human anthrax vaccine. *Eur J Epidemiol* **4**:12–19.

234. Garufi G, Wang Y, Oh SY, Maier H, Schneewind O, Missiakas D. 2012. Sortase-conjugation generates a capsule vaccine that protects guinea pigs against *Bacillus anthracis*. *Vaccine* **30**:3435–3444.

235. Cote C, Kaatz L, Reinhardt J, Bozue J, Tobery S, Bassett A, Sanz P, Darnell S, Alem F, O'Brien A, Welkos S. 2012. Characterization of a multi-component anthrax vaccine designed to target the initial stages of infection as well as toxemia. *J Med Microbiol* **61**: 1393–1400.

236. Glomski IJ, Corre JP, Mock M, Goossens PL. 2007. Cutting Edge: IFN-gamma-producing CD4 T lymphocytes mediate spore-induced immunity to capsulated *Bacillus anthracis*. *J Immunol* **178**:2646–2650.

237. Scorpio A, Tobery SA, Ribot WJ, Friedlander AM. 2008. Treatment of experimental anthrax with recombinant capsule depolymerase. *Antimicrob Agents Chemother* **52**:1014–1020.

238. Gill DM. 1982. Bacterial toxins: a table of lethal amounts. *Microbiol Rev* **46**:86–94.

239. Jones WI Jr, Klein F, Walker JS, Mahlandt BG, Dobbs JP, Lincoln RE. 1967. In vivo growth and distribution of anthrax bacilli in resistant, susceptible, and immunized hosts. *J Bacteriol* **94**:600–608.

240. Day J, Friedman A, Schlesinger LS. 2011. Modeling the host response to inhalation anthrax. *J Theor Biol* **276**: 199–208.

241. Popov S. 2008. New candidate anthrax pathogenic factors, p 25–35. *In* Georgiev V, Western KA, McGowan JJ (ed), National Institute of Allergy and Infectious Diseases, NIH, vol I. *Frontiers in Research*. Humana Press, Inc., Totowa, NJ.

242. Moayeri M, Leppla SH. 2004. The roles of anthrax toxin in pathogenesis. *Curr Opin Microbiol* **7**:19–24.

243. Moayeri M, Crown D, Dorward DW, Gardner D, Ward JM, Li Y, Cui X, Eichacker P, Leppla SH. 2009. The heart is an early target of anthrax lethal toxin in mice: a protective role for neuronal nitric oxide synthase (nNOS). *PLoS Pathog* **5**:e1000456. doi:10.1371/journal.ppat.1000456.

244. Duverger A, Jackson RJ, van Ginkel FW, Fischer R, Tafaro A, Leppla SH, Fujihashi K, Kiyono H, McGhee JR, Boyaka PN. 2006. Bacillus anthracis edema toxin acts as an adjuvant for mucosal immune responses to nasally administered vaccine antigens. *J Immunol* **176**: 1776–1783.

245. Firoved AM, Miller GF, Moayeri M, Kakkar R, Shen Y, Wiggins JF, McNally EM, Tang WJ, Leppla SH. 2005. *Bacillus anthracis* edema toxin causes extensive tissue lesions and rapid lethality in mice. *Am J Pathol* **167**: 1309–1320.

246. Crowe SR, Garman L, Engler RJ, Farris AD, Ballard JD, Harley JB, James JA. 2011. Anthrax vaccination induced anti-lethal factor IgG: fine specificity and neutralizing capacity. *Vaccine* **29**:3670–3678.

247. Aulinger BA, Roehrl MH, Mekalanos JJ, Collier RJ, Wang JY. 2005. Combining anthrax vaccine and therapy: a dominant-negative inhibitor of anthrax toxin is also a potent and safe immunogen for vaccines. *Infect Immun* **73**:3408–3414.

248. Chen Z, Moayeri M, Zhao H, Crown D, Leppla SH, Purcell RH. 2009. Potent neutralization of anthrax edema toxin by a humanized monoclonal antibody that competes with calmodulin for edema factor binding. *Proc Natl Acad Sci USA* **106**:13487–13492.

249. Kulshreshtha P, Bhatnagar R. 2011. Inhibition of anthrax toxins with a bispecific monoclonal antibody that cross reacts with edema factor as well as lethal factor of *Bacillus anthracis*. *Mol Immunol* **48**:1958–1965.

250. vor dem Esche U, Huber M, Zgaga-Griesz A, Grunow R, Beyer W, Hahn U, Bessler WG. 2011. Passive vaccination with a human monoclonal antibody: generation of antibodies and studies for efficacy in *Bacillus anthracis* infections. *Immunobiology* **216**:847–853.

251. Cromartie WJ, Bloom WL, Watson DS. 1947. Studies on infection with *Bacillus anthracis*. *J Infect Dis* **80**: 1–52.

252. Cromartie WJ, Bloom WL, Watson DW. 1947. Studies on infection with *Bacillus anthracis*. II. A histopathological study of skin lesions produced by *B. anthracis* in susceptible and resistant animal species. *J Infect Dis* **80**: 1–13.

253. Cromartie WJ, Watson DW, Bloom WL, Heckly RJ. 1947. Studies on infection with *Bacillus anthracis*. II. The immunological and tissue damaging properties of extracts prepared from lesions of *B. anthracis* infections. *J Infect Dis* **80**:14–27.

254. Smith H, Keppie J, Stanley JL. 1954. Observations on the cause of death in experimental anthrax. *Lancet* **267**: 474–476.

255. Smith H, Keppie J, Stanley JL, Harris-Smith PW. 1955. The chemical basis of the virulence of *Bacillus anthracis*. IV. Secondary shock as the major factor in death of guinea-pigs from anthrax. *Br J Exp Pathol* **36**: 323–335.

256. Ward MK, McGann VG, Hogge AL Jr, Huff ML, Kanode RG Jr, Roberts EO. 1965. Studies on anthrax infections in immunized guinea pigs. *J Infect Dis* **115**: 59–67.

257. Walker JS, Klein F, Lincoln RE, Fernelius AL. 1967. Temperature response in animals infected with *Bacillus anthracis*. *J Bacteriol* **94**:552–556.

258. Lincoln RE, Rhian MM, Klein F, Fernelius A. 1961. *Pathogenesis as Related to Physiological State of Anthrax Spore and Cell*. Burgess Publishing Company, Minneapolis, MN.

259. Jenkins A, Cote C, Twenhafel N, Merkel T, Bozue J, Welkos S. 2011. Role of purine biosynthesis in *Bacillus anthracis* pathogenesis and virulence. *Infect Immun* **79**: 153–166.

260. Coker PR, Smith KL, Fellows PF, Rybachuck G, Kousoulas KG, Hugh-Jones ME. 2003. *Bacillus anthracis* virulence in guinea pigs vaccinated with anthrax vaccine adsorbed is linked to plasmid quantities and clonality. *J Clin Microbiol* **41**:1212–1218.

261. Chitlaru T, Zaide G, Gat O, Ariel N, Zvi A, Grosfeld H, Cohen O, Shafferman A. 2011. Novel essential virulence determinants of *Bacillus anthracis* identified by various complementary high throughput analyses, abstr *Bacillus*-ACT: The International Conference on *Bacillus anthracis*, *B. cereus, and B. thuringiensis*, Bruges, Belgium, 7 to 11 August 2011.

262. Bozue JA, Parthasarathy N, Phillips LR, Cote CK, Fellows PF, Mendelson I, Shafferman A, Friedlander AM. 2005. Construction of a rhamnose mutation in *Bacillus anthracis* affects adherence to macrophages but not virulence in guinea pigs. *Microb Pathog* **38**: 1–12.

263. Chitlaru T, Gat O, Grosfeld H, Inbar I, Gozlan Y, Shafferman A. 2007. Identification of in vivo-expressed immunogenic proteins by serological proteome analysis of the *Bacillus anthracis* secretome. *Infect Immun* **75**: 2841–2852.

264. Cote CK, Bozue J, Moody KL, DiMezzo TL, Chapman CE, Welkos SL. 2008. Analysis of a novel spore antigen in *Bacillus anthracis* that contributes to spore opsonization. *Microbiology* **154**:619–632.

265. Scorpio A, Chabot DJ, Day WA, Hoover TA, Friedlander AM. Capsule depolymerase overexpression reduces *Bacillus anthracis* virulence. *Microbiology* **156**:1459–1467.

266. Ivins BE, Fellows PF, Nelson GO. 1994. Efficacy of a standard human anthrax vaccine against Bacillus anthracis spore challenge in guinea-pigs. *Vaccine* **12**: 872–874.

267. Turnbull PC, Broster MG, Carman JA, Manchee RJ, Melling J. 1986. Development of antibodies to protective antigen and lethal factor components of anthrax toxin in humans and guinea pigs and their relevance to protective immunity. *Infect Immun* **52**:356–363.

268. Kobiler D, Gozes Y, Rosenberg H, Marcus D, Reuveny S, Altboum Z. 2002. Efficiency of protection of guinea pigs against infection with *Bacillus anthracis* spores by passive immunization. *Infect Immun* **70**:544–560.

269. Little SF, Ivins BE, Fellows PF, Friedlander AM. 1997. Passive protection by polyclonal antibodies against *Bacillus anthracis* infection in guinea pigs. *Infect Immun* **65**:5171–5175.

270. Reuveny S, White MD, Adar YY, Kafri Y, Altboum Z, Gozes Y, Kobiler D, Shafferman A, Velan B. 2001. Search for correlates of protective immunity conferred by anthrax vaccine. *Infect Immun* **69**:2888–2893.

271. Lee DY, Chun JH, Ha HJ, Park J, Kim BS, Oh HB, Rhie GE. 2009. Poly-gamma-d-glutamic acid and protective antigen conjugate vaccines induce functional antibodies against the protective antigen and capsule of *Bacillus anthracis* in guinea-pigs and rabbits. *FEMS Immunol Med Microbiol* **57**:165–172.

272. Skoble J, Beaber JW, Gao Y, Lovchik JA, Sower LE, Liu W, Luckett W, Peterson JW, Calendar R, Portnoy DA, Lyons CR, Dubensky TW Jr. 2009. Killed but metabolically active *Bacillus anthracis* vaccines induce broad and protective immunity against anthrax. *Infect Immun* **77**:1649–1663.

273. Savransky V, Austin J, Tordoff K, Sanford D, Lemiale L, Park S, Nabors GS, Ionin B, Skiadopoulos MH. 2011. Protective efficacy of a novel anthrax vaccine against inhalational anthrax in guinea pigs is associated with induction of robust immune response, abstr *Bacillus*-ACT: The International Conference on *Bacillus anthracis, B. cereus, and B. thuringiensis*, Bruge, Belgium, 7 to 11 August 2011.

274. Twenhafel NA. 2010. Pathology of inhalational anthrax animal models. *Vet Pathol* **47**:819–830.

275. Comer JE, Noffsinger DM, McHenry DJ, Weisbaum DM, Chatuev BM, Chopra AK, Peterson JW. 2006. Evaluation of the protective effects of quinacrine against *Bacillus anthracis* Ames. *J Toxicol Environ Health A* **69**:1083–1095.

276. Lawrence WS, Hardcastle JM, Brining DL, Weaver LE, Ponce C, Whorton EB, Peterson JW. 2009. The physiologic responses of Dutch belted rabbits infected with inhalational anthrax. *Comp Med* **59**:257–265.

277. Weiss S, Kobiler D, Levy H, Marcus H, Pass A, Rothschild N, Altboum Z. 2006. Immunological correlates for protection against intranasal challenge of *Bacillus anthracis* spores conferred by a protective antigen-based vaccine in rabbits. *Infect Immun* **74**: 394–398.

278. Palmer JR, Albrecht M, Groen H, Elberson M, Fonseca M, Barnwell R, Pesce J, Mateczun A, Kean-Myers A. 2011. Protective efficacy of candidate anthrax vaccines and medical countermeasures against an aerosol challenge with the novel *Bacillus cereus* G9241 strain, abstr *Bacillus*-ACT: The International Conference on *Bacillus anthracis, B. cereus, and B. thuringiensis*, Bruge, Belgium, 7 to 11 August 2011.

279. Burrows TW, Gillett WA. 1971. Host specificity of Brazilian strains of *Pasteurella pestis*. *Nature* **229**: 51–52.

280. Meyer KF, Smith G, Foster L, Brookman M, Sung M. 1974. Live, attenuated *Yersinia pestis* vaccine: virulent in nonhuman primates, harmless to guinea pigs. *J Infect Dis* **129**(Suppl):S85–S12.

281. Oyston PC, Russell P, Williamson ED, Titball RW. 1996. An aroA mutant of *Yersinia pestis* is attenuated in guinea-pigs, but virulent in mice. *Microbiology* **142** (Pt 7):1847–1853.

282. Levy H, Weiss S, Altboum Z, Schlomovitz J, Rothschild N, Blachinsky E, Kobiler D. 2011. Lethal factor is not required for *Bacillus anthracis* virulence in guinea pigs and rabbits. *Microb Pathog* **51**:345–351.

283. Yee SB, Hatkin JM, Dyer DN, Orr SA, Pitt ML. 2010. Aerosolized *Bacillus anthracis* infection in New Zealand white rabbits: natural history and intravenous levofloxacin treatment. *Comp Med* **60**:461–468.

284. Wright GG, Hedberg MA, Feinberg RJ. 1951. Studies on immunity in anthrax. II. In vitro elaboration of protective antigen by non-proteolytic mutants of *Bacillus anthracis*. *J Exp Med* **93**:523–527.

285. Fasanella A, Scasciamacchia S, Garofolo G. 2009. The behaviour of virulent *Bacillus anthracis* strain AO843 in rabbits. *Vet Microbiol* **133**:208–209.

286. Yee SB, Dyer DN, Pitt MLM. 2011. Bacterial lipopolysaccharide-induced resistance to inhalational anthracis infection in New Zealand white rabbits, abstr *Bacillus*-ACT: The International Conference on *Bacillus anthracis, B. cereus, and B. thuringiensis*, Bruges, Belgium, 7 to 11 August 2011.

287. Lawrence WS, Marshall JR, Zavala DL, Weaver LE, Baze WB, Moen ST, Whorton EB, Gourley RL, Peterson

JW. 2011. Hemodynamic effects of anthrax toxins in the rabbit model and the cardiac pathology induced by lethal toxin. *Toxins* **3**:721–736.

288. **Nordberg BK, Schmiterlow CG, Hansen HJ.** 1961. Pathophysiological investigations into the terminal course of experimental anthrax in the rabbit. *Acta Pathol Microbiol Scand* **53**:295–318.

289. **Kobiler D, Weiss S, Levy H, Fisher M, Mechaly A, Pass A, Altboum Z.** 2006. Protective antigen as a correlative marker for anthrax in animal models. *Infect Immun* **74**:5871–5876.

290. **Sela-Abramovich S, Chitlaru T, Gat O, Grosfeld H, Cohen O, Shafferman A.** 2009. Novel and unique diagnostic biomarkers for *Bacillus anthracis* infection. *Appl Environ Microbiol* **75**:6157–6167.

291. **Wilson MK, Vergis JM, Alem F, Palmer JR, Keane-Myers AM, Brahmbhatt TN, Ventura CL, O'Brien AD.** 2011. Bacillus cereus G9241 makes anthrax toxin and capsule like highly virulent B. anthracis Ames but behaves like attenuated toxigenic nonencapsulated *B. anthracis* Sterne in rabbits and mice. *Infect Immun* **79**:3012–3019.

292. **Hoffmaster AR, Ravel J, Rasko DA, Chapman GD, Chute JD, Chung KM, De BK, Sacchi CT, Fitzgerald C, Mauer LW, Maiden MCJ, Priest FG, Barker M, Jiang L, Cer RZ, Rilstone J, Peterson SN, Weyant RS, Galloway DR, Read TD, Popovic T, Fraser CM.** 2004. Identification of anthrax toxin genes in *Bacillus cereus* associated with an illness resembling inhalation anthrax. *Proc Natl Acad Sci USA* **101**:8449–8454.

293. **Little SF, Ivins BE, Webster WM, Fellows PF, Pitt ML, Norris SL, Andrews GP.** 2006. Duration of protection of rabbits after vaccination with *Bacillus anthracis* recombinant protective antigen vaccine. *Vaccine* **24**:2530–2536.

294. **Little SF, Ivins BE, Webster WM, Norris SL, Andrews GP.** 2007. Effect of aluminum hydroxide adjuvant and formaldehyde in the formulation of rPA anthrax vaccine. *Vaccine* **25**:2771–2777.

295. **Ribot WJ, Powell BS, Ivins BE, Little SF, Johnson WM, Hoover TA, Norris SL, Adamovicz JJ, Friedlander AM, Andrews GP.** 2006. Comparative vaccine efficacy of different isoforms of recombinant protective antigen against *Bacillus anthracis* spore challenge in rabbits. *Vaccine* **24**:3469–3476.

296. **Auerbach S, Wright GG.** 1955. Studies on immunity in anthrax. VI. Immunizing activity of protective antigen against various strains of *Bacillus anthracis*. *J Immunol* **75**:129–133.

297. **Chawla A, Midha S, Bhatnagar R.** 2009. Efficacy of recombinant anthrax vaccine against *Bacillus anthracis* aerosol spore challenge: preclinical evaluation in rabbits and Rhesus monkeys. *Biotechnol J* **4**:391–399.

298. **Klas SD, Petrie CR, Warwood SJ, Williams MS, Olds CL, Stenz JP, Cheff AM, Hinchcliffe M, Richardson C, Wimer S.** 2008. A single immunization with a dry powder anthrax vaccine protects rabbits against lethal aerosol challenge. *Vaccine* **26**:5494–5502.

299. **Ren J, Dong D, Zhang J, Liu S, Li B, Fu L, Xu J, Yu C, Hou L, Li J, Chen W.** 2009. Protection against anthrax and plague by a combined vaccine in mice and rabbits. *Vaccine* **27**:7436–7441.

300. **Wimer-Mackin S, Hinchcliffe M, Petrie CR, Warwood SJ, Tino WT, Williams MS, Stenz JP, Cheff A, Richardson C.** 2006. An intranasal vaccine targeting both the *Bacillus anthracis* toxin and bacterium provides protection against aerosol spore challenge in rabbits. *Vaccine* **24**:3953–3963.

301. **Galloway DR, Baillie L.** 2004. DNA vaccines against anthrax. *Expert Opin Biol Ther* **4**:1661–1667.

302. **Keitel WA, Treanor JJ, El Sahly HM, Evans TG, Kopper S, Whitlow V, Selinsky C, Kaslow DC, Rolland A, Smith LR, Lalor PA.** 2009. Evaluation of a plasmid DNA-based anthrax vaccine in rabbits, nonhuman primates and healthy adults. *Hum Vaccin* **5**:536–544.

303. **Hermanson G, Whitlow V, Parker S, Tonsky K, Rusalov D, Ferrari M, Lalor P, Komai M, Mere R, Bell M, Brenneman K, Mateczun A, Evans T, Kaslow D, Galloway D, Hobart P.** 2004. A cationic lipid-formulated plasmid DNA vaccine confers sustained antibody-mediated protection against aerosolized anthrax spores. *Proc Natl Acad Sci USA* **101**:13601–13606.

304. **Migone TS, Subramanian GM, Zhong J, Healey LM, Corey A, Devalaraja M, Lo L, Ullrich S, Zimmerman J, Chen A, Lewis M, Meister G, Gillum K, Sanford D, Mott J, Bolmer SD.** 2009. Raxibacumab for the treatment of inhalational anthrax. *N Engl J Med* **361**:135–144.

305. **Mohamed N, Clagett M, Li J, Jones S, Pincus S, D'Alia G, Nardone L, Babin M, Spitalny G, Casey L.** 2005. A high-affinity monoclonal antibody to anthrax protective antigen passively protects rabbits before and after aerosolized *Bacillus anthracis* spore challenge. *Infect Immun* **73**:795–802.

306. **Peterson JW, Comer JE, Baze WB, Noffsinger DM, Wenglikowski A, Walberg KG, Hardcastle J, Pawlik J, Bush K, Taormina J, Moen S, Thomas J, Chatuev BM, Sower L, Chopra AK, Stanberry LR, Sawada R, Scholz WW, Sircar J.** 2007. Human monoclonal antibody AVP-21D9 to protective antigen reduces dissemination of the *Bacillus anthracis* Ames strain from the lungs in a rabbit model. *Infect Immun* **75**:3414–3424.

307. **Wycoff KL, Belle A, Deppe D, Schaefer L, Maclean JM, Haase S, Trilling AK, Liu S, Leppla SH, Geren IN, Pawlik J, Peterson JW.** Recombinant anthrax toxin receptor-Fc fusion proteins produced in plants protect rabbits against inhalational anthrax. *Antimicrob Agents Chemother* **55**:132–139.

308. **Hering D, Thompson W, Hewetson J, Little S, Norris S, Pace-Templeton J.** 2004. Validation of the anthrax lethal toxin neutralization assay. *Biologicals* **32**:17–27.

309. **Little SF, Ivins BE, Fellows PF, Pitt ML, Norris SL, Andrews GP.** 2004. Defining a serological correlate of protection in rabbits for a recombinant anthrax vaccine. *Vaccine* **22**:422–430.

310. **Pitt ML, Little SF, Ivins BE, Fellows P, Barth J, Hewetson J, Gibbs P, Dertzbaugh M, Friedlander AM.**

2001. In vitro correlate of immunity in a rabbit model of inhalational anthrax. *Vaccine* 19:4768–4773.

311. Pittman PR, Norris SL, Barrera Oro JG, Bedwell D, Cannon TL, McKee KT Jr. 2006. Patterns of antibody response in humans to the anthrax vaccine adsorbed (AVA) primary (six-dose) series. *Vaccine* 24:3654–3660.

312. Williamson ED, Hodgson I, Walker NJ, Topping AW, Duchars MG, Mott JM, Estep J, Lebutt C, Flick-Smith HC, Jones HE, Li H, Quinn CP. 2005. Immunogenicity of recombinant protective antigen and efficacy against aerosol challenge with anthrax. *Infect Immun* 73: 5978–5987.

313. Ingram RJ, Chu KK, Metan G, Maillere B, Doganay M, Ozkul Y, Dyson H, Williamson ED, Baillie L, Kim LU, Ascough S, Sriskandan S, Altmann DM. 2010. An epitope of *Bacillus anthracis* protective antigen that is cryptic in rabbits may be immunodominant in humans. *Infect Immun* 78:2353; author reply 2353–2354.

314. Oscherwitz J, Yu F, Cease KB. 2009. A heterologous helper T-cell epitope enhances the immunogenicity of a multiple-antigenic-peptide vaccine targeting the cryptic loop-neutralizing determinant of *Bacillus anthracis* protective antigen. *Infect Immun* 77:5509–5518.

315. Oscherwitz J, Yu F, Jacobs JL, Liu TH, Johnson PR, Cease KB. 2009. Synthetic peptide vaccine targeting a cryptic neutralizing epitope in domain 2 of *Bacillus anthracis* protective antigen. *Infect Immun* 77:3380–3388.

316. Albrink WS. 1961. Pathogenesis of inhalational anthrax. *Bacteriol Rev* 25:268–273.

317. Gleiser CA. 1967. Pathology of anthrax infection in animal hosts. *Fed Proc* 26:1518–1521.

318. Lever MS, Stagg AJ, Nelson M, Pearce P, Stevens DJ, Scott EA, Simpson AJ, Fulop MJ. 2008. Experimental respiratory anthrax infection in the common marmoset (*Callithrix jacchus*). *Int J Exp Pathol* 89:171–179.

319. Dabisch PA, Kline J, Lewis C, Yeager J, Pitt ML. 2010. Characterization of a head-only aerosol exposure system for nonhuman primates. *Inhal Toxicol* 22: 224–233.

320. Roy CJ, Reed DS, Hutt JA. 2010. Aerobiology and inhalation exposure to biological select agents and toxins. *Vet Pathol* 47:779–789.

321. Passalacqua KD, Bergman NH. 2006. *Bacillus anthracis*: interactions with the host and establishment of inhalational anthrax. *Future Microbiol* 1:397–415.

322. Allerberger F, Grif K, Dierich MP, Wimmer A, Plicka H. 2002. Anthrax inhalation and lethal human infection. *Lancet* 359:711. doi:10.1016/S0140-6736(02)07793-0.

323. Brachman PS. 1965. Human anthrax in the United States. *Antimicrobial Agents Chemother* (Bethesda) 5: 111–114.

324. Brachman PS. 1970. Anthrax. *Ann N Y Acad Sci* 174: 577–582.

325. Brachman PS. 1980. Inhalation anthrax. *Ann N Y Acad Sci* 353:83–93.

326. Brachman PS, Fekety FR. 1958. Industrial anthrax. *Ann N Y Acad Sci* 70:574–584.

327. Friedlander AM. 1999. Clinical aspects, diagnosis and treatment of anthrax. *J Appl Microbiol* 87:303. doi: 10.1046/j.1365-2672.1999.00896.x.

328. Jernigan DB, Raghunathan PL, Bell BP, Brechner R, Bresnitz EA, Butler JC, Cetron M, Cohen M, Doyle T, Fischer M, Greene C, Griffith KS, Guarner J, Hadler JL, Hayslett JA, Meyer R, Petersen LR, Phillips M, Pinner R, Popovic T, Quinn CP, Reefhuis J, Reissman D, Rosenstein N, Schuchat A, Shieh WJ, Siegal L, Swerdlow DL, Tenover FC, Traeger M, Ward JW, Weisfuse I, Wiersma S, Yeskey K, Zaki S, Ashford DA, Perkins BA, Ostroff S, Hughes J, Fleming D, Koplan JP, Gerberding JL. 2002. Investigation of bioterrorism-related anthrax, United States, 2001: epidemiologic findings. *Emerg Infect Dis* 8:1019–1028.

329. Stern EJ, Uhde KB, Shadomy SV, Messonnier N. 2008. Conference report on public health and clinical guidelines for anthrax. *Emerg Infect Dis* 14. doi:10.3201/eid1404.070969.

330. Wright M, Team NCAS. 2011. Long-term sequelae in a subset of 2001 anthrax survivors, abstr *Bacillus*-ACT: The International Conference on *Bacillus anthracis, B. cereus, and B. thuringiensis*, Bruges, Belgium, 7 to 11 August 2011.

331. Rossi CA, Ulrich M, Norris S, Reed DS, Pitt LM, Leffel EK. 2008. Identification of a surrogate marker for infection in the African green monkey model of inhalation anthrax. *Infect Immun* 76:5790–5801.

332. Boyer AE, Quinn CP, Hoffmaster AR, Kozel TR, Saile E, Marston CK, Percival A, Plikaytis BD, Woolfitt AR, Gallegos M, Sabourin P, McWilliams LG, Pirkle JL, Barr JR. 2009. Kinetics of lethal factor and poly-D-glutamic acid antigenemia during inhalation anthrax in rhesus macaques. *Infect Immun* 77:3432–3441.

333. Hoffmaster AR, Meyer RF, Bowen MD, Marston CK, Weyant RS, Thurman K, Messenger SL, Minor EE, Winchell JM, Rassmussen MV, Newton BR, Parker JT, Morrill WE, McKinney N, Barnett GA, Sejvar JJ, Jernigan JA, Perkins BA, Popovic T. 2002. Evaluation and validation of a real-time polymerase chain reaction assay for rapid identification of *Bacillus anthracis*. *Emerg Infect Dis* 8:1178–1182.

334. Tang S, Moayeri M, Chen Z, Harma H, Zhao J, Hu H, Purcell RH, Leppla SH, Hewlett IK. 2009. Detection of anthrax toxin by an ultrasensitive immunoassay using europium nanoparticles. *Clin Vaccine Immunol* 16: 408–413.

335. Saile E, Boons GJ, Buskas T, Carlson RW, Kannenberg EL, Barr JR, Boyer AE, Gallegos-Candela M, Quinn CP. 2011. Antibody responses to a spore carbohydrate antigen as a marker of nonfatal inhalation anthrax in rhesus macaques. *Clin Vaccine Immunol* 18:743–748.

336. Mabry R, Brasky K, Geiger R, Carrion R Jr, Hubbard GB, Leppla S, Patterson JL, Georgiou G, Iverson BL. 2006. Detection of anthrax toxin in the serum of animals infected with *Bacillus anthracis* by using engineered immunoassays. *Clin Vaccine Immunol* 13:671–677.

337. Quinn CP, Dull PM, Semenova V, Li H, Crotty S, Taylor TH, Steward-Clark E, Stamey KL, Schmidt DS,

Stinson KW, Freeman AE, Elie CM, Martin SK, Greene C, Aubert RD, Glidewell J, Perkins BA, Ahmed R, Stephens DS. 2004. Immune responses to Bacillus anthracis protective antigen in patients with bioterrorism-related cutaneous or inhalation anthrax. *J Infect Dis* 190:1228–1236.

338. Modlin JF, Advisory Committee on Immunization Practices. 2000. Use of anthrax vaccine in the United States. *Morb Mortal Wkly Rep* 49(RR15):1–20.

339. Galen JE, Chinchilla M, Pasetti MF, Wang JY, Zhao L, Arciniega-Martinez I, Silverman DJ, Levine MM. 2009. Mucosal immunization with attenuated Salmonella enterica serovar Typhi expressing protective antigen of anthrax toxin (PA83) primes monkeys for accelerated serum antibody responses to parenteral PA83 vaccine. *J Infect Dis* 199:326–335.

340. Gochenour WS Jr, Gleiser CA, Tigertt WD. 1962. Observations on penicillin prophylaxis of experimental inhalation anthrax in the monkey. *J Hyg* (Lond) 60:29–33.

341. Ivins BE, Pitt ML, Fellows PF, Farchaus JW, Benner GE, Waag DM, Little SF, Anderson GW Jr, Gibbs PH, Friedlander AM. 1998. Comparative efficacy of experimental anthrax vaccine candidates against inhalation anthrax in rhesus macaques. *Vaccine* 16:1141–1148.

342. Kao LM, Bush K, Barnewall R, Estep J, Thalacker FW, Olson PH, Drusano GL, Minton N, Chien S, Hemeryck A, Kelley MF. 2006. Pharmacokinetic considerations and efficacy of levofloxacin in an inhalational anthrax (postexposure) rhesus monkey model. *Antimicrob Agents Chemother* 50:3535–3542.

343. Kelly DJ, Chulay JD, Mikesell P, Friedlander AM. 1992. Serum concentrations of penicillin, doxycycline, and ciprofloxacin during prolonged therapy in rhesus monkeys. *J Infect Dis* 166:1184–1187.

344. Vietri NJ, Purcell BK, Lawler JV, Leffel EK, Rico P, Gamble CS, Twenhafel NA, Ivins BE, Heine HS, Sheeler R, Wright ME, Friedlander AM. 2006. Short-course postexposure antibiotic prophylaxis combined with vaccination protects against experimental inhalational anthrax. *Proc Natl Acad Sci USA* 103:7813–7816.

345. Vietri NJ, Purcell BK, Tobery SA, Rasmussen SL, Leffel EK, Twenhafel NA, Ivins BE, Kellogg MD, Webster WM, Wright ME, Friedlander AM. 2009. A short course of antibiotic treatment is effective in preventing death from experimental inhalational anthrax after discontinuing antibiotics. *J Infect Dis* 199:336–341.

346. Vitale L, Blanset D, Lowy I, O'Neill T, Goldstein J, Little SF, Andrews GP, Dorough G, Taylor RK, Keler T. 2006. Prophylaxis and therapy of inhalational anthrax by a novel monoclonal antibody to protective antigen that mimics vaccine-induced immunity. *Infect Immun* 74:5840–5847.

347. Ivins BE, Fellows P, Pitt LM, Estep J, Welkos SL, Worsham PL. 1996. Efficacy of a standard human anthrax vaccine against Bacillus anthracis spore challenge in rhesus monkeys. *Salisbury Med Bull* Suppl:125–126.

348. Pitt LM, Ivins BE, Estep J, Farchaus JW, Friedlander A. 1996. Comparative efficacy of a recombinant protective antigen vaccine against inhalational anthrax in guinea pigs, rabbits, and rhesus monkeys, abstr 96th Annual meeting of the American Society of Microbiology, New Orleans, LA, 19 to 23 May, 1996.

349. Ivins BE, Pitt MLM, Fellows PF, Farchaus JW, Benner GE, Waag DM, Little SF, Anderson GW Jr, Gibbs PH, Friedlander AM. 1998. Comparative efficacy of experimental anthrax vaccine candidates against inhalational anthrax in rhesus macaques. *Vaccine* 16:1141–1148.

350. Klinman DM, Klaschik S, Sato T, Tross D. 2009. CpG oligonucleotides as adjuvants for vaccines targeting infectious diseases. *Adv Drug Deliv Rev* 61:248–255.

351. Nguyen ML, Crowe SR, Kurella S, Teryzan S, Cao B, Ballard JD, James JA, Farris AD. 2009. Sequential B-cell epitopes of Bacillus anthracis lethal factor bind lethal toxin-neutralizing antibodies. *Infect Immun* 77:162–169.

352. Winterroth L, Rivera J, Nakouzi AS, Dadachova E, Casadevall A. 2010. Neutralizing monoclonal antibody to edema toxin and its effect on murine anthrax. *Infect Immun* 78:2890–2898.

353. Chabot D, Joyce JG, Caulfield M, Wang S, Vietri NJ, Leffel E, Pitt ML, Cook C, Helper W, Friedlander AM. 2009. Efficacy of a poly-gamma-D-Glutamic acid capsule conjugate vaccine against inhalational anthrax in rabbits and non-human primates, abstr *Bacillus*-ACT: The International Conference on *Bacillus anthracis, B. cereus, and B. thuringiensis*, Santa Fe, New Mexico, 30 August to 3 September 2009.

354. Lincoln RE, Klein F, Walker JS, Haines BW, Jones WI, Mahlandt BG, Friedman RH. 1964. Successful treatment of Rhesus Monkeys for septicemia anthrax. *Antimicrob Agents Chemother* (Bethesda) 10:759–763.

355. Pitt ML, Dyer D, Washington R, Blanset D, Riddle V, Meldorf M. 2009. Thereapeutic efficacy of Valortim, an anti-toxin monoclonal antibody, in the African green monkey model of inhalational anthrax, abstr *Bacillus*-ACT: The International Conference on *Bacillus anthracis, B. cereus, and B. thuringiensis*, Santa Fe, New Mexico, 30 August to 3 September 2009.

356. Dyer D, Leffel E, Washington R, Pitt ML. 2009. Development of a therapeutic African Green monkey model for inhalational anthrax to demonstrate added value of adjunct therapies, *Bacillus*-ACT: The International Conference on *Bacillus anthracis, B. cereus, and B. thuringiensis*, Santa Fe, New Mexico, 30 August to 3 September 2009.

357. Henning LN, Comer JE, Stark GV, Ray BD, Tordoff KP, Knostman KA, Meister GT. 2012. Development of an inhalational *Bacillus anthracis* exposure therapeutic model in cynomolgus macaques. *Clin Vaccine Immunol* 19:1765–1775.

358. Snoy PJ. 2010. Establishing efficacy of human products using animals: the US food and drug administration's "animal rule." *Vet Pathol* 47:774–778.

359. Watson LE, Kuo SR, Katki K, Dang T, Park SK, Dostal DE, Tang WJ, Leppla SH, Frankel AE. 2007. Anthrax toxins induce shock in rats by depressed cardiac ventric-

ular function. *PLoS One* **2**:e466. doi:10.1371/journal.pone.0000466.

360. Gupta PK, Moayeri M, Crown D, Fattah RJ, Leppla SH. 2008. Role of N-terminal amino acids in the potency of anthrax lethal factor. *PLoS One* **3**:e3130. doi:10.1371/journal.pone.0003130.

361. Beall FA, Dalldorf FG. 1966. The pathogenesis of the lethal effect of anthrax toxin in the rat. *J Infect Dis* **116**:377–389.

362. Beall FA, Taylor MJ, Thorne CB. 1962. Rapid lethal effect in rats of a third component found upon fractionating the toxin of *Bacillus anthracis*. *J Bacteriol* **83**:1274–1280.

363. Ezzell JW, Ivins BE, Leppla SH. 1984. Immunoelectrophoretic analysis, toxicity, and kinetics of in vitro production of the protective antigen and lethal factor components of *Bacillus anthracis* toxin. *Infect Immun* **45**:761–767.

364. Haines BW, Klein F, Lincoln RE. 1965. Quantitative assay for crude anthrax toxins. *J Bacteriol* **89**:74–83.

365. Nye SH, Wittenburg AL, Evans DL, O'Connor JA, Roman RJ, Jacob HJ. 2008. Rat survival to anthrax lethal toxin is likely controlled by a single gene. *Pharmacogenomics J* **8**:16–22.

366. Newman ZL, Printz MP, Liu S, Crown D, Breen L, Miller-Randolph S, Flodman P, Leppla SH, Moayeri M. 2010. Susceptibility to anthrax lethal toxin-induced rat death is controlled by a single chromosome 10 locus that includes rNlrp1. *PLoS Pathog* **6**:e1000906. doi:10.1371/journal.ppat.1000906.

367. Printz MP, Jirout M, Jaworski R, Alemayehu A, Kren V. 2003. Genetic Models in Applied Physiology. HXB/BXH rat recombinant inbred strain platform: a newly enhanced tool for cardiovascular, behavioral, and developmental genetics and genomics. *J Appl Physiol* **94**:2510–2522.

368. Chen Z, Moayeri M, Crown D, Emerson S, Gorshkova I, Schuck P, Leppla SH, Purcell RH. 2009. Novel chimpanzee/human monoclonal antibodies that neutralize anthrax lethal factor, and evidence for possible synergy with anti-protective antigen antibody. *Infect Immun* **77**:3902–3908.

369. Cui X, Li Y, Moayeri M, Choi GH, Subramanian GM, Li X, Haley M, Fitz Y, Feng J, Banks SM, Leppla SH, Eichacker PQ. 2005. Late treatment with a protective antigen-directed monoclonal antibody improves hemodynamic function and survival in a lethal toxin-infused rat model of anthrax sepsis. *J Infect Dis* **191**:422–434.

370. Joshi A, Kate S, Poon V, Mondal D, Boggara MB, Saraph A, Martin JT, McAlpine R, Day R, Garcia AE, Mogridge J, Kane RS. 2011. Structure-based design of a heptavalent anthrax toxin inhibitor. *Biomacromolecules* **12**:791–796.

371. Scobie HM, Thomas D, Marlett JM, Destito G, Wigelsworth DJ, Collier RJ, Young JA, Manchester M. 2005. A soluble receptor decoy protects rats against anthrax lethal toxin challenge. *J Infect Dis* **192**:1047–1051.

372. Sharma M, Swain PK, Chopra AP, Chaudhary VK, Singh Y. 1996. Expression and purification of anthrax

toxin protective antigen from *Escherichia coli*. *Protein Expr Purif* **7**:33–38.

373. Sharma S, Thomas D, Marlett J, Manchester M, Young JA. 2009. Efficient neutralization of antibody-resistant forms of anthrax toxin by a soluble receptor decoy inhibitor. *Antimicrob Agents Chemother* **53**:1210–1212.

374. Manayani DJ, Thomas D, Dryden KA, Reddy V, Siladi ME, Marlett JM, Rainey GJ, Pique ME, Scobie HM, Yeager M, Young JA, Manchester M, Schneemann A. 2007. A viral nanoparticle with dual function as an anthrax antitoxin and vaccine. *PLoS Pathog* **3**:1422–1431.

375. Little SF, Webster WM, Fisher DE. 2011. Monoclonal antibodies directed against protective antigen of Bacillus anthracis enhance lethal toxin activity in vivo. *FEMS Immunol Med Microbiol* **62**:11–22.

376. Goossens PL. 2009. Animal models of human anthrax: the Quest for the Holy Grail. *Mol Aspects Med* **30**:467–480.

377. Bolcome RE 3rd, Sullivan SE, Zeller R, Barker AP, Collier RJ, Chan J. 2008. Anthrax lethal toxin induces cell death-independent permeability in zebrafish vasculature. *Proc Natl Acad Sci USA* **105**:2439–2444.

378. Fellows PF, Linscott MK, Little SF, Gibbs P, Ivins BE. 2002. Anthrax vaccine efficacy in golden Syrian hamsters. *Vaccine* **20**:1421–1424.

379. Pomerantsev AP, Staritsin NA, Mockov Yu V, Marinin LI. 1997. Expression of cereolysine AB genes in *Bacillus anthracis* vaccine strain ensures protection against experimental hemolytic anthrax infection. *Vaccine* **15**:1846–1850.

380. Stepanov AV, Marinin LI, Pomerantsev AP, Staritsin NA. 1996. Development of novel vaccines against anthrax in man. *J Biotechnol* **44**:155–160.

381. Hoffmaster AR, Hill KK, Gee JE, Marston CK, De BK, Popovic T, Sue D, Wilkins PP, Avashia SB, Drumgoole R, Helma CH, Ticknor LO, Okinaka RT, Jackson PJ. 2006. Characterization of *Bacillus cereus* isolates associated with fatal pneumonias: strains are closely related to Bacillus anthracis and harbor B. anthracis virulence genes. *J Clin Microbiol* **44**:3352–3360.

382. Saile E, Koehler TM. 2006. Bacillus anthracis multiplication, persistence, and genetic exchange in the rhizosphere of grass plants. *Appl Environ Microbiol* **72**:3168–3174.

383. Weiner Z, Glomski I. 2012. Updating perspectives on the initiation of *Bacillus anthracis* growth and dissemination through its host. *Infect Immun* **80**:1626–1644.

384. Candela T, Dumetz F, Tosi-Couture E, Mock M, Goossens PL, Fouet A. 2012. Cell-wall preparation containing poly-gamma-d-glutamate covalently linked to peptidoglycan, a straightforward extractable molecule, protects mice against experimental anthrax infection. *Vaccine* **31**:171–175.

385. Uchida M, Harada T, Enkhtuya J, Kusumoto A, Kobayashi Y, Chiba S, Shyaka A, Kawamoto K. 2012. Protective effect of *Bacillus anthracis* surface protein EA1 against anthrax in mice. *Biochem Biophys Res Commun* **421**:323–328.

386. Levy H, Weiss S, Altboum Z, Schlomovitz J, Rothschild N, Glinert I, Sittner A, Kobiler D. 2012. The effect of deletion of the edema factor on Bacillus anthracis pathogenicity in guinea pigs and rabbits. *Microb Pathog* **52**: 55–50.

387. Levy H, Weiss S, Altboum Z, Schlomovitz J, Glinert I, Sittner A, Shafferman A, Kobiler D. 2012. Differential contribution of *Bacillus anthracis* toxins to pathogenicity in two animal models. *Infect Immun* **80**: 2623–2631.

388. Levy H, Glinert I, Weiss S, Sittner A, Schlomovitz J, Altboum Z, Kobiler D. 2014. Toxin-independent virulence of *Bacillus anthracis* in rabbits. *PLoS One* **9**: e84947. doi:10.1371/journal.pone.0084947.

389. Peachman KK, Li Q, Matyas GR, Shivachandra SB, Lovchik J, Lyons RC, Alving CR, Rao VB, Rao M. 2012. Anthrax vaccine antigen-adjuvant formulations completely protect New Zealand white rabbits against challenge with *Bacillus anthracis* Ames strain spores. *Clin Vaccine Immunol* **19**:11–16.

390. Lovchik J, Drysdale M, Koehler T, Hutt J, Lyons CR. 2012. Expression of either lethal toxin or edema toxin by *Bacillus anthracis* is sufficient for virulence in a rabbit model of inhalational anthrax. *Infect Immun* **80**:2414–2425.

391. Oscherwitz J, Yu F, Cease KB. 2010. A synthetic peptide vaccine directed against the 2β2-2β3 loop of domain 2 of protective antigen protects rabbits from inhalation anthrax. *J Immunol* **185**:3661–3668.

392. Bozue J, Welkos S, Cote CK. The *Bacillus anthracis* exosporium: what's the big hairy deal? *In* Eichenberger P, Driks A (ed), *The Bacterial Spore: From Molecules to Systems*. ASM Press, Washington, DC, in press.

393. Shah A, Liu MC, Vaughan D, Heller AH. 1999. Oral bioequivalence of three ciprofloxacin formulations following single-dose administration: 500 mg tablet compared with 500 mg/10 mL or 500 mg/5 mL suspension and the effect of food on the absorption of ciprofloxacin oral suspension. *J Antimicrob Chemother* **43**(Suppl A):49–54.

394. Tsai C-W, Morris S. 2015. Approval of raxibacumab for the treatment of inhalation anthrax under the US Food and Drug Administration "Animal Rule." *Front Microbiol* **6**:1320. doi:10.3389/fmicb.2015.01320.

The Clostridia

IV

The Bacterial Spore: From Molecules to Systems
Edited by P. Eichenberger and A. Driks
© 2016 American Society for Microbiology, Washington, DC
doi:10.1128/microbiolspec.TBS-0010-2012

Peter Dürre[1]

Physiology and Sporulation in *Clostridium*

15

Clostridia are anaerobic bacteria, although many species can tolerate oxygen to various extents. They are able to form endospores and are not capable of dissimilatory sulfate reduction. Most of them show a positive Gram reaction. These criteria have been used in the past for classification. However, phylogenetic analyses based on 16S rRNA sequences led to reattribution of many former clostridia to numerous other and also novel genera, such as *Blautia, Butyrivibrio, Caloramator, Cellulosilyticum, Dendrosporobacter, Eubacterium, Filifactor, Flavonifractor, Moorella, Oxalophagus, Oxobacter, Paenibacillus, Thermoanaerobacter, Thermoanaerobacterium, Sedimentibacter, Sporohalobacter, Syntrophomonas, Syntrophospora,* and *Tissierella* (J. P.Euzéby, List of prokaryotic names with standing in nomenclature – genus *Clostridium,* http://www.bacterio.cict.fr/c/clostridium.html). For the genus *Clostridium,* approximately 180 species have been validly described, rendering it one of the largest bacterial genera. Only a few of these species are pathogenic, however, involving microbes producing very dangerous toxins. On the other hand, a large number of species are used in biotechnological applications (enzyme, bulk chemicals, and biofuels production) and tested for use in cancer therapy. This is due to the enormous metabolic diversity within the clostridia, rendering them the avant-garde of biotechnologically

exploited microorganisms. During past years, techniques have been developed that allowed establishment of genetic systems for many clostridia. Thus, the tools are at hand for further elucidation and exploitation. Due to the limited space of this article, many aspects cannot be presented in detail. Thus, the interested reader is referred to recent references for additional information (1–6).

PHYSIOLOGY OF *CLOSTRIDIUM*

Clostridia are often differentiated by performing a saccharolytic or a proteolytic metabolism. However, such traits can also be present in a single organism. For example, *Clostridium tetani,* originally considered to be a paradigm of a proteolytic microbe, was reported to ferment glucose (7). Genome sequencing confirmed that this organism contains all genes required for glycolysis. Only polysaccharides cannot be degraded (8). In clostridia, there are a number of different pathways for substrate utilization, yielding one or several dominant fermentation products (Table 1).

Homoacetate Fermentation

This metabolic pathway has been mainly elucidated with the former *Clostridium thermoaceticum,* now reclassified as *Moorella thermoacetica.* However, there

[1]Institut für Mikrobiologie und Biotechnologie, Universität Ulm, 89069 Ulm, Germany.

Table 1 Major metabolic features of *Clostridium*

Fermentation pathway	Substrate(s)	Products	Representative species
Homoacetate (Wood-Ljungdahl)	Sugars (e.g., fructose), $CO_2 + H_2$, CO	Acetate (under special conditions also ethanol)	*C. aceticum*, *C. ljungdahlii*
Propionate			
Acrylyl-CoA	Lactate	Propionate, acetate, CO_2	*C. propionicum*
Succinate decarboxylation	Succinate	Propionate, CO_2	*C. mayombei*
Butyrate	Sugars (e.g., glucose)	Butyrate, acetate, CO_2, H_2	*C. butyricum*
Acetone/butanol	Sugars (e.g., glucose)	Butanol, acetone, CO_2, H_2, butyrate, acetate	*C. acetobutylicum*
Alcoholotrophic	Various alcohols, alcohol + acid	Various acids and solvents / Various acids	*C. aceticum* / *C. kluyveri*
Proteolytic, peptolytic, use of amino acids	Proteins, peptides, amino acids	Acetate, propionate, butyrate, CO_2, NH_4^+	*C. propionicum*, *C. tetanomorphum*
Stickland reaction	Pairs of amino acids	Acetate, CO_2, NH_4^+	*C. sticklandii*
Purinolytic	Purines	Acetate, formate, CO_2, NH_4^+	*C. purinilyticum*
Pyrimidinolytic	Pyrimidines	Acids, CO_2, NH_4^+	*C. oroticum*

are true clostridia that use this type of fermentation. It functions during the degradation of sugars as well as under autotrophic conditions. The sequence of enzymatic reactions leading to acetate formation from CO_2 plus H_2 or from CO is called the Wood-Ljungdahl pathway and the organisms using it are called acetogens, respectively. Clostridia, able to grow heterotrophically on sugars and autotrophically on CO_2/H_2 gas mixtures or CO, are, e.g., *C. aceticum* and *C. ljungdahlii*. *C. aceticum* was originally isolated in 1936 (9) and thought to be lost during World War II. It was rediscovered in 1981 as an old spore preparation in the laboratory of Horace A. Barker at the University of California in Berkeley (10). As *M. thermoacetica*, *C. aceticum* contains cytochromes (11). It is thus highly likely that energy conservation is achieved by generating a proton gradient over the membrane and using it for ATP synthesis via an F_1F_0-type ATPase. A model, showing a possible electron transfer chain with involvement of cytochrome, menaquinone, flavoproteins, flavodoxin, ferredoxin, and rubredoxin has been proposed by Das and Ljungdahl (12) for *M. thermoacetica* and might reflect as well the metabolism of *C. aceticum*. A recent study showed that in *M. thermoacetica* the NADH generated during sugar fermentation is converted into the NADPH required for acetogenesis from CO_2 by means of an electron-bifurcating NADH-dependent reduced ferredoxin: $NADP^+$ oxidoreductase (NfnAB), a cytoplasmic iron-sulfur-flavoprotein (13).

A different type of energy conservation is found in anaerobic autotrophs harboring the so-called Rnf system. The paradigm of this metabolism is *Acetobacterium woodii*, a nonsporulating bacterium (14). The membrane-bound Rnf complex containing iron-sulfur clusters and flavins was first detected in *Rhodobacter capsulatus* and found to play a role in nitrogen fixation (therefore the designation Rnf, from *Rhodobacter* nitrogen fixation) (15). In *A. woodii*, an Rnf complex was detected and analyzed that coupled electron transfer from reduced ferredoxin to NAD^+ to the export of sodium cations (16). Genome sequencing revealed that this is the only Na^+ pump in *A. woodii* (17). Thus, the long-standing previous assumption that a methyltransferase, transferring the methyl moiety from methyltetrahydrofolate (THF) to a corrinoid-iron sulfur protein, might be membrane bound and, in analogy to methanogens, exports sodium cations (reviewed in reference 18) turned out to be wrong. Energy conservation is now thought to start with hydrogen oxidation by an electron-bifurcating [FeFe]-hydrogenase, resulting in equal amounts of reduced ferredoxin and NADH. The reduced ferredoxin is oxidized at the Rnf complex, the electrons are transferred to NAD^+, yielding NADH, and, during this process, sodium cations are exported. The thus established sodium gradient is used by a Na^+-dependent F_1F_0-ATPase to generate ATP. NADH is used for the reduction of methenyl-THF via methylene-THF to methyl-THF in the methyl branch of the Wood-Ljungdahl pathway. Additional reduced ferredoxin required for CO_2 reduction to enzyme-bound CO in the carbonyl branch is thought to stem from the exergonic reduction of methylene-THF to methyl-THF either by electron bifurcation or by direct coupling to the Rnf complex (17). The ATP generated in the acetate kinase reaction is used for the activation of formate to formyl-THF, leaving the sodium cation gradient the only 1source for ATP formation. A similar mechanism is probably used by *C. ljungdahlii*, an autotrophic acetogen growing on CO, synthesis gas, and CO_2/H_2 mixtures, because this bacterium does not harbor genes for cytochromes and quinones, but for an Rnf complex (19). A major difference from *A. woodii*, however, is that *C. ljungdahlii* is not sodium dependent. Thus, the coupling ions in this case are probably protons (19). The autotrophic clostridia have gained a lot of industrial interest recently, as they can convert greenhouse gases into valuable bulk chemicals (20, 21). With the use of special media composition, it is also possible to turn, e.g., *C. ljungdahlii* into an ethanol producer when growing on synthesis gas (20).

Propionate Fermentation

The paradigm of a clostridial propionate fermentor is probably *C. propionicum*. It does not use the methylmalonyl-coenzyme A (CoA) pathway, known from propionibacteria, but uses instead the acrylyl-CoA pathway, branched into oxidative and reductive sequences. Lactate is one of the preferred substrates of *C. propionicum*. One molecule is oxidized first to pyruvate and then to acetyl-CoA and CO_2. Acetyl-CoA is converted to acetyl phosphate and then to acetate, thereby generating ATP in the acetate kinase reaction. The reducing equivalents obtained in the oxidation steps are used to reduce two molecules of acrylyl-CoA to two propionyl-CoA, which by means of a CoA-transferase are converted into the final product propionate. The two acrylyl-CoAs stem from another two lactate molecules, activated by the CoA-transferase and then dehydrated (3).

Another pathway, i.e., succinate decarboxylation, might be used by *C. mayombei*. This type of metabolism was first discovered in *Propionigenium modestum* (22). Succinate is converted into succinyl-CoA by transfer of the CoA-moiety from propionyl-CoA and subsequent release of propionate. Succinyl-CoA is first rearranged into methylmalonyl-CoA, which is then decarboxylated by a membrane-bound, Na^+-exporting decarboylase into propionyl-CoA and CO_2. Propionyl-CoA is used to restart the cycle. The generated sodium cation gradient drives ATP generation by a sodium-dependent F_1F_0-ATPase. A similar mechanism might be used by *C. mayombei*, because this organism ferments sodium succinate to propionate and CO_2 (23). However, this has not yet been experimentally confirmed.

Butyrate/Butanol Fermentations

Many clostridia form butyrate as a major fermentation product. The Embden-Meyerhof-Parnas pathway allows degradation of hexoses to pyruvate, which is converted to acetyl-CoA, CO_2, and reduced ferredoxin by pyruvate: ferredoxin-oxidoreductase. Part of the acetyl-CoA might be further metabolized to acetate, yielding ATP in the acetate kinase reaction. For butyrate formation, two acetyl-CoAs are combined into acetoacetyl-CoA, which is further converted via 3-hydroxyl-CoA, crotonyl-CoA, and butyryl-CoA into butyrate. Some clostridia use specific phosphotransbutyrylase and butyrate kinase for the last two steps, generating ATP in the kinase reaction. However, the majority use a CoA-transferase and acetate to convert butyryl-CoA into butyrate and acetyl-CoA, the latter of which is then used for acetate and ATP formation (24, 25). However, not only substrate-level phosphorylation during glycolysis and kinase reactions leads to ATP synthesis. Buckel and coworkers provided evidence for a ferredoxin reduction coupled to crotonyl-CoA reduction (26). The butyryl-CoA dehydrogenase/electron transfer flavoprotein (EtfAB) complex catalyzes the exergonic reduction of crotonyl-CoA to butyryl-CoA by NADH and concomitantly drives the endergonic reduction of ferredoxin by another NADH (electron bifurcation). The reduced ferredoxin can then transfer the electrons to NAD^+ at the Rnf complex, thus generating an ion gradient, which can be used for ATP synthesis by ATPases.

Not all clostridia do contain an Rnf complex. One of the few exceptions is *C. acetobutylicum*, an organism well known and extensively used for its ability to form the solvents butanol and acetone. Butanol is made by the action of butyraldehyde and butanol dehydrogenases from butyryl-CoA, while acetone is formed by decarboxylation of acetoacetate, being derived from acetoacetyl-CoA by a CoA-transferase. In *C. beijerinckii*, acetone can be further reduced to isopropanol. However, all these reactions are only performed at the end of exponential growth, which is fueled by a typical butyrate fermentation. The advantage of switching from an acidogenic to a solventogenic fermentation lies in avoiding life-threatening low pH caused by the acidic fermentation products. As undissociated acids at low pH, they can diffuse across the cytoplasmic membrane and lead to a collapse of the transmembrane proton gradient, as they dissociate in the more alkaline cytoplasm (27). Thus, conversion of acids into solvents allows the cells to stay metabolically active for a longer period and thus gives them an ecological advantage. As solvents become toxic in the long run as well, the induction of sporulation is coupled to the induction of solventogenesis by sharing the same master regulator Spo0A in phosphorylated form (8, 28, 29). For almost one hundred years, the acetone-butanol

(AB) fermentation has been a major biotechnological enterprise worldwide. It was second in size to ethanol fermentation. Low crude oil prices and thus cheap chemical synthesis led to a decline, a trend, which is currently reversed. The use of substrates with no competition to human nutrition, especially, will render the fermentation economically advantageous (30–33).

Fermentation of Alcohols

A great variety of mono-, di-, and polyols can be fermented by clostridia. Therefore, only a few examples are provided. Methanol is a substrate of *C. formicoaceticum*; ethanol is a substrate of *C. aceticum* (10); ethanol and propanol are substrates of *C. kluyveri*, but only in combination with acetate or succinate (34); butanol is a substrate of *C. ljungdahlii*, but only as a cosubstrate (19); 1,2-ethanediol (ethylene glycol) and 1,2-propanediol are fermented by *C. glycolicum* (35); 2,3-butanediol is fermented by *C. aceticum* and *C. magnum* (36); and glycerol is fermented by, e.g., *C. acetobutylicum*, *C. beijerinckii*, and *C. butyricum* (37).

The two-substrate fermentation by *C. kluyveri* deserves a more detailed description, because the energetics of this process remained mysterious and were only elucidated a few years ago. Typically, 6 ethanol and 3 acetate (molar ratios are provided) are fermented to 3 butyrate, 1 caproate, and 2 hydrogen. Only 1 ATP is gained in the acetate kinase reaction. It was with this organism that W. Buckel and coworkers discovered the coupling of the exergonic reduction of crotonyl-CoA to butyryl-CoA with NADH by the butyryl-CoA dehydrogenase/EtfAB complex to the endergonic reduction of ferredoxin with NADH (electron bifurcation). Thus, 2 NADH are consumed to generate butyryl-CoA and reduced ferredoxin (38). The reduced ferredoxin can be used for hydrogen formation as well as for additional ATP generation via the Rnf system. There, reduced ferredoxin is oxidized, transferring the electrons to NAD^+, and the protons are exported. This proton gradient can then be used by the F_1F_0-type ATPase. It is assumed that an additional 1.5 ATP are generated this way (39).

Fermentation of Amino Acids

Proteins, peptides, and single amino acids are substrates for all proteolytic clostridia. The former two are degraded to monomers by proteases, leaving the amino acids as the starting point for further utilization. Numerous pathways for the various compounds have been elucidated, sometimes even different ones for the same amino acid. Thus, the limited space of this chapter does not allow detailed descriptions of the various degrada-

tion sequences. General strategies, however, have been described (40–42). One of the first reactions always is removal of the amino group. This can be achieved by transamination, oxidative deamination, or elimination. In many cases, conversion into the corresponding 2-oxo acids follows. Degradation is often split into oxidative and reductive branches. Typical examples are the utilization of glycine and the degradation of amino acid pairs, the so-called Stickland reaction. Glycine can be used as the sole carbon source by *C. purinilyticum*, an organism specialized on purines and their degradation products (such as glycine) as substrates (43, 44). In the oxidative branch, 1 molecule of glycine is degraded to CO_2 via conversion into methylene-THF, with concomitant release of CO_2 and NH_4^+, methenyl-THF, formyl-THF, formate, and finally CO_2. The reducing equivalents generated during these reactions are used for reduction of another 3 molecules of glycine to acetyl phosphate and ammonia by the selenocysteine-containing glycine reductase complex. Acetyl phosphate is then converted into acetate. Thus, from degradation of every glycine molecule, 1 ATP can be formed (via formyl-THF synthetase or acetate kinase, respectively) (44). Similarly, *C. sticklandii* oxidizes 1 molecule of alanine to acetate, CO_2, and ammonia to yield sufficient reducing equivalents to convert 2 molecules of glycine to acetate and ammonia via the glycine reductase reaction.

Other reactions used for amino acid degradation involve *S*-adenosylmethionine-dependent carbon-nitrogen or vitamin B_{12}-dependent carbon-carbon rearrangements. Typical examples are lysine and glutamate fermentations (40, 42).

Fermentation of Purines and Pyrimidines

C. acidurici, *C. cylindrosporum*, and *C. purinilyticum* are specialized on decomposition of purines. In *C. purinilyticum*, a selenium-dependent degradation starts by cleaving the pyrimidine moiety of the heteroaromatic ring system. Products are CO_2, NH_4^+, and imidazole derivatives that are further degraded to formiminoglycine. This compound is cleaved into formimino-THF and glycine, both converted into formate (via methenyl-THF and formyl-THF) and acetate (via glycine reductase and acetate kinase). Thus, from every molecule of formiminoglycine, 2 molecules of ATP can be generated. Depending on the requirement of reducing equivalents, formate can be further oxidized to CO_2 (45). A selenium-independent pathway is possible, but it does not allow good growth. In this case, ring cleavage obviously starts at the imidazole moiety, yielding pyrimidines (46).

Pyrimidine degradation is performed by other specialized clostridia, e.g., *C. glycolicum* and *C. oroticum* (47, 48).

Other Fermentations

Clostridia are also able to ferment various organic acids, polymers, halogenated compounds, and aromatic compounds. Some are also able to fix nitrogen (e.g., *C. pasteurianum*). More details can be found in books completely devoted to the genus *Clostridium* (1, 2, 49).

SPORULATION IN *CLOSTRIDIUM*

Morphology of Spores, Spore Contents, and Spore Properties

Like *Bacillus*, *Clostridium* species are able to form endospores. Their shape can vary from coccoidal to cylindrical. A series of electron microscopy images taken from sporulating *C. formicoaceticum* cells resembles, in general, the stages found in *Bacillus subtilis* (50). Some clostridial spores (e.g., from *C. bifermentans*) were found to be associated with different appendages, ranging from long, tubular structures to ribbon-like structures or even capped, pin-like protrusions projecting from the spore surface (reviewed in reference 51). Another morphological difference to the archetype of

Bacillus sporulation is within *Clostridium* the formation of so-called clostridial forms at the beginning of sporulation. These cells are swollen, cigar-like, and accumulate storage components such as polysaccharides or polyhydroxybutyrate (Fig. 1). The polysaccharide has been designated granulose and determined in *C. saccharobutylicum* to consist of an $\alpha(1\rightarrow4)$-polyglucan of high molecular weight (52). In *C. botulinum*, *C. butyricum*, and *C. pasteurianum*, similar polymers have been detected, which are, in part, branched and sometimes identical to amylopectin or glycogen (53–57). The existence of this storage material was observed early in the last century, made visible by iodine staining, and used for differentiation (58, 59). Clostridia also produce dipicolinic acid, which complexes Ca^{2+} ions and thus adds to the thermoresistance of spores. However, several species (including *C. acetobutylicum*, *C. beijerinckii*, *C. botulinum*, *C. perfringens*, and *C. tetani*) lack the genes *spoVFA/B*, whose products catalyze the oxidation of 2,3-dihydrodipicolinic acid (DHDPA) (an early intermediate in bacterial lysine biosynthesis) to dipicolinic acid in *Bacillus* species. In *C. perfringens*, the EtfA protein produces dipicolinic acid from DHDPA (60), indicating an additional role for this protein to the one in electron bifurcation during the reduction of crotonyl-CoA to butyryl-CoA. In *Bacillus*, small, acid-soluble spore proteins (SASPs)

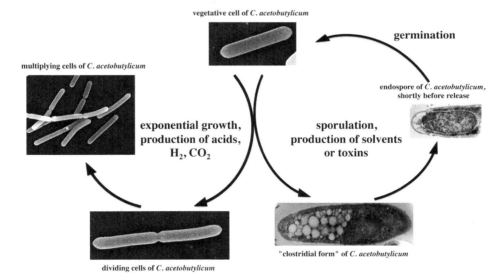

Figure 1 Vegetative growth and sporulation/germination cycle in clostridia. During normal growth, *Clostridium* cells (shown is *C. acetobutylicum*) divide and multiply (left). During this period, acids, carbon dioxide, and hydrogen are formed. Upon signals not yet determined, cells start to differentiate into "clostridial forms" with granulose as a storage material (solventogenic species produce at this time acetone, butanol, or isopropanol), then form endospores, which finally will germinate into vegetative cells again.

exert a protective effect against heat, peroxides, as well as UV radiation and also change the DNA conformation into an A-helix. Such proteins have also been detected in *C. bifermentans* and *C. perfringens* (61, 62) and since then in the sequenced genomes of clostridia. SASPs are classified into α/β and γ types, of which *Clostridium*, in contrast to *Bacillus*, obviously does not contain genes encoding γ-type SASPs (63). In *C. perfringens*, analyses of respective mutants and by antisense RNA showed that α/β-SASPs contribute significantly to spore resistance against moist heat, UV radiation, hydrogen peroxide, hydrochloric acid, nitrous acid, and formaldehyde (64–66).

Because bacterial endospores represent the most resistant cell type known (highly resistant to heat, chemicals, desiccation, and radiation), the presence of spores of pathogenic clostridia in food would be a serious problem. Therefore, measures are taken to prevent contamination in the first place (hygienic conditions, vacuum packing). Although the shelf life of food can be increased by a number of treatments, this does not per se result in the inactivation of spores. Rather, it prevents germination. Thus, the addition of preservatives such as nitrite has proved to be useful. Nitrite destroys redox-active iron-sulfur clusters, which are parts of essential enzymes in the vegetative cells of clostridia (67, 68). Not only do the spores of the well-known pathogens such as *C. botulinum*, *C. difficile*, *C. perfringens*, and *C. tetani* represent a threat, but also the spores of species that spoil food so that it can no longer be sold and used. Such an example is *C. tyrobutyricum* in the dairy industry. Because the organism converts lactate and acetate into butyrate and CO_2, it outcompetes the *Propionibacterium* used in cheese making and changes the taste and structure of the product (so-called late blowing).

Spores might also exhibit some metabolic activities without starting to germinate. In the spore coat of a number of *Firmicutes*, among them *C. tetani* and *C. thermocellum*, a new family of bacterial spore kinases has been detected; their physiological function is still unknown (69). *C. acetobutylicum* spores are obviously able to reduce uranyl acetate (U(VI)) to a mixture of UO_2 and other U(IV) products, when hydrogen is provided as an electron donor. The U(IV) products form a precipitate around the spores (70). Purified spores of *C. difficile* have been shown to exhibit catalase, peroxiredoxin, and chitinase activity, catalyzed by CotCB, CotD, and CotE, respectively. CotE is bifunctional, the latter two enzymatic activities are located at the N-terminal and C-terminal domains, respectively (71).

Mechanism of Sporulation

In *B. subtilis*, sporulation is initiated by phosphorylation of SpoOF, the first component of the so-called phosphorelay, by one of five different interacting histidine kinases. The His to Asp phosphotransfer is as in a classical two-component system and is repeated with the following proteins. SpoOB accepts the phosphoryl group from SpoOF~P at a histidine residue and then transfers it to SpoOA, again to an aspartate residue. SpoOA~P then becomes the master regulator of sporulation. This phosphorelay allows the integration of signals not only at SpoOF, but also at SpoOB or SpoOA. So, it came somewhat as a surprise when genome sequencing revealed that no *spoOF* and *spoOB* genes were found within the genus *Clostridium* (72–74). Thus, clostridia do not possess a phosphorelay. Considering the evolutionary relationship between *Clostridium* and *Bacillus* (clostridia appeared as a separate class approximately 2.7 billion years ago, bacilli approximately 2.3 billion years ago), this makes sense, because the adaptation to an aerobic lifestyle required enhanced sensing and regulating capabilities, which probably resulted in development of the phosphorelay (75, 76). On the other hand, SpoOA is indeed the master regulator of sporulation, and the respective gene has been found in all endospore formers (77). Studies on the molecular evolutionary history of this protein in all respective genera have revealed a division into two clades, mostly representing *Bacillus* and *Clostridium* species (78).

Phosphorylated SpoOA initiates sporulation in *Clostridium* as in *Bacillus* (79). In concert with σ^H, a number of operons are transcribed. Then, in the forespore, σ^F becomes active, when the anti-sigma factor SpoIIAB (ADP form) binds to the anti-anti-sigma factor SpoIIAA in its unphosphorylated form. The phosphorylated form of SpoIIAA, found predominantly in the mother cell, binds σ^F and thus prevents its transcriptional activity. SpoIIAB (ATP form) is responsible for both, the binding of σ^F and the phosphorylation of SpoIIAA, while SpoIIE catalyzes dephosphorylation. σ^F-dependent transcription leads to the formation of SpoIIR, which interacts with the membrane-bound protease SpoIIGA. SpoIIGA is responsible for the cleavage of pro-σ^E in the mother cell, yielding active σ^E. This process is supported by higher concentrations of SpoOA~P in the mother cell (as a consequence, higher concentrations of pro-σ^E in the mother cell) and a degradation of pro-σ^E in the forespore. After σ^F and σ^E become specifically active in the forespore or mother cell, respectively, the mother cell engulfs the forespore. Then, a second set of sigma factors enters the stage: σ^G in the forespore and σ^K in the mother cell. The anti-sigma factor SpoIIAB also acts on σ^G in the forespore, but additional components are required for activation (including SpoIIIJ, triggered by σ^E-dependent SpoIIIA from the mother cell). σ^K in the mother cell is first synthesized as an inactive precursor (similar to σ^E) and is activated by proteolysis. Again, help from the forespore is required for this process in the form of SpoIVB. The *spoIVB* gene is under control of the forespore-specific sigma factor G, and the protein is assumed to be inserted into the inner forespore membrane and to undergo autoproteolysis. Thus, released fragments serve as signals and interact with the SpoIVFA/FB-BofA complex in the outer forespore membrane. All these proteins stem from σ^E-dependent expression in the mother cell. The complex, upon interacting with fragments of SpoIVB, then cleaves pro-σ^K, yielding active σ^K. σ^G and σ^K allow expression of those operons, whose products are required for spore maturation. This whole process seems to be identical to what is happening in *C. acetobutylicum* during sporulation, the best-studied *Clostridium* in this respect. Genes encoding σ^F, σ^E, σ^G, and σ^K have been identified (80, 81), and the time course of expression of σ^E, σ^G, and σ^K as judged by Northern blots followed the scheme identified for *Bacillus* (82). However, closer inspection of regulons and their expression has revealed several differences. With the use of the *C. acetobutylicum sigF* mutant and complemented strain, it was found that σ^F was the first sporulation-related sigma factor to become active and that it is needed for the expression of σ^E and σ^G (83). This is in line with the *Bacillus* model and also true for *C. perfringens* (84). However, while *sigG* expression was found to be σ^F dependent, this was not the case for the genes *gpr*, *lonB*, *spoIIB*, and *spoIIR*, clearly a difference from *B. subtilis* (83). Reporter gene analyses led to the conclusion that SpoIIE controls sporulation in a manner similar to that in *Bacillus* (85), while investigations with a *spoIIE* mutant of *C. acetobutylicum* showed that SpoIIE affected *sigF* transcription and indicated an autostimulatory role for σ^F (86). A *sigE* mutant of *C. acetobutylicum* seems to be stalled in sporulation before stage II (87), which is in contrast to *B. subtilis* and also *C. perfringens sigE* mutants, stalled at early stage II (88). σ^E activity in *C. acetobutylicum* is also required for granulose synthesis and morphology of the "clostridial form." A *C. acetobutylicum sigG* mutant seemed to advance further in sporulation than a respective *Bacillus* mutant (87). The most important difference, however, seems to be the role of σ^K. A genome comparison of *Bacillus* with various clostridia has already revealed a lack of conserved genes beyond the developmental stage of *spoIVB* (75). This was followed by a detailed transcriptional study in *C. acetobutylicum*, providing the most comprehensive roadmap of clostridial sporulation up to now (89). The gene, originally identified as *sigK* (81), did not show increased expression, and the transcript level was well below those of the other sigma factor genes. Expression of two genes, known to be σ^K controlled in *Bacillus*, was also tested, but the data did not allow an unambiguous decision of whether there is a functional σ^K or not (89). On the other hand, the study confirmed the sequential order of σ^H, Spo0A, σ^F, σ^E, and σ^G in *C. acetobutylicum*, thus being identical to the scheme known for *Bacillus*. Genome sequencing revealed that *sigK* genes in most clostridia are intact as in most bacilli. *C. tetani* resembles *B. subtilis* by carrying a *skin* (*sigK* intervening sequence) element within the *sigK* gene. In *C. difficile*, a smaller *skin* element with divergent orientation was detected. *In vivo* excision of this element was found to be essential for sporulation (90). In *C. botulinum*, a *sigK* gene was disrupted (91). As a consequence, sporulation was blocked at an early stage, rather than at late stage as is the case in *Bacillus*. Thus, the function of σ^K (or a similar sigma factor) might be somewhat different in clostridia. In this respect, it has also been shown that σ^H in *C. difficile* governs the expression of many more genes than its counterpart in *Bacillus* (92).

Sporulation frequency in clostridia is not type associated, but rather is strain dependent, as shown for *C. difficile* strains (93, 94).

Regulation of Sporulation

The first experiments that indicated the close relationship of *Bacillus* and *Clostridium* at the molecular level were the discovery of the common presence of Spo0A in both genera and the detection of sporulation-specific sigma factor homologs in *C. acetobutylicum* (77, 81). Since then, it has been an open question concerning how Spo0A becomes phosphorylated to start the sporulation process. A comparative genomic study revealed a number of orphan histidine kinases in *C. acetobutylicum*, *C. perfringens*, and *C. tetani* (95). A transcriptional expression profile further narrowed down the number of potential candidates (75). Recent experiments based on newly developed mutation techniques for *Clostridium* finally provided compelling evidence for the situation in *C. acetobutylicum* (96). By insertional inactivation of all histidine kinase candidates (single and double mutants), Young and his coworkers could show that two different pathways exist for Spo0A phosphorylation in *C. acetobutylicum* (Fig. 2). One way of signal transduction occurred by concerted

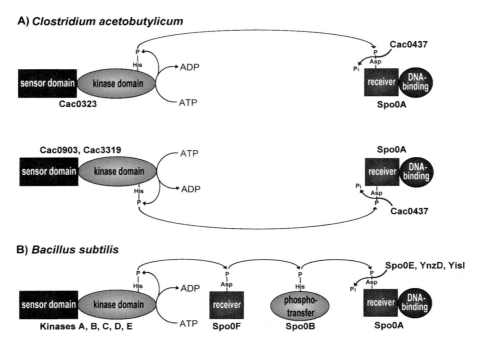

Figure 2 Signal transduction in *C. acetobutylicum* and *B. subtilis* leading to the onset of sporulation. (A) In *C. acetobutylicum*, three sensor kinases autophosphorylate at a histidine residue and transfer the phosphoryl group to an aspartate residue of the response regulator SpoOA. Cac0323 acts alone, while Cac0903 and Cac3319 act in concert. Cac0437 is a protein masquerading as a kinase, but acting as a phosphatase. (B) In *B. subtilis*, five different kinases channel the phosphoryl group to SpoOA via a phosphorelay consisting of SpoOF and SpoOB.

action of the kinases Cac0903 and Cac3319. Cac0903 showed the dominant activity at the transcript level. In addition, *spo0A* expression was found to be Cac0903 dependent, indicating that SpoOA~P induces *spo0A* expression as also found in *Bacillus*. The other pathway was governed by Cac0323 kinase alone. A fourth kinase, Cac0437, probably acted as a phosphatase of SpoOA~P, thus fulfilling the roles that SpoOE, YnzD, and YisI play in *Bacillus*. Interestingly, this reaction was ATP dependent and its mechanism still awaits elucidation. Steiner et al. (96) pointed out some similarity to the *Myxococcus xanthus* RedCDEF system, in which RedE becomes phosphorylated and then dephosphorylates RedF~P. However, there might also be the possibility of different sensor kinases forming heterodimers, leading to a different signal output (97).

The kinases found to be interaction partners of SpoOA in *C. acetobutylicum* must not be the same as in other clostridia. Genome comparisons revealed that an ortholog of Cac3319 is present in *C. tetani*, but not in *C. perfringens*. Similarly, orthologs of Cac0323 and Cac0903 can be found in *C. perfringens*, but not in *C. tetani* (97). This somewhat reflects the situation

in *Bacillus*. While *B. subtilis* possesses 5 kinases that are able to phosphorylate SpoOF, *Bacillus anthracis* contains 8 and *Bacillus cereus* even 11 (76). A reason might be that the bacteria use SpoOA to control additional networks to sporulation. In *C. acetobutylicum*, SpoOA~P also serves as the master regulator of solventogenesis, being responsible for the onset of acetone and butanol production (28, 29, 31, 98–100). In *C. perfringens*, production of the enterotoxin (CPE) is coupled to sporulation. CPE formation is σ^E and σ^K dependent (88). Another *C. perfringens* toxin, TpeL, belongs to the family of large clostridial cytotoxins, which use UDP-glucose as a substrate and glycosylate G-proteins such as Rac, Ras, and Rho (with some diversity among different toxins with respect to target proteins). TpeL is also expressed during sporulation, being controlled by SpoOA and σ^E (101). The well-known members of the same toxin family, toxin A and toxin B of *C. difficile* (encoded by *tcdA* and *tcdB*, respectively), are also massively produced during the stationary growth phase (102). They are under control of the alternative sigma factor TcdR and the anti-sigma factor TcdC (103). A link to sporulation obviously exists, but the data reported are somewhat contradictory. Using the new mutation techniques already mentioned, the *spo0A* gene and the gene of one of the orphan kinase interaction candidates were insertionally inactivated (104). The loss of Spo0A abolished sporulation completely as expected. In parallel, the concentration of toxin A, both intracellularly as well as extracellularly, was dramatically reduced. This was also the case, however, to a much lower extent, with the CD2492 kinase mutant. The latter result would be consistent with this kinase just being one of several that interact with Spo0A. However, it must be kept in mind that no Spo0A binding sites (so-called 0A boxes) have been found upstream of *tcdA* and *tcdB* (104). So, an indirect link must be assumed. On the other hand, a recent publication confirmed an influence of Spo0A on toxin production, but in a negative way. Again, a *spo0A* mutant of *C. difficile* was used. In this study, inactivation of *spo0A* led to increased production of toxins A and B (105). A possible reason might be the use of different strains, as was probably the case with two studies reporting conflicting data that only toxin B is essential for virulence of *C. difficile* (106) versus cytotoxic activity and corresponding virulence exerted by single action of toxin A as well as toxin B (107). In *C. botulinum*, production of the binary toxin C2 is linked to sporulation (108); however, the molecular details are not yet known.

A recent finding is that sporulation in pathogenic as well as apathogenic clostridia is regulated via the *agr* quorum-sensing system of Gram-positive bacteria. The membrane-associated AgrB protein processes and exports the autoinducer peptide, which is encoded by *agrD*. Sensing and regulation is achieved by the sensor kinase AgrC and the response regulator AgrA. An *agrD* mutant of *C. botulinum* showed a drastically reduced sporulation frequency and also reduced formation of neurotoxin. In *C. sporogenes*, AgrB activity was reduced by an antisense-RNA approach, which also led to an approximately 100-fold reduced sporulation frequency (109). An *agrB* mutant of *C. perfringens* could no longer produce spores and was blocked in enterotoxin (CPE) and beta2 toxin formation. Synthesis of alpha toxin and perfringolysin O was also reduced. Western blots revealed reduced levels of Spo0A as well as σ^F and the absence of σ^G in the mutant (110). In *C. acetobutylicum*, *agrA*, *agrB*, and *agrC* mutants could be constructed. All mutants no longer produced granulose. On solidified media, *agrA* and *agrC* mutants showed a significantly reduced spore formation, 5 orders of magnitude lower than the wild type. Solvent formation was not affected in all mutants. Mutations

could be restored by adding culture supernatant or synthetic autoinducer peptide (111). The *agr* system is obviously present in all clostridia as judged from genome data. As the response regulator AgrA controls directly a number of operons and also induces expression of the small, noncoding RNA RNAIII, it will be interesting to see whether such a sRNA is involved in control of sporulation. There is a recent report on CsfG, a sporulation-specific, small, noncoding RNA, which is highly conserved in endospore formers (112). It is synthesized exclusively in the forespore, but its physiological function seems to be in supporting later germination rather than the onset of sporulation (112).

Germination of Clostridial Spores

Under appropriate environmental conditions, a spore can germinate and thus convert back into a fully active vegetative cell. This can be achieved by inducing compounds, designated germinants, and physical factors. Interaction of germinants with specific receptors in the inner spore membrane first stimulate the release of H^+, K^+, and Na^+. Then, dipicolinate and Ca^{2+} ions as well as other divalent cations are set free, accompanied by imbibition of water. Potassium ions are reabsorbed by specific transporters. Hydrolysis of the spore's peptidoglycan cortex by spore cortex lytic enzymes (SCLEs) allow further uptake of water and swelling of the cell. SASPs are degraded by germination proteases; the resulting amino acids are then used for protein synthesis and energy metabolism.

L-Alanine has been identified as a germinant for *C. botulinum* and *C. sporogenes* spores (113); L-asparagine, a mixture of KCl and L-asparagine, L-alanine, and L-valine, and Na^+ ions together with inorganic phosphate, all to various degrees, for *C. perfringens* spores (114, 115); and inorganic phosphate, bile salts, and glycine for *C. difficile* spores, although with substantial diversity among different strains (116–118).

Germinant receptors belong to the GerA family and are organized in *B. subtilis* in three tricistronic operons (*gerA*, *gerB*, and *gerK*). Organization in clostridia is somewhat different and also strain dependent. The products of the *gerK* operons of *C. perfringens*, organized in a monocistronic *gerKB* and a bicistronic *gerKA/C* transcriptional unit, were found to be essential for L-asparagine-dependent germination and also to participate in KCl-induced germination (114). GerKA and GerKC seem to be the major players, because GerKB has an auxiliary role in spore germination, but is required for spore viability and outgrowth (119). On the other hand, GerA, being the product of the monocistronic *gerA* operon, obviously had only little effect on germination, as determined by mutant analyses (114).

Ion transport during *C. perfringens* spore germination might in part be catalyzed by GerO, a putative antiporter for Na⁺/H⁺-K⁺. This protein was found to be essential for germination, but might also play a role in spore formation (120). In *C. perfringens*, there are different spore-cortex lytic enzymes: an N-acetylmuramyl-L-alanine amidase, acting on the intact spore cortex and designated SCLE (encoding gene is *sleC*), and an N-acetylmuramidase, acting on the disrupted peptidoglycan fragments and designated CFLE (cortical fragment-lytic enzyme, encoding gene is *sleM*) (121). SleM and SleC have been localized on the outside of *C. perfringens* spore cortex (122). SleC probably also exerts transglycosylase activity (123) and was found to be essential for germination of *C. perfringens* spores, in contrast to SleM (124). SleC is also present in *C. difficile* and also essential for germination (125). A genome comparison between 12 *Bacillus* and 24 *Clostridium* strains revealed two major groups of germination-specific lytic enzymes in *Clostridium* (126). *C. beijerinckii* and *C. botulinum* Eklund 178 carried genes with homology to *C. perfringens sleC*, *sleM*, *gpr* (encoding a germination-specific protease degrading SASPs), *C. acetobutylicum*, *C. difficile*, and *C. botulinum* Alaska E43 homologs of *sleC* and *gpr*. These species and strains thus group with *C. perfringens*. On the other hand, *C cellulolyticum*, *C. kluyveri*, *C. novyi*, *C. tetani*, and most *C. botulinum* strains carry genes with homology to *Bacillus cwlJ*, *sleB*, and *ypeB*.

Industrial and Medical Applications of Clostridial Spores

An industrial application for clostridial spores is the use of *C. butyricum* spores as a probiotic (127). A general advantage of sporeformers as probiotics is that the number of spores added to food can be reduced by about 100-fold in comparison with vegetative cells. *C. butyricum* strain Miyairi 588 spore preparations have been commercially available since 1968 in Japan, China, and Korea and are used as a *C. difficile* prophylaxis and as treatment against non-antimicrobial-induced as well as antimicrobial-associated diarrhea. Spore preparations are produced by Miyarisan Pharmaceutical Co. Ltd., Tokyo, Japan. In 2012, approval as a novel food supplement within the European Union was requested (Miya-Pro tablets, http://www.food.gov.uk/multimedia/pdfs/clostridiumbutyricumdossier.pdf). These tablets contain a concentrate of *C. butyricum* Miyairi 588 spores. The history of the strain dates back

to 1933, when it was isolated by C. Miyairi. The 588 strain (the 588th isolate) was isolated in 1963. *C. butyricum* Miyairi 588 was shown to have therapeutic efficiency against inflammatory bowel disease because of its butyrate formation (128) and, as an antagonist, to have preventive and therapeutic effects on a number of pathogens, among them *C. difficile* and enterohemorrhagic *Escherichia coli* (EHEC) O157:H7 infection in a gnotobiotic mouse model (129).

An important medical application is the use of recombinant clostridial spores for cancer treatment (130–132). Because clostridia are anaerobic bacteria, their spores will only germinate in anaerobic environments. Normally, the tissues of humans and mammals in general are well oxygenated. Only within a fast-growing tumor and its surroundings does oxygen become limited and the tissue thus hypoxic. Thus, in healthy mammals, clostridial spores are removed quickly and without any deleterious effects (133, 134). However, when carrying a tumor, the mammal is colonized at and in the tumor. Clostridial spores recognize the hypoxic conditions, germinate, and the vegetative cells proliferate at the expense of the necrotic tissue. Thus, targeting of the tumor by clostridial spores is very specific, and the multiplication of bacterial cells at the target takes place. If the clostridia are now engineered to carry a gene whose gene product is toxic against tumor cells, a perfect and safe delivery vehicle has been developed. Safety is guaranteed because antibiotic treatment can kill and remove the microbes at any time. Three principal strategies are currently being tested: (i) the introduction into *Clostridium* of a gene whose gene product acts directly against the tumor (e.g., tumor necrosis factor, interleukin-2, *C. perfringens* enterotoxin) (3, 135, 136); (ii) introduction into *Clostridium* of a gene whose gene product converts an innocuous prodrug, injected separately after successful colonization, into a cytotoxic drug (e.g., cytosine deaminase and 5-fluorocytosine/5-fluorouracil; other enzymes are carboxypeptidase and nitroreductase) (reviewed in reference 132); and (iii) intravenous injection of nontoxic *C. novyi* together with conventional chemotherapeutic drugs (137). An important future step will be to seek and win FDA approval for respective clinical studies.

Acknowledgments. I thank Thiemo Standfest and Paul Walther for preparation of the electron microscopy photographs. Work in my laboratory was supported by grants from the BMBF GenoMikPlus project (Competence Network Göttingen) and the transnational BMBF SysMO2 project COSMIC2 (0315782A) (http://www.sysmo.net).

Citation. Dürre P. 2014. Physiology and sporulation in *Clostridium*. Microbiol Spectrum 2(4):TBS-0010-2012.

References

1. Bahl H, Dürre P. 2001. *Clostridia: Biotechnology and Medical Applications.* Wiley-VCH, Weinheim, Germany.

2. Dürre P. 2005. *Handbook on Clostridia.* CRC Press-Taylor and Francis Group, Boca Raton, FL.

3. Dürre P. 2007. *Clostridia. Encyclopedia of Life Sciences.* doi:10.1002/9780470015902.a0020370.

4. Dürre P. 2009. The genus *Clostridium*, p 339–353. *In* Goldman E, Green LH (ed), *Practical Handbook of Microbiology*, 2nd ed. CRC Press-Taylor and Francis Group, Boca Raton, FL.

5. Rood JI, McClane BA, Songer JG, Titball RW. 1997. *The Clostridia: Molecular Biology and Pathogenesis.* Academic Press, San Diego, CA.

6. Schiel B, Dürre P. 2010. *Clostridium*, p 1701–1715. *In* Flickinger MC (ed), *Encyclopedia of Industrial Biotechnology, Bioprocess, Bioseparation, and Cell Technology*, **vol. 3.** John Wiley & Sons, Hoboken, NJ. doi:10.1002/9780470054581.eib236

7. Wilde E, Hippe H, Tosunoglu N, Schallehn G, Herwig K, Gottschalk G. 1989. *Clostridium tetanomorphum* sp. nov., nom. rev. *Int J Syst Bacteriol* **39:**127–134.

8. Brüggemann H, Gottschalk G. 2004. Insights in metabolism and toxin production from the complete genome sequence of *Clostridium tetani. Anaerobe* **10:**53–68.

9. Wieringa KT. 1936. Over het verdwijnen van waterstof en koolzuur onder anaerobe voorwaarden. *Antonie van Leeuwenhoek* **3:**263–273.

10. Braun M, Mayer F, Gottschalk G. 1981. *Clostridium aceticum* (Wieringa), a microorganism producing acetic acid from molecular hydrogen and carbon dioxide. *Arch Microbiol* **128:**288–293.

11. Braun M. 1981. *Charakterisierung von anaeroben autotrophen Essigsäurebildnern und Untersuchungen zur Essigsäurebildung aus Wasserstoff and Kohlendioxid durch* Clostridium aceticum. Ph.D. dissertation. University of Göttingen, Germany.

12. Das A, Ljungdahl LG. 2003. Electron-transport system in acetogens, p 191–204. *In* Ljungdahl LG, Adams MW, Barton LL, Ferry JG (ed), *Biochemistry and Physiology of Anaerobic Bacteria.* Springer, New York, NY.

13. Huang H, Wang S, Moll J, Thauer RK. 2012. Electron bifurcation involved in the energy metabolism of the acetogenic bacterium *Moorella thermoacetica* growing on glucose or H_2 plus CO_2. *J Bacteriol* **194:**3689–3699.

14. Balch WE, Schoberth S, Tanner RS, Wolfe RS. 1977. *Acetobacterium*, a new genus of hydrogen-oxidizing, carbon dioxide-reducing anaerobic bacteria. *Int J Syst Bacteriol* **27:**355–361.

15. Schmehl M, Jahn A, Meyer zu Vilsendorf A, Hennecke S, Masepohl B, Schuppler M, Marxer M, Oelze J, Klipp W. 1993. Identification of a new class of nitrogen fixation genes in *Rhodobacter capsulatus*: a putative membrane complex involved in electron transport to nitrogenase. *Mol Gen Genet* **241:**602–615.

16. Biegel E, Müller V. 2010. Bacterial Na^+-translocating ferredoxin: NAD^+ oxidoreductase. *Proc Natl Acad Sci USA* **107:**18138–18142.

17. Poehlein A, Schmidt S, Kaster A-K, Goenreich M, Vollmers J, Thürmer A, Bertsch J, Schuchmann K, Voigt B, Hecker M, Daniel R, Thauer RK, Gottschalk G, Müller V. 2012. An ancient pathway combining carbon dioxide fixation with the generation and utilization of a sodium ion gradient for ATP synthesis. *PLoS One.* **7**(3): e33439. doi:10.1371/journal.pone.0033439

18. Müller V. 2003. Energy conservation in acetogenic bacteria. *Appl Environ Microbiol* **69:**6345–6353.

19. Köpke M, Held C, Hujer S, Liesegang H, Wiezer A, Wollherr A, Ehrenreich A, Liebl W, Gottschalk G, Dürre P. 2010. *Clostridium ljungdahlii* represents a microbial production platform based on syngas. *Proc Natl Acad Sci USA* **107:**13087–13092.

20. Köpke M, Mihalcea C, Bromley JC, Simpson SD. 2011. Fermentative production of ethanol from carbon monoxide. *Curr Opin Biotechnol* **22:**320–325.

21. Schiel-Bengelsdorf B, Dürre P. 2012. Pathway engineering and synthetic biology using acetogenic clostridia. *FEBS Lett* **586:**2191–2198.

22. Hilpert W, Schink B, Dimroth P. 1984. Life by a new decarboxylation-dependent energy conservation mechanism with Na^+ as coupling ion. *EMBO J* **3:**1665–1670.

23. Kane MB, Brauman A, Breznak JA. 1991. *Clostridium mayombei* sp. nov., an H_2/CO_2 acetogenic bacterium from the gut of the African soil-feeding termite, *Cubitermes speciosus. Arch Microbiol* **156:**99–104.

24. Charrier C, Duncan GJ, Reid MD, Rucklidge GJ, Henderson D, Young P, Russell VJ, Aminov RI, Flint HJ, Louis P. 2006. A novel class of CoA-transferase involved in shortchain fatty acid metabolism in butyrate-producing human colonic bacteria. *Microbiology* **152:**179–182.

25. Louis P, Duncan SH, McCrae SI, Millar J, Jackson MS, Flint HJ. 2004. Restricted distribution of the butyrate kinase pathway among butyrate-producing bacteria from the human colon. *J Bacteriol* **186:**2099–2106.

26. Herrmann G, Jayamani E, Mai G, Buckel W. 2008. Energy conservation via electron-transferring flavoprotein in anaerobic bacteria. *J Bacteriol* **190:**784–791.

27. Dürre P, Bahl H, Gottschalk G.. 1988. Membrane processes and product formation in anaerobes, p 187–220. *In* Erickson LE, Fung DY-C (ed), *Handbook on Anaerobic Fermentations.* Marcel Dekker Inc., New York, NY.

28. Dürre P. 2005. Formation of solvents in clostridia, p 671–693. *In* Dürre P (ed), *Handbook on Clostridia.* CRC Press, Taylor & Francis Group, Boca Raton, FL.

29. Dürre P. 2009. Metabolic networks in *Clostridium acetobutylicum*: interaction of sporulation, solventogenesis and toxin formation, p 215–227. *In* Brüggemann H, Gottschalk G (ed), *Clostridia. Molecular Biology in the Post-genomic Era.* Caister Academic Press, Norfolk, United Kingdom.

30. Dürre P. 2007. Biobutanol: an attractive biofuel. *Biotechnol J* **2:**1525–1534.

31. Dürre P. 2011. Fermentative production of butanol – the academic perspective. *Curr Opin Biotechnol* **22:** 331–336.

32. **Green EM**. 2011. Fermentative production of butanol – the industrial perspective. *Curr Opin Biotechnol* **22**: 337–343.

33. **Lütke-Eversloh T, Bahl H**. 2011. Metabolic engineering of *Clostridium acetobutylicum*: recent advances to improve butanol production. *Curr Opin Biotechnol* **22**: 634–647.

34. **Kenealy WR, Waselefsky DM**. 1985. Studies on the substrate range of *Clostridium kluyveri*: the use of propanol and succinate. *Arch Microbiol* **141**:187–194.

35. **Gaston LW, Stadtman ER**. 1963. Fermentation of ethylene glycol by *Clostridium glycolicum*, sp. n. *J Bacteriol* **85**:356–362.

36. **Schink B**. 1984. *Clostridium magnum* sp. nov., a nonautotrophic homoacetogenic bacterium. *Arch Microbiol* **137**:250–255.

37. **Forsberg CW**. 1987. Production of 1,3-propanediol from glycerol by *Clostridium acetobutylicum* and other *Clostridium* species. *Appl Environ Microbiol* **53**:639–643.

38. **Li F, Hinderberger J, Seedorf H, Zhang J, Buckel W, Thauer RK**. 2008. Coupled ferredoxin and crotonyl coenzyme A (CoA) reduction with NADH catalyzed by the butyryl-CoA dehydrogenase/Etf complex from *Clostridium kluyveri*. *J Bacteriol* **190**:843–850.

39. **Seedorf H, Fricke WF, Veith B, Brüggemann H, Liesegang H, Strittmatter A, Miethke M, Buckel W, Hinderberger J, Li F, Hagemeier C, Thauer RK, Gottschalk G**. 2008. The genome of *Clostridium kluyveri*, a strict anaerobe with unique metabolic features. *Proc Natl Acad Sci USA* **105**:2128–2133.

40. **Buckel W**. 1990. Amino acid fermentations: coenzyme B_{12}-dependent and –independent pathways, p 21–30. *In* Hauska G, Thauer RK (ed), *The Molecular Basis of Bacterial Metabolism*. Springer-Verlag, Heidelberg, Germany.

41. **Buckel W**. 1991. Ungewöhnliche Chemie bei der Fermentation von Aminosäuren durch anaerobe Bakterien. *Bioforum* **14**:7–19.

42. **Buckel W**. 2005. Special clostridial enzymes and fermentation pathways, p 177–220. *In* Dürre P (ed), *Handbook on Clostridia*. CRC Press, Taylor & Francis Group, Boca Raton, FL.

43. **Dürre P, Andersch W, Andreesen JR**. 1981. Isolation and characterization of an adenine-utilizing, anaerobic sporeformer, *Clostridium purinolyticum* sp. nov. *Int J Syst Bacteriol* **31**:184–194.

44. **Dürre P, Andreesen JR**. 1982. Selenium-dependent growth and glycine fermentation by *Clostridium purinolyticum*. *J Gen Microbiol* **128**:1457–1466.

45. **Dürre P, Andreesen JR**. 1983. Purine and glycine metabolism by purinolytic clostridia. *J Bacteriol* **154**: 192–199.

46. **Dürre P, Andreesen JR**. 1982. Anaerobic degradation of uric acid via pyrimidine derivatives by selenium-starved cells of *Clostridium purinolyticum*. *Arch Microbiol* **131**: 255–260.

47. **Andreesen JR**. 2005. Degradation of heterocyclic compounds, p 221–237. *In* Dürre P (ed), *Handbook on Clostridia*. CRC Press, Taylor & Francis Group, Boca Raton, FL.

48. **Vogels GD, van der Drift C**. 1976. Degradation of purines and pyrimidines by microorganisms. *Bacteriol Rev* **40**:403–468.

49. **Minton NP, Clarke DJ**. 1989. *Clostridia*. Plenum Press, New York, NY.

50. **Bahl H, Dürre P**. 1993. *Clostridia*, p 285–323. *In* Rehm H-J, Reed G, Pühler A, Stadler P (ed), Biotechnology, **vol 1**. Sahm H (ed), *Biological Fundamentals*. VCH Verlagsgesellschaft mbH, Weinheim, Germany.

51. **Labbé RG**. 2005. Sporulation (morphology) of clostridia, p 647–658. *In* Dürre P (ed), *Handbook on Clostridia*. CRC Press, Taylor & Francis Group, Boca Raton, FL.

52. **Reysenbach AL, Ravenscroft N, Long S, Jones DT, Woods DR**. 1986. Characterization, biosynthesis, and regulation of granulose in *Clostridium acetobutylicum*. *Appl Environ Microbiol* **52**:185–190.

53. **Bergère JL, Rousseau M, Mercier C**. 1975. Polyoside intracellulaire impliqué dans la sporulation de *Clostridium butyricum*. I. Cytologie, production et analyse enzymatique préliminaire. *Ann Inst Pasteur Microbiol* **126**:295–314.

54. **Brown RG, Lindberg B, Laishley EJ**. 1975. Characterization of two reserve glucans from *Clostridium pasteurianum*. *Can J Microbiol* **21**:1136–1138.

55. **Darvill AG, Hall MA, Fish JP, Morris JG**. 1977. The intracellular reserve polysaccharide of *Clostridium pasteurianum*. *Can J Microbiol* **23**:947–953.

56. **Hobson PN, Nasr H**. 1951. An amylopectin-type polysaccharide synthesized from sucrose by *C. butyricum*. *J Chem Soc* **1951**:1855–1857.

57. **Whyte JNC, Strasdine GA**. 1972. An intracellular α-D-glucan from *Clostridium botulinum*, type E. *Carbohydr Res* **25**:435–441.

58. **McCoy E, Fred EB, Peterson WH, Hastings EG**. 1926. A cultural study of the acetone butyl alcohol organism. *J Infect Dis* **39**:457–483.

59. **Spray RS**. 1948. The granulose reaction of certain anaerobes of the "butyric" group. *J Bacteriol* **55**:79–84.

60. **Orsburn BC, Melville SB, Popham DL**. 2010. EtfA catalyses the formation of dipicolinic acid in *Clostridium perfringens*. *Mol Microbiol* **75**:178–186.

61. **Cabrera-Martinez RM, Mason JM, Setlow B, Waites WM, Setlow P**. 1989. Purification and amino acid sequence of two small, acid-soluble proteins from *Clostridium bifermentans* spores. *FEMS Microbiol Lett* **52**: 139–143.

62. **Cabrera-Martinez RM, Setlow P**. 1991. Cloning and nucleotide sequence of three genes coding for small, acid-soluble proteins of *Clostridium perfringens* spores. *FEMS Microbiol Lett* **61**:127–131.

63. **Huang I-H, Raju D, Paredes-Sabja D, Sarker MR**. 2007. *Clostridium perfringens*: sporulation, spore resistance and germination. *Bangladesh J Microbiol* **24**:1–8.

64. **Paredes-Sabja D, Raju D, Torres JA, Sarker MR**. 2008. Role of small, acid-soluble spore proteins in the

resistance of *Clostridium perfringens* spores to chemicals. *Int J Food Microbiol* **122**:333–335.

65. Raju D, Waters M, Setlow P, Sarker MR. 2006. Investigating the role of small, acid-soluble spore proteins (SASPs) in the resistance of *Clostridium perfringens* spores to heat. *BMC Microbiol* **6**:50. doi:10.1186/1471-2180-6-50

66. Raju D, Setlow P, Sarker MR. 2007. Antisense-RNA-mediated decreased synthesis of small, acid-soluble spore proteins leads to decreased resistance of *Clostridium perfringens* spores to moist heat and UV radiation. *Appl Environ Microbiol* **73**:2048–2053.

67. Carpenter CE, Reddy DSA, Cornforth DP. 1987. Inactivation of clostridial ferredoxin and pyruvate-ferredoxin oxidoreductase by sodium nitrite. *Appl Environ Microbiol* **53**:549–552.

68. Reddy D, Lancaster LR Jr, Cornforth DP. 1983. Nitrite inhibition of *Clostridium botulinum*: electron spin resonance detection of iron-nitric oxide complexes. *Science* **221**:769–770.

69. Scheeff ED, Axelrod HL, Miller MD, Chiu H-J, Deacon AM, Wilson IA, Manning G. 2010. Genomics, evolution, and crystal structure of a new family of bacterial spore kinases. *Proteins* **78**:1470–1482.

70. Dalla Vecchia E, Veeramani H, Suvorova EI, Wigginton NS, Bargar JR, Bernier-Latmani R. 2010. U(VI) reduction by spores of *Clostridium acetobutylicum*. *Res Microbiol* **161**:765–771.

71. Permpoonpattana P, Tolls EH, Nadem R, Tan S, Brisson A, Cutting SM. 2011. Surface layers of *Clostridium difficile* endospores. *J Bacteriol* **193**:6461–6470.

72. Dürre P, Hollergschwandner C. 2004. Initiation of endospore formation in *Clostridium acetobutylicum*. *Anaerobe* **10**:69–74.

73. Dürre P. 2005. Sporulation in clostridia (genetics), p 659–669. *In* Dürre P (ed), *Handbook on Clostridia*. CRC Press, Taylor & Francis Group, Boca Raton, FL.

74. Stragier P. 2002. A gene odyssey: exploring the genomes of endospore-forming bacteria, p 519–525. *In* Sonenshein AL, Hoch JA, Losick R (ed), Bacillus subtilis *and its closest relatives: From Genes to Cells*. American Society for Microbiology, Washington, DC.

75. Paredes CJ, Alsaker KV, Papoutsakis ET. 2005. A comparative genomic view of clostridial sporulation and physiology. *Nat Rev Microbiol* **3**:969–978.

76. Stephenson K, Lewis RJ. 2005. Molecular insights into the initiation of sporulation in Gram-positive bacteria: new technologies for an old phenomenon. *FEMS Microbiol Lett* **29**:281–301.

77. Brown DP, Ganova-Raeva L, Green BD, Wilkinson SR, Young M, Youngman P. 1994. Characterization of *spo0A* homologs in diverse *Bacillus* and *Clostridium* species reveals regions of high conservation within the effector domain. *Mol Microbiol* **14**:411–426.

78. Gutierrez Escobar AJ, Montoya Castaño D. 2009. Evolutionary analysis for the functional divergence of the Spo0A protein: the key sporulation control element. *In Silico Biol* **9**:149–162.

79. Hilbert DW, Piggot PJ. 2004. Compartmentalization of gene expression during *Bacillus subtilis* spore formation. *Microbiol Mol Biol Rev* **68**:234–262.

80. Nölling J, Breton G, Omelchenko MV, Makarova KS, Zeng Q, Gibson R, Lee HM, Dubois J, Qiu D, Hitti J, GTC Sequencing Center Production, Finishing, and Bioinformatics Teams, Wolf YI, Tatusov RL, Sabathe F, Doucette-Stamm L, Soucaille P, Daly MJ, Bennett GN, Koonin EV, Smith DR. 2001. Genome sequence and comparative analysis of the solvent-producing bacterium *Clostridium acetobutylicum*. *J Bacteriol* **183**:4823–4838.

81. Sauer U, Treuner A, Buchholz M, Santangelo JD, Dürre P. 1994. Sporulation and primary sigma factor homologous genes in *Clostridium acetobutylicum*. *J Bacteriol* **176**:6572–6582.

82. Santangelo JD, Kuhn A, Treuner-Lange A, Dürre P. 1998. Sporulation and time course expression of sigma-factor homologous genes in *Clostridium acetobutylicum*. *FEMS Microbiol Lett* **161**:157–164.

83. Jones SW, Tracy BP, Gaida SM, Papoutsakis ET. 2011. Inactivation of σ^F in *Clostridium acetobutylicum* ATCC 824 blocks sporulation prior to asymmetric division and abolishes σ^E and σ^G protein expression but does not block solvent formation. *J Bacteriol* **193**:242–2440.

84. Li J, McClane BA. 2010. Evaluating the involvement of alternative sigma factors SigF and SigG in *Clostridium perfringens* sporulation and enterotoxin synthesis. *Infect Immun* **78**:4286–4293.

85. Scotcher MC, Bennett GN. 2005. SpoIIE regulates sporulation but does not directly affect solventogenesis in *Clostridium acetobutylicum* ATCC 824. *J Bacteriol* **187**:1930–1936.

86. Bi C, Jones SW, Hess DR, Tracy BP, Papoutsakis ET. 2011. SpoIIE is necessary for asymmetric division, sporulation, and expression of σ^F, σ^E, and σ^G but does not control solvent production in *Clostridium acetobutylicum* ATCC 824. *J Bacteriol* **193**:5130–5137.

87. Tracy BP, Jones AW, Papoutsakis ET. 2011. Inactivation of σ^E and σ^G in *Clostridium acetobutylicum* illuminates their roles in clostridial-cell-form biogenesis, granulose synthesis, solventogenesis, and spore morphogenesis. *J Bacteriol* **193**:1414–1426.

88. Harry KH, Zhou R, Kroos L, Melville SB. 2009. Sporulation and enterotoxin (CPE) synthesis are controlled by the sporulation-specific sigma factors SigE and SigK in *Clostridium perfringens*. *J Bacteriol* **191**:2728–2742.

89. Jones SW, Paredes CJ, Tracy B, Cheng N, Sillers R, Senger RS, Papoutsakis ET. 2008. The transcriptional program underlying the physiology of clostridial sporulation. *Genome Biol* **9**:R114. doi:10.1186/gb-2008-9-7-r114.

90. Haraldsen JD, Sonenshein AL. 2003. Efficient sporulation in *Clostridium difficile* requires disruption of the σ^K gene. *Mol Microbiol* **48**:811–821.

91. Kirk DG, Dahlsten E, Zhang Z, Korkeala H, Lindström M. 2012. *Clostridium botulinum* ATCC 3502 sigma factor K is involved with early stage sporulation. *Appl Environ Microbiol* **78**:4590–4596.

92. Saujet L, Monot M, Dupuy B, Soutourina O, Martin-Verstraete I. 2011. The key sigma factor of transition phase, SigH, controls sporulation, metabolism, and virulence factor expression in *Clostridium difficile*. *J Bacteriol* 193:3186–3196.

93. Burns DA, Minton NP. 2011. Sporulation studies in *Clostridium difficile*. *J Microbiol Methods* 87:133–138.

94. Burns DA, Heeg D, Cartman ST, Minton NP. 2011. Reconsidering the sporulation characteristics of hypervirulent *Clostridium difficile* BI/NAP1/027. *PLoS One*. 6(9):e24894. doi:10.1371/journal.pone.0024894.

95. Doß S, Gröger C, Knauber T, Whitworth DE, Treuner-Lange A. 2005. Comparative genomic analysis of signal transduction proteins in clostridia, p 561–582. *In* Dürre P (ed), *Handbook on Clostridia*. CRC Press, Taylor & Francis Group, Boca Raton, FL.

96. Steiner E, Dago AE, Young DI, Heap JT, Minton NP, Hoch JA, Young M. 2011. Multiple orphan histidine kinases interact directly with Spo0A to control initiation of endospore formation in *Clostridium acetobutylicum*. *Mol Microbiol* 80:641–654.

97. Dürre P. 2011. Ancestral sporulation initiation. *Mol Microbiol* 80:584–587.

98. Harris LM, Welker NE, Papoutsakis ET. 2002. Northern, morphological, and fermentation analysis of *spo0A* inactivation and overexpression in *Clostridium acetobutylicum* ATCC 824. *J Bacteriol* 184:3586–3597.

99. Köpke M, Dürre P.. 2010. Biochemical production of biobutanol, p 221–257. *In* Luque R, Campelo J, Clark J (ed), *Handbook of Biofuels Production*. Woodhead Publishing Ltd, Abington, Cambridge, UK.

100. Ravagnani A, Jennert KCB, Steiner E, Grünberg R, Jefferies JR, Wilkinson SR, Young DI, Tidswell EC, Brown DP, Youngman P, Morris JG, Young M. 2000. Spo0A directly controls the switch from acid to solvent production in solvent-forming clostridia. *Mol Microbiol* 37:1172–1185.

101. Paredes-Sabja D, Sarker N, Sarker MR. 2011. *Clostridium perfringen stpeL* is expressed during sporulation. *Microbial Pathogen* 51:384–388.

102. Popoff MR, Stiles BG. 2005. Clostridial toxins vs. other bacterial toxins, p 323–383. *In* Dürre P (ed), *Handbook on Clostridia*. CRC Press, Taylor & Francis Group, Boca Raton, FL.

103. Carter GP, Douce GR, Govind R, Howarth PM, Mackin KE, Spencer J, Buckley AM, Antunes A, Kotsanas D, Jenkin GA, Dupuy B, Rood JI, Lyras D. 2011. The anti-sigma factor TcdC modulates hypervirulence in an epidemic BI/NAP1/027 clinical isolate of *Clostridium difficile*. *PLoS Pathog* 7(10):e1002317. doi:10.1371/journal.ppat.1002317.

104. Underwood S, Guan S, Vijayasubhash V, Baines SD, Graham L, Lewis RJ, Wilcox MH, Stephenson K. 2009. Characterization of the sporulation initiation pathway of *Clostridium difficile* and its role in toxin production. *J Bacteriol* 191:7296–7305.

105. Deakin LJ, Clare S, Fagan RP, Dawson LF, Pickard DJ, West MR, Wren BW, Fairweather NF, Dougan G, Lawley TD. 2012. *Clostridium difficile spo0A* gene is a persistence and transmission factor. *Infect Immun* 80:2704–2711.

106. Lyras D, O'Connor JR, Howarth PM, Sambol SP, Carter GP, Phumoonna T, Poon R, Adams V, Vedantam G, Johnson S, Gerding DN, Rood JI. 2009. Toxin B is essential for virulence of *Clostridium difficile*. *Nature* 458:1176–1179.

107. Kuehne SA, Cartman ST, Heap JT, Kelly ML, Cockayne A, Minton NP. 2010. The role of toxin A and toxin B in *Clostridium difficile* infection. *Nature* 467:711–713.

108. Nakamura S, Serikawa T, Yamakawa K, Nishida S, Kozaki S, Sakaguchi G. 1978. Sporulation and C2 toxin production by *Clostridium botulinum* type C strains producing no C1 toxin. *Microbiol Immunol* 22:591–596.

109. Cooksley CM, Davis IJ, Winzer K, Chan WC, Peck MW, Minton NP. 2010. Regulation of neurotoxin production and sporulation by a putative *agrBD* signaling system in proteolytic *Clostridium botulinum*. *Appl Environ Microbiol* 76:4448–4460.

110. Li J, Chen J, Vidal JE, McClane BA. 2011. The Agr-like quorum-sensing system regulates sporulation and production of enterotoxin and beta2 toxin by *Clostridium perfringens* type A non-food-borne human gastrointestinal disease strain F5603. *Infect Immun* 79:2451–2459.

111. Steiner E, Scott J, Minton NP, Winzer K. 2012. An *agr* quorum sensing system that regulates granulose formation and sporulation in *Clostridium acetobutylicum*. *Appl Environ Microbiol* 78:1113–1122.

112. Marchais A, Duperrier S, Durand S, Gautheret D, Stragier P. 2011. CsfG, a sporulation-specific, small non-coding RNA highly conserved in endospore formers. *RNA Biol* 8:358–364.

113. Brousolle V, Alberto F, Shearman CA, Mason DR, Botella L, Nguyen-The C, Peck MW, Carlin F. 2002. Molecular and physiological characterization of spore germination in *Clostridium botulinum* and *C. sporogenes*. *Anaerobe* 8:89–100.

114. Paredes-Sabja D, Torres JA, Setlow P, Sarker MR. 2008. *Clostridium perfringens* spore germination: characterization of germinants and their receptors. *J Bacteriol* 190:1190–1201.

115. Paredes-Sabja D, Udompijitkul P, Sarker MR. 2009. Inorganic phosphate and sodium ions are cogerminants for spores of *Clostridium perfringens* type A food poisoning-related isolates. *Appl Environ Microbiol* 75:6299–6305.

116. Heeg D, Burns DA, Cartman ST, Minton NP. 2012. Spores of *Clostridium difficile* clinical isolates display a diverse germination response to bile salts. *PLoS One* 7(2):e32381. doi:10.1371/journal.pone.0032381.

117. Paredes-Sabja D, Bond C, Carman RJ, Setlow P, Sarker MR. 2008. Germination of spores of *Clostridium difficile* strains, including isolates from a hospital outbreak of *Clostridium difficile*-associated disease (CDAD). *Microbiology* 154:2241–2250.

118. Sorg JA, Sonenshein AL. 2008. Bile salts and glycine as cogerminants for *Clostridium difficile* spores. *J Bacteriol* 190:2505–2512.

119. Paredes-Sabja D, Setlow P, Sarker MR. 2009. Role of GerKB in germination and outgrowth of *Clostridium perfringens* spores. *Appl Environ Microbiol* **75**:3813–3817.

120. Paredes-Sabja D, Setlow P, Sarker MR. 2009. GerO, a putative Na$^+$/H$^+$-K$^+$ antiporter, is essential for normal germination of spores of the pathogenic bacterium *Clostridium perfringens*. *J Bacteriol* **191**:3822–3831.

121. Makino S, Moriyama R. 2002. Hydrolysis of cortex peptidoglycan during bacterial spore germination. *Med Sci Monit* **8**:RA119–127.

122. Miyata S, Kozuka S, Yasuda Y, Chen Y, Moriyama R, Tochikubo K, Makino S. 1997. Localization of germination-specific spore-lytic enzymes in *Clostridium perfringens* S40 spores detected by immunoelectron microscopy. *FEMS Microbiol Lett* **152**:243–247.

123. Kumazawa T, Masayama A, Fukuoka S, Makino S, Yoshimura T, Moriyama R. 2007. Mode of action of a germination-specific cortex-lytic enzyme, SleC, of *Clostridium perfringens* S40. *Biosci Biotechnol Biochem* **71**:884–892.

124. Paredes-Sabja D, Setlow P, Sarker MR. 2009. SleC is essential for cortex peptidoglycan hydrolysis during germination of spores of the pathogenic bacterium *Clostridium perfringens*. *J Bacteriol* **191**:2711–2720.

125. Burns DA, Heap JT, Minton NP. 2010. SleC is essential for germination of *Clostridium difficile* spores in nutrient-rich medium supplemented with the bile salt taurocholate. *J Bacteriol* **192**:657–664.

126. Xiao Y, Francke C, Abee T, Wells-Bennik MHJ. 2011. Clostridial spore germination versus bacilli: genome mining and current insights. *Food Microbiol* **28**:266–274.

127. Bader J, Albin A, Stahl U. 2012. Spore-forming bacteria and their utilisation as probiotics. *Benef Microbes* **3**:67–75.

128. Okamoto T, Sasaki M, Tsujikawa T, Fujiyama Y, Bamba T, Kusunoki M. 2000. Preventive efficacy of butyrate enemas and oral administration of *Clostridium butyricum* M588 in dextran sodium sulfate-induced colitis in rats. *J Gastroenterol* **35**:341–346.

129. Takahashi M, Taguchi H, Yamaguchi H, Osaki T, Komatsu A, Kamiya S. 2004. The effect of probiotic treatment with *Clostridium butyricum* on enterohemorrhagic *Escherichia coli* O157:H7 infection in mice. *FEMS Immunol Med Microbiol* **41**:219–226.

130. Mengesha A, Wie JZ, Zhou S-F, Wie MQ. 2010. Clostridial spores to treat solid tumours – potential for a new therapeutic model. *Curr Gene Ther* **10**:15–26.

131. Minton NP, Mauchline ML, Lemmon MJ, Brehm JK, Fox M, Michael NP, Giaccia A, Brown JM. 1995. Chemotherapeutic tumour targeting using clostridial spores. *FEMS Microbiol Rev* **17**:357–364.

132. Minton NP. 2003. Clostridia in cancer therapy. *Nat Rev Microbiol* **1**:237–242.

133. Brown JM, Liu S-C. 2004. Use of anaerobic bacteria for cancer therapy, p 211–219. *In* Nakano MM, Zuber P (ed), *Strict and Facultative Anaerobes. Medical and Environmental Aspects.* Horizon Bioscience, Wymondham, UK.

134. Lambin P, Theys J, Landuyt W, Rijken P, van der Kogel A, van der Schueren E, Hodgkiss R, Fowler J, Nuyts S, de Bruijn E, van Mellaert L, Anné J. 1998. Colonization of *Clostridium* in the body is restricted to hypoxic and necrotic areas of tumours. *Anaerobe* **4**:183–188.

135. Barbé S, van Mellaert L, Theys J, Geukens N, Lammertyn E, Lambin P, Anné J. 2005. Secretory production of biologically active rat interleukin-2 by *Clostridium acetobutylicum* DSM792 as a tool for anti-tumor treatment. *FEMS Microbiol Lett* **246**:67–73.

136. Theys J, Nuyts S, Landuyt W, van Mellaert L, Dillen C, Böhringer M, Dürre P, Lambin P, Anné J. 1999. Stable *Escherichia coli-Clostridium acetobutylicum* shuttle vector for secretion of murine tumor necrosis factor alpha. *Appl Environ Microbiol* **65**:4295–4300.

137. Dang LH, Bettegowda C, Huso DL, Kinzler KW, Vogelstein B. 2001. Combination bacteriolytic therapy for the treatment of experimental tumors. *Proc Natl Acad Sci USA* **98**:15155–15160.

The Bacterial Spore: From Molecules to Systems
Edited by P. Eichenberger and A. Driks
© 2016 American Society for Microbiology, Washington, DC
doi:10.1128/microbiolspec.TBS-0022-2015

Jihong Li,[1] Daniel Paredes-Sabja,[2]
Mahfuzur R. Sarker,[3] and Bruce A. McClane[1]

Clostridium perfringens Sporulation and Sporulation-Associated Toxin Production

16

The ability of the Gram-positive, anaerobic rod *Clostridium perfringens* to form resistant spores contributes to its survival in many environmental niches, including soil, waste water, feces, and foods (1). In addition, sporulation and germination play a significant role when this important pathogen causes disease (2, 3). As introduced in the next section of this review, spores often facilitate the transmission of *C. perfringens* to hosts and then germinate *in vivo* to cause disease.

Toxin production is well appreciated as a critical factor for the pathogenicity of *C. perfringens* (1). At least 17 different *C. perfringens* toxins have been described in the literature; however, individual isolates produce only portions of this impressive toxin arsenal. Consequently, *C. perfringens* strains are commonly classified into one of five types (A to E) based on their ability to produce four "typing" toxins, i.e., alpha-, beta-, epsilon-, and iota-toxins. While all isolates produce alpha-toxin, type B strains also express beta- and epsilon-toxins, type C isolates also make beta-toxin, type D strains also produce epsilon-toxin, and type E isolates also express iota-toxin. Besides producing one or more of the typing toxins, sporulating cells of some *C. perfringens* strains produce additional toxins such as *C. perfringens* enterotoxin (CPE), or a recently identified toxin named TpeL (1, 4). The connection between CPE production and sporulation has disease relevance, as introduced below.

THE IMPORTANCE OF SPORES FOR *C. PERFRINGENS* DISEASE

In humans and several important livestock species, *C. perfringens* causes a spectrum of diseases that remain important medical and veterinary concerns. The most notable of those *C. perfringens* diseases are as follows: (i) histotoxic infections such as clostridial myonecrosis, also known as traumatic gas gangrene (5), and (ii) diseases such as enteritis or enterotoxemias that originate in the intestinal tract (1, 2). As will now be described, spores can play an important role in the transmission of all these illnesses.

C. perfringens is the most common cause of traumatic human gas gangrene, which remains challenging

[1]Department of Microbiology and Molecular Genetics, University of Pittsburgh School of Medicine, Pittsburgh, PA 15219; [2]Departamento de Ciencias Biológicas, Universidad Andrés Bello, Santiago, 920-8640, Chile; [3]Department of Biomedical Sciences, College of Veterinary Medicine, Department of Microbiology, College of Science, Oregon State University, Corvallis, OR 15219.

to treat even by using modern medical approaches (5). *C. perfringens* type A causes clostridial myonecrosis when spores or vegetative cells gain entry into muscle tissue via a wound. Spores can germinate if low oxidation-reduction (Redox) conditions are present in the muscle tissue; the resultant vegetative cells then grow rapidly to further reduce tissue Redox conditions, promoting additional bacterial growth. The growing *C. perfringens* vegetative cells produce alpha-toxin and perfringolysin O, which cause local and regional necrosis in muscle, allowing rapid and progressive spread of the infection. In addition, these toxins can enter the systemic circulation to induce organ damage, circulatory problems, and death (5).

Many cases of human or animal enteritis and enterotoxemia (i.e., absorption of toxins from the intestines into the circulation, from where they damage nonintestinal organs) are also caused by *C. perfringens* (2). Spores often play a critical role in transmission of the *C. perfringens* illnesses originating in the intestines, particularly during two human food-borne illnesses. The first of those diseases, i.e., *C. perfringens* type A food poisoning, is caused by CPE-producing type A strains and ranks as the second most prevalent bacterial food-borne illness in the United States at 1 million cases/year (6). While the enterotoxin (*cpe*) gene can be either chromosomal or plasmid borne, ~75% of all *C. perfringens* food poisoning cases are caused by type A strains carrying a chromosomal *cpe* gene (1). Food poisoning typically occurs when type A chromosomal *cpe*-positive strains are ingested with foods and then sporulate in the small intestine, where they produce CPE (further discussion later) (1). The second *C. perfringens* foodborne illness of humans is enteritis necroticans, which is caused by beta-toxin-producing type C strains (7). Historically, enteritis necroticans was first observed in post-World War II Germany, where it was known as Darmbrand (7). However, enteritis necroticans is most often associated with childhood infections in Papua New Guinea (where the disease is known locally as PigBel because it often follows ingestion of contaminated pork), although it occasionally occurs in developed countries (7). In both *C. perfringens* type A and type C foodborne diseases, improper cooking or holding of foods plays a critical role in transmission. This temperature abuse facilitates the survival of resistant *C. perfringens* spores present in foods (as discussed later in this review); those spores later germinate and cause illness when the food is ingested.

CPE-producing type A strains carrying a plasmid *cpe* gene are responsible for 2 to 15% of all cases of non-foodborne human gastrointestinal (GI) diseases, such as

antibiotic-associated diarrhea (8). These illnesses are primarily acquired by ingesting spores that are present in the environment. Although less studied, spores could also contribute to *C. perfringens* enteritis and enterotoxemias in livestock, which can be caused by all types (A to E) of this bacterium.

After introducing *C. perfringens* spore ultrastructure and describing the basic processes of sporulation and germination in this bacterium, the remainder of this review will focus on recent insights into (i) the resistance properties that allow spores to contribute to *C. perfringens* disease transmission and (ii) the molecular basis for the expression of CPE and TpeL by sporulating cells.

THE ULTRASTRUCTURE OF *C. PERFRINGENS* SPORES

C. perfringens spores (Fig. 1) contain several different structural layers, all of which contribute to spore resistance properties (9, 10). Unlike some spore-forming species (11), *C. perfringens* does not possess an exosporium. Instead, the outermost layer of the *C. perfringens* spore is the spore coat. The composition and

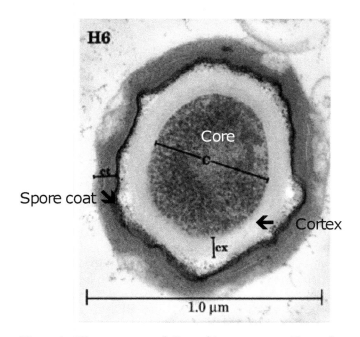

Figure 1 Ultrastructure of *C. perfringens* spores. Transmission electron micrograph of a spore from *C. perfringens* strain H-6, a food poisoning strain. Components of spore shown include proteinaceous spore coat layers, the cortex region, and the core with ribosomes giving a granular appearance. The bar represents 1.0 M. Reproduced with permission from reference 9.

function of the spore coat in *C. perfringens* have not yet been carefully studied; however, in other Gram-positive sporeformers, the spore coat is thought to comprise >50 spore-specific proteins and provides protection to the spore from reactive chemicals and lytic enzymes (12–14). In *C. perfringens*, a small fraction of the spore population (∼5%) is defective in spore coats (15). Those spores are permeable to lysozyme (15) so they can germinate inside the host under specific conditions.

The layers that underlie the coat are common for all known Gram-positive sporeformers, including *C. perfringens*. Beneath the coat is the outer membrane, which does not provide protection to dormant spores, but it is essential for spore formation and is presumably lost by shearing forces after short periods. The spore peptidoglycan cortex underlies the outer membrane, with a structure similar to that of peptidoglycan in a growing cell wall (16). The cortex plays an essential role in spore core dehydration and therefore directly contributes to spore resistance to environmental stress and chemicals. In studied sporeformers, and presumably also in *C. perfringens*, the spore peptidoglycan has three novel structural modifications that contribute to its resistance to cell wall hydrolases typically found in growing cells: (i) only one-quarter of cortex *N*-acetylmuramic acid (NAM) residues are substituted with short peptides, giving the cortex a lower degree of cross-linking than the germ cell wall; (ii) about one-quarter of the NAM residues carry a single L-alanine modification not present in the glycan strands of the germ cell wall; and (iii) nearly every second muramic acid residue in the cortex peptidoglycan is converted to muramic-δ-lactam (MAL) (17, 18), which seems to be the recognition substrate element for the cortex lytic enzymes (CLEs) which uniquely hydrolyze the peptidoglycan cortex, but not the germ cell wall, during spore germination (19, 20). The germ cell wall underlies the spore peptidoglycan cortex, has no role in spore resistance, and is converted into the growing cell wall during spore outgrowth. The spore inner membrane underlies the germ cell wall and is the last layer of protection of the spore core. The spore inner membrane is significantly compressed, resulting in highly immobile lipids (12), which results in a low permeability to small molecules including water and DNA-damaging chemicals (12, 21).

The core is the innermost layer of the *C. perfringens* spore and contains the spore DNA, RNA, and most enzymes. Three major factors, including the low water content of the core (20 to 50% of wet weight), its high levels of Ca-dipicolinic acid (Ca-DPA) (25% of core dry weight), and the saturation of DNA with small acid-soluble proteins (SASPs, discussed further below), together contribute to the resistance properties of these spores (22, 23).

SPORULATION OF *C. PERFRINGENS*

To survive unfavorable conditions, *C. perfringens* initiates the process of sporulation by undergoing an asymmetrical division of its cytoplasm membrane. This process gives rise to two compartments, i.e., a small compartment (termed the forespore), and a large compartment (termed the mother cell), each with a complete genome. As sporulation progresses through a series of morphological and biochemical changes, the forespore becomes the mature *C. perfringens* spore that is eventually released to the environment upon lysis of the mother cell (24).

The sporulation process in spore-forming bacteria, including *C. perfringens*, is initiated by the integration of a wide range of environmental and physiological signals induced from changes in cell density, the Krebs cycle, and nutrient starvation (25). In *C. perfringens*, initiation of sporulation requires the presence of inorganic phosphate (P_i) in the environment (26). In contrast, in sporulation medium containing P_i, *C. perfringens* was blocked at a very early stage of sporulation (i.e., the absence of polar septation and DNA partitioning) in cells reaching the stationary phase of growth (26). Importantly, P_i can neutralize the inhibitory effect of glucose at the onset of sporulation and induces *spo0A* expression, indicating that P_i acts as a key signal triggering spore formation in *C. perfringens* (26). As introduced earlier, *C. perfringens* sporulation directly contributes to pathogenesis since sporulation leads to the synthesis of CPE and consequently to intestinal damage of epithelial cells (27). Coupling this finding with the fact that P_i is normally present in the GI tract of humans and animals, it appears that *C. perfringens* has efficiently adapted to sporulating in the GI tract. By extension, it can be speculated that P_i directly contributes to pathogenesis and survival of the progeny of *C. perfringens* type A food poisoning strains, i.e., it induces the production of CPE to cause diarrhea that disseminates metabolically dormant spores into the environment. These spores are able to withstand unfavorable conditions and remain viable for long periods of time.

Global regulation between the transition state and sporulation has not been studied in as great detail in *C. perfringens* as for the model sporulation system, *Bacillus subtilis*. However, several studies have shown significant differences in the molecular regulation of sporulation between *C. perfringens* and *B. subtilis* (28, 29). As for *B. subtilis*, glucose has been found to act as a catabolic repressor of sporulation in *C. perfringens* (30). The transcriptional regulator carbon catabolite protein (CcpA) of the LacI/GalR family of repressor in *C. perfringens* (31) regulates many catabolite repressor effects from glucose. Interestingly, and in contrast to *B. subtilis*, CcpA is required for the efficient sporulation of *C. perfringens* (31). Initiation of sporulation in *B. subtilis* is mediated by the phosphorylation state of the master regulator of sporulation, i.e., the transcriptional factor SpoOA, which is present in all sequenced *Clostridium* species, including *C. perfringens* (32–35). Although the genome of CPE-negative *C. perfringens* strain 13 has a premature stop codon in *spoOA* (36), other *C. perfringens* isolates, including SM101 (a CPE-positive transformable derivative of a type A food poisoning strain), possess an intact *spoOA* gene (37). Evidence of a master regulatory role for SpoOA in *C. perfringens* sporulation was provided by a study showing that a *spoOA* knockout mutant of SM101 was unable to form spores. This *spoOA* phenotype was restored upon complementing the mutant with a recombinant plasmid carrying a wild-type *spoOA* copy (37). Interestingly, complementing the SM101 *spoOA* mutant with wild-type *spoOA* from other clostridial species revealed that SpoOA homologues can also induce the initiation of sporulation in enterotoxigenic *C. perfringens* (38).

The environmental signals that drive the initiation of sporulation are sensed by sporulation-specific orphan histidine kinases, which have not yet been identified in *C. perfringens*. Those histidine kinases integrate the sporulation signals and trigger a complex phosphorelay that increases the concentration of SpoOA in a phosphorylated state (SpoOA~P) (39, 40). Once threshold levels of SpoOA~P are reached, many genes (including those required for polar septum formation) become up- or downregulated, leading to a series of biochemical and morphological events (41). For example, expression of small acid-soluble protein-4 (SASP4, discussed later) and CPE by *C. perfringens* is dependent on SpoOA (37, 42).

In *C. perfringens*, the morphological events during sporulation are divided into seven stages (I to VII), resembling those of sporulating *B. subtilis* cells. In *B. subtilis*, four major sporulation-specific sigma (σ) factors (σ^F, σ^E, σ^G, and σ^K) regulate this sporulation process. The homologues of these genes are present in *C. perfringens*, where two recent studies (28, 29) demonstrated the expression and function of these sigma factors (Fig. 2). Those studies also presented evidence for both similarities and differences between the sporulation of *C. perfringens* and that of *B. subtilis*. Sporulation similarities between these bacteria include the following: (i) SpoOA and SigF mediate control of expression of the other sporulation-associated sigma factors; (ii) SigG is expressed in the late sporulation stage; and (iii) sporulation requires the production of all four sporulation-associated alternative sigma factors. Differences in sporulation include (i) *C. perfringens* lacks a *B. subtilis*-like phosphorelay and (ii)

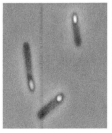  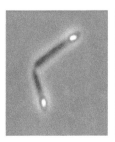 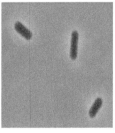 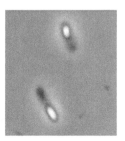

SM101   SM101::*sigF* SM101*sigF*comp   SM101::*sigG*   SM101*sigG*comp

Figure 2 Sporulation-associated sigma factors are required for *C. perfringens* sporulation. Shown are photomicrographs of sporulating cultures of SM101, a transformable derivative of a food poisoning strain, after growth for 8 h in Duncan-Strong sporulation medium. Also shown is the absence of sporulating cells in similar Duncan-Strong cultures of a *sigF* or *sigG* null mutant of SM101 (SM101::*sigF* or SM101::*sigG*). This loss of sporulation was specifically due to inactivation of the *sigF* or *sigG* genes in those mutants since the effect was reversible by complementation, i.e., by adding back a wild-type *sigF* or *sigG* gene, respectively, to those mutants (SM101::*sigF*Comp or SM101::*sigG*Comp). Reproduced with permission from reference 28. Similar loss of sporulation was observed with *sigE* or *sigK* mutants of SM101 (29).

the expression of a key mother cell transcription factor (SpoIIID) depends on σ^E-associated RNA polymerase in *B. subtilis*, but not in *C. perfringens*.

In detail (Fig. 3), *C. perfringens* transcribes *sigF* as part of a *spoIIA* tricistronic operon containing the *sigF*, *spoIIAA*, and *spoIIAB* genes in the early sporulation stage (28). The bacterium then uses SigF to regulate the production of other sporulation-associated sigma factors (28). SigE and SigK, but not SigF or SigG, are initially made as inactive proproteins and then proteolytically activated to their mature form (28, 29). Formation of *C. perfringens* mature spores requires expression of all four sigma factors.

Most, if not all, *C. perfringens* strains possess (43–45) an Agr-like quorum-sensing (QS) system involving proteins homologous to (i) AgrD, which is the putative precursor signaling peptide of this Agr system, and (ii) AgrB, a membrane protein that is thought to modify AgrD to the active form. A recent study (44) used an *agrB* mutant to demonstrate that the Agr-like QS system is also required for the initiation of sporulation in *C. perfringens*. Specifically, inactivating the *agrB* gene in *C. perfringens* strain F5603 decreased sporulation by ~1,500-fold. Furthermore, inactivation of the *agrB* gene in F5603 results in reduced or lost expression of SigF and SigG, respectively, which are needed for sporulation. In addition, this *agrB* mutant produced less Spo0A, which is also necessary for *C. perfringens* sporulation; in fact, this reduced Spo0A production may explain why the *agrB* mutant produced reduced amounts of alternative sigma factors and sporulates poorly. Collectively, these results indicate that, in *C. perfringens*,

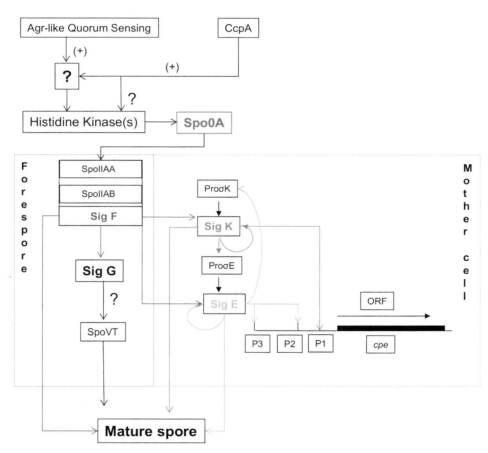

Figure 3 Sporulation in *C. perfringens*. Working through unidentified intermediates, the Agr QS system and CcpA affect Spo0A expression or, possibly, phosphorylation to initiate sporulation. This triggers a cascade of sigma factors where SigF controls production of the three other sporulation-associated sigma factors. Two of these sigma factors (SigE and SigK) then regulate CPE production during sporulation. Compiled from references 28, 29, 31, and 44. Not shown in this drawing, SigE (and possibly SigK) can also regulate production of TpeL toxin (97).

the Agr-like QS system regulates sporulation at an early stage, i.e., by controlling SpoOA and SigF synthesis.

Recent studies identified two sporulation repressors, named *virX* mRNA and CodY protein (46, 47). The global regulator CodY repressed sporulation in a *C. perfringens* type D strain, and resulted in spores with lower germination ability (46). The *virX* gene encodes a regulatory RNA that significantly inhibits sporulation and CPE production in the type A SM101. Because transcription levels of *sigE*, *sigF*, and *sigK* were higher in an isogenic *virX*-null mutant compared with wild-type SM101, it appears that *virX* RNA negatively regulates spore formation through the sporulation-specific sigma factors (47).

A wide variety of genes are expressed during sporulation in *C. perfringens*. Whole-genome expression profiling of the sporulation process in type A food poisoning strain SM101 was recently performed using DNA microarrays (48). This analysis revealed that a large number of genes showing sporulation-associated upregulated expression are homologues of known *Bacillus* genes involved in sporulation or germination. Expression levels of 106 SM101 genes exceeded 5 $\log_2$-fold increases during sporulation. Similarly, 294 SM101 genes showed upregulation between 3 and 5 $\log_2$-fold increases and 451 genes were upregulated at a lower level (2 to 3 $\log_2$-fold increases) in SM101 sporulation cultures.

GERMINATION OF
C. PERFRINGENS SPORES

C. perfringens spores can remain in dormancy for extended periods of time and survive extreme environmental conditions (see next section). However, in the presence of favorable conditions they can return to life, with outgrowth in less than 20 min (19). Spore germination is also an important factor for *C. perfringens* foodborne disease transmission. From a practical food safety perspective, the process of germination is of considerable interest because (i) germination of *C. perfringens* spores in food stuffs can lead to food poisoning; and (ii) upon germination, these spores lose their resistance and become susceptible to mild decontamination treatments. Therefore, understanding the molecular mechanism of *C. perfringens* spore germination might allow modulation of the germination process in foods by either inhibitors or artificial germinants that could allow the control of spore contamination loads with milder treatment conditions.

Germination is also important for transmission of other *C. perfringens* diseases. As already mentioned,

spore germination in wounds can lead to clostridial myonecrosis. Furthermore, germination of *C. perfringens* spores in the intestines is presumably important during CPE-associated non-foodborne human GI diseases, since these illnesses are thought to be transmitted by the ingestion of environmental spores. Those spores would need to germinate in the human intestinal tract before they could colonize and then cause disease.

Germination does not require metabolism and is initiated by small molecules called germinants, which can include amino acids, sugars, purines, nucleosides, and salts (19, 49, 50). Germinant specificity can vary significantly between bacterial species and strains, and this selectivity is likely to be influenced by adaptation to specific environmental niches. Indeed, significant differences in specificity of germinants are observed between spores of *C. perfringens* food poisoning strains versus non-foodborne GI disease isolates. Spores of *C. perfringens* food poisoning isolates are able to germinate in the presence of KCl, NaPi (pH 6.0), L-asparagine, or the exogenous 1:1 chelate of Ca^{2+} and Ca-DPA, while spores of non-foodborne GI disease isolates initiate germination in the presence of L-alanine, L-valine, and with the mixture of KCl and L-asparagine (51, 52). Also, spores of a non-foodborne GI isolate germinated to a greater extent than spores of a food poisoning isolate in the presence of cultured intestinal epithelial cells (53). These results support the hypothesis that spores of food poisoning isolates have adapted to food niches (i.e., processed meat products) where nutrients like KCl and NaPi are highly abundant, while spores of non-foodborne GI disease isolates are better adapted to germinate in the host's intestinal epithelium environment.

In addition to recognizing the aforementioned germinants, bacterial spores in the host encounter several host-derived components that are capable of inducing germination. Among these are lysozyme, which is released by Peyer patches in the small intestine (54), present in the serum (55), and composes part of the antibacterial arsenal of phagocytic cells. Lysozyme can trigger germination of spores of *C. perfringens* strain SM101 by directly degrading the spore peptidoglycan cortex (15). This germination pathway might have implications for the pathogenesis of *C. perfringens* (15), especially for superdormant spores that are extremely slow to germinate (56).

The germinant receptors (GRs) that recognize these nutrient germinants are relatively low-abundance proteins that localize to the spore inner membrane and belong to the GerA family of GRs (19, 21, 49). In *B. subtilis* spores, three tricistronic operons (*gerA*, *gerB*, and *gerK*) encode the three major GRs, with different receptors responding to different germinants (57). In contrast, *C. perfringens* has no tricistronic *gerA*-like operon and only a monocistronic *gerAA* that is far from the *gerK* locus. This *gerK* locus contains a bicistronic *gerKA-KC* operon and a monocistronic *gerKB* upstream of, and in the opposite orientation from, *gerKA-KC* (58, 59). Interestingly, the tricistronic *gerA* operons found in *B. subtilis* account for ~50% of GRs, while the *gerK* locus found in *C. perfringens* accounts for nearly 5% of the GRs present in sequenced genomes of endospore-forming members of *Bacillales* and *Clostridiales* (49). In *C. perfringens* strain SM101, gene knockout studies have identified the main receptors for L-asparagine, KCl, AK, and NaPi as the products of the bicistronic operon *gerKA-KC*, while the products of *gerAA* and *gerKB* play auxiliary roles in germination (51, 52, 60). Further gene knockout and protein localization studies demonstrated that GerKC is the essential GR for germination of *C. perfringens* spores and also that GerKC is located in spore inner membrane (58).

GerKA-KC and GerKB receptors are also required for viability and outgrowth of *C. perfringens* spores (51, 60). Binding of nutrient germinants to GRs located in the spore's inner membrane triggers the release of monovalent ions (Na$^+$ and K$^+$) and the spore core's depot of dipicolinic acid as a 1:1 chelate with Ca^{2+} (Ca-DPA) is replaced by water (49). Although the precise mechanism of monovalent ion release remains unknown, GrmA-like antiporter homologues (named GerO and GerQ, which are not to be confused with the coat protein GerQ in *B. subtilis* and the germination receptor GerQ in *Bacillus anthracis*) in *C. perfringens* strain SM101 were shown to be involved in transport of K$^+$ and/or Na$^+$, and GerO was also required for normal germination (61). However, because both GerO and GerQ are expressed in the mother cell compartment during sporulation, it is likely that their effect on spore germination is primarily during spore formation (61).

A major event after germinant nutrient binding to GRs is the release of the large deposit of Ca-DPA in the spore core (19). The precise mechanism of Ca-DPA release remains to be fully understood, although proteins encoded by the *spoVA* operon are thought to be involved in Ca-DPA movement (49). Instead of a hexacistronic *spoVA* operon as in *B. subtilis*, *C. perfringens* carries a tricistronic (*spoVAC*, *spoVAD*, and *spoVAE*) *spoVA* operon (49). Interestingly, *C. perfringens* SM101 *spoVA* null mutant spores lacking DPA are stable and germinate well, suggesting that Ca-DPA is not required for either *C. perfringens* spore stability or signal transduction from GRs to downstream effectors as

is the case in *B. subtilis* (49). This release of ions allows a slight hydration of the spore core that does not restore enzymatic activity but does lead to a decrease in spore wet-heat resistance (19). Taken together, the aforementioned events constitute an initial stage of germination known as Stage I (19).

Once Ca-DPA is released from the spore core (signaling the start of Stage II of germination), a series of biochemical events take place, with the hallmark being the hydrolysis of the peptidoglycan cortex of the spore (19). The cortex, which acts as a strait jacket restricting spore core hydration and therefore expansion, is hydrolyzed by CLEs that will specifically degrade the cortex (19). Two CLEs are present in the *C. perfringens* spore; one of these, SleC, is synthesized as an inactive zymogen and is the sole essential CLE for cortex hydrolysis of *C. perfringens* food poisoning isolates. SleC is a bifunctional enzyme with lytic transglycosylase and N-acetylmuramoyl-L-alanine amidase activity on crosslinked peptide moieties in the cortex (62). In contrast to the case of *B. subtilis* CLE CwlJ, which is activated by Ca-DPA released from the spore core (63), *C. perfringens* SleC is controlled by the Csp proteins (64) that belong to the subtilisin family of serine proteases. Csp proteins are localized in the cortex and activate cortex hydrolysis by converting pro-SleC to active SleC (64).

In contrast to SleC, the second CLE, SleM, is synthesized in a mature form with N-acetylmuramidase activity (65) and has little role in cortex hydrolysis of spores made by chromosomal *cpe* food poisoning isolates (66). Interestingly, complementation of a *sleC* mutant of a food poisoning isolate with wild-type *sleC* from a non-foodborne human GI disease isolate only partially restored the germination phenotype, suggesting that the precise role of both CLEs (i.e., SleC and SleM) in nonfoodborne GI disease isolates might be different from that in chromosomal *cpe* food poisoning isolates (67). Significant differences between spores of both food poisoning and non-foodborne GI disease isolates also exist in regard to Csp proteins. While spores of chromosomal *cpe* food poisoning isolates possess only one Csp protein, CspB (64), spores of non-foodborne isolates have three Csp proteins (i.e., CspA, CspB, and CspC) encoded by a tricistronic operon (68). Studies with *C. perfringens* chromosomal *cpe* strain SM101 demonstrated that CspB alone is localized to the spore coat and alone is sufficient for converting pro-SleC to active SleC and activate cortex hydrolysis (58, 64). Degradation of the spore peptidoglycan cortex allows full core hydration, remodeling of the germ cell wall, resumption of metabolism, degradation of SASPs, and complete loss of spore resistance properties (19).

RESISTANCE PROPERTIES OF C. PERFRINGENS SPORES

One reason why C. perfringens is such a successful foodborne pathogen is because it can form resistant spores that allow survival in improperly held or incompletely cooked foods. Specifically, spores provide C. perfringens with resistance against such common food environment stresses as low or high temperatures, osmotic pressure, chemical preservatives, and pH. For example, while vegetative cells of this bacterium cannot survive even brief exposure to 55°C, spores of some C. perfringens strains can survive boiling for an hour or longer (7, 69).

Importantly, C. perfringens spores exhibit significant strain-to-strain differences in their food environment stress resistance properties (7, 69–71). As mentioned earlier in this review, type A strains with a chromosomal cpe gene are strongly associated with food poisoning; it is those isolates that also typically form the most resistant spores, regardless of their geographic origin, date of isolation, or isolation source (72). For example, in terms of decimal reduction values (D_{100} value or the time that a culture must be kept at 100°C to obtain a one log reduction in viable spore numbers), the spores of type A chromosomal cpe food poisoning isolates are, on average, ~60-fold higher than the D_{100} values for spores of type A isolates carrying a plasmid cpe gene or cpe-negative type A strains. Notably, spores produced by some chromosomal cpe food poisoning isolates have D_{100} values exceeding 2 h (69).

Of epidemiologic significance, a survey detected both spores and vegetative cells of cpe-positive type A isolates in nonoutbreak raw meats and seafood sold retail in the United States (73). Importantly, those cpe-positive type A retail food isolates all carried a chromosomal cpe gene. Furthermore, the spores made by each of these raw food isolates exhibited exceptionally strong heat resistance, indicating that the spore heat-resistant phenotype is an intrinsic trait of most type A chromosomal cpe isolates, rather than a survivor trait selected by cooking. This spore resistance phenotype should be an important virulence determinant since it likely favors survival of type A chromosomal cpe isolates in improperly warmed or incompletely cooked foods.

Storage of foods at low temperatures (in refrigerators or freezers) is another important food safety approach. The spores of most chromosomal cpe strains also show exceptional cold resistance compared with the spores of type A plasmid cpe strains or cpe-negative strains. For example, after a 6-month storage at 4°C or −20°C, the average log reduction in viability for spores

of plasmid cpe or cpe-negative strains was about 3- to 4-fold greater, respectively, compared with the average log reduction in viability of spores made by chromosomal cpe strains (71). These results suggest that the chromosomal cpe strains are strongly associated with food poisoning, not only because of their exceptional spore heat resistance properties, but also because their spores are unusually tolerant of storage at low temperature (71).

Other factors besides temperature are also used to control the presence of pathogens in foods. For example, commercial curing of meats often involves use of sodium nitrite, which can inhibit outgrowth of clostridial spores or, at high concentrations, kill bacterial spores. One study (70) showed that the spores of type A chromosomal cpe isolates exhibit significantly better tolerance of, and survival against, nitrite-induced stress compared with the spores of other type A isolates. This nitrite resistance should further facilitate the ability of the chromosomal cpe, type A strains to cause foodborne illness.

MECHANISMS OF C. PERFRINGENS SPORE RESISTANCE

C. perfringens spore resistance depends on a synergistic interplay of multiple factors, which include sporulation temperature, mineralization of the core with DPA and its cations, binding of α/β-type small acid-soluble proteins (SASPs) to spore DNA, and core water content. Spore core water content is directly affected by sporulation temperature; a higher sporulating temperature produces C. perfringens spores with higher heat resistance. Spores of an spmA/B null mutant have more core water content, which directly reduces heat resistance by 50% (74). A spoVA null mutant makes spores with 2-fold more core water than wild-type spores, and those mutant spores also exhibit lower resistance to moist heat, UV radiation, and chemical treatment (75). A possible explanation for this decreased resistance is that increased hydration decreases SASP binding to spore DNA (75). The degree of cross-linking of the spore peptidoglycan cortex also plays a major role in C. perfringens spore resistance since dacF/B null mutant strains, which have significantly increased cross-linking of muropeptides, make spores with decreased heat resistance (74). However, the degree of cross-linking of muropeptides does not affect the core water content of C. perfringens spores (74).

SASPs bind to and saturate spore DNA, providing protection from various environmental stresses. The C. perfringens genome encodes four major α/β-type SASPs (i.e., SASP1, SASP2, SASP3, and SASP4). When SASP1,

SASP2, and SASP3 levels in sporulating cells were reduced by >90% using antisense-RNA-mediated downregulation approaches, a 5-fold reduction in spore heat resistance was observed (73, 74, 76). Those spores also showed greater sensitivity to chemicals (i.e., nitrous acid, hydrogen peroxide, formaldehyde, and HCl) and UV radiation. However, SASP1, SASP2, and SASP3 levels could not explain the resistance differences observed between spores of type A chromosomal *cpe* isolates versus spores of other *C. perfringens* since all strains produce similar amounts of SASP1, SASP2, and SASP3 and no consistent sequence variation in these proteins occurs among *C. perfringens* strains (73, 74, 76, 77).

However, an Asp is found at residue 36 of the SASP4 made by most, if not all, of the type A chromosomal *cpe* isolates forming highly resistant spores (78). In contrast, Gly is consistently present at this SASP4 residue in those *C. perfringens* strains producing more sensitive spores. An important contribution of the Asp36 SASP4 variant to the exceptional heat, sodium nitrite, and cold resistance properties of spores made by most chromosomal *cpe* food poisoning strains has been directly demonstrated using *ssp4* null mutants (78). Furthermore, electrophoretic mobility shift assays (EMSAs) and DNA binding studies showed that SASP4 variants with an Asp at residue 36 bind DNA more efficiently and tightly than do SASP4 variants with a Gly at residue 36 (Fig. 4A). Results from saturation mutagenesis experiments (42) indicated that both amino acid size and charge at SASP4 residue 36 are important for tight DNA binding and for spore resistance properties. It was also shown that *C. perfringens* SASP4 binds preferentially to AT-rich DNA sequences, while SASP2 binds better to GC-rich DNA sequences (Fig. 4B). Since the *C. perfringens* genome is more than 70% AT rich, these binding preferences may help to explain why SASP4 plays such an important role in providing cold, heat, and nitrite resistance by protecting spore DNA. However, maximal spore resistance requires production of all four *C. perfringens* SASPs.

While SASP4 variations are clearly a major determinant of relative stress resistance for *C. perfringens* spores, the *ssp4* null mutant of a type A chromosomal *cpe* strain still showed somewhat more resistance than did wild-type spores of other type A isolates, indicating that other factors also contribute to the exceptional spore resistance associated with chromosomal *cpe* strains. Several studies (9, 10) have attempted to correlate structural features of *C. perfringens* spores with the striking heat resistance of chromosomal *cpe*-positive isolates. Factors analyzed included the core, cortex, coat, and total spore size (9, 10), but the most significant

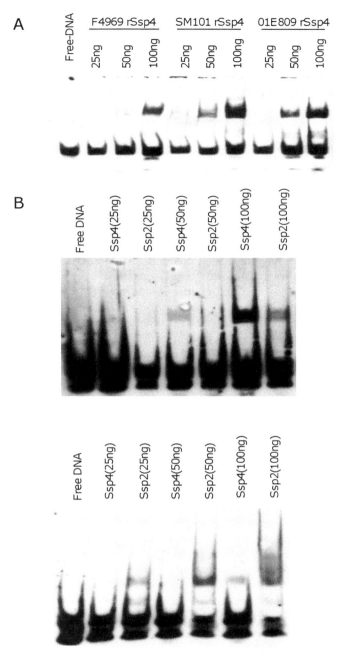

Figure 4 DNA binding properties of recombinant His6-tagged SASP4. (A) Electromobility shift assays (EMSAs) showing binding to biotin-labeled *C. perfringens* DNA by purified rSASP4 from F4969 (a CPE-positive non-foodborne human GI disease strain that forms sensitive spores and produces an SASP4 variant with a Gly at residue 36), SM101 or 01E809 (two CPE-positive food poisoning isolates that form resistant spores and produce an SASP4 variant with Asp at residue 36). (B) EMSAs showing binding by purified SM101 rSASP4 or rSASP2 to (left) *C. perfringens* AT-rich biotin-labeled DNA or (right) biotin-labeled *C. perfringens* GC-rich DNA. Reproduced with permission from references 42 and 78.

correlation noted between *C. perfringens* spore structure and heat resistance was for the ratio of core volume and core plus peptidoglycan layer, with a lower ratio giving higher heat resistance (10). However, in general, the ultrastructural features of *C. perfringens* spores are similar to those of other sporeformer species, so it is likely that the main differences involved in spore resistance differences between *C. perfringens* spores action at the molecular, rather than structural, level.

Multilocus Sequence Typing (MLST) analyses (79) of eight housekeeping genes determined that chromosomal *cpe* isolates represent a distinct genetic cluster within the global *C. perfringens* population, i.e., these studies (72) identified many genetic differences in typical type A chromosomal *cpe* isolates besides their carriage of a chromosomal *cpe* gene, a variant *ssp4* gene, and ability to produce highly resistant spores. These type A chromosomal *cpe* isolates apparently have now evolved to excel at causing foodborne disease.

Since those initial MLST studies, molecular analyses have also been performed on type C strains (7), the only non-type-A isolates that cause human enteritis necroticans. In post-World War II Germany, this disease was named Darmbrand, and recent molecular analyses (7) of these Darmbrand isolates showed they carry both beta (*cpb*) and *cpe* genes on large plasmids. Even though these Darmbrand isolates carry a plasmid-borne *cpe*, they produce highly heat-resistant spores. Interestingly, these type C Darmbrand strains produce the same variant Ssp4 made by chromosomal *cpe* strains. MLST analysis indicated that these type C Darmbrand strains and type A chromosomal *cpe* strains also share a similar genetic background. This helps to explain why both Darmbrand strains and type A chromosomal *cpe* strains are so well suited to cause human foodborne illness: i.e., they both produce spores that are highly resistant to food environment stresses—which allows their spores to germinate for entry of vegetative cells into the human GI tract by ingestion in food—and they both produce toxins that are highly active in the intestines. In terms of evolution, it is likely that these strains emerged by entry of different mobile genetic elements (conjugative plasmids and/or transposons) into *C. perfringens* strains with a similar background, including production of the Asp36 SASP variant.

SPORULATION-ASSOCIATED TOXINS

C. perfringens Enterotoxin
As already introduced, CPE-producing type A strains of *C. perfringens* cause the second most common bacterial foodborne illness in the United States, along with many cases of non-foodborne human GI diseases such as antibiotic-associated diarrhea. During this food poisoning (1), spores often germinate in temperature-abused foods, followed by rapid multiplication of the resultant vegetative cells in those contaminated foods (note that *C. perfringens* has a doubling time of only ~10 min). After the food is ingested, many vegetative cells are killed by exposure to the low pH of the stomach. However, when a food was sufficiently contaminated, some ingested bacteria survive and pass into the small intestines. Initially these bacteria multiply, but they soon commit to *in vivo* sporulation, possibly when they encounter P_i in the intestinal tract. It is during this sporulation in the small intestines that CPE is produced during *C. perfringens* type A food poisoning.

Considerable evidence implicates CPE as the toxin responsible for the GI symptoms that characterize both *C. perfringens* type A food poisoning and CPE-associated non-foodborne human GI diseases (1). For example, ingestion of purified CPE plus bicarbonate was shown to be sufficient to induce diarrhea and cramping in human volunteers (1). Additionally, studies fulfilling Koch's molecular postulates demonstrated that CPE production is essential for the GI pathogenicity of CPE-positive type A human food poisoning and non-foodborne human GI disease isolates in animal models (27).

CPE can induce substantial small intestinal histologic damage, which includes villus blunting along with epithelial necrosis and desquamation (1). This damage is thought to cause intestinal fluid and ion loss, effects that manifest clinically as diarrhea (1). Evidence with experimental animals suggests that CPE can sometimes be absorbed into the systemic circulation, where it can bind to and damage the liver and other organs. These effects can lead to a lethal increase in serum potassium levels, which could explain some deaths associated with *C. perfringens* type A food poisoning (80).

The histologic damage that occurs in the CPE-treated small intestine is a consequence of the cellular action of this toxin (Fig. 5). This action starts with CPE binding to receptors that include certain members of the claudin family of tight junction proteins (1). There are ~27 members of the claudin family, but only some claudins can serve as CPE receptors (1). An Asn residue located in the middle of the second extracellular loop of receptor claudins is necessary for a claudin to bind CPE (81) and amino acid residues near this Asn residue modulate the affinity of CPE binding properties (82). Once bound to a claudin receptor, the toxin becomes localized in a small complex of ~90 kDa that can contain, at minimum, a CPE, claudin receptor, and a claudin nonreceptor (1).

At 37°C, several small complexes rapidly oligomerize to form a larger CPE complex named CH-1 (for CPE hexamer-1). CH-1 contains, at minimum, six CPE molecules as well as both receptor claudins and claudin nonreceptors (83). Formation of CH-1, which is ~450 kDa in size, occurs on the host cell surface; however, once formed, this CH-1 prepore soon inserts into the plasma membrane to form a pore.

CPE pore formation increases the permeability of mammalian cells for the second messenger calcium, among other ions. Calcium influx plays a critical role in CPE-induced cell death (84). At low CPE doses, where small amounts of the CH-1 pore form, there is a modest calcium influx, and host cells die by a classical caspase-3-mediated apoptosis. At high CPE doses, where

more CH-1 pores form, a massive calcium influx triggers mammalian cell death via oncosis.

Substantial morphologic damage, such as cell rounding, develops in CPE-treated host cells. This effect damages the tight junction and exposes the basolateral surface of the cell, which allows formation of a second large CPE complex named CH-2. CH-2 is ~600 kDa in size; like CH-1, it contains six CPE molecules and both receptor and nonreceptor claudins. However, CH-2 uniquely contains another tight junction protein named occludin (83). The consequences of CH-2 formation are less clear than for CH-1 formation but may include formation of additional pores and further damage to the tight junction by sequestering even more tight-junction proteins in CPE complexes and inducing the internalization of tight-junction proteins into the host cell cytoplasm.

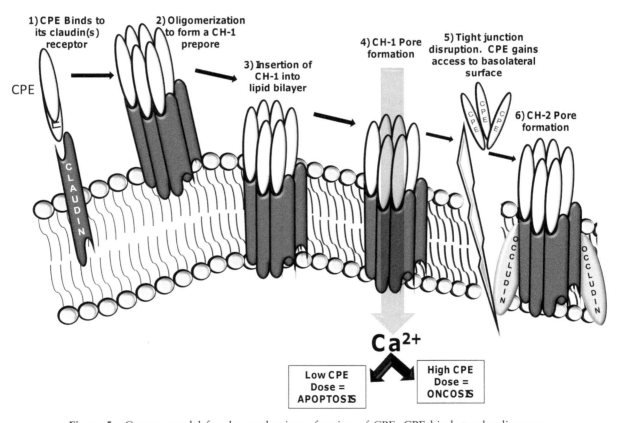

Figure 5 Current model for the mechanism of action of CPE. CPE binds to claudin receptors to form small complexes. Those small complexes then oligomerize on the host cell surface to form an ~450-kDa prepore known as CH-1. The prepore inserts into the membrane to form an active pore that alters host plasma membrane permeability for small molecules. As a result, calcium enters the cytoplasm and triggers either apoptosis (caused by low CPE doses, where there is a modest calcium influx) or oncosis (caused by high CPE doses, where there is a strong calcium influx). Reproduced with permission from reference 1.

The CPE protein is a single polypeptide consisting of 319 amino acids, with a molecular weight of ~35 kDa. The structure of the toxin was recently solved and shown to consist of two major domains with a resemblance to the aerolysin-like pore forming toxin family (85, 86). When this structural information is collated with results from structure/function mutagenesis studies (87), it revealed that claudin binding activity is mediated by the C-terminal domain of CPE. This binding involves several tyrosine residues located near the extreme C terminus of the toxin that interact with the ECL-2 loop of receptor claudin. In contrast, the N-terminal domain of the toxin, which consists of two halves, mediates both CPE oligomerization and pore formation. The extreme N-terminal sequences of the toxin are susceptible to cleavage by intestinal proteases such as trypsin or chymotrypsin and thus may be removed in the intestines during disease. Proteolytic removal of these sequences increases cytotoxicity by about 2- to 3-fold, as it exposes the CPE residues (notably residue D48) to promote toxin oligomerization (2).

During sporulation, some *C. perfringens* strains produce very large amounts of CPE, which can comprise up to 20% of the total protein present in a sporulating cell (2). The *cpe* gene is transcribed as a message of ~1.2 kb, beginning ~3 h after inoculation into Duncan-Strong sporulation medium (2). The CPE protein becomes detectable by Western blotting at 4 to 5 h post-inoculation into Duncan-Strong medium. CPE is not secreted from sporulating cells, but instead accumulates in the cytoplasm until the mother cell lyses to release the mature spore (2).

Regardless of whether an isolate carries a chromosomal *cpe* gene or a plasmid-borne *cpe* gene, CPE production is strictly sporulation associated. Therefore, it is not surprising that a *spo0A* null mutant of SM101, which cannot sporulate, is unable to produce CPE (37). Similarly, the Agr-like QS system, which regulates sporulation in *C. perfringens* by reducing production of Spo0A, is also required for wild-type production levels of CPE (44). In contrast, *virX* RNA inhibits CPE production by repressing sporulation (47).

One reason for the strong, sporulation-associated expression of CPE by *cpe*-positive type A strains is the presence of three promoters upstream of the *cpe* open reading frame (Fig. 3). Two of those *cpe* promoters (named P2 and P3) are similar to consensus SigE-dependent promoters that drive mother cell gene expression, while the other *cpe* promoter (named P1) resembles a consensus SigK-dependent promoter. Consistent with *cpe* transcription being dependent on these two sporulation-associated sigma factors, *sigK*- and

sigE-null mutants of strain SM101 failed to drive beta-glucuronidase production when transformed with a plasmid carrying the *cpe* promoter region fused to the *Escherichia coli* reporter gene *gusA* (29). Another study (28) investigated the role of SigF and SigG in CPE production and reported that *cpe* transcription is also blocked in an SM101 *sigF* null mutant, but not in a SM101 *sigG* null mutant. The role of SigF in controlling CPE production can be explained by the dependence of SigE and SigK upon SigF expression. The ability of a *sigG*-null mutant to produce CPE indicates that, while all four sporulation-associated sigma factors are needed for *C. perfringens* sporulation, those sigma factors are not all necessary for *cpe* transcription and CPE production (28, 29).

TpeL Toxin

Many *C. perfringens* isolates encode a novel toxin named TpeL. TpeL was initially identified in the supernatant of *C. perfringens* strain CP4 and shown to be cytotoxic to Vero cells by causing cell rounding (4). TpeL has a molecular mass of ~205 kDa (88) and belongs to the family of large clostridial toxins (LCTs) that encompass *Clostridium difficile* toxin A (TcdA) and B (TcdB), along with similar toxins such as *Clostridium sordellii* lethal toxin (TcsL) made by other clostridial species (89). TpeL has no signal peptide region within the open reading frame (4).

Classically, large clostridial toxins were thought to contain four domains: (i) A domain, involved in N-terminal biological activity; (ii) B domain, C-terminal carbohydrate-binding repeats, often assumed to be involved in receptor binding; (iii) C domain, autoproteolytic cleavage during toxin-processing; and (iv) D domain, delivery of the A domain into cytosol (89). LCTs enter host cells via endocytosis and insert via the D domain into the endosome membrane. The protease C domain is activated by intracellular inositol hexaphosphate (InsP$_6$), resulting in toxin cleavage and release of the A domain to the cytosol, where it glycosylates small GTPases, inactivating their cellular functions.

Notably, TpeL has a shorter amino acid sequence than other LCTs, with homology encompassing to the N-terminal domain, including the DXD motif (essential for glycosyltransferase activity) and a conserved W102 (essential for enzymatic activity) (4, 89–91). Interestingly, despite the dogma that LCTs use their B domain to bind to host cell membrane receptor(s) (89, 92), the carbohydrate binding repeats present in the C-terminal domain of other LCTs are absent from TpeL (88). Recently, this puzzle was resolved when a receptor binding domain was identified in the C-terminal half of the TpeL D domain (93). This region allows TpeL to bind to low-density lipoprotein receptor-related protein 1 (LRP1) as a receptor (93).

The *tpeL* gene is present ~3 kb downstream of the beta-toxin-encoding gene (*cpb*) on large plasmids in many *C. perfringens* type B and C strains (94, 95). The *tpeL* gene is also present in some type A isolates (96), although its location (plasmid versus chromosomal) has not been determined in those strains. Studies have shown that *tpeL* is present (i) in ~18% of type A necrotic enteritic outbreak isolates (96); (ii) in ~2% of *C. perfringens* isolates from retail chicken samples (96); and (iii) in 100% and 75%, respectively, of type B and type C isolates (94, 95). The contribution of TpeL, when produced, to *C. perfringens* pathogenesis remains unclear.

Bioinformatic analysis of the *tpeL* promoter region on the 65-kb plasmid of *C. perfringens* type B strain ATCC 3626 revealed the presence of σ^E- and σ^K-dependent promoter sequences (97). In contrast, no sequence with similarity to the consensus OA box was found (97). Evidence that *tpeL* can be expressed during sporulation came from fusion of the *tpeL* promoter region with the *E. coli gusA* reporter gene; when that construct was introduced into *C. perfringens* strain SM101, no significant beta-glucuronidase (GUS) activity was observed in vegetative growth of SM101 transformants. However, GUS activity became significant in sporulating cultures of those SM101 transformants within as little as 4 h after initiation of sporulation (97). Sporulation-specific expression of *tpeL* was confirmed by introducing plasmids carrying the *tpeL*-*gusA* fusions into an SM101 *spo0A* mutant (37); no GUS specific activity was observed when this *spo0A* mutant carrying the *tpeL*-*gusA* fusion construct was grown under sporulation conditions, indicating that sporulation-regulated *tpeL* expression is dependent upon *spo0A* expression.

Evidence of σ^E-dependent expression of *tpeL* came from experiments that measured GUS specific activity in vegetative and sporulation cultures of an SM101 *sigE* mutant carrying the *tpeL*-*gusA* fusion construct (97). No significant GUS specific activity was observed during vegetative or sporulation growth of *sigE* mutant carrying the *tpeL*-*gusA* fusion construct (97). The expression of *tpeL* was only detectable in the mother cell compartment and at ~80- to 150-fold lower levels than *cpe* expression (97). Given the presence of the putative σ^K-dependent promoter upstream of *tpeL*, it is possible that other σ factors such as σ^K might also be involved in regulating *tpeL* expression (97). Further detailed

studies on *tpeL* promoter binding with Spo0A, σ^E, and σ^K should clarify the mechanism of sporulation-regulated *tpeL* expression (97).

While the above evidence supports *tpeL* expression during sporulation, TpeL production by *C. perfringens* isolates during vegetative growth has also been reported (98, 99). Furthermore, those studies found repression of TpeL production mediated by glucose in a similar manner to LCT production by *C. difficile*. During vegetative growth, the regulator TpeR is critical for TpeL production, similar to the cases of *C. difficile* TcdR and *C. sordellii* TcsR.

SUMMARY

The important pathogen *C. perfringens* utilizes its spores to survive in harsh environments. Spores are also important for transmission of this foodborne disease pathogen, particularly where type A chromosomal *cpe* food poisoning isolates often produce highly resistant spores that can survive in temperature-abused foods. Another linkage between sporulation and virulence for *C. perfringens* is the expression of two *C. perfringens* toxins from promoters recognized by sporulation-associated sigma factors. The production of one of those toxins, CPE, is essential for the pathogenesis of the second most common bacterial foodborne disease in the United States. Interestingly, another Gram-positive sporeformer, *Bacillus thuringiensis*, also produces sporulation-associated, pore-forming toxins that affect the gastrointestinal tract, albeit in insects rather than humans. The similarities between the pathogenesis of *C. perfringens* and *B. thuringiensis* suggest that these bacteria show a common strategy of producing toxins when sporulating in the GI tract of their host to induce diarrhea, which may then facilitate transmission of their spores back into the environment so they can be picked up by additional hosts. The prevalence of *C. perfringens* type A food poisoning suggests this strategy has been highly successful.

Acknowledgments. Preparation of this review was supported in part by R01 AI19844-32 from the National Institute of Allergy and Infectious Diseases (to B.McC.) and by Department of Defense Multi-disciplinary University Research Initiative (MURI) award through the U.S. Army Research Laboratory and the U. S. Army Research Office under contract number W911NF-09-1-0286 (to M.R.S.); and by grants from MECESUP UAB0802, the Fondo Nacional de Ciencia y Tecnología de Chile (FONDECYT Grant 1110569) and from the Research Office of Universidad Andres Bello (DI-35-11/R) (to D.P.-S).

Citation. Li J, Paredes-Sabja D, Sarker MR, McClane BA. 2016. Clostridium perfringens sporulation and sporulation-associated toxin production. Microbiol Spectrum 4(1):TBS-0022-2015.

References

1. **McClane BA, Robertson SL, Li J.** 2013. *Clostridium perfringens*, p. 465–489. *In* Doyle MP, Buchanan RL (ed), *Food Microbiology: Fundamentals and Frontiers*, 4th ed. ASM Press, Washington, DC.

2. **McClane BA, Uzal FA, Miyakawa MF, Lyerly D, Wilkins TD.** 2006. The enterotoxic clostridia, p. 688–752. *In* Dworkin M, Falkow S, Rosenburg E, Schleifer H, Stackebrand E (ed), *The Prokaryotes*, 3rd ed. Springer-Verlag, New York, NY.

3. **Mallozzi M, Viswanathan VK, Vedantam G.** 2010. Spore-forming bacilli and clostridia in human disease. *Future Microbiol* **5**:1109–1123.

4. **Amimoto K, Noro T, Oishi E, Shimizu M.** 2007. A novel toxin homologous to large clostridial cytotoxins found in culture supernatant of *Clostridium perfringens* type C. *Microbiology* **153**:1198–1206.

5. **Stevens DL, Rood JI.** 2006. Histotoxic clostridia, p. 715–725. *In* Fischetti VA, Novick RP, Ferretti JJ, Portnoy DA, Rood JI (ed), *Gram-Positive Pathogens*, 2nd ed. ASM Press, Washington, DC.

6. **Scallan E, Hoekstra RM, Angulo FJ, Tauxe RV, Widdowson MA, Roy SL, Jones JL, Griffin PM.** 2011. Foodborne illness acquired in the United States–major pathogens. *Emerg Infect Dis* **17**:7–15.

7. **Ma M, Li J, McClane BA.** 2012. Genotypic and phenotypic characterization of *Clostridium perfringens* isolates from Darmbrand cases in post-World War II Germany. *Infect Immun* **80**:4354–4363.

8. **Carman RJ.** 1997. *Clostridium perfringens* in spontaneous and antibiotic-associated diarrhoea of man and other animals. *Rev Med Microbiol* **8**(Suppl 1):S43–S5.

9. **Novak JS, Juneja VK, McClane BA.** 2003. An ultrastructural comparison of spores from various strains of *Clostridium perfringens* and correlations with heat resistance parameters. *Int J Food Microbiol* **86**:239–247.

10. **Orsburn B, Melville SB, Popham DL.** 2008. Factors contributing to heat resistance of *Clostridium perfringens* endospores. *Appl Environ Microbiol* **74**:3328–3335.

11. **Henriques AO, Moran CP Jr.** 2007. Structure, assembly, and function of the spore surface layers. *Annu Rev Microbiol* **61** 555–588.

12. **Setlow P.** 2006. Spores of *Bacillus subtilis*: their resistance to and killing by radiation, heat and chemicals. *J Appl Microbiol* **101**:514–525.

13. **Klobutcher LA, Ragkousi K, Setlow P.** 2006. The *Bacillus subtilis* spore coat provides "eat resistance" during phagocytic predation by the protozoan *Tetrahymena thermophila*. *Proc Natl Acad Sci USA* **103**:165–170.

14. **Laaberki MH, Dworkin J.** 2008. Role of spore coat proteins in the resistance of *Bacillus subtilis* spores to *Caenorhabditis elegans* predation. *J Bacteriol* **190**:6197–6203.

15. **Paredes-Sabja D, Sarker MR.** 2011. Host serum factor triggers germination of *Clostridium perfringens* spores lacking the cortex hydrolysis machinery. *J Med Microbiol* **60**:1734–1741.

16. **Popham DB, Bernhards CB.** Spore peptidoglycan. *In* Eichenberger P, Driks A (ed), *The Bacterial Spore*. ASM Press, Washington, DC, in press.

17. **Warth AD, Strominger JL.** 1969. Structure of the peptidoglycan of bacterial spores: occurrence of the lactam of muramic acid. *Proc Natl Acad Sci USA* **64**:528–535.

18. **Warth AD, Strominger JL.** 1972. Structure of the peptidoglycan from spores of *Bacillus subtilis*. *Biochemistry* **11**:1389–1396.

19. **Setlow P.** 2003. Spore germination. *Curr Opin Microbiol* **6**:550–556.

20. **Makino S, Moriyama R.** 2002. Hydrolysis of cortex peptidoglycan during bacterial spore germination. *Med Sci Monit* **8**:RA119–RA127.

21. **Moir A, Corfe BM, Behravan J.** 2002. Spore germination. *Cell Mol Life Sci* **59**:403–409.

22. **Setlow P.** 2007. I will survive: DNA protection in bacterial spores. *Trends Microbiol* **15**:172–180.

23. **Setlow P.** 2014. Spore resistance properties. *Microbiol Spectrum* **2**(4):TBS-0003-2012. doi:10.1128/microbiolspec.TBS-0003-2012

24. **Labbe RG.** 2005. Sporulation (morphology) of *Clostridium*, p 647–658. *In* Durre P (ed), *Handbook on Clostridia*. CRC Press, Boca Raton, FL.

25. **Stragier P.** 2002. A gene odyssey: exploring the genomes of endospore-forming bacteria, p. 519–525. *In* Sonenshein AL, Hoch JA, Losick R (ed), *Bacillus subtilis and Its Closest Relatives: From Genes to Cells*. ASM Press, Washington, DC.

26. **Philippe VA, Méndez MB, Huang IH, Orsaria LM, Sarker MR, Grau RR.** 2006. Inorganic phosphate induces spore morphogenesis and enterotoxin production in the intestinal pathogen *Clostridium perfringens*. *Infect Immun* **74**:3651–3656.

27. **Sarker MR, Carman RJ, McClane BA.** 1999. Inactivation of the gene (*cpe*) encoding *Clostridium perfringens* enterotoxin eliminates the ability of two *cpe*-positive *C. perfringens* type A human gastrointestinal disease isolates to affect rabbit ileal loops. *Mol Microbiol* **33**:946–958.

28. **Li J, McClane BA.** 2010. Evaluating the involvement of alternative sigma factors SigF and SigG in *Clostridium perfringens* sporulation and enterotoxin synthesis. *Infect Immun* **78**:4286–4293.

29. **Harry KH, Zhou R, Kroos L, Melville SB.** 2009. Sporulation and enterotoxin (CPE) synthesis are controlled by the sporulation-specific sigma factors SigE and SigK in *Clostridium perfringens*. *J Bacteriol* **191**:2728–2742.

30. **Shih N-J, Labbé RG.** 1994. Effect of glucose on sporulation and extracellular amylase production by *Clostridium perfringens* type A in a defined medium. *Curr Microbiol* **29**:163–169.

31. **Varga J, Stirewalt VL, Melville SB.** 2004. The CcpA protein is necessary for efficient sporulation and enterotoxin gene (*cpe*) regulation in *Clostridium perfringens*. *J Bacteriol* **186**:5221–5229.

32. Bettegowda C, Huang X, Lin J, Cheong I, Kohli M, Szabo SA, Zhang X, Diaz LA Jr, Velculescu VE, Parmigiani G, Kinzler KW, Vogelstein B, Zhou S. 2006. The genome and transcriptomes of the anti-tumor agent *Clostridium novyi*-NT. *Nat Biotechnol* 24:1573–1580.

33. Bruggemann H, Baumer S, Fricke WF, Wiezer A, Liesegang H, Decker I, Herzberg C, Martinez-Arias R, Merkl R, Henne A, Gottschalk G. 2003. The genome sequence of *Clostridium tetani*, the causative agent of tetanus disease. *Proc Natl Acad Sci USA* 100:1316–1321.

34. Myers GS, Rasko DA, Cheung JK, Ravel J, Seshadri R, DeBoy RT, Ren Q, Varga J, Awad MM, Brinkac LM, Daugherty SC, Haft DH, Dodson RJ, Madupu R, Nelson WC, Rosovitz MJ, Sullivan SA, Khouri H, Dimitrov GI, Watkins KL, Mulligan S, Benton J, Radune D, Fisher DJ, Atkins HS, Hiscox T, Jost BH, Billington SJ, Songer JG, McClane BA, Titball RW, Rood JI, Melville SB, Paulsen IT. 2006. Skewed genomic variability in strains of the toxigenic bacterial pathogen, *Clostridium perfringens*. *Genome Res* 16:1031–1040.

35. Nölling J, Breton G, Omelchenko MV, Makarova KS, Zeng Q, Gibson R, Lee HM, Dubois J, Qiu D, Hitti J, Wolf YI, Tatusov RL, Sabathe F, Doucette-Stamm L, Soucaille P, Daly MJ, Bennett GN, Koonin EV, Smith DR. 2001. Genome sequence and comparative analysis of the solvent-producing bacterium *Clostridium acetobutylicum*. *J Bacteriol* 183:4823–4838.

36. Shimizu T, Ohtani K, Hirakawa H, Ohshima K, Yamashita A, Shiba T, Ogasawara N, Hattori M, Kuhara S, Hayashi H. 2002. Complete genome sequence of *Clostridium perfringens*, an anaerobic flesh-eater. *Proc Natl Acad Sci USA* 99:996–1001.

37. Huang IH, Waters M, Grau RR, Sarker MR. 2004. Disruption of the gene (*spo0A*) encoding sporulation transcription factor blocks endospore formation and enterotoxin production in enterotoxigenic *Clostridium perfringens* type A. *FEMS Microbiol Lett* 233:233–240.

38. Huang IH, Sarker MR. 2006. Complementation of a *Clostridium perfringens spo0A* mutant with wild-type *spo0A* from other *Clostridium* species. *Appl Environ Microbiol* 72:6388–6393.

39. Sonenshein AL. 2000. Control of sporulation initiation in *Bacillus subtilis*. *Curr Opin Microbiol* 3:561–566.

40. Kroos L. 2007. The *Bacillus* and *Myxococcus* developmental networks and their transcriptional regulators. *Annu Rev Genet* 41:13–39.

41. Molle V, Fujita M, Jensen ST, Eichenberger P, González-Pastor JE, Liu JS, Losick R. 2003. The Spo0A regulon of *Bacillus subtilis*. *Mol Microbiol* 50:1683–1701.

42. Li J, Paredes-Sabja D, Sarker MR, McClane BA. 2009. Further characterization of *Clostridium perfringens* small acid soluble protein-4 (Ssp4) properties and expression. *PLoS One* 4:e6249. doi:10.1371/journal.pone.0006249.

43. Vidal JE, Chen J, Li J, McClane BA. 2009. Use of an EZ-Tn5-based random mutagenesis system to identify a novel toxin regulatory locus in *Clostridium perfringens* strain 13. *PLoS One* 4:e6232. doi:10.1371/journal.pone.0006232.

44. Li J, Chen J, Vidal JE, McClane BA. 2011. The Agr-like quorum-sensing system regulates sporulation and production of enterotoxin and beta2 toxin by *Clostridium perfringens* type A non-food-borne human gastrointestinal disease strain F5603. *Infect Immun* 79:2451–2459.

45. Ohtani K, Yuan Y, Hassan S, Wang R, Wang Y, Shimizu T. 2009. Virulence gene regulation by the *agr* system in *Clostridium perfringens*. *J Bacteriol* 191:3919–3927.

46. Li J, Ma M, Sarker MR, McClane BA. 2013. CodY is a global regulator of virulence-associated properties for *Clostridium perfringens* type D strain CN3718. *MBio* 4:e00770-13. doi:10.1128/mBio.00770-13.

47. Ohtani K, Hirakawa H, Paredes-Sabja D, Tashiro K, Kuhara S, Sarker MR, Shimizu T. 2013. Unique regulatory mechanism of sporulation and enterotoxin production in *Clostridium perfringens*. *J Bacteriol* 195:2931–2936.

48. Xiao Y, van Hijum SA, Abee T, Wells-Bennik MH. 2015. Genome-wide transcriptional profiling of *Clostridium perfringens* SM101 during sporulation extends the core of putative sporulation genes and genes determining spore properties and germination characteristics. *PLoS One* 10:e0127036. doi:10.1371/journal.pone.0127036.

49. Paredes-Sabja D, Setlow P, Sarker MR. 2011. Germination of spores of *Bacillales* and *Clostridiales* species: mechanisms and proteins involved. *Trends Microbiol* 19:85–94.

50. Moir A, Cooper G. 2014. Spore germination. *Microbiol Spectrum* 3(5):TBS-0014-2012. doi:10.1128/microbiol-spec.TBS-0014-2012.

51. Paredes-Sabja D, Torres JA, Setlow P, Sarker MR. 2008. *Clostridium perfringens* spore germination: characterization of germinants and their receptors. *J Bacteriol* 190:1190–1201.

52. Paredes-Sabja D, Udompijitkul P, Sarker MR. 2009. Inorganic phosphate and sodium ions are cogerminants for spores of *Clostridium perfringens* type A food poisoning-related isolates. *Appl Environ Microbiol* 75:6299–6305.

53. Paredes-Sabja D, Sarker MR. 2011. Germination response of spores of the pathogenic bacterium *Clostridium perfringens* and *Clostridium difficile* to cultured human epithelial cells. *Anaerobe* 17:78–84.

54. Mercado-Lubo R, McCormick BA. 2010. A unique subset of Peyer's patches express lysozyme. *Gastroenterology* 138:36–39.

55. Reitamo S, Klockars M, Adinolfi M, Osserman EF. 1978. Human lysozyme (origin and distribution in health and disease). *Ric Clin Lab* 8:211–231.

56. Ghosh S, Zhang P, Li YQ, Setlow P. 2009. Super-dormant spores of *Bacillus* species have elevated wet-heat resistance and temperature requirements for heat activation. *J Bacteriol* 191:5584–5591.

57. Moir A, Smith DA. 1990. The genetics of bacterial spore germination. *Annu Rev Microbiol* 44:531–553.

58. Banawas S, Paredes-Sabja D, Korza G, Li Y, Hao B, Setlow P, Sarker MR. 2013. The *Clostridium perfringens* germinant receptor protein GerKC is located in the spore

inner membrane and is crucial for spore germination. *J Bacteriol* 195:5084–5091.

59. **Udompijitkul P, Alnoman M, Banawas S, Paredes-Sabja D, Sarker MR.** 2014. New amino acid germinants for spores of the enterotoxigenic *Clostridium perfringens* type A isolates. *Food Microbiol* 44:24–33.

60. **Paredes-Sabja D, Setlow P, Sarker MR.** 2009. Role of GerKB in germination and outgrowth of *Clostridium perfringens* spores. *Appl Environ Microbiol* 75:3813–3817.

61. **Paredes-Sabja D, Setlow P, Sarker MR.** 2009. GerO, a putative Na+/H+-K+ antiporter, is essential for normal germination of spores of the pathogenic bacterium *Clostridium perfringens*. *J Bacteriol* 191:3822–3831.

62. **Kumazawa T, Masayama A, Fukuoka S, Makino S, Yoshimura T, Moriyama R.** 2007. Mode of action of a germination-specific cortex-lytic enzyme, SleC, of *Clostridium perfringens* S40. *Biosci Biotechnol Biochem* 71: 884–892.

63. **Paidhungat M, Ragkousi K, Setlow P.** 2001. Genetic requirements for induction of germination of spores of *Bacillus subtilis* by Ca(2+)-dipicolinate. *J Bacteriol* 183: 4886–4893.

64. **Paredes-Sabja D, Setlow P, Sarker MR.** 2009. The protease CspB is essential for initiation of cortex hydrolysis and dipicolinic acid (DPA) release during germination of spores of *Clostridium perfringens* type A food poisoning isolates. *Microbiology* 155:3464–3472.

65. **Chen Y, Miyata S, Makino S, Moriyama R.** 1997. Molecular characterization of a germination-specific muramidase from *Clostridium perfringens* S40 spores and nucleotide sequence of the corresponding gene. *J Bacteriol* 179:3181–3187.

66. **Paredes-Sabja D, Setlow P, Sarker MR.** 2009. SleC is essential for cortex peptidoglycan hydrolysis during germination of spores of the pathogenic bacterium *Clostridium perfringens*. *J Bacteriol* 191:2711–2720.

67. **Paredes-Sabja D, Sarker MR.** 2010. Effect of the cortex-lytic enzyme SleC from non-food-borne *Clostridium perfringens* on the germination properties of SleC-lacking spores of a food poisoning isolate. *Can J Microbiol* 56: 952–958.

68. **Shimamoto S, Moriyama R, Sugimoto K, Miyata S, Makino S.** 2001. Partial characterization of an enzyme fraction with protease activity which converts the spore peptidoglycan hydrolase (SleC) precursor to an active enzyme during germination of *Clostridium perfringens* S40 spores and analysis of a gene cluster involved in the activity. *J Bacteriol* 183:3742–3751.

69. **Sarker MR, Shivers RP, Sparks SG, Juneja VK, McClane BA.** 2000. Comparative experiments to examine the effects of heating on vegetative cells and spores of *Clostridium perfringens* isolates carrying plasmid genes versus chromosomal enterotoxin genes. *Appl Environ Microbiol* 66:3234–3240.

70. **Li J, McClane BA.** 2006. Comparative effects of osmotic, sodium nitrite-induced, and pH-induced stress on growth and survival of *Clostridium perfringens* type A isolates carrying chromosomal or plasmid-borne enterotoxin genes. *Appl Environ Microbiol* 72:7620–7625.

71. **Li J, McClane BA.** 2006. Further comparison of temperature effects on growth and survival of *Clostridium perfringens* type A isolates carrying a chromosomal or plasmid-borne enterotoxin gene. *Appl Environ Microbiol* 72:4561–4568.

72. **Deguchi A, Miyamoto K, Kuwahara T, Miki Y, Kaneko I, Li J, McClane BA, Akimoto S.** 2009. Genetic characterization of type A enterotoxigenic *Clostridium perfringens* strains. *PLoS One* 4:e5598. doi:10.1371/journal.pone.0005598.

73. **Wen Q, McClane BA.** 2004. Detection of enterotoxigenic *Clostridium perfringens* type A isolates in American retail foods. *Appl Environ Microbiol* 70:2685–2691.

74. **Paredes-Sabja D, Sarker N, Setlow B, Setlow P, Sarker MR.** 2008. Roles of DacB and Spm proteins in *Clostridium perfringens* spore resistance to moist heat, chemicals, and UV radiation. *Appl Environ Microbiol* 74:3730–3738.

75. **Paredes-Sabja D, Setlow B, Setlow P, Sarker MR.** 2008. Characterization of *Clostridium perfringens* spores that lack SpoVA proteins and dipicolinic acid. *J Bacteriol* 190: 4648–4659.

76. **Raju D, Setlow P, Sarker MR.** 2007. Antisense-RNA-mediated decreased synthesis of small, acid-soluble spore proteins leads to decreased resistance of *Clostridium perfringens* spores to moist heat and UV radiation. *Appl Environ Microbiol* 73:2048–2053.

77. **Raju D, Waters M, Setlow P, Sarker MR.** 2006. Investigating the role of small, acid-soluble spore proteins (SASPs) in the resistance of *Clostridium perfringens* spores to heat. *BMC Microbiol* 6:50. doi:10.1186/1471-2180-6-50.

78. **Li J, McClane BA.** 2008. A novel small acid soluble protein variant is important for spore resistance of most *Clostridium perfringens* food poisoning isolates. *PLoS Pathog* 4:e1000056. doi:10.1371/journal.ppat.1000056.

79. **Miyamoto K, Li J, Akimoto S, McClane BA.** 2009. Molecular approaches for detecting enterotoxigenic *Clostridium perfringens*. *Res Adv Appl Environ Microbiol* 2: 1–7.

80. **Caserta JARS, Robertson SL, Saputo J, Shrestha A, McClane BA, Uzal FA.** 2011. Development and application of a mouse intestinal loop model to study the *in vivo* action of *Clostridium perfringens* enterotoxin. *Infect Immun* 79:3020–3027.

81. **Robertson SL, Smedley JG III, McClane BA.** 2010. Identification of a claudin-4 residue important for mediating the host cell binding and action of *Clostridium perfringens* enterotoxin. *Infect Immun* 78:505–517.

82. **Shrestha A, McClane BA.** 2013. Human claudin-8 and -14 are receptors capable of conveying the cytotoxic effects of *Clostridium perfringens* enterotoxin. *MBio* 4: e00594-12. doi:10.1128/mBio.00594-12.

83. **Robertson SL, Smedley JG III, Singh U, Chakrabarti G, Van Itallie CM, Anderson JM, McClane BA.** 2007. Compositional and stoichiometric analysis of *Clostridium perfringens* enterotoxin complexes in Caco-2 cells and claudin 4 fibroblast transfectants. *Cell Microbiol* 9: 2734–2755.

84. **Chakrabarti G, McClane BA.** 2005. The importance of calcium influx, calpain and calmodulin for the activation of CaCo-2 cell death pathways by *Clostridium perfringens* enterotoxin. *Cell Microbiol* **7:**129–146.

85. **Kitadokoro K, Nishimura K, Kamitani S, Fukui-Miyazaki A, Toshima H, Abe H, Kamata Y, Sugita-Konishi Y, Yamamoto S, Karatani H, Horiguchi Y.** 2011. Crystal structure of *Clostridium perfringens* enterotoxin displays features of beta-pore-forming toxins. *J Biol Chem* **286:**19549–19555.

86. **Briggs DC, Naylor CE, Smedley JG III, Lukoyanova N, Robertson S, Moss DS, McClane BA, Basak AK.** 2011. Structure of the food-poisoning *Clostridium perfringens* enterotoxin reveals similarity to the aerolysin-like pore-forming toxins. *J Mol Biol* **413:**138–149.

87. **Gao Z, McClane BA.** 2012. Use of *Clostridium perfringens* enterotoxin and the enterotoxin receptor-binding domain (C-CPE) for cancer treatment: opportunities and challenges. *J Toxicol* **2012:**981626. doi:10.1155/2012/981626.

88. **Guttenberg G, Hornei S, Jank T, Schwan C, Lü W, Einsle O, Papatheodorou P, Aktories K.** 2012. Molecular characteristics of *Clostridium perfringens* TpeL toxin and consequences of mono-O-GlcNAcylation of Ras in living cells. *J Biol Chem* **287:**24929–24940.

89. **Belyi Y, Aktories K.** 2010. Bacterial toxin and effector glycosyltransferases. *Biochim Biophys Acta* **1800:**134–143.

90. **Busch C, Hofmann F, Gerhard R, Aktories K.** 2000. Involvement of a conserved tryptophan residue in the UDP-glucose binding of large clostridial cytotoxin glycosyltransferases. *J Biol Chem* **275:**13228–13234.

91. **Busch C, Hofmann F, Selzer J, Munro S, Jeckel D, Aktories K.** 1998. A common motif of eukaryotic glyco-

syltransferases is essential for the enzyme activity of large clostridial cytotoxins. *J Biol Chem* **273:**19566–19572.

92. **Voth DE, Ballard JD.** 2005. *Clostridium difficile* toxins: mechanism of action and role in disease. *Clin Microbiol Rev* **18:**247–263.

93. **Schorch B, Song S, van Diemen FR, Bock HH, May P, Herz J, Brummelkamp TR, Papatheodorou P, Aktories K.** 2014. LRP1 is a receptor for *Clostridium perfringens* TpeL toxin indicating a two-receptor model of clostridial glycosylating toxins. *Proc Natl Acad Sci USA* **111:** 6431–6436.

94. **Gurjar A, Li J, McClane BA.** 2010. Characterization of toxin plasmids in *Clostridium perfringens* type C isolates. *Infect Immun* **78:**4860–4869.

95. **Sayeed S, Li J, McClane BA.** 2010. Characterization of virulence plasmid diversity among *Clostridium perfringens* type B isolates. *Infect Immun* **78:**495–504.

96. **Chalmers G, Bruce HL, Hunter DB, Parreira VR, Kulkarni RR, Jiang YF, Prescott JF, Boerlin P.** 2008. Multilocus sequence typing analysis of *Clostridium perfringens* isolates from necrotic enteritis outbreaks in broiler chicken populations. *J Clin Microbiol* **46:**3957–3964.

97. **Paredes-Sabja D, Sarker N, Sarker MR.** 2011. *Clostridium perfringens tpeL* is expressed during sporulation. *Microb Pathog* **51:**384–388.

98. **Chen J, McClane BA.** 2015. Characterization of *Clostridium perfringens* TpeL toxin gene carriage, production, cytotoxic contributions, and trypsin sensitivity. *Infect Immun* **83:**2369–2381.

99. **Carter GP, Larcombe S, Li L, Jayawardena D, Awad MM, Songer JG, Lyras D.** 2014. Expression of the large clostridial toxins is controlled by conserved regulatory mechanisms. *Int J Med Microbiol* **304:**1147–1159.

Practical
Technologies

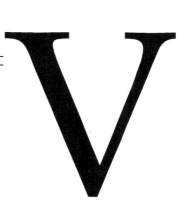

The Bacterial Spore: From Molecules to Systems
Edited by P. Eichenberger and A. Driks
© 2016 American Society for Microbiology, Washington, DC
doi:10.1128/microbiolspec.TBS-0011-2012

Rachele Isticato[1]
Ezio Ricca[1]

Spore Surface Display

17

SURFACE DISPLAY ON BACTERIAL CELLS OR PHAGES

Display systems that present biologically active molecules on the surface of microorganisms have become increasingly used to address environmental and biomedical issues (1–3). Strategies using environmentally relevant proteins or peptides for display on the surface of phages or bacterial cells have been extensively reviewed by Wu et al. (4). Examples include proteins able to bind metal ions that can be used as bioadsorbents or biocatalysts, including cysteine-rich metallothioneins (MTs) or Cys-His rich synthetic peptides, known to bind Cd^{2+} and Hg^{2+} with a very high affinity. Eukaryotic MTs have been expressed on the surface of *Escherichia coli* cells through fusion to the porin LamB, with a 20-fold increased ability of Cd^{2+} accumulation of the recombinant cell with respect to its parental strain (5, 6). In addition, metal-binding peptides have also been expressed on the surface of soil bacteria known to survive in contaminated environments. The mouse MT was displayed on the surface of *Pseudomonas putida* (7) and *Ralstonia metallidurans* CH34 (8), resulting in a 3-fold increase in binding and removal of Cd^{2+}, sufficient to improve plant growth in a contaminated soil (8). Synthetic phytochelatins (ECn) with the repetitive metal-binding motif (Glu-Cys)nGly were displayed on the surface of *Moraxella* sp. cells causing a 10-fold improvement in Hg^{2+} intracellular accumulation (9–11).

In addition to heavy metals, organic contaminants can be removed from the environment by the use of microbial cells displaying heterologous enzymes. Examples include organophosphorus hydrolases (OPHs). These bacterial enzymes are able to degrade organophosphates, which are toxic compounds widely used as pesticides. *E. coli* cells expressing OPH on their surface via the Lpp-OmpA fusion system were able to degrade parathion and paraoxon 7-fold faster than cells expressing OPH intracellularly (12). Surface display approaches have also been used to develop whole-cell diagnostic tools and vaccine delivery systems. Functional single-chain antibody fragments have been expressed on bacterial cells and used as diagnostic devices in immunological tests. In the first report of an antibody fragment expressed in an active form on a bacterial surface, the murine anti-human-IgE scFv antibody fragment was exposed on the surface of *Staphylococcus xylosus* and *S. carnosus* cells (13). More recently, the oral commensal bacterium *Streptococcus gordonii* was engineered to display a single-chain Fv (scFv) antibody fragment, derived from a monoclonal antibody raised against the major adhesin of the dental caries-producing bacterium *Streptococcus mutans* (streptococcal antigen I/II or SA I/II). Recombinant *S. gordonii* was found to specifically bind to immobilized SA I/II and represents the first step toward the development of a stable system for the delivery of recombinant antibodies (14).

[1]Department of Biology, Federico II University of Naples, Naples, 80126 Italy.

Surface Display for Vaccine Delivery

The most common application of surface display systems has been the development of new vaccines. While the display of peptides on phages has been used mostly as a tool for epitope discovery, the display on bacterial surfaces usually takes advantage of previously identified vaccine candidates (immunogenic peptides or proteins) to obtain a recombinant bacterium that can then be used as a live carrier for vaccine delivery (15). The possibility of using phage display for the identification of neutralizing viral epitopes was first demonstrated by inserting peptide sequences from the V3 loop of gp120 from HIV-1 into the N-terminal region of the major coat protein (pVIII) of the filamentous bacteriophage fd, which led to their display in multiple copies on the surface of the virion (16). While most of the initial studies were carried out using filamentous phage fd, other phage systems were later developed for similar purposes (17–20). Initial attempts at using bacterial display for the development of new vaccines were focused on Gram-positive bacteria (21, 22) and *E. coli*, *Salmonella*, and other Gram-negative bacteria were considered only later (23). Studies on *Salmonella* provided clear evidence that attenuated strains can be effectively used as a general vaccine vehicle to deliver antiphagocytic virulence determinants of unrelated bacteria (24, 25). To avoid the use of engineered pathogens for vaccine delivery, systems that allow the expression of heterologous antigens on the surface of commensal bacteria were also developed (26).

Strategies of Surface Display

Strategies to display heterologous proteins on the bacterial surface are generally based on a cell surface molecule (outer membrane proteins, lipoproteins, subunits of cell appendages, S-layer proteins, cell wall and cell membrane proteins) acting as a carrier and able to anchor an heterologous passenger protein on the cell surface (1) (Fig. 1). Several properties of carrier proteins can affect the efficiency of surface display and have different effects on the stability and integrity of the host. A successful carrier protein should have a strong anchoring motif to avoid detachment from the surface and should be resistant to proteases present in the extracellular medium or in the periplasmic space (27). The site of insertion of the heterologous part into the carrier is another important factor, since it can influence the stability, activity, and posttranslational modification of the fusion protein. Therefore, fusions at the N or C terminus of the carrier protein, as well as sandwich fusions (insertions), have been constructed (Fig. 2). In some instances, a linker peptide (spacer) consisting of repeats of 5 to 10 aliphatic amino acid residues is inserted between the carrier and the passenger to minimize potential steric effects disturbing the correct folding of either protein (Fig. 2). Some characteristics of the passenger protein can also influence the translocation process and efficiency of surface display. The folded structure of the passenger protein, such as the formation of disulfide bridges in the periplasm or the presence of many charged or

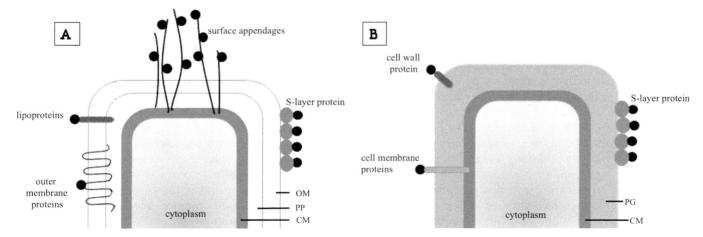

Figure 1 Surface proteins used as carriers for surface display systems in Gram-negative (**A**) and Gram-positive (**B**) bacteria. CM, cytoplasmic membrane; OM, outer membrane; PP, periplasm; PG, peptidoglycan. Black circles indicate the heterologous proteins used as passengers.

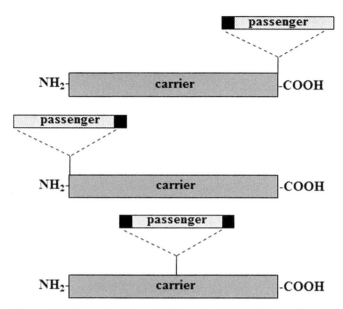

Figure 2 C-terminal (top), N-terminal (middle), and sandwich (bottom) fusions. Black squares indicate the linker sequence, in some cases used to separate carrier and passenger.

hydrophobic residues, can affect translocation through the cell membranes (27). In some cases, surface display of heterologous proteins has been taking advantage of the autotransport properties of certain proteins (28). In these cases, the C-terminal end of the autotransporter (the carrier) is anchored at the cell membrane and exposes its own N-terminal end (the passenger) on the external surface of the membrane. Replacement of the passenger domain with a heterologous protein results in the autodisplay of N-terminally fused passengers (3, 29). Autodisplay technology has been used in several biotechnological applications, including the construction of a variety of whole-cell biocatalytic systems (3).

SURFACE DISPLAY ON *BACILLUS SUBTILIS* SPORES

As described in detail elsewhere, *Bacillus* spores are encased in a coat, a protein structure protecting the spore from toxic chemicals, lytic enzymes (30), and phagocytic predation by protozoans (31). The rigidity and compactness of the spore coat immediately suggest the possibility of using its structural components as anchoring motifs for the expression of heterologous polypeptides on the spore surface. A genetic system to engineer the coat of *B. subtilis* spores has been developed and a model passenger has been efficiently displayed (32). The spore-based approach provides several advantages

over other display systems, such as high stability even after prolonged storage, the possibility of displaying large multimeric proteins, and safety for human use (see "Advantages of the Spore Surface Display Systems," below). The spore coat components of *B. subtilis* and other *Bacillus* species that have been used as carriers, the heterologous proteins used as passengers, and their proposed potential applications are listed in Table 1. Because of limited information on the mechanisms of protein incorporation into the coat, the structure of the proteins forming the spore surface, and the existence of anchoring motifs, initial attempts to expose heterologous proteins on the spore surface were focused on two coat components selected for their surface location (CotB; 32) or for their abundance (CotC; 33). Subsequently, additional coat proteins, as well as coat-associated proteins, have been proposed as carriers (Table 1). A review of the various examples of spore surface display is presented in the following subsections organized on the basis of the carrier utilized.

CotB as a Carrier Protein

CotB was the first coat protein to be used as a carrier to target a heterologous protein to the spore surface (32). Expression of the *cotB* gene is under the control of the mother cell-specific sigma factor, σ^K, and of the transcriptional regulators GerE and GerR (34). CotB has a strongly hydrophilic C-terminal half formed by three serine-rich repeats (Fig. 3). Serine residues account for over 50% of the CotB C-terminal half. CotB has a deduced molecular mass of 46 kDa, but migrates on SDS-PAGE as a 66-kDa polypeptide. The discrepancy between measured and deduced molecular mass has been explained by showing that CotB is initially synthesized as a 46-kDa species (CotB-46) and converted into a 66-kDa species (CotB-66), presumably a CotB homodimer (35). Conversion of the CotB-46 form into CotB-66 requires expression of *cotG* and *cotH*. In *cotG* mutants, the CotB-46 species accumulates and undergoes assembly (35). The exact role of CotG in formation of CotB-66 is not known. However, CotG was found to be present in complexes with mainly the CotB-46 form at the time of coat assembly. Moreover, CotG and CotB were found to interact directly in a yeast two-hybrid assay (35). Presumably, the interaction of CotG with CotB-46 is essential for the formation of the CotB-66 species. The requirement of *cotH* expression for the assembly of both CotG and CotB-66 is in part explained by the observation that CotG does not accumulate in cells mutant for *cotH*. This is reminiscent of the situation with CotC, whose various

Table 1   List of carriers, passenger proteins, and potential applications described for spore surface display systems

Bacterium	Carrier	Passenger	Potential application	Ref.
B. subtilis	CotB	TTFC of C. tetani	Vaccine	32
		LTB of E. coli	Vaccine	33
		PA of B. anthracis	Vaccine	44
		CPA of C. perfringens	Vaccine	45
		UreA of H. acinonychis	Vaccine	46
		TcdA-TcdB of C. difficile	Vaccine	47
		VP28 of WSSV	Vaccine	48
		18xHis	Bioremediation	49
	CotC	TTFC of C. tetani	Vaccine	32
		LTB of E. coli	Vaccine	33
		PA of B. anthracis	Vaccine	44
		Pep23 of HIV	Vaccine	56
		TP22.3 of C. sinensis	Vaccine	57
		UreA of H. acinonychis	Vaccine	46
		GP64 of B. mori	Antiviral	58
		TcdA-TcdB of C. difficile	Vaccine	47
		HAS	Clinical use	55
		ADH of B. mori	Biocatalysis	59
	CotG	Streptavidin	Diagnostic tool	64
		GFP$_{UV}$	Display system	65
		ω-Transaminase	Biocatalysis	66
		β-Galactosidase	Biocatalysis	67
		Neu5Ac aldolase	Biocatalysis	68
		Phytase	Animal probiotic	69
		UreA of H. acinonychis	Vaccine	46
	OxdD	Phytase	Animal probiotic	69
		β-Glu	Animal probiotic	69
	SpsC	pIII coat protein of M13	Display system	75
	CotA		Autodisplay	78
	CotZ	UreA of H. acinonychis	Vaccine	106
		FliD of C. difficile	Vaccine	107
	CgeA	CagA of H. pylori	Vaccine	108
B. thuringiensis	Cry1Ac	GFP	Display system	80
		Anti-PhOx scFv	Diagnostic tool	80
	InhA	β-Gal	Display system	81
B. anthracis	BlcA/B	GFP	Display system	82

forms (see below) also do not accumulate in a *cotH* mutant (36). These observations have led to the proposal that CotH, or a CotH-controlled protein, protects CotC and CotG from proteolysis prior to their assembly. CotH or a factor under its control could directly interact with CotC and CotG in a chaperone-like manner, or otherwise with a protease that uses CotC and CotG as a substrate (35–37). The possibility of using CotB as a carrier protein was based on its surface localization, as shown by a fluorescence-activated cell sorter (FACS) analysis with anti-CotB antibody (32). To verify that a heterologous protein fused to CotB could be displayed on the spore surface, two antigens were initially selected as model passenger proteins:

(i) the nontoxic 459-amino-acid C-terminal fragment of the tetanus toxin (TTFC), a well-characterized and highly immunogenic peptide of 51.8 kDa encoded by the *tetC* gene of *Clostridium tetani* (38); and (ii) the 103-amino-acid B subunit of the heat-labile toxin (LTB) of enterotoxigenic strains of *E. coli*, a 12-kDa peptide encoded by the *eltB* gene (39). The strategy to obtain recombinant *B. subtilis* spores expressing CotB-TTFC or CotB-LTB on their surface was based on (i) use of the *cotB* gene and its promoter for the construction of translational fusions and (ii) chromosomal integration of the *cotB-tetC* and *cotB-eltB* gene fusions into the nonessential gene locus *amyE* (32). Placing the fusion proteins under the control of the *cotB* promoter

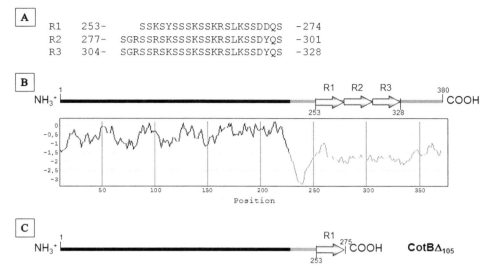

Figure 3 (A) Amino acid sequences of the three repeats (R1, R2, R3) present in the C-terminal half of CotB. (B) Hydrophobic (black) and hydrophilic (gray) regions of CotB as deduced by a Kyte-Doolittle plot (ProtScale software on ExPASy). (C) Truncated form of CotB used as a carrier for surface display.

and ribosome-binding site ensured correct timing of expression during sporulation, while its chromosomal integration guaranteed the genetic stability of the construct. When TTFC and LTB were fused to the C-terminal end of CotB, the chimeric proteins failed to correctly and efficiently assemble on the spore surface (27). In order to bypass such problems, TTFC and LTB were fused to the C-terminal end of a truncated CotB, deleted of 105 C-terminal amino acids (CotBΔ$_{105}$), thus removing two of the three amino acid repeats (Fig. 3C). The CotBΔ$_{105}$-TTFC chimeric protein was efficiently assembled and exposed on the spore surface (32). A quantitative dot blot showed that each recombinant spore exposed an amount of CotBΔ$_{105}$-TTFC fusion protein equal to 0.00022 pg and made it possible to conclude that 1.5×10^3 chimeric molecules are present on the surface of each recombinant spore (32). Unlike CotBΔ$_{105}$-TTFC, CotBΔ$_{105}$-LTB was not properly assembled. The strain expressing this chimera showed reduced sporulation and germination efficiencies and its spores were not resistant to lysozyme. These observations, together with the SDS-PAGE analysis of the released coat proteins, suggested that the presence of CotBΔ$_{105}$-LTB strongly altered the integrity of the spore coat. An *in silico* analysis (27) showed some identity between the chimeric product (in the fusion region) and LytF, a cell wall-associated hydrolase produced by *B. subtilis* during vegetative growth and involved in daughter cell separation (40). When cells are no longer growing, LytF is degraded by the cell

surface protease WprA and the extracellular protease Epr (40). This raised the possibility that the chimeric product CotBΔ$_{105}$-LTB could be degraded by those proteases during sporulation (27). Exposure of CotBΔ$_{105}$-TTFC on the spore surface required the expression of wild-type alleles of *cotH*, *cotG*, and *cotB* (32). As discussed above, these three gene products act in a hierarchical way with CotH to assemble CotG which, in turn, interacts with CotB. This interaction is essential for the formation of the CotB homodimer of 66 kDa (35). This would suggest that a wild-type copy of CotB forms a heterodimer with a copy of CotBΔ$_{105}$-TTFC, allowing the surface display of the fusion. The model passenger protein TTFC has been also fused at the N-terminal and inserted in the middle of CotB (32) (Fig. 2). In both cases the CotBΔ$_{105}$ form of CotB was used. Both the N-terminal and the sandwich fusions were properly assembled in the coat structure (32). At least in the CotB-TTFC case, it was then possible to conclude that the site of insertion of the passenger protein in the carrier does not affect display on the spore surface. Western blot and FACS analysis were used to show that recombinant spores containing either one of the three types of fusions exposed TTFC at their surface, while *in vivo* experiments showed that spore-displayed TTFC was able to induce a TTFC-specific immune response in a murine model and, therefore, was exposed in a biologically active form (32). Surface display of TTFC fused to CotB provided validation that the spore coat of *B. subtilis* can be engineered to incorporate bioactive molecules of large molecular mass (TTFC is 51.8 kDa). TTFC-specific immune response induced by spores displaying the CotBΔ₁₀₅-TTFC chimera (C-terminal fusion) was analyzed in more detail in subsequent studies. Mice orally or nasally immunized with recombinant spores induced fecal sIgA and serum IgG at protective levels, and mice were protected when challenged with purified tetanus toxin (41). Similar spores were used in a mucosal-priming–parenteral-boosting vaccination strategy (42). Recombinant spores were used to orally prime BALB/c and C57BL/6 mice that were then subcutaneously boosted with soluble TTFC (without adjuvant). Protective levels of TTFC-specific IgA and IgG were observed also in this case (42). In a different study, recombinant spores displaying the CotBΔ₁₀₅-TTFC chimera (C-terminal fusion) were shown to induce spleen and mesenteric lymph node cell proliferation as well as production of gamma interferon (IFNγ), but not of interleukin 4 (IL-4) and IL-10, in both locations, indicating that recombinant spores preferentially induce a strong cell-mediated immune response with a Th1 phenotype independently from their ability to germinate in the gastrointestinal tract (43). In more recent studies, CotB has been used as a carrier to display other antigens: the domains 1b-3 and 4 of the protective antigen (PA) of *Bacillus anthracis* (44), the C-terminal part of the alpha toxin (CPA) of *Clostridium perfringens* (45), the UreA protein of *Helicobacter acinonychis* (46), the carboxy-terminal repeat domains of toxins A and B of *Clostridium difficile* (47), and the VP28 protein of the white spot syndrome virus (WSSV) (48) (Table 1). In all cases, recombinant spores were shown to induce antigen-specific and protective immune responses. While the display of *H. acinonychis* and WSSV antigens was achieved by a C-terminal fusion to the truncated form of the carrier, CotBΔ₁₀₅, display of *B. anthracis*, *C. perfringens*, and *C. difficile* antigens was obtained by fusing the heterologous molecules to the C terminus of a full-length version of CotB. The truncated form of CotB has also been used to express a peptide formed by 18 histidine (18×His) residues and recombinant spores used to adsorb nickel ions (49). Metal binding was not affected by either pH or temperature; the number of spores present was the only factor influencing nickel adsorption. In addition, approximately 40% of bound nickel could be recovered from the spore surface (49). The efficiency of metal binding, the robustness of the spore, and the possibility of recovering bound nickel suggest that recombinant spores are a potentially powerful bioremediation tool for the treatment of contaminated ecosystems (49).

CotC as a Carrier Protein

CotC is a 12-kDa protein in the *B. subtilis* coat, initially identified by reverse genetics (50) and then shown to be close to the surface based on genetic evidence (51). CotC was selected as a carrier candidate for its relative abundance in the coat. CotG, CotD, and CotC represent about 50% of the total solubilized coat proteins by mass. Such relatively high amounts could allow for the assembly of a significant number of CotC-based fusions on the coat, thus ensuring an efficient heterologous display. Expression of the *cotC* gene is under the control of the mother cell-specific σ factor, σ^K, and the transcriptional regulators GerE and SpoIIID. *cotC* expression is also positively controlled by GerR in an indirect way, through the activation of SpoVIF, which stabilizes the transcriptional activator GerE and consequently induces the expression of the GerE-dependent genes (34). The primary product of the *cotC* gene is a 66-amino-acid polypeptide extremely rich in tyrosine (30.3%) and lysine (28.8%) residues (50). However, it has been shown that CotC is assembled into at least five distinct protein forms, ranging in size from 12 to 30 kDa (36, 52). Two of these, having molecular masses of 12 and 21 kDa and corresponding most likely to a monomeric and homodimeric form of CotC, respectively, are assembled on the spore surface right after their synthesis 8 h after the onset of sporulation. The other two forms, 12.5 and 30 kDa, are probably the products of posttranslational modifications of the two early forms, occurring directly on the coat surface during spore maturation (36). The 23-kDa CotC species has been shown to be a CotC-CotU heterodimer (52), formed under the control of the spore coat protein CotE (53). As for CotB, CotC strictly requires *cotH* for its assembly, and none of the CotC forms is assembled in the coat of *cotH* spores. In a wild-type strain, CotC does not accumulate in the mother cell compartment, while in a mutant strain that overproduces CotH, at least two CotC forms, of 12 and 21 kDa, were found in the mother cell compartment. This supports the hypothesis that CotH, or a CotH-controlled protein, protects CotC from proteolysis prior to its assembly (37). In the case of CotC, only C-terminal fusions have been constructed with the two model antigens used with CotB, TTFC, and LTB (33). Both CotC-TTFC and CotC-LTB gene fusions were obtained by cloning *tetC* or *eltB* in frame with the last *cotC* codon under the transcriptional and translational control of the *cotC* promoter region. The gene fusion was then integrated into the *B. subtilis* chromosome at the *amyE* locus by double crossover recombination (33). Both of these fusion proteins were assembled on the coat of recombinant spores without major effect on the spore structure and/or function. A quantitative determination of recombinant proteins exposed on *B. subtilis* spores revealed that ca. 9.7×10^2 and 2.7×10^3 molecules of CotC-TTFC and CotC-LTB, respectively, were extracted from each spore. Although CotC appears more abundant than CotB within the coat, comparable amounts of heterologous proteins are exposed by the CotC- and CotBΔ_{105}-based systems. A possible explanation for this unexpected result comes from the finding that the C-terminal end of CotC is essential for the interaction of a CotC molecule with other CotC molecules and with other coat components (36). Support for this explanation also comes from the observation that when a heterologous peptide is fused at the N-terminal end of CotC, a normal pattern (four protein forms) of CotC assembly is observed and the efficiency of display is 5-fold increased (54). As in the case of CotB, CotC has been used as a carrier for a variety of heterologous proteins (Table 1). In addition to TTFC and LTB (33, 54), other antigens such as the domains 1b-3 and 4 of the PA of *B. anthracis* (44), the C-terminal repeats of toxins A and B of *C. difficile* (47), the VP28 protein of WSSV (48), and the UreA protein of *H. acinonychis* (46), described above as passengers fused to CotB, were also fused to the C terminus of CotC (Table 1). Human serum albumin (HSA) (55), the T-helper cell epitope Pep23 (KDSWTVNDIQKLVGK, amino acid residues 249 to 263) of the HIV1-RT (56), the tegumental protein TP22.3 of the human parasite *Clonorchis sinensis* (57), the envelope glycoprotein GP64 of the nucleopolyhedrovirus of the model lepidopteran *Bombyx mori* (BmNPV), and the alcohol dehydrogenase of *B. mori* (58, 59) have also been displayed on the spore surface by using CotC as a carrier (Table 1). With the use of UreA as a passenger, CotB and CotC were compared for their respective efficiency in anchoring the passenger to the spore surface (34). Although CotC was about 10-fold more efficient than CotB in expressing the passenger within the coat, it was not as efficient as CotB in displaying UreA on the spore surface, suggesting that CotB is an optimal carrier when external exposure is required, whereas CotC is better suited when a higher amount of passenger protein is needed (46).

CotG as a Carrier Protein

CotG is a 24-kDa protein produced in the mother cell compartment of the sporangium around hour 8 of sporulation under the control of the mother cell-specific σ factor, σ^K, and of transcriptional regulator GerE. Like *cotC*, *cotG* expression is indirectly controlled by GerR, through the activation of SpoVIF, which positively acts on GerE and on GerE-dependent genes (34). CotG is assembled on the forespore surface as two main forms of about 32 and 36 kDa. The 32-kDa form most likely represents the unmodified product of the *cotG* gene (24 kDa) whose abnormal migration may be attributed to its unusual primary structure, characterized by the presence of 9 repeats of 13 amino acids (60) or, alternatively, as 7 tandem repeats of 7 and 6 amino acids, followed by 5 repeats of 7 amino acids (61). At the DNA level, this section of the *cotG* sequence consists of 19 likely paralogous segments of two different sizes: 12 21-bp-length copies and 7 18-bp-length copies (61). It has been proposed that the modular structure of *cotG* is the outcome of several rounds of gene elongation events of an ancestral module (61). It is interesting to note that in all CotG-containing species, CotG preserves a modular structure, although the number and the length of the repeats differ (61). The 36-kDa form of CotG could be generated by extensive cross-linking of the protein as it is assembled into the spore coat. The view that CotG is able to form cross-links is suggested by analysis of the coat structure of *sodA* mutant spores (62). Spores produced by *cotG* mutants are not affected in their germination properties or resistance to lysozyme (60). As in the cases of CotB and CotC, CotG strictly requires *cotH* expression for its assembly (63). CotG controls the conversion of the CotB-46 form into the form of 66 kDa (CotB-66) found in mature spores (35). In the first attempt to use CotG as an anchoring motif, the biotin-binding protein streptavidin was fused to the C-terminal end of CotG (62) (Table 1). The coding sequence for the 24-amino-acid streptavidin secretion signal was not included in the gene fusion, and a linker (Gly-Gly-Gly-Gly-Ser) was inserted between carrier and passenger (64). The gene fusion was expressed under the control of *cotG* promoter but, in contrast to the previously discussed cases of spore display, it was present on an extrachromosomal plasmid (62). Although the genetic stability of the gene fusion was not directly assessed, recombinant spores displayed active streptavidin on their surface, as demonstrated by FACS and fluorescence microscopy experiments (62). A similar plasmid-based approach was used to express a fusion of the C-terminal end of CotG to a variant form of the GFP protein of *Aequorea victoria*, GFP$_{UV}$ (65), a cofactor containing ω-transaminase of *Vibrio fluvialis* (66) and a N-acetyl-D-neuraminic acid (Neu5Ac) aldolase (67). In these cases, the same Gly(4)-Ser linker was inserted between CotG and the heterologous part. Neu5Ac aldolase was expressed using both a low- and a high-copy-number plasmid. An about 5-fold increase of the spore-associated enzymatic activity was obtained with the high-copy-number plasmid (67). The β-galactosidase of *E. coli* (68), the phytase of *B. subtilis* (69), and the UreA of *H. acinonychis* (46) were also fused to the C terminus of CotG, but those chimeras were encoded by recombinant DNA integrated into the *B. subtilis* chromosome. In the case of UreA a dot-blot analysis determined that about 5.3×10^3 UreA molecules were present in the spore coat (46). As summarized in Table 1, CotG has been mainly used as a carrier of passengers with enzymatic activity. Some of the CotG-displayed enzymes are either active as a dimer (ω-transaminase; 66) or a tetramer (streptavidin; 62), thus suggesting that the elimination of the cell wall translocation step for surface display is a clear advantage of spore-based over cell-based systems.

OxdD as a Carrier Protein

OxdD is a 43-kDa minor component of the spore coat with oxalate decarboxylase activity, therefore capable of converting oxalate into formate and CO_2 (70). The *oxdD* gene is transcribed during sporulation from a promoter recognized by σ^K and is negatively regulated by GerE (71). Therefore, OxdD is produced in the mother cell compartment of the sporangium, and its assembly within the coat occurs in a *safA*-dependent manner (72). Genetic and cell biological analyses suggest that OxdD is at an internal location within the coat, not at the surface. Because of its internal location, OxdD has been proposed as a carrier able to provide a high degree of protection to the passenger protein by burying the passenger protein below the surface (69). OxdD has been tested as a carrier by fusing to its C terminus a phytase (Phy) and a β-glucuronidase (β-Glu) (69). Phytases are monomeric enzymes widely used in animal nutrition. A *B. subtilis* phytase and an *E. coli* β-glucuronidase, encoded by the *uidA* gene, have been used as model enzymes for spore coat expression (69). In both cases a 10-amino-acid (10×Ala) linker was inserted between carrier and passenger (69). Recombinant spores carrying either the *cotG::phy* or the *cotG::uidA* gene fusions showed phytase or β-glucuronidase activity, indicating that OxdD is a suitable carrier for coat expression (69). In the case of spores displaying OxdD-Phy, a specific activity of 2.7×10^3 U/g (dry weight) was measured, indicating that 0.4 g of spores (dry weight) would represent 1,000 U of phytase, an amount of enzyme that, if added to 1 kg of feed in a daily swine diet, would replace 1 g of inorganic phosphorus supplementation (69). Spores expressing the phytase fused to OxdD or to CotG were then compared for their respective efficiency of surface expression. The phytase fused to OxdD was less efficiently

displayed but more enzymatically active than the phytase fused to CotG, consistent with the possibility that, when fused to OxdD, the enzyme is covered by external coat layers (69). Protection of heterologous enzymes is an additional feature of the display system that could be relevant for some applications. For example, OxdD-Phy-expressing spores have been proposed as probiotics for animal use (69). Orally ingested spores of *B. subtilis* are known to perform an entire cycle of sporulation, germination, growth, and sporulation in the mammalian gastrointestinal tract, transiting unaffected through the highly acidic environment of the stomach, germinating in the upper part of the intestine, and resporulating in the lower intestinal region (73). In this context, the use of spores expressing OxdD-Phy in animal probiotic products could provide the further advantage of supply of usable phosphorus due to the continuous presence of active enzyme in the intestine (69).

SpsC as a Spore-Anchoring Signal

SpsC is a 389-amino-acid protein encoded by the third of the 11-gene *sps* operon. The operon is transcribed by a σ^K-controlled promoter and is positively regulated by the GerE transcriptional factor (74). Expression of the *sps* operon is involved in the biosynthesis of spore coat polysaccharides (74, 75). At its N terminus, at positions 6 to 10, SpsC contains the sequence Asn-His-Phe-Leu-Pro, proposed as a spore-anchoring motif (76). That sequence is preceded by two basic amino acids (Lys-Arg), a possible cleavage site for a trypsin-like protease. This observation has provided support for the hypothesis that during sporulation, proteolytic cleavage of a cytoplasmic form of SpsC is a necessary step (76). The sequence Asn-His-Phe-Leu-Pro was identified by screening phage display peptide libraries for peptides able to bind tightly to spores of *B. subtilis* (76). All the peptides isolated in this study contained the sequence Asn-His-Phe-Leu at the N terminus and exhibited clear preferences for Pro, at position 5. The peptide Asn-His-Phe-Leu-Pro was then shown to be sufficient for tight binding to spores of *B. subtilis* and of the closely related species *B. amyloliquefaciens* and *B. globigii* (76). Although this study was not specifically aimed at developing a surface display system, it highlighted that the SpsC protein or just the N-terminal peptide Asn-His-Phe-Leu-Pro can be used to display a passenger protein on the spore surface.

CotA as a Display System

CotA is a 65-kDa component of the outer layer of the *B. subtilis* coat (50, 51). CotA is responsible for the brown pigmentation that characterizes mature spores of *B. subtilis* when cultured on plates and which acts as a melanin-like pigment that protects the spore against UV radiation (77). CotA shows a high degree of sequence similarity with fungal and bacterial copper-dependent oxidases and was found to be a copper-dependent laccase that is active within the coat (77, 78). When assembled into the coat, CotA showed a half-life of inactivation at 80°C of about 4 h, indicating that it is a highly stable enzyme, naturally immobilized on the surface of the spore (78). The level of CotA at the spore surface could be increased with no detrimental effect on spore assembly or properties. Indeed, a strain bearing *cotA* on a multicopy plasmid formed colonies that were more pigmented than those of the wild type, and the spores had increased amounts of CotA and increased laccase activity (78). In a recent study (79), the laccase activity of CotA was used to perform directed evolution experiments on the spore coat. A plasmid-generated mutagenesis library of *cotA* was transformed into a *B. subtilis* mutant strain lacking the *cotA* gene, and spores of control and transformed strains were assayed for CotA activity with 2,2-azinobis (3-ethylbenzothiazoline-6-sulfonate) (ABTS) and 4-hydroxy-3,5-dimethoxybenzaldehydeazine (SGZ) as substrates. Spores of one mutant had a 120-fold increase in laccase activity (79).

DISPLAY ON SPORES OF BACILLI OTHER THAN *B. SUBTILIS*

In addition to *B. subtilis*, other *Bacillus* species have been proposed as systems for the surface display of heterologous proteins. The 130-kDa Cry1Ac protoxin of *B. thuringiensis* is a major component of the spore coat, and its N terminus is found to be spore surface exposed (80). Based on this, Cry1Ac has been suggested as a suitable carrier for surface display on spores of *B. thuringiensis* (81). The system was initially developed by using the green fluorescent protein (GFP) as a model passenger. This identified the minimal region of the protoxin required for surface display. DNA coding for the carrier region of the protoxin was carried on a plasmid and, when fused to GFP, used to transform a Cry⁻ mutant strain, thus avoiding competition with the native protoxin (81). The system was used to successfully display a single-chain antibody (scFv) recognizing the hapten 4-ethoxylmethylene-2-phenyl-2-oxazolin-5-one (phOx) (81). Fluorescence microscopy experiments showed that the anti-phOx antibody was expressed and displayed on the *B. thuringiensis* spore surface (81). InhA, an exosporium protein of *B. thuringiensis*, was shown to be able to display proteins active either as a

monomer (GFP) or as a tetramer (β-Gal) (82). This study then proposed that the exosporium, the surface structure surrounding the spore of some *Bacillus* species, would be a suitable surface for the display of heterologous proteins. Two components of the exosporium of *B. anthracis*, BclA and BclB, have also been proposed as carriers for spore surface display (83). BclA and BclB first assemble at the spore surface and then are proteolytically cleaved. Finally, the mature proteins become stably attached to the spore surface via their N termini (83). The initial N-terminal 19 residues (amino acids 2 to 24) of BclA, including the proteolytic cleavage site, targeted GFP to the spore surface. However, because released spores did not show fluorescence, the 19-amino-acid motif alone was not sufficient for attachment of the fusion protein to the mature exosporium (83).

SURFACE DISPLAY ON NONRECOMBINANT SPORES

All display systems based on the use of bacterial cells or bacteriophages, as well as the spore-based systems summarized in the previous sections, rely on the genetic engineering of the host. This is a major drawback when the application of the display system involves the release into nature of the recombinant host (field applications) and, in particular, when the display system is designed for human or animal use (delivery of antigens or enzymes to mucosal surfaces). Serious concerns over the use of live genetically modified microorganisms, their release into nature, and their clearance from the host following oral delivery have been raised (84). To overcome this obstacle, nonrecombinant approaches are highly desirable and their development is strongly encouraged by control agencies (84). In this context, various nonrecombinant approaches to display heterologous proteins on the spore surface have been recently proposed. In the first study suggesting that heterologous proteins can be adsorbed on the spore surface, the gene encoding the NADPH-cytochrome P450 reductase (CPR), a diflavin-containing enzyme, was overexpressed in *B. subtilis* cells by using the isopropyl-β-D-thiogalactopyranoside-inducible vegetative promoters P*groE* and P*tac* (85). The expressed CPR was released into the culture medium after sporulation by autolysis of the mother cell and was found associated to the spore surfaces. Purified spores showed CPR activity, and the enzyme was accessible to anti-CPR antibodies in FACS experiments (85). It is noteworthy that the enzyme used in this study contains two flavin cofactors (FMN and FAD) and, for this reason, cannot be produced or secreted by standard expression systems (85). Although this first study did not address the mechanisms involved in the spontaneous adhesion of the enzyme to the spore surface and did not investigate whether other enzymes had a similar behavior, it clearly suggested a new approach for spore display. A similar result was obtained using spores of *B. thuringiensis*. The endo-β-N-acetylglucosaminidase (Mbg) of *B. thuringiensis* is a putative peptidoglycan hydrolase containing two LysM domains at its N terminus and had been used to display GFP on the surface of *B. thuringiensis* cells (86). Fusions of Mbg to GFP and a bacterial laccase (WlacD) were found to associate with the spore surface, suggesting that Mbg is also a possible carrier for spore surface display (87).

Nonrecombinant Display of Heterologous Antigens and Enzymes

A different approach for spore adsorption was followed by Huang et al. (88). Previously purified proteins were mixed with purified spores of *B. subtilis* and adsorption conditions were developed. A collection of antigens (the TTFC of *C. tetani*, PA of *B. anthracis*, Cpa of *C. perfringens* described above, and glutathione S-transferase (Sj26GST) from *Schistosoma japonicum* [89]) were expressed in *E. coli* and adsorbed to spores. Adsorbed spores were able to induce specific and protective immune responses in mice immunized mucosally (88). Spore adsorption was more efficient when the pH of the binding buffer was acidic (pH 4) and less efficient or totally inhibited at pH values of 7 or 10 (88). A combination of electrostatic and hydrophobic interactions between spores and antigen were suggested to drive adsorption. Interestingly, adsorption was not dependent on any specific coat proteins but, rather, was due to the negatively charged and hydrophobic spore surface (88). In addition, the same study showed that killed or inactivated spores were equally effective as live spores in adsorbing the various antigens (88). A similar approach was also used to adsorb the phytase of *E. coli* to spores of a probiotic strain of *B. polyfermenticus* (90) and the β-galactosidase of *Alicyclobacillus acidocaldarius* to *B. subtilis* spores (91). In the case of the phytase, immobilization on the spore surface stabilized the enzyme by increasing its half-life at temperatures ranging between 60 and 90°C, but also caused a loss of activity of about 30% (91). The β-galactosidase of *A. acidocaldaricus* was more stable when adsorbed to spores than the free enzyme at high temperatures and at low pH values (91). This study also reported that spores of mutant strains having an altered (or totally lacking) protein in the outer layer of the coat adsorbed enzymes much more efficiently than wild-type spores (91). This observation is consistent with results obtained with *B. subtilis* and *B. anthracis* spores (92) and suggests a negative effect of the spore outermost structures on adhesion to surface or molecules.

Spore Surface Properties and Adsorption

The physicochemical properties of the spore surface have been addressed in different studies with different approaches. An early study showed that spores of *B. subtilis* are negatively charged by time-resolved micropotentiometry (93). In an aqueous environment the spore behaves like an almost infinite ionic reservoir and can accumulate billions of protons (approximately 2×10^{10} per spore) (93). The carboxyl groups were identified as the major ionizable groups in the spore and, on the basis of the diffusion time analysis, it was found that proton diffusion is much lower in the spore core than within the coat and cortex (93). This, then, suggested that the inner membrane, separating core from cortex and coat in a dormant spore, is probably a major permeability barrier for protons (93). The role of electrostatic forces in spore adhesion to a planar surface has been also addressed by studying spores of *B. thuringiensis* (94). The surface potentials of spore and mica surfaces were experimentally obtained using a combined atomic force microscopy (AFM)-scanning surface potential microscopy technique (94). By these techniques, the surface charge density of the spores was estimated at 0.03 μC/cm² at 20% relative humidity and decreased with increasing humidity. This work showed that the electrostatic force can be an important component in the adhesion between the spore and a planar surface (94). However, the interaction between spores and a protein in an aqueous environment may involve additional forces, as suggested by the observation that the enzyme β-galactosidase of *A. acidocaldaricus* binds more efficiently to modestly negatively charged mutant spores than to highly negatively charged wild-type spores (91). A recent study showed that the electric charge of wild-type and mutant spores can be measured at the single-spore level by using optical tweezers, allowing for a detailed analysis of the physical properties of the spore surface (95). Adhesion of *B. subtilis* and *B. anthracis* spores to abiotic surfaces has also been shown to be largely due to negative charges on the surfaces of spores of both species (which is notable given that, unlike *B. subtilis* spores, *B. anthracis* spores have an exosporium surrounding the coat). In both species, the spore surface thermodynamic properties changed in similar ways when the spore surface was altered due to mutations, and in general, adhesion increased in the mutants (96).

ADVANTAGES OF THE SPORE SURFACE DISPLAY SYSTEMS

Spore-based surface display systems provide several advantages with respect to other display approaches. A first advantage comes from the well-documented robustness of the *Bacillus* spore, which grants high stability to the display system even after a prolonged storage. This aspect has been tested with spores displaying CotBΔ$_{105}$-TTFC. Aliquots of purified recombinant spores were stored at −80°C, −20°C, +4°C, and at room temperatures and assayed for the amount of heterologous protein present on the spore after different storage times. A dot-blot analysis with anti-TTFC antibody showed identical amounts of displayed TTFC at all time points (27). The stability of the display system is an extremely useful property for a variety of biotechnological applications. Heat stability, for example, is a stringent requirement in developing new mucosal vaccine delivery systems, especially in developing countries, where poor distribution and storage conditions are major limitations. High stability is also extremely useful in industrial and environmental applications, where (nonspore) cell-based systems are limited because most cells have poor long-term survival, especially under extreme conditions (97). The safety record of several *Bacillus* species (98) is another important advantage of spores over other systems. Several *Bacillus* species, including *B. subtilis*, are widely used as probiotics and have been on the market for human or animal use for decades in many countries (98). Although most studies in humans or animals have been performed with laboratory strains of *B. subtilis*, in some cases, intestinal isolates and strains with probiotic properties have also been used to display heterologous proteins (99). The safety of the live host is obviously an essential requirement if the display system is intended for human or animal use, such as delivery of vaccines or therapeutic molecules to mucosal surfaces. A limitation of cell- and phage-based display systems is the size of the heterologous protein to be exposed, since it may affect the structure of a cell membrane-anchoring protein or of a viral capsid. In addition, cell-based systems require a membrane translocation step in order to externally expose a protein produced in their cytoplasm (97). With the spore-based system, heterologous proteins are produced in the mother cell and their assembly on the spore surface does not involve membrane translocation (Fig. 4). Therefore, this system has fewer size limitations and, as described above, large proteins (for example, TTFC of 51.8 kDa) have been successfully displayed on the spore surface.

Spore surface display by the nonrecombinant approach has an additional advantage. A recent publication showed that the B subunit of the heat-labile toxin of *E. coli* (LTB) is exposed on the spore surface in its native pentameric form (100). This is an important observation because in other display systems LTB could only be displayed as a monomer. Because only the pentameric form of LTB binds its receptor, the ganglioside GM1, spore surface display allows enhanced activation of the immune system using this molecule (100).

FUTURE DEVELOPMENTS

A better understanding of coat protein assembly mechanisms should permit the design of improved strategies for the display of fusion proteins at the spore surface. An example is CotB: although it has been used to display several passenger proteins, the presence of a wild-type copy of the *cotB* gene is needed for display to be efficient (32). Assembly of CotB involves a complex pattern of protein-protein interactions (35), leading to the hypothesis that a wild-type copy of CotB is able to recruit the CotB-based chimera to the coat. The identification of the CotB regions that undergo covalent modifications during assembly would be an important step in designing novel fusions that, as required for each application, would or would not be cross-linked to the spore. Some coat components have an enzymatic activity. In addition to CotA and OxdD discussed above, Tgl, a 24-kDa coat component, has been shown to have a transglutaminase activity (101, 102). In a mutant strain lacking *tgl*, some coat components are more easily extractable, suggesting that Tgl is involved in the cross-linking (and therefore in the insolubilization) of some coat proteins (101–103). Only some Tgl substrates have been identified (101, 102), and, also in those cases, the amino acid residues involved in the cross-linking reaction are not known. A better understanding of this process may be important for surface display. A passenger containing the appropriate signal could be covalently bound to a coat component (and therefore to the coat) in a Tgl-dependent way, allowing its controlled display. It has been recently discovered that the *B. subtilis* spore has an additional structure surrounding the outer coat layer. This outermost layer has been named the crust, and has been shown to be formed by proteins, probably in a glycosylated state (104, 105). The *cgeA* gene and genes of the

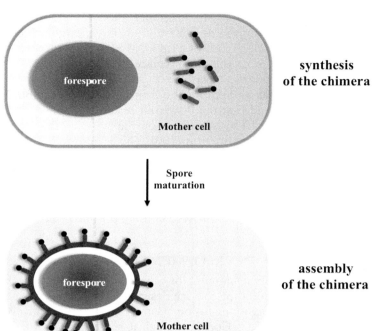

synthesis
of the chimera

Figure 4 Gene fusions are synthesized into the mother cell cytoplasm due to mother cell-specific transcription signals. Chimera are then assembled around the forming spore during the spore maturation process and released by autolysis of the mother cell. Dark gray cylinders represent the coat components used as carriers, and black circles indicate the heterologous passengers.

assembly
of the chimera

cotVWXWZ cluster have been shown to be involved in crust formation, and CotZ and CgeA have been identified as essential crust components (105). Identification of the crust has been extremely relevant for the development of future spore display systems. Crust components, being on the surface of the spore, are obvious carrier candidates. Recently, CotZ has been fused to UreA of *H. acinonychis* (106) and to FliD of *C. difficile* (107), resulting in fusions that were efficiently exposed on the spore surface. Vectors to target heterologous proteins to the spore surface by using CgeA as a carrier have also been developed and shown to successfully display a *Helicobacter pylori* antigen (108).

Future developments of spore display systems will include the approach involving protein adsorption on nonrecombinant spores. Any future development of this technique must necessarily be based on a better understanding of the mechanisms involved in spore adsorption. A combination of electrostatic and hydrophobic interactions between spores and passengers has been suggested as being responsible for adhesion (88, 90). However, at least in the case of β-galactosidase adhesion to *B. subtilis* spores, the electrostatic force does not seem to play a predominant role (91). Whether other factors (e.g., van der Waals and capillary forces) are also involved, and how the involved forces are affected by external factors such as humidity or by properties of the passenger protein, are all relevant questions that still need to be addressed.

A future and exciting extension of recombinant and nonrecombinant spore display systems comes from the recent observation that spores of *B. subtilis* can be used as building blocks for stimuli-responsive new biomaterials (109). Spores displaying heterologous molecules can then be used to self-assemble submicrometer-thick monolayers and build functionalized surfaces able to catalyze a chemical reaction or sense a specific molecule.

Note in Proof. Since this chapter was published in Microbiology Spectrum [2(5):TBS-0011-2012, doi:10.1128/microbiolspec. TBS-0011-2012, 2014], several additional papers on spore display have appeared; see references 110–115.

Acknowledgments. *Work in the authors' laboratory is supported by the European Union (VII Framework contract number 613703 and 614088) to E.R.*

Citation. Isticato R, Ricca E. 2014. Spore surface display. Microbiol Spectrum 2(5):TBS-0011-2012.

References

1. **Lee SY, Choi JH, Xu Z.** 2003. Microbial cell-surface display. *Trends Biotechnol* **21:**45–52.

2. **Kronqvist N, Malm M, Rockberg J, Hjelm B, Uhlén M, Ståhl S, Löfblom J.** 2010. Staphylococcal surface display in combinatorial protein engineering and epitope mapping of antibodies. *Recent Pat Biotechnol* **4:**171–182.

3. **van Bloois E, Winter RT, Kolmar H, Fraaije MW.** 2011. Decorating microbes: surface display of proteins on *Escherichia coli*. *Trends Biotechnol* **29:**79–86.

4. **Wu CH, Mulchandani A, Chen W.** 2008. Versatile microbial surface-display for environmental remediation and biofuels production. *Trends Microbiol* **16**: 181–188.

5. **Sousa C, Kotrba P, Ruml T, Cebolla A, De Lorenzo V.** 1998. Metalloadsorption by *Escherichia coli* cells displaying yeast and mammalian metallothioneins anchored to the outer membrane protein LamB. *J Bacteriol* **180**:2280–2284.

6. **Sousa C, Cebolla A, de Lorenzo V.** 1996. Enhanced metalloadsorption of bacterial cells displaying poly-His peptides. *Nat. Biotechnol.* **14**:1017–1020.

7. **Valls M, de Lorenzo V, Gonzalez-Duarte R, Atrian S.** 2000. Engineering outer-membrane proteins in *Pseudomonas putida* for enhanced heavy metal bioadsorption. *J Inorg Biochem* **79**:219–223.

8. **Valls M, Atrian S, de Lorenzo V, Fernandez LA.** 2000. Engineering a mouse metallothionein on the cell surface of *Ralstonia eutropha* CH34 for immobilization of heavy metals in soil. *Nat Biotechol* **18**:661–665.

9. **Bae W, Chen W, Mulchandani A, Mehra R.** 2000. Enhanced bioaccumulation of heavy metals by bacterial cells displaying synthetic phytochelatins. *Biotechnol Bioeng* **70**:518–524.

10. **Bae W, Mehra R, Mulchandani A, Chen W.** 2001. Genetic engineering of *Escherichia coli* for enhanced uptake and bioaccumulation of mercury. *Appl Environ Microbiol* **67**:5335–5338.

11. **Bae W, Mulchandani A, Chen W.** 2002. Cell surface display of synthetic phytochelatins using ice nucleation protein for enhanced heavy metal bioaccumulation. *J Inorg Biochem* **88**:223–227.

12. **Richins R, Kaneva I, Mulchandani A, Chen W.** 1997. Biodegradation of organophosphorus pesticides by surface-expressed organophosphorus hydrolase. *Nature Biotechnol* **15**:984–987.

13. **Gunneriusson E, Samuelson P, Uhlen M, Nygren PA, Stahl S.** 1996. Surface display of a functional single-chain Fv antibody on staphylococci. *J Bacteriol* **178**: 1341–1346.

14. **Giomarelli B, Maggi T, Younson J, Kelly C, Pozzi G.** 2004. Expression of a functional single-chain Fv antibody on the surface of *Streptococcus gordonii*. *Mol Biotechnol* **28**:105–112.

15. **Benhar I.** 2001. Biotechnological applications of phage and cell display. *Biotechnol Adv* **19**:1–33.

16. **di Marzo Veronese F, Willis AE, Boyer-Thompson C, Appella E, Perham RN.** 1994. Structural mimicry and enhanced immunogenicity of peptide epitopes displayed on filamentous bacteriophage. The V3 loop of HIV-1 gp120. *J. Mol Biol* **243**:167–172.

17. **Jiang J, Abu-Shilbayeh L, Rao VB.** 1997. Display of a PorA peptide from *Neisseria meningitidis* on the bacteriophage T4 capsid surface. *Infect Immun* **65**:4770–4777.

18. **Felici F, Castagnoli L, Musacchio A, Jappelli R, Cesareni G.** 1991. Selection of antibody ligands from a large library of oligopeptides expressed on a multivalent exposition vector. *J Mol Biol* **222**:301–310.

19. **Felici F, Luzzago A, Monaci P, Nicosia A, Sollazzo M, Traboni C.** 1995. Peptide and protein display on the surface of filamentous bacteriophage. *Biotechnol Annu Rev* **1**:149–183.

20. **Folgori A, Tafi R, Meola A, Felici F, Galfre G, Cortese R, Monaci P, Nicosia A.** 1994. A general strategy to identify mimotopes of pathological antigens using only random peptide libraries and human sera. *EMBO J* **13**: 2236–2243.

21. **Nguyen TN, Hansson M, Stahl S, Bachi T, Robert A, Domzig W, Binz H, Uhlen M.** 1993. Cell-surface display of heterologous epitopes on *Staphylococcus xylosus* as a potential delivery system for oral vaccination. *Gene* **128**:89–94.

22. **Nguyen TN, Gourdon MH, Hansson M, Robert A, Samuelson P, Libon C, Andreoni C, Nygren PA, Binz H, Uhlen M, Stahl S.** 1995. Hydrophobicity engineering to facilitate surface display of heterologous gene products on *Staphylococcus xylosus*. *J Biotechnol* **42**:207–219.

23. **Wu JY, Newton S, Judd A, Stocker B, Robinson WS.** 1989. Expression of immunogenic epitopes of hepatitis B surface antigen with hybrid flagellin proteins by a vaccine strain of *Salmonella*. *Proc Natl Acad Sci USA* **86**:4726–4730.

24. **Newton SM, Jacob CO, Stocker BA.** 1989. Immune response to cholera toxin epitope inserted in *Salmonella* flagellin. *Science* **244**:70–72.

25. **Schorr J, Knapp B, Hundt E, Kupper HA, Amann E.** 1991. Surface expression of malarial antigens in *Salmonella typhimurium*: induction of serum antibody response upon oral vaccination of mice. *Vaccine* **9**:675–681.

26. **Fischetti VA, Medaglini D, Pozzi G.** 1996. Gram-positive commensal bacteria for mucosal vaccine delivery. *Curr Opin Biotechnol* **7**:659–666.

27. **Isticato R, Cangiano G, De Felice M, Ricca E.** 2004. Display of molecules on the spore surface. p 193–200. *In* E. Ricca, Henriques AO, Cutting SM (ed), *Bacterial Spore Formers: Probiotics and Emerging Applications*. Horizon Biosciences, Norfolk, UK.

28. **Jose J, Meyer TF.** 2007. The autodisplay story, from discovery to biotechnical and biomedical applications. *Microbiol Mol Biol Rev* **71**:600–619.

29. **Maurer J, Jose J, Meyer TF.** 1997. Autodisplay: one-component system for efficient surface display and release of soluble recombinant proteins in *Escherichia coli*. *J Bacteriol* **179**:794–804.

30. **Henriques AO, Moran CP Jr.** 2007. Structure, assembly and function of the spore surface layers. *Annu Rev Microbiol* **61**:555–588.

31. **Klobutcher LA, Ragkousi K, Setlow P.** 2006. The *Bacillus subtilis* spore coat provides "eat resistance" during phagocytic predation by the protozoan *Tetrahymena thermophila*. *Proc Natl Acad. Sci USA* **103**:165–170.

32. **Isticato R, Cangiano G, Tran H-T, Ciabattini A, Medaglini D, Oggioni MR, De Felice M, Pozzi G, Ricca E.** 2001. Surface display of recombinant proteins on *Bacillus subtilis* spores. *J Bacteriol* **183**:6294–6301.

33. **Mauriello EMF, Duc LH, Isticato R, Cangiano G, Hong HA, De Felice M, Ricca E, Cutting SM.** 2004. Display

of heterologous antigens on the *Bacillus subtilis* spore coat using CotC as a fusion partner. *Vaccine* 22:1177–1187.

34. Cangiano G, Mazzone A, Baccigalupi L, Isticato R, Eichenberger P, De Felice M, Ricca E. 2010. Direct and indirect control of late sporulation genes by GerR of *Bacillus subtilis*. *J Bacteriol* 192:3406–3413.

35. Zilhao R, Isticato R, Ozin A, Serrano M, Moran CP, Ricca E, Henriques OA. 2004. Interactions among CotB, CotG, and CotH during assembly of the *Bacillus subtilis* spore coat. *J Bacteriol* 186:1110–1119.

36. Isticato R, Esposito G, Zilhao R, Nolasco S, Cangiano G, De Felice M, Henriques OA, Ricca E. 2004. Assembly of multiple CotC forms into the *Bacillus subtilis* spore coat. *J Bacteriol* 186:1129–1135.

37. Baccigalupi L, Castaldo G, Cangiano G, Isticato R, Marasco R, De Felice M, Ricca E. 2004. GerE-independent expression of *cotH* leads to CotC accumulation in the mother cell compartment during *Bacillus subtilis* sporulation. *Microbiol UK* 150:3441–3449.

38. Helting TB, Zwisler O. 1977. Structure of tetanus toxin. I. Breakdown of the toxin molecule and discrimination between polypeptide fragments. *J Biol Chem* 252:194–198.

39. Douce G, Turcotte C, Cropley I, Roberts M, Pizza M, Domenghini M, Rappuoli R, Dougan G. 1995. Mutants of *Escherichia coli* heat-labile toxin lacking ADP-ribosyltransferase activity act as nontoxic, mucosal adjuvants. *Proc Natl Acad Sci USA* 92:1644–1648.

40. Yamamoto H, Kurosawa S-I, Sekigushi J. 2003. Localization of the vegetative cell wall hydrolases LytC, LytE, and LytF on the *Bacillus subtilis* cell wall surface and stability of these enzymes to cell wall-bound or extracellular proteases. *J Bacteriol* 22:6666–6677.

41. Duc LH, Hong HA, Fairweather N, Ricca E, Cutting SM. 2003. Bacterial spores as vaccine vehicles. *Infect Immun* 71:2810–2818.

42. Ciabattini A, Parigi R, Isticato R, Oggioni MR, Pozzi G. 2004. Oral priming of mice by recombinant spores of *Bacillus subtilis*. *Vaccine* 22:4139–4143.

43. Mauriello EMF, Cangiano G, Maurano F, Saggese V, De Felice M, Rossi M, Ricca E. 2007. Germination-independent induction of cellular immune response by *Bacillus subtilis* spores displaying the C fragment of the tetanus toxin. *Vaccine* 25:788–793.

44. Duc LH, Hong HA, Atkins HS, Flick-Smith HC, Durrani Z, Rijpkema S, Titball RW, Cutting SM. 2007. Immunization against anthrax using *Bacillus subtilis* spores expressing the anthrax protective antigen. *Vaccine* 25:346–355.

45. Hoang TH, Hong HA, Clark GC, Titball RW, Cutting SM. 2008. Recombinant *Bacillus subtilis* expressing the *Clostridium perfringens* alpha toxoid is a candidate orally delivered vaccine against necrotic enteritis. *Infect Immun* 76:5257–5265.

46. Hinc K, Isticato R, Dembek M, Karczewska J, Iwanicki A, Peszyńska-Sularz G, De Felice M, Obukowski M, Ricca E. 2010. Expression and display of UreA of *Helicobacter acinonychis* on the surface of *Bacillus subtilis* spores. *Microb Cell Fact* 9:2. doi:10.1186/1475-2859-9-2.

47. Permpoonpattana P, Hong HA, Phetcharaburanin J, Huang JM, Cook J, Fairweather N, Cutting SM. 2011. Immunization with *Bacillus* spores expressing toxin A peptide repeats protects against infection with *Clostridium difficile*. *Infect Immun* 79:2295–2302.

48. Ning D, Leng X, Li Q, Xu W. 2011. Surface-displayed VP28 on *Bacillus subtilis* spores induces protection against infection by white spot syndrome virus in crayfish by oral administration. *J Appl Microbiol* 111:1327–1336.

49. Hinc K, Ghandili S, Karbalaee G, Shali A, Noghabi K, Ricca E, Ahmadian G. 2010. Efficient binding of nickel ions to recombinant *Bacillus subtilis* spores. *Res Microbiol* 161:757–764.

50. Donovan W, Zheng L, Sandman K, Losick R. 1987. Genes encoding spore coat polypeptides from *Bacillus subtilis*. *J Mol Biol* 196:1–10.

51. Zheng L, Donovan WP, Fitz-James PC, Losick R. 1988. Gene encoding a morphogenic protein required in the assembly of the outer coat of the *Bacillus subtilis* endospore. *Genes Dev* 2:1047–1054.

52. Isticato R, Pelosi A, Zilhao R, Baccigalupi L, Henriques AO, De Felice M, Ricca E. 2008. CotC-CotU heterodimerization during assembly of the *Bacillus subtilis* spore coat. *J Bacteriol* 190:1267–1275.

53. Isticato R, Pelosi A, De Felice M, Ricca E. 2010. CotE binds to CotC and CotU and mediates their interaction during spore coat formation in *Bacillus subtilis*. *J Bacteriol* 192:949–954.

54. Isticato R, Scotto Di Mase D, Mauriello EMF, De Felice M, Ricca E. 2007. Amino terminal fusion of heterologous proteins to CotC increases display efficiencies in the *Bacillus subtilis* spore system. *BioTechniques* 42:151–156.

55. Mao L, Jiang S, Li G, He Y, Chen L, Yao Q, Chen K. 2012. Surface display of human serum albumin on *Bacillus subtilis* spores for oral administration. *Curr Microbiol* 64:545–551.

56. D'Apice L, Sartorius R, Caivano A, Mascolo D, Del Pozzo G, Scotto Di Mase D, Ricca E, Li Pira G, Manca F, Malanga D, De Palma R, De Berardinis P. 2007. Comparative analysis of new innovative vaccine formulations based on the use of procaryotic display systems. *Vaccine* 25:1993–2000.

57. Zhou Z, Xia H, Hu X, Huang Y, Li Y, Li L, Ma C, Chen X, Hu F, Xu J, Lu F, Wu Z, Yu X. 2008. Oral administration of a *Bacillus subtilis* spore-based vaccine expressing *Clonorchis sinensis* tegumental protein 22.3 kDa confers protection against *Clonorchis sinensis*. *Vaccine* 26:1817–1825.

58. Li G, Tang Q, Chen H, Yao Q, Ning D, Chen K. 2011. Display of *Bombyx mori* nucleopolyhedrovirus GP64 on the *Bacillus subtilis* spore coat. *Curr Microbiol* 62:1368–1373.

59. Wang N, Chang C, Yao Q, Li G, Qin L, Chen L, Chen K. 2011. Display of *Bombyx mori* alcohol dehydrogenases on the *Bacillus subtilis* spore surface to enhance enzymatic activity under adverse conditions. *PLoS ONE* 6:e21454. doi:10.1371/journal.pone.0021454.

60. Sacco M, Ricca E, Losick R, Cutting S. 1995. An additional GerE-controlled gene encoding an abundant spore coat protein from *Bacillus subtilis*. *J Bacteriol* **177**:372–377.

61. Giglio R, Fani R, Isticato R, De Felice M, Ricca E, Baccigalupi L. 2011. Organization and evolution of the *cotG* and *cotH* genes of *Bacillus subtilis*. *J Bacteriol* **193**:6664–6673.

62. Henriques AO, Melsen LR, Moran CP. 1998. Involvement of superoxide dismutase in spore coat assembly in *Bacillus subtilis*. *J Bacteriol* **180**:2285–2291.

63. Naclerio G, Baccigalupi L, Zilhao R, De Felice M, Ricca E. 1996. *Bacillus subtilis* spore coat assembly requires *cotH* gene expression. *J Bacteriol* **178**:4375–4380.

64. Kim J-H, Lee C-S, Kim B-G. 2005. Spore-displayed streptavidin: a live diagnostic tool in biotechnology. *Biochem Biophys Res Commun* **331**:210–214.

65. Kim J-H, Roh C, Lee C-W, Kyung D, Choi S-K, Jung H-C, Pan J-G, Kim B-G. 2007. Bacterial surface display of GFP$_{UV}$ on *Bacillus subtilis* spores. *J Microb Biotechnol* **17**:677–680.

66. Hwang B-Y, Kim B-G, Kim J-H. 2011. Bacterial surface display of a co-factor containing enzyme ω-transaminase from *Vibrio fluvialis* using *Bacillus subtilis* spore display system. *Biosci Biotechnol Biochem* **75**:1862–1865.

67. Xu X, Gao C, Zhang X, Che B, Ma C, Qiu J, Tao F, Xu P. 2011. Production of *N*-acetyl-D-neuraminic acid by use of an efficient spore surface display system. *Appl Environ Microbiol* **77**:3197–3201.

68. Kwon SJ, Jung H-C, Pan J-G. 2007. Transgalactosylation in a water-solvent biphasic reaction system with β-galactosidase displayed on the surfaces of *Bacillus subtilis* spores. *Appl Environ Microbiol* **73**:2251–2256.

69. Potot S, Serra CR, Henriques AO, Schyns G. 2010. Display of recombinant proteins on *Bacillus subtilis* spores using a coat-associated enzyme as the carrier. *Appl Environ Microbiol* **76**:5926–5933.

70. Tanner A, Bowater L, Fairhurst A, Bornemann S. 2001. Oxalate decarboxylase requires manganese and dioxygen for activity: overexpression and characterization of *Bacillus subtilis* YvrK and YoaN. *J Biol Chem* **276**:43627–43634.

71. Costa T, Steil L, Martins LO, Volker U, Henriques AO. 2004. Assembly of an oxalate decaroxylase produced under the sigma k control into the *Bacillus subtilis* spore coat. *J Bacteriol* **186**:1462–1474.

72. Ozin AJ, Samford CS, Henriques AO, Moran CP. 2001. SpoVID guides SafA to the spore coat in *Bacillus subtilis*. *J Bacteriol* **183**:3041–3049.

73. Tam NK, Uyen NQ, Hong HA, Duc LH, Hoa TT, Serra CR, Henriques AO, Cutting SM. 2006. The intestinal life cycle of *Bacillus subtilis* and close relatives. *J Bacteriol* **188**:2692–2700.

74. Eichenberger P, Fujita M, Jensen ST, Conlon EM, Rudner DZ, Wang ST, Ferguson C, Haga K, Sato T, Liu JS, Losick R. 2004. The program of gene transcription for a single differentiating cell type during sporulation in *Bacillus subtilis*. *PLoS Biol* **2**:e328. doi:10.1371/journal.pbio.0020328.

75. Plata G, Fuhrer T, Hsiao T-L, Sauer U, Viktup D. 2012. Global probabilistic annotation of metabolic networks enables enzyme discovery. *Nat Chem Biol* **8**:848–854.

76. Knurr J, Benedek O, Heslop J, Vinson RB, Boydston JA, McAndrew J, Kearney JF, Turnbough CL. 2003. Peptide ligands that bind selectively to spores of *Bacillus subtilis* and closely related species. *Appl Environ Microbiol* **69**:6841–6847.

77. Hullo MF, Moszer I, Danchin A, Martin-Verstraete I. 2001. CotA of *Bacillus subtilis* is a copper-dependent laccase. *J Bacteriol* **183**:5426–5430.

78. Martins LO, Soares CM, Pereira MM, Teixeira M, Costa T, Jones GH, Henriques AO. 2002. Molecular and biochemical characterization of a highly stable bacterial laccase that occurs as a structural component of the *Bacillus subtilis* endospore coat. *J Biol Chem* **277**:1849–1859.

79. Gupta N, Farinas ET. 2010. Directed evolution of CotA laccase for increased substrate specificity using *Bacillus subtilis* spores. *Protein Eng Des Sel* **23**:679–682.

80. Du C, Nickerson KW. 1996. *Bacillus thuringiensis* HD-73 spores have surface-localized Cry1Ac toxin: physiological and pathogenic consequences. *Appl Environ Microbiol* **62**:3722–3726.

81. Du C, Chan WC, McKeithan W, Nickerson KW. 2005. Surface display of recombinant proteins on *Bacillus thuringiensis* spores. *Appl Environ Microbiol* **71**:3337–3341.

82. Park TJ, Choi S-K, Jung H-C, Lee SY, Pan J-G. 2009. Spore display using *Bacillus thuringiensis* exosporium protein InhA. *J Microbiol Biotechnol* **19**:495–501.

83. Thompson BM, Stewart GC. 2008. Targeting of the BclA and BclB proteins to the *Bacillus anthracis* spore surface. *Mol Microbiol* **70**:421–434.

84. Detmer A, Glenting J. 2006. Live bacterial vaccines—a review and identification of potential hazards. *Microb Cell Fact* **5**:23. doi:10.1186/1475-2859-5-23.

85. Yim S-K, Jung H-C, Yun C-H, Pan J-G. 2009. Functional expression in *Bacillus subtilis* of mammalian NADPH-cytochrome P450 oxidoreductase and its spore-display. *Protein Expr Purif* **63**:5–11.

86. Shao X, Jiang M, Yu Z, Cai H, Li L. 2009. Surface display of heterologous proteins in *Bacillus thuringiensis* using a peptidoglycan hydrolase anchor *Microb Cell Fact* **8**:48. doi:10.1186/1475-2859-8-48.

87. Jiang M, Shao X, Ni H, Yu Z, Li L. 2011. In vivo and in vitro surface display of heterologous proteins on *Bacillus thuringiensis* vegetative cells and spores. *Process Biochem.* **46**:1861–1866.

88. Huang JM, Hong HA, Van Tong H, Hoang TH, Brisson A, Cutting SM. 2010. Mucosal delivery of antigens using adsorption to bacterial spores. *Vaccine* **28**:1021–1030.

89. Walker J, Crowley P, Moreman AD, Barrett J. 1993. Biochemical properties of cloned glutathione S-transferases from *Schistosoma mansoni* and *Schistosoma japonicum*. *Mol Biochem Parasitol* **61**:255–264.

90. Cho EA, Kim EJ, Pan JG. 2011. Adsorption immobilization of *Escherichia coli* phytase on probiotic *Bacillus polyfermenticus* spores. *Enzyme Microb Technol* **49**: 66–71.

91. Sirec T, Strazzulli A, Isticato R, De Felice M, Moracci M, Ricca E. 2012. Adsorption of β-galactosidase of *Alicyclobacillus acidocaldaricus* on wild type and mutant spores of *Bacillus subtilis*. *Microb Cell Fact* **11**:100. doi:10.1186/1475-2859-11-100.

92. Chen G, Driks A, Tawfiq K, Mallozzi M, Patil S. 2010. *Bacillus anthracis* and *Bacillus subtilis* spore surface properties and transport. *Colloids Surf B Biointerfaces* **76**:512–518.

93. Kazakov S, Bonvouloir E, Gazaryan I. 2008. Physicochemical characterization of natural ionic microreservoirs: *Bacillus subtilis* dormant spores. *J Phys Chem* **112**:2233–2244.

94. Chung EA, Iacoumi S, Lee I, Tsouris C. 2010. The role of the electrostatic force in spore adhesion. *Environ Sci Technol* **44**:6209–6214.

95. Pesce G, Rusciano G, Sasso A, Isticato R, Sirec T, Ricca E. 2014. Surface charge and hydrodynamic coefficient measurements of *Bacillus subtilis* spore by optical tweezers. *Colloids Surf B Biointerfaces* **116**:568–575.

96. Chen G, Driks A, Tawfiq K, Mallozzi M, Patil S. 2010. *Bacillus anthracis* and *Bacillus subtilis* spores surface properties and transport. *Colloids Surf B Biointerfaces* **76**:512–518.

97. Knecht LD, Pasini P, Daunert S. 2011. Bacterial spores as platforms for bioanalytical and biomedical applications. *Anal Bioanal Chem* **400**:977–989.

98. Cutting SM. 2011. *Bacillus* probiotics. *Food Microbiol* **28**:214–220.

99. Cutting SM, Hong HA, Baccigalupi L, Ricca E. 2009. Oral vaccine delivery by recombinant spore probiotics. *Intern Rev Immunol* **28**:487–505.

100. Isticato R, Sirec T, Treppiccione L, Maurano F, De Felice M, Rossi M, Ricca E. 2013. Non-recombinant display of the B subunit of the heat labile toxin of *Escherichia coli* on wild type and mutant spores of *Bacillus subtilis*. *Microb Cell Fact* **12**:98. doi:10.1186/1475-2859-12-98.

101. Ragkousi K, Setlow P. 2004. Transglutaminase-mediated cross-linking of GerQ in the coats of *Bacillus subtilis* spores. *J Bacteriol* **186**:5567–5575.

102. Zilhão R, Isticato R, Martins LO, Steil L, Völker U, Ricca E, Moran CP, Henriques AO. 2005. Assembly and function of a transglutaminase at the surface of the developing spore in *Bacillus subtilis*. *J Bacteriol* **187**: 7753–7764.

103. Kobayashi K, Kumazawa Y, Miwa K, Yamanaka S. 1996. ε-(γ-glutamyl)Lysine crosslinks of spore coat proteins and transglutaminase activity in *Bacillus subtilis*. *FEMS Microbiol Lett* **144**:157–160.

104. McKenney PT, Driks A, Eskandarian HA, Grabowski P, Guberman J, Wang KH, Gitai Z, Eichenberger P. 2010. A distance-weighted interaction map reveals a previously uncharacterized layer of the *Bacillus subtilis* spore coat. *Curr Biol* **20**:934–938.

105. Imamura D, Kuwana R, Takamatsu H, Watabe K. 2011. Proteins involved in formation of the outermost layer of *Bacillus subtilis* spores. *J Bacteriol* **193**:4075–4080.

106. Hinc K, Iwanicki A, Obuchowski M. 2013. New stable anchor protein and peptide linker suitable for successful spore surface display in *B. subtilis*. *Microb Cell Fact* **12**: 22. doi:10.1186/1475-2859-12-22.

107. Negri A, Potocki W, Iwanicki A, Obuchowski M, Hinc K. 2013. Expression and display of *Clostridium difficile* protein FliD on the surface of *Bacillus subtilis* spores. *J Med Microbiol* **62**:1379–1385.

108. Iwanicki A, Piatek I, Stasilojć M, Grela A, Lega T, Obuchowski M, Hinc K. 2014. A system of vectors for *Bacillus subtilis* spore surface display. *Microb Cell Fact* **13**:30. doi:10.1186/1475-2859-13-30.

109. Chen X, Mahadevan L, Driks A, Sahin O. 2014. *Bacillus* spores as building blocks for stimuli-responsive materials and nanogenerators. *Nat Nanotechnol* **9**:137–141.

110. Hwang BY, Pan JG, Kim BG, Kim JH. 2013. Functional display of active tetrameric beta-galactosidase using *Bacillus subtilis* spore display system. *J Nanosci Nanotechnol* **13**:2313–2319.

111. Ricca E, Baccigalupi L, Cangiano G, De Felice M, Isticato R. 2014. Mucosal vaccine delivery by non-recombinant spores of *Bacillus subtilis*. *Microb Cell Fact* **13**:115.

112. Sirec T, Cangiano G, Baccigalupi L, Ricca E, Isticato R. 2014. The spore surface of intestinal isolates of *Bacillus subtilis*. *FEMS Microbiol Lett* **358**:194–201.

113. Bonavita R, Isticato R, Maurano F, Ricca E, Rossi M. 2015. Mucosal immunity induced by gliadin-presenting spores of *Bacillus subtilis* in HLA-DQ8-transgenic mice. *Immunol Lett* **165**:84–89.

114. Wu IL, Narayan K, Castaing JP, Tian F, Subramaniam S, Ramamurthi KS. 2015. A versatile nano display platform from bacterial spore coat proteins. *Nat Commun* **6**:6777.

115. Stasiłojć M, Hinc K, Peszyńska-Sularz G, Obuchowski M, Iwanicki A. 2015. Recombinant *Bacillus subtilis* spores elicit Th1/Th17-polarized immune response in a murine model of *Helicobacter pylori* vaccination. *Mol Biotechnol* **57**:685–691.

The Bacterial Spore: From Molecules to Systems
Edited by P. Eichenberger and A. Driks
© 2016 American Society for Microbiology, Washington, DC
doi:10.1128/microbiolspec.TBS-0018-2013

Cristina N. Butterfield
Sung-Woo Lee
Bradley M. Tebo

The Role of Bacterial Spores in Metal Cycling and Their Potential Application in Metal Contaminant Bioremediation

18

BIOGEOCHEMISTRY AND BACTERIAL METAL TRANSFORMATIONS

Life and elemental cycles are intertwined through biogeochemistry. Organisms not only order atoms into dynamic molecules, they also help control the composition of their natural environments along with chemical, physical, and geological processes. Elements such as C, H, O, N, P, and S make up the backbone of life on earth. These, combined with a suite of trace nutrients including metals such as Fe, Cu, and Mn, compose all the structural, mechanical, and messaging components of the cell. They are fixed from the environment and cycled through metabolic transformations. Eukaryotic and prokaryotic microorganisms are abundant and perform many geochemical cycling processes including biotransformation, mineral dissolution, and biomineralization. This review focuses on the contribution of bacteria and, more specifically, bacterial spores to metal speciation in the environment. Many of these metal transformations are required for cellular metabolism and are facilitated by metals via electron transfer in metal-protein centers.

Metal-binding proteins make up 40% of all proteins (1); 75% of these proteins utilize metals to catalyze diverse essential processes. Nutritionally essential metals (Na, Mg, Ca, V, Cr, Mo, Cu, Zn, Ni, Co, Fe, and Mn) have the propensity to bind to certain protein ligands in a rather predictable manner. The chemical theory regarding hard and soft Lewis acids and bases asserts that hard metals (Mn, Ca, Fe^{3+}, Co) tend to bind hard ligands (OH^-, PO_4^{3-}, NH_3, CO_3^{2-}), while soft metals (Cu^{1+}, Cd) tend to bind soft ligands (RSH, RS^-). Proteins evolved to carry these metals in such a way as to optimize their redox potential to drive kinetically slow reactions. They may do this by stabilizing the intermediate or product state of their redox active metals, e.g.,

Division of Environmental and Biomolecular Systems, Institute of Environmental Health, Oregon Health & Science University, Portland, OR 97239.

$Mn^{2+}/Mn^{3+}/Mn^{4+}$, $Fe^{2+}/Fe^{3+}/Fe^{4+}$, and Cu^{+}/Cu^{2+}, to lower the energy barrier to electron transfer. Such electron transfer reactions are at the center of important metabolic processes such as photosynthesis, oxidative energy generation, and respiration, which have important impacts on environmental elemental cycling.

Although many metals are essential, they can often be toxic at higher concentrations, so bacteria rely on highly regulated mechanisms for metal homeostasis. Metals enter the cell by selective import through transporters and receptors like P1B-type ATPases that are tuned to the specific import and export of Cu, Cd, Zn, Pb, Co, and/or Ag. When essential metals are not abundant in their soluble, bioavailable form, they are scavenged in the environment by small organic molecules, like siderophores in the case of Fe(III). With their high affinity for Fe(III), siderophores are able to capture the metal ion and deposit it onto cell surface receptors. Metabolically active bacteria are able to regulate their metal uptake by regenerating transporter machinery after damage and subsequently modulating the production of metal-scavenging small organic molecules. Once transported in, the metal is chelated, or sequestered, by organic ligands like thiols. An intermediate protein synthesis partner, or chaperone, will then shuttle the metal to the required protein ligand set.

Once the metal quota is met, the cell will sequester superfluous metal ions so that harmful redox products, the most dangerous being reactive oxygen species generated by Fenton chemistry, and improper metal placement are minimized. Sequestering metals provides a reversible mode of metal storage to manage metals over a cell's lifetime. Reversible binding allows proteins to deliver metals as needed, while hyper-phosphate-accumulating organisms mineralize numerous hard metals with phosphate (2). The ubiquitous iron storage protein, ferritin, is a highly dynamic and flexible protein that mineralizes iron in its hollow core to sequester it from harmful side reactions in the cell but keeps the Fe available for use in Fe-S clusters and heme-containing proteins. In the *Escherichia coli* periplasm Cu(I) is oxidized to Cu(II) to prevent passive transport of Cu(I) into the cytoplasm where Cu is not needed (3, 4). In fact, metal homeostasis is so tightly regulated that there are essentially no free Cu atoms in the *E. coli* cytoplasm (5), so copper must be either compartmentalized, bound in proteins, or exported through an efflux mechanism (6, 7). External, semispecific metal control is provided by biosorption to the cell's surface: surface carboxylates, sulfhydryls, and phosphates are all available for nonspecific metal sequestration (8–11).

Another trick up the microbe's sleeve is the ability to detoxify metals by decreasing the bioavailability of the metal through metal precipitation reactions, effectively decreasing the availability of unwanted metals to metal binding sites of proteins. Redox active soluble metals and metalloids can be dangerous, not only because they generate radicals such as reactive oxygen species, but they also compete with metals for protein metal binding sites based on similar Lewis acid base ligand affinities. Especially problematic is when an improper metal has a higher affinity than the natural metal for the metal binding site, so binding is irreversible and the protein loses function. Metal precipitation may result from either formation of insoluble metal species (e.g., through production of a metabolite such as hydrogen sulfide or phosphate) or through a change in oxidation state of the metal (most frequently through reduction). Many different bacteria are capable of these reactions, including some bacterial spores (Table 1). For example, various organisms like some *Pseudomonas* or *Bacillus* species are capable of reducing metals such as Cr(VI) to Cr(III) and Se(IV) and/or Se(VI) to Se(0), where the lower oxidation states of these metal(loid)s are less soluble and hence tend to precipitate (12–14). Obviously, the ability to detoxify metals is not rare in bacteria; microbes have encountered changing and unpredictable levels of metals in their environment throughout the history of the Earth.

Remediation Through Biogeochemistry

The ability of microbes to detoxify elements has attracted the attention of those interested in removing, or remediating, toxic elements from contaminated environments. Metals may reach toxic levels in the environment through natural biogeochemical processes or from anthropogenic sources such as metal ore processing, coal burning at power plants, or preparation of nuclear fuels. Understanding biogeochemical cycling in relation to the removal of toxic metals could be the solution to saving vulnerable environments. Following is a discussion of remediation challenges and strategies to introduce bioremediation as an eco-friendly and sustainable remediation regime.

When designing remediation methods to remove contaminants like metals in groundwater to protect the health of humans and our environments, there are several important considerations: soil and water chemistry and the form of the metal (e.g., pH, metal concentration, redox conditions), soil or sediment permeability, and geological properties that affect movement of materials, microbial community composition,

Table 1 Examples of elements precipitated by bacteria

Product	Organism (*indicates activity from spore)	Reference
Precipitation via oxidation of metal/metalloid/nonmetal		
Mn(II) → Mn(IV)	*Pseudomonas putida* GB-1	173
	Bacillus sp. SG-1*	68
	Aurantimonas manganoxydans SI85-9A1	174
	Leptothrix discophora	175
	Pedomicrobium sp. ACM 3067	176
	Roseobacter sp. AzwK-3b	177
Fe(II) → Fe(III)	*Mariprofundus ferrooxydans*	178
	Sideroxydans sp. ES-1	179
	Gallionella ferruginea	180
	Thiobacillus denitrificans	181
Precipitation via reduction of metal/metalloid/nonmetal		
Se(VI) → Se(0)	*Sulfurospirillum barnesii*	182
	Bacillus selenitireducens	13
Se(IV)/Se(VI) → Se(0)	*Thauera selenatis*	183
	Pseudomonas stutzeri	14
	Wolinella succinogenes	184
U(VI) → U(IV)	*Shewanella oneidensis*	185
	Anaeromyxobacter dehalogenans	186
	Cellulomonas sp. ES6	157
	Desulfotomaculum reducens MI-1*	152
	Clostridium acetobutylicum DSM 792*	152
	S. oneidensis	152
	Desulfitobacterium spp.	122
Cr(VI) → Cr(III)	*Pseudomonas putida*	12
	Pseudomonas ambigua	187
	Agrobacterium radiobacter	188
	Enterobacter cloacae	189
U(VI) → U(IV), Cr(VI) → Cr(III)	*Desulfotomaculum reducens* MI-1	123
Tc(VII) → Tc-oxide, Tc-sulfide	*Desulfovibrio desulfuricans*	190, 191
Precipitation via formation of phosphate		
U(VI)-phosphate	*Bacillus* sp. Y9-2	192
	Rahnella sp. Y9602	192
Precipitation via formation of sulfide		
Hg-sulfide	*Clostridium cochlearium* T-2	193
As(III)-sulfide	*Desulfotomaculum auripigmentum* sp. nov	147

and cost of treatment and maintenance. Soils and water compositions and environment are highly variable, so one technique optimized for a certain site may not be compatible for other sites.

Metal remediation methods can be split up into three types of methods including chemical, physicochemical, and biological (as reviewed in reference 15). Chemical methods involve ensuring that the contaminant is in a suitable form to be contained (e.g., through *in situ* precipitation using chemical amendments) or removed (e.g., through facilitating mobilization and flushing using strong chelators to complex the metals). Physicochemical methods use a similar approach to adjust the chemistry but with the addition of a physical treatment process (e.g., pump and treat, filtration, electrokinetic processing). Since these topics are beyond the scope of this review, we will focus on biological methods for metal remediation.

Natural microbial and chemical processes could remediate a given site over time through a process of natural attenuation brought about by (bio)sorption and biotransformations. Natural attenuation can be hastened by stimulating the microbial communities by providing more favorable environmental conditions. For example, biostimulation can occur by enriching the soil with exogenous nutrients or aerating the water to promote growth and speed metabolism of vegetative cells. Biological remediation technologies range from this *in situ* approach to *ex situ* treatment with bacteria engineered to be optimized for metal recovery (Fig. 1). These approaches for removing contaminants exploiting the capabilities of microorganisms are attractive for remediation because they can be less disruptive of the soil environment than chemical or physicochemical methods, and they require no toxic additives and create little or no waste. Biosorption and biochemical remediation techniques are the most recent and promising modes of efficient heavy metal containment removal or transformation to nontoxic states. What follows are a just a few examples of bacterial metal homeostasis strategies being applied to heavy metal remediation.

METAL TRANSFORMATIONS WITH VEGETATIVE CELLS

Mercury

Hg and compounds that bind Hg are highly toxic even at low concentrations because Hg competes with natural metals like Cu for thiols in proteins. If redox-inactive Hg displaces a native metal in an electron-transferring protein, the functionality of the protein is destroyed. Further, Hg is often methylated so that direct sequestration is limited by its altered size and chemistry, so Hg(II) must be removed from the methyl groups by reduction or oxidation before it can be transformed into the less toxic metallic mercury [Hg(0)] state.

Hg is transformed by proteins encoded by the *mer* gene cluster. The *mer* genes are present in Gram-positive and Gram-negative bacteria and are tightly regulated and sensitive to even low concentrations of Hg. The structure of the *mer* gene operon varies between bacteria, but, at the core, there are *merT*, *merP*, *merA*, and *merD*. MerD is a *mer* operon regulator protein. MerP is a periplasmic protein with two cysteine residues that bind Hg(II) during entry and interacts with the inner membrane protein, MerT. MerT has two cysteine residue pairs and translocates Hg(II) across the inner membrane. Once in the cytoplasm, Hg can be complexed by sulfhydryl agents or reduced by mercuric reductase (MerA), a flavin-adenine dinucleotide (FAD)-dependent pyridine nucleotide-oxidoreductase requiring NADPH and sulfhydryl compounds for activity (16–18). MerA transfers electrons from NADPH via FAD to Hg(II) through another pair of redox-active Cys to produce Hg(0) gas, removing it from the cell. Volatilized Hg(0) is released into the atmosphere, natural waters, and soils where it is oxidized.

Understanding Hg resistance in bacteria has led to some intriguing Hg remediation strategies, including the introduction of the *mer* genes into *E. coli* (19, 20), radiation-resistant *Deinococcus* (21, 22), and plants (23, 24). One of the best developed systems for Hg remediation is a packed bed bioreactor containing a gel-encapsulated bacterial biofilm of a Hg-resistant strain of genetically engineered *Pseudomonas putida* (25). The Mer enzymes engineered into the bacteria in the biofilm volatilize Hg, and the Hg precipitates into droplets through fast oxidative absorption. The Hg droplets are then removed by distillation or containment of the bioreactor contents. This technique improves on prior methods, because it removes all of the Hg in the gas phase and does not require regeneration of the biocatalyst.

Chromium

Bacterial metal resistance is also being employed for the remediation of Cr. Cr precipitation through reduction of Cr(VI) to Cr(III) effectively detoxifies it, because the insoluble form cannot easily be taken up by the cell. Cr(VI)-reducing bacteria are widespread as evidenced by their isolation from many contaminated sites including near a chromate-producing chemical factory, a nuclear production complex, and an alkaline lake (26–28). These bacteria reduce Cr with NADH-dependent Cr reductases (29–31) in aerobic conditions

and cytochromes *b*, *c*, and *d* in anaerobic conditions (32, 33).

Jeyasingh and coworkers (34) demonstrated a Cr remediation strategy that employed a combination of biological and physical methods. They introduced a thick biobarrier (0.44 mg Cr reducer biomass/g sand mixture) or reactive zone (injection of 100 g wet weight Cr reducer biomass) of Cr-reducing isolates into the soil. (See *in situ* remediation methods for setup in Fig. 1.) Molasses and other nutrients were injected to sustain the biomass while Cr(III) and Cr(VI) levels were monitored downstream of the biobarrier. The Cr-reducing bacteria successfully removed Cr from the bench-scale contaminant plume. These results were used to evaluate a two-dimensional transport and transformation computer model for simulating bioremediation of contaminated aquifers. Their model agreed well with the transport and transformation of the Cr(VI), save for one case where they did not expect to see bacterial growth upstream of the plume because they assumed the bacteria were immobile. Clearly, engineering, chemical, and microbiological intricacies need to be considered when optimizing remediation strategies.

METAL TRANSFORMATIONS WITH SPORES

Although spores are dormant and not metabolically active, they still may serve as catalysts for certain reactions and thus have an overlooked role in geochemical processes such as metal sorption and transformation. Because of this, spores carrying out these processes may provide promising avenues in certain treatment strategies for heavy metal remediation. While microbial transformations carried out by vegetative cells may offer an environmentally friendly option for remediation technologies, these organisms are limited by their resistance to harsh conditions (including high metal concentrations), and their growth, or at least activity, must be sustained. Under extreme pH and temperature, for example, the vegetative cell population could collapse and fail to recover. Dormant spores with appropriate activities or surface properties may provide a good alternative to vegetative cells. Some microbial spores can carry out certain metal transformations that are beneficial to human and environmental health, even under harsh conditions.

The spore's resilience and coat protein composition are attractive qualities for remediation with spores. The spore is heat and drought resistant, allowing it to persist in the environment long after conditions for microbial growth expire (35–37). The outer layers of the spore, the spore coat and the exosporium, provide a

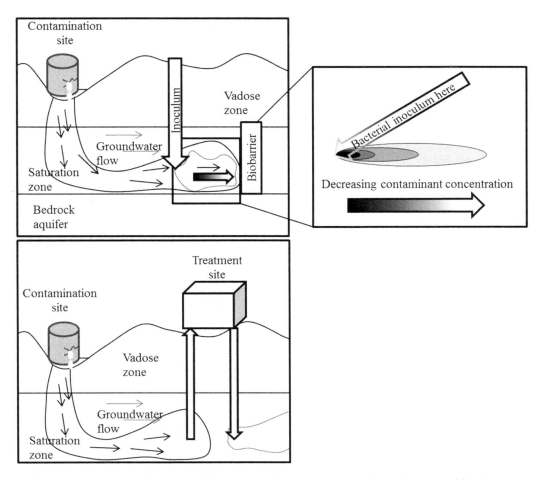

Figure 1 Schematic diagrams of *in situ* remediation by bacterial inoculation and biobarrier installation (top) and *ex situ* remediation by pump and treat method (bottom).

highly sorptive surface for metal sequestration and subsequent removal from the environment (38). For example, spores bind Cu(II) 2 to 4 orders of magnitude greater than that determined for fungi, bacteria, and algae (38). Spores are dormant and resistant to environmental extremes, so most of the complications of high metal toxicity are moot, making them ideal in relatively acidic and radioactive environments. Following are examples of how spores participate in biogeochemical cycling of Mn and U processes that may serve as a cost-effective and safe alternative to remediation with vegetative cells.

Manganese

Mn is ubiquitous in the environment and abundant in the Earth's crust, comprising about 0.1% of average crustal rocks (39). Mn cycles primarily between soluble Mn(II) and insoluble Mn(III,IV) oxides (collectively referred to as "Mn oxides") with a complex-stabilized

soluble Mn(III) intermediate occurring as a transient state. Mn(II) is released from igneous and metamorphic rocks by interactions with surface water and groundwater, either through reductive dissolution of Mn(III,IV) oxides or leaching of Mn(II) from rocks.

Mn oxides have high surface area and are abundant in deposits such as nodules, microconcretions, coatings, and crusts in soil as Mn hydrous oxides and on the ocean floor in ferromanganese nodules (40). They form about 30 different types of oxide/hydroxide minerals with tunnel and layered structures that contain vacancies in which other metals can sorb, such as Ra, Pb, and Po (41). Mn and Fe oxides readily incorporate radionuclides including ^{234}Th, ^{228}Th, ^{228}Ra, and ^{226}Ra (42, 43). These properties inspired their description as "scavengers of the sea" (44).

Mn(III,IV) oxides are some of nature's strongest oxidants, which make them attractive for removing contaminants that are immobilized by oxidation. In fact, they are already used in drinking water treatment facilities (45). They can oxidize both simple organic matter including phenols and quinones (46–48) and recalcitrant organic matter, forming low-molecular-weight organic compounds (49) that could be used as microbial food. They also oxidize metals such as Se(IV) to Se(VI) (50), Cr(III) to Cr(VI) (51), and As(III) to As(V) (52). Additionally, synthetic contaminants like Bisphenol A (53), polychlorinated biphenyls, chlorinated anilines, and atrazine are all degraded by Mn oxides (47, 54, 55). A review by Tebo et al. (56) highlights the mechanisms by which these compounds are degraded: free-radical oxidation; nucleophilic addition of the substrate to o-quinones (57); oxidation and release of CO_2 without organic intermediates (58, 59); and dealkylation at Mn oxide surfaces (60). At low pH these mechanisms are enhanced (53, 61, 62).

Bacterial manganese(II) oxidation

Chemical Mn(II) oxidation is thermodynamically favorable but kinetically very slow at pH < ~8.5 (63). It is likely that the environmentally relevant mechanism of mineralization is through microbial (bacterial and fungal) activities that are several orders of magnitude faster than abiotic autocatalysis on the surface of Mn oxides (64). Low G + C *Firmicutes*, high G + C Gram-positives, and *Alphaproteobacteria*, *Betaproteobacteria*, and *Gammaproteobacteria* all have Mn(II)-oxidizing members, many of whom catalyze this oxidative reaction in their vegetative form including *Pseudomonas*, *Aurantimonas*, *Leptothrix*, *Pedomicrobium*, and *Erythrobacter* (56).

Biogenic oxides tend to localize to the outer layers of the cell, often with the extracellular polysaccharide matrix. Mn oxides likely template on the organic matter by direct and indirect processes (65, 66) forming predominately birnessite minerals (56). The evolutionary purpose of Mn(II) oxidation remains a mystery. Bacteria do not require the process for survival, yet one can imagine that a Mn mineral coat may deter predation, act as UV protection, or degrade recalcitrant organic matter for nutrition for the microbe. That dormant spores from some *Bacillus* species can also oxidize Mn(II) to form Mn oxides, makes the role of Mn(II) oxidation even more inexplicable for these organisms.

Bacillus spores and Mn(II) oxidation

Bacteria have been linked to Mn(II) oxidation since the beginning of the 20th century, but it was not until the 1970s that a spore had been linked to this important process. Transmission electron microscopy captured the

first images of a bacterial spore, *Bacillus* sp. SG-1, isolated from an enrichment culture of near-shore sediments from San Diego and encased in a web of Mn oxides (Fig. 2) (67). Rosson and Nealson proposed that an enzyme on the spore surface was the Mn oxide deposition culprit since azide and cyanide inhibited Mn(II) binding and oxidation (68). Since then, work has continued to determine how the bacteria generate these oxides and what kind of effect these Mn oxides have on geochemical cycling of Mn and other elements.

Mn(II)-oxidizing spores are not uncommon. Although it may be difficult to analyze the abundance of spores and their activity in the environment, our laboratory has demonstrated the diversity and pervasiveness of Mn(II)-oxidizing *Bacillus* spores through isolation from a variety of environments (69–72). Mn(II)-oxidizing spores are isolated by diluting a soil or water sample, heating it to 80°C for 10 to 20 min, and spreading it onto Mn(II)-containing solid media. Colonies that sporulate and become encrusted with

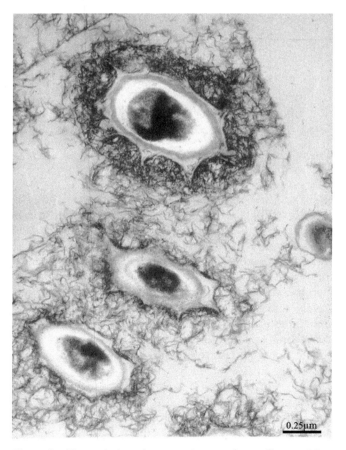

Figure 2 Transmission electron micrograph *Bacillus* sp. SG-1 spore with spiny MnO_2 oxides localized to the exosporium from reference 68 (bar = 0.25 m).

brown Mn oxides are cultured for DNA extraction and 16S rRNA gene sequence analysis. In 2002, Francis and Tebo reported the isolation of 13 different species in the shore sediments around San Diego, CA that grouped into three phylogenetically distinct groupings (70). These included isolates from various environments grouping with strain SG-1, the MB-7 cluster that was similar to moderately halophilic or halotolerant species, and the PL-12 cluster. The PL-12 cluster was similar to strains isolated from a variety of sources including a Korean traditional fermented seafood; a uranium mine tailings pile near Dresden, Germany; a hydrocarbon seep; and rice paddy-associated anoxic bulk soil. Clearly Mn(II)-oxidizing activity of *Bacillus* spores is present in phylogenetically and environmentally diverse *Bacillus* species, suggesting that Mn(II)-oxidizing *Bacillus* spores are common and widespread in the environment.

If Mn oxides are the "scavengers of the sea" (44) and Mn(II)-oxidizing spores are widespread, then Mn(II)-oxidizing spores should impact the Mn biogeochemical cycle in the oceans. Dick and colleagues were also able to map out the diversity of 20 more isolates of Mn(II)-oxidizing *Bacillus* spores falling into two clusters in the water column of the Guaymas Basin, a semienclosed basin with a sediment-covered hydrothermal system (73). Mn(II)-oxidizing activity was associated with the spores and could occur at high temperatures, as high as 70°C. Their study suggested that bacterial spores play a role in the short Mn residence time in the Guaymas Basin and that rapid and stable Mn(II)-oxidizing spores at hydrothermal sediments and plumes could be the mechanism of elemental scavenging as put forth by Cowen (74–76).

Metals like Cr(III), Co(II), and U(IV) are oxidized in the presence of Mn(II)-oxidizing spores. An exciting prospect was the possibility that the spores were directly oxidizing these metals, but alas these metals were not oxidized by stringently washed *Bacillus* sp. SG-1 spores without added Mn (77) and do not compete for sites within the same enzyme as Mn(II) (77, 78). Cr(III), Co(II), and U(IV) have been shown to be oxidized abiotically by synthetic Mn oxides, δ-MnO_2 (77, 79, 80), leading to the speculation that the Mn oxides are the oxidant. Indeed, newly formed biogenic MnO_2 oxidizes these metals faster than synthetic oxides: seven times faster for Cr(III) (78), two to five times faster for Co(II) (77), and almost two times faster for U(IV) (81). The higher reactivity of the biogenic nanoparticulate MnO_2 oxides can be attributed to their mixed valence Mn(II, III, IV) (82) and vacancies within the structure (56). Clearly, determining the mechanism of generating reactive biogenic MnO_2 oxides is linked with understanding the cycling of other metals.

mnx genes and Mn(II) oxidation

Developing a system in *Bacillus* sp. SG-1 to alter its genomic DNA was not trivial. van Waasbergen and coworkers attempted many transformation techniques with several plasmids (83). She finally conceived a protoplast preparation transformation of the pLTV1 plasmid carrying a modified *Streptococcus faecalis* transposon to disrupt the gene responsible for Mn(II) oxidation. van Waasbergen identified the seven-gene operon called *mnxA-mnxG* (83, 84). Twenty years later, and still only *Bacillus* SG-1 MnxC and MnxG have enough amino acid sequence similarity to known proteins to predict function. MnxC shows similarity to a Cu-loading protein-folding chaperone, Sco, that is thought to be responsible for Cu loading in the synthesis of cytochrome *c* oxidase. MnxG is similar to another Cu-containing protein, cerruloplasmin, a large six-domain multicopper oxidase that oxidizes Fe(II) to Fe(III) and is involved in Fe transport in the bloodstream (84).

Further evidence that the *mnx* gene products are linked to spore Mn(II) oxidation activity followed in the next decade. The *mnxD-lac* gene reporter activity assay localized gene expression to around stage III of sporulation, peaking between stages IV and V, suggesting regulation of *mnx* genes by late sporulation sigma factor K (84). The Cu binding regions in the *mnxG* gene were used to design degenerate primers to amplify a region of DNA to be used in phylogenetic analysis of environmental isolates, because the rest of the gene was too divergent from one isolate to the next (70). *mnxG* genes clustered with their respective 16S rRNA genes, indicating little horizontal gene transfer of *mnxG*. Further, detection of *mnxG* correlated to Mn(II)-oxidizing activity in the endospore and, because the exosporium retains Mn(II)-oxidizing activity after being removed by high pressure with the French press, the activity was localized to this layer (70).

Direct evidence of Mnx proteins catalyzing Mn(II) oxidation came after isolating activity from the exosporium to a single protein band in nondenaturing sodium dodecyl sulfide-polyacrylamide gel electrophoresis (SDS-PAGE) and analyzing the peptides by tandem mass spectrometry. Only peptides matching MnxF and MnxG amino acid sequences were identified from the exosporium preparation of *Bacillus* sp. MB-7, whose *mnxD*, *mnxE*, *mnxF*, and *mnxG* genes are well conserved among the sequenced Mn(II)-oxidizing *Bacillus* species. The Mn(II) oxidation reaction was inactivated by multicopper oxidase inhibitors, azide and cyanide, further implicating a Cu protein in catalysis (85). Thus, it appeared that multicopper oxidase, MnxG, was the Mn(II) oxidase.

There is precedence for multicopper oxidases to be localized to the outer layer of the *Bacillus* spore. *Bacillus subtilis* coat protein, CotA, is a well-studied multicopper oxidase located in the thick, striated outer coat capable of generating a brown pigment (not Mn oxides) around the spore which it uses for UV protection (86). CotA is a 67-kDa three-domain protein encoded by a single gene and is most similar to *E. coli*'s Cu(I)-oxidizing CueO multicopper oxidase. It is also heat resistant and even fully active at 80°C (87). It has a flexible, lid-like region near the substrate binding region that can play a role in binding larger substrates than the metal-oxidizing multicopper oxidases.

Multicopper oxidases contain at least four Cu atoms that facilitate the single-electron oxidation of a reduced metal or phenolic compound with the reduction of O_2 to H_2O. The four Cu sites are made up of one type 1 center consisting of two S ligands from Cys and Met, and two imidazole ligands from His, one type 2 center consisting of two imidazole ligands, and two type 3 centers consisting of three imidazole ligands each. The type 1 Cu center's Cu-S (Cys) charge transfer gives the protein its characteristic blue color. It also contains enough redox potential to rip a single electron off the substrate and send it down a proton bonding network through the protein approximately 10 Å to the trinuclear center where the remaining Cu centers lie and are ready to transfer the electron to O_2 creating H_2O (88).

To investigate how MnxG could oxidize Mn(II) to Mn(IV) minerals via a multicopper mechanism, the protein needed to be isolated to high purity and quantity. Attempts were made at expressing *mnxF* and *mnxG* in *E. coli* but none resulted in soluble, active protein. Eventually, all of the highly conserved *Bacillus mnx* genes (*mnxCEFG*) were amplified together as a single fragment, cloned, and coexpressed in an *E. coli* protein expression strain to successfully produce a Mn(II)-oxidizing protein (89). After purification, the resulting indigo blue protein, with an absorbance maximum of 580 nm, was analyzed by mass spectrometry to conclude that not only was MnxG present, but also were MnxE and MnxF. This complex, now named Mnx, was approximately 230 kDa in mass and contained between 6 and 10 Cu per mol of protein depending on the purification and dialysis conditions (90). This was the first observation of a multisubunit multicopper oxidase.

Suddenly, the proteins of unknown function, MnxE and MnxF, became much more interesting to the story of Mn oxidation in *Bacillus* and their isolation soon followed. The same strategy of coexpressing *mnxEF* in *E. coli* was employed and the metal binding properties of the protein were determined. MnxEF forms a hexamer that binds type 2 Cu and low-spin heme upon loading with excess heme (91). This heme/Cu binding is notable because only cytochrome *c* oxidase is known to bind type 2 Cu and heme. These results are intriguing because they hint at a possible role for the MnxEF subunits in metal-ligand-mediated redox and Mn oxidation in the Mnx protein complex.

Since multicopper oxidases catalyze single-electron transfers, the role of MnxG in the two-electron oxidation of Mn(II) to Mn(IV) oxides is an alluring conundrum. Several studies have theorized thermodynamically feasible Mn(II) oxidation mechanisms (92–94). In order for Mn(II) to convert directly to Mn(IV) via a single two-electron step, a multinuclear complex is necessary to bridge O_2, resulting in H_2O_2, which is thermodynamically favorable but kinetically inhibited by the difficulty of forcing electrons into H_2O_2 π* orbital (93). It is also difficult to imagine a multinuclear complex within a multicopper oxidase. Thus the two single step oxidation mechanism is more probable (94). This scheme fits with the single electron transfer mechanism of multicopper oxidases, but how MnxG doubles this reaction to oxidize Mn(II) to Mn(IV) is a mystery.

Mn(II) oxidation mechanism and implications of Mn(III) intermediate

X-ray absorption spectroscopy techniques determined that the primary product of Mn(II) oxidation and the most environmentally relevant mineral produced by *Bacillus* sp. SG-1 [and other Mn(II)-oxidizing bacteria] is most closely similar to hexagonal birnessite [Mn(IV)O_2] (82, 95, 96). Experiments spectrophotometrically measuring the trapping of Mn(III) with pyrophosphate using purified exosporium from *Bacillus* sp. SG-1 and purified Mn oxidase from *Bacillus* sp. PL-12 confirmed the two single-electron sequential oxidation steps by the native exosporium and the MnxG complex, respectively (97, 98). The reaction can be initiated with Mn(II) or Mn(III)-pyrophosphate to produce brown Mn(IV) oxides, so the enzyme may have two binding sites for different Mn species.

Soldatova and coauthors put forward a mechanistic model to rationalize these results within the thermodynamic limitations (Fig. 3) (98). They postulate that Mn(II) is oxidized at the substrate site proximal to the type 1 Cu just as in other multicopper oxidases. The

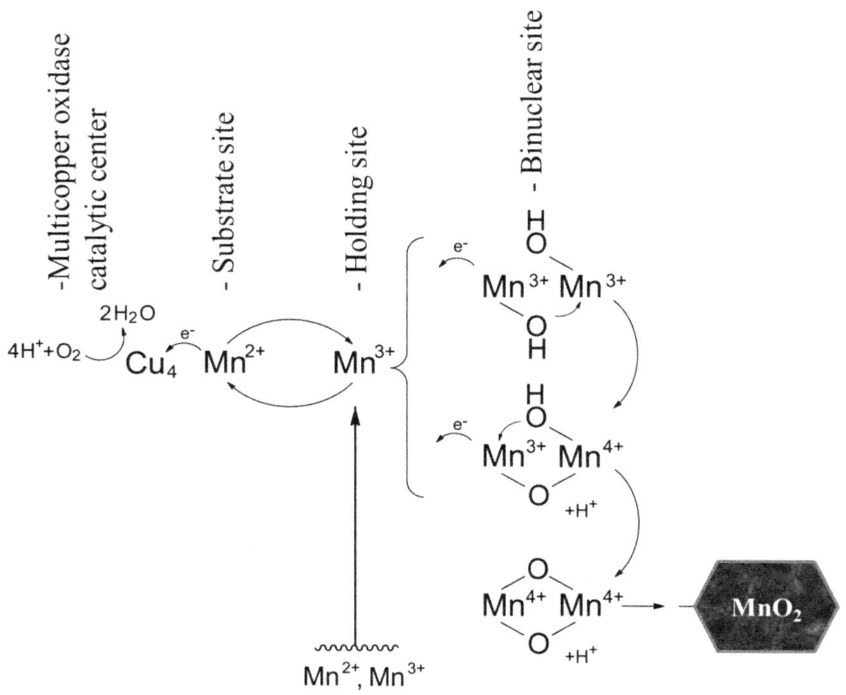

Figure 3 Proposed Mn(II) oxidation mechanism by *Bacillus* spp. multicopper oxidase MnxG (adapted from reference 98).

Mn(III) product is then held in a holding site away from the substrate site, akin to ceruloplasmin's Fe(III) holding site, before it is transferred to a Mn(III) binuclear site. Newly oxidized Mn(III) drives the formation of oxo bridges in the binuclear site to form Mn(IV)O_2. The binuclear site may nucleate MnO_2 formation by following one of the mechanisms put forward for the ferritin Fe(II) to Fe(III) oxidation (99), such as stabilizing the binuclear Mn(IV) until a new Mn(III) ion triggers sequential displacement of Mn(IV) and oxidation of the displacing Mn(III).

To test the model, Mnx was loaded with Mn(II) and oxidation was allowed to proceed at room temperature. Aqueous Mn(II), mononuclear Mn(II) binding to a nitrogenous ligand, and a binuclear Mn(II) binding were observed through continuous-wave electron paramagnetic resonance and electron spin-echo envelope modulation spectroscopy (100). The first two species agree with the above model, but the binuclear center was more reduced than predicted. Instead of three separate Mn binding sites, there could just be two: a mononuclear and a binuclear Mn(II). Perhaps the oxidation at the binuclear site was too fast to be observed by the methods above, and rapid freeze-quench experiments will have to be performed to capture Mn(III) and Mn(IV).

As an intermediate of oxidation of Mn(II) to Mn(IV), Mn(III) is a powerful oxidant and its lability in this reaction could have interesting environmental implications. In the absence of chelator, Mn(III) is rapidly disproportionated to Mn(II) and Mn(IV) oxides. In the environment, however, there are other Mn(III) chelators available, like oxalate, citrate, pyrophosphate, or even organic matter. It is possible that Mn(III) could be hijacked by a natural chelator for other purposes like competing with Fe(III) for complexation with siderophores (101), degrading lignin (102–105), or oxidizing sulfur and nitrogen compounds (106–109). Thus, Mn(II) oxidation by "dormant" spores can impact Earth's biogeochemistry by having a catalytic effect on a variety of other major elemental cycles.

URANIUM AND URANIUM REDUCTION

While U is an element that is not too abundant, it contaminates certain environments by various mechanisms (110). For example, the U contamination at the Department of Energy (DOE) site at Rifle, CO is primarily due to the mill tailings from a U ore-processing facility on site leaching into the subsurface (111). Contamination at the DOE site at Oak Ridge, TN and Hanford, near Richland, WA is attributed to U-containing nuclear munitions waste entering the environment through leaking underground storage tanks (112, 113). In addition to U mining and weapons manufacturing, it can also be released from combustion of coal and fertilizers

(114). Because U can cause liver, kidney, brain, and heart problems (115), strategies to remediate U-contaminated sites have been sought.

U is generally found in two oxidation states in the environment, U(IV) and U(VI), with U(VI) being more mobile and more toxic than U(IV). Therefore, in U-contaminated sites, immobilization of U via reduction of U(VI) to U(IV) is an intriguing strategy for remediation. For a while, reduction of U in the environment was considered primarily an abiotic reaction where a strong reductant was attributed to carry out the process (116). However, studies in the early 1990s started showing the capability of dissimilatory metal-reducing bacteria and sulfate-reducing bacteria to enzymatically reduce U(VI) to U(IV) (117–123). Indeed, microbial community analyses indicated the stimulation of dissimilatory metal-reducing bacteria and sulfate-reducing bacteria during bioreduction of U in the subsurface via injection of electron donors such as acetate or ethanol (111, 124–126).

Most of the pure culture studies on U reduction have been focused on vegetative cells, but as we have seen from *Bacillus* spores, biogeochemical cycles are not exclusive to active bacteria. While vegetative cells require costly nutrients, bioremediation with the following U-reducing spores could be an attractive strategy to explore.

Clostridium acetobutylicum

Clostridia are Gram-positive anaerobes that form endospores like *Bacillus* sp. in response to stressful environmental conditions (127). Previous studies on the vegetative cells have shown that *Clostridium* species can be used for bioremediation of various contaminants in the environment; these contaminants include 2,4,6-trinitrotoluene (TNT) (128–130) radionuclides such as U and Tc (131, 132), pesticides (133), herbicides (134), and insecticides (135). Microbial community analyses on environmental samples stimulated for U reduction have also reported observation of *Clostridium* spp. suggesting *Clostridium* spp. may have great environmental relevance regarding bioremediation of U (125, 136, 137). Vegetative *Clostridium* spp. have been shown to be capable of U reduction at low pH (131, 132). To augment metal reduction *Clostridium* spp. spores have been used as an inoculum to maintain the activity of the vegetative cells after various storage methods (130). Thus, in this case, the spores are being used as an agent to ultimately deliver active vegetative cells.

Recently, a study has reported that spores of *C. acetobutylicum* DSM 792 can carry out reduction of U (138). This study reported that spores of *C. acetobutylicum* were able to reduce U in spent growth media but not in fresh growth media, suggesting a role for a *C. acetobutylicum*-derived compound(s) in reducing U. The study suggested that some form of (genus specific) compound present in the spent medium serves as an electron shuttle between H_2 and U(VI). As in the case of MnO_2 forming spores of *Bacillus* species, the product of U reduction was primarily found to be associated with the exosporium. Based on these findings, it was hypothesized that either (i) a hydrogenase on the exosporium oxidizes H_2 and then the consequent electrons reduce the compound which subsequently reduces U(VI) or (ii) the hydrogenase forms electrons that are shuttled by the compound to a reductase also on the exosporium to catalyze U reduction (Fig. 4). Using X-ray absorption near edge structure (XANES), Vecchia et al. (138) verified that the product of U reduction was U(IV) instead of U removal being due to U(VI) precipitation or adsorption.

Compared with what is known about Mn(II) oxidation by spores of *Bacillus* sp. SG-1, little is known about the mechanism of U(VI) reduction by spores of *C. acetobutylicum*. The recent discovery of U reduction by *C. acetobutylicum* spores opens up questions as to the roles of a hydrogenase and/or reductase in the outer spore layers and thus further biochemical work is needed to verify the mechanism(s). Within the species *C. acetobutylicum*, only strains ATCC 824 (139), EA 2018 (140), and DSM 1731 (141) have their genomes sequenced, which does not include *C. acetobutylicum* DSM 792. Among these strains, vegetative cells of *C. acetobutylicum* ATCC 824 (131) have shown the ability to reduce U(VI), while both vegetative cells and spores of *C. acetobutylicum* DSM 792 have been shown to reduce U(VI). For both of these strains genetic systems have been developed (142–145) that should facilitate identification of the genes involved in U(VI) reduction.

Desulfotomaculum reducens

Similar to *C. acetobutylicum*, *Desulfotomaculum reducens* belongs to the *Clostridiales* family of the *Firmicutes* (138). Although both are part of Clostridiales, *D. reducens* can utilize metal reduction and sulfate reduction for growth (123, 146). Species of *Desulfotomaculum* have been shown to play important roles in bioremediation by not only primarily reducing oxidized metals and/or metalloids, but also by forming sulfides that can precipitate metals. For example, *D. reducens* reduces Cr(VI) and U(VI) (123), *D. auripigmentum* reduces As(V) (147), *D. nigrificans* removes

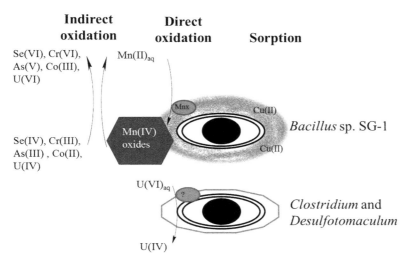

Indirect oxidation

Direct oxidation

Sorption

Se(VI), Cr(VI), As(V), Co(III), U(VI)

Mn(II)$_{aq}$

Mnx

Mn(IV) oxides

Cu(II)

Cu(II)

Bacillus sp. SG-1

Se(IV), Cr(III), As(III), Co(II), U(IV)

U(VI)$_{aq}$

?

U(IV)

Clostridium and *Desulfotomaculum*

Figure 4   Examples of oxidation and sorption of metals by bacterial spores.

zinc by precipitation of ZnS (148), and *Desulfotomaculum* sp. DF-1 forms NiS to remove Ni (149). Microbial community analyses of sediments actively reducing U(VI) identified closely related *Desulfosporosinus* species, suggesting a possible role of *Desulfosporosinus* in U(VI) reduction (125, 150, 151). However, as was the case for *Clostridium* and most studies with *Desulfotomaculum*, this study focused on vegetative cell-mediated and not spore-mediated bioremediation.

More recent studies verifying formation of U(IV) in the solid phase showed that the spores of *D. reducens* MI-1 can also reduce U(VI) (146, 152). Interestingly, the spores of *D. reducens* MI-1, similar to *C. acetobutylicum*, could only reduce U(VI) in the presence of H$_2$ and spent growth media. The study showed it was not the fermentation products of pyruvate in the spent medium; instead, it may be some form of stimulatory factor: something smaller than 3 kDa and produced by vegetative cells of *D. reducens* MI-1. Possible candidates for the small compound could be anthraquinone-2,6-disulfonic acid and riboflavin, which have been shown to function as electron shuttles for metal reduction (153–159). However, neither were effective in assisting U(VI) reduction mediated by spores of *D. reducens* MI-1. Nonetheless, the requirement of a stimulatory factor in U(VI) reduction by *C. acetobutylicum* and *D. reducens* spores is a significant difference to the Mn(II) oxidation mechanism by *Bacillus* sp. SG-1 spores. Also similar to the *C. acetobutylicum* U reduction story is the mystery as to which *D. reducens* spore enzymes are reducing U. The genome of *D. reducens* MI-1 has been sequenced recently (160) enabling *in silico* analyses on putative genes that may be involved in U(VI) reduction. A transcriptomic study was also performed to identify changes in gene expression upon exposure to U(VI) (161). It identified several upregulated genes involved in energy metabolism, metal detoxification, and iron metabolism, some of which might be used in U(VI) reduction. Since no genetic system has been described to perform mutagenesis studies on *D. reducens* MI-1 and no enzymes have been implicated in function, there is only speculation on which enzymes are responsible for U(VI) reduction. Based on what was observed with the addition of H$_2$, it was suggested that a hydrogenase functions in the absence of a reductase to transfer electrons to a stimulatory factor which then shuttles the electrons directly to U. Future work on identifying the enzymes and associated factors should elucidate the mechanism of U reduction by these spores.

FUTURE WORK

Spores in Biogeochemistry

Spores have long been known to be relatively abundant in the environment (162, 163). For example, in San Diego, CA coastal sediments they have been found to represent 0.085% to 0.29% of the total viable bacteria (71). Of these, 17% to 30% of the viable spore-forming bacteria oxidized Mn(II). Recent studies using dipicolinic acid as a marker specific for endospores showed that spores can be present in significant numbers, and, in certain environments, spores can be as equally abundant as vegetative cells (164, 165). Given the recent recognition that spores can carry out chemical reactions that affect a variety of elements and chemicals, the role of bacterial spores in biogeochemistry and contaminant remediation needs to be reevaluated. The challenge will be to quantitatively assess the impact spores have.

The current approach of extracting nucleic acids from the environment to investigate which microbes and genes are important is insufficient in addressing the spore influence. Progress in sequencing technologies has enabled environmental genomic and transcriptomic studies to help researchers understand which microbes are responsible for certain functions (166). DNA-based techniques provide information on which microbes or genes are present in the system, while mRNA-based techniques suggest which genes are actively being expressed. Also, some studies have made a correlation between gene expression levels to specific activities (e.g., references 136, 167, and 168). When it is the function of vegetative cells that is under investigation, transcript levels can be sufficient in elucidating which enzymes are active in the environment. When spores are taken into consideration, however, determining which enzymes are active by looking at transcripts can provide limited information because dormant cells do not transcribe DNA. Therefore, transcription levels will fail to take into account input from spores and overestimate the role of enzymes of vegetative cells regarding specific activity. To fully capture which enzymes are active and attribute them to the environment's dynamics, directly monitoring proteins would provide a more accurate representation. Although obtaining sufficient amount of protein to perform proteomic studies is challenging, it has been applied to study various aspects of biogeochemical cycling or bioremediation, such as Mn(II) oxidation (169) and reduction (170). But as techniques for quantifying and identifying proteins in environmental samples, proteomic approaches may be useful to analyze whether spore proteins are at all environmentally significant.

We are left with lingering questions that, if answered, may propel remediation efforts with spores and their vegetative cell counterparts. First, why would spores participate in functions such as MnO_2 or U(IV) formation? Are there any benefits for the spores to carry out such processes? At least vegetative cells have been shown to utilize U(VI) as an electron acceptor (121), while the advantages of bacterial Mn(II) oxidation are yet to be elucidated (171). Also, is there a dual ability (in both vegetative and spore life stage) of strains to reduce U(VI), and is it widespread among spore-forming U(VI)-reducing bacteria? Because both vegetative cells and spores of *C. acetobutylicum* DSM 792 are capable of reducing U(VI) (138), it would be interesting to see if both the vegetative cells and spores share the same mechanism.

Spores in Remediation

Like remediation with vegetative cells, spore genetic engineering would allow remediation projects to reach a higher level of efficiency while maintaining the eco-friendly appeal. While the deployment of genetically modified dormant bacteria is not ethical because of the possible horizontal transmission of derived sequences, *ex situ* remediation would benefit from a more reactive agent like genetically engineered spores. On the other hand, it may be possible to use spores that are modified so they don't germinate (e.g., by killing them before introducing them). For example, *Bacillus* sp. SG-1 spores fixed with glutaraldehyde are still capable of oxidizing Mn(II) (68). Such treatments may render the spores environmentally friendly, even if genetically modified.

An emerging strategy is the engineering of a microbe that remediates in both the vegetative and dormant forms. One may either insert the gene for expression of a membrane-bound enzyme to perform remediation during the active stage or utilize spore display to tether a remediation enzyme to the spore surface of a microbe that has remediation activity during its active phase. An example of the spore surface-tethering mechanism is one devised by the Jae-Gu Pan laboratory (172) where they fused the *Bacillus thuringiensis* exosporium protein InhA to green fluorescent protein so that it was displayed during sporulation. This technique may work in order to optimize the activity of interest, for example, Mn(II) oxidation, on the spore surface. Another strategy could be to increase the capacity of *Bacillus* SG-1 to oxidize Mn(II) by fusing the *mnx* genes to *inhA*, creating extra enzyme on the spore surface and increasing activity without using additional spores.

Bioremediation is particularly attractive with robust spores but obviously suffers from the real possibility of spore germination in the soil environment and loss of the desired activity. Perhaps the engineering of a germination receptor loss-of-function mutant or knocking out other genes involved in activation may result in a permanent spore form. If spores could be engineered to never germinate, one might easily imagine that a remediation scheme utilizing a filter with immobilized spores to act as reactive mineral media would treat water in an eco-friendly and efficient manner.

SUMMARY

Bacteria play a key role in geochemical cycling of several metals. Essential metals are transformed through normal metabolic processes and utilized for enzyme catalysis and electron transfer. But too much of a good thing can be harmful in the case of these metals.

Bacteria must also regulate metal uptake and speciation within the cell to minimize improper placement in enzymes and devastating reactive oxygen species-mediated damage.

Metal homeostasis mechanisms are capable of effectively pulling out metals from the surrounding environment, so they have been an attractive method for developing remediation strategies for heavy-metal-contaminated environments. Great interest in vegetative cell bioremediation has led to pilot scale studies using genetically engineered microbes and isolates from contaminated environments to some promising results.

Even though the dormant spore form does not have all of the homeostasis mechanisms of vegetative cells, they have been found to carry out activities with remediation potential and may be a more cost-effective technique. Mn oxides are highly reactive and sorptive so they are an attractive tool for remediation of heavy metals. Marine *Bacillus* spores of many different lineages are able to mineralize soluble Mn(II) and Mn(III) to Mn(IV) oxides through the use of an exosporium multicopper oxidase complex involving MnxE, MnxF, and MnxG (18). This process is not only interesting in relation to the global cycling of Mn, but also because of its unique mechanism of a two-step electron oxidation and as a potential bioremediation agent. Another harmful metal, U, may be able to be removed by *D. reducens* MI-1 and *C. acetobutylicum* DSM 792 spores. Although the mechanism of this reaction is not yet understood, these spores are likely candidates for *in situ* U remediation technologies. There are probably many more spore-driven metal removal processes than those extant in the literature. Further investigation of environmental spore-forming isolates should yield additional spore-mediated metal transformation methods useful for bioremediation.

Utilizing spores in bioremediation overcomes many of the shortcomings of vegetative cell schemes. Spores offer an immobile, robust platform for enzymatic metal transformations without the worry of bacterial population and activity controls like the addition of nutrients and monitoring of a constant, nontoxic environment. If spore remediation is to be implemented, more research is needed on the mechanisms of metal removal to ensure harmful side reactions do not mobilize other harmful metals in the process.

Acknowledgments. *We gratefully acknowledge the agencies funding the research in our laboratory over the years including the National Science Foundation (OCE-1129553 and 1154307 and CHE-1410688) and the Department of Energy (Subsurface Biogeochemical Research Program, DE-SC0005324). We also thank Kati Geszvain and Matthew Jones for their thoughtful comments during the preparation of this manuscript.*

Citation. Butterfield CN, Lee S-W, Tebo BM. 2016. The role of bacterial spores in metal cycling and their potential application in metal contaminant bioremediation. Microbiol Spectrum 4(2):TBS-0018-2012.

References

1. Gray HB. 2003. Biological inorganic chemistry at the beginning of the 21st century. *Proc Natl Acad Sci USA* **100:**3563–3568.

2. Bond PL, Hugenholtz P, Keller J, Blackall LL. 1995. Bacterial community structures of phosphate-removing and non-phosphate-removing activated sludges from sequencing batch reactors. *Appl Environ Microbiol* **61:** 1910–1916.

3. Beswick PH, Hall GH, Hook AJ, Little K, McBrien DC, Lott KA. 1976. Copper toxicity: evidence for the conversion of cupric to cuprous copper in vivo under anaerobic conditions. *Chem Biol Interact* **14:**347–356.

4. Outten FW, Huffman DL, Hale JA, O'Halloran TV. 2001. The independent cue and cus systems confer copper tolerance during aerobic and anaerobic growth in *Escherichia coli. J Biol Chem* **276:**30670–30677.

5. Changela A, Chen K, Xue Y, Holschen J, Outten CE, O'Halloran TV, Mondragón A. 2003. Molecular basis of metal-ion selectivity and zeptomolar sensitivity by CueR. *Science* **301:**1383–1387.

6. Chacón KN, Mealman TD, McEvoy MM, Blackburn NJ. 2014. Tracking metal ions through a Cu/Ag efflux pump assigns the functional roles of the periplasmic proteins. *Proc Natl Acad Sci USA* **111:**15373–15378.

7. Bagai I, Rensing C, Blackburn NJ, McEvoy MM. 2008. Direct metal transfer between periplasmic proteins identifies a bacterial copper chaperone. *Biochemistry* **47:** 11408–11414.

8. Andrès Y, Redercher S, Gerente C, Thouand G. 2001. Contribution of biosorption to the behavior of radionuclides in the environment. *J Radioanal Nucl Chem* **247:**89–93.

9. Song Z, Kenney JPL, Fein JB, Bunker BA. 2012. An X-ray absorption fine structure study of Au adsorbed onto the non-metabolizing cells of two soil bacterial species. *Geochim Cosmochim Acta* **86:**103–117.

10. Beveridge TJ, Murray RGE. 1976. Uptake and retention of metals by cell walls of *Bacillus subtilis. J Bacteriol* **127:**1502–1518.

11. Fein JB, Daughney CJ, Yee N, Davis TA. 1997. A chemical equilibrium model for metal adsorption onto bacterial surfaces. *Geochim Cosmochim Acta* **61:**3319–3328.

12. Ishibashi Y, Cervantes C, Silver S. 1990. Chromium reduction in *Pseudomonas putida. Appl Environ Microbiol* **56:**2268–2270.

13. Switzer Blum J, Burns Bindi A, Buzzelli J, Stolz JF, Oremland RS. 1998. *Bacillus arsenicoselenatis*, sp. nov., and *Bacillus selenitireducens*, sp. nov.: two haloalkaliphiles from Mono Lake, California that respire oxyanions of selenium and arsenic. *Arch Microbiol* **171:**19–30.

14. Lortie L, Gould WD, Rajan S, McCready RGL, Cheng KJ. 1992. Reduction of selenate and selenite to elemental selenium by a *Pseudomonas stutzeri* isolate. *Appl Environ Microbiol* 58:4042–4044.

15. Hashim MA, Mukhopadhyay S, Sahu JN, Sengupta B. 2011. Remediation technologies for heavy metal contaminated groundwater. *J Environ Manage* 92:2355–2388.

16. Izaki K, Tashiro Y, Funaba T. 1974. Mechanism of mercuric chloride resistance in microorganisms. 3. Purification and properties of a mercuric ion reducing enzyme from *Escherichia coli* bearing R factor. *J Biochem* 75:591–599.

17. Schottel JL. 1978. The mercuric and organomercurial detoxifying enzymes from a plasmid-bearing strain of *Escherichia coli*. *J Biol Chem* 253:4341–4349.

18. Furukawa K, Tonomura K. 1972. Induction of metallic mercury-releasing enzyme in mercury-resistant *Pseudomonas*. *Agric Biol Chem* 36:2441–2448.

19. Chen S, Wilson DB. 1997. Construction and characterization of *Escherichia coli* genetically engineered for bioremediation of Hg(2+)-contaminated environments. *Appl Environ Microbiol* 63:2442–2445.

20. Deng X, Wilson DB. 2001. Bioaccumulation of mercury from wastewater by genetically engineered *Escherichia coli*. *Appl Microbiol Biotechnol* 56:276–279.

21. Brim H, McFarlan SC, Fredrickson JK, Minton KW, Zhai M, Wackett LP, Daly MJ. 2000. Engineering *Deinococcus radiodurans* for metal remediation in radioactive mixed waste environments. *Nat Biotechnol* 18:85–90.

22. Brim H, Venkateswaran A, Kostandarithes HM, Fredrickson JK, Daly MJ. 2003. Engineering *Deinococcus geothermalis* for bioremediation of high-temperature radioactive waste environments. *Appl Environ Microbiol* 69:4575–4582.

23. Bizily SP, Rugh CL, Meagher RB. 2000. Phytodetoxification of hazardous organomercurials by genetically engineered plants. *Nat Biotechnol* 18:213–217.

24. Bizily SP, Rugh CL, Summers AO, Meagher RB. 1999. Phytoremediation of methylmercury pollution: merB expression in *Arabidopsis thaliana* confers resistance to organomercurials. *Proc Natl Acad Sci USA* 96:6808–6813.

25. Deckwer WD, Becker FU, Ledakowicz S, Wagner-Döbler I. 2004. Microbial removal of ionic mercury in a three-phase fluidized bed reactor. *Environ Sci Technol* 38:1858–1865.

26. Zhang K, Li F. 2011. Isolation and characterization of a chromium-resistant bacterium *Serratia* sp. Cr-10 from a chromate-contaminated site. *Appl Microbiol Biotechnol* 90:1163–1169.

27. Bowen De León K, Young ML, Camilleri LB, Brown SD, Skerker JM, Deutschbauer AM, Arkin AP, Fields MW. 2012. Draft genome sequence of *Pelosinus fermentans* JBW45, isolated during in situ stimulation for Cr(VI) reduction. *J Bacteriol* 194:5456–5457.

28. VanEngelen MR, Peyton BM, Mormile MR, Pinkart HC. 2008. Fe(III), Cr(VI), and Fe(III) mediated Cr(VI) reduction in alkaline media using a Halomonas isolate from Soap Lake, Washington. *Biodegradation* 19:841–850.

29. Suzuki T, Miyata N, Horitsu H, Kawai K, Takamizawa K, Tai Y, Okazaki M. 1992. NAD(P)H-dependent chromium (VI) reductase of *Pseudomonas ambigua* G-1: a Cr(V) intermediate is formed during the reduction of Cr(VI) to Cr(III). *J Bacteriol* 174:5340–5345.

30. Park CH, Keyhan M, Wielinga B, Fendorf S, Matin A. 2000. Purification to homogeneity and characterization of a novel *Pseudomonas putida* chromate reductase. *Appl Environ Microbiol* 66:1788–1795.

31. Bae W, Chen W, Mulchandani A, Mehra RK. 2000. Enhanced bioaccumulation of heavy metals by bacterial cells displaying synthetic phytochelatins. *Biotechnol Bioeng* 70:518–524.

32. Bopp LH, Ehrlich HL. 1988. Chromate resistance and reduction in *Pseudomonas fluorescens* strain LB300. *Arch Microbiol* 150:426–431.

33. Lovley DR, Phillips EJP. 1994. Reduction of chromate by *Desulfovibrio vulgaris* and its C3 cytochrome. *Appl Environ Microbiol* 60:726–728.

34. Jeyasingh J, Somasundaram V, Philip L, Bhallamudi SM. 2011. Pilot scale studies on the remediation of chromium contaminated aquifer using bio-barrier and reactive zone technologies. *Chem Eng J* 167:206–214.

35. Kennedy MJ, Reader SL, Swierczynski LM. 1994. Preservation records of micro-organisms: evidence of the tenacity of life. *Microbiology* 140:2513–2529.

36. Nicholson WL, Munakata N, Horneck G, Melosh HJ, Setlow P. 2000. Resistance of *Bacillus* endospores to extreme terrestrial and extraterrestrial environments. *Microbiol Mol Biol Rev* 64:548–572.

37. Sneath PHA. 1962. Longevity of micro-organisms. *Nature* 195:643–646.

38. He LM, Tebo BM. 1998. Surface charge properties of and Cu(II) adsorption by spores of the marine *Bacillus* sp. strain SG-1. *Appl Environ Microbiol* 64:1123–1129.

39. Turekian KK, Wedepohl KH. 1961. Distribution of the elements in some major units of the earth's crust. *Bull Geol Soc Am* 72:175–192.

40. Post JE. 1999. Manganese oxide minerals: crystal structures and economic and environmental significance. *Proc Natl Acad Sci USA* 96:3447–3454.

41. Towler PH, Smith JD, Dixon DR. 1996. Magnetic recovery of radium, lead and polonium from seawater samples after preconcentration on a magnetic adsorbent of manganese dioxide coated magnetite. *Anal Chim Acta* 328:53–59.

42. Todd JF, Elsinger RJ, Moore WS. 1988. The distributions of uranium, radium and thorium isotopes in two anoxic fjords: Framvaren Fjord (Norway) and Saanich Inlet (British Columbia). *Mar Chem* 23:393–415.

43. Wei C-L, Murray JW. 1991. ^{234}Th/^{238}U disequilibria in the Black Sea. *Deep-Sea Res* 38:S855–S873.

44. Goldberg ED. 1954. Marine geochemistry I. Chemical scavengers of the sea. *J Geol* 62:249–265.

45. Prasad VS, Chaudhuri M. 1995. Removal of bacteria and turbidity from water by chemically treated manganese and iron ores. *Aqua Lond* **44**:80–82.

46. Stone AT, Morgan JJ. 1984. Reduction and dissolution of manganese(III) and manganese(IV) oxides by organics: 2. Survey of the reactivity of organics. *Environ Sci Technol* **18**:617–624.

47. Stone AT, Morgan JJ. 1984. Reduction and dissolution of manganese(III) and manganese(IV) oxides by organics. 1. Reaction with hydroquinone. *Environ Sci Technol* **18**:450–456.

48. Lehmann RG, Cheng HH, Harsh JB. 1987. Oxidation of phenolic acids by soil iron and manganese oxides. *Soil Sci Soc Am J* **51**:352–356.

49. Sunda WG, Kieber DJ. 1994. Oxidation of humic substances by manganese oxides yields low-molecular-weight organic substrates. *Nature* **367**:62–64.

50. Scott MJ, Morgan JJ. 1996. Reactions at oxide surfaces. 2. Oxidation of Se(IV) by synthetic birnessite. *Environ Sci Technol* **30**:1990–1996.

51. Manceau A, Charlet L. 1992. X-ray absorption spectroscopic study of the sorption of Cr(III) at the oxide-water interface. I. Molecular mechanism of Cr(III) oxidation on Mn oxides. *J Colloid Interface Sci* **148**:425–442.

52. Huang PM. 1991. Kinetics of redox reactions on manganese oxides and its impact on environmental quality, p 191–230. *In* Sparks DL, Suarez DL (ed), *Rates of Soil Chemical Processes*. Soil Science Society of America, Inc, Madison, WI.

53. Lin K, Liu W, Gant J. 2009. Oxidative removal of bisphenol A by manganese dioxide: efficacy, products, and pathways. *Environ Sci Technol* **43**:3860–3864.

54. Stone AT. 1987. Reductive dissolution of manganese (III/IV) oxides by substituted phenols. *Environ Sci Technol* **21**:979–988.

55. Ulrich HJ, Stone AT. 1989. Oxidation of chlorophenols adsorbed to manganese oxide surfaces. *Environ Sci Technol* **23**:421–428.

56. Tebo BM, Bargar JR, Clement BG, Dick GJ, Murray KJ, Parker D, Verity R, Webb SM. 2004. Biogenic manganese oxides: properties and mechanisms of formation. *Annu Rev Earth Planet Sci* **32**:287–328.

57. Park JW, Dec J, Kim JE, Bollag JM. 1999. Effect of humic constituents on the transformation of chlorinated phenols and anilines in the presence of oxidoreductive enzymes or birnessite. *Environ Sci Technol* **33**:2028–2034.

58. Cheney MA, Sposito G, McGrath AE, Criddle RS. 1996. Abiotic degradation of 2,4-D (dichlorophenoxyacetic acid) on synthetic birnessite: a calorespirometric method. *Colloids Surf A Physicochem Eng Asp* **107**:131–140.

59. Nasser A, Sposito G, Cheney MA. 2000. Mechanochemical degradation of 2,4-D adsorbed on synthetic birnessite. *Colloids Surf A Physicochem Eng Asp* **163**:117–123.

60. Cheney MA, Shin JY, Crowley DE, Alvey S, Malengreau N, Sposito G. 1998. Atrazine dealkylation on a manganese oxide surface. *Colloids Surf A Physicochem Eng Asp* **137**:267–273.

61. Stone AT. 1987. Microbial metabolites and the reductive dissolution of manganese oxides: oxalate and pyruvate. *Geochim Cosmochim Acta* **51**:919–925.

62. Laha S, Luthy RG. 1990. Oxidation of aniline and other primary aromatic amines by manganese dioxide. *Environ Sci Technol* **24**:363–373.

63. Morgan JJ. 2000. Manganese in natural waters and earth's crust: Its availability to organisms, p 1–33. *In* Sigel A, Sigel H (ed), Metal Ions in Biological Systems, **vol 37**. *Manganese and Its Role in Biological Processes*. Marcel Dekker, New York, NY.

64. Hastings D, Emerson S. 1986. Oxidation of manganese by spores of a marine bacillus: kinetic and thermodynamic considerations. *Geochim Cosmochim Acta* **50**:1819–1824.

65. Richardson LL, Aguilar C, Nealson KH. 1988. Manganese oxidation in pH and O2 microenvironments produced by phytoplankton. *Limnol Oceanogr* **33**:352–363.

66. Hullo MF, Moszer I, Danchin A, Martin-Verstraete I. 2001. CotA of *Bacillus subtilis* is a copper-dependent laccase. *J Bacteriol* **183**:5426–5430.

67. Nealson KH. 1978. The isolation and characterization of marine bacteria which catalyze manganese oxidation, p 847–858. *In* Krumbein WE (ed), Environmental Biogeochemistry and Geomicrobiology. **vol 3**: *Methods, Metals and Assessment*. Ann Arbor Science Publishers Inc., Ann Arbor, MI.

68. Rosson RA, Nealson KH. 1982. Manganese binding and oxidation by spores of a marine bacillus. *J Bacteriol* **151**:1027–1034.

69. Dick GJ. 2006. *Molecular Biogeochemistry of Mn(II) Oxidation: Deep-sea Hydrothermal Plumes, Enzymes, and Genomes*. Ph.D. dissertation. University of California, San Diego.

70. Francis CA, Tebo BM. 2002. Enzymatic manganese(II) oxidation by metabolically dormant spores of diverse Bacillus species. *Appl Environ Microbiol* **68**:874–880.

71. Lee Y. 1994. *Microbial Oxidation of Cobalt: Characterization and Its Significance in Marine Environments*. Ph.D. dissertation. University of California, San Diego.

72. Templeton AS, Staudigel H, Tebo BM. 2005. Diverse Mn(II) oxidizing bacteria isolated from submarine basalts at Loihi Seamount. *Geomicrobiol J* **22**:127–139.

73. Dick GJ, Lee YE, Tebo BM. 2006. Manganese(II)-oxidizing *Bacillus* spores in Guaymas Basin hydrothermal sediments and plumes. *Appl Environ Microbiol* **72**:3184–3190.

74. Cowen JP, Hui Li Y. 1991. The influence of a changing bacterial community on trace metal scavenging in a deep-sea particle plume. *J Mar Res* **49**:517–542.

75. Cowen JP, Massoth GJ, Baker ET. 1986. Bacterial scavenging of Mn and Fe in a mid- to far-field hydrothermal particle plume. *Nature* **322**:169–171.

76. Cowen JP, Massoth GJ, Feely RA. 1990. Scavenging rates of dissolved manganese in a hydrothermal vent plume. *Deep-Sea Res A, Oceanogr Res Pap* **37**:1619–1637.

77. Murray KJ, Webb SM, Bargar JR, Tebo BM. 2007. Indirect oxidation of Co(II) in the presence of the marine

Mn(II)-oxidizing bacterium *Bacillus* sp. strain SG-1. *Appl Environ Microbiol* 73:6905–6909.

78. Murray KJ, Tebo BM. 2007. Cr(III) is indirectly oxidized by the Mn(II)-oxidizing bacterium *Bacillus* sp. strain SG-1. *Environ Sci Technol* 41:528–533.

79. Fredrickson JK, Zachara JM, Kennedy DW, Liu C, Duff MC, Hunter DB, Dohnalkova A. 2002. Influence of Mn oxides on the reduction of uranium(VI) by the metal-reducing bacterium *Shewanella putrefaciens*. *Geochim Cosmochim Acta* 66:3247–3262.

80. Fendorf SE, Zasoski RJ. 1992. Chromium(III) oxidation by δ-MnO2. 1. Characterization. *Environ Sci Technol* 26:79–85.

81. Chinni S, Anderson CR, Ulrich KU, Giammar DE, Tebo BM. 2008. Indirect UO2 oxidation by Mn(II)-oxidizing spores of *Bacillus* sp. strain SG-1 and the effect of U and Mn concentrations. *Environ Sci Technol* 42:8709–8714.

82. Bargar JR, Tebo BM, Villinski JE. 2000. In situ characterization of Mn(II) oxidation by spores of the marine *Bacillus* sp. strain SG-1. *Geochim Cosmochim Acta* 64:2775–2778.

83. van Waasbergen LG, Hoch JA, Tebo BM. 1993. Genetic analysis of the marine manganese-oxidizing *Bacillus* sp. strain SG-1: protoplast transformation, Tn917 mutagenesis, and identification of chromosomal loci involved in manganese oxidation. *J Bacteriol* 175:7594–7603.

84. van Waasbergen LG, Hildebrand M, Tebo BM. 1996. Identification and characterization of a gene cluster involved in manganese oxidation by spores of the marine *Bacillus* sp. strain SG-1. *J Bacteriol* 178:3517–3530.

85. Dick GJ, Torpey JW, Beveridge TJ, Tebo BM. 2008. Direct identification of a bacterial manganese(II) oxidase, the multicopper oxidase MnxG, from spores of several different marine Bacillus species. *Appl Environ Microbiol* 74:1527–1534.

86. Moeller R, Schuerger AC, Reitz G, Nicholson WL. 2012. Protective role of spore structural components in determining *Bacillus subtilis* spore resistance to simulated mars surface conditions. *Appl Environ Microbiol* 78:8849–8853.

87. Martins LO, Soares CM, Pereira MM, Teixeira M, Costa T, Jones GH, Henriques AO. 2002. Molecular and biochemical characterization of a highly stable bacterial laccase that occurs as a structural component of the *Bacillus subtilis* endospore coat. *J Biol Chem* 277:18849–18859.

88. Solomon EI, Sundaram UM, Machonkin TE. 1996. Multicopper oxidases and oxygenases. *Chem Rev* 96:2563–2606.

89. Butterfield CN, Soldatova AV, Lee SW, Spiro TG, Tebo BM. 2013. Mn(II,III) oxidation and MnO2 mineralization by an expressed bacterial multicopper oxidase. *Proc Natl Acad Sci USA* 110:11731–11735.

90. Butterfield CN. 2014. *Characterizing the Mn(II) oxidizing enzyme from the marine* Bacillus *sp. PL-12 spore. Dissertation.* Oregon Health & Science University, Portland, OR.

91. Butterfield CN, Tao L, Chacón KN, Spiro TG, Blackburn NJ, Casey WH, Britt RD, Tebo BM. 2015. Multicopper manganese oxidase accessory proteins bind Cu and Heme. *Biochim Biophys Acta* 1854:1853–1859.

92. Learman DR, Wankel SD, Webb SM, Martinez N, Madden AS, Hansel CM. 2011. Coupled biotic-abiotic Mn(II) oxidation pathway mediates the formation and structural evolution of biogenic Mn oxides. *Geochim Cosmochim Acta* 75:6048–6063.

93. Luther GW III. 2010. The role of one- and two-electron transfer reactions in forming thermodynamically unstable intermediates as barriers in multi-electron redox reactions. *Aquat Geochem* 16:395–420.

94. Luther GW III. 2005. Manganese(II) oxidation and Mn(IV) reduction in the environment - Two one-electron transfer steps versus a single two-electron step. *Geomicrobiol J* 22:195–203.

95. Bargar JR, Tebo BM, Bergmann U, Webb SM, Glatzel P, Chiu VQ, Villalobos M. 2005. Biotic and abiotic products of Mn(II) oxidation by spores of the marine *Bacillus* sp. strain SG-1. *Am Mineral* 90:143–154.

96. Villalobos M, Toner B, Bargar J, Sposito G. 2003. Characterization of the manganese oxide produced by *Pseudomonas putida* strain MnB1. *Geochim Cosmochim Acta* 67:2649–2662.

97. Webb SM, Dick GJ, Bargar JR, Tebo BM. 2005. Evidence for the presence of Mn(III) intermediates in the bacterial oxidation of Mn(II). *Proc Natl Acad Sci USA* 102:5558–5563.

98. Soldatova AV, Butterfield C, Oyerinde OF, Tebo BM, Spiro TG. 2012. Multicopper oxidase involvement in both Mn(II) and Mn(III) oxidation during bacterial formation of MnO(2). *J Biol Inorg Chem* 17:1151–1158.

99. Honarmand Ebrahimi K, Bill E, Hagedoorn PL, Hagen WR. 2012. The catalytic center of ferritin regulates iron storage via Fe(II)-Fe(III) displacement. *Nat Chem Biol* 8:941–948.

100. Tao L, Stich TA, Butterfield CN, Romano CA, Spiro TG, Tebo BM, Casey WH, Britt RD. 2015. Mn(II) binding and subsequent oxidation by the multicopper oxidase mnxg investigated by electron paramagnetic resonance spectroscopy. *J Am Chem Soc* 137:10563–10575.

101. Parker DL, Sposito G, Tebo BM. 2004. Manganese(III) binding to a pyoverdine siderophore produced by a manganese(II)-oxidizing bacterium. *Geochim Cosmochim Acta* 68:4809–4820.

102. Glenn JK, Akileswaran L, Gold MH. 1986. Mn(II) oxidation is the principal function of the extracellular Mn-peroxidase from *Phanerochaete chrysosporium*. *Arch Biochem Biophys* 251:688–696.

103. Perez J, Jeffries TW. 1992. Roles of manganese and organic acid chelators in regulating lignin degradation and biosynthesis of peroxidases by *Phanerochaete chrysosporium*. *Appl Environ Microbiol* 58:2402–2409.

104. Höfer C, Schlosser D. 1999. Novel enzymatic oxidation of Mn2+ to Mn3+ catalyzed by a fungal laccase. *FEBS Lett* 451:186–190.

105. Schlosser D, Höfer C. 2002. Laccase-catalyzed oxidation of Mn(2+) in the presence of natural Mn(3+)

chelators as a novel source of extracellular H(2)O(2) production and its impact on manganese peroxidase. *Appl Environ Microbiol* 68:3514–3521.

106. Kostka JE, Luther GW III, Nealson KH. 1995. Chemical and biological reduction of Mn(III)-pyrophosphate complexes: potential importance of dissolved Mn(III) as an environmental oxidant. *Geochim Cosmochim Acta* 59:885–894.

107. Luther GW III, Ruppel DT, Burkhard C. 1998. Reactivity of dissolved Mn(III) complexes and Mn(IV) species with reductants: Mn redox chemistry without a dissolution step, p 265–280. *In* Sparks DL, Grundl TJ (ed), *Mineral-Water Interfacial Reactions: Kinetics and Mechanisms*, ACS Symposium Series, vol 715. American Chemical Society, Washington, DC.

108. Klewicki JK, Morgan JJ. 1999. Dissolution of β-MnOOH particles by ligands: pyrophosphate, ethylenediaminetetraacetate, and citrate. *Geochim Cosmochim Acta* 63:3017–3024.

109. Klewicki JK, Morgan JJ. 1998. Kinetic behavior of Mn(III) complexes of pyrophosphate, EDTA, and citrate. *Environ Sci Technol* 32:2916–2922.

110. Wall JD, Krumholz LR. 2006. Uranium reduction. *Annu Rev Microbiol* 60:149–166.

111. Anderson RT, Vrionis HA, Ortiz-Bernad I, Resch CT, Long PE, Dayvault R, Karp K, Marutzky S, Metzler DR, Peacock A, White DC, Lowe M, Lovley DR. 2003. Stimulating the in situ activity of *Geobacter* species to remove uranium from the groundwater of a uranium-contaminated aquifer. *Appl Environ Microbiol* 69:5884–5891.

112. Wu WM, Carley J, Fienen M, Mehlhorn T, Lowe K, Nyman J, Luo J, Gentile ME, Rajan R, Wagner D, Hickey RF, Gu B, Watson D, Cirpka OA, Kitanidis PK, Jardine PM, Criddle CS. 2006. Pilot-scale in situ bioremediation of uranium in a highly contaminated aquifer. 1. Conditioning of a treatment zone. *Environ Sci Technol* 40:3978–3985.

113. Fredrickson JK, Zachara JM, Balkwill DL, Kennedy D, Li SMW, Kostandarithes HM, Daly MJ, Romine MF, Brockman FJ. 2004. Geomicrobiology of high-level nuclear waste-contaminated vadose sediments at the Hanford site, Washington state. *Appl Environ Microbiol* 70:4230–4241.

114. Markich SJ. 2002. Uranium speciation and bioavailability in aquatic systems: an overview. *Scientific World Journal* 2:707–729.

115. Craft E, Abu-Qare A, Flaherty M, Garofolo M, Rincavage H, Abou-Donia M. 2004. Depleted and natural uranium: chemistry and toxicological effects. *J Toxicol Environ Health B Crit Rev* 7:297–317.

116. Kauffman JW, Laughlin WC, Baldwin RA. 1986. Microbiological treatment of uranium mine waters. *Environ Sci Technol* 20:243–248.

117. Gorby YA, Lovley DR. 1992. Enzymatic uranium precipitation. *Environ Sci Technol* 26:205–207.

118. Lovley DR, Phillips EJP. 1992. Reduction of uranium by *Desulfovibrio desulfuricans*. *Appl Environ Microbiol* 58:850–856.

119. Lovley DR, Phillips EJP, Gorby YA, Landa ER. 1991. Microbial reduction of uranium. *Nature* 350:413–416.

120. Marshall MJ, Beliaev AS, Dohnalkova AC, Kennedy DW, Shi L, Wang Z, Boyanov MI, Lai B, Kemner KM, McLean JS, Reed SB, Culley DE, Bailey VL, Simonson CJ, Saffarini DA, Romine MF, Zachara JM, Fredrickson JK. 2006. c-Type cytochrome-dependent formation of U(IV) nanoparticles by *Shewanella oneidensis*. *PLoS Biol* 4:e268.

121. Sanford RA, Wu Q, Sung Y, Thomas SH, Amos BK, Prince EK, Löffler FE. 2007. Hexavalent uranium supports growth of *Anaeromyxobacter dehalogenans* and *Geobacter* spp. with lower than predicted biomass yields. *Environ Microbiol* 9:2885–2893.

122. Fletcher KE, Boyanov MI, Thomas SH, Wu Q, Kemner KM, Löffler FE. 2010. U(VI) reduction to mononuclear U(IV) by *Desulfitobacterium* species. *Environ Sci Technol* 44:4705–4709.

123. Tebo BM, Obraztsova AY. 1998. Sulfate-reducing bacterium grows with Cr(VI), U(VI), Mn(IV), and Fe(III) as electron acceptors. *FEMS Microbiol Lett* 162:193–198.

124. Brodie EL, Desantis TZ, Joyner DC, Baek SM, Larsen JT, Andersen GL, Hazen TC, Richardson PM, Herman DJ, Tokunaga TK, Wan JM, Firestone MK. 2006. Application of a high-density oligonucleotide microarray approach to study bacterial population dynamics during uranium reduction and reoxidation. *Appl Environ Microbiol* 72:6288–6298.

125. Suzuki Y, Kelly SD, Kemner KM, Banfield JF. 2003. Microbial populations stimulated for hexavalent uranium reduction in uranium mine sediment. *Appl Environ Microbiol* 69:1337–1346.

126. Wu WM, Carley J, Luo J, Ginder-Vogel MA, Cardenas E, Leigh MB, Hwang C, Kelly SD, Ruan C, Wu L, Van Nostrand J, Gentry T, Lowe K, Mehlhorn T, Carroll S, Luo W, Fields MW, Gu B, Watson D, Kemner KM, Marsh T, Tiedje J, Zhou J, Fendorf S, Kitanidis PK, Jardine PM, Criddle CS. 2007. In situ bioreduction of uranium (VI) to submicromolar levels and reoxidation by dissolved oxygen. *Environ Sci Technol* 41:5716–5723.

127. Paredes CJ, Alsaker KV, Papoutsakis ET. 2005. A comparative genomic view of clostridial sporulation and physiology. *Nat Rev Microbiol* 3:969–978.

128. Preuss A, Fimpel J, Diekert G. 1993. Anaerobic transformation of 2,4,6-trinitrotoluene (TNT). *Arch Microbiol* 159:345–353.

129. Lewis TA, Goszczynski S, Crawford RL, Korus RA, Admassu W. 1996. Products of anaerobic 2,4,6-trinitrotoluene (TNT) transformation by *Clostridium bifermentans*. *Appl Environ Microbiol* 62:4669–4674.

130. Sembries S, Crawford RL. 1997. Production of *Clostridium bifermentans* spores as inoculum for bioremediation of nitroaromatic contaminants. *Appl Environ Microbiol* 63:2100–2104.

131. Gao W, Francis AJ. 2008. Reduction of uranium(VI) to uranium(IV) by clostridia. *Appl Environ Microbiol* 74:4580–4584.

132. Francis AJ, Dodge CJ, Lu F, Halada GP, Clayton CR. 1994. XPS and XANES studies of uranium reduction by *Clostridium* sp. *Environ Sci Technol* 28:636–639.

133. Zhang S, Yin L, Liu Y, Zhang D, Luo X, Cheng J, Cheng F, Dai J. 2011. Cometabolic biotransformation of fenpropathrin by *Clostridium* species strain ZP3. *Biodegradation* 22:869–875.

134. Hammill TB, Crawford RL. 1996. Degradation of 2-sec-butyl-4,6-dinitrophenol (dinoseb) by *Clostridium bifermentans* KMR-1. *Appl Environ Microbiol* 62: 1842–1846.

135. Lal R, Pandey G, Sharma P, Kumari K, Malhotra S, Pandey R, Raina V, Kohler HPE, Holliger C, Jackson C, Oakeshott JG. 2010. Biochemistry of microbial degradation of hexachlorocyclohexane and prospects for bioremediation. *Microbiol Mol Biol Rev* 74:58–80.

136. N'Guessan AL, Vrionis HA, Resch CT, Long PE, Lovley DR. 2008. Sustained removal of uranium from contaminated groundwater following stimulation of dissimilatory metal reduction. *Environ Sci Technol* 42: 2999–3004.

137. Tapia-Rodriguez A, Tordable-Martinez V, Sun W, Field JA, Sierra-Alvarez R. 2011. Uranium bioremediation in continuously fed upflow sand columns inoculated with anaerobic granules. *Biotechnol Bioeng* 108:2583–2591.

138. Vecchia ED, Veeramani H, Suvorova EI, Wigginton NS, Bargar JR, Bernier-Latmani R. 2010. U(VI) reduction by spores of *Clostridium acetobutylicum*. *Res Microbiol* 161:765–771.

139. Nölling J, Breton G, Omelchenko MV, Makarova KS, Zeng Q, Gibson R, Lee HM, Dubois J, Qiu D, Hitti J, Wolf YI, Tatusov RL, Sabathe F, Doucette-Stamm L, Soucaille P, Daly MJ, Bennett GN, Koonin EV, Smith DR, GTC Sequencing Center Production, Finishing, and Bioinformatics Teams. 2001. Genome sequence and comparative analysis of the solvent-producing bacterium *Clostridium acetobutylicum*. *J Bacteriol* 183: 4823–4838.

140. Hu S, Zheng H, Gu Y, Zhao J, Zhang W, Yang Y, Wang S, Zhao G, Yang S, Jiang W. 2011. Comparative genomic and transcriptomic analysis revealed genetic characteristics related to solvent formation and xylose utilization in *Clostridium acetobutylicum* EA 2018. *BMC Genomics* 12:93.

141. Bao G, Wang R, Zhu Y, Dong H, Mao S, Zhang Y, Chen Z, Li Y, Ma Y. 2011. Complete genome sequence of *Clostridium acetobutylicum* DSM 1731, a solvent-producing strain with multireplicon genome architecture. *J Bacteriol* 193:5007–5008.

142. Nakotte S, Schaffer S, Böhringer M, Dürre P. 1998. Electroporation of, plasmid isolation from and plasmid conservation in *Clostridium acetobutylicum* DSM 792. *Appl Microbiol Biotechnol* 50:564–567.

143. Feustel L, Nakotte S, Dürre P. 2004. Characterization and development of two reporter gene systems for *Clostridium acetobutylicum*. *Appl Environ Microbiol* 70:798–803.

144. Inui M, Suda M, Kimura S, Yasuda K, Suzuki H, Toda H, Yamamoto S, Okino S, Suzuki N, Yukawa H. 2008.

Expression of *Clostridium acetobutylicum* butanol synthetic genes in *Escherichia coli*. *Appl Microbiol Biotechnol* 77:1305–1316.

145. Green EM, Boynton ZL, Harris LM, Rudolph FB, Papoutsakis ET, Bennett GN. 1996. Genetic manipulation of acid formation pathways by gene inactivation in *Clostridium acetobutylicum* ATCC 824. *Microbiology* 142:2079–2086.

146. Junier P, Frutschi M, Wigginton NS, Schofield EJ, Bargar JR, Bernier-Latmani R. 2009. Metal reduction by spores of *Desulfotomaculum reducens*. *Environ Microbiol* 11:3007–3017.

147. Newman DK, Kennedy EK, Coates JD, Ahmann D, Ellis DJ, Lovley DR, Morel FMM. 1997. Dissimilatory arsenate and sulfate reduction in *Desulfotomaculum auripigmentum* sp. nov. *Arch Microbiol* 168:380–388.

148. Radhika V, Subramanian S, Natarajan KA. 2006. Bioremediation of zinc using *Desulfotomaculum nigrificans*: bioprecipitation and characterization studies. *Water Res* 40:3628–3636.

149. Fortin D, Southam G, Beveridge TJ. 1994. Nickel sulfide, iron-nickel sulfide and iron sulfide precipitation by a newly isolated Desulfotomaculum species and its relation to nickel resistance. *FEMS Microbiol Ecol* 14: 121–132.

150. Chang YJ, Peacock AD, Long PE, Stephen JR, McKinley JP, Macnaughton SJ, Hussain AK, Saxton AM, White DC. 2001. Diversity and characterization of sulfate-reducing bacteria in groundwater at a uranium mill tailings site. *Appl Environ Microbiol* 67:3149–3160.

151. Nevin KP, Finneran KT, Lovley DR. 2003. Microorganisms associated with uranium bioremediation in a high-salinity subsurface sediment. *Appl Environ Microbiol* 69:3672–3675.

152. Bernier-Latmani R, Veeramani H, Vecchia ED, Junier P, Lezama-Pacheco JS, Suvorova EI, Sharp JO, Wigginton NS, Bargar JR. 2010. Non-uraninite products of microbial U(VI) reduction. *Environ Sci Technol* 44:9456–9462.

153. Hernandez ME, Newman DK. 2001. Extracellular electron transfer. *Cell Mol Life Sci* 58:1562–1571.

154. Finneran KT, Anderson RT, Nevin KP, Lovley DR. 2002. Potential for bioremediation of uranium-contaminated aquifers with microbial U(VI) reduction. *Soil Sediment Contam* 11:339–357.

155. von Canstein H, Ogawa J, Shimizu S, Lloyd JR. 2008. Secretion of flavins by *Shewanella* species and their role in extracellular electron transfer. *Appl Environ Microbiol* 74:615–623.

156. Marsili E, Baron DB, Shikhare ID, Coursolle D, Gralnick JA, Bond DR. 2008. *Shewanella* secretes flavins that mediate extracellular electron transfer. *Proc Natl Acad Sci USA* 105:3968–3973.

157. Sivaswamy V, Boyanov MI, Peyton BM, Viamajala S, Gerlach R, Apel WA, Sani RK, Dohnalkova A, Kemner KM, Borch T. 2011. Multiple mechanisms of uranium immobilization by *Cellulomonas* sp. strain ES6. *Biotechnol Bioeng* 108:264–276.

158. Lovley DR, Coates JD, Blunt-Harris EL, Phillips EJP, Woodward JC. 1996. Humic substances as electron acceptors for microbial respiration. *Nature* 382:445–448.

159. Ahmed B, Cao B, McLean JS, Ica T, Dohnalkova A, Istanbullu O, Paksoy A, Fredrickson JK, Beyenal H. 2012. Fe(III) reduction and U(VI) immobilization by *Paenibacillus* sp. strain 300A, isolated from Hanford 300A subsurface sediments. *Appl Environ Microbiol* 78:8001–8009.

160. Junier P, Junier T, Podell S, Sims DR, Detter JC, Lykidis A, Han CS, Wigginton NS, Gaasterland T, Bernier-Latmani R. 2010. The genome of the Gram-positive metal- and sulfate-reducing bacterium *Desulfotomaculum reducens* strain MI-1. *Environ Microbiol* 12:2738–2754.

161. Junier P, Vecchia ED, Bernier-Latmani R. 2011. The response of desulfotomaculum reducens MI-1 to U(VI) exposure: a transcriptomic study. *Geomicrobiol J* 28:483–496.

162. Bonde GJ. 1981. *Bacillus* from marine habitats: allocation to phena established by numerical techniques, p 181–215. *In* Berkeley RCW, Goodfellow M (ed), *The Aerobic Endospore-Forming Bacteria: Classification and Identification.* Academic Press, New York, NY.

163. Hong HA, To E, Fakhry S, Baccigalupi L, Ricca E, Cutting SM. 2009. Defining the natural habitat of *Bacillus* spore-formers. *Res Microbiol* 160:375–379.

164. Langerhuus AT, Røy H, Lever MA, Morono Y, Inagaki F, Jørgensen BB, Lomstein BA. 2012. Endospore abundance and D:L-amino acid modeling of bacterial turnover in holocene marine sediment (Aarhus Bay). *Geochim Cosmochim Acta* 99:87–99.

165. Lomstein BA, Langerhuus AT, D'Hondt S, Jørgensen BB, Spivack AJ. 2012. Endospore abundance, microbial growth and necromass turnover in deep sub-seafloor sediment. *Nature* 484:101–104.

166. Tyson GW, Chapman J, Hugenholtz P, Allen EE, Ram RJ, Richardson PM, Solovyev VV, Rubin EM, Rokhsar DS, Banfield JF. 2004. Community structure and metabolism through reconstruction of microbial genomes from the environment. *Nature* 428:37–43.

167. Rahm BG, Richardson RE. 2008. Dehalococcoides' gene transcripts as quantitative bioindicators of tetrachloroethene, trichloroethene, and cis-1,2-dichloroethene dehalorespiration rates. *Environ Sci Technol* 42:5099–5105.

168. Han JI, Semrau JD. 2004. Quantification of gene expression in methanotrophs by competitive reverse transcription-polymerase chain reaction. *Environ Microbiol* 6:388–399.

169. Anderson CR, Davis RE, Bandolin NS, Baptista AM, Tebo BM. 2011. Analysis of in situ manganese(II) oxidation in the Columbia River and offshore plume: linking Aurantimonas and the associated microbial community to an active biogeochemical cycle. *Environ Microbiol* 13:1561–1576.

170. Wilkins MJ, Verberkmoes NC, Williams KH, Callister SJ, Mouser PJ, Elifantz H, N'guessan AL, Thomas BC, Nicora CD, Shah MB, Abraham P, Lipton MS, Lovley DR, Hettich RL, Long PE, Banfield JF. 2009. Proteogenomic monitoring of *Geobacter* physiology during stimulated uranium bioremediation. *Appl Environ Microbiol* 75:6591–6599.

171. Tebo BM, Ghiorse WC, van Waasbergen LG, Siering PL, Caspi R. 1997. Bacterially-mediated mineral formation: Insights into manganese(II) oxidation from molecular genetic and biochemical studies, p 225–266. *In* Banfield JF, Nealson KH (ed), *Geomicrobiology: Interactions Between Microbes and Minerals*, vol 35. Mineralogical Society of America, Washington, D.C.

172. Park TJ, Choi SK, Jung HC, Lee SY, Pan JG. 2009. Spore display using *Bacillus thuringiensis* exosporium protein InhA. *J Microbiol Biotechnol* 19:495–501.

173. Okazaki M, Sugita T, Shimizu M, Ohode Y, Iwamoto K, de Vrind-de Jong EW, de Vrind JPM, Corstjens PLAM. 1997. Partial purification and characterization of manganese-oxidizing factors of Pseudomonas fluorescens GB-1. *Appl Environ Microbiol* 63:4793–4799.

174. Caspi R, Haygood MG, Tebo BM. 1996. Unusual ribulose-1,5-bisphosphate carboxylase/oxygenase genes from a marine manganese-oxidizing bacterium. *Microbiology* 142:2549–2559.

175. Boogerd FC, de Vrind JPM. 1987. Manganese oxidation by *Leptothrix discophora*. *J Bacteriol* 169:489–494.

176. Ridge JP, Lin M, Larsen EI, Fegan M, McEwan AG, Sly LI. 2007. A multicopper oxidase is essential for manganese oxidation and laccase-like activity in *Pedomicrobium* sp. ACM 3067. *Environ Microbiol* 9:944–953.

177. Hansel CM, Francis CA. 2006. Coupled photochemical and enzymatic Mn(II) oxidation pathways of a planktonic Roseobacter-Like bacterium. *Appl Environ Microbiol* 72:3543–3549.

178. Emerson D, Rentz JA, Lilburn TG, Davis RE, Aldrich H, Chan C, Moyer CL. 2007. A novel lineage of proteobacteria involved in formation of marine Fe-oxidizing microbial mat communities. *PLoS One* 2: e667. http://dx.doi.org/10.1371/journal.pone.0000667.

179. Emerson D, Moyer C. 1997. Isolation and characterization of novel iron-oxidizing bacteria that grow at circumneutral pH. *Appl Environ Microbiol* 63:4784–4792.

180. Hallbeck L, Pedersen K. 1991. Autotrophic and mixotrophic growth of *Gallionella ferruginea*. *J Gen Microbiol* 137:2657–2661.

181. Straub KL, Benz M, Schink B, Widdel F. 1996. Anaerobic, nitrate-dependent microbial oxidation of ferrous iron. *Appl Environ Microbiol* 62:1458–1460.

182. Oremland RS, Blum JS, Culbertson CW, Visscher PT, Miller LG, Dowdle P, Strohmaier FE. 1994. Isolation, growth, and metabolism of an obligately anaerobic, selenate-respiring bacterium, strain SES-3. *Appl Environ Microbiol* 60:3011–3019.

183. Debieux CM, Dridge EJ, Mueller CM, Splatt P, Paszkiewicz K, Knight I, Florance H, Love J, Titball RW, Lewis RJ, Richardson DJ, Butler CS. 2011. A bacterial

process for selenium nanosphere assembly. *Proc Natl Acad Sci USA* **108**:13480–13485.

184. **Tomei FA, Barton LL, Lemanski CL, Zocco TG.** 1992. Reduction of selenate and selenite to elemental selenium by *Wolinella succinogenes*. *Can J Microbiol* **38**:1328–1333.

185. **Burgos WD, McDonough JT, Senko JM, Zhang G, Dohnalkova AC, Kelly SD, Gorby Y, Kemner KM.** 2008. Characterization of uraninite nanoparticles produced by *Shewanella oneidensis* MR-1. *Geochim Cosmochim Acta* **72**:4901–4915.

186. **Wu Q, Sanford RA, Löffler FE.** 2006. Uranium(VI) reduction by *Anaeromyxobacter dehalogenans* strain 2CP-C. *Appl Environ Microbiol* **72**:3608–3614.

187. **Horitsu H, Futo S, Miyazawa Y, Ogai S, Kawai K.** 1987. Enzymatic reduction of hexavalent chromium by hexavalent chromium tolerant *Pseudomonas ambigua* G-1. *Agric Biol Chem* **51**:2417–2420.

188. **Llovera S, Bonet R, Simon-Pujol MD, Congregado F.** 1993. Chromate reduction by resting cells of Agrobacterium radiobacter EPS-916. *Appl Environ Microbiol* **59**:3516–3518.

189. **Wang PC, Mori T, Komori K, Sasatsu M, Toda K, Ohtake H.** 1989. Isolation and characterization of an Enterobacter cloacae strain that reduces hexavalent chromium under anaerobic conditions. *Appl Environ Microbiol* **55**:1665–1669.

190. **Lloyd JR, Ridley J, Khizniak T, Lyalikova NN, Macaskie LE.** 1999. Reduction of technetium by *Desulfovibrio desulfuricans*: biocatalyst characterization and use in a flowthrough bioreactor. *Appl Environ Microbiol* **65**:2691–2696.

191. **Lloyd JR, Nolting HF, Solé VA, Bosecker K, Macaskie LE.** 1998. Technetium reduction and precipitation by sulfate-reducing bacteria. *Geomicrobiol J* **15**:45–58.

192. **Martinez RJ, Beazley MJ, Taillefert M, Arakaki AK, Skolnick J, Sobecky PA.** 2007. Aerobic uranium (VI) bioprecipitation by metal-resistant bacteria isolated from radionuclide- and metal-contaminated subsurface soils. *Environ Microbiol* **9**:3122–3133.

193. **Pan-Hou HSK, Imura N.** 1981. Role of hydrogen sulfide in mercury resistance determined by plasmid of *Clostridium cochlearium* T-2. *Arch Microbiol* **129**:49–52.

Index

Spore surface display
  advantages of, 361
  applications for, 354
  on *Bacillus subtilis* spores, 353–359
  on bacterial cells or phages, 351–353
  CotA as display system, 354, 358–359
  CotB as carrier protein, 353–356
  CotC as carrier protein, 354, 356–357
  CotG as carrier protein, 354, 357–358
  future developments with, 361–362
  heterologous antigens and enzymes in
    nonrecombinant, 360
  list of carriers for, 354
  mother cell-specific transcription signals
    with, 362
  N and C terminus sandwich fusions
    with, 353
  nonrecombinant spores in, 359–361
  OxdD as carrier protein, 354, 358
  passenger proteins for, 354
  spores of bacilli other than *B. subtilis*
    in, 359
  spore surface properties in, 360–361
  SpsC as spore-anchoring signal, 354, 358
  strategies of, 352–353
  surface proteins used as carriers for, 352
  for vaccine delivery, 352
Sporobacter termitidis, 6
Sporolactobacillaceae, sporeformer
  distribution with, 6
Sporolactobacillus inulinus, 6
Sporomusa ovata, genomes available for, 9
Sporomusa paucivorans, 8
Sporomusa sphaeroides, sporulation in, 8
Sporosarcina globispora, renaming of, 7
Sporulation in bacteria, standard model
  comparative genomics perspective on,
    95–97
  endospore formation in, 87–88
    more than two endospores, 90–92
    twin endospore in *Bacillus subtilis*,
      89–90
    unusual nonmodel
      endosporeformers, 95
  nondormant internal offspring production
    in, 92–95
  value of comparative approaches for, 97
spoVAC gene, 11
SpoVB, synthesis of spore peptidoglycan
  in, 164
SpoVD
  localization in spore coat of, 152
  in synthesis of spore peptidoglycan, 164
SpoVD/SpoVE/SpoVB/YkvU, protein
  asymmetry in sporulation, 152–153
SpoVE, localization in spore coat of, 152
spoVG gene, 11
SpoVM
  localization in spore coat of, 152
  protein asymmetry in sporulation,
    151–152
spoVR gene, 11
spoVS gene, 11
SpsC, spore-anchoring signal with, 354, 358
sspF gene, 11
Staphylococcaceae, sporeformer distribution
  with, 6
Streptococcaceae, sporeformer distribution
  with, 6

Streptohalobacillus, environmental habitats
  distribution for, 63
Streptohalobacillus salinus, 30
SubtiWiki, 138
Sulfobacillus acidophilus, genome
  sequencing for, 10
Sulfurospirillum barnesii, metal
  transformations with, 369
Superoxide dismutases (SODs), *Bacillus
  anthracis* exosporium with, 256
Symbiobacterium thermophilum, 3, 6
Synthetic genetic arrays (SGAs), *Bacillus
  subtilis* with, 134–135
Syntrophomonadaceae, sporeformer
  distribution with, 6
Syntrophomonas wolfei, sporulation genes
  in, 11
σ^D. *See* Alternative sigma factor

T

Temperature gradient gel electrophoresis
  (TGGE), 69
Tenericutes, 5
Tenuibacillus, environmental habitats
  distribution for, 63
Tenuibacillus multivorans, 30
Terminal restriction fragment length
  polymorphism (T-RFLP), 69
Terribacillus
  environmental habitats distribution for, 63
  species, 30
Terribacillus aidingensis, 30
Terribacillus goriensis, 30
Terribacillus halophilus, 30
Terribacillus saccharophilus, 30
Texcoconibacillus, environmental habitats
  distribution for, 63
TGGE. *See* Temperature gradient gel
  electrophoresis
Thalassobacillus
  environmental habitats distribution for, 63
  species, 30
Thalassobacillus cyri, 30
Thalassobacillus devorans, 30
Thalassobacillus hwangdonensis, 30
Thalassobacillus pellis, 30
Thauera selenatis, metal transformations
  with, 369
Thermaerobacter marianensis, sporulation
  genes in, 11
Thermincola carboxydiphila, 8
Thermincola ferriacetica, 8
Thermincola potens, 8
Thermoactinomycetaceae, sporeformer
  distribution with, 6
Thermoanaerobacteraceae, sporeformer
  distribution with, 6
Thermoanaerobacterales
  sporeformer distribution with, 6
  sporulation genes in, 11
Thermoanaerobacter mathranii
  evolution of sporulation with, 13
  genome sequencing for, 10
Thermodesulfobiaceae, sporeformer
  distribution with, 6
Thermolithobacteraceae, sporeformer
  distribution with, 6
Thermolithobacterales, sporeformer
  distribution with, 6

Thermolithobacteria, 5
  sporeformer distribution with, 6
Thermolongibacillus, environmental habitats
  distribution for, 63
Thermophiles, genome sequencing for, 10
Thermosinus carboxydivorans, genomes
  sequencing for, 9
Thermotogae
  Firmicutes gene families shared with, 5
  sporulation genes in, 11
Thiobacillus denitrificans, metal
  transformations with, 369
TpeL toxin, 342–343
Transtracheal injection, *Bacillus anthracis*
  delivery using, 271
T-RFLP. *See* Terminal restriction fragment
  length polymorphism
Tuberibacillus calidus, 6, 30
Tumebacillus, environmental habitats
  distribution for, 63

U

Uranium metal transformations, 375–377
  Clostridium acetobutylicum in, 376
  Desulfotomaculum reducens in, 376–377
U.S. Food and Drug Administration (FDA),
  animal rule, 272
UV radiation, spore resistance with, 203,
  205–206
UV resistance
  directed evolution to, 39
  genome sequencing for strains with, 10

V

Veillonellaceae, 5
  sporeformer distribution with, 6
  sporulation in, 8
Veillonella parvula, genomes available for, 9
Vibrio subtilis, 3
Virgibacillus
  environmental habitats distribution for,
    63
  species, 30
Virgibacillus alimentarius, 30
Virgibacillus arcticus, 31
Virgibacillus byunsanensis, 31
Virgibacillus campisalis, 31
Virgibacillus carmonensis, 31
Virgibacillus chiguensis, 31
Virgibacillus dokdonensis, 31
Virgibacillus halodenitrificans, 31
  genome sequencing for, 10
Virgibacillus halophilus, 31
Virgibacillus kekensis, 31
Virgibacillus koreensis, 31
Virgibacillus marismortui, 31
Virgibacillus necropolis, 31
Virgibacillus olivae, 31
Virgibacillus pantothenticus, 31
  renaming of, 7
Virgibacillus proomii, 31
Virgibacillus salarius, 31
Virgibacillus salexigens, 31
Virgibacillus salinus, 31
Virgibacillus sediminis, 31
Virgibacillus siamensis, 31
Virgibacillus soli, 31
Virgibacillus subterraneus, 31
Virgibacillus xinjiangensis, 31